UNDERSTANDING
SOLID STATE PHYSICS

UNDERSTANDING SOLID STATE PHYSICS

PROBLEMS AND SOLUTIONS

Jacques Cazaux

PAN STANFORD PUBLISHING

Published by

Pan Stanford Publishing Pte. Ltd.
Penthouse Level, Suntec Tower 3
8 Temasek Boulevard
Singapore 038988

Email: editorial@panstanford.com
Web: www.panstanford.com

British Library Cataloguing-in-Publication Data
A catalogue record for this book is available from the British Library.

Understanding Solid State Physics: Problems and Solutions

ISBN 978-981-4267-89-2 (Hardcover)
ISBN 978-981-4267-90-8 (eBook)

Translated to English by Laurie E. Calvet, University of Paris-Sud, France

Printed in the USA

"I stand upon my desk to remind myself that we must constantly look at things in a different way. You see the world looks very different from up here. Don't believe me? Come see for yourself. Come on. Just when you think you know something you have to look at it in another way even though it may seem silly or wrong. You must try . . . "

—John Keating, in the *Dead Poets Society*, a movie by Peter Weir

Contents

Preface

There are excellent text books available in solid state physics (see bibliography, a non-exhaustive list), and the better the books, the easier looks solid state physics. But even clear explanations do not necessarily involve a full understanding of this field of physics. The best test for the reader is to try an alternative point of view: solve exercises or problems, and this is the assigned goal of this book. This book provides a stimulating challenge for the grey cells as it gives the opportunity to make useful (for the understanding) errors and also to draw curves and figures or to deal with orders of magnitude in the numerical applications. The questions at the end of each chapter permit to have another proper view as they are free of mathematical calculations and can be solved using common sense.

This book is the deeply revised version of the French book *Initiation à la physique du solide: exercices comméntes avec rappels de cours*, which was written more than 20 years ago and was the result of years of teaching solid state physics at the Reims University, France. Its initial idea was just a compilation of problems given for examinations at the end of each year to test the young students on their understanding of lectures based upon the first ten chapters of the well-known book of C. Kittel with complements from books of pioneers of this subject. Next a sabbatical stay in a laboratory involved in microelectronics (lab directed by Dr. A. Frederic at Thomson CSF, presently Thales, Orsay) opened my eyes on the fascinating samples that were elaborated and operated: nearly academic samples involving the physics of surfaces and of 1- or 2D samples, such as atom aggregates, multilayers, and semiconducting devices, full of promising applications in nanotechnologies. Finally came the idea of adding exercises and problems inspired from a dozen of Nobel Prize–winning works on solid state physics, including the most recent on graphene and carbon nanotubes.

The prerequisites to use this book are only the mathematics of differential equations involved in Maxwell's and Schrödinger's equations in their simplest form. The detailed statements permit to overcome other mathematical difficulties by suggesting practical simplifications in order to lead the reader step by step up to the end. I hope the readers will find the same pleasure as I had while writing this book and also making errors.

I am greatly indebted to my colleagues of various French universities: M. Gerl, B. Gruzza, J. Philibert, Y. Queré, C. Colliex, and O. Jbara for their comments and constructive remarks. I also wish to thank again A. Friederich, J. Massies, A. Fert, C. Weisbuch, and their colleagues for their stimulating hospitality in the Thomson-CSF group during the years 1987–1990. Last but not at least, I am very grateful to the hundreds of students in physics at the University of Reims who, by their errors, unintentionally acted as guinea pigs in testing this book and thus improving it.

Jacques Cazaux
Reims, France
Winter 2014

Note to the Reader

This book is divided into five chapters numbered I to V. Each chapter is composed of a course summary, exercises with solutions, and then problems that are also followed by solutions. Simple questions are given at the end of each chapter.

At the end of some questions one may find, between parentheses, indications such as e, m, k_B, etc. to denote that the numerical value of the electron charge, e, the electron mass, m, and the Boltzmann constant, k_B, etc. are needed to solve them. These numerical values are given in Table I at the beginning of the book. Table II lists some useful numerical values that would be required to solve the exercises and problems. SI units have been used in both tables. Table III presents the periodic table of elements for easy reference.

At the right end of some solutions one may find between parentheses indications such as Chapter IV, Ex. 5, or Chapter V, Pb. 6. These indications refer to other exercises, Ex., or problems, Pb., being in continuity to the exercise or problem under consideration. The roman capital number of the chapter is omitted when the correlated exercises or problems belong to the same chapter. This arrangement permits the reader to gather various exercises and problems that have a common aspect. For instance, an extended investigation of the various properties of graphene may be performed from the addition of Chapter I, Ex. 17 (Crystallography); Chapter III, Pb. 9 (Phonons); Chapter IV, Ex. 14b; and Chapter V, Pb. 11 (π electrons). Similar correlations may be established between the phonon and the electron dispersion curves in 1D, 2D, or 3D.

Tables

Table I Physical Constants

Charge of 1 electron	e or q	1.60×10^{-19} C
Mass of 1 electron	m	9.1×10^{-31} kg
Planck's constant	H $\hbar$	6.62×10^{-34} J·s 1.05×10^{-34} J·s
Boltzmann constant	k_B	1.38×10^{-23} J/°K
Avogadro's number	N	6.02×10^{23} mol^{-1}
Dielectric constant of vacuum	ε_0	8.85×10^{-12} F/m
Magnetic permeability of vacuum	μ_0	1.257×10^{-6} N/A^2 (H/n)
Ionization energy of hydrogen atom	R_H	13.6 eV
Bohr radius	$\hbar^2 4\pi\varepsilon_0/me^2 = r_B$	0.53 Å
Bohr magneton	μ_B $\mu_0\mu_B$	0.93×10^{-23} A·m^2 1.165×10^{-29} (MKSA)
Ideal gas constant	$R = N \cdot k_B$	8.314 J/° K·mole
Speed of light	c	$\sim 3 \times 10^8$ m·s^{-1}

Table II Some Useful Numerical Values

- $(4\pi\varepsilon_0)^{-1} = 9 \times 10^9$ N·m^2/C^2 (F/m)
- 1 eV : Temperature: ≈ 12000 °K ; Frequency: ≈ 2.4×10^{14} Hz; Molar energy: ≈ 23 Kcal/mole; Wavenumber: ≈ 8000 cm^{-1}; Wavelength: ≈ 12400 Å
- $h/k_B = 4.8 \times 10^{-1}1$ sec·°K
- $e^2/h = 3.874 \times 10^{-5}\ \Omega^{-1}$
- $h/e^2 = 25812\ \Omega$
- $hc/e\lambda = 12400$ eV for $g = 1$ Å
- $h/(2\ meV)^{1/2} = 12.26$ Å for $V = 1$ volt
- $k_BT = 1/40$ eV = 25 meV for T (ambient) = 290 °K
- $\hbar^2/2m = 3.8$ eV·Å^2
- $h^3(2\pi mk_B)^{-3/2} = 4.2 \times 10^{-22}$m^3 (deg)$^{3/2}$
- $e/4\pi\varepsilon_0 r_0 = 14.4$ volts for $r_0 = 1$ Å

Table III Periodic Table of Elements

I	II												III	IV	V	VI	VII	VIII
hydrogen 1 **H** 1.0079																		helium 2 **He** 4.0026
lithium 3 **Li** 6.941	beryllium 4 **Be** 9.0122												boron 5 **B** 10.811	carbon 6 **C** 12.011	nitrogen 7 **N** 14.007	oxygen 8 **O** 15.999	fluorine 9 **F** 18.998	neon 10 **Ne** 20.180
sodium 11 **Na** 22.990	magnesium 12 **Mg** 24.305												aluminium 13 **Al** 26.982	silicon 14 **Si** 28.086	phosphorus 15 **P** 30.974	sulfur 16 **S** 32.065	chlorine 17 **Cl** 35.453	argon 18 **Ar** 39.948
potassium 19 **K** 39.098	calcium 20 **Ca** 40.078		scandium 21 **Sc** 44.956	titanium 22 **Ti** 47.867	vanadium 23 **V** 50.942	chromium 24 **Cr** 51.996	manganese 25 **Mn** 54.938	iron 26 **Fe** 55.845	cobalt 27 **Co** 58.933	nickel 28 **Ni** 58.693	copper 29 **Cu** 63.546	zinc 30 **Zn** 65.39	gallium 31 **Ga** 69.723	germanium 32 **Ge** 72.61	arsenic 33 **As** 74.922	selenium 34 **Se** 78.96	bromine 35 **Br** 79.904	krypton 36 **Kr** 83.80
rubidium 37 **Rb** 85.468	strontium 38 **Sr** 87.62		yttrium 39 **Y** 88.906	zirconium 40 **Zr** 91.224	niobium 41 **Nb** 92.906	molybdenum 42 **Mo** 95.94	technetium 43 **Tc** [98]	ruthenium 44 **Ru** 101.07	rhodium 45 **Rh** 102.91	palladium 46 **Pd** 106.42	silver 47 **Ag** 107.87	cadmium 48 **Cd** 112.41	indium 49 **In** 114.82	tin 50 **Sn** 118.71	antimony 51 **Sb** 121.76	tellurium 52 **Te** 127.60	iodine 53 **I** 126.90	xenon 54 **Xe** 131.29
caesium 55 **Cs** 132.91	barium 56 **Ba** 137.33	57-70 *	lutetium 71 **Lu** 174.97	hafnium 72 **Hf** 178.49	tantalum 73 **Ta** 180.95	tungsten 74 **W** 183.84	rhenium 75 **Re** 186.21	osmium 76 **Os** 190.23	iridium 77 **Ir** 192.22	platinum 78 **Pt** 195.08	gold 79 **Au** 196.97	mercury 80 **Hg** 200.59	thallium 81 **Tl** 204.38	lead 82 **Pb** 207.2	bismuth 83 **Bi** 208.98	polonium 84 **Po** [209]	astatine 85 **At** [210]	radon 86 **Rn** [222]
francium 87 **Fr** [223]	radium 88 **Ra** [226]	89-102 **	lawrencium 103 **Lr** [262]	rutherfordium 104 **Rf** [261]	dubnium 105 **Db** [262]	seaborgium 106 **Sg** [266]	bohrium 107 **Bh** [264]	hassium 108 **Hs** [269]	meitnerium 109 **Mt** [268]	ununnilium 110 **Uun** [271]	unununium 111 **Uuu** [272]	ununbium 112 **Uub** [277]		ununquadium 114 **Uuq** [289]				

*Lanthanide series	lanthanum 57 **La** 138.91	cerium 58 **Ce** 140.12	praseodymium 59 **Pr** 140.91	neodmium 60 **Nd** 144.24	promethium 61 **Pm** [145]	samarium 62 **Sm** 150.36	europium 63 **Eu** 151.96	gadolinium 64 **Gd** 157.25	terbium 65 **Tb** 158.93	dysprosium 66 **Dy** 162.50	holmium 67 **Ho** 164.93	erbium 68 **Er** 167.26	thulium 69 **Tm** 168.93	ytterbium 70 **Yb** 173.04
**Actinide series	actinium 89 **Ac** [227]	thorium 90 **Th** 232.04	protactinium 91 **Pa** 231.04	uranium 92 **U** 238.03	neptunium 93 **Np** [237]	plutonium 94 **Pu** [244]	americium 95 **Am** [243]	curium 96 **Cm** [247]	berkelium 97 **Bk** [247]	californium 98 **Cf** [251]	einsteinium 99 **Es** [252]	fermium 100 **Fm** [257]	mendelevium 101 **Md** [258]	nobelium 102 **No** [259]

Chapter I

Crystal Structure and Crystal Diffraction

Course Summary

A. Crystal Structure

1. *Definitions*

Crystal: An ideal crystal consists of a regular repetition in space of atoms or groups of atoms.

Lattice: The group of points from which the observed atomic environment is the same as at the origin. This identity concerns both the chemical nature of the atoms and their orientation.

Basis: The atoms or group of atoms which constitute the structural unit applied to each points of the lattice.

Primitive vectors of the lattice: To go from one point of the crystal to an equivalent point, in which the arrangement and orientation appear exactly the same, it is sufficient to effectuate a translation of the form $\mathbf{T} = m\vec{a} + n\vec{b} + p\vec{c}$ in which $\vec{a}, \vec{b}, \vec{c}$ are the primitive vectors of the matrix and m, n, p are integers (positive, zero, or negative).

Crystal structure = Lattice + Basis

Understanding Solid State Physics: Problems and Solutions
Jacques Cazaux

ISBN 978-981-4267-89-2 (Hardcover), 978-981-4267-90-8 (eBook)
www.panstanford.com

2. *Simple and Multiple Lattices*

The choice of vectors $\vec{a}, \vec{b}, \vec{c}$ is not unique. The vectors $\vec{a}, \vec{b}, \vec{c}$ are said to be primitive if the translations allow a complete description of all the points of the crystal lattice. The parallelepiped $\vec{a}(\vec{b} \times \vec{c})$ corresponds to a simple (or primitive unit) cell of the lattice.

A multiple cell is constructed from the non-fundamental vectors $\vec{a}', \vec{b}', \vec{c}'$, where the parallelepiped contains n equivalent points ($n > 1$): the corresponding non-primitive cells, therefore, consists of n primitive cells and the order of the lattice cell will be n, where n = 2 or $n = 4$.

3. *Lattice Rows and Miller Indices*

To describe the crystal structure it is sufficient to state the choice of lattice vectors, and the nature and position of atoms that make up the basis. These positions are expressed with the help of the vectors $\vec{a}, \vec{b}, \vec{c}$, considered as the unit vectors: $\vec{r}_j = u_j\vec{a} + v_j\vec{b} + w_j\vec{c}$.

The lattice points are arranged along various rows and planes. When the rows and the planes are parallel and equidistant to each other they are equivalent and they are represented with the same symbols.

A series of parallel rows is represented by (m, n, p) where $\vec{\rho} = m\vec{a} + n\vec{b} + p\vec{c}$ when the row is parallel to the line that connects the origin to the lattice point m, n, p.

A series of parallel planes can be represented all by Miller indices (h, k, l), which describe the equation of the form $hx + ky + lz = 1$ of plane nearest the origin and using a, b, c units (see Ex. 8). This definition means that the intersections of the plane (h, k, l) with axes x, y, and z are $1/h$, $1/k$, $1/l$, respectively. The atoms chosen to define a plane must not be collinear.

4. *Point Symmetry*

In nearly all crystals one or several directions are equivalent. This orientation or point symmetry of a crystal can be represented by the symmetry of the figure formed by the group of half-lines which, emanating from the same point 0, are parallel to the directions from which all the properties of the crystal are identical. The point symmetries that are encountered include the rotations of order n

around an axis (the angle of rotation is $2\pi/n$ with n = 1, 2, 3, 4, 6) and the rotation-inversions, written as $\bar{n}$ ($\bar{1} \equiv$ inversion with respect to 0, $\bar{2} \equiv$ m: mirror symmetry, $\bar{3}, \bar{4}, \bar{6}$). Several symmetry elements can be associated around a point but the number of distinct combinations and possibilities is limited to 32. That is, there are 32 symmetry point groups which result in a classification of 32 crystal classes.

5. *The 7 Crystallographic Systems and the 14 Bravais Lattices*

Limited to the symmetry of the lattice (and not that of the crystal = lattice + basis), there are only seven possibilities, corresponding to the seven crystallographic systems shown in Table 1.

Table 1 The 7 crystallographic systems and the 14 Bravais lattices

System	Number of lattices	Nature of axis and angles	Lengths and angles to be defined	Symmetry
Triclinic	1 P	$a \neq b \neq c$ $\alpha \neq \beta \neq \gamma$	a, b, c α, β, γ	$\bar{1}$
Monoclinic	2 P,C	$a \neq b \neq c$ $\alpha = \gamma = 90° \neq \beta$	a, b, c β	2/m
Orthorhombic	4 P,C,I,F	$a \neq b \neq c$ $\alpha = \beta = \gamma = 90°$	a, b, c	mmm
Tetragonal	2 P,I	$a = b \neq c$ $\alpha = \beta = \gamma = 90°$	a, c	4/mmm
Cubic	3 P,I,F	$a = b = c$ $\alpha = \beta = \gamma = 90°$	a	m3m
Trigonal (Rhombohedral)	1 P (R)	$a = b = c$ $\alpha = \beta = \gamma \neq 90°$ $\neq 120°$	a α	$\bar{3}$m
Hexagonal	1 P	$a = b \neq c$ $\alpha = \beta = 90°$; $\gamma = 120°$	a, c	6/mm

(P = primitive, C = base-centered, I = body-centered, F = face-centered, R = rhombohedral)

Besides the primitive cells, P, if one includes in this classification certain multiple lattices because they render the symmetry of the lattice more clearly, 14 possible Bravais lattices are obtained from the 7 crystallographic systems in 3D: they are labeled C, I, F. (In 2D there are five Bravais lattices, see Ex. 17.)

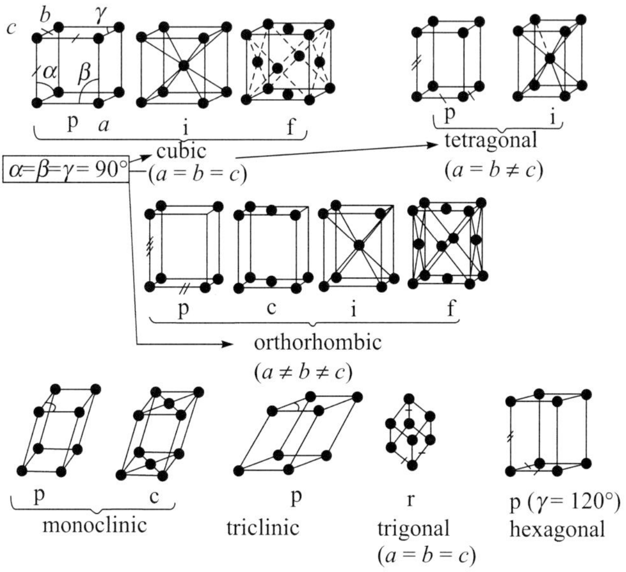

Figure 1 The 14 Bravais lattices.

6. *Space Symmetry*

If one sought to study nature at the atomic scale and enumerated all the possible symmetry operations, keeping unchanged the positions of atoms (position symmetry), the number of possibilities is 230, making up what are known as the space groups.

These symmetry operations involve placing the point symmetries into the Bravais lattices of the system.

B. Diffraction and the Reciprocal Lattice

1. *Bragg's Law*

One can use the interferences of waves scattered by atoms of a crystal to deduce its underlying crystal structure. The Bragg's law,

$2d_{h,k,l}\sin\theta = \lambda$, is satisfied when the wavelength of the radiation, λ is smaller than the interatomic distance (say <3 Å).

For such a purpose, three types of radiation are used in general:

(a) Photons or electromagnetic radiations with very short wavelength X-rays so that, 3 Å > λ > 0.1 Å and $\lambda = hc/\mathrm{E}$, where λ(Å) = 12.4/E (keV) (see Exs. 15 and 20; Pbs. 1, 5, and 6).

(b) Fast or slow electrons (Pbs. 3 and 4):

$$\lambda = h/(mv) = h\Bigg/\left[2m_0\mathrm{eV}_0\left(1+\frac{\mathrm{eV}_0}{2m_0c^2}\right)\right]^{1/2}$$

giving in the non-relativistic case, $\lambda(\text{Å}) = 12.26/\sqrt{\mathrm{V}}\,(\text{volts})$.

(c) Neutrons (Pbs. 7 and 11):

$\lambda = h/Mv$

where $\lambda(\text{Å}) = 0.286/\sqrt{\mathrm{E}}\,(\mathrm{eV})$

2. *X-Rays*

In laboratory diffraction experiments, X-ray radiation is obtained by bombarding a metallic anode by a beam of high-energy electrons, eV_0 (~10s of keV). The obtained radiation spectra (see Fig. 2) consist of a broad distribution of photon energies ranging from $h\nu$~0 to $h\nu$~eV_0 (continuous radiation or bremsstrahlung) superimposed with some very intense characteristic lines, which correspond to transitions between electronic shells in the target metal of atomic number, Z. Schematically, the energetic order of magnitude of the X-rays can be evaluated using the Bohr model applied to an atom, Z where, $h\nu_{n,n'} = R_{\mathrm{H}}Z^2\left(\frac{1}{n^2}-\frac{1}{n'^2}\right)$, leading to the Moseley law, $\sqrt{\nu} = \sqrt{cR_{\mathrm{H}}}\,(Z-\sigma)$, where n = 1, n' = ∞, σ is related to the screening and R_{H} = 13.6 eV. The nomenclature of these X-rays summarizes the atomic number Z of the metallic anode: the quantum number of the electron vacancy induced with the incident electrons (K: n = 1, L: n = 2, M: n = 3); the quantum number of the atomic electron filling this vacancy ($n' - n = 1$: α, $n' - n = 2$: β, etc.) and finally, in subscript the sub-shells concerned (taking into account the selection rules: $l' = l \pm 1$). For instance, if the target metal is Mo, the most intense characteristic line is noted as Mo Kα_1.

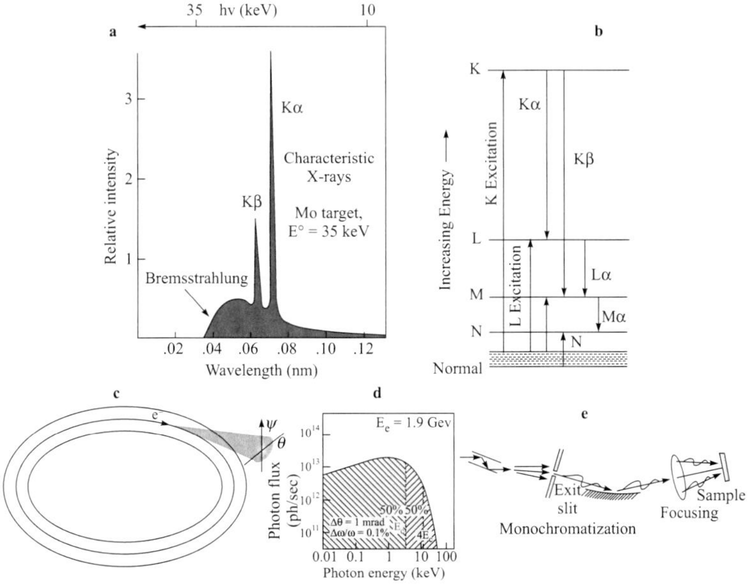

Figure 2 **X-rays sources. (Top) Conventional**: (a) Example of X-ray spectrum issued from a conventional X-ray tube. (b) Electronic transitions involved in the production of characteristic X-rays. **(Bottom) Synchrotron radiation**: (c) Source. (d) Spectral distribution of X-rays. (e) Monochromatization and focussing on the sample.

X-rays are also emitted when ultrafast charged particles experience radial accelerations. Such a radiation, called synchrotron radiation, is produced in synchrotrons using bending magnets, undulators, and/or wigglers. This radiation is tunable, linear polarized, left and right, in the soft X-ray range. In solid state physics, the main applications concern not only X-ray diffraction but also photoelectron spectroscopy and X-ray absorption spectroscopy (Chapter IV).

3. *Reciprocal Lattice (Exs. 12, 13, and 19)*

If $\vec{a}, \vec{b}$, and $\vec{c}$ are the vectors of the primitive (or fundamental) crystal lattice, then the reciprocal lattice of the system is constructed from the vectors $\vec{A}, \vec{B}$, and $\vec{C}$ such that:

$$\vec{A} = \sigma^2 \frac{\vec{b} \wedge \vec{c}}{\vec{a}\left(\vec{b} \wedge \vec{c}\right)},\ \vec{B} = \sigma^2 \frac{\vec{c} \wedge \vec{a}}{\vec{b}\left(\vec{c} \wedge \vec{a}\right)},\ \vec{C} = \sigma^2 \frac{\vec{a} \wedge \vec{b}}{\vec{c}\left(\vec{a} \wedge \vec{b}\right)}$$

Then,

$$\vec{A}\cdot\vec{b}=\vec{A}\cdot\vec{c}=\vec{B}\cdot\vec{a}=\vec{B}\cdot\vec{c}=\vec{C}\cdot\vec{a}=\vec{C}\cdot\vec{b}=0$$

$$\vec{A}\cdot\vec{a}=\vec{B}\cdot\vec{b}=\vec{C}\cdot\vec{c}=\sigma^2$$

In general, σ^2 is defined as either $\sigma^2 = 1$ or $\sigma^2 = 2\pi$. In the following we adopt the latter choice, which is the metric used for representing reciprocal and vector pace $(2\pi/\lambda)$, which is convenient for the Ewald construction.

Property Exs. 10a and b

The vector $\vec{G}$ $(h, k, l) = h\vec{A} + k\vec{B} + l\vec{C}$ of the reciprocal space is perpendicular to the planes of the same indices (h, k, l) of the direct lattice such that $\left|\vec{G}(h,k,l)\right|\cdot d_{h,k,l}=2\pi$, where $d_{h,k,l}$ is the distance between two adjacent (h, k, l) planes.

4. *More Detailed Analysis of Diffraction*

- *Phase difference*: The phase difference between two scattered waves, O and M is:

  $$\delta=\vec{r}(\vec{s}-\vec{s}_0),$$

  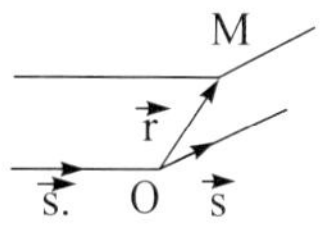

  $$\varphi=2\pi\frac{\delta}{\lambda}=(\vec{k}-\vec{k}_0)\vec{r}=\Delta\vec{k}\cdot\vec{r}$$

  where $\vec{k}_0=\dfrac{2\pi}{\lambda}\vec{s}_0$

  For a simple lattice of identical atoms, the constructive interference conditions are reduced to:

  $\varphi=\Delta\vec{k}\cdot\vec{r}=N\cdot 2\pi=2\pi(\alpha m+\beta n+\gamma p)$, where $\Delta\vec{k}=\vec{G}$.

  One can then deduce the Ewald construction (see Exs. 18 and 19). The diffraction condition can also be written as $\vec{k}_0\cdot\vec{G}/2=(G/2)^2$, which imposes that the incident wave vector, k_0, extend from the origin, Γ, to a point of any plane of symmetry between the origin and any point (h, k, l) of the reciprocal lattice space. The smallest possible volume surrounding the origin and limited by such symmetry planes is known as the first Brillouin zone (BZ; Ex. 14).

- *Structure factor:* When several atoms make up the basis of a crystal, the scattered wave amplitudes of each atom f_j will contribute a *structure factor* to the total amplitude:

$$A = \sum_{\rho}\sum_{j} f_j \exp - i\Delta\vec{k}\cdot(\vec{\rho}+\vec{r}_j)$$

where $\vec{\rho} = m\vec{a} + n\vec{b} + p\vec{c}$ and $\vec{r}_j = u_j\vec{a} + v_j\vec{b} + w_j\vec{c}$.

Furthermore, when the Bragg condition is met so that $\Delta\vec{k} = \vec{G}$, the contribution, from all the atoms of the basis is:

$$F(h,k,l) = \sum_{j} f_j \exp - i\vec{G}\cdot\vec{r}_j = \sum_{j} f_j \exp - i2\pi(hu_j + kv_j + lw_j)$$

For applications we can refer, for example, the distinction between ordered alloys and disordered alloys in Pb. 5.

- *Atomic form factor:* X-rays striking an atom "see" the electrons surrounding the nucleus. These atoms are not points compared to λ and the corresponding atomic form factor, f, includes the electronic distribution into the atom, $\rho(\mathrm{r})$, and the phase difference between the origin and any point of the electron cloud:

$$f \propto \int_{\text{atom}} \rho(\vec{r})\exp - i(\Delta\vec{k}\cdot\vec{r}) \quad \text{(see Ex. 22)}$$

Exercises 1 to 9 are meant to familiarize the reader with the crystal lattices in 1-, 2-, and 3D. Similarly, Exs. 10 to 18 are meant to introduce the basic notions of the reciprocal lattice and the BZ. The knowledge of these crystallographic elements is indispensable for what follows in Chapters III to V, because this book concerns essentially the crystal and the behavior of the atoms and electrons from which it is constructed. Thus, the effects of diffraction concern both external electrons in the crystal and also the internal valence and conduction electrons that lead to the theory of bands in Chapter V. In this Chapter, external radiation is used to characterize periodic structures such as crystal surfaces and their growth (with a preference for incident electrons, Pbs. 3 and 4) and magnetic structures (preferably with neutrons, Pb. 11). The more general method for investigating crystal structures in 3D consists of using X-rays to characterize alloys, for instance (Pbs. 5 and 6) and superlattices (Pb. 10) and finally, structures such as those found in Ex. 20.

Exercises

Exercise 1: Description of some crystal structures

Describe the crystal structures represented in Fig. 3. For each case state the Bravais lattice, the position of the atoms making up the basis (Hint: The crystals in "a" to "f" belong to the cubic system). In the case of multiple lattices, it is best to describe the atoms and the atom fractions in each lattice, in order to deduce the chemical formula attributed to the basis. In the case of multiple lattices, state the primitive lattice (of order 1).

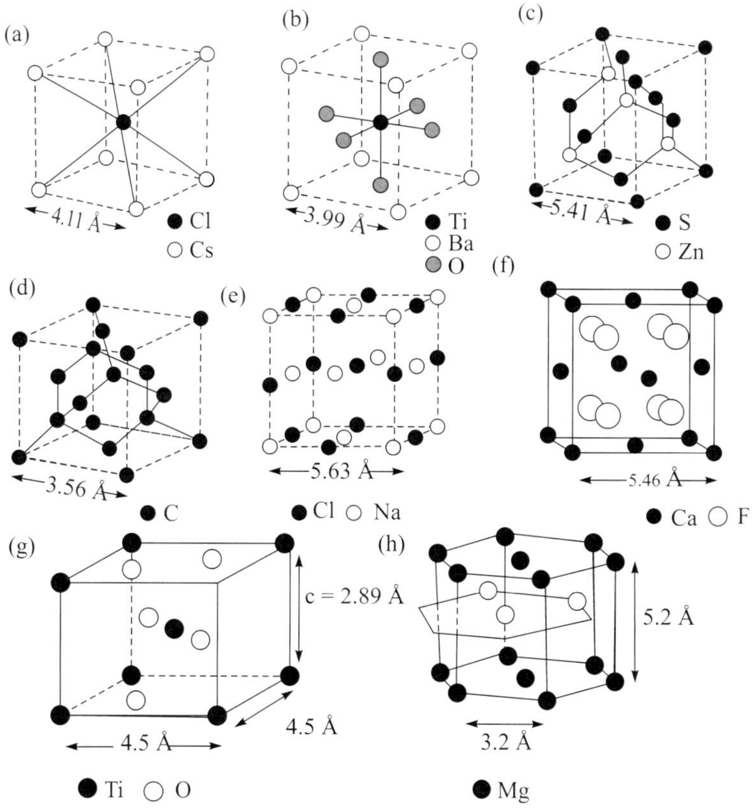

Figure 3

Solution:

One must first characterize the crystalline lattice by searching for the places of the points where one observes the same chemical environment (1°) as at the origin with the same orientation (2°) (For definition of the lattice points, see Course Summary).

If one atom of species A is taken as the origin, then condition 1 excludes all other atoms of species B ≠ A. Among atoms of species A, the second condition excludes certain others: (Fig. 3d) for instance, the atoms at the same positions as the white atoms in Fig. 3c, would be excluded; (Fig. 3g) the central atom would be excluded; (Fig. 3h) the atoms located in the intermediate plane are excluded.

The lattice points thus determined, one has to connect two identical points by translations of type: $\vec{T} = m\vec{a} + n\vec{b} + p\vec{c}$ (where, m, n, and p are integers). The vectors $\vec{a}, \vec{b}$, and $\vec{c}$ form the basis vectors of the primitive lattice cell if such a translation allows an identical description of all the lattice points.

For the crystals represented in Figs. 3a and 3b, there are three orthogonal vectors of length "a" which extend to the edges of the cube.

For Fig. 3g, the lattice is tetragonal.

For Fig. 3h, the structure contains in fact three juxtaposed lattices of order 1. The lattice vectors in the base plane are represented by bold lines in Fig. 4 (where there is a projection on the plane of atoms situated in the intermediate plane).

The lattices shown in the crystals c, d, e, and f are multiple of order 4 (face-centered cubic). The vectors of the fundamental lattice (order 1) are represented in Fig. 5. It is clear that such a lattice choice (and the corresponding vectors) masks the obvious symmetries of the lattice and that it is better to choose multiple lattices which do represent these symmetries (e.g., Bravais lattices).

In order to characterize the basis, one should first determine the number and the nature of the constituent atoms. It is sufficient to evaluate the total number of atoms (and atom fractions) of each species situated in the interior of the lattice of order n and to divide this number by n. One then obtains the position of the basis atoms and uses the basis vectors of the lattice of order n as the metric. Applying these principles to the lattice structures, we find the following results:

(a) Bravais lattice: cubic simple; a = 4.11 Å, Basis: 1 Cs at (0,0,0) and 1 Cl at (½,½,½)

(b) Bravais lattice: cubic simple; a = 3.99 Å, Basis: 1 Ba at (0,0,0) and 1 Ti at (½,½,½); 3 O at (½,½,0), (½,0,½), and (0,½,½). The structure is $BaTiO_3$, known as a perovskite.

(c) Bravais lattice: face-centered cubic; a = 5.41 Å, Basis: 1 S at (0,0,0) and 1 Zn at (¼,¼,¼). The structure is known as zinc blende.

(d) Bravais lattice: face-centered cubic; a = 3.56 Å, Basis: 1 C at (0,0,0) and 1 C at (¼,¼,¼). The structure is known as diamond structure. Also it is that of Si, Ge, and of many other binary semiconductors.

(e) Bravais lattice: face-centered cubic; a = 5.63 Å, Basis: 1 Na at (0,0,0) and 1 Cl at (½,0,0). The structure is known as NaCl.

(f) Bravais lattice: face-centered cubic; a = 5.46 Å, Basis: 1 Ca at (0,0,0) and 2 F Zn at (¼,¼,¼) and (¾,¼,¼ (or at (−¼,¼,¼). The structure is known as fluorine, CaF_2.

(g) Bravais lattice: tetragonal; a = 4.5 Å, c = 2.89 Å; Basis: 2 Ti at (0,0,0) and (½,½,½); 4 O at ≈ (⅓,⅓,0), (⅔,⅔,0), (⅔,⅓,½), and (⅓,⅔,½).
The structure is known as rutile (TiO_2).

(h) Bravais lattice: hexagonal; a = 3.2 Å, c = 5.2 Å; Basis: 2 Mg at (0,0,0) and (⅔,⅓,½). The structure is hexagonal close packed (hcp), which is not a Bravais lattice.

It can be verified that the above descriptions are sufficient to reconstruct the different structures. Exercise 3 shows the determination of the structures from their descriptions. Note that the meaning of the word “structure” is less precise than “Bravais lattice”.

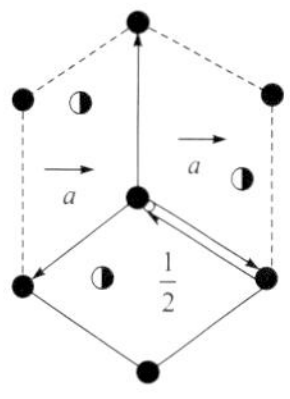

Figure 4

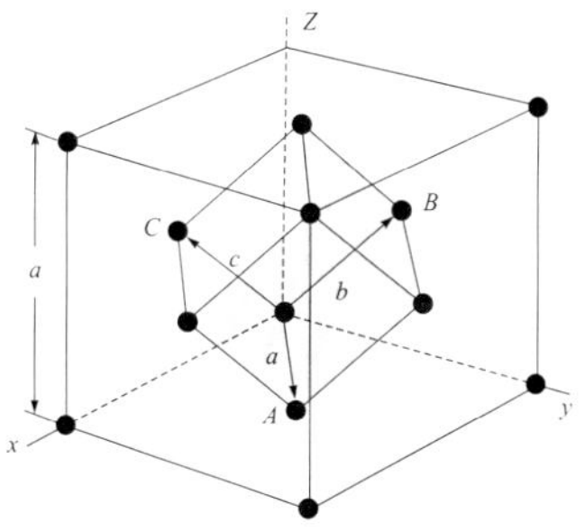

Figure 5

Exercise 2: Mass per unit volume of crystals

With the help of Table III (periodic table at the beginning of the book), find the mass per unit volume (V.M.) of the crystals shown in Fig. 3.

Solution:

It is easy to see that V.M. = $\sum_i p_i A_i / NV$, where V is the volume of the lattice containing the equivalent of p atoms of species with atomic mass A and N is Avogardo's number. We thus find:

$$\text{V.M. (CsCl)} = (132.9 + 35.45)\ /N\, a^3 = 4\ \text{g·cm}^{-3}$$

$$\text{V.M. (Diamond)} = 8 \times 12\ /N\, a^3 = 3.53\ \text{g·cm}^{-3}$$

$$\text{V.M. (CaF}_2\text{)} = 4(40 + 2.19)\ /\ N\, a^3 = 3.18\ \text{g·cm}^{-3}$$

Exercise 3: Construction of various crystal structures

Construct the corresponding crystal structure from the following descriptions of the Bravais lattices and the bases:

(1) One-dimension
Bravais lattice: a line of length a. Basis: Two atoms of species A spaced by 0 and ¼.

(2) Two-dimensions

(a) Bravais lattice: simple rectangle with a = 4 Å, b = 3 Å; Basis: one atom of species A located at (0,0) and another atom of species B located at (½,½).

(b) The same parameters as in (a) but now using atoms A and B that are chemically the same (A = B). In addition to the crystal structure, state the basis vectors of the primitive lattice and specify the nature of the new Bravais lattice.

(c) Bravais lattice: hexagonal (a = b, γ = 120°). Basis: two atoms of C, one located at (0,0) and the other at (⅓,⅔).

(3) Three-dimensions
The Bravais lattice of (a) silicon, (b) GaAs, and (c) Mg_2Si are face-centered cubic. Their basis vectors are respectively located at:

(a) Two atoms of Si; one at (0,0,0) and the other at (¼,¼,¼)
(b) One atom of As located at (0,0,0) and one atom of Ga located at (¼,¼,¼)
(c) One atom of Si located at (0,0,0); two atoms of Mg located at (¼,¼,¼) and (¾,¼,¼)

For the sake of simplicity, limit the representation of the atomic projection on the base plane, specifying the relevant height.

Solution:

This is a complementary exercise to Ex. 1

(1) See Fig. 13, Ex. 15.
(2) (a) See Fig. 6a.

(b) See Fig. 6b. Each atom is situated on a lattice junction. The base vectors of the primitive lattice are given by: a' = b' = 2.5 Å; γ = 73°6. Translations of the type $m\vec{a}' + n\vec{b}'$, allow the crystal to be constructed but do not evidence the rectangular symmetry of the lattice. The Bravais lattice, better adapted to show this symmetry, is rectangular centered. There exist only five 2D Bravais lattices of order 1 (see Ex. 17).

(c) This is the graphite structure (see Fig. 16, Ex. 17).

(3) After substituting C atoms for Si, the crystal structure of Si is represented by Fig. 3d. After substituting the atoms of S by the atoms of As and the atoms of Zn by the atoms of Ga, the crystal structure of GaAs is represented in Fig. 3c (Ex. 1).

The crystal structure of Mg_2Si is analogous to that of CaF_2 with the atoms of Mg substituted for F and those of Ca substituted for Si (Fig. 3f).

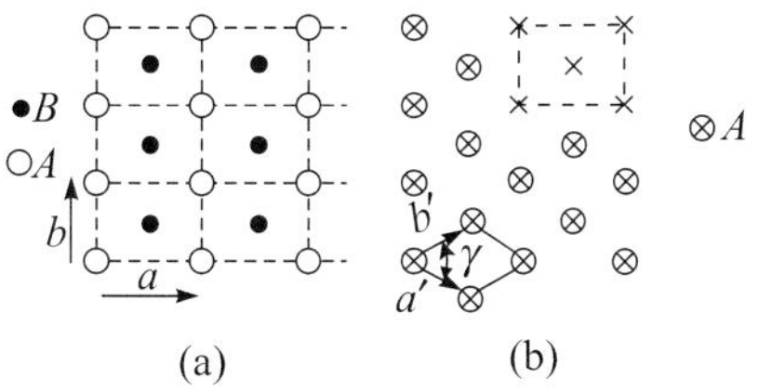

Figure 6

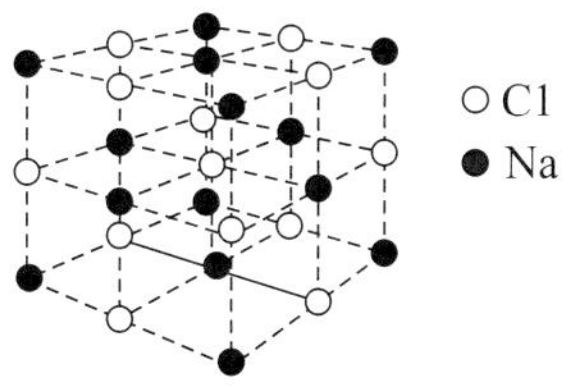

Figure 7

In order to avoid the difficulty of representing 3D structures, one can project the atoms on the base plane, while specifying their side. Figure 7 shows this representation for the case of silicon (or equivalently diamond).

Exercise 4: Lattice rows

Determine the lattice rows noted by the indices [m, n, p], from the two points in the lattice m_1, n_1, p_1 and m_2, n_2, p_2.
(a) 321 and $2\bar{4}0$; (b) 321 and $33\bar{1}$; (c) $\bar{1}21$ and 111; (d) $\bar{1}21$ and $\bar{2}12$.

Solution:

One must determine the translation between the origin and the line formed between the two points so that the lattice direction can be determined (see definition of lattice planes in Course Summary, Section 3). Note that the direction [m, n, p] is confused with the direction [$\bar{m}$, $\bar{n}$, $\bar{p}$].

We thus use: [m_2-m_1, n_2-n_1, p_2-p_1] and find:
(a) [1 $\bar{2}$ 1]; (b) [0 1 $\bar{2}$]; (c) [$\bar{2}$ 1 0]; and (d)[1 1 $\bar{1}$]

Exercise 5a: Lattice rows and reticular planes

Using the property of the reciprocal lattice, find the condition that permits the lattice direction [m, n, p] to be found in the lattice plane (h, k, l).

Is the lattice direction [$\bar{2}$ 1 0] contained in the plane (1 2 3)?

Solution:

$\vec{G}(h, k, l)$ is perpendicular to the plane (h, k, l). For the lattice row [m, n, p] to be contained in this plane it is sufficient for the scalar product: $\vec{G}(h,k,l)\cdot\left(m\vec{a}+n\vec{b}+p\vec{c}\right)=0$. We thus arrive at the condition, $hm + kn + lp = 0$.

The lattice direction [$\bar{2}$, 1, 0] is thus contained in the plane (1, 2, 3).

Exercise 5b: Lattice rows and reticular planes (continued)

Using the properties of the reciprocal lattice, determine the indices of the plane (h, k, l) which contains the lattice directions $[m_1, n_1, p_1]$ and $[m_2, n_2, p_2]$.

Do this for the lattice directions [1, 1, 1] and [3, 2, 1].

Solution:

The vector product of two vectors in the plane is a vector perpendicular to this plane. It is therefore parallel and proportional to $G(h, k, l)$.

$$h = \alpha(n_1p_2 - p_1n_2);\ k = \alpha(p_1m_2 - m_1p_2);\ l = \alpha(m_1n_2 - m_2n_1)$$

The coefficients α are chosen so that h, k, l are of lowest order and so that the planes (h, k, l) and $(\bar{h}, \bar{k}, \bar{l})$ are equivalent.

For the given numbers we find $(\bar{1}\,2\,\bar{1})$ or equivalently $(1\,\bar{2}\,1)$.

Exercise 6: Intersection of two reticular planes

Using the properties of the reciprocal lattice, determine the indices $[m, n, p]$ of lattice direction defined by the intersection of the planes (h_1, k_1, l_1) and (h_2, k_2, l_2).

Do this for the lattice planes (3 2 1) and (1 2 3).

Solution:

The vector product of $\vec{G}\,(h_1, k_1, l_1)$ and $\vec{G}\,(h_2, k_2, l_2)$ defines a vector that is parallel to both planes (h_1, k_1, l_1) and (h_2, k_2, l_2) and is thus, parallel to their intersection, which is the lattice direction $[m, n, p]$. The procedure to follow is thus analogous, to that in the previous exercise:

$$m = \alpha(k_1l_2 - k_2l_1);\ n = \alpha(l_1h_2 - l_2h_1);\ p = \alpha(h_1k_2 - h_2k_1)$$

For the given numbers we thus find $[1, \bar{2}, 1]$.

Exercise 7: Lattice points, rows, and planes

Determine the indices $[m, n, p]$ of the row that results from the intersection of two planes where one passes through the lattice points 3 2 1, $2\,\bar{4}\,0$, and $3\,3\,\bar{1}$ and the other through the lattice points $\bar{1}\,2\,1$, 1 1 1, $\bar{2}\,1\,2$.

Solution:

The solution is the synthesis of Exs. 4, 5b, and 6, using the appropriate numerical applications. The solution is thus $[1, \bar{2}, 1]$.

One observes that the technique used in the Exs. 4 to 7 is applicable to all the crystalline systems, including the triclinic one.

Exercise 8: Atomic planes and Miller indices: application to lithium

The Bravais lattice of lithium is simple cubic with lattice parameter a = 3.48 Å.

(a) Suppose that the atoms (assumed to be spheres) are placed along the [111], represent the atomic distribution along the following faces (100), (110), (111), and (201).

(b) For each such 2D structure, state the direction and basis vectors $\vec{a}$ and $\vec{b}$ of the elementary lattice as well as the value of the angle, γ.

(c) Find the atomic concentration and the mass density of lithium ($A \approx 7$).

Solution:

(a) By definition, the intersection of P, A, and R in the plane (h, k, l) with the base vectors $\vec{a}, \vec{b}, \vec{c}$ from the direct lattice are such that $\overrightarrow{OP} = \frac{\vec{a}}{h}$; $\overrightarrow{OQ} = \frac{\vec{b}}{k}$; $\overrightarrow{OR} = \frac{\vec{c}}{l}$ (see Fig. 11 and Ex. 10a). When one indice is zero; the intersection with the corresponding axis is at infinity. The planes in this exercise are shown on the top of Fig. 8.

The atoms are in contact along the direction [1,1,1], their radius is $r = \frac{a\sqrt{3}}{4}$. It is easy to find the atomic distribution, shown in the bottom of Fig. 6.

(b) The base vectors of the elementary lattice relative to the different sides are respectively:

$$(100): |\vec{a}| = |\vec{b}| = a; \hat{\gamma} = 90°; (110): |\vec{a}| = |\vec{b}| = a\sqrt{3}/2; \hat{\gamma} \approx 70°;$$

$$(111): |\vec{a}| = |\vec{b}| = a\sqrt{2}; \hat{\gamma} = 60° \text{ and } 120°;$$

$$(201): |\vec{a}| = a; |\vec{b}| = a\sqrt{5}; \hat{\gamma} = 90°.$$

We observe that the atomic density decreases when the indices h, k, l increase (see also Chapter IV, Ex. 15).

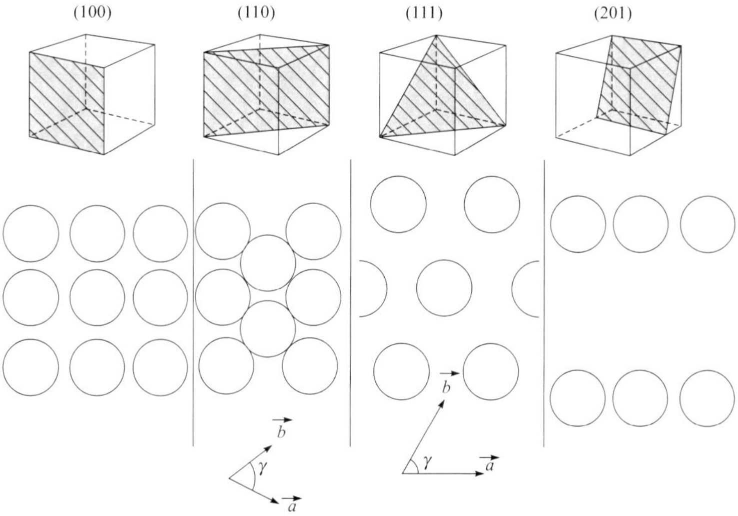

Figure 8

The packing rate (t = 0.68) is found in the following exercise (Ex. 9).

(c) There are two atoms (c.c) of lithium at each cube (within a volume a^3).
N (Li) $\approx 4.7 \times 10^{28}$ atoms/m^3.
The mass density is found using $\rho = \frac{NA}{N}$, where $\rho \approx 546$ kg/m^3.

Exercise 9: Packing

Assume that the elemental atoms are hard spheres of radius r. Calculate the maximum packing rate t obtained when this element crystallizes into the following structures:

(a) simple cubic (sc)
(b) body-centered cubic (bcc)
(c) face-centered cubic (fcc)
(d) diamond
(e) hexagonal close packed (hcp) (first calculate the optimal c/a relation)

Using these results, determine the value of the lattice parameter(s) of the following real crystal systems where d is distance between nearest neighbors:

(f) magnesium (hcp), $d = 3.20$ Å

(g) aluminum (fcc), $d = 2.86$ Å

(h) silicon (diamond), $d = 2.35$ Å

Solution:

We denote r as the radius of the hard spheres and s as the side of the cube.

(a) The cube occupies $8 \times \frac{1}{8} = 1$ sphere so that $s = 2r$ and $t = \pi/6 = 0.524$.

(b) The spheres touch along the diagonal of the cube, thus $\sqrt{3}s = 4r$ so that the cube contains two spheres. We thus have: $t = \sqrt{3}\pi/8 = 0.680$.

(c) The spheres touch along the diagonals of the cube faces, thus $\sqrt{2}s = 4r$ so that the cube contains $6 \times \frac{1}{2} + 8 \times \frac{1}{8}$ spheres, totaling four spheres per cube. We thus find: $t = \sqrt{2}\pi/6 = 0.740$.

(d) The distance between the two atoms (at 000 and ¼ ¼ ¼) forms the basis so that $\sqrt{3}s/4 = 2r$ and 8 atoms are contained in a volume of s^3 (order 4 with two atoms per basis), where $t = \sqrt{2}\pi/16 = 0.340$.

(e) In the elementary lattice (Fig. 9) the atoms at 000(O), 100(A), 110(B), and 2/3; 1/3; 1/2 (C) are placed at the summits of a tetrahedron with side a, which is regular when the relation c/a is optimal.

In this case $\mathrm{AH} = \frac{2}{3}\mathrm{AK} = \frac{2}{3}a\sin 60^\circ = \frac{s\sqrt{3}}{3}$; the height $\mathrm{CH} = \frac{s\sqrt{2}}{3}$. We thus have $h = \frac{c}{2}$, so that the optimal $\frac{c}{a}$ relation for the hexagonal close-packed structure is:

$$\frac{c}{a} = \sqrt{\frac{8}{3}} = 1.63.$$

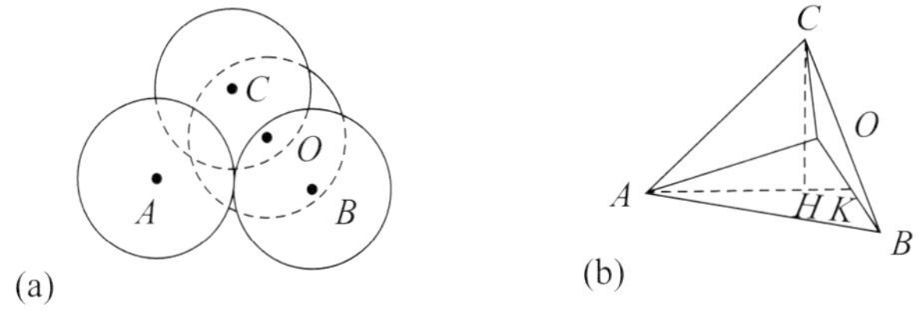

Figure 9

The lattice parameter in Fig. 3h contains $\frac{12}{6}+3+\frac{2}{2}=6$ atoms and $a = 2r$. The packing is thus:

$t = \pi/3\sqrt{2} = 0.74$.

We have thus verified that there are two ways to minimize the volume occupied by hard spheres: the face-centered cubic and the hexagonal close packed structures. As shown in Fig. 10, both result in the same packing ($t = 0.74$) because they represent, in fact, the same method of placing spheres in a third layer.

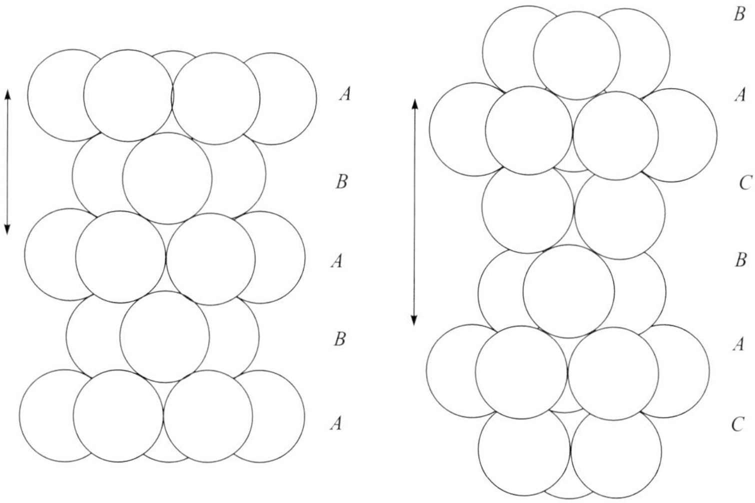

Figure 10 Three-dimensional filling of the hexagonal close packed (left) and face-centered cubic (right) structures.

(f) Mg: $d = 2r = a = 3.2$ Å, $c = 3.2\sqrt{8/3} = 5.23$ Å

(g) Al: $d = 2r = \sqrt{2}a/2$, thus $a = 4.04$ Å

(h) Si: $d = 2r = \sqrt{3}a/4$, thus $a = 5.43$ Å

Exercise 10a: Properties of the reciprocal lattice

(a) Show that all reciprocal lattice vectors of the form $\vec{G} = h\vec{A} + k\vec{B} + l\vec{C}$ are perpendicular to the planes of the same indices (h, k, l) in real space.

(b) Show that the distance $d_{h,k,l}$ between two consecutive planes (h, k, l) is inversely proportional to $\left|\vec{G}_{h,k,l}\right|$.

(c) Find $d_{h,k,l}$ for:

(i) A simple cubic lattice

(ii) A orthorhombic lattice ($a \neq b \neq c$, $\alpha = \beta = \gamma = \pi/2$)

Solution:

(a) By definition, the intersections of plane (h, k, l) with the basis vectors of the lattice (see Fig. 11) are:

$$P\left(\frac{a}{h},0,0\right),Q\left(0,\frac{b}{k},0\right),R(0,0,\frac{c}{l}).$$

The vector $\overrightarrow{PQ}$ containing the plane (h, k, l) obeys the relation

$\overrightarrow{PQ} = \frac{\vec{b}}{k} - \frac{\vec{a}}{h}$ and the product $\vec{G}_{h,k,l} \cdot \overrightarrow{PQ} = \left(h\vec{A} + k\vec{B} + l\vec{C}\right)\left(\frac{\vec{b}}{k} - \frac{\vec{a}}{h}\right)$

$= 0$ because by the definition of $\vec{G}_{h,k,l}$:

$$\vec{A}\cdot\vec{a} = \vec{B}\cdot\vec{b} = \vec{C}\cdot\vec{c} = 2\pi \text{ and } \vec{A}\cdot\vec{b} = \vec{A}\cdot\vec{c} = \vec{B}\cdot\vec{a} = 0.$$

$\vec{G}_{h,k,l}$ is therefore perpendicular to $\overrightarrow{PQ}$ and we may also show similarly that $\vec{G}_{h,k,l}$ is perpendicular to $\overrightarrow{PR}$: it is thus perpendicular to the plane (h, k, l).

(b) The distance $d_{h,k,l}$ is represented by the length OH in Fig. 11. $\overline{OH}$ is perpendicular to the plane (h, k, l) and is therefore parallel to $\vec{G}_{h,k,l}$. Consider $\overrightarrow{OH}$ as the projection of $\overrightarrow{OP}$ on $\vec{G}_{h,k,l}$. We obtain

$$\left|\vec{G}_{h,k,l}\right| \cdot d_{h,k,l} = \vec{G}_{h,k,l} \cdot \overrightarrow{OP} = \left(h\vec{A} + k\vec{B} + l\vec{C}\right)\left(\frac{\vec{a}}{h}\right) = 2\pi .$$

Thus, $d_{h,k,l} = \frac{2\pi}{\left|\vec{G}_{h,k,l}\right|}$.

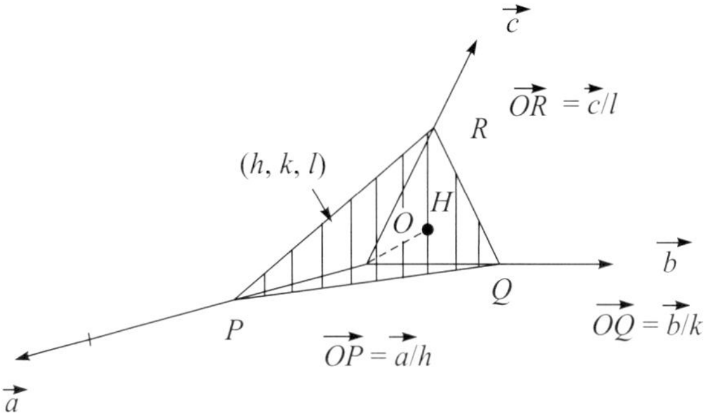

Figure 11

(c) From the definition of a reciprocal lattice

$$\left(\vec{A} = 2\pi\frac{\vec{b}\Lambda\vec{c}}{(\vec{a},\vec{b},\vec{c})},\ \vec{B} = \frac{2\pi(\vec{c}\Lambda\vec{a})}{(\vec{a},\vec{b},\vec{c})},\ \vec{C} = \frac{2\pi(\vec{a}\Lambda\vec{b})}{(\vec{a},\vec{b},\vec{c})}\right)$$

in the rectangular coordinate system, we obtain the following relations:

$$\vec{A} = \frac{2\pi\vec{a}}{a^2},\ \vec{B} = \frac{2\pi\vec{b}}{b^2},\ \vec{C} = \frac{2\pi\vec{c}}{c^2}$$

from which we obtain

$$G^2_{h,k,l} = (h\vec{A} + k\vec{B} + l\vec{C})^2 = 4\pi^2\left(\frac{h^2}{a^2} + \frac{k^2}{b^2} + \frac{l^2}{c^2}\right)$$

If the lattice is orthorhombic

$$d_{h,k,l} = \frac{2\pi}{|G_{h,k,l}|} = \frac{1}{\left(\frac{h^2}{a^2} + \frac{k^2}{b^2} + \frac{l^2}{c^2}\right)^{1/2}}$$

If the lattice is cubic ($a^2 = b^2 = c^2$)

$$(a^2 = b^2 = c^2) \rightarrow d_{h,k,l} = \frac{a}{(h^2 + k^2 + l^2)^{1/2}}$$

Exercise 10b: Distances between reticular planes

(1) Starting from (a) the definition and (b) a property of the reciprocal lattice, give an expression for the distance $d_{h,k,l}$ between inter-reticular planes (h, k, l) of a cubic array with edges a and next an orthorhombic array ($a \neq b \neq c$).

(2) In each of the two cases, state the eight first distances in decreasing order. For the cubic array consider only the non-equivalent planes. For the orthorhombic array use a = 3Å, b = 4 Å, and c = 5 Å.

Solution:

(1) The distance between two parallel planes with equation $Ax+By+Cz+D_1=0$ and $Ax+By+Cz+D_2=0$ is $\delta=\dfrac{|D_1-D_2|}{\sqrt{A^2+B^2+C^2}}$.

(a) The plane equation (h, k, l), the nearest to the origin in a orthorhombic matrix is (see Course Summary, Section 3):

$$\left(\frac{h}{a}\right)x+\left(\frac{k}{b}\right)y+\left(\frac{l}{c}\right)z-1=0\,.$$

Thus, $d(h,k,l)=1/[(h^2/a^2+k^2/b^2+l^2/c^2)]^{1/2}$ for the orthorhombic case; $d(h,k,l)=a/(h^2+k^2+l^2)^{1/2}$ for the cubic case.

(b) On can also obtain easily obtain these results using the relation $\vec{\mathrm{d}}(h,k,l)\cdot\vec{\mathrm{G}}(h,k,l)=2\pi$ where $\vec{\mathrm{G}}(h,k,l)=\left(h\vec{\mathrm{A}}+k\vec{\mathrm{B}}+l\vec{\mathrm{C}}\right)$ (see Ex. 10a for further details).

(2) For the cubic lattice, it is sufficient to classify the quantity $h^2+k^2+l^2$ by increasing order. We thus find: (100); (110); (111); (200); (210); (211); (220); (300) and (221).

For the orthorhombic lattice: (001); (010); (011); (100); (101); (002); (110); (111). This series was found from the increasing values of $|\vec{\mathrm{G}}|$.

Exercise 11: Angles between the reticular planes

In a cubic lattice, determine the angle Φ between the planes (h_1, k_1, l_1) and (h_2, k_2, l_2). Verify the result for (100) and (110) planes.

Solution:

The angle between the planes (h_1, k_1, l_1) and (h_2, k_2, l_2) is in reciprocal space the angle Φ between $G_1(h_1, k_1, l_1)$ and $G_2(h_2, k_2, l_2)$. (see Course Summary of Chapter II, Section 3).

It is therefore sufficient to determine the scalar product, $\vec{G}_1 \cdot \vec{G}_2 = G_1 G_2 \cos\Phi$.

Using $\vec{G}_1 = h_1\vec{A} + k_1\vec{B} + l_1\vec{C}$ and $\vec{G}_2 = h_2\vec{A} + k_2\vec{B} + l_2\vec{C}$, we obtain:

$$\cos\Phi = \frac{\left(h_1 h_2 + k_1 k_2 + l_1 l_2\right)}{\left(h_1^2 + k_1^2 + l_1^2\right)^{\frac{1}{2}} \left(h_2^2 + k_2^2 + l_2^2\right)^{\frac{1}{2}}}$$

Applied to planes (100) and (110), we find $\cos\Phi = \sqrt{2}/2$, or $\Phi = 45°$.

Exercise 12: Volume of reciprocal space

Consider successively the direct (cubic) simple cubic, body-centered cubic, and face-centered cubic lattices, each with lattice parameter a and their associated reciprocal lattices. In reciprocal space, find the corresponding volume of the primitive lattice (order 1).

Solution:

Simple cubic reciprocal lattice has a volume $V = (2\pi/a)^3$.

Body-centered cubic: In real space, the cubic lattice is of order 2. Its volume is therefore (see Course Summary) $v = a^3/2$. The reciprocal lattice of the bcc is fcc (order 4) with a cubic parameter $2\left(\frac{2\pi}{a}\right)$, which gives $V = \left(\frac{1}{4}\right)\left(\frac{4\pi}{a}\right)^3 = 2(2\pi/a)^3$

Face-centered cubic: In real space, the cubic lattice is of order 4: $v = a^3/4$. In reciprocal space, the lattice is of order 2 (bcc) with a cubic parameter $2\left(\frac{2\pi}{a}\right)$ so that the volume is $V = 4\left(\frac{2\pi}{a}\right)^3$.

Note: In these special cases, one can verify the general relation: $v \cdot V = (2\pi)^3$.

Exercise 13: Reciprocal lattice of a face-centered cubic structure

Construct the reciprocal lattice of a face-centered cubic structure using:

(a) The definitions

(b) The structure factor:

$$F(h,k,l) = \sum_j f_j \exp[-i2\pi(hu_j + kv_j + lw_j)]$$

Solution:

(a) Starting from the definitions, one must consider the primitive rhombohedral (or trigonal) lattice from which the basis vectors a′, b′ and c′ obey the following relations (see Ex. 1, Fig. 5):

$$\vec{a}' = (\vec{x}+\vec{y})(a/2);\ \vec{b}' = (\vec{y}+\vec{z})(a/2); \vec{c}' = (\vec{z}+\vec{x})(a/2).$$

The volume of this primitive lattice is such that $V = a^3/4$, because the fcc lattice is of order 4. The translation vectors of the reciprocal lattice are given by $(A = 2\pi b' \times c'/V)$:

$$\vec{A} = \frac{2\pi}{a}(\vec{x}+\vec{y}-\vec{z}),\ \vec{B} = \frac{2\pi}{a}(-\vec{x}+\vec{y}+\vec{z}),\ \vec{C} = \frac{2\pi}{a}(\vec{x}-\vec{y}+\vec{z})$$

These are the primitive lattice vectors of a bcc lattice.

(b) The diffraction conditions of a direct lattice impose the diffusion vector $\Delta\vec{k}$ from being equal to the reciprocal lattice vector $\vec{G}(h,k,l)$ and the presence of identical atoms in the interior of the lattice (by the intermediate of the structure facture) "erases" certain of these reflections.

To construct a reciprocal lattice for a given Bravais lattice, one must therefore consider the multiple reciprocal lattices chosen, erasing in the reciprocal space the points corresponding to a forbidden reflection (associated to the presence of additional points in real space). Thus, in the case of the fcc lattice $F(h, k, l)$ is:

$$F(h,k,l) = f\{1+\exp-i\pi(k+1)+\exp-i\pi(k+h)+\exp-i\pi(h+1)\}$$

$F(h, k, l)$ is zero when h, k, and l have different parity. Reflections of type 100 and 110 are forbidden and in reciprocal space erase the corresponding points, which are only present in a cubic-centered lattice with edge $4\pi/a$.

Remark: We are concerned with deleting the reflections related to the existence of points of the primitive lattice contained in the multiple lattice considered here and not with deleting all of the forbidden reflections. For instance, the reciprocal lattice of diamond remains bcc (direct lattice is fcc), even though the position of an atom at ¼, ¼, ¼ is chemically identical, but not equivalent, to that found at 0, 0, 0, forbids reflections at which $h + k + l = 4n + 2$ (such as the reflection 200; see Pb. 1).

Exercise 14: Reciprocal lattice of body-centered and face-centered cubic structures

Construct the BZs for these two lattices.

Solution:

- The reciprocal lattice of the bcc structure is the fcc lattice with a lattice edge of $2 \cdot 2\pi/a$ and points positioned at the center of the faces. This structure takes into account only the reflections of the simple cubic lattice such as the $h + k + l$ = an even number: 110, 200, etc. The perpendicular bisector (or symmetry) planes between the center $\Gamma(0\ 0\ 0)$ and the reciprocal points of the type 110 (represented by a black point on Fig. 12a) are sufficient to define the first BZ. These planes pass through the points of type N with coordinates of $\frac{\pi}{a}, \frac{\pi}{a}, 0$. The same bisector planes pass through points of the type H and P. Note that at points H there are also perpendicular bisector planes between Γ and points of the type 200. All these planes delimit the cube in Fig. 12a. The points P are half-distance between Γ and the points 111, and do not appear in the reciprocal lattice. We thus fine that the first BZ consists of a regular rhombic dodecahedron.

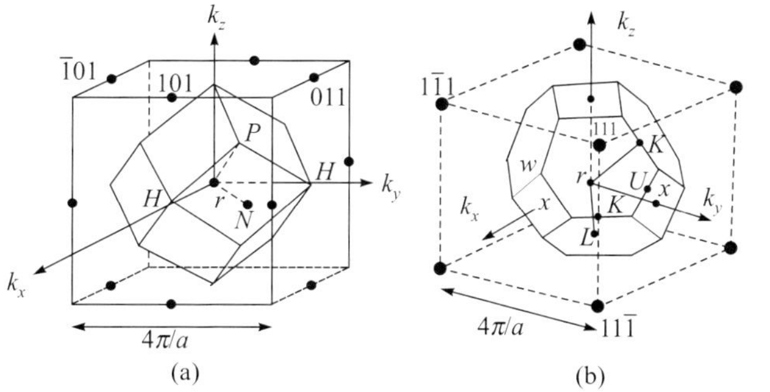

Figure 12

- The reciprocal lattice of the fcc structure is the bcc lattice with lattice edge of $2 \cdot 2\pi/a$, where the only allowed reflections are those where h, k, l have the same parity, such as 111, 200, etc. The points of type 111 allow the construction of the cube (indicated by dashed lines in Fig. 12b), which, with the Γ point in the center, constitutes a bcc lattice. The intersections of the eight symmetry planes between the origin and the points of type 111 construct eight regular hexagons centered on the points of type L because these median planes are truncated by the six median planes between Γ and the points of type 200, from which appear the squares centered at X. The first BZ is thus a truncated octahedron.

Exercise 15: X-ray diffraction by a row of identical atoms

Consider the linear chain of carbon atoms shown in Fig. 13. This structure could easily be a chain of hydrocarbons with alternating single and double bonds such as –C=C–C=C–C=C–C=.

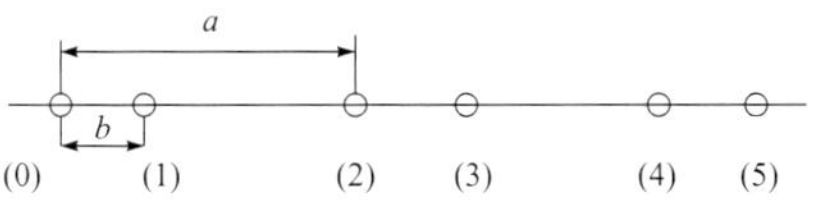

Figure 13

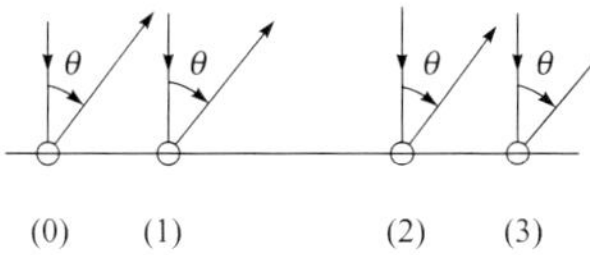

Figure 14

(a) What is the lattice vector? What is the basis? Specify using the conventional notation with $b = \frac{a}{4}$.

(b) Monochromatic X-rays of wavelength λ illuminate the chain.

(i) Evaluate the path difference between the waves diffused in the angle θ by the atom positioned at the origin (0) and the atom placed in position (2). Indicate the possible values of θ (or of one of its trigonometric function) observed by diffraction assuming that the chain consists only of atom pairs (crystal diffraction). Show that the addition of odd atoms accentuates the diffracted intensity in certain directions while diminishing it in others (always assuming that $b = \frac{a}{4}$). Find the result using the structure factor.

(ii) *Numerical application*: λ = 0.5 Å, a = 0.5 Å, determine the table of increasing values from 0, $0 \leq \theta \leq \frac{\pi}{2}$, for which the diffraction conditions are satisfied. State the values corresponding to the intensities I_T/I_f in which I_T is the diffracted intensity of the group of atoms of the chain and I_f is the diffracted intensity of the atoms only situated at the lattice points.

Solution:

(a) The lattice vector is $\vec{a}$ and the basis consists of two atoms of carbon and one atom of 0 such that $\frac{b}{a} = \frac{1}{4}$.

(b) *Crystal diffraction*:

The path difference δ_r between waves diffused by two consecutive points of the basis is such that, $\delta_r = a\sin\theta$; the diffraction condition of the corresponding matrix is thus $\delta_r = a\sin\theta = n\lambda$.

Note: Beware of simply writing the Bragg diffraction condition with the usual form $2d\sin\theta = n\lambda$ without considering the problem at hand.

The diffraction waves emitted by the crystal are thus of the form:

$$\theta = \text{Arcsin}\left(\frac{n\lambda}{a}\right)$$

Diffraction by a basis:

The step difference δ_m between waves emitted by the two different atoms in the same lattice cell is such that

$$\delta_m = b\sin\theta = \frac{a}{4}\sin\theta.$$

(i) The second interference system combines with the first such that when $\delta_m = n'\lambda\left(\sin\theta = \frac{4n'\lambda}{a}\right)$ the diffraction is enhanced (by a factor of 4) by the diffraction of the basis but the system of interferences destroys the values of θ corresponding to $\delta_m = \frac{(2n'+1)}{2}\lambda$. In other cases (where n is odd), the resultant intensity will be double because of the phase difference between the diffracted waves is $\pm\pi/2$.

One can obtain this result from the structure factor:

$$F(h,k,l) = \sum_j f_j \exp[-i2\pi(hx_j + ky_j + lz_j)],$$

which here reduces to (1D where, $x_j = 0.1/4$):

$$F(h) = f_0 + f_{1/4}\exp{-i}\left(\frac{\pi}{2}h\right)$$

With $f_0 = f_{1/4}$ (identical atoms)
For $h(n) = 4k$; $F(h) = 2f_0$, $I_T = 4I_f$
For $h(n) = (2k+1)\cdot 2$; $F(h) = 0$, $I_T = 0$
For $h(n) = $ odd; $F(h) = f_0(1 \pm i)$, $I_T = 2I_f$

(ii) *Numerical application*: $\lambda = 0.5$ Å, $\lambda/a = 0.1$

$n(h)$	1	2	3	4	5	6	7	8	9	10
$\sin\theta = \dfrac{n\lambda}{a}$	0.1	0.2	0.3	0.4	0.5	0.6	0.7	0.8	0.9	1
θ	5°.74	11°.53	17°.45	25°.50	30°	36°.86	44°.42	53°.13	64°.16	90°
I_T/I_r*	2	0	2	4	2	0	2	4	2	0

*In reality, the absolute values of the intensities I_T and I_F will vary with θ because they depend on the polarization and Lorentz factors and on the form factor (see Ex. 22). The evaluation of their ratio related to a given diffraction angle, however, is exact (see Chapter III, Ex. 1).

Exercise 16: X-ray diffraction by a row of atoms with a finite length

We continue with the same structure in Ex. 15, but now assume that the linear chain of atoms has a finite length with the basis repeated N times.

(a) Find the expression for the intensity due only to the lattice as a function of θ.

(b) Find the expression for the intensity due to the crystal structure (lattice + basis) assuming that $b = a/4$.

(c) Apply the result to the case when $\lambda = 0.5$ Å, $a = 5$ Å and $N = 10$. What is the angular width of the first reflection ($n = 1$)?

Solution:

(a) As previously, if we neglect polarization and Lorentz diffusion factors (see pp. 134 and 242 of Ref. [15a] and the atomic form factor (see. Ex. 22), we find the classic optics problem consisting of N waves of the same amplitude. Taking the phase origin to be the initial atom, we obtain the resultant amplitude:

$$A = \sum_{0}^{N-1} f \exp - i\frac{2\pi n\delta_r}{\lambda}, \text{ with } \delta_r = a\sin\theta.$$

This leads to the determination of the geometric progression $\exp(-i\pi\delta_r/\lambda)$ that is of the form $\exp(-i\phi)$ where $\phi = 2\pi a\sin\theta/\lambda$. The intensity is thus:

$$I_R = AA^x = ff^x \cdot \frac{\sin^2\frac{N\varphi}{2}}{\sin^2\frac{\varphi}{2}} = ff^x \frac{\sin^2\left(\pi Na\ \sin\frac{\theta}{\lambda}\right)}{\sin^2\left(\frac{\pi a\ \sin\theta}{\lambda}\right)}$$

The main maxima correspond to the values of θ which simultaneously cancel the numerator and the denominator:

$$\frac{\pi Na\ \sin\theta}{\lambda} = n'\pi \text{ and } \frac{\pi a\ \sin\theta}{\lambda} = n\pi$$

This last condition, which is more restrictive, is just the diffraction condition ($n\lambda = a\sin\theta$) and the associated intensities are given by $I_r = ff^x N^2$ (and are therefore proportional to the square of the number of scattering atoms).

However, the diffused intensities are strictly zero when $\sin\theta = n'\lambda/Na \neq n\lambda/a$. The characteristic of the variation of the diffused intensity by the lattice as a function of $\sin\theta$ is shown in Fig. 15a.

Note that the reflection order n is given mostly by the angular interval such that $\left(\frac{nN-1}{N}\right)\frac{\lambda}{a} < \sin\theta_n < \left(\frac{nN+1}{N}\right)\frac{\lambda}{a}$ and the associated angular width of diffraction is inversely proportional to the number N of irradiated bases.

(b) If we consider the waves diffused by the crystalline structure (lattice + basis), we obtain the following resultant amplitude:

$$A = f\exp - i\left(2\pi\cdot 0\frac{\delta_r}{\lambda}\right) + f\exp - i\left(2\pi\frac{\delta_m}{\lambda}\right) + f\exp - i\left(2\pi\cdot 1\frac{\delta_r}{\lambda}\right)$$
$$+ f\exp - i\left[\frac{2\pi(\delta_r+\delta_m)}{\lambda}\right] + f\exp - i\left[\frac{2\pi(N-1)\delta_r}{\lambda}\right]0$$
$$+ f\exp - i\left[\frac{2\pi(N-1)(\delta_r+\delta_m)}{\lambda}\right].$$

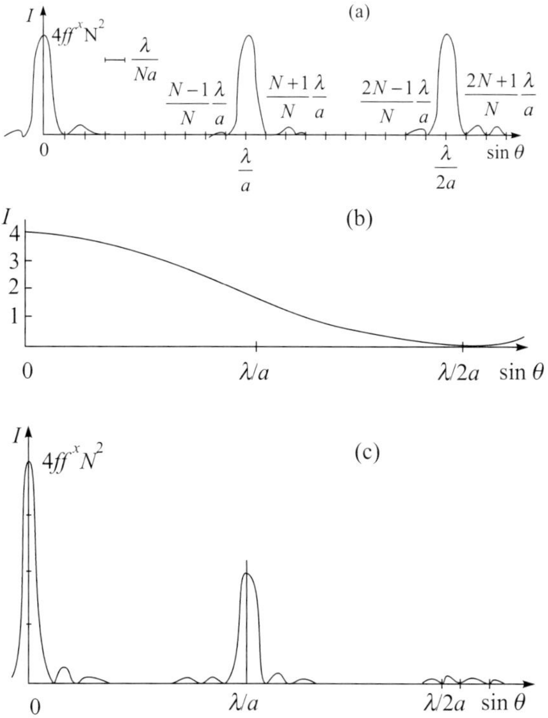

Figure 15 (a) Intensity diffracted by the lattice; (b) modulation induced by the basis, $4\cos\left(\dfrac{\pi a \sin\theta}{4\lambda}\right)^2$; (c) resulting intensity: lattice + basis.

This expression is of the form:

$$A = f\left[1+\exp\left(-i\frac{2\pi\delta_m}{\lambda}\right)\right]\sum_0^N \exp\left(-i\frac{2\pi n\delta_r}{\lambda}\right)$$

where the term between brackets is just the structure factor $[F(h) = f + f\exp - ih\pi/2]$ when the Bragg condition for the crystal is satisfied: $\delta_m = (a/4)\sin\theta$ and $a\sin\theta = h\lambda$.

The resultant intensity, I is:

$$I = AA^x = ff^x \cdot 4\cos^2\left(\frac{\pi a\sin\theta}{4\lambda}\right)\frac{\sin^2\left(\dfrac{\pi Na\sin\theta}{\lambda}\right)}{\sin^2\left(\dfrac{\pi a\sin\theta}{\lambda}\right)}$$

The intensity of the lattice (Fig. 15a) is modulated by the crystal structure via the term $4\cos^2\left(\dfrac{\pi a \sin\theta}{4\lambda}\right)$, represented in Fig. 15b, with the final result shown in Fig. 15c.

(c) For the reflection, $n = 1$ and neglecting the slow variations introduced by the modulation of the basis, we note that the corresponding reflection half-width is of the order λ/Na or $\Delta(\sin\theta) \approx \Delta\theta = 10^{-2}$ radian.

The angular widths of the diffraction are inversely proportional to the number of basis atoms in the linear crystal.

Exercise 17: Bravais lattices in 2D: application to a graphite layer (graphene)

(a) Starting from basic elementary considerations, list the five possible Bravais lattices in 2D.

(b) Graphite is lamellar crystal in which single layers of carbon atoms or graphene are distributed at the points of regular hexagons (with sides d) to form a honeycomb pattern.

(c) Characterize this structure by its Bravais lattice and basis (see also the definition given in Ex. 3, Q. 2c).

(d) Sketch the corresponding reciprocal lattice and the first BZ.

(e) Give the expression of the structure factor $F(h, k)$ and then state the different values for graphite.

Solution:

(a) In 2D the primitive Bravais lattices are characterized by two vectors $\vec{a}$ and $\vec{b}$ and the angle γ formed between them. The different possibilities are thus:

$a \neq b$; $\gamma \neq 90°$; oblique system

$a \neq b$; $\gamma = 90°$; rectangular system

$a = b$; $\gamma = 120°$; hexagonal system

$a = b$; $\gamma = 90°$; square system

All lattices can be only centered, and thus, of order 2 but the centered hexagonal lattice can be described as a simple hexagon and the centered square as a simple square because they involve no reduction in symmetry. Consequently, in 2D the only multiple lattice is thus the centered rectangular lattice (see note at the end of problem).

(b) The carbon atoms in a graphite monolayer (or graphene) are depicted in Fig. 16 (see also Chapter III, Pb. 9, and Chapter V, Pb. 11).

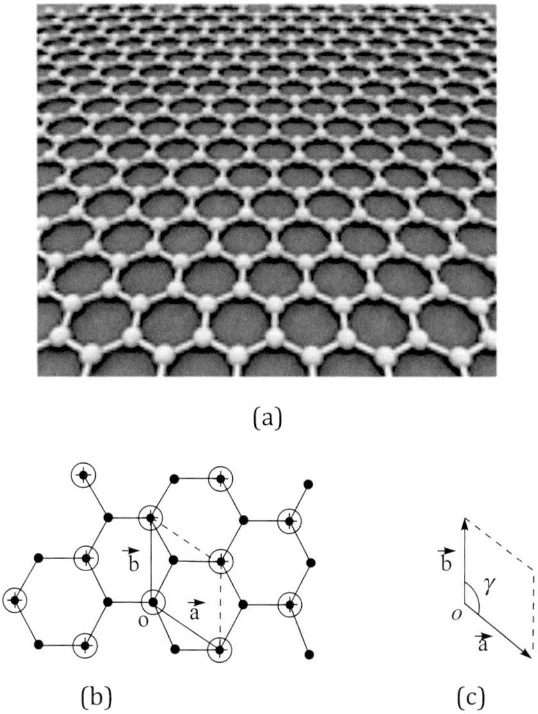

Figure 16

After having arbitrarily chosen the origin on the atom O, it can be noted that the other lattice points (where the chemical environment is indistinguishable from the origin *with the same orientation*) should not be confused with the geometric position of all the atoms but only the atomic position of every other atom (see the points indicated in Fig. 16). The primitive lattice forms a rhombic unit cell ($a = b = d\sqrt{3}$, $\gamma = 120°$) which belongs to the hexagonal system and the basis consists of two atoms of carbon situated at 0,0 and at 1/2, 2/3 (or 2/3, 1/3).

(c) The reciprocal lattice is defined by the following relations given by the definition in Chapter II, Section 3: $\vec{a}\cdot\vec{A}=\vec{b}\cdot\vec{B}=2\pi;\ \vec{a}\cdot\vec{B}=\vec{b}\cdot\vec{A}=0$.

Thus, $|\vec{A}| = |\vec{B}| = 2\pi/(a\cos 30°)$ because $\vec{A}$ is not parallel to $\vec{a}$ as $\gamma = 60°$ (or 120°):

$$F(h,k) = f_0\left[1 + \exp - i2\pi\left(\frac{h}{3} + \frac{2k}{3}\right)\right]$$

(d) The factor F(h, k) is complex, but the intensity (ff^*) is real.
If $h + 2k = 3n$, $F(h, k) = 2$.
If $h + 2k = 3n \pm 1$, $F(h, k) = 1 + \exp \pm i2\pi/3$.
(See Chapter IV, Pb. 12 and Ex. 14b)

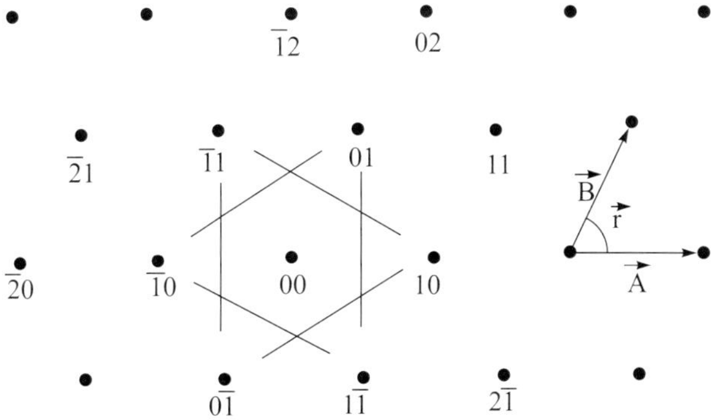

Figure 17

Notes: (1) The lattice of order 1 in a centered rectangular lattice (order 2) is a variation of the oblique lattice ($a \neq b$, $\gamma \neq 90°$) called the lozenge lattice ($a = b$, $\gamma \neq 90°$ and 120°). It is represented in Ex. 3, Fig. 7, and Ex. 8, Fig. 8, face (110).

(2) A single monolayer of graphite is also called "graphene". It is a very important material in particular for its electronic properties and in the form of carbon nanotubes (see the corresponding exercises in Chapter V) but also for its specific atomic vibrations (see Chapter III, Pb. 9, and Chapter V, Pb. 11).

Exercise 18a: Ewald construction and structure factor of a diatomic row

Consider an infinite row of atoms placed at equidistant positions d with alternating species Z_1 and Z_2. A radiation of wavelength λ irradiates this row at normal incidence. As shown in the figure below, the diffracted rays are detected using a photographic plate (with a hole) that is placed perpendicular to the incident beam.

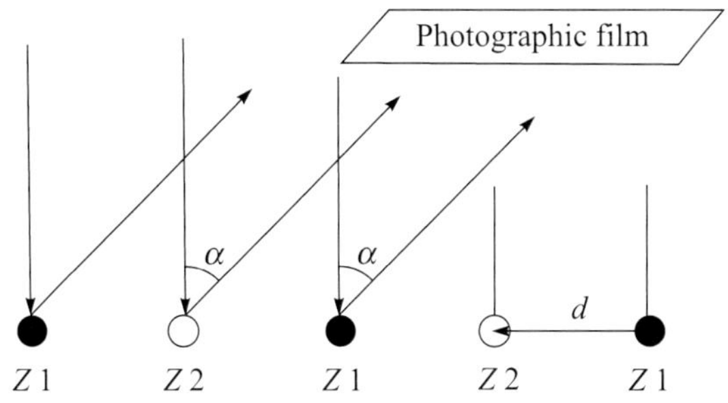

Figure 18

(1) From simple geometric considerations, numerically determine the angles $\alpha_1, \alpha_2, \alpha_3, \ldots$, which corresponds to the constructive interference (evaluated from α increasing but limited to the plane of incidence). Give the expressions of the corresponding amplitudes A_1, A_2, A_3. Take $\lambda = 2d/3$.

(2) Find the above results using the graphic construction of Ewald in the reciprocal space. Note that the reciprocal lattice of a lattice in 1D in the direct space reduces to a lattice of planes (not points) with abscissa $n\vec{A}$ (n is an integer not equal to zero). In short, represent the planes and their intersection with the Ewald sphere ($\lambda = 2d/3$), determine the direction of the diffracted rays (and thus $\alpha_1, \alpha_2, \alpha_3, \ldots$), and evaluate the amplitude using the structure factor.

On the same construction, show the Bragg angles θ and the vectors $\Delta\vec{k}$.

(3) Sketch what is observed on the photographic plate.

(4) Answer all the questions above for the case when all the atoms are identical.

Solution:

(1) We are considering a linear lattice with $a = 2d$ and basis Z_1 at 0 and Z_2 at ½.

The interferences are constructive when the ray $IK = n\lambda$, equivalent to $2d\sin\alpha = n\lambda$ (note that it is purely accidental that this expression corresponds to the usual Bragg formula).

We thus have α_1=Arc sin$(\lambda/2d)$ = 19.5°. For n = 1, and more generally for any odd n), the step difference between waves diffracted by consecutive atoms Z_1 and Z_2 is $(2n+1)\lambda/2$ [with phase difference $(2n + 1)\pi$] and amplitude $A_1 \propto Z_1 - Z_2$. The other numerical values are α = 41.8° ($A_2 \propto Z_1 + Z_2$); α_3 = 90° ($A_3 \propto Z_1 - Z_2$).

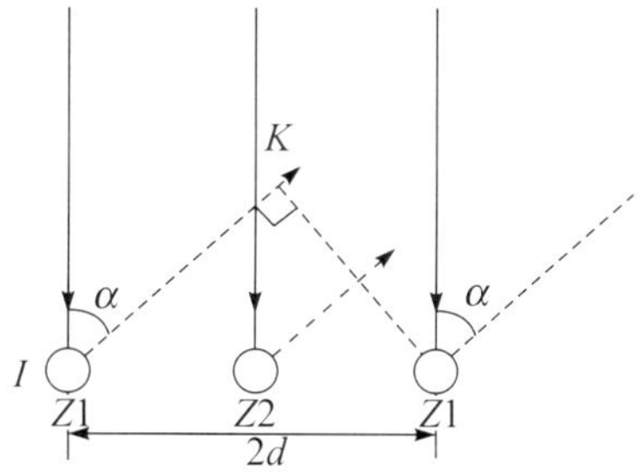

Figure 19

(2) The Ewald sphere is defined by the ray k_0, where $k_0 = 2\pi/\lambda = 3\pi/d$ and the position of its center C, which is deduced from the origin of the reciprocal space knowing that $\vec{k}_0$ is parallel to the direction of the incident rays. The points of the reciprocal space (in 3d) are reduced here (1d) to planes characterized by the translations $\vec{T} = n\vec{A}$ with $\vec{A} = \dfrac{2\pi}{a}\vec{u}_a$. (To show this result, consider the given definition in the Course Summary and extend $\vec{b}$ and $\vec{c}$ to infinity in order to observe that $\vec{B}$ and $\vec{C}$ go to zero; the points of the reciprocal lattice are thus, located on planes perpendicular to $\vec{A}$.)

After noting that $|\vec{A}| = 2\pi/2d$, it is easy to obtain the construction shown in Fig. 20a. One observes that, because the reciprocal lattice is formed on planes, the intersection of the Ewald sphere with these planes is necessary (and henceforth that $\lambda < 2d$). The diffraction conditions are always fulfilled on the contrary to the situation where the reciprocal lattice is made up of points (in 3d for monochromatic radiations and regardless of the incident angle). We also note the existence of symmetrical reflections $(\bar{1}, \bar{2}, \bar{3})$ and the observable diffractions by transmission (when $\vec{k}$ is pointing down.)

The amplitude of the interference is deduced from the structure factor in 1d:

$$F(n) = f_{z_1} + f_{z_2} e^{-i\pi n}$$

Planes of odd n correspond to phases in opposition (destructive interferences) and thus, the diffracted intensity is reduced.

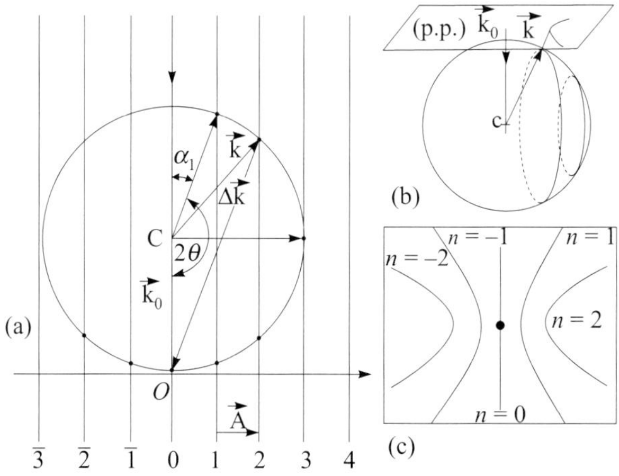

Figure 20

(3) The intersections of the Ewald sphere with the planes of the reciprocal lattice result in circumferences. The extremities of vectors $\vec{k}$ (extending from the origin C in Fig. 20a) are therefore, that form these circumferences satisfy the Bragg condition (Fig. 20b). The diagram observed in the photographic plate, shown in Fig. 20c, indicates the intersection of the different cones formed by $\vec{k}$ and the plate. The diagram is formed of conics, in this case parabolas, with the intensity of the even ones ($n = \pm 2$ in Fig. 18c) greater than the odd ones ($n = \pm 1$ in Fig. 20c).

(4) If one imagines the atoms of species Z_2 become progressively identical to those of species Z_1, one expects that the above odd reflections will be diminished and next vanish, leaving the even reflections from the previous parts of this question. This intuitive result is coherent with the Ewald construction because the direct lattice has lattice constant d (and not 2d) and the only planes in the reciprocal lattice are those equidistant

at $2\pi/d$ (and not π/d). See the even planes indicated in Fig. 20a.

Only the parabolas at n = 2 and n = −2 will appear on the photographic plate, along with the line at n = 0.

Exercise 18b: Structure factor for a tri-atomic basis; Ewald construction at oblique incidence (variation of Ex. 18a)

Consider a linear crystal with lattice constant a in which the basis consists of atom of species A at 0 and two atoms of species B at 1/3 and 2/3. Find the structure factor of such a crystal.

X-rays of wavelength λ (= $a/4$) irradiate this crystal at an oblique incidence (45°). Using the Ewald construction in the plane of incidence, show graphically, the diffraction directions and indicate the relative weights of the different diffracted intensities. State the physical significance of the vector $\vec{\mathrm{k}}$ from which the extremity is on a plane passing through the origin of the reciprocal lattice.

Solution:

The linear crystal is shown below:

$$F(h) = f_\mathrm{A} + f_\mathrm{B}e^{-i2\pi h/3} + f_\mathrm{B}e^{i2\pi h/3} \text{ because } e^{-i4\pi h/3} = e^{i2\pi h/3}$$

$$F(h) = f_\mathrm{A} + 2f_\mathrm{B}\cos\left(\frac{2\pi h}{3}\right)$$

$$F(h) = f_\mathrm{A} + 2f_\mathrm{B} \text{ for the reflections such that } h = 3n$$

$$F(h) = f_\mathrm{A} - f_\mathrm{B} \text{ for the others}$$

Keeping in mind that $I(h) = F(h) \times F^{x}(h)$, Fig. 21 shows the reciprocal space, where the planes corresponding to the strong reflections ($h = 3n$) are represented by solid lines and the those corresponding to weak reflections ($h \neq 3n$) by dashed lines.

The vectors $\vec{\mathrm{k}}$ extending from C to the plane passing by origin O, such as k_S, correspond zero path difference between the rays scattered from all the atoms in the and the constructive interferences exist even for randomly distributed atoms along the row. In the plane of the incident beam, the corresponding reflected beam is symmetric to the incident beam and it is called specular reflection. In addition to the $\vec{\mathrm{k}}$ vectors diffracted up (reflections), there are also those being diffracted down, corresponding to transmission diffraction (see Chapter II, Ex. 2b , and Chapter III, Ex. 2b).

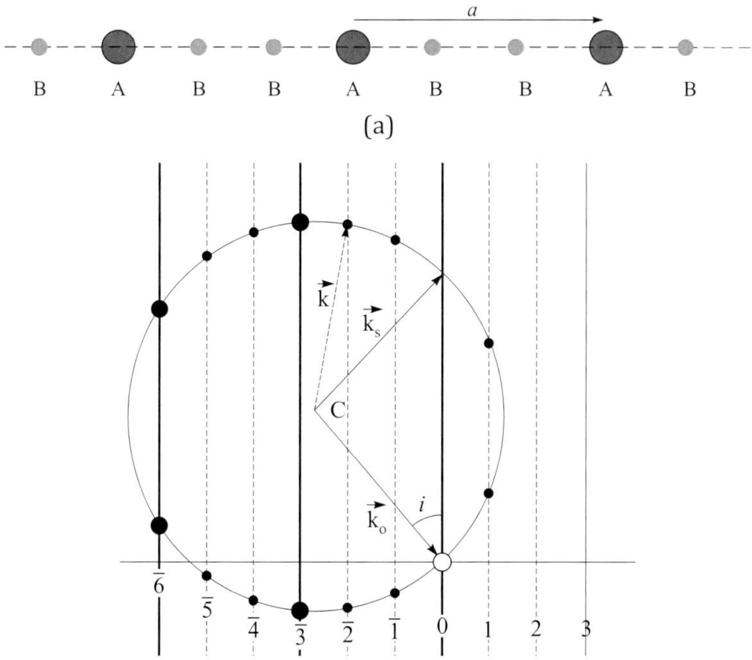

Figure 21

Exercise 19: Reciprocal lattice, BZs, and Ewald construction of a 2D crystal

Consider a 2D crystalline structure characterized by a simple rectangular lattice (a = 3 Å, b = 4 Å) and a basis consisting of one atom Z_1 located at (0,0) and another atom Z_2 located at (1/2,1/2) (see Ex. 3, Q. 2).

Starting from an orthorhombic simple crystal structure ($a \neq b \neq c$; $\alpha = \beta = \gamma = 90°$), show that the reciprocal lattice in 3D is transformed into a series of parallel lines when parameter c decreases down to zero. Build the 2D reciprocal lattice from the vectors $\vec{A}, \vec{B}$. Construct the first BZ (all the following considerations are limited to the AB plane).

A radiation of wavelength λ is directed along the row $[\bar{1}, 0]$ in direct space. Use the Ewald construction to determine the reflection indices that will be observed and indicate the numerical value of the first Bragg angle. Use $\lambda = 1.8$ Å.

In the general case for radiations at any angle, list the six first non-equivalent reflections (with increase in Bragg angles) and determine the corresponding amplitudes. On the figure drawn in part (2) draw the extremities of the diffracted vectors corresponding to the reflection $(\bar{1}, 1)$ and relative to the incident wave vector $\vec{k}_0$, which has an extremity at (0, 0).

Atoms of species Z_1 and Z_2 are identical. Show the reciprocal lattice of a centered rectangular lattice is a centered rectangular lattice. Represent the first BZ and give the list of the first five non-equivalent reflections being allowed.

Solution:

(1) Applying the definition of the reciprocal basis vectors to the simple orthorhombic lattice (see Course Summary), we find that the vectors are $\vec{A}=\frac{2\pi}{a}\vec{u}_a$; $\vec{B}=\frac{2\pi}{b}\vec{u}_b$; $\vec{C}=\frac{2\pi}{c}\vec{u}_c$. When $c \to \infty$ in direct space the planes $(0,0,n)$ separate from each other to infinity, leaving only the 2D lattice. In reciprocal space $\vec{C}$ goes to zero and the points of the type $(0, 0, n)$ are transformed into lines normal to the plane defined by $\vec{A}$ and $\vec{B}$. In this plane, the reciprocal lattice is drawn from translations of the type $h\vec{A} + k\vec{B}$ from the origin at (0, 0) where each point is indexed by integer values of h and k. The result of this operation leads to the scheme shown in Fig. 22.

The limits of the BZs are the symmetry planes (lines in $2d$) between the origin (0, 0) and the various (h, k) points and the first BZ corresponds to the minimum surface defined by these symmetry lines and surrounding the origin (area of the rectangle with hatched contour in Fig. 22.

(2) The points intercepting the Ewald sphere are reflections located at $(1, 2)$; $(1, \bar{2})$; $(3, 1)$; $(3, \bar{1})$, where we have considered only the waves diffracted in the plane $(\vec{A}, \vec{B})$. In $3d$ the Ewald sphere intercepts in fact all lines normal to $\vec{A}, \vec{B}$, where the points h, k are in the interior of the circumference shown in Fig. 22.

The angle $(\vec{k}_0, \vec{k}) = 2\theta$ is such that:

$$\sin\theta_{12} = \frac{OA}{2k_0} = \frac{\lambda}{2}\left(\frac{1}{a^2} + \frac{4}{b^2}\right)^{1/2} = 32°7$$

(3) The sequences of increasing angles θ corresponding to the Bragg's law may be evaluated numerically from the increasing values of $\dfrac{1}{d(h,k)} = \left(\dfrac{h^2}{a^2} + \dfrac{k^2}{b^2}\right)^{\frac{1}{2}}$. They can also be directly determined from measurements of $|G|$ in Fig. 22 because $|G(h,k)| = \dfrac{2\pi}{d(h,k)}$ (see Ex. 10b).

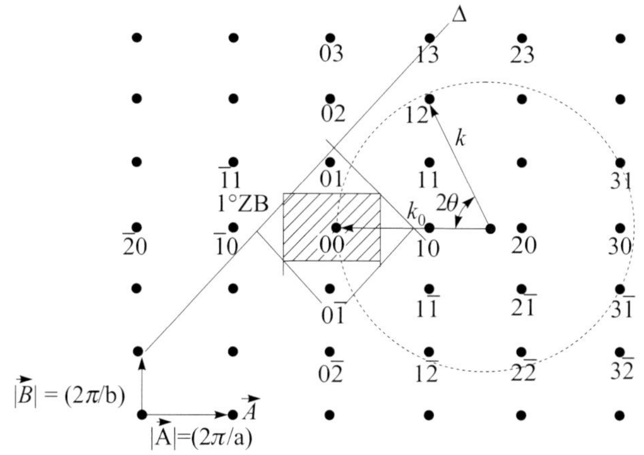

Figure 22

Using either method we find the following sequence (0,1); (1,0); (1,1); (0,2); (1,2); (2,0); (2,1); (0,3). The amplitude of the reflections is given by the structure factor which here is:

$$F(h,k) = f_{Z_1} + f_{Z_2}e^{-i\pi(h+k)}$$

When $h + k$ is even, $F(h,k) = f_{Z_1} + f_{Z_2}$ (Ex. 1, 3)

When $h + k$ is odd, $F(h,k) = f_{Z_1} + f_{Z_2}$ (Ex. 1, 2)

Finally, the line labeled by Δ in Fig. 22 (median plane between the origin and the point $(\bar{1},1)$ represents the end of all incident vectors $\vec{k}_0$ which can be excited by the reflection at $(\bar{1},1)$. Such vectors obey either $\vec{k}_0 \cdot \dfrac{\vec{G}}{2}(\bar{1}1) = |G(\bar{1}1)|/4$ or $\vec{k}_0 - \vec{k} = \Delta\vec{k} = \vec{G}(\bar{1}1)$.

(4) When $Z_1 = Z_2$ the direct lattice is a centered rectangle (see the solution to Ex. 3, Q. 2) and the reciprocal lattice must be constructed from the vectors $\vec{a}'$ and $\vec{b}'$ of the lattice of order 1 in the form of a lozenge. It is simpler to describe a centered rectangle structure as a simple rectangle with two identical atoms at 0,0 and ½,½. The presence of this basis leads to forbidden reflections of the type $h + k$ = odd number. Erasing the corresponding points on Fig. 22, we obtain Fig. 23, where the forbidden reflections are marked by an ×).

The reciprocal lattice thus obtained corresponds to a centered rectangle lattice with $\vec{A}' = 2\vec{A}$ and $\vec{B}' = 2\vec{B}$.

The first five allowed reflections that are non-equivalent are (1, 1); (0, 2); (2, 0); (1, 3); and (2, 2).

In Fig. 23, the hatched area indicates the contour of the first BZ when $Z_1 = Z_2$.

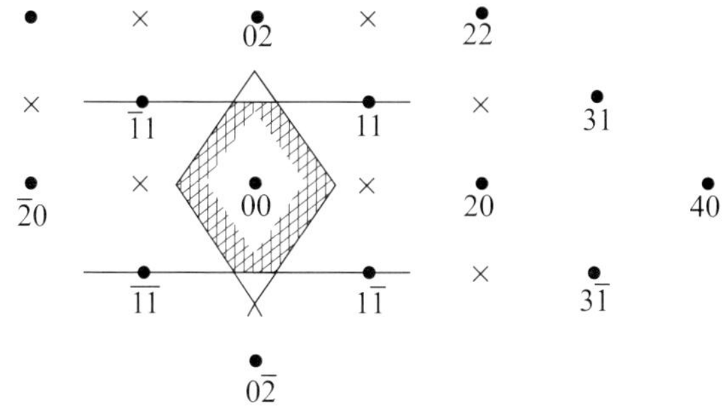

Figure 23

Exercise 20: X-ray diffraction patterns and the Ewald construction

The main experimental methods of X-ray diffraction are:

(a) the Laüe method
(b) the crystal rotation method
(c) the powder method (or the Debye–Scherrer method)

Describe the corresponding experimental set-up and with the help of the Ewald construction in reciprocal space, deduce the observed diffraction patterns in each case.

Solution:

Figure 22 shows a schematic experimental set-up and an example of X-ray spectrum is given in the Course Summary with the correlated electron transitions.

In the Laüe method, the sample is a fixed crystal and the useful part of incident X-ray spectrum is the continuous radiation spectrally varying between λ_m and λ_M. These two extremes correspond to Ewald spheres with respective radii of $k_M = 2\pi/\lambda_m$ (from center C_M) and $k_m = 2\pi/\lambda_m$ (from center C_m). All the points of the reciprocal lattice included between these two extreme spheres can be cut by an intermediate sphere of intermediate radius k (from center, C) and the Bragg condition, $\Delta\vec{k} = \vec{G}$, are satisfied for each of these points. The diffraction pattern observed on a photographic plate will be formed by a limited number of spots from the transmission (if the angle2θ between $\vec{k}_0$ and $\vec{k}$ is such that $|2\theta| \leq \pi/2$ and from the reflection $|2\theta| > \pi/2$.

Max T. F. von Laüe won the Nobel Prize in physics, 1914, for his discovery of the diffraction of X-rays by crystals. For the first time his experiments demonstrate the periodic structure of crystals (that was suspected from a long time but not evidenced) and they established also the wave nature of X-rays. Bragg formulation of X-ray diffraction was first proposed by William Lawrence Bragg and his father William Henry Bragg. They were awarded with the Nobel Prize in physics, 1915.

In the crystal rotation method, the useful part of incident X-ray spectrum is the characteristic radiation that is monochromatic. The crystal rotates around an axis which effectively rotates the reciprocal axis about the same axis. Each point in the reciprocal lattice describes a circumference which will intersect the Ewald sphere in two symmetrical points. These intersections define the diffraction directions and the pattern is formed with the corresponding points.

In the powder method, the radiation is also monochromatic. The sample is grinded into a fine powder and each grain has a random direction compared with its neighbors. This random distribution of directions results in a reciprocal space in which the points (h, k, l) will be distributed on a sphere of radius $|\vec{G}(h,k,l)|$. The intersection of these spheres with the Ewald sphere corresponds to a series of circumferences where the vectors $\vec{k}$ emanating from the center C whose end intersects these circumferences and defines the diffraction direction. The resulting pattern diagram is formed of concentric rings.

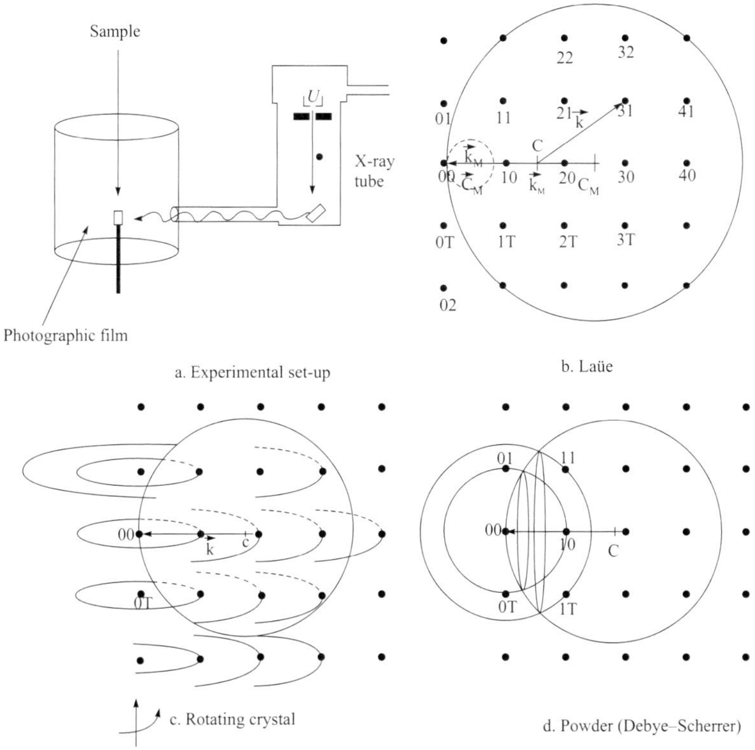

Figure 24

Exercise 21a: Resolution sphere

(a) A crystal is irradiated with a beam of monochromatic X-rays with wavelength, λ.
Using the Ewald construction, show that when the crystal can take any possible orientation, the only possible reflections (h, k, l) correspond to the points in reciprocal space that are included in sphere (resolution sphere). Determine the radius of this sphere.

Numerical application: Consider a cubic crystal with a = 3 Å and an incident wavelength of λ = 0.5 Å. Find the maximal value of $h^2 + k^2 + l^2$.

Consider a crystal with a fixed orientation. Show that the divergence angle of the incident beam and its variation from monochromaticity as well as small crystallite sizes or slight crystal misorientations enhances the probability of observation of Bragg reflections.

Contrarily to X-ray diffraction, diffraction of mono-energetic fast electrons by transmission through thin films ($t \approx 1000$ Å) gives rise quite always to diffraction patterns. In order to explain this phenomenon, evaluate the associated wavelength of incident electrons of 100 keV kinetic energy.

(h, m, e)

Solution:

When the radiation is monochromatic and the crystal orientation is fixed, the probability for a crystal plane satisfying the Bragg condition is exceptional (i.e., to find that the Ewald sphere passes by a point corresponding to the reciprocal lattice). Exploring all the angular possibilities between the incident radiation and the crystal orientation is equivalent to turning the reciprocal lattice around its center and in all directions, with the Ewald sphere remaining fixed (or equivalently the reciprocal lattice is fixed and the Ewald sphere is rotated around O). As shown in Fig. 25, this operation defines a sphere of radius $2k_0 = 4\pi/\lambda$ that allows the exploration of all Bragg conditions containing the points h, k, l in the interior of the sphere.

We note that the geometry of the diffraction pattern reflects that of the reciprocal lattice and that the amplitudes of each point correspond to the Fourier components of the electronic periodicity in real space.

Unfortunately, the diffraction experiments permit to obtain the intensities of the waves but losing their phase and they do not allow observation of all the points in reciprocal space. There is thus, a loss of information in the reflections that are not accessible.

(a) The only accessible reflections are those that:

$$G^2(h,k,l) = (h^2 + k^2 + l^2)\left(\frac{2\pi}{a}\right)^2 \leq \left(\frac{4\pi}{\lambda^2}\right) \text{ or } \sqrt{h^2 + k^2 + l^2} \leq \frac{2a}{\lambda}$$

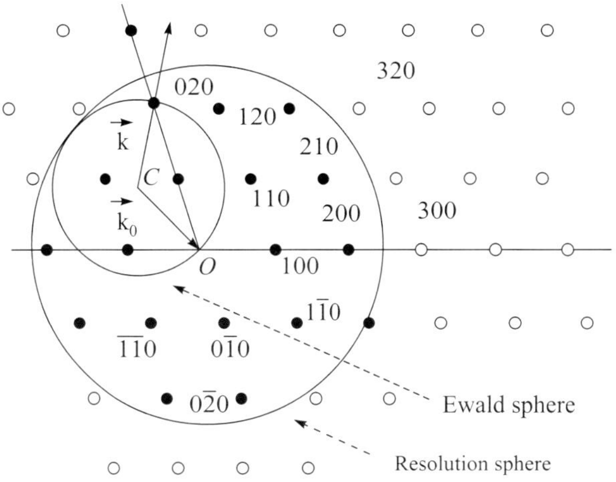

Figure 25

The influence of these parameters is easily determined from the analysis shown in this exercise and the previous one. As shown in Fig. 26a, non-monochromatic radiations may be treated as in the Laüe method with two external Ewald spheres with interval k_m and k_M. A finite divergence of a beam can be treated as a reduced rotation of the Ewald sphere around the origin of the reciprocal lattice. A sample composed of small crystallites corresponds to small pivots of the reciprocal lattice around its origin, as seen in Fig. 26b.

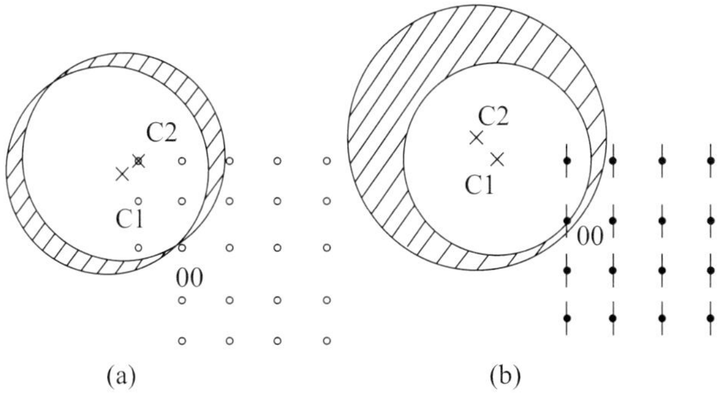

Figure 26

The size of the crystallites relaxes the angular constraints induced by the Bragg condition $\Delta\theta \propto N^{-1}$ where N is the number of diffusing bases (see Ex. 16). This influence can also be treated as a broadening of the points of the reciprocal lattice.

$$\lambda = \frac{h}{\left[2m_0eV_0\left(1+\frac{eV_0}{2m_0c^2}\right)\right]^{1/2}} = \frac{12.26}{\left[V_0\left(1+\frac{eV_0}{2m_0c^2}\right)\right]^{1/2}}$$

λ = 0.037 Å. The vector $\vec{k}_0$ will be much larger ($2\pi/\lambda \approx 100$ Å^{-1}) than the basis vectors of the reciprocal lattice $2\lambda/a \approx 1$ Å. The probability that the Ewald sphere will pass through a point of the reciprocal lattice is thus increased.

As shown in Fig. 27, in transmission experiments, one can associate this sphere to the tangent passing through the origin of the reciprocal lattice. In addition the diffraction conditions are relaxed by the fact that the number of participating atoms is reduced ($N \approx$ 100s) in the direction of the film thickness which is equivalent to increasing the points of the reciprocal lattice in the corresponding direction. The combination of these two factors (in addition to the dynamic theories of the diffraction of electrons) results in the realization of a diagram consisting of a number of points for a single crystal or rings in the case of a polycrystalline material.

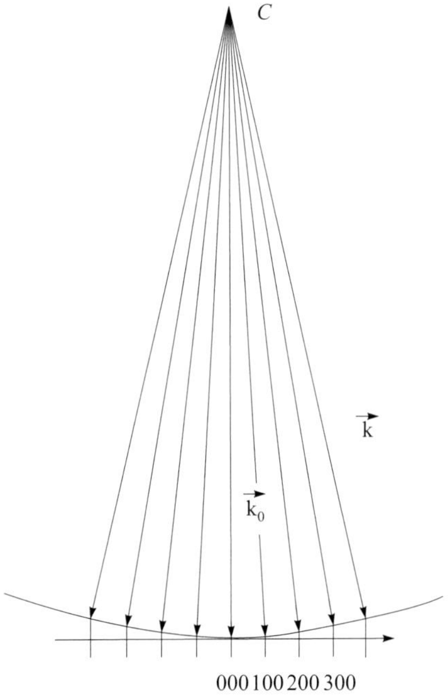

Figure 27

We note that the diffraction of slow electrons by reflection at the surface of single crystals also results in diffraction patterns, but for different reasons (see Pb. 3).

Exercise 21b: Crystal diffraction with diverging beams (electron backscattered diffraction, EBSD)

Mono-energetic electrons, energy $E°$, are issued isotropically from a point source C. This point source is situated into a micro-crystal but near from one of its plane surfaces in front of which a fluorescent screen is set at distance, D. The corresponding diffraction patterns are displayed on the screen that is parallel to the exit crystal surface.

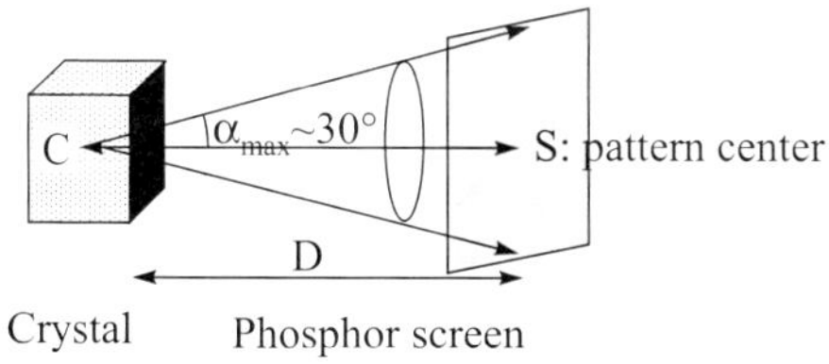

Figure 28

(a) Evaluate the associated wave length λ of the diffracted electrons for $E°$= 20 keV. Evaluate the first Bragg angles, θ_{100}; θ_{110}; θ_{111}; for a simple cubic crystal of parameter, a = 4 Å.

(b) Again from the Bragg's law, describe the expected shape of the diffraction patterns. With D = 0.2 m, represent them for the following crystal orientations:

  (1) The crystalline plane parallel to the screen is the (100) plane formed from the lattice rows [010], horizontal, and [001], vertical.

  (2) The crystalline plane parallel to the screen is the (111) plane formed from the lattice rows $[11\bar{1}]$ and $[\bar{1}11]$, horizontal and vertical respect.

  (3) The crystalline plane parallel to the screen is the (110) plane formed from the lattice rows [110] and [001] horizontal and vertical respect.

(c) In reality in the corresponding experimental arrangement (EBSD), a 20 keV incident electron beam is nearly parallel to the screen and it irradiates the sample surface with an incident angle of ~70° but the crystalline orientations are unchanged.

Are there changes in the observed diffraction patterns? The atoms in materials scatter inelastically, a fraction of the electrons with small energy losses to form a divergent source of electrons close to the surface of the sample. Thus, this fraction of incident electrons is backscattered into the vacuum after having experienced diffraction effects on the lattice planes. The diffracted electrons are not mono-energetic electrons, strictly speaking. What is the consequence on the diffraction patterns?

Solution:

(a) From the solution (d) of Ex. 21, above the associated wave length of the diffracted electrons λ is ~0.086 Å for $E°$= 20 keV.

(b) $\sin \theta_{100} = \lambda/ 2a$= 10.8 mrad.~0.6° ~$\theta_{100}$
$\sin \theta_{110} = \lambda\sqrt{2}/2a$= 15 mrad. ~$\theta_{110}$~0.86°
$\sin\theta_{111} = \lambda\sqrt{3}/2a$= 18.6 mrad. ~$\theta_{111}$~1.07°
As shown in Fig. 29, the Bragg conditions are satisfied for electron beams describing conical surfaces of apex C' (symmetric to C with respect to the diffracting lattice planes) and the diffraction patterns correspond to the intersections of the various cones with the screen. In theory such intersections are parabolic, elliptic, or hyperbolic but they are nearly straight lines as a consequence of the small values of the Bragg's angles d′.

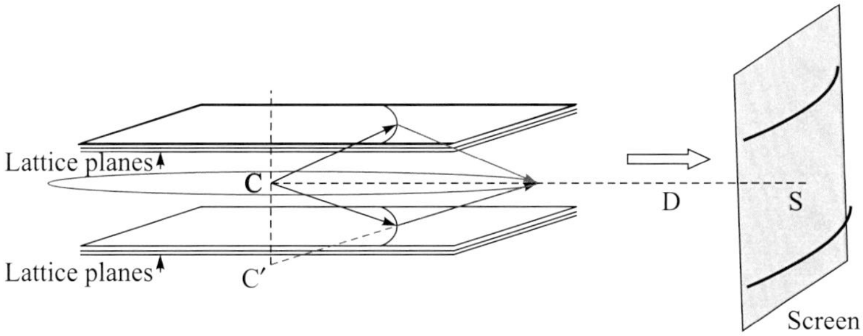

Figure 29

(1) When the plane parallel to the screen is defined by the lattice rows [010] and [001], the lattice row [100] is perpendicular to the screen. The first lines of the

diffraction pattern correspond to the planes (010) and (0$\underline{1}$0), and also to the planes (001) and (00$\underline{1}$). Being perpendicular to the [100] direction the conical surfaces relative to the (100) reflections cannot intersect the screen. Similarly, the conical surfaces of axes making an angle α larger than α_{max} to the [100] direction cannot intercept a screen of limited dimension. The observable reflections may be derived from the solution of Ex. 11 where the angle Φ between two lattice planes (h_1, k_1, l_1) and (h_2, k_2, l_2) obeys to:

$$\cos\Phi = \frac{h_1h_2 + k_1k_2 + l_1l_2}{\left(h_1^2 + k_1^2 + l_1^2\right)^{1/2}\left(h_2^2 + k_2^2 + l_2^2\right)^{1/2}}$$

Applied to the (010) planes this equation leads to $\cos\Phi = k_2/\left(h_2^2 + k_2^2 + l_2^2\right)^{1/2} > \cos\alpha_{max}$. A more precise evaluation may take into account the deviation induced by the small Bragg angles: $\Phi - \theta$. In addition there are the reflections of order n = 2, 3, 4 such as (020) of Bragg angle nearly twice that of the (010) reflection. From sin θ_{010} $= \lambda/2a$ = 10 mrad. ~ tgθ_{100} = SA/D. One obtains SA = 2 mm. Then, the parallel lines corresponding to (001) and (00$\underline{1}$) are distant from 4 mm.

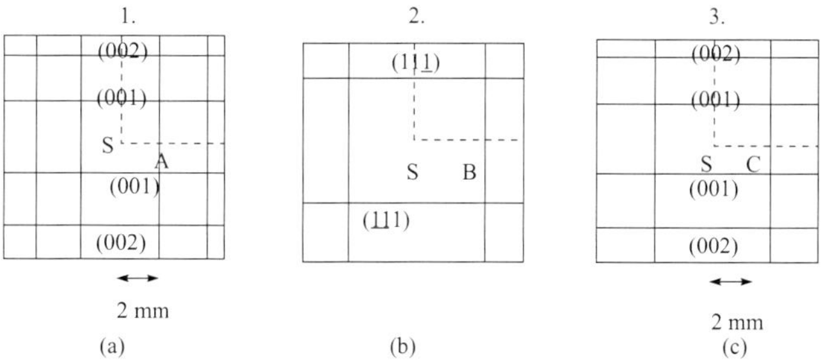

Figure 30

(2) When the plane parallel to the screen is the (111) plane formed from the lattice rows [11$\underline{1}$] and [$\underline{1}$11], the lattice row [111] is thus perpendicular to the screen. The diffraction pattern is homothetic to that of situation 1 with a multiplication factor of √3.

From $\sin\theta_{111} = \lambda\sqrt{3}/2a$ = 18.6 mrad.~θ_{111} one obtains SB~3.5 mm and the parallel lines are distant from 7 mm.

(3) When the plane parallel to the screen is the (110) plane formed from the lattice rows [$\bar{1}$10] and [001] horizontal and vertical respect, the lattice row [110] is perpendicular to the screen. The horizontal lattice planes of type (001) lead to diffraction lines similar to those obtained for situation 1. The vertical lattice planes of type (011) lead to vertical lines distant from each other of 2 SC~6 mm.

(c) The crystalline orientations being unchanged the main features of the diffraction patterns remain unchanged when the surface is not geometrically parallel to the screen. Due to the energy losses experienced by the BSEs, their spectral distribution is composed of a band of wavelengths slightly larger than the value evaluated for $E°$. The corresponding Bragg angles are slightly increased and broadened: the lines of the patterns are broadened into bands.

Note that the geometry of patterns looks similar to a magnified projection of the planes with only a small shift due to the small values of the Bragg angles. Thus, this geometry is nearly independent from λ or $E°$ but not the intensity of the lines.

Comments: Kikuchi and Kossel lines; texture analysis with EBSD

The lines obtained in the diffraction pattern are named Kikuchi lines (S. Kikuchi (1928), Diffraction of cathode rays by mica. *Japanese Journal of Physics* **5**, 83). The bombardment of a crystal target with 10 keV incident electrons generates also characteristic X-rays of well-defined wavelengths and the Bragg diffraction of these X-rays leads to patterns composed of fine lines that are referred to as Kossel lines (named after Walther Kossel).

The goal of this exercise is to demonstrate the possibility to identify the orientations of small crystals from the use of diverging electron or X-ray beams. The old Kikuchi experiments have been the subject of a renewed interest very recently with the digital acquisition of the diffraction patterns via the use of electronic cameras back to the phosphor screen (instead of looking at photographic films) and the automatic identification of the crystal orientations with specific software. Thus, the corresponding ESBD experiments are conducted using a scanning electron microscope (SEM) equipped with an EBSD detector containing at least a phosphor screen, compact lens and low-light, charge-coupled device (CCD) camera chip. The typical dimension of the irradiated volume is of a few microns. Next, the incident electron beam may be scanned across the surfaces of polycrystalline samples and the crystal orientation may be measured at each point micro-crystals leading to maps revealing the constituent grain morphology, orientations, and boundaries.

As illustrated in the images, the applications concern metallurgy and semiconductors and also geological samples and nano-devices.

Among many others, additional details may be founded in articles (F. J. Humphreys, P. S. Bate, P. J. Hurley (2001), *Journal of Microscopy* **201**, 50; J. Basinger, D. Fullwood, J. Kacher, B. Adams (2011), *Microscopy Microanalysis* **17**, 330) or books (W. Zhou and Z. L. Wang, Eds. (2007) *Scanning Microscopy for Nanotechnology, Techniques and Applications*. Berlin, NY: Springer, particularly the contribution of T. Maitland and S. Sitzman, Electron Backscattered Diffraction Technique and Materials Characterization, Section 2, p. 41. See also A. J. Schwartz, M. Kumar, B. L. Adams, D. J. Klumer, Eds. (2000) *Electron Backscattered Diffraction*. New York: Materials Science Academic/Plenum Press).

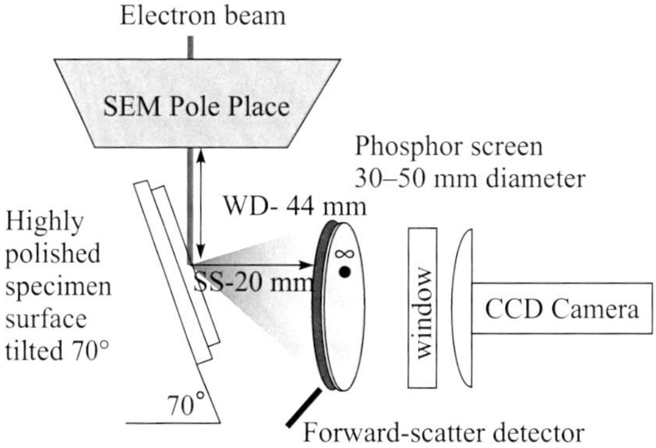

Figure 31 Geometry of an EBSD system.

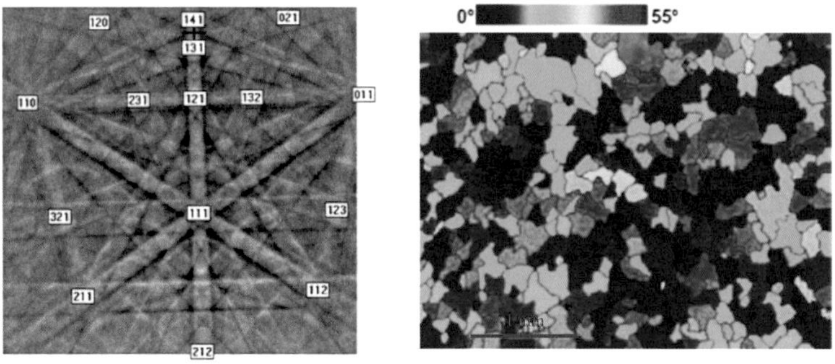

(a) (b)

Figure 32 (a) Example of an indexed EBSD pattern of a Si crystal (E° = 40 keV). (b) Crystalline orientation map of a Ni-base alloy. The scale is colored considering the misorientation between each point and the <001> direction oriented along the normal.

Exercise 22: Atomic form factor

A beam of monochromatic X-rays with wavelength λ propagates in a vacuum and encounters a spherical atom of radius R. The incident wave with wave vector $\vec{k}_0 = \left(2\pi/\lambda\right)\vec{u}_0$ is partially diffused by the Z electrons of the atom, assumed to have a density distribution of ρ.

(a) Find the amplitude of a wave diffused in the direction $\vec{u}_0$ by the electrons contained in the volume dV described by the radius vector $\vec{r}$ (with respect to the wave diffused by a point electron placed at the center 0 of the atom).
In the form of an integral, deduce the expression for the form factor of an electron distribution with spherical symmetry.

(b) Find the atomic diffusion coefficient of a uniform electron distribution at the interior of a sphere of radius R. These results should be expressed as a function of $\vec{k} = \vec{k} - \vec{k}_0$ where, $\vec{k}_0 = (2\pi/\lambda)\,\vec{u}_0$ and then, state the results in terms of the parameters $\sin\theta/\lambda$ which are more directly related to experiments.

Solution:

As shown in Fig. 26, the phase φ between the wave diffused at point M in the direction $\vec{u}$ and the wave diffused at point O in the same direction is such that

$$\varphi = 2\pi\delta/\lambda = 2\pi(\overline{OK} - \overline{MH})/\lambda = (2\pi/\lambda)\vec{r}(\vec{u} - \vec{u}_0) = (\vec{k} - \vec{k}_0)\vec{r} = \Delta\vec{k}\cdot\vec{r}$$

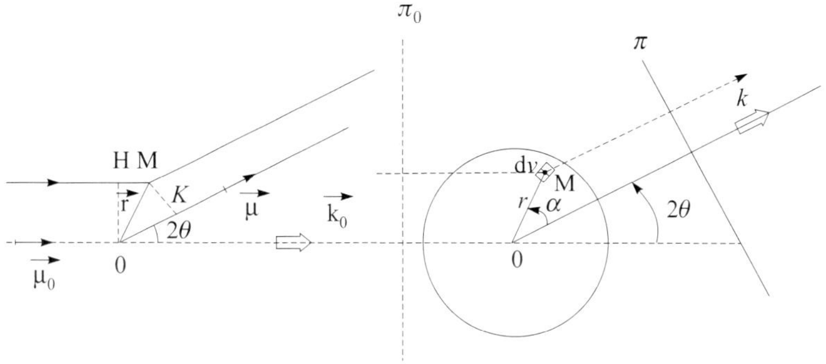

Figure 33

Compared to the diffused amplitude by an electron at 0, that diffused at M is proportional to the number of electrons ρdv contained in the volume dv. Then, df = ρ dv $\exp(-i\Delta\vec{k}\cdot\vec{r})$.

The diffusing power of an atom with respect to that of an electron is:

$$f = \int_{\substack{\text{Volume}\\ \text{of atom}}} \rho\cdot\exp - i(\Delta\vec{k}\cdot\vec{r}).dv$$

For a spherical distribution $\rho = \rho\ (r)$, the axis symmetry around k permits to select dv limited between two spheres of radiuses r and $r + dr$ and semi-apex angles α and $d\alpha$:

$$dv = 2\pi r^2 \sin\alpha\ dr\ d\alpha \text{ (see Fig. 33)}$$

The form factor becomes $2\pi\int_0^R \rho(r) r^2 dr \int_0^R [\exp(-i|\Delta\vec{k}| r\cos\alpha)]\cdot\sin\alpha\ d\alpha$

$$f = \frac{4\pi}{|\Delta k|}\int_0^R \rho(r)\cdot\sin(|\Delta\vec{k}|\cdot r)\cdot r\ dr$$

When the electronic distribution is uniform $\left(\frac{4\pi \boldsymbol{R}^3}{3}\rho = Z\right)$, f can be integrated by parts to obtain:

$$f = \frac{4\pi\rho}{|\Delta k|}\left[\frac{\sin(|\Delta k|\cdot r)}{|\Delta k|^2} - \frac{r\cos(|\Delta k|\cdot r)}{|\Delta k|}\right]_0^R$$

$$= \frac{3z}{(|\Delta k| R)^3}[\sin(|\Delta k| R) - |\Delta k|\cdot R\cdot\cos(|\Delta k| R)]$$

If $\left|\Delta\vec{k}\right|$ is replaced by $4\pi\sin\theta/\lambda$, as is accessible by experiment, the form factor becomes:

$$f = \frac{3Z}{\left(4\pi\sin\theta\cdot\frac{R}{\lambda}\right)^3}\left[\sin\left(\frac{4\pi\sin\theta}{\lambda}R\right) - \frac{4\pi\sin\theta}{\lambda}R\cdot\cos\left(\frac{4\pi\sin\theta}{\lambda}R\right)\right]$$

Notes:
- By modifying the integration limits, from the atom to the entire crystal, we may obtain the amplitude of the wave diffused by the entire crystal:

  $$A(\Delta\vec{k}) = \int_{\substack{\text{Crystal}\\\text{volume}}} \rho(r)\exp(-i\Delta\vec{k}\cdot r)dv$$

  This is just the Fourier transform in reciprocal space of the electronic distribution of the crystal under investigation.
- Note that if the incident beam is extended θ = 0, $\left|\Delta\vec{k}\right|$ = 0, all the diffused waves are in phase and the integral expression for f reduces to:

  $$f(\Delta\vec{k} = 0) = \int_{\substack{\text{Atom}\\\text{volume}}} \rho(r)dv$$

  f is thus, equal to Z. Starting from this initial value the evolution of f as a function of θ will depend on the way in which the Z electrons are distributed around the atom. Observe in particular that $f(\theta)$ is constant if all the electronic charge is concentrated at O. This fact explains why f is called the form factor (see also Pb. 11).

Exercise 23: X-ray diffusion by an electron (Thomson)

A beam of X-rays with angular frequency ω is polarized along the x-axis and propagates along the z-axis, irradiating a free electron situated at the origin. Under the electric field $E_0e^{i\omega t}$ of this electromagnetic wave the electron vibrates at the same frequency ω and therefore, acts like an oscillating dipole, $P_0e^{i\omega t}$.

(a) Knowing that the electric field emitted by such a dipole at a distance, r = OM ($r >> 2\pi c/\omega$) is of the form ($\alpha = \overrightarrow{\text{Ox}}, \overrightarrow{\text{OM}}$):

$$E_r = \frac{-1}{4\pi\varepsilon_0}\frac{\omega^2\sin\alpha}{c^2 r}P_0\exp i\omega\left(t - \frac{r}{c}\right)$$

show that the ratio between the amplitudes of the diffused and incident electric fields is of the form $E_r/E_0 = r_e \sin\alpha/r$, where, r_e is a constant to evaluate.

(b) Find the relation σ_e between the intensity diffused over 4π rad. and the incident intensity.

(c) In theory, the same mechanism also applies to the nucleus with atomic charge, Z_e. Why can this effect be neglected?

(d) The electron is bonded to the atom and its movement must include a friction force, which is opposite its velocity $(-mv/\tau)$ and a restoring force (of the form $-m\omega_0^2 x$, where x is the elongation). Show that the relation E_t/E_0 resembles that found in (a) when r_e is weighted by a complex factor $\vec{f}$ that should be explicitly stated. (e, e_0, m, m_0)

Solution:

(a) $P_0 = -ex_0$ and $m\gamma = -m_0\omega^2 x_0 = -eE_0$. Making this substitution, we find:

$$E_r/E_0 = \frac{r_e}{r}\sin\alpha\,,$$

where r_e is the classical radius of an electron, $r_e = e^2/(4\pi\varepsilon_0 mc^2) = 2.82 \times 10^{-15}$ m.

(b) $$\sigma_e = \frac{I}{I_0} = r_e^2 \int_0^{\pi} 2\pi \sin^3\alpha\, d\alpha = 8\pi r_e^2/3 = 66.6 \times 10^{-30}\ \mathrm{m}^2$$

σ_e = 0.666 barn (10^{-24} cm^2). This is the coherent diffusion cross-section of an electron.

(c) The calculation is formally the same as replacing e by Z_e and m_0 by M but in the most favorable case (the hydrogen atom) the amplitude diffused by the nucleus is 1840 times weaker than that diffused by an electron (because of the relation between m/M) and the corresponding intensity will be negligible.

(d) The equation of motion of an electron is:

$$m_0\frac{d^2x}{dt^2} + \frac{m_0 dx}{\tau dt} + m_0\omega_0^2 x = -\mathrm{e}E_0\mathrm{e}^{i\omega t}$$

So that:

$$x_0 = \frac{-\mathrm{e}}{m_0}\frac{E_0\mathrm{e}^{i\omega t}}{\left(\omega_0^2 - \omega^2\right) + i\omega/\tau}$$

Compared to the free electron treated in (a), it is sufficient to introduce here an anomalous diffusion factor $\tilde{f}$ that relates the different accelerations obtained in the two situations:

$$\tilde{f} = \frac{\omega^2}{(\omega^2 - \omega_0^2) - i\omega/\tau}$$

This diffusion factor is takes a complex form and on can easily separate the real and imaginary parts by multiplying by the complex conjugate of the denominator.

Comments: The interactions between X-rays and atoms

In coherent diffusion, the X-rays interact essentially with the electrons of the atom or of the solid. The weight of this interaction is relatively weak compared to the dominant interaction which is the absorption of a photon X by an atomic electron (the principle of radiography). The absorption cross section of an X-ray photon is several orders larger than the diffusion cross-section σ_e evaluated here in b).

Diffusion will only acquire a significant amplitude when the X-rays are reflected by ~10^{23}–10^{24} electrons contained in 1 cm^3 of a crystal are in phase so that Bragg diffraction or small angle diffusion in almost the same direction as the incidence occurs.

The calculation done here is not applicable to polarized incident X-rays, which is the case when using synchrotron radiation but not when using conventional sources of X-rays. When the incident radiation is not polarized, the weighted intensity is evaluated in (b) by the factor $\left(\frac{1+\cos^2 2\theta}{2}\right)$ in which the angle 2θ is such that $2\theta = \overrightarrow{Oz}, \overrightarrow{OM}$. (For further details see Ref. [9].)

Finally, the atomic electrons are not free and their binding energy $\hbar\omega_0$ represents the energy necessary to remove them from their initial state (1s, 2s, 2p, etc.) The corrective term $\tilde{f}$ is nearly equal to 1 when $\omega >> \omega_0$, corresponding to incident photons of energy $h\nu$ greater than the binding energy of the atomic electrons. In this case, the atomic electrons diffuse the X-rays like free electrons. Conversely when $\omega = \omega_0 + \varepsilon$ (with $\varepsilon > 0$) an anomalous diffusion related to the real part of f) and an absorption effect is observed, related to the imaginary part of f and directly correlated with the friction term,

$-mv/\tau$. For values $\omega < \omega_0$, the classical approach seen here must be replaced with a quantum mechanical one. We will consider the influence of these phenomena in the propagation of X-rays in materials (see Chapter V, Ex. 27).

Problems

Problem 1: X-ray diffraction by cubic crystals

The powder method is used to record the diffraction pattern of the three cubic cyrstals, A, B, and C which crystallize in the body-centered cubic, face-centered cubic, or diamond structures.

The measured diffusion angles Φ (between the directly transmitted rays and those relative to the two first rings) give the following results:

Crystal A: 42.2° and 49.2°
Crystal B: 28.8° and 41°
Crystal C: 42.8° and 73.2°

- Determine the corresponding structure (bcc, fcc, diamond, etc.) of each crystal.
- Find the lattice parameter a in each case using the wavelength of $\lambda = 1.5$ Å.
- Show that the information on the Φ value is insufficient when one of the crystals is simple cubic possibly.

Solution:

The measure angles Φ correspond to 2θ.

If the crystal is body-centered cubic, the forbidden reflections are such that $h + k + l = 2n + l$. The two first allowed reflections are (110) and (200). The relation between the corresponding $\sin\theta$ will be such that:

$$\sin\theta_{200}/\sin\theta_{110} = d_{110}/d_{200} = 2/\sqrt{2} = 1.414$$

If the crystal is face-centered cubic, the allowed reflections are indices with the same parity. The two first ones are (111) and (200). The relation with $\sin\theta$ is:

$$\sin\theta_{200}/\sin\theta_{111} = 2/\sqrt{3} = 1.154$$

If the crystal is a diamond structure in addition to the forbidden reflections of the fcc lattice, one must consider those introduced by the two-atom basis of identical atoms at (000) and (1/4, 1/4, 1/4). The corresponding forbidden reflections are such that $h + k + l = 2n+2 = 2, 6, 10, \ldots$. The two allowed reflections are (111) and (220). The relation between the $\sin\theta$ is:

$$\sin\theta_{220}/\sin\theta_{111} = \sqrt{8/3} = 1.633$$

The structures are thus the following: A is fcc, B is bcc, and C is diamond.

Next, the lattice parameters can be determined using:

$$2\frac{a}{\sqrt{h^2+k^2+l^2}}\sin\theta = \lambda$$

a(A) = 3.608 Å; a(B) = 4.265 Å; a(C) = 3.56 Å which is clearly diamond, see Fig. 3d.

If one of the crystals was cubic simple, the relation of the two first allowed reflections (100) and (110) would be equal to $\sqrt{2}$ and one could not distinguish between body-centered cubic without taking into account the angles related to the seventh reflection, that is the (220) in the case of a simple cubic and the (321) in the case of a body-centered cubic.

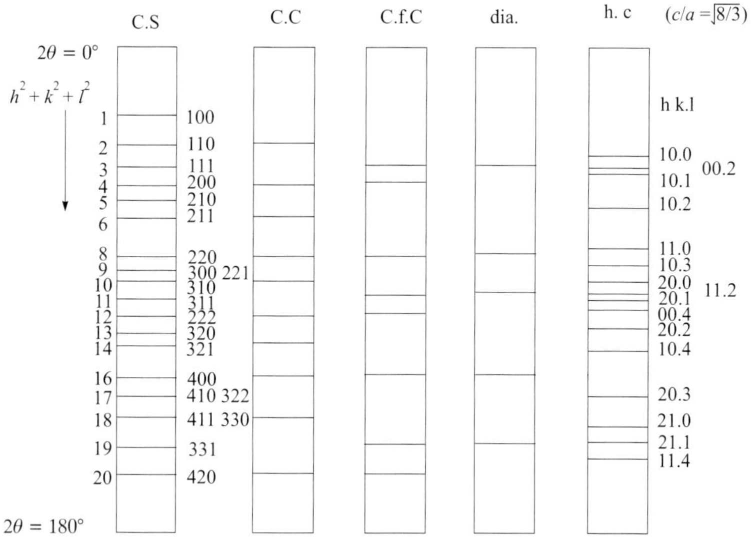

Figure 34

Figure 34 summarizes the different allowed reflections in the different cubic structures and in the hcp structure with an ideal c/a ratio. (This figure is inspired by Ref. [7] and the description by Ref. [1].

Problem 2: Analysis of an X-ray diffraction diagram

Figure 35 shows the intensity profile of a powder diagram obtained from X-ray diffraction by a cubic crystal.

- Find the Bravais lattice of the corresponding crystal and index the observed reflections.
- Determine the lattice parameter a knowing that the K_α radiation of copper was used (λ = 1.54 Å).
- Taking a two-atom basis with Z_1 at (000) and Z_2 at (1/2,0,0), find the approximate ratio of atomic numbers Z_1/Z_2.

The material shown in Fig. 28 is in fact KBr (Z_1 = 19; Z_2 = 35). What are some of the possible differences between the theoretical and experimental determination of the ratio of atomic numbers?

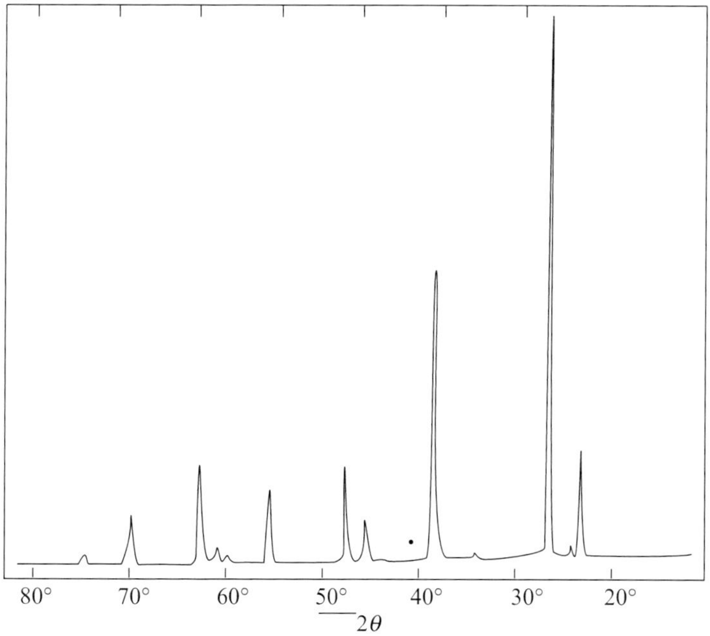

Figure 35

Solution:

The two first reflections are located at $2\theta \approx 23.6°$ and $27.1°$. The relation between sinus of the angles θ is ~1.145. It is very close to the expected result for a fcc Bravais lattice (1.154) (see Pb. 1: X-ray diffraction by cubic crystals). The allowed reflections have indices of the same parity. The sequence of reflections on the diagram is (111); (200); (220); (311); (222); (400); (331); (420); (422); and (333). The lattice parameter a can be found from the measure of the angle $2\theta_{422} \approx 70°$, which results in $\frac{2a}{\sqrt{24}}\sin\theta = 1.54$ Å and $a = 6.57$ Å. The structure factor of the basis is $f_{Z_1} + f_{Z_2}e^{-i\pi h}$ which implies that the intensities, I of the reflections relative to the values of h (k and l) odd are weaker than the ones corresponding to the even indices.

To the first approximation the relation $I(200)/I(111)$ corresponds to:

$$(f_{Z_1} + f_{Z_2})^2/(f_{Z_1} - f_{Z_2})^2 \approx (Z_1 + Z_2)^2/(Z_1 - Z_2)^2.$$

The measure of $I(200)/I(111)$ (≈ 4.9) leads to an estimation of Z_1/Z_2 of order 2.56, which is rather far from the expected value of 1.84. In fact the value must take into account that the crystal is ionic (K^+ and Br^-). The expected theoretical relation is thus ~2. This charge transfer explains the fact that crystals of K^+Cl^- (19–1 = 17 + 1) show odd reflections $h + k + l$ which are quite forbidden.

Other causes of this disparity are due to:

(i) Atomic diffusion factors are proportional to the atomic number of the corresponding atoms (Z_1 and Z_2) only for a diffusion angle of zero (which is the direction of the transmitted wave). Outside of this direction, the atomic diffusion factor decreases when the angle of diffusion increases. This change is a result of the non-negligible size of the atoms compared to the wavelength λ, which introduces a phase difference between the various diffused waves by the Z electrons (see Ex. 22).

(ii) Thermal vibrations of atoms induce a decrease in the diffused intensity as the order of the reflections increases.

(iii) In the powder method the reflections intensity is proportional to the number of equivalent planes responsible of the corresponding reflection. There are eight for the 111 and six for the 200.

All of these considerations explain the decrease with 2θ of the reflection intensities of the same parity. It is also important to consider how the intensities are taken: profile along a radius or integrated intensity along the perimeter of each ring.

Problem 3: Low-energy electron diffraction (LEED) by a crystalline surface: absorption of oxygen

Accelerated by a voltage V of ~156 volts, a beam of mono-energetic electrons strikes a single-crystalline surface. As shown in Fig. 36 the diffraction diagram is obtained by reflection.

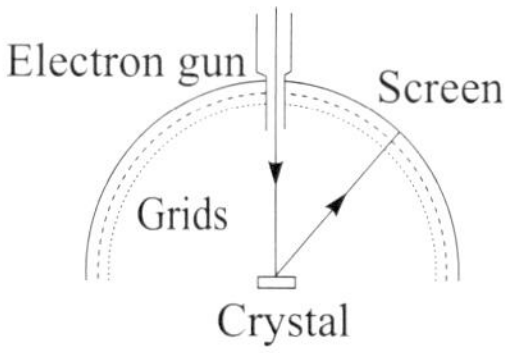

Figure 36

(1) Show that the reciprocal lattice of such a surface is composed of rods. Using the Ewald construction and after having determined the associated wavelength of incident electrons, show that the diffraction diagram is always formed of points independently of the incidence angle.

(2) (a) Consider a normal incidence and a crystalline surface such as the (100) face of Ni with a fcc Bravais lattice and with lattice parameter, c = 3.52 Å.

Represent the corresponding reciprocal space in the plane of vectors $\vec{A}, \vec{B}$ containing the origin 00.

Using a graphic construction, state the order of the different reflections, which are accessible in an experiment where a hemispherical screen is centered on the crystal.

(b) Same as in part (a) but now consider the (111) face.

(3) The (111) and (100) faces are exposed to a partial pressure of oxygen and the oxygen atoms form a lattice superimposed on the lattice of the crystalline nickel. The diffraction diagrams undergo the modifications shown in Figs. 37 and 38.

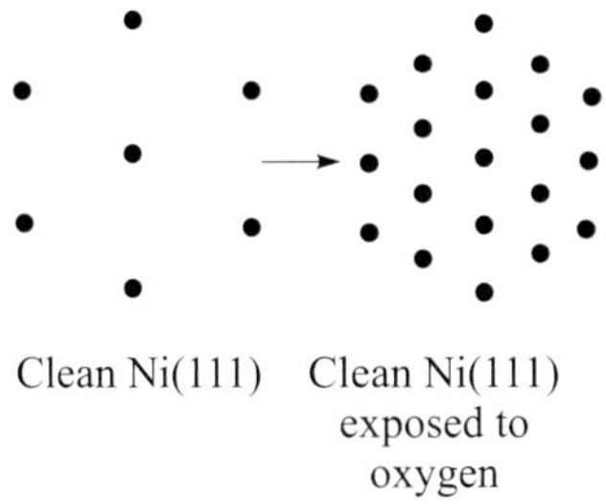

Figure 37

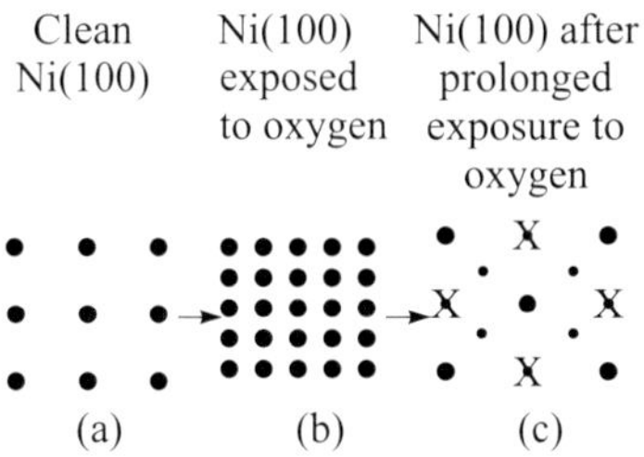

Figure 38

Considering only the geometry of these diagrams, characterize the new elemental cell relative to the (111) face and the two successive lattices relative to the (100) face.

(4) We would like to determine the position of the oxygen atoms relative to the nickel atoms taking into account that both are regularly distributed. Restricting the study to the (100) face that is exposed to oxygen for the longest time, there are two possible hypotheses:

(a) The oxygen atoms are on the top of some of the nickel atoms.

(b) The oxygen atoms are located at the center of certain squares formed by the atoms of nickel.

Determine the corresponding structure factor. To simplify the problem, assume that the presence of the oxygen atoms does not change the amplitude diffused by the nickel atoms.

Considering the intensities shown in Fig. 38c, which model is the best, a or b? (h, e, m_0)

Solution

(1) In the case of a 2D lattice (see Ex. 19) the reciprocal lattice will consist of lines perpendicular to the plane $(\vec{a}, \vec{b})$ (which is also the plane $\vec{A}, \vec{B}$) and the traces of the lines in the plane are determined from the origin with the help of the translations vectors $\vec{A}$ and $\vec{B}$.

One can also note that the reciprocal space is the Fourier space (in 3D) of the direct space. In direct space, the surface results in a Dirac function in the direction normal to it. Thus, the corresponding direction in reciprocal space is a constant. The intersections of the lines of the reciprocal lattice with the Ewald sphere (such as D in Fig. 39a) determine the direction of the diffused beams giving rise to constructive interferences, that is, to say to points on the diffraction diagram. In the $\vec{A}, \vec{B}$ plane, the observed reflections are contained in the circle of radius $|k_0| = 2\pi/\lambda$ and are shown as lines in the plane (Fig. 39b).

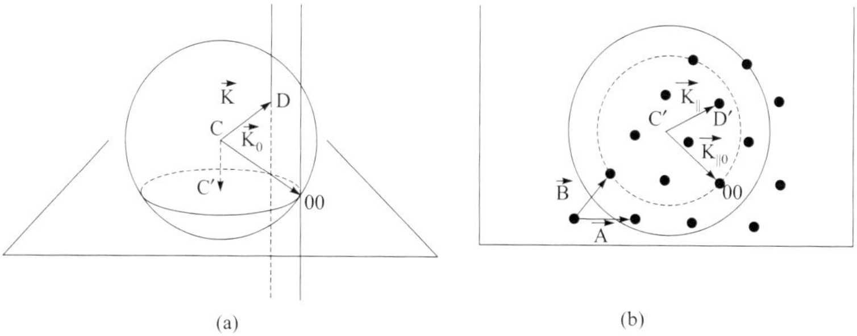

Figure 39

$$\lambda = h/[2m_0eV_0(1+eV_0/2m_0c^2)]^{1/2}$$

$$\lambda(\text{Å}) \approx 12.26/\sqrt{V} \approx 0.98\ \text{Å}$$

The last result holds because the relativistic effects are negligible.

(2) (a) The (100) face of a fcc lattice is a square lattice with constant $a = c/\sqrt{2} = 2.49$ Å because the square centered lattice is not a Bravais lattice in 2D (see Ex. 17). The length of the basis vectors $\vec{A}, \vec{B}$ of the reciprocal lattice is such that $|\vec{A}| = \dfrac{2\pi}{a} = 2.52\ \text{Å}^{-1}$.

The experimentally accessible reflections (h, k) correspond to the points of the reciprocal lattice included in the projection of the Ewald sphere on the surface:

$$(h^2+k^2)(2\pi/a)^2 \le (2\pi/\lambda)^2$$

In total, we can only visualize the nine reflections of the type 01 and 11 because $h^2 + k^2 \le 2.5$: 10, $\bar{1}0$, 01, $0\bar{1}$, 11, $1\bar{1}$, $\bar{1}1$, $\bar{1}\bar{1}$ including the specular reflection at 00.

(b) See Fig. 40.

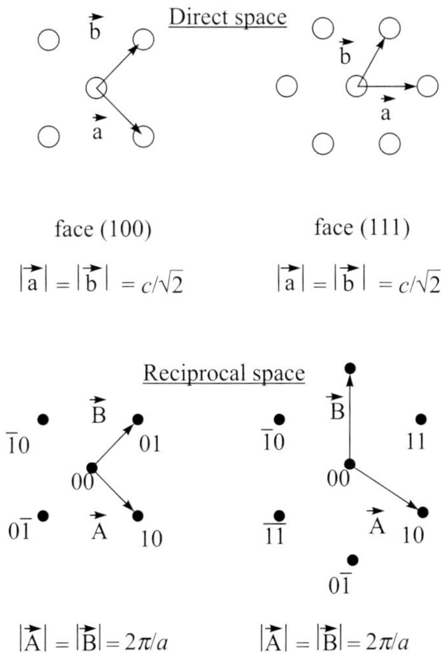

Figure 40

(3) After examining the change of the diffraction pattern that results by the contribution of oxygen, one can say that the some of the oxygen atoms are regularly distributed on the surface because the diagram is made up of points and not of rings. When the (111) or the (100) faces of nickel transform from being free of oxygen to being covered with oxygen (first stage for the (100) face, the basis vectors of the reciprocal lattice (that is to say the diffraction diagram) are reduced by 2. This implies that the inter-reticular distances are doubled in each direction of the direct lattice and that there is one oxygen atom for four atoms of nickel. In the case of prolonged exposure in the Ni(100) face, each reciprocal basis vector has been reduced by $1/\sqrt{2}$ (or equivalently that the simple square lattice as become a centered square lattice). This implies that the number of oxygen to Ni is 1 to 2 and that the translation vectors of the direct lattice are multiplied by a factor $\sqrt{2}$ This result is expected: when the direct lattice increases,

the reciprocal lattice decreases proportionately. (This is a property of Fourier space.)

(4) To decide between the two hypotheses, it is sufficient to consider the structure factor for each case, as shown in Fig. 41 (a and b). For the situation where Ni(100)—p($\sqrt{2}\times\sqrt{2}$)45°—0, for example, and taking $\vec{a}_s$ and $\vec{b}_s$ as the primitive vectors, hypothesis (a) corresponds to a centered lattice with a basis of one oxygen atom at 00 and two nickel atoms at 0,0 and ½,½.

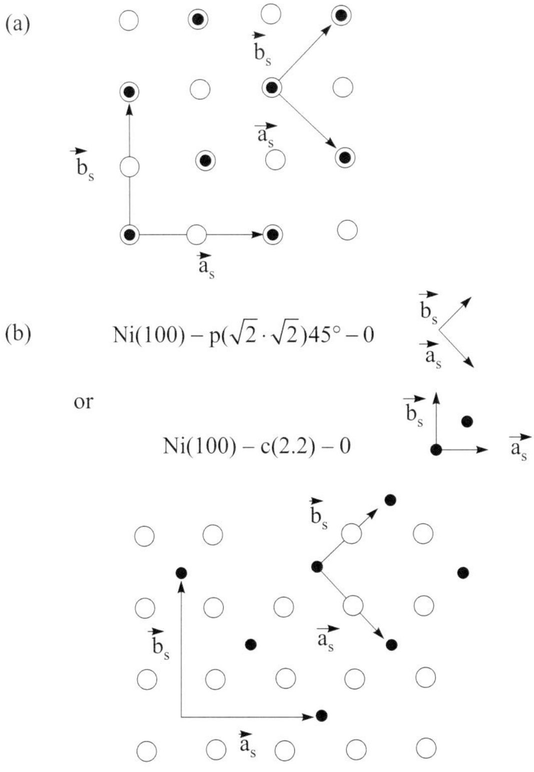

Figure 41

The factor of structure will be such that:

$$F(h,k) = f_0 + f_{\mathrm{Ni}}(1 + e^{-i\pi(h+k)})$$

This leads to two types of diffractions spots with inequal intensities: those corresponding to $h + k$ even which are strong

$\approx (f_0 + 2f_{Ni})^2$ and those corresponding to $h + k$ odd which are weak $\approx f_0^2$.

In the hypothesis (b), the lattice is the same as in (a) but the basis consists of an atom of oxygen at 00 and two atoms of Ni at ½,0, and 0,½. The corresponding structure factor is:

$F(h,k) = f_0 + f_{Ni}(e^{-i\pi h} + e^{-i\pi k})$

This structure factor leads to three types of intensities:

(i) h and k are odd: $I \approx (f_0 - 2f_{Ni})^2$

(ii) h and k are even: $I \approx (f_0 + 2f_{Ni})^2$ where the intensity is the strongest

(iii) h and k have opposite parity, $I \approx f_0^2$

Taking into account the diagrams shown in Fig. 38c, it is the last hypothesis that explains the data with (i) x, (ii) ●, (iii) •.

Comments: LEED and crystallography of surfaces

(1) Incident electrons are charged particles and therefore, are sensitive to the crystalline potential of the crystal periodicity. The first experiment was conducted on Ni by C. Davisson and L. Germer in 1927, which confirmed the de Broglie hypothesis saying that particles of matter such as electrons have wave-like properties.

Having a relatively weak kinetic energy, the penetration depth of incident electrons is very weak and they can only penetrate the first few atomic layers of the surface. The diffraction diagram is thus related to the surface layer and can be very different from that obtained by transmission (see Fig. 27, Ex. 21).

As a result of this sensitivity, it is necessary to maintain an ultra-high vacuum pressure (lower than 10^{-10} torr or 1.33×10^{-8} Pa) in the diffraction apparatus. The product pressure $\times$ exposure time defines the unit of a Langmuir (10^{-6} torr $\times$ second), which corresponds in general to the coverage of a monoatomic layer of residual gas on all surfaces and in particular on that of the investigated sample when the sticking coefficient for atoms and molecules of the specific gas is unitary.

The controlled admission of oxygen in the apparatus relies on this mechanism and is easily explained by the kinetic theory of gases. The obtained covering rate (θ = ¼ or ½) corresponds therefore sensitively to the exposure of ¼ or ½ Langmuir.

(2) The surface system the most widely investigated in the past was that of sulfur on nickel because of the surface migration (or segregation) of impurities of sulfur when nickel is raised to temperatures greater than 800°C. The detailed analysis of the diffused intensities has established that the sulfur atoms are regularly and spontaneously interposed between nickel atoms (situation b).

The system analyzed here has been studied to elucidate the position of the oxygen sites at the first steps of the oxidation process of nickel.

Even if in the present problem the qualitative interpretation of intensities was sufficient to resolve the present problem, in general the use of just the structure factor is insufficient. Opposite to X-ray diffraction, the analysis of the diffraction intensity of electrons needs to include dynamic theories, not just kinetic theories, because of the strong interaction between electrons and matter.

(3) To represent a system corresponding to a substrate in the presence of superstructures created by the adsorbed atoms, the following notation is often used:

$$M(h,k,l)-\alpha\left(\frac{a_S}{a}\times\frac{b_S}{b}\right)\zeta-S$$

where, M(h, k, l) represents the face h k l of the substrate M, S is the chemical symbol of the adsorbed atoms, α indicates if the lattice is primitive (p) or centered (c); $(a_s/a)\times(b_S/b)$ is the relation between the unitary vectors of the adsorbed atom lattice and that of the substrate, and ζ is the angle between $\vec{a}_S$ and $\vec{a}$, which is omitted when the angle is zero.

As an example, the legend in Fig. 41 uses this notation. Nevertheless, it may be confusing in certain cases and we refer to reader to the lecture by E. A. Wood (1964), *Journal of Applied Physics* **35**, 1306. For slow electron diffraction in general, see R. L. Park and H. H. Madden (1968) *Surface Science* **11**, 188 or L. J. Clarke (1985), in: *An Introduction to LEED*, John Wiley.

Problem 4: Reflection high-energy electron diffraction (RHEED) applied to epitaxy and to surface reconstruction

The Bravais lattice of gallium arsenide and that of aluminum is face-centered cubic. The crystalline parameters are a (GaAs) = 5.635 Å and a (Al) = 4.05 Å, respectively.

(1) *Surface crystallography*: Adopt a common scale and represent the distribution of arsenic atoms (in GaAs) and thus, also of aluminum atoms (in Al) in the (001) plane. The basis of GaAs consists of an atom of As in the 000 and an atom of Ga in the ¼, ¼, ¼.

(2) *Lattice matching*: Consider the epitaxy of Al on GaAs, that is to say the growth of single crystal of Al on a GaAs single crystal. To realize such an operation successfully, it is necessary that the two crystals have a lattice mismatch, $\Delta a/a$, smaller than a few percent. Show graphically that the epitaxy of Al (001) on GaAs (001) is nevertheless possible if the [100] axis of GaAs and that of Al form an angle of 45°.

Because of the difference in size of the different atoms, the atomic sites of Al must not coincide with those of As and the surface density of Al atoms must be close to that bulk Al. Thus, the fit of the Al lattice to that of AsGa leads to its slight contraction or dilation. Evaluate the corresponding relative variation Δa (Al)/a (Al).

(3) *Diffraction*: Epitaxy is realized by sublimating metallic aluminum under ultra-high vacuum conditions on a single crystalline substrate of GaAs at an appropriate temperature, using the molecular beam epitaxy (MBE) technique.

The operation is controlled by irradiating the sample at a grazing angle ($\theta \sim$ a few degrees) with a beam of fast electrons (~10 keV). The diffraction pattern of reflected electrons is observed on a screen perpendicular to the plane of the sample and to the plane of incidence. Taking into account the weak electron penetration in the material, this diagram indicates a perfectly flat surface or a surface with roughness.

(a) Using the Ewald construction (take λ = 0.12 Å), explain why the surface diffraction diagram (2D lattice), and in particular the specular reflection at (0,0), consists of elongated spots. Give the direction of the elongated spots.

(b) The plane of incidence, which is vertical, contains the row [100] of the (001) plane, which is horizontal, of GaAs. Show in the sample plane the different points of reciprocal space of GaAs (001) and those of Al (100) when this last has been epitaxial (see part 2 of this problem). In the horizontal plane of the sample, deduce the angular separation between the specular reflection point and the first diffraction spot in the case of the initial virgin surface of GaAs and when this surface is covered by a single layer of Al.

(4) *Reconstruction*: The surfaces considered up to now are ideal surfaces in which the atoms conserve their ideal positions from the bulk crystal. However, such surfaces are $\overline{1}$ unstable and their energy can be diminished when they are rearranged (or reconstructed) at the surface. Compared to an ideal surface of GaAs, the atomic sites are often practically unchanged in the surface plane, but certain atoms have moved slightly deeper in the bulk, while others appear to be raised. In the case of the (100) surface of GaAs, lightly doped with arsenic for stability, the As atoms emerge $\overline{1}$ under certain conditions to form a periodic over-layer having a lattice four times larger in the [110] direction and two times larger in the [1 $\vec{a}_S$ 0] direction compared with the ideal primitive surface lattice. This type of reconstruction is called (2×4). (For the notation in the general case see commentary no. 3 of the preceding problem).

Show the 2×4 reconstruction of GaAs (001) in direct and reciprocal space. State the influence of this reconstruction on the diffraction diagram taking into account that the emerging atoms will have an atomic diffusion factor slightly greater than the atoms that are located further down from the ideal.

(5) *Intensity oscillations*: When the epitaxy of aluminum is formed layer-by-layer (Franck, Van der Merwe model as opposed to growth by pyramidal islands), the intensity of the central spot (specular reflection) evolves sinusoidally as a function of the coverage rate, θ, of the substrate. It is maximal when the number n of deposed single layers is an integer and minimal when $\theta = n + ½$ (see Fig. 42). Can you explain simply why this result is true regardless of the incident angle α with respect to the surface?

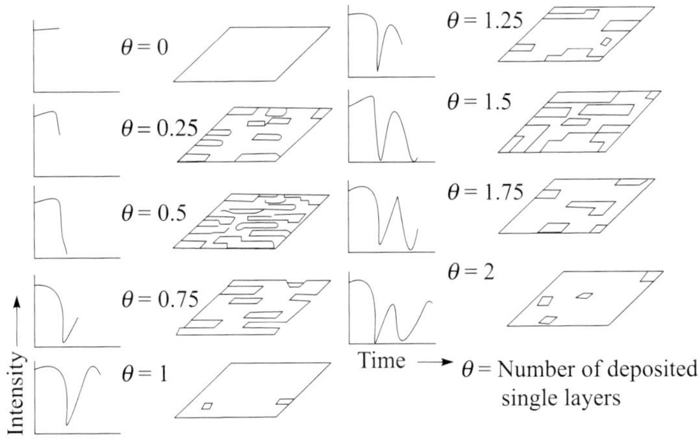

Figure 42

Solution:

(1) See Fig. 43, left panel. Note that the (001) face of a fcc crystal has points distributed on a square centered lattice of side *a*, but that such a lattice is not a 2D Bravais lattice (see Ex. 17). The Bravais lattice is a simple square with parameter $a/\sqrt{2}$ of which the sides, for GaAs, are parallel to the [110] and [$\bar{1}$10].

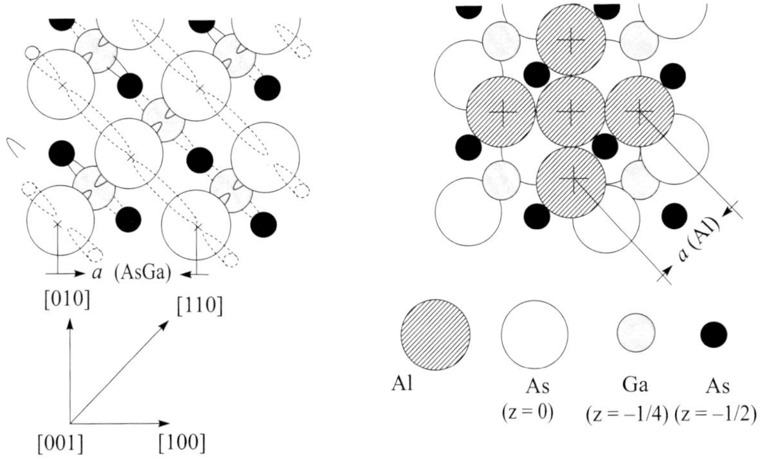

Figure 43

(2) (See Fig. 43, right panel) The atoms of Al must be inserted between the atoms of As such that the surface density of Al and As are nearly conserved $n_s(\text{Al})/n_s(\text{As}) = a^2(\text{GaAs})/a^2(\text{Al}) = 1.936 \sim 2$. The Al atoms on the (001) GaAs face form a simple square lattice with parameter $a\,(\text{Al})/\sqrt{2}$. The edges of this lattice are parallel to the [100] and [010] axis of GaAs and therefore, form a 45° angle with their equivalents in the arsenic lattice.

In bulk aluminum the distance between two atoms in contact is $a(\text{Al})/\sqrt{2} = 2.864$ Å; however, in the epitaxial layer $a(\text{Al}) = a(\text{GaAs})/2 = 2.817$ Å. The epitaxied Al layer is thus trained by $\Delta a/a \approx -1.6\%$.

(3a) The reciprocal lattice of a 2D lattice is a lattice formed with vertical rods perpendicular to the surface (see Ex. 19 and previous problem). In the surface plane, the reciprocal square lattice is characterized by $|\vec{\text{A}}| = |\vec{\text{B}}| = 2\pi/d$. For the (001) face of GaAs $|\vec{\text{A}}| = \dfrac{2\pi}{a(\text{As Ga})/\sqrt{2}} \approx 1.57\,\text{Å}^{-1}$; its direction is parallel to the [110] row. For epitaxied aluminum on this face, $|\vec{\text{A}'}| = 2.23\ \text{Å}^{-1}$ ($\vec{\text{A}'} \parallel [100]$). The radius of the Ewald sphere $|\vec{\text{k}_0}| = |k| = \dfrac{2\pi}{\lambda} = 52\ \text{Å}^{-1}$, is more than 20 times larger than the primitive vectors of the reciprocal lattice.

Taking into account this different dimension, at grazing incidence the Ewald sphere will be practically tangent to the rod passing through the origin of the reciprocal lattice. The spot, 00, of specular reflection will therefore be an elongated form with its principal axis being perpendicular to the sample plane. (See the representation in Fig. 44). If there is surface roughness, the grazing electrons crossing these prismatic crystals will give a diffraction pattern formed with points. In this case the reciprocal lattice will be formed of points and not rods because it is relative to a 3D lattice. In particular, the points situated normal to the surface at point 00 (Fig. 44) will be equidistance from $2\pi/c$ and will not form a continuous line.

(3b) Figure 45 shows the different points of the reciprocal lattice in the plane of the sample.

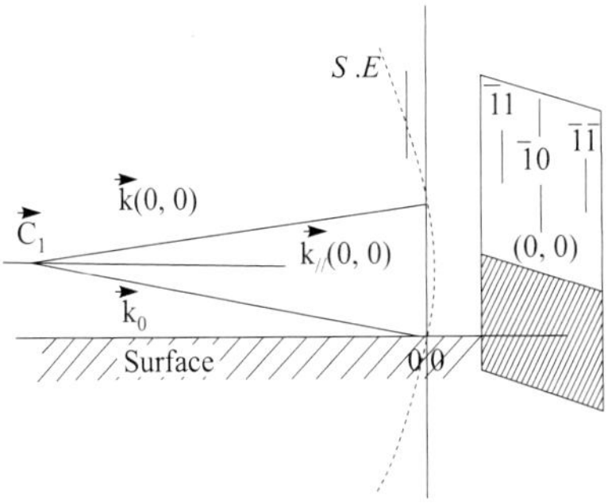

Figure 44

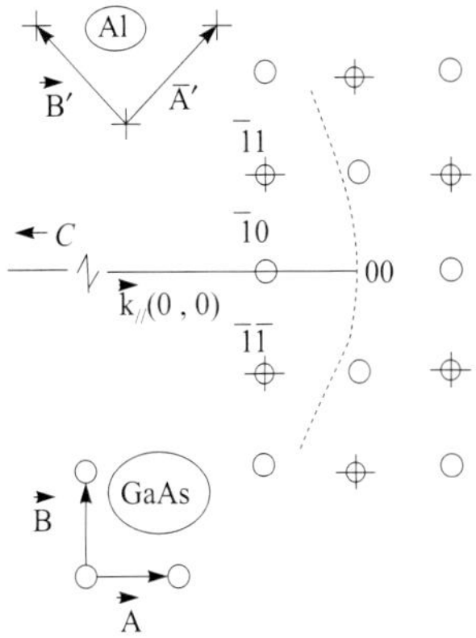

Figure 45

The screen displays the intersections of the Ewald sphere with the different rods of the reciprocal lattice postulating that the directions thus, defined are contained within the useful solid angle defined by the point source C and the screen. For arsenic and for aluminum, there are the two symmetric lines at $\bar{1}1$ and $\bar{1}\bar{1}$ (using the vectors $\vec{A}, \vec{B}$ of the 2D lattice of As on the (001) face of GaAs). Moreover, for As the $\bar{1}0$ spot situated below the 00 spot appears—although it is forbidden for aluminum. Relative to the direction of $\vec{k}$ (0,0), the angular separation β between these two symmetric spots (common in the two diagrams) will be given by $\beta \sim \tan\beta = \dfrac{|\vec{A}|}{|\vec{k}_0|} = 0.03 \text{ rad} = 1.73°$.

(4) Figure 46a shows the direct lattice and the reciprocal lattice of the reconstructed (001) face of GaAs. Conditioned by the reciprocal lattice, the form of the diffraction diagram presents additional spots related to the reconstruction. The incident beam containing the axis [110], is essentially of periodicity of order 2 along the axis $[1\bar{1}0]$ and it appears as two satellite spots with a smaller intensity and situated between the spot associated with $(\bar{1}1)$ and $(\bar{1}0)$ for one case and $(\bar{1}0)$ and $(\bar{1}\bar{1})$ for the other case: spots marked with a B in Fig. 46b. An azimuth rotation of the sample of 90° would lead to the

appearance of three equidistant satellite spots between the two principle spots $(\bar{1}1)$ and $(\bar{1}0)$ instead of a single one.

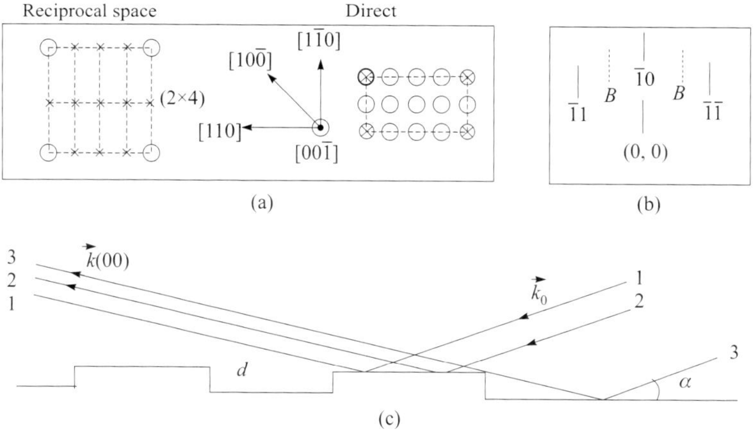

Figure 46

(5) The specular reflection of electron waves on terraces are all in phase because they all have a zero phase difference between them: Rays 1 and 2 in Fig. 46c. Conversely, the phase difference between the electron waves reflected on a terrace and the valley of another region (Rays 1 and 3) is $2d \sin\alpha$ where the grazing angle α is not a Bragg angle except accidentally. Then, the interference between these waves is destructive and the effects more pronounced when d approaches the condition $2d \sin\alpha = (n + ½)\lambda$. Consequently, the specular spot, 00, is bright initially when the initial surface is atomically flat. Its intensity decreases with the increase in the areas of terraces during atom deposition by MBE. Its intensity vanishes when the area of terraces is equal to the area of valleys $\theta = 1/2$. Its intensity starts to increase again when the valleys are progressively filled with evaporated atoms and the specular spot is bright again when the coverage θ is unity. During a layer-by-layer growth the specular reflection oscillates in intensity and the number of oscillations is equals to the number of atomic layers deposited on the initial surface. Finally, the automatic recording of the oscillations permits to monitor the number of atomic layers being deposited and it

permits to stop the evaporation when the required number of layers is attained. This simplified explanation does not take into account refraction effects due to the inner potential that have to be combined to purely geometrical shadowing effects. The key result is that the intensity oscillations of the specular spot are characteristic of layer-by-layer growth: Franck, Van der Merwe model (see *Surface Science* **168**, 1986, 423, for more details concerning these oscillations as well as Turco et al. *Revue de Physique Appliquée* **22**, 1987, 827). Besides its great interest for fundamental surface science, epitaxy is also of interest in applied science, most notably for the semiconductor industry. It allows the realization of most devices used in microelectronics and optoelectronics, ranging from metal-semiconductor contacts (ohmic or Schottky) to more complicated structures such as superlattices and quantum wells (see Pb. 10 and Chapter V, Pb. 9).

***Additional Remark*:** In fact even in the case of a surface plane at the atomic scale all the electrons are not susceptible to interference. Their spatial and temporal coherence is limited by the energetic dispersion of the incident beam $\Delta E/E$ and by the beam divergence on the sample β. For the spatial coherence Δx one obtains $\Delta x = \lambda/2\beta\,[1 + \Delta E/2E]^{1/2}$ which is typically of order 1000 Å.

Comments: Surface reconstruction and stepped surfaces

(1) *Surface reconstruction:* The presence of the surface modifies the position normally occupied by the atoms in a 3D crystal. One distinguishes in fact two types of displacements illustrated in Figs. 47–49.

(a) A surface relaxation is characterized by a variation of the spacing between the last atomic plane and the penultimate one. In most of the cases the spacing, $d_{1\text{-}2}$, between first and second atomic layers contracts relative to bulk spacing, d_{bulk}. There is no change in the periodicity parallel to surface.

This situation happens mainly in metals where the density of the conduction electron gas, $n(z)$, in the bulk decreases suddenly when approaching the surface (see Chapter IV, Pb. 2). Then the surface atoms have lost electrons and they tend to move in the direction of optimum density. In some case positive ions are attracted toward the bulk region of large electron density. However, in some cases, for example, Al(100) and (111) surfaces, the most densely packed planes undergo no contraction or a slight expansion. For ionic crystals, the relaxation effects are easier to evaluate from the calculation of the Madelung constant (see for instance Chapter II, Ex. 7).

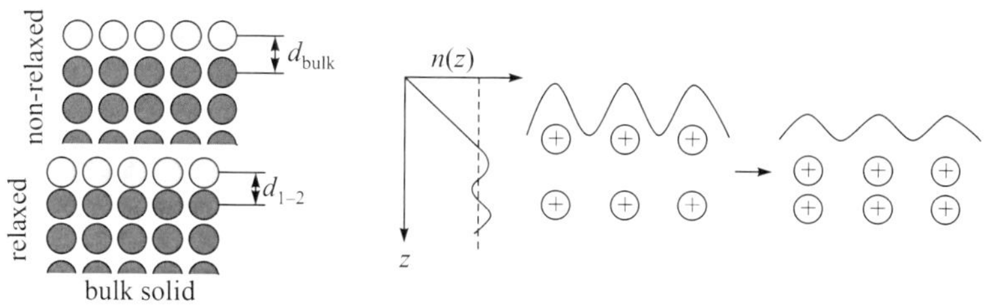

Figure 47 Side view of relaxation (left), driving force (right), resulting from the in-depth evolution of the electron density, $n(z)$.

The surface undulation can occur in ionic crystals as in ionocovalent materials like GaAs when the anions and the cations are subject to an alternating vertical displacement (here, As goes toward the exterior and Ga toward the interior. The driving forces are Coulomb forces where the surface ions of a given sign are attracted toward the bulk by ions in excess of opposite sign or are pushed toward the exterior by ions in excess of the same sign.

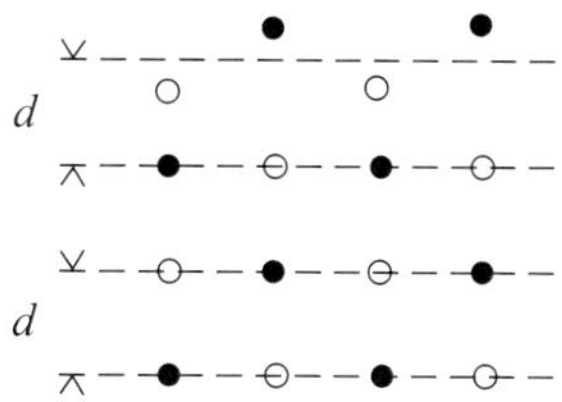

Figure 48 Side view of a surface undulation.

The situation described in Pb. 4 above combines in fact reconstruction and undulation. For details on this point, see Ref. [26].

(b) A surface reconstruction occurs when two neighboring surface atoms move closer to form a "dimer" bond. This reconstruction may occur on metal surface but its mechanism is understood more easily in covalent crystals. In such crystals, diamond or Si, the atoms on an ideal surface would have two dangling bonds because of the missing atoms in the vacuum side. In reconstructed surfaces each atoms has one dangling bond instead of two as it may be seen in Fig. 49 for Si(001) 2×1 surface reconstruction.

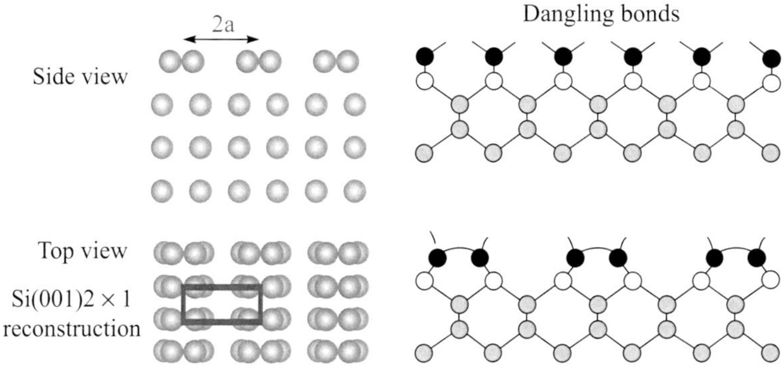

Figure 49 Side view and front view of Si(001) 2×1 surface reconstruction (left). Its driving force (right) results from the formation of dimer bonds.

(2) *Surface defects: Stepped surfaces*

There are many deviations to ideally flat surface such as terraces, steps, kinks, and adatoms. At the atomic scale, the names of some of them are indicated in Fig. 50 for cubic crystals. The investigation of these surface defects are of prime importance in surface science topics such as crystal growth, surface diffusion, roughening, and vaporization as well as in catalysis where the reactivity of foreign atoms on a crystalline substrate is a function on their position. The thermodynamics of crystal surface formation, transformation, and reactivity is describes simply with the terrace ledge kink (TLK) model. This model is based upon the idea that the energy of an atom's position on a crystal surface is determined by its bonding to neighboring atoms and where transitions simply involve the counting of broken and formed bonds. For instance, the layer-by-layer homo-epitaxy is easily understood from the thermally assisted migration of a weakly bounded adatom toward monoatomic steps or kinks.

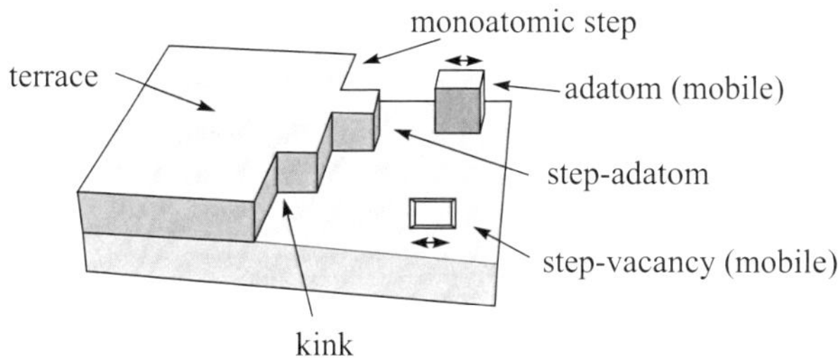

Figure 50 Names for the various atomic positions in the TLK model.

A correlated topic is that of stepped surfaces and their description. Example of such a stepped surface is shown in the top of Fig. 42. Illustrated in the bottom of Fig. 51, this description is often of the form S-$[m(h, k, l) \times n(h', k', l')]$, where S denotes a stepped surface, $m(h, k, l)$ denotes a h, k, l terrace m atoms width and $n(h', k', l')$ denotes the step height (omitted if $n = 1$). For additional details the readers are referred to Refs. [27] and [28].

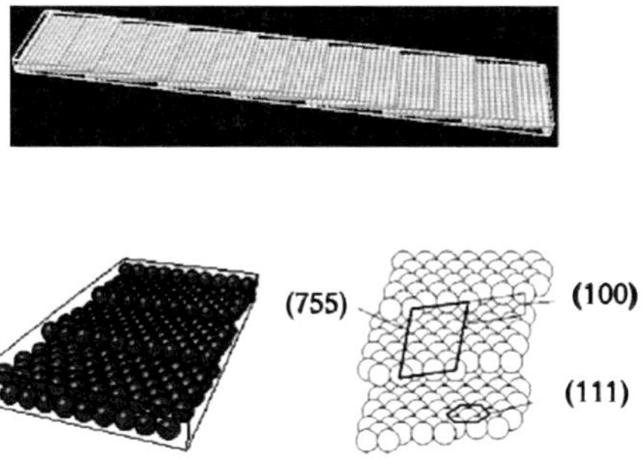

Figure 51 Top: View of a stepped surface. Bottom: Stepped surface of platinum described by Pt S-[7(111) × (100)] for a (111) terrace seven atoms wide and a (100) step, one atom high. Another description would be Pt (755).

Comment: Surface imaging (LEEM and STM)

It has been seen that diffraction experiments permit to access directly to the reciprocal space. The use of low-energy incident electrons at normal incidence or the use of high-energy incident electrons at grazing incidence permits to access to the 2D reciprocal space of surfaces because of the low penetration depth of these electrons. A direct visualization of surfaces at (or close to) the atomic resolution is possible from the use of microscopes based also on low-energy incidents electrons.

In particular there is the low-energy electron microscope (LEEM) which has been invented by Bauer (E. Bauer, *Surface Review and Letters* **5**, 1998, 1275, and *Reports on Progress in Physics* **57**, 1994, 895). LEEM is a technique used by surface scientists to image atomically clean surfaces, atom–surface interactions, and thin (crystalline) films. For instance its use permits to visualize terraces of monoatomic steps on a Si(100) surface from the constructive interferences of 6 eV electrons reflected from the two terraces: a mechanism similar to that giving rise to oscillations of the specularly reflected spot (0,0) as described above. An important aspect of this surface microscopy is that it is very fast. Videos of the evolution of the surface with nanometer resolution (~10 nm) can be acquired in real time to follow thin film growth such as, the growth of Co islands on Ru. Many other microscopy techniques have been developed based upon LEEM. These include photo-excitation electron spectroscopy (PEEM); mirror electron microscopy (MEM); reflectivity contrast imaging; spin-polarized LEEM (SPLEEM). In PEEM, upon exposure to electromagnetic radiation (photons), secondary electrons are excited from the surface and imaged. PEEM was first developed in the early 1930s, using ultraviolet (UV) light to induce photoemission of electrons (see Chapter IV, Pb. 4). However, since then, this technique has made many advances, the most important of which was the pairing of PEEM with synchrotron radiation.

With a slightly different performance, there is also the scanning tunneling microscope (STM) invented by G. Binning and H. Rohrer (Nobel Prize in physics, 1986). Its principle is the subject of Chapter IV, Pb. 5, where it is shown that the resolutions of this microscope are ~1 Å, lateral resolution, and ~0.1 Å depth resolution. With these resolutions, individual atoms on surfaces are routinely imaged and manipulated. The STM can be used not only in ultra-high vacuum but also in air, water, and various other liquid or gas ambient, and at temperatures ranging from near 0 K to a few hundred degrees centigrade (see R. V. Lapshin in H. S. Nalwa, ed., *Encyclopedia of Nanoscience and Nanotechnology* **14**, 1994, 105). Many other microscopy techniques have been developed based upon STM. There are considered briefly in Chapter IV, Pb. 5, and other STM methods involving manipulating the tip in order to change the topography of the sample or the atomic deposition of metals with any desired (pre-programmed) pattern are indicated in Chapter II, Ex. 11. The atomic force microscope (AFM) and the STM are two early versions of scanning probes that launched the wide field of nanotechnology. See also Chapter II, Ex. 11, and Chapter IV, Pb. 5.

Problem 5: Identification of ordered and disordered alloys

Following the elaboration conditions, the alloy Cu_3Au (type L12) can be obtained in an ordered or in a disordered phase. The ordered phase corresponds to a cubic crystal having an atom of gold at 000 and three atoms of copper at ½ ½ 0, ½ 0 ½, and 0 ½ ½. In the disordered phase all the preceding sites are randomly occupied by either Au (P_{Au} = 25%) or by copper atoms (P_{Cu} = 75%).

Determine the structure factors of each of these two phases. Identify the reflections relative to the different rings relative to in the powder diagrams shown in Fig. 52a and b below. Which pattern corresponds to the ordered phase?

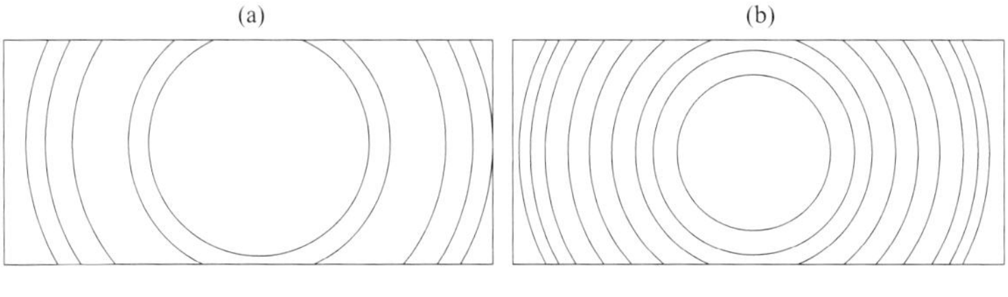

Figure 52 Powder patterns of Cu_3Au.

In reality, depending on the temperature of formation, the alloy can be obtained in intermediate phases between the two extremes just mentioned. Each phase is characterized by an order parameter at large distance, S, such that, $S = (P_A - N_A)/(1 - N_A)$, where N_A represents the fraction of atoms A of the alloy and P_A is the probability for this type of atom to occupy the expected sites of the ordered phase. State briefly how the X-ray diffraction experimentally allows the determination of the order parameter S and why the method will not work for the case of brass β (type L2) of which the ordered phase is simple cubic with one atom of Cu at 000 and one atom of Zn at ½ ½ ½.

Solution:

The structure factor of the alloy Cu_3Au in the ordered phase has the form:

$$F_0(h,k,l) = f_{Au} + f_{Cu}\{\exp - i\pi(h+k) + \exp - i\pi(h+l) + \exp - i\pi(k+l)\}$$

It is different from zero for all the reflections of simple cubic crystals. In the disordered phase, the structure can be considered as a fcc crystal having an average basis of one atom, resulting from the association of ¾ of Cu and ¼ of Au. The structure factor is thus:

$$F_D(h,k,l) = f_{(3Cu+Au)/4}\{1 + \exp - i\pi(h+k) + \exp - i\pi(h+l) + \exp - i\pi(k+l)\}$$

The reflections for which the indices h, k, and l are not the same parity will therefore be forbidden and their absence in the experimental diagram will therefore, characterize the disordered phase.

The expected intensities in the two extreme situations can easily be deduced from the table given by the structure factor for the different reflections.

Cu_3Au	Reflections	$F_0(h, k, l)$–Ordered alloy	$F_D(h, k, l)$–Disordered alloy
	(100)	$f_{Au}-f_{Cu}$ (SL)	0
	(110)	$f_{Au}-f_{Cu}$ (SL)	0
	(111)	$f_{Au}+3f_{Cu}$	$4f_{\frac{3Cu}{4}+\frac{1}{4}Au}$
	(200)	$f_{Au}+3f_{Cu}$	$4f_{\frac{3Cu}{4}+\frac{1}{4}Au}$
	(210)	$f_{Au}+f_{Cu}$ (SL)	0

Cu_3Au	**Reflections**	**$F_0(h, k, l)$-Ordered alloy**	**$F_D(h, k, l)$-Disordered alloy**
	(211)	$f_{Au}-f_{Cu}$ (SL)	0
	(220)	$f_{Au}+3f_{Cu}$	$4f_{\frac{3Cu}{4}+\frac{1}{4}Au}$
	(221) (300) Confound	$f_{Au}-f_{Cu}$ (SL) $f_{Au}-f_{Cu}$ (SL)	0 0

The measured intensities of the lines normally forbidden by the fully disordered structure (known as superstructure lines (SL) in the table) allow therefore to experimentally determine the order parameter S of Bragg and Williams. This parameter varies between 0 and 1 when going from the disordered to the ordered phases.

The X-ray diffraction is relatively ineffective in the case of Brass β because the lines of the superstructure (corresponding to $h + k + l$ odd) have a very weak intensity even for the ordered phase ($f_{Zn} - f_{Cu} \sim Z(Zn) - Z(Cu) = 30 - 29$). In such a case, already discussed for KCl (see Pb. 2), the use of neutron diffraction will be more precise (see Pb. 7).

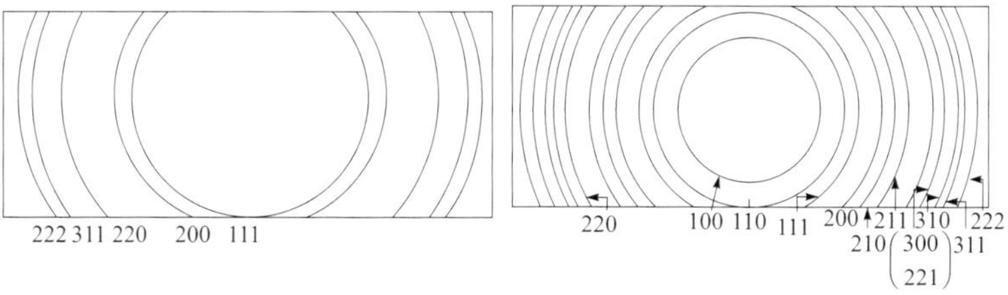

(a) Disordered Cu_3Au (b) Ordered Cu_3Au

Figure 53 Indexed diffraction rings.

Problem 6: X-ray diffraction study of an AuCu alloy

The crystal structure related to the ordered phase of AuCu (type L10) is shown in Fig. 54.

Describe (Bravais lattice and basis) this structure. State the sequence of first non-equivalent reflections relative to the lattice in the order corresponding to increasing Bragg angles. Determine the structure factor and the intensities of the different reflections (strong, weak, and forbidden).

The disordered phase of this alloy is cubic ($c = a$). The Cu and Au atoms are randomly placed on the different sites. Describe the new crystal structure. Eliminate the forbidden reflections and then list the order of the first non-equivalent allowed reflections.

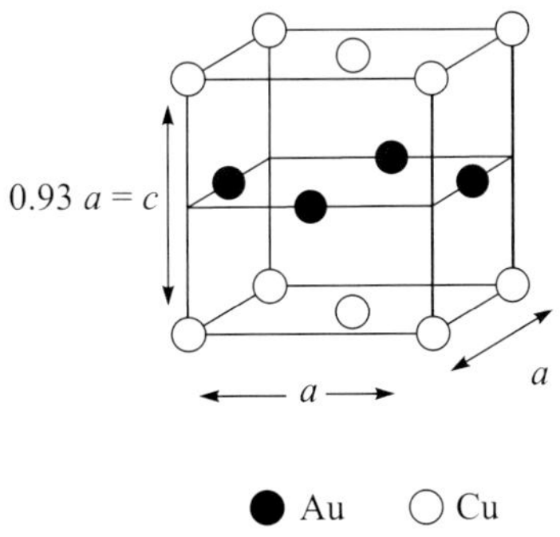

Figure 54

Solution:

The lattice shown in Fig. 54 is not a primitive cell. It is of order 2 as it may be seen by taking the origin on a Cu atom at a corner and by pointing out that all the other lattice points are on all the other Cu atoms including those situated at the center of the top and bottom squares. Like for a centered square lattice in 2D, a tetragonal Bravais lattice with one face centering does not exist. Thus, the correct description is simple tetragonal with $a' = a\sqrt{2}/2$, $c' = c$ with a basis of one Cu at 000 and one Au at ½ ½ ½.

The distance between the forbidden inter-reticular planes is given by:

$$d_{h,k,l} = \left[\frac{h^2}{a'^2} + \frac{k^2}{a'^2} + \frac{l^2}{c^2}\right]^{\frac{1}{2}}$$

It is used to establish the order of the non-equivalent reflections of increasing θ (corresponding to decreasing $d_{h,k,l}$: (001), (100), (101), (111), (002), etc.

All the reflections are allowed and those such that $h + k + l$ are odd will be less intense than the others. Nevertheless, anticipating the description of the disordered fcc phase, one can also consider that the represented structure is a non-primitive quadratic cell with $a' = a$, $c' = c$ with a basis of two Cu at 000 and ½ ½ ½ and two Au at ½ 0 ½ and 0 ½ ½.

The increasing sequence of non-equivalent reflections would then be (100), (001), (110), (101), (111), (200), (002), (210), (201), (102), (211), (112).

The structure factor is given by:

$$F(h,k,l) = f_{\text{Cu}}(1 + e^{-i\pi(h+k)}) + f_{\text{Au}}(e^{-i\pi(h+l)} + e^{i\pi(k+l)})$$

The forbidden reflections correspond to values of h and k having different parity. In addition the corresponding reflections with h, k, l of the same parity will be strong, $F_M = 2f_{\text{Cu}} + 2f_{\text{Au}}$, and those having h, k, and l of a different parity will be weak, $F_M = 2f_{\text{Cu}} - 2f_{\text{Au}}$). The sequence of allowed reflections are thus (001) and (110): F_m; next (111), (200), and 002): F_M; (201) (112): F_m; (220) and (202): F_M, etc.

One finds the same results to build the reciprocal space if one takes care to remove the forbidden reflections and the same intensities at the same Bragg angles. The final result is independent from the initial description: primitive or non-primitive cell. Only the notation of the points will be different because the vectors $\vec{a}$, $\vec{b}$ in the square plane will be changed and thus the orientations $\vec{A}$, $\vec{B}$ will change as a result.

The ordered phase is fcc with an atom Cu/2 + Au/2 at 000. The reflections for which h, k, and l do not have the same parity are forbidden. The sequence of allowed reflections is (111), (200), (220), (311), (222), etc. Compared to the description of reflections relative to the quadratic lattice of order 2, one observes that the weak reflections have disappeared and that two previously non-equivalent reflections (example, 200 and 002) have become equivalent.

Problem 7: Neutron diffraction of diamond

The figure below is adapted from a neutron diffraction pattern of diamond (fcc lattice with 2 C atoms at 000 and ¼ ¼ ¼; lattice a = 3.56 Å, see Fig. 3d). Corrected from some experimental factors such as the temperature effects, this figure represents the angular variation of the scattered intensities, by postulating that they are uniquely proportional to the square of the structure factor $F(h, k, l)$.

- Give the indices of the three reflections shown in the figure.
- Explain why they have unequal weight.
- Deduce the associated wavelength λ and the kinetic energy, E_k (in eV) of incident neutrons.

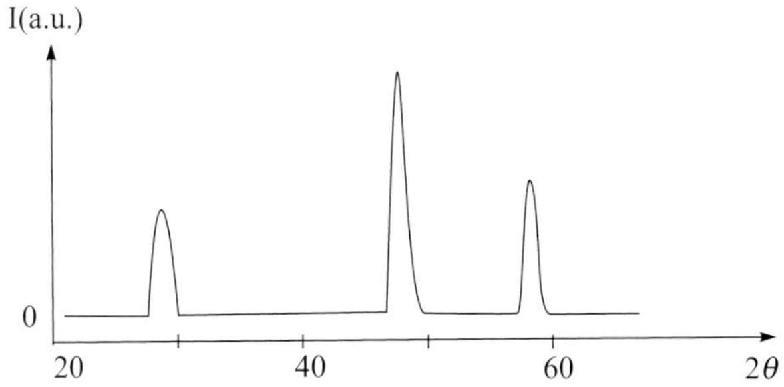

Figure 55

- With respect to X-rays with comparable wavelength value (see Pb. 2, Fig. 35), one can observe that the atomic form factor of neutrons $f(n)$ does not depend on θ. Why is this? (M, h, e)

Solution:

$$F(h,k,l) = f_c \left(\underbrace{1 + e^{-i\frac{\pi}{2}(h+k+l)}}_{\text{Basis}} \right) \left(\underbrace{1 + e^{-i\pi(h+k)} + e^{-i\pi(h+l)} + e^{-i\pi(k+l)}}_{\text{fcc lattice}} \right)$$

The first three allowed reflections are (111), (220), and (311) (see Fig. 34). In intensity their weight is given by $F(h, k, l) \times F^{\times}(h, k, l)$. The results are $2(4f_c)^2$; $8(4f_c)^2$, and $2(4f_c)^2$ where the coefficient of two results from $(1 + i)(1 - i)$. The angle $2\theta_{311}$ is exactly equal to 58°.

The Bragg's law for this reflection leads to λ = 1.04 Å from $\lambda = h/Mv$.

The kinetic energy of incident neutrons is:

$$E_k = (1/2)Mv^2 = h^2/2M\lambda^2 \text{ or } E_k(\text{eV}) = [0.28/\lambda(\text{Å})]^2$$

$$E_k = 72.5 \times 10^{-3} \text{ eV}$$

The atomic diffusion factor of neutrons is isotropic because the neutrons are neutral particles interacting with the atomic nucleus and not with the electrons, as it is the case for X-rays. The dimension of this nucleus is smaller than λ whereas, the electron cloud was comparable to λ. The phase composition that is taken into account in the atomic form factor for X-rays (see Pb. 2) and therefore does not need to be considered for thermal neutrons.

Comments: Neutron diffraction

The neutrons considered here are called thermal (or cold) neutrons because their energy is of the order $k_B T$ (around 25 meV at ambient conditions). They are obtained by nuclear reactors, such as the ISIS neutron source in England, or the at the Spallation neutron source at Oak Ridge national laboratory in the United States. Incident neutrons interact with the nucleus of atoms in the target either elastically or in elastically. Elastic collisions change the direction of incident neutrons without energy exchange between the projectile and the target atoms. The corresponding application is neutron diffraction or elastic neutron scattering and it is the subject of the present exercise. In addition to these elastic interactions (that is, λ is not changed and the process is coherent), there are also collisions that involve a change in energy with phonons and magnons. In inelastic collisions, the incident neutrons lose or win energy as a consequence of atomic vibrations (phonons) or of atomic spin (magnons): the corresponding application is neutron spectroscopy (discussed in Chapter III, Pb. 7: Phonons in germanium and neutron diffusion). C. G. Shulland and B. N. Brockhouse shared the Nobel Prize in physics in 1994 for pioneering contributions to the development of neutron scattering techniques for studies of condensed matter, in particular, "for the development of neutron spectroscopy".

Here, the technique of neutron diffraction is similar to X-ray diffraction but due to their different scattering properties, neutrons and X-rays provide complementary information. Whereas in crystalline diffraction using X-rays the amplitude is proportional to Z, the advantage of neutron diffraction is that the diffused amplitude has no correlation with the atomic number. Thus the diffused amplitude of light elements such as H and C is comparable to that of other elements, which is important when distinguishing two neighboring elements. The applications of neutron diffraction concern the structural analysis of materials such as polymers (C–H chains) or alloys of the type CuZn, which in X-ray diffraction give rise to certain reflections that are nearly forbidden in their ordered phase; f_{Cu}–f_{Zn} because $Z(\text{Cu}) = 28$ and $Z(\text{Zn}) = 29$ (see Pbs. 5 and 11 and also Pb. 2 when applied to KCl).

Although neutrons are uncharged, they carry a spin, and therefore interact with magnetic moments, including those arising from the electron cloud around an atom. Neutron diffraction can therefore reveal the microscopic magnetic structure (ferro, ferri, antiferro, etc.) of materials (http://en.wikipedia.org/wiki/Neutron_diffraction). In the reverse to non-magnetic materials, magnetic scattering does require an atomic form factor as it is caused by the much larger electron cloud around the tiny nucleus. The intensity of the magnetic contribution to the diffraction peaks will therefore, dwindle toward higher angles.

Problem 8: Diffraction of modulated structures: application to charge density waves

Consider a row of N atoms equally spaced by a (N is a very large even number). The electronic charge carried by each atom is the sum of a term common to all the atoms and of a term that varies sinusoidally with the position of the atom in the row. The periodicity of the modulation is d where $d > a$. For an atom in position ma, the atomic form factor is given by:

$$f(m\vec{a}) = f_1 + f_2 \cos \vec{q} \cdot m\vec{a}, \text{ where } |q| = 2\pi/d \text{ and } \vec{q} \parallel \vec{a}.$$

(1) Taking into account the term of phase difference, $e^{-i\Delta\vec{k}\cdot\vec{r}}$ (see Chapter II, Course Summary, Section 4) find the expression giving the amplitude diffused by N atoms in the direction $\vec{k}$ when these atoms are irradiated with an incident beam of wave vector $\vec{k}_0$ ($|\vec{k}_0| = |k|$ and $\Delta\vec{k} = \vec{k}_0 - \vec{k}$). Deduce the conditions imposed on $\Delta\vec{k}$ for the diffused amplitude-different from zero. Show these results in reciprocal space.
Verify that when the modulation is zero, the typical diffraction conditions are recovered, $\Delta\vec{k} \cdot \vec{a} = 2\pi n$, and show that the presence of this modulation results in satellite lines in the diffraction pattern. (Hint: Express the cosine using Euler notation.)

(2) Consider the hypothesis $d = 2a$. Where are the satellite lines located? What is their amplitude? What is the amplitude of the principle lines? Compare with the results established directly using the structure factor of arrow of alternating atoms. Consider in particular the results relative to $\left|\Delta\vec{k}\right| = 2\pi/a$ and $\left|\Delta\vec{k}\right| = \pi/a$.

(3) The sinusoidal modulation has an incommensurable period with that of the lattice, that is, to say that d/a is an irrational number. For the present case, give the position of the satellite lines compared to the principle lines for $d/a = 3.6$.

In a lamellar crystal such as graphite (see Ex. 17), the charge density wave is of the form $\Delta\rho(r) = \Delta\rho(\cos\vec{q}_1\vec{r} + \Delta\rho(\cos\vec{q}_3\vec{r})$ and it is incommensurable. Show schematically the form of the diffraction pattern (or the reciprocal lattice) obtained by the transmission electron experiments at normal incidence, knowing that $|q_1| = |q_2| = |q_3| = 2\pi/3.6a$; that these three vectors are parallel to the plane of the layer at 120° from one another and directed parallel to the principle axes of the lattice.

Solution:

(1) $A = \sum_{m=1}^{N} f(m\vec{a})\, e^{-i\Delta\vec{k}\cdot m\vec{a}}$

Expanding $f(ma)$, the first term $\sum_{m=1}^{N} f_1 e^{-i\Delta\vec{k}\cdot m\vec{a}}$ corresponds to the amplitude diffused by the unperturbed lattice. This amplitude will be non-zero (and equal to Nf_1 only when $\Delta\vec{k}\cdot\vec{a} = 2\pi n$ where n is a whole number. This is the usual condition: $\Delta\vec{k} = \vec{G}$ (see Course Summary).

Using Euler notation for trigometric functions, the second term reduces to:

$$\sum_{m=1}^{N} (f_2/2)(e^{i\vec{q}m\vec{a}} + e^{-i\vec{q}m\vec{a}})e^{-i\Delta\vec{k}m\vec{a}}$$

It will be nonzero for $(\Delta\vec{k} \pm \vec{q})\vec{a} = 2\pi n$. The amplitude of the corresponding satellite lines will be equal to $Nf_2/2$.

Figure 56 shows the electronic distribution in direct space (top of the figure) and the weighted amplitude of the diffraction spots in reciprocal space (bottom of the figure).

For the principle lines, $\Delta\vec{k}\cdot\vec{a} = 2\pi n$ corresponds to $\Delta k_{||} = 2\pi n/a$ (where $\Delta k_{||}$ is the projection of $\Delta\vec{k}$ on the axis $Ox \parallel \vec{a}$).

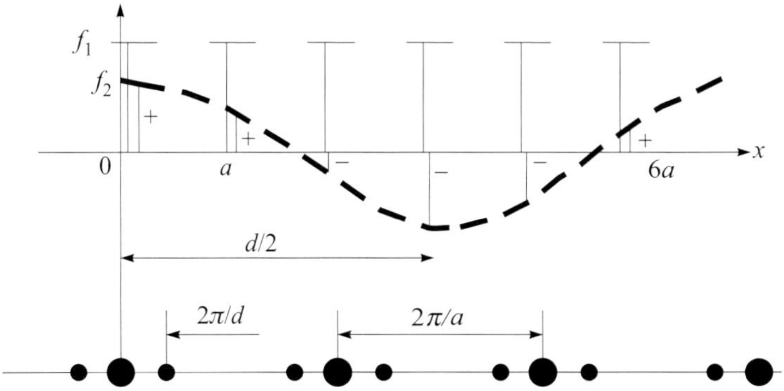

Figure 56

For the satellite lines $(\Delta\vec{k} \pm \vec{q})\vec{a} = 2\pi n$ leads to $\Delta k_{||}\cdot a = 2\pi n \mp 2\pi a/d$. These lines, weighted by $Nf_2/2$ disappear when f_2 is zero.

(2) When $d = 2a$, the adjacent satellite lines will merge and difficult to distinguish. Their amplitude will be Nf_2, and that of the principle lines will not change. This result can be directly obtained by considering a lattice with $d = 2a$ consisting of an atom Z_1 at 0 such that $f_{Z_1} = f_1 + f_2$ and an atom Z_2 at position $d/2$ such that $f_{Z_2} = f_1 - f_2$.
This is the same situation that was seen in Ex. 18. There are $N/2$ bases and the reciprocal lattice has periodic steps of n/a. The amplitude associated to the even points (such that $\Delta k_{||} = 2\pi/a$, for instance) will be A (even n) = $(N/2)$ $(f_{Z_1} - f_{Z_2})$ = Nf_1. The amplitude associated to the odd points (such that, e.g., $\Delta k_{||} = \pi/a$) will be A (odd n) = $(N/2)$ $(f_{Z_1} - f_{Z_2})$ = Nf_2. (See Ex. 18 for an easy calculation of the structure factor F(h)).

(3) The satellite lines are found at $\Delta\vec{k} = \vec{G} \pm \vec{q}$ with $|\vec{q}| = 2\pi/3.6a$.
The charge density waves can materialize on electron diffraction patterns. These patterns correspond most prominently to the reciprocal lattice (the large black dots in Fig. 57). The effect of the charge density waves correspond to smaller satellite spots (the small black dots). The observation of such a pattern shows the incommensurability of the

charge density waves (refer Ex. 17 for the construction of the reciprocal lattice of the graphite type).

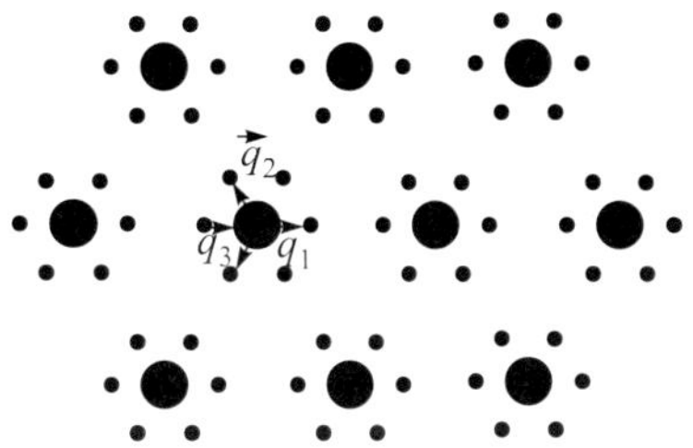

Figure 57

Remarks: There are numerous composed materials with reduced dimensions that have structural transitions at low temperature, known as Peierls transitions. These show up as a sinusoidal modulation of the crystal lattice which is related to a similar modulation of the electronic density (also known as a charge density wave). This incommensurability has been observed by electron microscopy of the trichalcogenides of transition metals, $NbSe_3$ and TaS_3.

This phenomenon of incommensurability can also be observed by diffraction in a number of other situations: condensation of rare gases on graphite or magnetism of $MnAu_2$ for instance. (For further information, see Société Française de Physique (1978) *Solid State Phase Transformations in Metals and Alloys*: Ecole d'été d'Aussois. Editions de Physique, p. 523; in particular, the article by Pouget on modulated structures).

Problem 9: Structure factor of $Ga_xAl_{1-x}As$

The Bravais lattice of GaAs, like that of AlAs is face-centered cubic with an atom of As at 000 and an atom of Ga (or Al) at ¼ ¼ ¼ (see Fig. 3c, Ex. 1, the equivalent structure of ZnS). The two lattices have practically the same lattice parameter a and starting from AlAs, it is possible to progressively substitute atoms of Al for those of Ga. On the initial Al sites, this substitution is random such that a ternary composition is formed that can be crystallographically defined as having x atoms of Ga and $(1-x)$ atoms of Al (with $0 \le x \le 1$) in the ¼ ¼ ¼ position.

(1) Find the structure factor $F(h, k, l)$ of AlAs ($x = 0$) and of GaAs ($x = 1$). Find the first five non-equivalent allowed reflections as a function of their atomic form factors f_{Ga}, f_{Al}, and f_{As}. Next, evaluate $F(h, k, l)$ numerically taking $f_Z \propto Z$, such that $f_{Al} \propto 13$, $f_{Ga} \propto 31$, and $f_{As} \propto 33$.
All the sites are now occupied by Si atoms: determine the forbidden reflections in silicon (or in diamond).

(2) In the case of $Ga_xAl_{1-x}As$, state the general expression of $F(h, k, l)$ and indicate the evolution of $F(2\ 0\ 0)$ as a function of x. Explain what is observed.

Solution:

(1) $$F(h,k,l) = \sum_j f_j \exp - i2\pi(hu_j + kv_j + lw_j)$$

It becomes

$$F(h,k,l) = \left(f_{As} + f_{Al} e^{-i\frac{\pi}{2}(h+k+l)} \right) \sum_{j'} \exp - i2\pi\left(hu_{j'} + kv_{j'} + lw_{j'}\right)$$

The first term between large parentheses is the contribution of the basis and the summation is that of the fcc lattice with: $u_{i'}, v_{i'}, w_{i'}, = (000), (1/2\ 1/2\ 0), (0\ 1/2\ 1/2)$, and $(1/2\ 0\ 1/2)$
The reflections for which h, k, and l do not have the same parity are forbidden and the following table is obtained:

hkl	(111)	(200)	(220)	(311)	(222)
$F(h, k, l)$	$f_{As}+if_{Al}$	$f_{As}-f_{Al}$	$f_{As}+f_{Al}$	$f_{As}-if_{Al}$	$f_{As}-f_{Al}$
AlAs	33 + i13	20	46	33–i13	20
GaAs	33 + i31	2	64	33–i31	2

For silicon instead, in addition to the forbidden reflections resulting from the fcc lattice, there are the reflections such that $h + k + l = 2 + 4n$ that are also forbidden because the two identical atoms form the basis (see Pb. 1).

(2) $$F(h,k,l) = \left(f_{As} + [xf_{Ga} + (1-x)f_{Al}] e^{-i\frac{\pi}{2}(h+k+l)} \right) \cdot \sum_{j'} (\text{cfc lattice})$$

$F(200) = f_{As} - xf_{Ga} - (1-x)\ f_{Al}$. When x changes from 0 to 1,

$F(200)$ decreases linearly from 20 to 2. The (200) reflection of GaAs becomes almost forbidden when the partial electron transfer from As to Ga atoms is taken into account. The measured intensity of this reflection (FF^x) therefore allows the determination of x and establish the composition of this ternary semiconductor (see also Chapter V, Ex. 30).

Problem 10: Structure factor of superlattices

Consider a 1D infinite crystal with lattice constant $d = Na$ of which the basis consists of N atoms chemically different but equidistant by a. (The atomic form factor of an atom at position j is f_j).

(1) Derive the structure factor $F(h)$ from the general expression of the structure factor $F(h\ k\ l)$?

(2) In fact the basis consists of N_1 atoms of species A ($f_j = f_A$) followed by N_2 atoms of species B ($f_j = f_B$) in such a manner that the crystal structure is now a 1D superlattice (with $N = N_1 + N_2$). From the resulting summation in $F(h)$, rewrite $F(h)$ in function of f_A, and f_B and $\alpha = \exp\text{-}i(h2\pi/N)$.

(3) Determine the expression of $F(h)$ for the following three hypothesis:

  (a) $h = nN$ (where, n is whole number)

  (b) $h = nN/N_1 \neq n'\ N$ (where n' is whole number)

  (c) $h = (2n + 1)\ N/2$ (where N is even and N_1 is odd)

(4) State the order (h) of the forbidden reflections when $N_1 = N_2$ and next when $N_1 = N_2/2$. Verify the validity of the results for the simple case where, $N_1 = N_2 = 1$. Show that it is possible to determine N_1 and N_2 from the results established in (3) and from the order of the maximal reflections and of the forbidden reflections.

(5) In the general case express the quantity FF^x, which governs the intensity of the X-ray diffraction of such a structure. To simplify the problem assume that f_A and f_B are real and state the results using multiples and sub-multiples of the trigonometric functions of the angle $2\pi/N$. In the hypothesis where, $N_1 = N_2 = N/2$, establish the law of the intensity variation of the first allowed reflections when N is very large compared to the order of the reflections.

(6) With $f_A = 3f_B$, represent the comparative table of the reflection intensities as a function of h ($1 \leq h \leq 13$) for the three following situations: (a) $N_1 = N_2 = 6$; (b) $N_1 = 5$; $N_2 = 7$; (c) $N_1 = 6$; $N_2 = 7$ with $f_A = 3f_B$. Comments are provided on the result.

(7) The crystalline structure, now a 3D structure, can be deduced from the row described before in (2) by perpendicular translations of this row of type $n\vec{a}+m\vec{b}$ (where $\vec{a}$ is in the y-direction and $\vec{b}$ is in the z-direction with $\vec{a}=\vec{b}$). Thus, the resulting crystal consists of N_1 atomic layers of element A (simple cubic with lattice parameter a) followed by N_2 layers of element B (also simple cubic with lattice parameter a) repeated infinitively (see Fig. 58). What is the general expression of the structure factor for this superlattice? Is there a significant change compared to the 1D description?

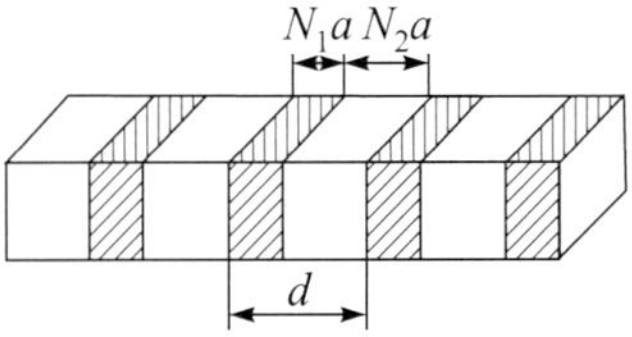

Figure 58

(8) Represent the diffraction pattern of the superlattice limited to the plane xOy when $N_1 = N_2 = 6$. (Hint: Starting from a cubic reciprocal lattice, associate to each point a spot size proportional to the intensity of the X-rays scattered by the reticular planes of the corresponding indices.)

Solution:

(1) The general 3D expression of the structure factor is:

$$F(h,k,l)=\sum f_j \exp -i2\pi(hu_j+kv_j+lw_j)$$

In 1D and using reduced coordinates (in unit of $d = Na$ where the position of an atom j is $u_j = j/N$), $F(h)$ takes the form:

$$F(h)=\sum_{j=0}^{N-1} f_j \exp -i2\pi(h_j/N)$$

(2) Applied to a basis composed of N_1 atoms A followed by N_2 atoms B, the expression of $F(h)$ becomes:

$$F(h) = f_A \sum_{j=0}^{N_1-1} \exp - i2\pi h(j/N) + f_B \sum_{j=N_1}^{N-1} \exp - i2\pi(j/N)$$

Writing $\alpha = e^{-ih\Phi}$ with $\Phi = 2\pi/N$, the sum of the two geometric progressions is:

$$F(h) = f_A \frac{1-\alpha^{N_1}}{1-\alpha} + f_B \alpha^{N_1} \frac{1-\alpha^{N_2}}{1-\alpha} \quad (1)$$

This expression is undetermined when 0/0, which occurs when $\alpha = 1$ [see 3(a) below], otherwise, since $\alpha^N = 1$, it can be written:

$$F(h) = (f_A - f_B)(1-\alpha^{N_1})/(1-\alpha) \quad (2)$$

(3) (a) When the order of reflections, h, is such that $h = nN$ (where n is an integer), Eq. (1) is undetermined. To find the result, we determine the relation between the derivatives with respect to n or h and find $F(h = nN) = N_1 f_A + N_2 f_B$. The corresponding amplitude is maximal when $h = N$).

(b) When $h = nN/N_1 \neq n'N$, the numerator goes to zero (Eq. 2) and the corresponding reflections are forbidden.

(c) When N_1 is odd and N is even, Eq. 2 can be simplified: $F(h) = f_A - f_B$ for $h = (2n + 1)\,N/2$ because the exponentials are equal to –1.

(4) When $N_1 = N_2$, the even reflections will be forbidden (see 3(b) above) except for the reflection $h = n(N_1 + N_2)$ which will be maximal [see 3(a) above].

When $N_1 = N_2/2 = N/3$, for which h is a multiple of 3 the reflections are forbidden with the exception of the multiple reflections of N.

When $N_1 = N_2 = 1$, we find the result established in Ex. 18 $F(h \text{ even}) = f_A + f_B$; $F(h \text{ odd}) = f_A - f_B$.

Consequently, the order of reflections giving the maximal intensity (and in particular the first ones) allows one to find the number N of atoms corresponding to the basis [see 3(a) above]. Next, the order of the forbidden reflections allows one to calculate N_1 [see 3(b) above]. If N_1 and N are not too large, it is possible to determine the geometry of the superlattice from a simple examination of the diffraction pattern. The analysis of the corresponding intensities also allows one in some cases to find the chemical identification of the superlattice from f_A and f_B which are sensitively proportional to $Z(A)$ and $Z(B)$ (see Pb. 9).

(5) By multiplying by the complex conjugates, functions of the type $(1-e^{i\Phi})$ become:

$$2-e^{i\Phi}-e^{i\Phi}=2-2\cos\Phi=4\sin^2(\Phi/2)$$

Thus

$$F(h)\cdot F^x(h)=\frac{(f_A-f_B)^2\sin^2(N_1h\pi/N)}{\sin^2(h\pi/N)}$$

This result only applies when there are no indeterminate values ($h = Nn$). Otherwise one obtains:

$$F(h)F^x(h=nN)=(N_1f_A+N_2f_B)^2$$

When the number of atoms forming the superlattice is large and when the different sub-layers have the same thickness, the intensity of the first reflections orders obeys:

$$F(h)\cdot F^x(h)\cong\frac{N^2}{h^2\pi^2}(f_A-f_B)^2$$

Only odd orders are allowed and their intensity decreases as l/h^2.

(6) See the table (in units of f_B^2).

h	1	2	3	4	5	6	7	8	9	10	11	12	13
$N_1=N_2$ =6	60	0	8	0	4.3	0	4.3	0	8	0	60	580	60
$N_1=5$ $N_2=7$	56	4	4	4	0.3	4	0.3	4	4	4	56	484	56
$N_1=6$ $N_2=7$	51	1.4	8	1.3	3	1.8	1.8	3	1.3	9	1.4	51	625

This numerical application confirms the general analysis done in point 3 above). A look at the diffracted intensities allows one to deduce the numbers N_1 and N_2 layers of the superlattice. In particular the order for which the intensity is maximal allows one to find $N = N_1 + N_2$ (h = 12 for the cases α and β; h = 13 for the case γ). Moreover, the even reflections (with the exception of $h = nN$) are forbidden when $N_1 = N_2$.

(7) From the crystallographic point of views the superlattice is quadratic ($d = Na$; $b = c = a$; $\alpha = \beta = \gamma = 90°$). In this super-cell the coordinates $v_j = w_j = 0$ so that $F(h, k, l) = F(h)$ regardless of the particular values taken by k and l.

(8) The reciprocal superlattice can be built from the three vectors $\vec{A}$, $\vec{B}$, $\vec{C}$ orthogonal such that $|B| = |C| = 2\pi/a$; $|A| = 2\pi/d$. Limited just to the plane (A, B), the successive points along the row (h, 0) will therefore, be N times closer than those of the direction (0, k).

Figure 59 shows a part of the corresponding reciprocal lattice. Having weighted the points of this lattice in function of the corresponding intensities this figure also represents the diffraction pattern of such a superlattice.

Note: We have adopted a methodology which starts directly from a superlattice to the diffraction pattern. Another approach would consist of starting from the diffraction pattern of the elementary lattice (here a simple cubic) and to exploring the appearance of the satellite lines associated with the super-periodicity created by the stratification A/B/A/B.

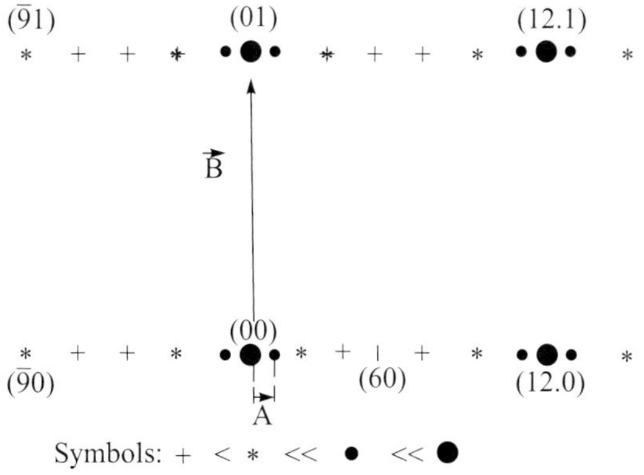

Figure 59

Comment: The characterization of the superlattices

Multilayer structures (N_1 layers of material A, N_2 layers of material B) are prepared by repeated sequential deposition of the constituents A and then B using MBE for instance (see Pb. 4).

The periodicity of these structures (from 10 to 2000 Å) makes these objects particularly interesting to realize monochromators and neutron polarizers as well as X-ray mirrors operating at normal incidence (as well as other systems; see Chapter IV, Ex. 30).

Finally, for solid state physics, semiconducting superlattices have been of wide interest from a fundamental point of view as well as applications since they have been first proposed by Esaki and Tsu in 1970. The effect of the superlattice periodicity on electronic states will be considered in Pbs. 9a and b of Chapter V.

Regardless of their constituents (magnetic–nonmagnetic such as Fe/Ag or Ni/Cu or two semiconductors), their characterization can be done during their elaboration using RHEED (see Pb. 4). Another method is based on the use of transmission electron microscopy. The purpose of the present exercise was to illustrate the power of diffraction of X-rays which allows the determination of the number of constituent layers and of their chemical composition from the inspection of their X-ray diffraction patterns. When the epitaxy conditions are not strictly satisfied, the deformations of the lattice imposed by the lattice matching can also be analyzed by this method. (See Chapter II, Pb. 5 for an elementary study of these deformations.)

Problem 11: Diffraction of X-rays and neutrons from vanadium

(I) Investigation by X-ray diffraction

(1) When a sample of powdered vanadium is exposed to monochromatic X-rays of wavelength $\lambda = 1.54$ Å, the characteristics of the resulting diffraction pattern are given in the following table:

Line number	1	2	3	4	5	6	7
Bragg angle	22.1°	30.5°	38.5°	46.0°	53.5°	61.7°	72.0°
Amplitude	32.6	30.2	23.6	22.4	19.2	17.8	16.6

Express the Bragg formula in terms of a and of the Miller indices (h, k, l) of each line for a cubic system.

Starting from these parameters, show that vanadium crystallizes in a bcc structure and calculate the parameters of the lattice and the index of the lines.

What are the values of f_x, the atomic structure factor, for each of the observed lines? Can you give qualitative reasons for which f_x is a function of the angle θ and λ?

(2) Establish the relation between f_x and the electronic density supposed to be of spherical symmetry. What is the phase difference between the diffused ray that which passes through the center O of the atom and that which passes by the point M with the coordinates r and α defined in Fig. 60.

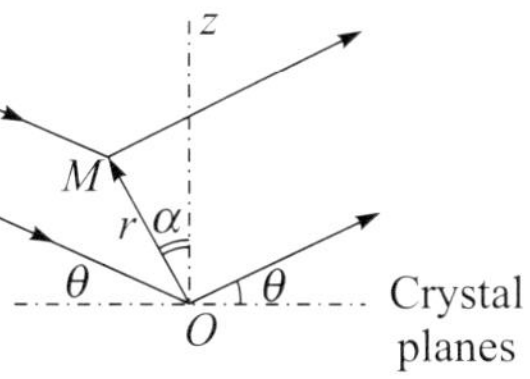

Figure 60

If $\rho(r)$ represents the electronic density at M, show that the expression of f_x, defined as the relation of the diffused amplitude by an atom and that by an electron, has the form:

$$f_x = \int_0^\infty U(r) \cdot \frac{\sin \mu r}{\mu \cdot r} \cdot dr \qquad (1)$$

$U(r)$ dr representing the number of electrons contained between the two spheres of radius r and r + dr, give the expression of μ as a function of θ and λ.

(II) Investigation by neutron diffraction

(1) The sample of vanadium is now irradiated with a beam of neutrons issued from a nuclear reactor where the internal temperature is 27°C. What will the associated wavelength be? (Hint: The neutron kinetic energy is $E_c = (3/2)\, k_B T$).

(2) The diffraction pattern of the neutron beam is very similar to that of Section I above, with the exception of two points: the lines are shifted toward weaker Bragg angles and an additional line appears. Comment on these results.

All these lines have the same amplitude equal to +0.35. We define f_N as being the nucleus form factor. What is its value? Why is it independent of λ and θ?

(3) From now, we consider that f_N is negative as would show considerations out of the present problem.

The beam of neutrons is polarized, that is to say that all their spins are parallel to each other. In addition the Vanadium sample is immersed into a magnetic field and a depolarization of the neutrons of the beam results. The depolarization factor R being defined as the ratio of the intensities of the two polarizations (parallel and antiparallel), the following values of R for the first lines are obtained:

Line number	1	2	3	4	5
R	0.9880	0.9920	0.9968	0.9980	0.9988

Show that the ratio R is equal to $1+4\frac{f_M}{f_N}$ where f_M is the magnetic form factor of the V electrons that is smaller than f_N, (which should be verified).

(4) Since f_x can be related to $U(r)$, state why f_M can also be represented by a formula analogous to Eq. (1) What is the new physical signification of $U(r)$? $U(r)$ can be expressed by a Fourier series using:

$$U(r)=r\cdot\sum_m A_m \sin 2\pi m x$$

In the expression for f_M, is it possible to replace $\sin\theta/\lambda$ by another value, taking into account the Fourier series?

State the value of x in the expression found for f_M. What does m correspond to in the transformed expression of $\sin\theta/\lambda$? Show that the replacement of $\sin\theta/\lambda$ suggests a reasonable upper limit for the integration. What is this limit and why? Calculate the value of A_m.

Among the observed lines, select the ones to be taken into account? Find the values of U(r) relative to the following points:

r = 0.2 Å, r = 0.4 Å, r = 0.8 Å, r = 1.0 Å.

Draw the curve U(r). (h, k_B, M)

Solution:

(I) *Investigation by X-ray diffraction*

(1) In the cubic system, the distance d between two reticular adjacent planes of index (h, k, l) is $d(h, k, l) = a/(h^2 + k^2 + l^2)^{1/2}$. The Bragg formula becomes $\sin\theta(h, k, l) = \lambda\,(h^2 + k^2 + l^2)^{1/2}/2a$.

The structure $F(h,k,l) = \sum_j f_j \exp - i2\pi(hx_j + ky_j + lz_j)$ factor, goes to zero for a bcc lattice. The reflections are such that $h + k + l$ = even. Compared to a cubic simple lattice the amplitude of the scattered waves is doubled for a bcc lattice ($F = 2f_j$). In these conditions, we find the following table:

Reflection n°	h, k, l	$\theta\,(h, k, l)$ in degrees	$\sin\theta\,(h,k,l)$	a(Å) (calculated)	f_x
1	1 1 0	21.1	0.360	3.025	16.3
2	2 0 0	30.5	0.507	3.034	15.1
3	2 1 1	38.5	0.622	3.030	11.8
4	2 2 0	46	0.719	3.028	11.2
5	3 1 0	53.5	0.804	3.029	9.6
6	2 2 2	61.7	0.880	3.029	8.9
7	3 2 1	72	0.951	3.029	8.3

The atomic form factor decreases with θ because the atoms cannot be considered as point atoms with respect to the wavelength of the incident X-rays (see Ex. 22).

(2) The phase difference between the diffused rays in O and M is:

$$\varphi = \Delta\vec{k}\cdot\vec{r} = \frac{2\pi}{\lambda}(\vec{u}_0 - \vec{u})\cdot\vec{r} = \frac{4\pi}{\lambda} r\sin\theta\cdot\cos\alpha$$

This phase difference is the same for all the points of a ring around the axis Oz and the elementary volume is equal to $dv = 2\pi r^2\cdot\sin\alpha\, d\alpha\cdot dr$.

The atomic form factor, f_x, corresponds to the integral over the volume of the atom:

$$f_x = \int \rho(r) \exp i\varphi . dv$$

Then

$$f_x = 2\pi \int_0^{\infty} \rho(r) r^2 dr \int_0^{\pi} \exp i \left(\frac{4\pi r \sin\theta \cos\alpha}{\lambda} \right) \cdot \sin\alpha d\alpha.$$

or

$$f_x = 4\pi \int_0^{\infty} \rho(r) \cdot r^2 \frac{\sin(4\pi r \sin\theta)/\lambda}{(4\pi r \sin r\theta)/\lambda} dr$$

which has the form of the proposed integral when $U(r)$ is chosen as:

$U(r) = 4\pi\rho(r) r^2$ and $\mu = 4\pi \sin\theta/\lambda$

(II) *Investigation by neutron diffraction*

(1) The wavelength associated with neutrons is $\lambda = h/p$ with $E_c = p^2/2M$, which gives λ (Å) $= 0.286/\sqrt{E_c(eV)}$. Here, $E_c = 3k_B T/2 \approx 40$ meV ($k_B T/2$ per degree of freedom). Then, $\lambda \approx 1.42$ Å.

(2) With respect to the investigation with X-rays, the lines bring nearer each other because the neutron wavelength is slightly shorter. Then the reflection (4 0 0) is also possible: $\sin\theta(400) = 2\lambda/a = 0.94 < 1$.

Excepting the sign, the nucleus form factor is equal to half the amplitude of the scattered waves from a bcc lattice, $f_N = -0.175$.

It is independent of λ and θ because if the electron cloud scatters the X-rays, the nucleuses scatter neutrons and the size of the nucleuses is small compared to the incident wavelengths. There is no phase difference between different scattering points in the same nucleus (the same atom).

(3) The polarized incident neutrons interact with atoms of vanadium which are, themselves, polarized in the magnetic field. The depolarization is related to the fact that the spin of the electrons of vanadium will be parallel or anti-parallel to the spin of the incident neutrons. If we call f_M the magnetic form factor of the V electrons the scattered amplitude is proportional to $f_N + f_M$ in one case to $f_N - f_M$ in the other case.

Thus, the ratio R of the intensities is:

$$R = \frac{I(\uparrow\uparrow)}{I(\downarrow\uparrow)} = \left(\frac{f_N + f_M}{f_N - f_M}\right)^2$$

When f_M is small compared to f_N: $R \approx 1 + 4f_M/f_N$ and the evolution of f_M for the various crystallographic planes is the following:

$h\,k\,l$	(1 1 0)	(2 0 0)	(2 1 1)	(2 2 0)	(3 1 0)
$\frac{f_M}{f_N}$	-3×10^{-3}	-2×10^{-3}	-0.8×10^{-3}	-0.5×10^{-3}	-0.3×10^{-3}
f_M	5.25×10^{-4}	3.5×10^{-4}	1.4×10^{-4}	0.875×10^{-4}	0.525×10^{-4}

One may observe that f_M is effectively small compare to f_N.

(4) Like f_x, f_M is decreasing because the spin of atoms of vanadium results from the combined spins of electrons in the V atom. The resulting electronic spin is due to the $3d$ electrons in the outer shell of vanadium. The magnetic diffusion is therefore, related to this diffusion of electrons which is represented by $U(r)$ from which:

$$f_M = \int_0^\infty U(r)\cdot(\sin \mu r/\mu r)\cdot dr.$$

On the one hand, one has (see Ex. 1):

$$\sin\theta/\lambda = \frac{(h^2 + k^2 + l^2)^{1/2}}{2a}$$

On the other hand the electronic distribution of the crystal may be re-written as:

$$\rho(r) = \sum_G \rho_G \exp i\vec{G}\cdot\vec{r} = \sum_m \rho_m \exp i\frac{2\pi m}{a}(h^2 + k^2 + l^2)^{1/2}\cdot r$$

Taking into account that $G(h, k, l)\cdot d(h, k, l) = 2\pi$, in the expression proposed for $U(r)$ we can identify x to be the dimensional quantity:

$$\frac{r}{d_{h,k,l}} = \frac{r(h^2 + k^2 + l^2)^{1/2}}{a}$$

In the sum, the whole number m would represent the order of the various reflections.

Finally, taking into account the fact that the integration over the volume of the atom can, to a first approximation be limited up to the inter-reticular distance along each of the directions (that is to say $d_{h,k,l}/2$), the following expression for f_M is obtained:

$$f_M = \int_0^{\frac{d}{2}} r \cdot \sum_m A_m \sin\frac{2\pi mr}{d} \cdot \frac{\sin 2\pi nr/d}{2\pi nr/d} . dr$$

$$= \frac{d}{2\pi n} \int_0^{\frac{d}{2}} \sum_m A_m \left[\cos\frac{2\pi(m-n)}{d} r - \cos\frac{2\pi(m+n)}{d} \cdot r \right] dr$$

The integrals lead to a null result except when $m = n$. Then:

$$f_M = \frac{A_m \cdot d^2}{8\pi n} = \frac{A_m \cdot a^2}{8\pi n(h^2+k^2+l^2)} \text{ or } A_m = \frac{8\pi f_M}{d^2} \cdot n$$

The (110) and (220) reflections permit to obtain the first two terms of the expansion of $U(r)$, instead of only one.
$A_1 = 2.87 \times 10^{-3}$ Å^{-2} and $A_2 = 0.958 \times 10^{-3}$ Å^{-2} in the development of $U(r)$ of the form:

$$U(r) = r\left(A_1 \sin\frac{2\pi r}{d} + A_2 \sin\frac{4\pi r}{d} + \ldots \right)$$

Finally, the following table and graph can be derived:

r (Å)	0	0.2	0.4	0.6	0.8	1
x	0	0.093	0.187	0.28	0.373	0.487
U(r) 10^{-3} Å^{-1}	0	0.494	1.36	1.49	0.874	0.026

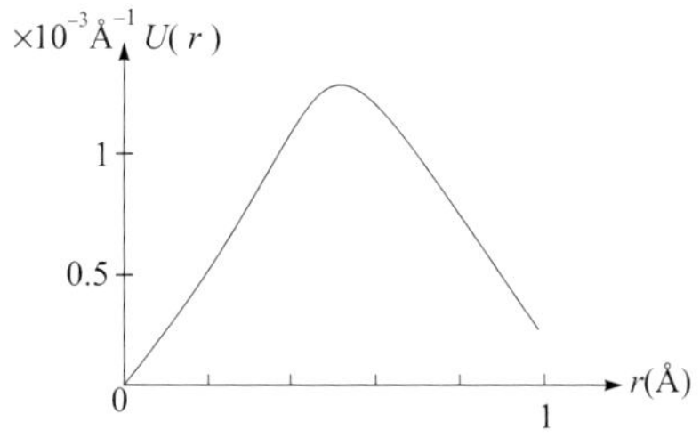

Figure 61

Remark: The probability of the presence of $\rho(r)$ of an electron between r and r + dr is proportional to U(r) where the coefficient of proportionality C can be deduced from $c\int U(r)\,dr = 3$ (electrons/atom). Here, $C = 3 \times 10^3$. See also Chapter III, Pb. 7.

Problem 12: X-ray diffraction of intercalated graphite

Graphite crystallizes in a hexagonal system with a = 2.455 Å; c = 6.7 Å and a basis consisting of four carbon atoms of which two are localized on the plane A at 000 and 2/3 1/3 0 (to create a 2D structure in the form of a honeycomb; see Ex. 17) and two C atoms are in the intermediary plane, B, at 0 0 1/2 and at 1/3 2/3 1/3 (see Fig. 62).

(1) Give the expression of the structure factor F (h, k, l) of graphite. Deduce the rule corresponding to the reflected amplitude on different planes of the lattice and state in particular the order of the forbidden reflections.

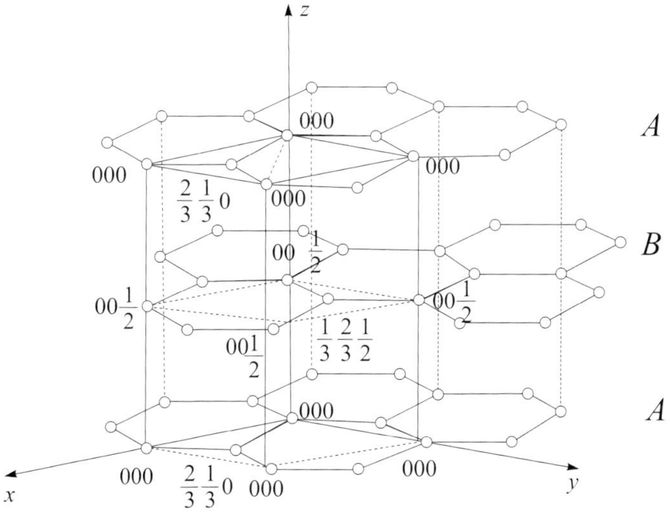

Figure 62

(2) Using the goniometric arrangement (combining a the rotation of θ for the sample and a rotation of 2θ of the detector) describe the diffraction diagram of a single crystal of graphite, limiting first to reflections of the type (001) to simplify the problem and then next to reflections of type (h00). State the

sequence of numerical values for the successive Bragg angles $\theta(001)$ and $\theta(h00)$ when the wavelength is $\lambda = 1.54$ Å.

(3) Potassium atoms are inserted into the leaflets of graphite to obtain the first insertion stage KC_8 (Fig. 63). The goal is now to explore the changes in the reflections in the (001) and (h00) considered before in (2). Find the structure factor $F(0, 0, 1)$ and $F(h, 0, 0)$ in order to determine the modifications to the Bragg angles and to the amplitudes of these reflections. Note in particular that the distance between the two successive planes, A and B, of carbon has changed from 3.35 Å to 5.51 Å and the two planes are now of the same type A. Also, the plane of the inserted K atoms is hexagonal with lattice parameter $a' = 2a$.

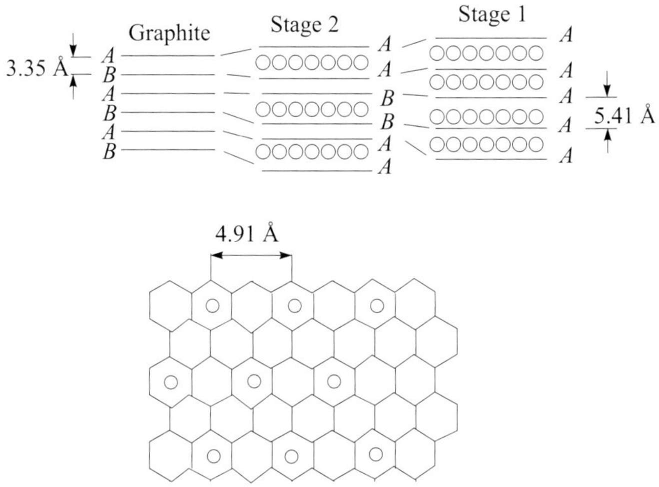

Figure 63 Intercalation of K into graphite (front and top views).

(4) For a synthesis of all these results, draw the diagrams of $F(001)$ and $F(h00)$ as a function of θ. Assume that the atomic form factors f_c and f_K are real and independent of θ.

(5) To study a laminar material is it necessary to consider all the points (h, k, l) accessible in reciprocal space?

Solution:

(1) Starting from the general expression of $F(h, k, l)$ one obtains:

$$F(h,k,l)/f_C = \left(1+e^{-i\frac{2\pi}{3}(2h+k)}\right) + e^{-i\pi l}\left(1+e^{-i\frac{2\pi}{3}(h+2k)}\right)$$

When l is even, the contribution of the intermediate plane is added to that of the base plane because $e^{-i\pi l} = 1$. The weight of these contributions will be maximal, $F(h, k, l) = 4f_c$, when $h + 2k$ (or $h-k$ or $k-h$ or $2h + k$) $= 3n$. It will be reduced to $F(h, k, l) = f_c\,(2 - 2\cos 60°) = f_c$ when $h - k = 3n \pm 1$.

When l is odd, the contribution of the intermediary plane subtracts from that of the base ($e^{-i\pi l} = -1$) and the reflections of type $h - k = 3n$ are forbidden. The reflections of type $h - k = 3n \pm 1$ correspond to the structure factor, such that $F(h, k)/f_c = 2i\sin(2\pi 3) = i\sqrt{3}$ and the corresponding intensity is real $I(h, k, l) = 3f_c f_c^x$.

(2) When considering the Bragg angles of type $\theta(0, 0, l)$, the reciprocal lattice along the $\vec{C}$ axis is investigated and thus the stratification of graphite has to be considered. The only allowed reflections are of the type (002), (004), (006), ... and their amplitude is maximal ($4f_c$) because for odd reflections there is destructive interference between the diffused waves of two adjacent planes. The application of the Bragg's law, $2c\sin\theta(001) = l\lambda$ results in $\theta(002) = 13.3°$; $\theta(004) = 27.4°$; $\theta(006) = 43.6°$. Explorations of Bragg angles of the type $\theta(h00)$ can be considered as explorations along the reciprocal axis $\vec{A}$ and thus, the distribution of atoms in a layer. All the reflections are allowed ($l = 0$ is even), but reflections where $h = 1, 2, 4, 5$ are weak ($F = f_c$) whereas, those of type $h = 3, 6, 9$ have the maximal amplitude $4f_C$. The corresponding angles $\theta(h00)$ obey the rule $2a\sin\theta(h00) = h\lambda$ or $\theta(100) = 18.28°$; $\theta(200) = 38.85°$; $\theta(300) = 70.2°$.

(3) The new crystalline structure is hexagonal with a =4.91 Å; and c = 5.41 Å. Choosing the origin at a K atom in the plane of carbon atoms, the basis can be described by 1K at $0\,0\,\frac{1}{2}$ and 8C at: $\frac{1}{8}\,\frac{1}{8}0$; $\frac{3}{8}\,\frac{3}{8}0$; $\frac{5}{8}\,\frac{1}{8}0$; $\frac{7}{8}\,\frac{3}{8}0$; $\frac{1}{8}\,\frac{5}{8}0$; $\frac{3}{8}\,\frac{7}{8}0$; $\frac{5}{8}\,\frac{5}{8}0$; and $\frac{7}{8}\,\frac{7}{8}0$.

The reflection amplitude of type 001 will be given by $F(001) = 8f_C + f_K\,e^{-i\pi l}$.

They will be strong for l even and weak for l odd, which is easily seen by simply considering that the phase difference of waves diffused by the atomic planes of K and those of C will be

π compared with waves diffused by two consecutive planes of C atoms that will be 2π.

The corresponding Bragg angles will be $\theta(001) = 8.18°$; $\theta(002) = 16.53°$; $\theta(003) = 25.27°$; $\theta(004) = 34.7°$.

As was obvious previously, the change in the interlayer distance due to K insertion narrows the sequence of Bragg angles.

The reflection amplitude of type $h00$ is given by:

$$F(h00) = f_K + 4f_C\left(\cos\frac{h\pi}{4} + \cos\frac{3h\pi}{4}\right)$$

The reflections such that $h = 4n$ will have the maximal amplitude ($F = f_k + 8f_C$) and all others will have the minimal amplitude ($F = f_K$). The lattice parameter a having doubled, the $\sin\theta$ values will be halved $\theta(100) = 9.02°$; $\theta(200) = 18.28°$; $\theta(300) = 28.6°$; $\theta(400) = 38.85°$; $\theta(500) = 51.64°$

(4) See Fig. 64

(5) Concerning the intercalation effect, the complete analysis of the diffraction diagram does not bring any new information compared to that following the $\vec{C}$ axis. Therefore, the study of the diagram may be limited to the exploration along the direction of diffraction spots that is perpendicular to the plane, a fact already formulated in Pb. 10.

Comments: Intercalations in layered materials

It is possible to insert some elements or molecules that are partially ionized between the different layers of lamellar crystals, for instance Li, K, H_2SO_4 in C; NH_3, I_2 in MoS_2. This is done essentially by chemical processes or by electrochemistry. The resulting composite modifies the physical properties of both the lamellar material and that inserted.

Besides the 2D behavior of these synthetic materials, they are interesting from the fundamental viewpoint because of their anisotropic character is adjustable depending on the species inserted and the insertion stage, corresponding to the number of lamellar crystals separating two layers of the inserted species (see Fig. 63).

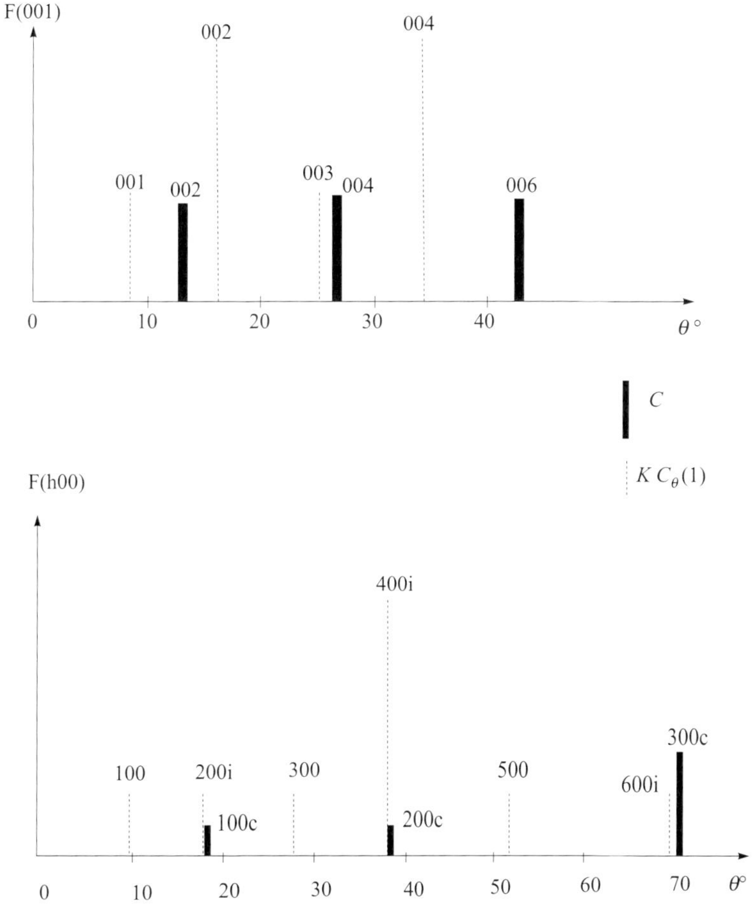

Figure 64

A certain number of applications can be envisioned based on the electronic particularities of these materials and in particular the excellent electric conductivity that runs parallel to the layers. Catalytic applications based on the selective modification of the chemical reactivity of the insertions are also considered as well as the storage of electrical energy associated with the insertion–extraction reactions.

Outside of the specific case of insertion, the present problem illustrates a common theme in materials science which is that the understanding of a new material is followed necessarily by its structural characterization. This characterization serves to control and refine the materials processing but is also indispensable for understanding the specific physical properties and deduce the atomic arrangements. In the arsenal of techniques available for structural characterization, X-ray diffraction holds a privileged place even if the prediction or the analysis of the results is not always as easy as in the simplified examples chosen here.

Questions

Q.1: Give the order of magnitude of the interatomic distance d in a solid, the atomic density of a crystal N_V, the surface density, N_S, and the linear density, N_L?

Q.2: In diffraction experiments of crystals what do the following techniques "see": (a) incident X-rays, (b) incident electrons, and (c) incident neutrons?

Q.3: Why, at normal incidence, does one see always a diffraction pattern with electrons that are "slow" by using reflection from and "fast" when using transmission through thin films?

Q.4: Monochromatic and parallel X-rays beam irradiated a perfect crystal at any incidence angle. What happens from the point of view of diffraction effects?

Q.5: In diffraction experiments, what does the 000 reflection correspond to?

Q.6: Are the diffraction conditions more favorable when the atomic planes are closer and closer to each other?

Q.7: What is epitaxy?

Q.8: What is the meaning of Ni (100) c (2×2)0?

Q.9: In surface science, what is a Langmuir?

Q.10: What is a reconstructed surface? Give one cause of such a reconstruction.

Q.11: What is a disordered alloy? How can a disordered and an ordered alloy be differentiated?

Q.12: What is a superlattice? Give some examples.

Q.13: What is the structure of a insertion compound in a lamellar crystal?

Q.14: What is the difference between the following two expressions: $2d_{300} \sin\theta = \lambda$ and $2d_{100} \sin\theta = 3\lambda$?

Q.15a: Not taking into account the different lattice parameter, what are the expected differences between an X-ray diffraction pattern of GaAs (zinc-blended structure as shown in Fig. 3c) and that of germanium (diamond structure shown in Fig. 3d)? [Z(Ga) = 31; Z(As) = 33; Z(Ge) = 32]

Q.15b: In the X-diffraction pattern of KCl (fcc structure with 1 K at 000 and 1 Cl at ½00, what are the quasi-forbidden reflections caused by the basis known that Z(K) = 19 and Z(Cl) = 17?

Q.16: Consider particles with wavelength λ = 1.54 Å. State in eV, their energy depending on whether they are associated with:

(a) Photons (X-rays); (b) electrons; (c) neutrons.

Q.17: Show the reciprocal lattice and the first BZ of CsCl, NaCl, and CaF_2 (Fig. 3a, e, and f, respectively). What is the difference between the last two?

Q.18: The following techniques are named often by their acronyms, what are their full name and their main characteristics:

(a) LEED; (b) RHEED; (c) EBSD.

Q.19: Give a definition for: (a) The first BZ; (b) the Kikuchi lines; (c) the Kossel lines.

Q.20: How to produce the X-rays?

Q.21: Why the results of the Laüe experiments were important when known in 1912?

Q.22: Why Davisson and Germer were important in 1927?

Answers at the end of the book

Chapter II

Crystal Binding and Elastic Constants

Course Summary

A. Crystal Binding

1. *Statement of the Problem*

QUESTIONS: It is natural to ask why a given solid chooses a particular crystal structure and what kinds of properties are connected with it. This topic is called "cohesion" and the questions are:

- What are the forces that result in the cohesive energy of a solid? What is their physical origin?
- Why do all materials (except He) become solids when the temperature is sufficiently lowered?

QUALITATIVE REPLY: Electronic forces are the main answer to these questions. They essentially affect the valence electrons and assure the crystalline cohesion. These forces overcome the repulsion between the overlap of electron clouds of neighbor atoms and the kinetic energy of the nucleus and of the electrons.

DEFINITION: The cohesive energy of a solid is the energy required to break the atoms of the solid into isolated atomic species, that is,

$$E_{\mathrm{c}} = E_{\mathrm{solid}} - \sum_{\mathrm{A}} E_{\mathrm{A}}^{\mathrm{isolated}}$$

Understanding Solid State Physics: Problems and Solutions
Jacques Cazaux

ISBN 978-981-4267-89-2 (Hardcover), 978-981-4267-90-8 (eBook)
www.panstanford.com

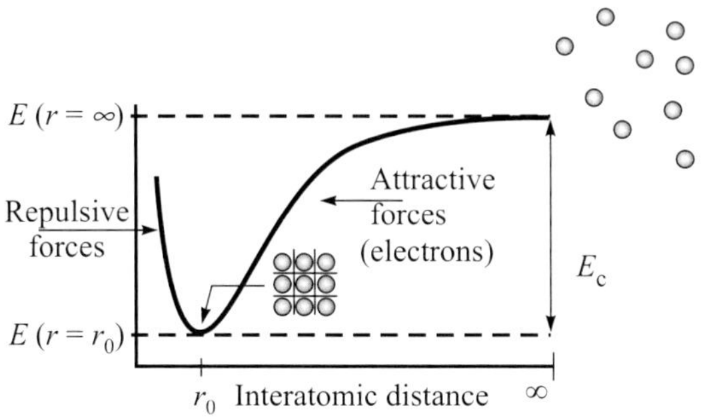

Figure 1 Dependence of potential energy on interatomic distance.

The calculated values of the cohesive energy are compared with experimental results, which can be obtained by measuring the latent heat of sublimation at various low temperatures, and extrapolating to zero Kelvin. The empirical parameters relative to the cohesive forces can be evaluated starting from the inter-reticular distance, r_0, and the compressibility, β, of the solid: $1/\beta = -Vdp/dV$.

The analysis of different types of solids can reveal the nature of different attractive forces.

Outside of these general considerations, detailed explorations of crystalline cohesion are generally complex and do not rely on a single rule for all solids. Usually the type of bonds may be classified into one of four general categories and each of them is treated by specific simplifications. We follow these general rules but it must be pointed out that the approach by Harrison provides a coherent theory taking into account covalent solids, crystals of rare gases, ionic crystals, and simple and transitions metals [12].

2. *Rare Gas Crystals*

In addition to chemical bonding between atoms, there is another type of attractive force that exists between atoms, ions, or molecules known as van der Waals forces. These forces represent the main contribution to the cohesive energy for rare gas crystals, from Ne to Rn with eight outershell electrons/atom.

The attractive energy between two atoms results in an induced dipole–dipole interaction (called the van der Waals–London interaction) which varies as r^{-6}:

$$W_p = -\vec{P}_B \cdot \vec{E}_A = -\varepsilon_0 \alpha_e E^2 \propto \alpha_e P_A^2 / r^6$$

where α is the polarizability of the atom.

- The repulsive energy due to the overlapping of orbital electron clouds varies even more abruptly as r^{-12} or as λ exp – (r/ρ).
- The total potential energy, neglecting the kinetic energy terms, is known as the Lennard-Jones energy (see Pb. 3) is:
  $W_{\mathrm{T}}(\text{atom–atom}) = 4\varepsilon[(\sigma/r)^{12} - (\sigma/r)^{0}]$

in which ε is the cohesive energy of a molecule and σ is related to the equilibrium distance r_0 between adjacent atoms by $r_0 = 1.12\ \sigma$ (see Ex. 10).

For N atoms in a crystal, the potential energy is the sum of the first j neighbors:

$$E_{\mathrm{e}} = \frac{N}{2} 4\varepsilon \left[\sum_j \left\{ \left(\frac{\sigma}{r_{0j}} \right)^{12} - \left(\frac{\sigma}{r_{0j}} \right)^{6} \right\} \right]$$

The cohesive energy is weak, 0.02 eV/atom or less, which results in a low melting temperature.

3. *Ionic Crystals*

The cohesive energy results essentially from the electrostatic attraction between ions of opposite sign. $U_{0\mathrm{n}} = -q^2\alpha/4\pi\varepsilon_0 r_0$ in which α, the Madelung constant, represents the alternate summation of different types of ions with respect to the reference ion ($q = 1.6 \cdot 10^{-19}$ C and r_0: distance between nearest neighbors at equilibrium); As with rare gas crystals, the electrostatic repulsive energy at short distance is due to the overlapping of electronic orbitals between nearest neighbors. It is given either by an expression of the form $\lambda e^{-r/\rho}$ (Born-Meyer type with constants λ, ρ, see Pbs. 1 and 2) or by a Lennard-Jones type potential Ar^{-p} (with $p \approx 9$ or 10) (see Exs. 1–7 and 10).

Excepting the ionization energy of the species and neglecting the kinetic energy terms, the cohesion energy of N pairs of ions takes one of the following forms:

$$E_{\mathrm{C}} = N[z\lambda e^{-r/\rho} - (\alpha q^2/4\pi\varepsilon_0 r)]$$

$$E_{\mathrm{C}} = N[zAr^{-p} - (\alpha q^2/4\pi\varepsilon_0 r)]$$

where z is the number nearest neighbors and r is the inter-atomic distance at equilibrium.

The equilibrium condition, $(\partial E_C/\partial r)_{r=r_0} = 0$ allows the calculation of r_0 as a function of A and p or ρ and λ. Reciprocally it permits to evaluate one of the two constants as a function of the measured crystalline parameter. If necessary the other constant can be deduced from a measurement of the compressibility β (see Pbs. 1 and 2).

The major result is that the binding energy results essentially (80–90%) from the Coulomb attraction:

$$E_C = -\frac{q^2\alpha}{4\pi\varepsilon_0 r_0}\left(1-\frac{\rho}{r_0}\right) \quad \text{or} \quad E_C = -\frac{q^2\alpha}{4\pi\varepsilon_0 r_0}\left(1-\frac{1}{p}\right)$$

in which the Madelung constant, α, depends on the crystalline structure considered.

Conventionally, here α is the multiplying factor of $q^2/4\pi\varepsilon_0 r$ in the expression for the ion potential energy with $q = 1.6\times10^{-19}$ C. In other conventions it is sometimes used as the multiplying factor of the quantity $Z_1Z_2q^2/4\pi\varepsilon_0 r$ in the expression of the attractive potential energy of ions Z_1q and Z_2q. This last convention leads to α(MgO) = α(NaCl), which will modify certain expressions but does not change the physical consequences mentioned above.

This ionic structure gives high bond strength that provides brittle substances with high melting points and low electrical conductivity.

4. *Metallic Bonds*

Metallic binding constitutes the electrostatic attractive forces between the delocalized electrons, called conduction electrons, gathered in a mobile electron cloud (or a Fermi sea) and the positively charged metal ions that are fixed and immersed in this cloud (or sea). Whereas most chemical bonds are localized between specific neighboring atoms, metallic bonds extend over the entire molecular structure. The resulting cohesion is essentially the reduction of the total energy of conduction electrons in the metal compared to a free atom. In a simple metal (but not in transition metals), it results in certain plasticity when defaults can be neglected with also relatively low melting points and easy reshaping (bending, flattening). The delocalized electrons provide high electrical conductivity. The detailed study of this cohesion energy necessitates first the examination of behavior of free electrons as considered in Chapter IV, Ex. 24 and Pb. 1, or of the band structure theory (Chapter V) for transition metals (see Chapter V, Ex. 8).

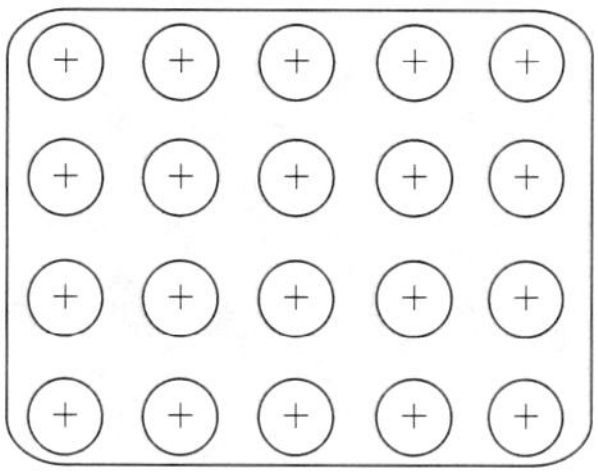

Figure 2 Model for simple metals: Ions lattice in a Fermi sea.

5. *Covalent Bonds*

Covalent chemical bonds involve the sharing of a pair of valence electrons by two atoms, in contrast to the transfer of electrons in ionic bonds. Such bonds lead to stable molecules if electrons are acted upon by the electrons and nuclei of adjacent atoms as atoms come within close proximity. For each individual atom there are discrete energy levels that may be occupied by electrons and this causes each distinct atomic state to split into a series of closely spaced electron states in the solid, to form what is termed electron energy bands. All the atoms are bonded together into a giant molecule.

Even though this study will not be covered until Chapter V, with the tight binding approximation (see Course Summary of Chapter V and the related exercises), one can observe that the covalent bond is a strong bond that accounts for the hardness of diamond (as well as Ge and Si). This occurs despite a weak filling rate where a tetrahedral bond allows only four nearest neighbors.

Covalent bonds are formed when non-metallic atoms approach and share valence electrons. These are the strongest of all bonds. Covalent networks are very hard to disrupt, giving these substances very high melting points and low conductivity in any state.

B. Elastic Constants

1. *Introduction*

This chapter is concerned with the elastic constants of single crystals. They are macroscopic parameters relating stress to strain in homogenous solids and they are of interest for their insight into the nature of the binding forces in solids. They are also of importance for the velocity of sound and for the thermal properties of solids as investigated in Chapter III from the microscopic point of view.

2. Stress

In continuum mechanics, stress is a measure of the internal forces per unit area, Pa or N/m^2, acting within a solid.

There are nine stress components. Three are directed along x, X_x, X_y, and X_z where the subscript indicates the normal to the plane to which the force X is applied (see Fig. 2). The six others are directed along y and z, respectively. In equilibrium one has $Y_z = Z_y$, $Z_x = X_z$, and $Y_x = Z_y$ then the stress exerted on a solid can be described by six independent parameters.

For the simple case of an uniaxial stress, for example, a bar subjected to tension or compression, the stress tensor reduces to

$$\begin{bmatrix} X_x & 0 & 0 \\ 0 & 0 & 0 \\ 0 & 0 & 0 \end{bmatrix}$$

For hydrostatic pressure, it reduces to

$$\begin{bmatrix} -p & 0 & 0 \\ 0 & -p & 0 \\ 0 & 0 & -p \end{bmatrix}$$

A different type of stress occurs when the non-diagonal components such as X_y or Y_z differ from 0. They correspond to shear forces.

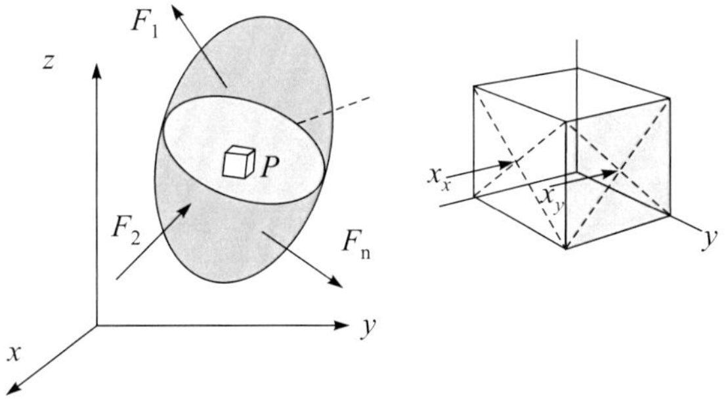

Figure 3 Example of stress components.

3. *Strain*

The application of stress results in deformation of the body's shape or strain. Strain is the observable deformation of a sample relative to its initial shape.

(a) **1D STRAIN**

In 1D, for example, a bar subjected to tension changes in length: elongation.

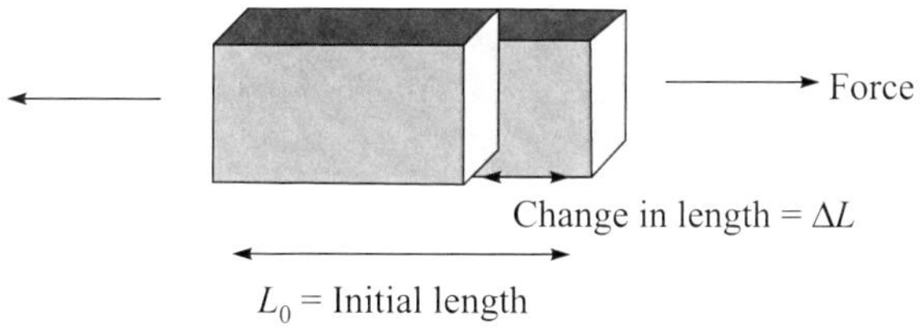

1D strain is illustrated in the following figure where the top figure shows the system before and after strain application.

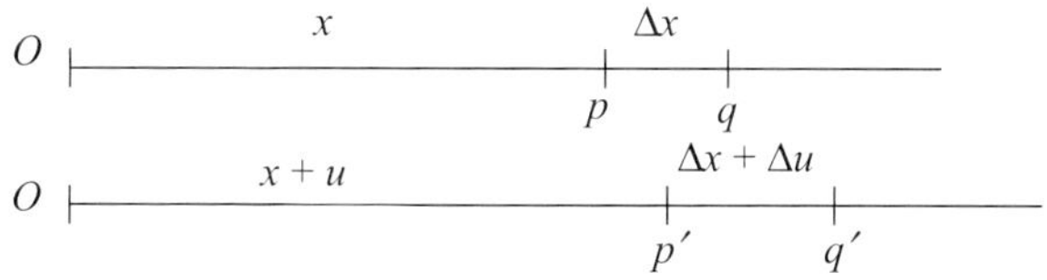

The strain of the segment $\overline{PQ}$ is defined by $\dfrac{\overline{P'Q'}-\overline{PQ}}{\overline{PA}}=\dfrac{\Delta u}{\Delta x}$

and the strain at point P by $e=\lim\limits_{\Delta x\to 0}\dfrac{\Delta u}{\Delta x}=\dfrac{\partial u}{\partial x}$.

Elongation along the axes x, y, or z are written, respectively, as

$e_{xx}=\partial u/\partial x$

$e_{yy}=\partial v/\partial xy$

$e_{zz}=\partial w/\partial z$

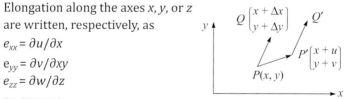

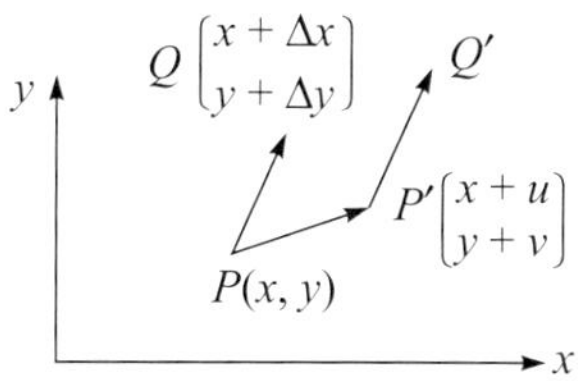

(b) **2D STRAIN**

Applied to the distortion of a vector PQ in a plane, a 2D strain corresponds to

$$\overrightarrow{P'Q'}=\overrightarrow{PQ}+\left(\frac{\partial u}{\partial x}+\frac{\partial v}{\partial x}\right)\overrightarrow{\Delta x}+\left(\frac{\partial u}{\partial y}+\frac{\partial v}{\partial y}\right)\overrightarrow{\Delta y}$$

The description of the deformation involves strain components of the form: $e_{xy} = (\partial v/\partial x) + (\partial u/\partial y)$.

(c) **3D STRAIN**

Extended to 3D, the six components of the strain are

$$e_{xx} = \frac{du}{dx}; e_{yy} = \frac{dv}{dy}; e_{zz} = \frac{dw}{dz};$$

$$e_{xy} = \frac{dv}{dx} + \frac{du}{dy}; e_{yz} = \frac{dw}{dy} + \frac{dv}{dz}; e_{zx} = \frac{du}{dz} + \frac{dw}{dx}.$$

Occasionally definitions of e_{xy}, e_{yz}, and e_{zx} are given which differ by a factor ½ from those given here. In the literature [Refs. 15 or 19] one finds also strain components expressed in a different form, $\varepsilon_{i,j}$. These values are related by $e_{ii} = \varepsilon_{ii}$ and by $e_{ij} = e_{ji} = 2 \cdot \varepsilon_{ij} = 2 \cdot \varepsilon_{ji}$ when $(j \neq i)$.

For example, $e_{xy} = \frac{\partial u}{\partial y} + \frac{\partial v}{\partial x} = 2\varepsilon_{xy}$

Although the table e_{ij} is not a tensor [19], we use this representation in subsequent exercises. This representation implies that the coefficients such as C_{44}, C_{55}, C_{66} in the Hooke's law are two times smaller here than those deduced from the tensor definition.

The existence of these two definitions makes the interpretation of different numerical values coming from different sources difficult. Physically the diagonal elements e_{ij} $(= \varepsilon_{ij})$ represent dilatation or linear strain. The off-diagonal elements e_{ij} where $i \neq j$ represent a shearing strain or the variation (half of ε_{ij}) of the angle made by two vectors, that were respectively parallel to two of the three previous axes.

4. *Hooke's Law*

The Hooke postulates that the strain is a linear function of the stress. Such approximation is only valid very below the limit of elasticity of the solid of interest applied in 1D, it is $X_x = c\frac{\partial u}{\partial x}$). In 3D it leads to the following series of linear equations:

$$e_{xx} = S_{11}X_x + S_{12}Y_y + S_{13}Z_z + S_{14}Y_z + S_{15}Z_x + S_{16}X_y;$$

$$e_{yy} = S_{21}X_x + S_{22}Y_y + S_{23}Z_z + S_{24}Y_z + S_{25}Z_x + S_{26}X_y;$$

$$e_{zz} = S_{31}X_x + S_{32}Y_y + S_{33}Z_z + S_{34}Y_z + S_{35}Z_x + S_{36}X_y;$$

$$e_{yz} = S_{41}X_x + S_{42}Y_y + S_{43}Z_z + S_{44}Y_z + S_{45}Z_x + S_{46}X_y;$$

$$e_{zx} = S_{51}X_x + S_{52}Y_y + S_{53}Z_z + S_{54}Y_z + S_{55}Z_x + S_{56}X_y;$$

$$e_{xy} = S_{61}X_x + S_{62}Y_y + S_{63}Z_z + S_{64}Y_z + S_{65}Z_x + S_{66}X_y;$$

or

$$X_x = C_{11}e_{xx} + C_{12}e_{yy} + C_{13}e_{zz} + C_{14}e_{yz} + C_{15}e_{zx} + C_{16}e_{xy};$$

$$Y_y = C_{21}e_{xx} + C_{22}e_{yy} + C_{23}e_{zz} + C_{24}e_{yz} + C_{25}e_{zx} + C_{26}e_{xy};$$

$$Z_z = C_{31}e_{xx} + C_{32}e_{yy} + C_{33}e_{zz} + C_{34}e_{yz} + C_{35}e_{zx} + C_{36}e_{xy};$$

$$Y_z = C_{41}e_{xx} + C_{42}e_{yy} + C_{43}e_{zz} + C_{44}e_{yz} + C_{45}e_{zx} + C_{46}e_{xy};$$

$$Z_x = C_{51}e_{xx} + C_{52}e_{yy} + C_{53}e_{zz} + C_{54}e_{yz} + C_{55}e_{zx} + C_{56}e_{xy};$$

$$X_y = C_{61}e_{xx} + C_{62}e_{yy} + C_{63}e_{zz} + C_{64}e_{yz} + C_{65}e_{zx} + C_{66}e_{xy}.$$

The quantities S_{ij} are called the elastic (compliance) constants and C_{ij} are known as the elastic stiffness constants (or moduli of elasticity). They are expressed respectively in m^2/N and in Pa (N/m^2). The corresponding numerical values characterize the mechanical properties of the materials. Certain coefficients (C_{ij} in particular) can be dependent on temperature as well as on residual impurities and dislocation densities. In fact, any thermal treatment can change these constants, most notably when the solid is metallic.

The matrices of S_{ij} and C_{ij} are symmetric. As a result, there are 21 independent coefficients for a triclinic crystal, 3 for a cubic crystal (C_{11}, C_{12}, and C_{44}), and 2 for an isotropic solid because $2C_{44} = C_{11} - C_{12}$.

In an isotropic solid, one can use coefficients which have a more direct physical meaning such as the Young modulus, E, and the Poisson coefficient, σ:

$$E = \frac{F/S}{\Delta l/l} = \frac{X_x}{e_{xx}} = C_{11}; \sigma = \frac{\Delta b/b}{\Delta l/l} = \frac{-e_{yy}}{e_{xx}} = \frac{C_{12}}{C_{11}+C_{12}} \quad \text{or} \quad 0 < \sigma < 0.5$$

One also uses the Lamé coefficients, λ and μ: $\lambda = C_{12}$; $\mu = C_{44}$ so that $C_{11} = \lambda + 2\mu$. The bulk modulus, $B = -V(\partial p/\partial V)$ with p as the pressure and V as the volume, and its reverse, the compressibility, $\beta = 1/B$ are also of frequent use. In cubic crystals one has $B = (C_{11} + 2C_{12})/3$.

The goal of this subsection, elastic constants, is to correlate the macroscopic properties of solids with their microscopic causes such as the force constant between atoms and the suggested exercises will be limited to cubic crystals and to isotropic solids (Exs. 13, 15, 16).

5. *Velocity of Elastic Waves*

(a) In 1D, the equation of motion applied to an object of length Δx is, in the O_x direction:

$$\rho \Delta x \frac{\partial^2 u}{\partial t^2} = \Delta F = c\Delta e = c\frac{\partial^2 u}{\partial x^2}\Delta x.$$

The general solution is of the form $u = f\left(t - \frac{x}{v_s}\right) + g\left(t + \frac{x}{v_s}\right)$.

The sinusoidal solutions are of form $u = u_0 \exp i(\omega t - kx)$ and the velocity is $v_s = \sqrt{\frac{c}{\rho}}$.

(b) In 3D, the motion in the x direction of an element of volume $\Delta x\, \Delta y\, \Delta z$ obeys the equation:

$$\rho \frac{\partial^2 u}{\partial t^2} = \frac{\partial X_x}{\partial x} + \frac{\partial X_y}{\partial y} + \frac{\partial x_z}{\partial z}$$

Using the stress/strain relation, Hooke's law, in the above equation and in the equations of motion relative to the Oy and Oz axes, one searches sinusoidal solutions of the form: $\vec{u} = \overrightarrow{u_0} \exp i(\omega t - k'\vec{r})$. These solutions correspond to waves, of wave vector $\vec{k}$ that may propagate along the different crystallographic axes. In the general case, the three normal modes of propagation are such that the polarization of these modes, $\overrightarrow{u_0}$ (the displacement direction of the particles) is not strictly parallel or perpendicular to $\vec{k}$. Nevertheless to simplify the analysis the present study would be limited to propagations along the axes such as the [100] or the [110] of the cubic crystals where the motion is either purely transversal ($\vec{u} \perp \vec{k}$) or purely longitudinal ($\vec{u} \parallel \vec{k}$).

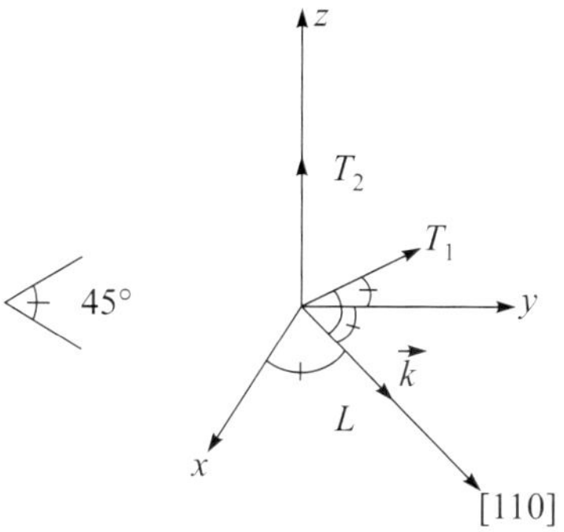

Thus in a cubic crystal when the longitudinal wave propagates along the [110] axis, one introduces solutions of the form $u = u_0 \exp i(\omega t - k_x x - k_y y)$ and $v = v_0 \exp i(\omega t - k_x x - k_y y)$. For a transverse wave polarized along the Oz direction and propagating in the [110] direction, the solution is of the form: $w = w_0 \exp i(\omega t - k_x x - k_y y)$. Taking into account that in each case $k_x = k_y = \frac{|k|}{\sqrt{2}}$, it can be shown that (see Pb. 4):

$$V_L[110] = \frac{(C_{11} + C_{12} + 2C_{44})^{1/2}}{2\rho}, V_{T_1}[110] = \left(\frac{C_{44}}{\rho}\right)^{1/2}$$

Pb. 5 explores the elementary deformations introduced by hetero-epitaxy of materials (of semiconductors here).

Exercises

Exercise 1: Compression of a ionic linear crystal

Consider a line of equidistant ions of R with alternative charges equal to ±q.

(a) Evaluate the electrostatic potential energy U_p and the repulsion energy U_r of an ion placed at the origin in the field of all the other ions. Assume that the repulsion energy between two atoms is of the form A/R^p.

Find the expression for the total energy of $2N$ ions in the chain and deduce the expression of A in equilibrium.

(b) A compression of the crystal transforms R_0 into $R_0(1-\delta)$: show that the compressive work per unit length is approximately equal to ½ $C\,\delta^2$ and find C.

Solution:

(a) The potential energy of an ion placed at the origin is

$$U_\mathrm{p} = \frac{q^2}{4\pi\varepsilon_0}\sum_i (\pm)\frac{1}{R_i} = \frac{2q^2}{4\pi\varepsilon_0 R}\left(-\frac{1}{1}+\frac{1}{2}-\frac{1}{3}+\frac{1}{4}-\ldots\right)$$

$$= (-2\log 2)\cdot\frac{q^2}{4\pi\varepsilon_0 R}$$

$2\log 2 = \alpha$, the Madelung constant of a linear ionic lattice. The total energy U_T of $2N$ ions will be

$$U_\mathrm{T}(R) = N\left(\frac{Az}{R^p} - \frac{q^2}{4\pi\varepsilon_0}2\log 2\cdot\frac{1}{R}\right)$$

with $z = 2$ (i.e., only nearest neighbors). (Note that we include the mutual potential and repulsive energy only once.)

At equilibrium:

$$\left(\frac{\partial U}{\partial R}T\right)_{R=R_0} = 0 = 2N\left(\frac{-pA}{R_0{}^{p+1}} + \frac{q^2}{4\pi\varepsilon_0}\frac{1}{R_0{}^2}\log 2\right)$$

which leads to

$$A = \frac{q^2\log 2}{p\cdot 4\pi\varepsilon_0}R_0{}^{p-1}$$

Substituting the initial expression for U_T, we find

$$U_\mathrm{T} = \frac{2Nq^2\log 2}{4\pi\varepsilon_0 R_0}\left(\frac{1}{p}-1\right) = \frac{N\alpha q^2}{4\pi\varepsilon_0 R_0}\left(\frac{1}{p}-1\right)$$

(b) When the crystal is compressed, the compressive work increases the total energy. For a single ion this energy is

$$u(R) = \frac{q^2\log 2}{4\pi\varepsilon_0}\left(\frac{R_0{}^{p-1}}{pR^p} - \frac{1}{R}\right)$$

Its increase corresponds to

$$u[R_0(1-\delta)]-u[R_0]=\frac{q^2\log 2}{4\pi\varepsilon_0 R_0}\left(\frac{p-1}{2}\right)\delta^2$$

The compressive work per unit length $(1/R_0)$ is therefore of the form (½) $C\delta^2$ where

$$C=\frac{q^2\log 2}{4\pi\varepsilon_0 R_0^{\,2}}(p-1)$$

Exercise 2a: Madelung constant for a row of divalent ions

Determine the Madelung constant for ions placed at the center and at the end of a very long row of equidistant ions with alternating charges of $\pm 2q$, where $q = 1.6 \times 10^{-19}$ C.

Solution:

By convention the Madelung constant is the multiplicative factor α of $-q^2/4\pi\varepsilon_0 r$ in the expression of potential energy of attraction where q is the electronic charge and r is the distance between neighbors. As a result for a $(2q)$ central ion, it will be four times larger than a $(1q)$ ion at the same place but in a row of monovalent ions (see previous Exercise). We thus have

$$\alpha = 4(2\log 2) = 1.386 \times 4 = 5.545.$$

If the ion $2q$ is located at the end of a chain, the Madelung constant will be half of this: $\alpha = 4(\log 2) = 2.773$.

It is thus easy to find the Madelung constant for a crystal of multiple charged ions from that of a crystal of single charged ions having the same structure. Consequently, assuming that the ionic radii are comparable, one may expect a significant increase in the cohesive energy and a less reduction in the inter-ionic distance at equilibrium. (See the case of MgO compared to NaCl in Ex. 8.)

An ion situated at the end of the chain will have a cohesive energy two times less than that of the central ion and it will be more easily removed from the row. Its equilibrium distance with its first neighbor will be larger and its vibration frequencies will be weaker compared to ions of the same type but located far from the ends (see Ex. 6).

Exercise 2b: Madelung constant of a row of ions $-2q$ and $+q$

(a) Consider a linear crystal with the following basis: at 0, an ion A with charge $-2q$ and two ions B with charge $+q$ symmetrically located relative to A at 1/3 and $-1/3$. Find the Madelung constant of each ion, $\alpha(\mathrm{A})$ and $\alpha(\mathrm{B})$.

(b) Find the Madelung constant for crystal similar as that in (a) but with the ions B now located at ¼ and −¼.

(c) What are the limiting values of $\alpha(\mathrm{A})$ and $\alpha(\mathrm{B})$ when the distance between the successive 'molecules' BAB increases? Comment on the result.

Note: The evaluation of the summations can be simplified by regrouping the groups of 3 ions so that the successive contributions of the 'molecules' of BAB appears.

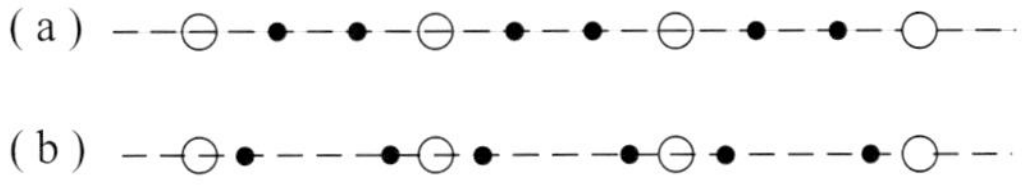

Solution:

(a) Taking into account the symmetry to the right and left, for the ion A of charge $-2q$, we find

$$\alpha(\mathrm{A}) = 2\cdot 2\left(1+\frac{1}{2}-\frac{2}{3}+\frac{1}{4}+\frac{1}{5}-\frac{2}{6}+\frac{1}{7}\cdots\right)$$

Regrouping the terms to find the contributions of type BAB we have

$$\frac{1}{n-1}-\frac{2}{n}+\frac{1}{n+1}=\frac{2}{(n-1)n(n+1)}$$

With n = 3, 6, 9 we have

$$1+\frac{2}{2\cdot 3\cdot 4}+\frac{2}{5\cdot 6\cdot 7}=1.093$$

so that $\alpha(\mathrm{A})$ = 4.37.

For ion B, the symmetry is broken and the left and right sum leads to

$$\alpha(B)=\frac{2-1}{1}+\frac{2-1}{2}-\frac{2}{3}+\frac{2-1}{4}+\cdots=1.093$$

We thus find that $\alpha(\mathrm{B})$ is four times weaker than $\alpha(\mathrm{A})$ within the convention given in Course Summary and in Ex. 8.

(b) For ion A, we obtain the series:

$$\alpha(\mathrm{A}) \approx 2 \cdot 2\left(1+\frac{1}{3}-\frac{2}{4}+\frac{1}{5}+\frac{1}{7}-\frac{2}{8}+\frac{1}{9}\cdots\right)$$

which may be evaluated as in a) taking $n = 4, 8, 12$. The series is equal to 1.037 and $\alpha(\mathrm{A}) = 4.15$.

For ion B, the left and right sum is:

$$\alpha(\mathrm{B}) \approx 2\left(1-\frac{1}{2}+\frac{1}{3}-\frac{1}{4}+\frac{1}{5}\cdots\right) = 2\log 2$$

$\alpha(\mathrm{B}) = 1.386$.

(c) In the isolated molecule BAB, the Madelung constants are $\alpha_{\mathrm{M}}(\mathrm{A}) = 4$ and $\alpha_{\mathrm{M}}(\mathrm{B}) = 1.5$, which are the asymptotic values obtained when the molecules are separated from each other.

Starting from these molecules, it is logical to find that $\alpha(\mathrm{A})$ increases when the second nearest neighbors with opposite sign approaches whereas $\alpha(\mathrm{B})$ decreases as the repulsion acting on an ion of the same sign becomes closer.

The sum $\alpha_{\mathrm{M}}(\mathrm{A}) + 2\alpha(\mathrm{B})$ is equal to 6.556 in (a) whereas it was 6.922 in (b). It approaches the asymptotic value of 7 as the molecules are pulled further apart. We can take that difference between 7 and this sum as a measure of the cohesive energy of the crystal and thus see that such a crystal will have the tendency to dissociate to create additional bonds (such as hydrogen bonds in $H_2^+O^{2-}$, for example). At minimum we would expect that the bonds A–B be shorter than the bonds B–B.

Exercise 3: Cohesive energy of an aggregate of ions

Consider that an aggregate of NaCl forms a cube of side equals to $a/2$ of the full fcc lattice, see Fig. 4.

Assume that the cohesion energy of an ion corresponds to the sum of the attractive Coulomb potential energy of form $V_1 = -\alpha q^2/4\pi\varepsilon_0 r$ and the repulsive potential energy (related to the overlap of the electron clouds between nearest neighbors) of form $V_2 = zA/r^p$ (where A and p are constants).

(a) Compare the cohesive energy of an ion in this aggregate $U(\mathrm{ag})$ and the equilibrium distance which separates two consecutive ions $r_0(\mathrm{ag})$ to those corresponding to a diatomic molecule $V(\mathrm{m})$ and $r_0(\mathrm{m})$ and those of a bulk crystal $U(\mathrm{c})$ and $r_0(\mathrm{c})$.

Express the results in the form of ratios such as $[r_0(\text{ag})/r_0(\text{c})]$ to emphasize the influence of the Madelung constant and the number z of neighbors. How the results change with the size of the aggregate?

(b) Give the numerical values of $r_0(\text{ag})$ and $U(\text{ag})$ taking p = 9 for all cases with $r_0(\text{c})$ = 2.814 Å and $\alpha(\text{c})$=1.747, $U(\text{c})$ = −3.95 eV/ion. Evaluate $r_0(\text{m})$ and compare with the experimental result $r(\text{m})$ = 2.51 Å.

Note: The cohesive energy also includes the ionization energy of the species but this term is the same for the three types of NaCl considered here and then it is not involved in the calculations of r_0.

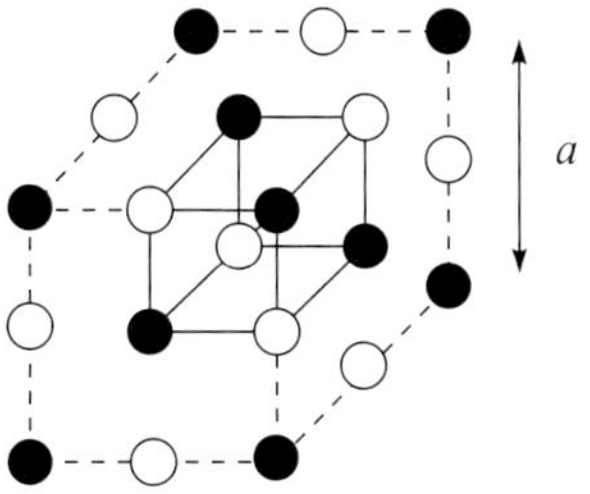

Figure 4

Solution:

(1) The Madelung constant will be the same for each of the eight ions of the aggregate thus the repulsive potential energy is only felt by the z (= 3) first neighbors.

$\alpha(\text{ag}) = 3 - (3/\sqrt{2}) + (1/\sqrt{3}) = 1.456$

The cohesive energy of an ion of the aggregate is

$$U(\text{ag}) = -\frac{q^2}{4\pi\varepsilon_0}\alpha(\text{ag}) + z(\text{ag})A/r^p$$

The energy of an ion located in the bulk crystal will be described by a similar expression with the substitution of $\alpha(\text{ag})$ by $\alpha(\text{c})$ and $z(\text{ag})$ by six (for the six nearest neighbors). In addition for the ion energy in a molecule α = 1 and z = 1.

At equilibrium, $\partial U/\partial r = 0$ one obtains

$$r_0 = [\alpha q^2/4\pi\varepsilon_0 pzA]^{\frac{1}{(1-p)}} \tag{1}$$

After substituting it into the equation for U one obtains:

$$U = (\alpha q^2/4\pi\varepsilon_0 r_0)(1 - \frac{1}{p}) \tag{2}$$

This expression applies to all ionic crystals (in the limit that the expression for U is applicable) with the appropriate choice of α and r_0. We thus have

$$r_0(\text{ag})/r_0(c) = [\alpha(\text{ag})/\alpha(c)]^{\frac{1}{1-p}}[z(\text{c})/z(\text{ag})]^{\frac{1}{1-p}}$$

Numerically we find $r_0(\text{ag})/r_0(c) = 0.938$ and $r_0(m)/r_0(c) = 0.857$, which results in r_0 = 2.412 Å. Compared to the experimental value, r = 2.51 Å, this result is satisfying and could be improved by including the kinetic energy of ions:

$U_0(\text{ag})/U_0(c) = 0.888$, which gives $U_0(\text{ag}) = -3.5$ eV/ion.

The calculation of the cohesive energy of an ion in the molecule is –2.6 eV, which is compatible with the result of a more detailed analysis (–3.47 eV/molecule by taking into account the formation energy of the ions which is U_F = 1.43 eV).

It is coherent to find that the cohesion energy of an ion in an aggregate is smaller than that of the ion in a crystal, otherwise the crystal would spontaneously form aggregates. It is also greater than that of a single molecule. The size effect is in fact more important in the cohesive energy than in the inter-ionic distance, because of the direct influence of α, compared to the inter-ionic distance where the exponent ~1/8 in Eq. 1 reduces its importance. When the size of the aggregate increases, the values of r_0(ag) and U_0(ag) move closer to that of the crystal. However, the calculation becomes more difficult when neither α nor the number first nearest neighbors is the same for all of the ions in the aggregate (see Ex. 5).

Exercise 4: Madelung constant of a 2D ionic lattice

Consider a lattice in 2D in which the ions $+q$ and $-q$ are distributed in the same manner as on the face of a (100) NaCl structure (see Chapter I, Fig. 3e).

(a) Sketch the distribution of ions. Determine the Madelung constant, α(2d) of such a lattice using a direct method in adding contributions of successive neighbors up to the 7th. Comment on the convergence of the series.

(b) Evaluate α(2d) using the Evjen method. This method consists of determining the contributions of ion fractions contained in each successive square in 2d (cubes in 3d) where the reference ion is placed in the center. Stop the summation when the precision of α reaches 10^{-2}. Comment on why the Evjen method allows a faster convergence than the direct method.

(c) Include the repulsive energy between two nearest neighbor ions of the form Ar^{-p}. Determine the distance r_0 that separates the ions in equilibrium.

(d) Noting r_0(3d) as the distance separating the two nearest neighbors in a cubic structure of type NaCl, find the value of the ratio $r_0(2d)/r_0(3d)$ with α(NaCl) $\approx$ 1.75 and assuming that the constant A and the exponent p are the same for the two types of structures ($p \approx 9$). Find the ratio $U_p(2d)/U_p(3d)$ between the two total potential energies.

Summarize these results in showing the curves of U_p (2d) and U_p (3d) as a function of r and comment on the result.

Solution:

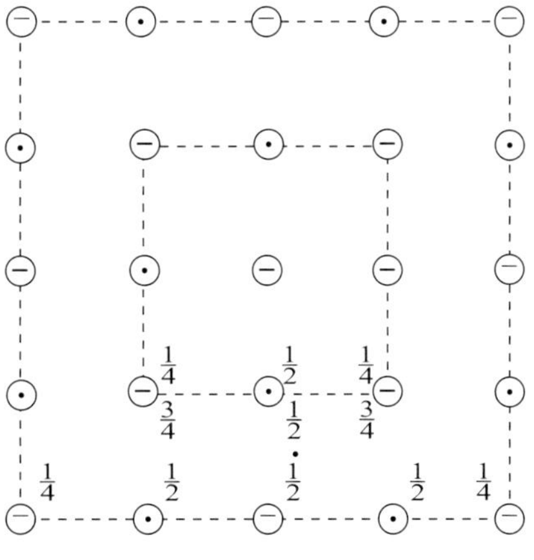

Figure 5

(a) Using the central ion (–) as the origin and noting r_0 as the distance separating two nearest neighbor ions (Fig. 5), the attractive potential energy has the form:

$U_p(\text{attr}) = -\alpha q^2/4\pi\varepsilon_0 r_0$

with $\alpha = 4 - (4/\sqrt{2}) - (4/2) + (8/\sqrt{5}) - (4/\sqrt{8}) + 4/3 - (8/\sqrt{10}) + \cdots$

in which the contributions have been classed by increasing distances (first neighbors, second neighbors, etc.)

It is clear that this series does not converge simply because the three first neighbors lead to a negative value of α (e.g., repulsion) and even the weight of the seventh neighbor (–2.53) is similar to that of the sixth preceding it (+2.75).

(b) Following the Evjen method, we obtain for the first square:

$\alpha_1 = (4/2) - (4/4\sqrt{2}) = 1.293$

The contribution α_2 for charges between the second square and the first square is:

$\alpha_2 = (4/2) - (3/\sqrt{2}) - 1 + (4/\sqrt{5}) - (1/\sqrt{8}) = 0.314$

Finally, the contribution α_3 for charges between the third square (not shown) and the second is:

$$\alpha_3 = -1 + (4/\sqrt{5}) - (3/\sqrt{8}) + (2/3) - (4/\sqrt{10}) + (4/\sqrt{13}) - (1/\sqrt{18}) = 3.6 \times 10^{-3}$$

From this we find $\alpha = \alpha_1 + \alpha_2 + \alpha_3 \approx 1.61$.

Even though we are considering the same summation as in (a), it is done in a different manner. This method leads to a faster convergence because the algebraic sum of charges contained between successive squares is zero and the polar contributions in r^{-1} are zero to first order with the exception of the first square. Therefore only dipolar contributions of order r^{-2}, associated with alternate dipoles, are present.

(c) Calling z the number of first neighbors, the repulsive energy of $2N$ ions in the lattice is $N\,U(r) = zNAr^{-p}$.

Their attractive Coulomb energy is $-N\alpha q^2/4\pi\varepsilon_0 r$.

Thus the total potential energy of an ion is given by

$U_p = [(zA/r^p)] - (\alpha/4\pi\varepsilon_0)\cdot(q^2/r)]$

In equilibrium we have $\left(\dfrac{\partial U}{\partial r}\right)_{r=r_0} = 0$, which leads to $r_0^{\,p-1} = 4\pi\varepsilon_0 \cdot pzA/\alpha q^2$.

(d) $$\frac{a_0(2d)}{a_0(3d)} = \frac{r_0(2d)}{r_0(3d)} = \left[\frac{z(2d)}{z(3d)}\cdot\frac{\alpha(3d)}{\alpha(3d)}\right]^{\frac{1}{1-p}} = \left[\frac{4}{6}\cdot\frac{1.75}{1.61}\right]^{\frac{1}{8}} = 0.96$$

We find that the lattice parameter in an ionic plane is 4% smaller than that of the same crystal in 3D.

If we replace A by $A=(\alpha q^2/4\pi\varepsilon_0)\cdot(r_0{}^{p-1}/pz)$ in the formula for the total energy U_p we find

$$U_\text{p}(r_0)=(\alpha q^2/4\pi\varepsilon_0 r_0)[(1/p)-1]$$

The ratio of energies is thus

$$\frac{U(2d)}{U(3d)}=\frac{\alpha(2d)}{\alpha(3d)}\cdot\frac{r_0(3d)}{r_0(2d)}=\left[\frac{\alpha(2d)}{\alpha(3d)}\right]^{\frac{p}{p-1}}\cdot\left[\frac{z(2d)}{z(3d)}\right]^{\frac{1}{1-p}}=0.957$$

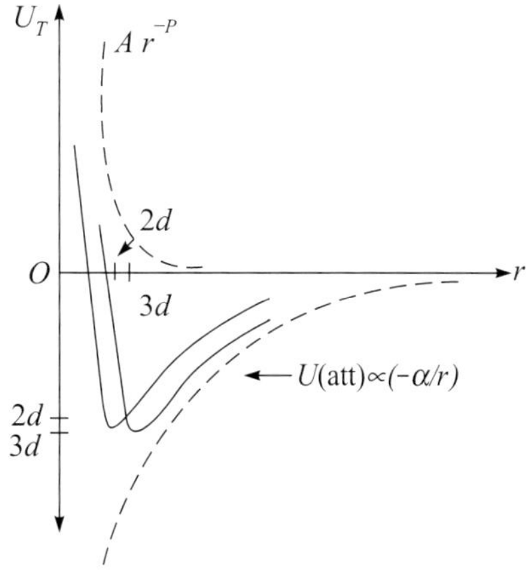

Figure 6

The cohesive energy of the lattice plane will thus be smaller by 4% than the 3D crystal lattice.

These results show that in equilibrium the attractive Coulomb energy contributes nearly 90% (ratio $1 - p^{-1}$ instead of 1) to the total energy in both 2D and 3D. If the total energy is logically smaller in 2D compared to 3D, their only slight difference is due to the different Madelung constants. In addition, the 2D lattice is smaller than that in 3D due to the smaller number of neighbors (four as compared with six) which proportionately reduces the repulsive energy.

Exercise 5: Madelung constant of ions on a surface, an edge, and a corner

In a cubic ionic crystal of NaCl type we consider an ion situated successively on: (A) the surface of the lattice but far from the edges of the finite crystal, (B) on the edge of the straight dihedral formed by two equivalent surfaces and (C) on the corner situated at the intersection of three faces (see Fig. 7).

Give the expression of the corresponding Madelung constants noted as $\alpha(s)$, $\alpha(e)$, and $\alpha(c)$ as a function of the Madelung constants, α_1, α_2, α_3 in 1D, 2D, and 3D, respectively, having ionic distributions analogous to that in a row, 2D plane, and crystal.

What are the numerical values for the (100) face and its intersections with the (010) and (001) faces of NaCl with $\alpha_3 = 1.747$; $\alpha_2 = 1.615$; $\alpha_1 = 1.386$.

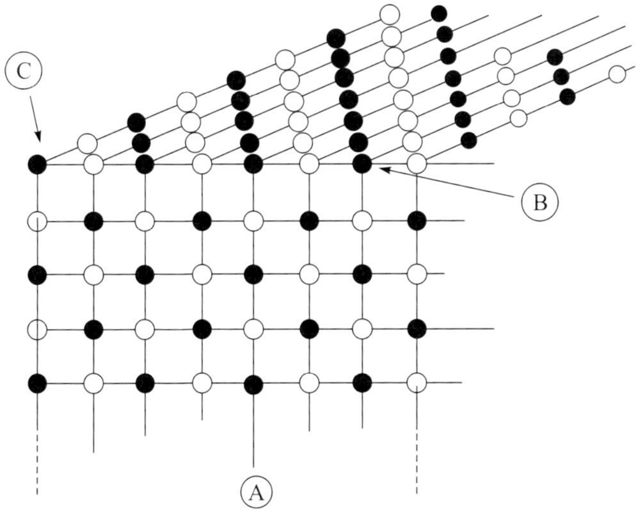

Figure 7

Solution:

When the ion is centered on the surface its Madelung constant is

$\alpha(s) = \alpha_2 + \alpha'$ with $\alpha_3 = \alpha_2 + 2\alpha'$ where α' is given by the contribution of continuous half-space in the plane containing the reference ion. We thus have:

$$\alpha(s) = (\alpha_2 + \alpha_3)/2 = 1.681$$

When the ion is centered on an edge, $\alpha(e)$ is given by: $\alpha(e) = \alpha_2 + \alpha''$ where α'' is the contribution of ions in the interior of the dihedral but excluding those which are situated on the two adjacent faces.

For the entire crystal one has: $\alpha_3 = 4\alpha'' + 2\alpha_2 - \alpha_1$.

Therefore $\alpha(e) = (\alpha_1 + 2\alpha_2 + \alpha_3)/4 = 1.591$.

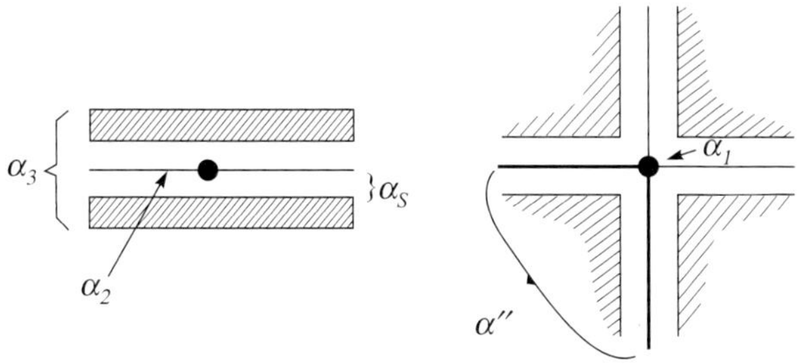

Figure 8

The Madelung constant of an ion situated at the intersection of three orthogonal planes $\alpha(c)$ is related to $\alpha(e)$ by $\alpha(e) = 2\alpha(c) - \alpha'''$ where α''' is the contribution of a quarter of the perpendicular plane of the dihedral that cuts through the reference ion: $\alpha''' = \alpha_2/4 + \alpha_1/2$

In total we have

$$\alpha(c) = \frac{3\alpha_1 + 3\alpha_2 + \alpha_3}{8} = 1.344.$$

Note: This type of calculation can be used to determine the energy necessary to cleave an ionic crystal following the different crystallographic planes. It also can be used to evaluate the decrease of the cohesive energy of an ion (and the increase of its reactivity) as it move away from the crystal. Note also that the Madelung constant involves $\varepsilon_0\ \varepsilon_r$ with a relative dielectric constant $\varepsilon_r \sim 80$ for ions into water leading to the decrease of the attractive energy in proportion and thus facilitating their dissolution.

Exercise 6: Madelung constant of an ion on top of a crystal surface

Figure 9 shows ions $+q$ located above a crystalline surface A, along a step B and along a discontinuity C but always far from the edges.

(1) Find the corresponding Madelung constants α(A), α(B), and α(C) as a function of their value in a row, in a 2D plane and in a 3D crystal, denoted as α_1, α_2, α_3. The nearest neighbors of the ion $+q$ have the opposite sign of the ion so that the system is electrically neutral. Assume that the ion has come from the interior of a crystal where it has left a vacancy.

Calculate the results for a (100) face of NaCl with $\alpha_3 = 1.747$, $\alpha_2 = 1.615$, $\alpha_1 = 1.386$.

(2) Taking the repulsive potential between two nearest neighbor ions to be of the form $U_2 = Ar^{-p}$ (where $p = 9$), find the cohesive energy of the ion A above the surface and compare it to that of an ion placed at the center of a 3D crystal $U(c)$. Find the ratio between the distance between the ion A at equilibrium and the surface, r_A compared to the distance $a/2$ between two neighbor ions in the crystal.
$U(c) = -3.95$ eV/ion and $a/2 = 2.814$ Å
Ion A is moved from its equilibrium position by a displace u normal to the surface such that $u \ll r_A$. Show that it feels a restoring force of the form $F = -\beta u$ and find the expression for β. Determine the frequency ν_0 of vibration of this ion, assuming for simplification that the surface ion to which it is attached is fixed.
Use $M(\text{Na}) = 23$ g for 6.02×10^{23} atoms. Comment the best choice of temperatures to use for epitaxy.

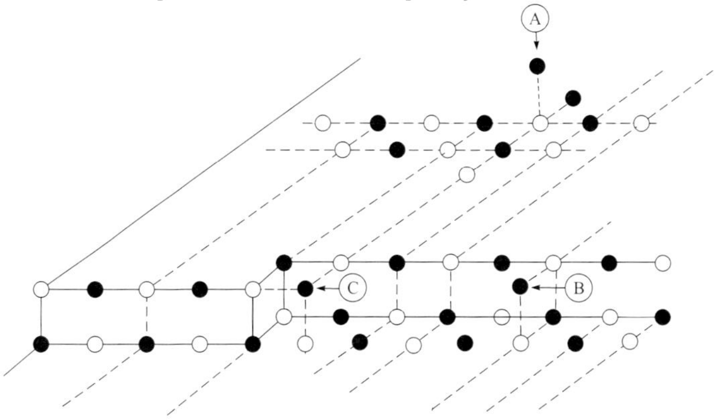

Figure 9

Solution:

(1) The evaluation is easiest to determine in the reverse order:
For symmetry reasons it is easy to see that: $\alpha(\text{C}) = \alpha_3/2$
$\alpha(\text{B}) = \alpha(\text{C}) - \alpha_1/2 = (\alpha_3 - \alpha_1)/2$
$\alpha(\text{A}) = \alpha(\text{B}) - \alpha_2/2 + \alpha_1/2 = (\alpha_3 - \alpha_2)/2$

We may verify this last result evaluating the Madelung constant of an ion included on the surface, denoted $\alpha(\text{s})$, see Ex. 5, to which one must subtract the contribution α_2:

$\alpha(\mathrm{C}) = 0.8735;\ \alpha(\mathrm{B}) = 0.1805;\ \alpha(\mathrm{A}) = 0.066.$

Following the procedure in Ex. 3 [see Eqs. (1) and (2)], we find

$U(\mathrm{A}) = [\alpha(\mathrm{A})q^2/4\pi\varepsilon_0 r_0][1-1/p]$

and

$r_\mathrm{A} = [\alpha(\mathrm{A})q^2/4\pi\varepsilon_0 r_0][1-1/p]$

because here $z = 1$.

$$\frac{r_\mathrm{A}}{a/2} = \left[\frac{\alpha_3}{6\alpha(\mathrm{A})}\right]^{1/8} \cong 1.20 \text{ and } \frac{U(\mathrm{A})}{U(\mathrm{C})} = 3.2\%$$

$U(\mathrm{A}) = 122$ meV and $r_\mathrm{A} = 3.39$ Å

The restoring force is

$$F = -\frac{\partial U}{\partial z} = \frac{pA}{r_\mathrm{A}^{p+1}}\left[1-(p+1)\frac{u}{r_\mathrm{A}}\right] - \frac{\alpha q^2}{4\pi\varepsilon_0 r_\mathrm{A}^2}\left(1-\frac{2u}{r_\mathrm{A}}\right)$$

where $z = r_\mathrm{A} + u$ and $F(r_\mathrm{A}) = 0$.

$$F = -\beta u \text{ with } \beta = \frac{\alpha q^2}{4\pi\varepsilon_0 r_\mathrm{A}^3}(p-1)$$

$$\beta = 3.74 \text{ N/m}; \omega_0\sqrt{\frac{\beta}{m}}; \nu_0 = 1.57\times 10^{12}\text{ c/s}$$

With each particular value of α, the vibration frequency of the ion depends on the site (A, B, or C) that it occupies. This vibration frequency is experimentally accessible by spectroscopy of low-energy electrons.

Comments: Epitaxy

An epitaxial deposition on a single crystal surface is operated generally under high vacuum conditions. If the temperature of the single crystal substrate is too low, the deposited atoms will be distributed randomly on the surface as a function of their point of impact. The crystallization seeds will thus create a polycrystalline coating. If the temperature of the substrate is too high, the atoms such as A that are very weakly bounded to the surface will re-evaporate. The best strategy consists in choosing intermediate temperature to facilitate a surface diffusion of the deposited atoms in such a way that those on sites A can move towards B and next C where their cohesive energy is larger. Favorable conditions for epitaxy, that is mono-crystalline growth, are found if the flux of incident atoms is optimized and in the case of hetero-epitaxy, especially if the natural lattice parameter of the deposition matches that of the substrate (see Chapter I, Pb. 4, and in particular Fig. 50 and related comments).

Beside this specific calculation which concerns ionic crystals, the considerations above are very general and it is particularly obvious that the atomic bonds on the sites such as C, with two neighbors, will be stronger than those on sites such as A. One can thus 'decorate' the crystalline steps of a surface by a simple vaporization of Au atoms (see Y. Quéré, **21**, p. 143.)

In the specific case of the alkali halides, the fact that the Madelung constant of the surface ions is less than unity implies that ions of opposite signs deposited simultaneously tend to form molecules (where $\alpha = 1$) rather than remaining on their initial sites. The diffusion of molecules toward discontinuities assures the homoepitaxy. The present exercise allows the evaluation of the Madelung constant and of the cohesion energy of the molecules of the NaCl type on surfaces.

Exercise 7: Madelung constant of parallel ionic layers

(1) One considers two ionic monolayers, P_1 and P_2, parallel to each other. In a layer the ion arrangement is that of the (100) face of a NaCl crystal and each ion of the first layer is straight above the ion of opposite sign in the second layer. The distance between the two is $a/2$ (see Fig. 10). Find the Madelung constant of an ion in P_1. Use the Evjen method two times to find the contribution of the plane containing the ion, $\alpha(P_1)$, and then of the adjacent plane, $\alpha(P_2)$.

(2) Same question when three consecutive planes are considered in distinguishing two possibilities: an ion placed on the middle plane $\alpha(M)$ and ion placed on an exterior plane $\alpha(E)$. Describe qualitatively the changes in $\alpha(M)$ and $\alpha(E)$ with the increases of the number of ionic monolayers.

Note: The Madelung constant of an ion in the center of a face is $\alpha(f) = 1.681$ as compared $\alpha(v) = 1.747$ when it is in an infinite crystal (see Ex. 4).

(3) For a semi-infinite crystal determine how α changes when the reference ion is on the surface (first layer) and then on subsequent layers (the second, third, etc.) of the crystal. Suggest an explanation for the surface relaxation (in-depth variation of the monolayer spacing) in NaCl.

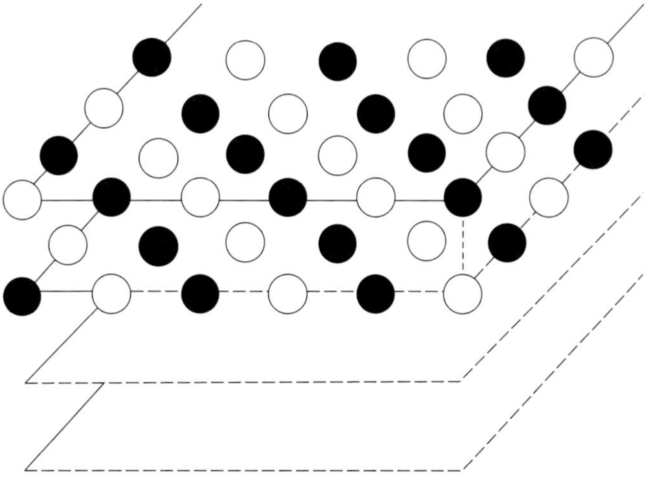

Figure 10

Solution:

(1) The contribution to the crystalline plane P_1 containing the ion was discussed already in Ex. 4. Limited to the first two squares this contribution is

$$\alpha(P_1) = 4 - \frac{4}{\sqrt{2}} - \frac{4}{2\cdot 2} + \frac{8}{2\sqrt{5}} - \frac{4}{4\sqrt{8}} = 1.607$$

The contribution of the homologous ions in the plane P_2 appears as a change in sign and a unitary increase of the measured distance in z:

$$\alpha(P_2) = -\frac{4}{\sqrt{2}} + \frac{4}{\sqrt{3}} + \frac{4}{2\sqrt{5}} - \frac{8}{2\sqrt{6}} + \frac{4}{4\sqrt{9}} \quad (+1)$$

The addition of +1 takes into account the ion of opposite sign straight below the reference ion to obtain a global neutrality of the charges.

$\alpha(P) = 0.076$ or $\alpha = 1.683$

(2) $\alpha(M) = \alpha(P_1) + 2\alpha(P_2) = 1.759$

To evaluate $\alpha(E)$ one must include the contribution from the third plane following an analogous procedure used for $\alpha(P_2)$:

$$\alpha(P_3) = +\frac{4}{\sqrt{5}} - \frac{4}{\sqrt{6}} - \frac{4}{2\sqrt{8}} + \frac{8}{2\sqrt{9}} - \frac{4}{4\sqrt{12}}\left(-\frac{1}{2}\right) = -7\times 10^{-3}$$

$$\alpha(E) = \alpha(P_1) + \alpha(P_2) + \alpha(P_3) = 1.676$$

when there are four layers and the reference ion is found in the two intermediate layers, its Madelung constant will be such that

$\alpha(\mathrm{M}) = \alpha(\mathrm{P}_1) + 2\alpha(\mathrm{P}_2) + (\mathrm{P}_3) = 1.752.$

This constant becomes smaller for N = 5 because in addition to the two negative contributions of the external planes of type P_3: $\alpha(\mathrm{M}) = \alpha(\mathrm{P}_1) + 2\alpha(\mathrm{P}_2) + 2\alpha(\mathrm{P}_3) = 1.745$.

The general tendency is shown in Fig. 11. The contribution of every even neighboring planes (second, fourth, sixth, etc.) is negative because the most strongly effect of the equivalent ion of the same sign as the reference ion. But the effect of successive planes, regardless of whether their contribution is positive or negative, decreases very quickly as distance increases. The asymptotic value $\alpha(\mathrm{s})$ for ions in E and $\alpha(\mathrm{v})$ for ions in M are practically reached for the first layers (N = 3, 4, 5), depending on the desired precision. On must nevertheless, note that the Madelung constant of an ion in a double layer (α = 1.681), as well as that located in the intermediate plane of a triple layer ($\alpha \sim 1.759$) is greater than that in the bulk crystal [$\alpha(v)$ = 1.747].

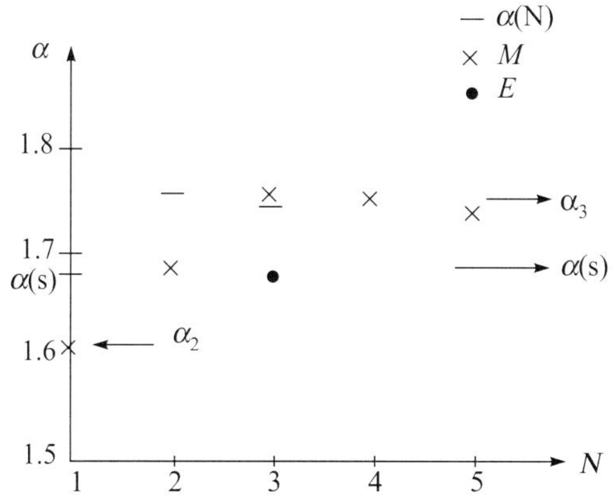

Figure 11

(3) Starting from the surface (N = 1), we find successively:
$\alpha(N=1) = \alpha(\mathrm{s}) = 1.681; \alpha(N=2) = \alpha(\mathrm{s}) + \alpha(\mathrm{P}_2) = 1.757;$
$\alpha(N=3) = \alpha(\mathrm{s}) + \alpha(\mathrm{P}_2) + \alpha(\mathrm{P}_3) = 1.750.$

The corresponding evolution is represented by dashed lines in Fig. 11 with the analogous remarks to those in the preceding section: the rapid change from $\alpha(\mathrm{s})$ to $\alpha(\mathrm{v})$ where the value of α exceeds that of $\alpha(\mathrm{v})$ for the nearest layer situated above the surface layer.

Comment

The significant variation of α associated with changing number of nearest neighbors (from five to six) when considering the surface of an ionic layer and that of the next one, suggests the existence of surface constraints when assuming that the ions occupy ideal sites of the 3D crystal because these sites do not strictly correspond to the minimum of potential energy.

Even though the calculation of the exact configuration taken at equilibrium is complex (the existence of the surface imposes a different calculation for the displacements parallel or perpendicular to the surface), one can nevertheless consider that the top surface layer is going to move slightly from its ideal position (when α decreases, r_0 increases) and thus give rise to a tetragonal type first layer (where $a = b$ and $c > a$) so that the surface tension is relaxed. Other configurations are possible also, such as the surface reconstructions see Chapter I, Pb. 4, Fig. 49.

The above considerations may be applied to other types of crystals. If these perturbations affect only the first few atomic layers they can still greatly influence the physical properties of the materials (such as their catalytic activity for instance.)

Exercise 8: Cohesive energy of a MgO crystal

The crystalline structure of MgO is identical to that of NaCl (fcc with an ion at 0 0 0 and one of opposite sign at ½ 0 0) but the charge carried by each ion is $+2q$, where q is the electronic charge. Find the Madelung constant of an ion $2q$ and its cohesive energy U_0, which is the sum of the Coulomb attractive energy and the repulsive energy between two nearest neighbors of the form $\mathrm{A}r^{-\mathrm{p}}$. Find the expected ratio between these parameters and those of NaCl assuming that A takes the same value in both crystals.

Give the numerical values with:

$$\alpha(\mathrm{NaCl}) = 1.747;\, a(\mathrm{NaCl}) = 5.63\ \text{Å};\, U_0(\mathrm{NaCl}) = (-)7.9\ \mathrm{eV/mol\ and}\ p = 9.$$

From X-ray diffraction, a_0 (MgO) is found to be 4.21 Å.

Solution:

By convention here α is the coefficient in the quantity $q^2/4\pi\varepsilon_0 r$ in the expression of the attractive potential energy of ions. Since the ions have a double charge ($Mg^{2+}O^{2-}$), the Madelung constant will be four times that of NaCl which consists of ions with a single charge (see Ex. 2 and comments at the end of this exercise):

$$\alpha(\mathrm{MgO}) = 4\alpha(\mathrm{NaCl})$$

Using equations (1) and (2) of Ex. 3, we find

$$a_0(\mathrm{MgO})/a_0(\mathrm{NaCl}) = [\alpha(\mathrm{NaCl})/\alpha(\mathrm{MgO})]^{\frac{1}{p-1}}$$

and

$$U(\mathrm{MgO})/U(\mathrm{NaCl}) = [\alpha(\mathrm{MgO})/\alpha(\mathrm{NaCl})][a_0(\mathrm{NaCl})/a_0(\mathrm{MgO})]$$

Numerically one obtains

$$a_0(\mathrm{MgO})/a_0(\mathrm{NaCl}) = 0.841 \quad \text{and} \quad U(\mathrm{MgO})/U(\mathrm{NaCl}) = 4.76$$

which corresponds to

$$U_0(\mathrm{MgO}) = -37.5\ \mathrm{eV/molecule}.$$

The expected lattice parameter would be $a_0(\mathrm{MgO})$ = 4.73 Å; a value relatively far off from the real value of a_0(= 4.21 Å) because of the difference of the ionic radius (see next exercise) but the general tendency is correct due to the contraction of the lattice when the ions increase in charge.

Comments

Despite the obvious simplifications this exercise shows relatively well the influence of charge ions on inter-ionic distances, which decrease when the charges increase, and also on the cohesive energy which increases in proportion to the product of the charges carried by each ion species. This exercise explains how crystals of type MX^2 (where M is a divalent metal and X is a halogen) can have values of α greater than four (see Ex. 2b).

In this exercise the numerical values of the cohesion energy (–37.5 eV/molecule) is remarkably close to the generally accepted value (–40.8 eV/molecule; see Ref. [18]) even though the accepted value for p (which describes the interionic repulsion at short distances: Lennard-Jones) would be 4.82.

Finally, the interested reader may try doing the exercises in this chapter that use parameters of NaCl using MgO instead. For certain systems (such as a homogenous chain or cubic aggregates), one can compare the results obtained to those of Moukouri and Noguera (*Z. Phys. D* **24**, 1992, p. 71).

Exercise 9: Ionic radii and the stability of crystals

In the alkali halides one can assume that the ions are hard spheres of radius r_+ (cation) and r_- (anion).

(a) What inequality should the ratio r_-/r_+ satisfy so that in a simple cubic structure of CsCl type (see Fig. 3a in Chapter I, Ex. 1) the + and – ions are in contact along the diagonal without the overlapping of the largest ions of the same sign at the corners of the cube? What is the optimal packing parameter, *t*?

(b) What inequality should the ratio r_-/r_+ satisfy so that in a fcc lattice of the NaCl type (Fig. 3e, Chapter I, Ex. 1) the ions of opposite signs located along the [100] rows are in contact without the overlap of the biggest ions of the same sign located along the [110] rows? What is the optimal packing parameter?

(c) Suppose that CsCl can crystallize in either a simple cubic lattice either or a fcc lattice. For both hypotheses, find the distance r_0 between nearest neighbors of opposite signs, the lattice parameter *a* of the cube and the corresponding packing parameter. Same question for NaCl. Comment on the results. Use r_+ (Na^+) = 0.98 Å; r_+ (Cs^+) = 1.67 Å; r_- (Cl^-) = 1.81 Å.

(d) In the alkali halides the cohesive energy E_c can be approximately described by (see Course Summary in Chapter I, Section 3):

$$E_c = (\alpha e^2/4\pi\varepsilon_0 r_0)\left(1-\frac{\rho}{r_0}\right),$$ where α is the Madelung constant and ρ = 0.345 Å.

For CsCl, find the value of the ratio of cohesive energies, $E_c(\text{sc})/E_c$ (fcc), for the two hypotheses considered in (c). Same question for NaCl, α (sc) = 1.7626; α(fcc) = 1.7476.

Solution:

(a) $2(r_+ + r_-) = a\sqrt{3}$ and $2r_+, 2r_- \le a$.

Thus $\sqrt{3}-1=0.732 \le \dfrac{r_-}{r_+} \le \dfrac{1}{\sqrt{3}-1}=1.366$

$$t=\frac{(4\pi/3)(r_+^3+r_-^3)}{(8/3\sqrt{3})(r_++r_-)^3}=\frac{\pi\sqrt{3}}{2}\cdot[1-3p/(1+p)^2] \quad \text{with} \quad p=\frac{r_-}{r_+}$$

- t is minimal when $r_-=r_+$. It takes the value $t_m=\dfrac{\pi\sqrt{3}}{8}=0.68$, corresponding to a bcc lattice (see Chapter I, Ex. 9). It is maximal for the limiting values of p so that:

  $t_M=\pi[\sqrt{3}-(3/2)]=0.729$

(b) $2(r_++r_-)=a$ and $4r_+, 4r_- \le a\sqrt{2}$ which leads to

$\sqrt{2}-1=0.414 \le r_-/r_+ \le 1/(\sqrt{2}-1)=2.414$

$$t=(1/2)(4\pi/3)(r_+^3+r_-^3)/(r_++r_-)^3=\frac{2\pi}{3}[1-3p/(1+p)^2]$$

- t is minimal for $r_+ = r_-$ where it is $t_m=\dfrac{\pi}{6}=0.524$

  This minimal value corresponds to a simple cubic lattice (see Chapter I, Ex. 9a).

  It is maximal for the limiting values of p so that t_M = 0.793.

(c) • $\dfrac{r(Cl^-)}{r(Cs^+)}=1.08$

CsCl must obey the two inequalities in (a) and (b). It can crystallize in the two lattices with the overlap of the nearest neighbors of opposite signs which leads to

$r_0=r_-(Cl^-)+r_+(Cs^+)=3.48$ Å and

$a=4.02$ Å (sc) or $a=6.96$ Å (fcc); $\dfrac{t(\text{sc})}{t(\text{fcc})}=\dfrac{3\sqrt{3}}{4}\approx 1.30$

- $\dfrac{r(Cl^-)}{r(Na^+)}\approx 1.85$

NaCl obeys the inequality in (b) but not that in (a). When crystallized in a simple cubic system, the neighboring ions of opposite sign cannot be in contact and the lattice parameter will be determined by the contact of the Cl^- ions (the largest ones) along the rows [100]:

sc: $a = 2r_- = 3.62$ Å, $r_0 = a\sqrt{2}/2 = 3.14$ Å, $t = 0.606$
fcc: $a_0 = 5.58$ Å, $r_0 = 2.79$ Å, $t = 0.662$
which results in

$$\frac{t(\mathrm{sc})}{t(\mathrm{fcc})} = 0.915$$

Remark: Under the action of Coulomb forces, ions of opposite signs tend to move nearer to each other and the most stable equilibrium corresponds to the most compact configuration taking into account the ion volume. From this simple model, one can therefore predict that NaCl will crystallize in a fcc lattice and that CsCl will crystallize in a simple cubic lattice.

(d) For CsCl, r_0 is the same for the two structures so the ratio of cohesion energy is directly related to the ratio of the Madelung constants:

$$\frac{E(\mathrm{sc})}{E(\mathrm{fcc})} = 1.0086$$

For NaCl on the other hand we find:

$$\frac{E(\mathrm{cs})}{E(\mathrm{fcc})} = \frac{\alpha(\mathrm{sc})}{\alpha(\mathrm{fcc})} \cdot \frac{r_0^2(\mathrm{fcc})}{r_0^2(\mathrm{sc})} \cdot \frac{r_0(\mathrm{sc}) - \rho}{r_0(\mathrm{fcc}) - \rho} = 0.91$$

Comments

One of the most important (and most difficult) problems in physics of crystals is the determination of the relative stability of different types of structures. If it is easy to determine that the most stable lattice is that which has the smallest free energy, its evaluation in the different situations is very delicate even for the alkali halides. The energy taken into account in the present exercise is valid at 0 K and in addition to the kinetic energy terms, it neglects corrections due to van der Waals bonds (see Pb. 3) which lead to uncertain results if the two possible structures are energetically very close to each other.

The approach to ionic radii is still rather coarse because the repulsive term due to the overlapping of electronic orbitals (in exp $(-r/p)$ is based on a model of hard spheres. This explains the differences between the values of a thus calculated and those determined experimentally (CsCl : $a = 4.11$ Å, NaCl: $a = 5.63$ Å).

Although schematic, this exercise gives nevertheless the correct reply to the following question: "Why does sodium chloride crystallize in a face-centered cubic lattice when cesium chloride crystallizes in a simple cubic lattice?" We refer the reader to Ref. [25], Chapter 2, p. 83 for further details on the physics of alkali halides and also the excellent article of J. C. Phillips (*Rev. Mod. Phys.* **42,** 1970, p. 317) on the "Spectroscopy theory of the chemical bond" relative to compounds of type $A^N B^{8-N}$. See also Ref. [12].

Exercise 10: Lennard-Jones potential of rare gas crystals

The potential energy of attraction between two atoms of rare gas separated by a distance r is of the form A/r^6, van der Waals attraction, and the repulsive energy due to the overlapping of electronic orbitals is of the form B/r^{12}. The Lennard-Jones potential is of the form:

$$U = 4\varepsilon\left[\left(\frac{\sigma}{r}\right)^{12} - \left(\frac{\sigma}{r}\right)^{6}\right]$$

(a) Show that one can express A and B as a function of ε and σ so that these expressions are equivalent.

(b) Find the physical meaning of the parameters ε and σ by expressing the distance r_0 separating two atoms in equilibrium as a function of σ and by expressing the cohesive energy which results as a function of ε.

(c) The Bravais lattice of rare gas crystals is fcc with lattice parameter a = 4.46 Å(Ne); a = 5.31 Å(Ar); a = 5.64 Å(Kr) and a = 6.13 Å(Xe). The respective cohesive energies are 20 meV (Ne), 80 meV (Ar), 116 meV (Xe). Find the values of σ and ε and state the relative errors of ε that arise because second neighbors have been neglected.

Solution:

(a) $B/r^{12} - A/r^6 = (4\varepsilon\sigma^{12}/r^{12}) - (4\varepsilon\sigma^6/r^6)$ so that $B = 4\varepsilon\sigma^{12}$ and $A = 4\varepsilon\sigma^6$.

(b) $(\partial U/\partial r)r_0 = 0$, which results in $12\sigma^{12}r_0^{-13} = 6\sigma^6 r_0^{-7}$ or $r_0 = 2^{1/6}\sigma = 1.12\sigma$

$$E_c = 4\varepsilon\left[\frac{1}{4} - \frac{1}{2}\right] = |\varepsilon|$$

The Lennard-Jones parameter ε has the advantage of directly expressing the cohesive energy and σ is proportional to the distance r_0 at equilibrium. In addition $E_c = 0$ when $r = \sigma$.

(c) In fcc crystals $r_0 = a/\sqrt{2} = 1.12\,\sigma$ or $\sigma = 0.631a$ and each atom has 12 nearest neighbors. The cohesive energy of an atom in the crystal is thus 6ε because the Lennard-Jones formula is in fact the mutual energy between two atoms, thus the factor $N/2$, and not the factor N for the cohesive energy of N atoms (see Course Summary). We find the following table:

	Ne	Ar	Kr	Xe
σ(Å)	2.81	3.35	3.56	3.87
ε(meV)	3.33	13.33	19.33	28.33

The number of second neighbors is six at a distance a.
The relative error of ε (and E_c) essentially involves the r^6 term and its weight is $1/2^4 \approx 6\%$.
For more details see Pb. 2.

Exercise 11: Chemisorption on a metallic surface

Under certain conditions the binding energy of an atom on the surface of a metal (essentially transition metals) may take the form:

$$E_L = zAe^{-pr} - \sqrt{z}Be^{-qr}$$

in which the first terms describes repulsion (of the Born–Mayer type) due to the overlapping of electronic orbitals and the second term is related to interatomic attraction (in the tight-binding approximation); z is the number of nearest neighbors (or coordination number) between the adsorbed atom and the atoms of the substrate. The parameters A, B, p, and q are positive.

(1) What inequality between the exponents p and q must hold so that the adsorbed atom can take a stable equilibrium position r_0 with respect to its z neighbors. Find an expression for the equilibrium distance r_0 as a function of p, q, A, B, and z.

(2) Show that the binding energy E_L can take the form $E_L = Cz^{\alpha}$. Find explicitly the range of variation of the exponent α when p varies between $2q$ and $4q$.

Assume that the nature of the atom is such that $p = 4q$ and that the bond energy is ~7 eV when it occupies the most stable site on the (100) face of a bcc metal. Find the atomic distribution on the (100) and (110) faces of this metal. Determine the most favorable site (point C). Also find other possible sites and the associated binding energies.

(3) In the hypothesis that $p = 2q$, indicate if there is preferred site and if so, which one?

(4) Determine the angular vibration ω of the ad-atom on the surface as a function of its equilibrium bond energy $E_L(r_0)$, knowing that the force constant β is such that

$$\beta = \left(\frac{d^2E}{dr^2}\right)_{r=r_0}$$

Using $E_L = 7$ eV, $p = 8$ Å $= 4q$, find the energy of the quantum vibration $\hbar\omega$ for an oxygen atom ($A = 16$). ($\hbar$, N)

Solution:

(1) The equilibrium condition is obtained by evaluating dE_L/dr which must vanish at a distance r_0 positive. In addition $(d^2E_L/dr^2)_{r=r_0}$ must be positive in order for the equilibrium to be stable.

One obtains successively:
$zpAe^{-pr_0} = \sqrt{z}qBe^{-qr_0}$ which leads to $e^{(p-q)r_0} = \sqrt{z}\dfrac{pA}{qB}$ from which we find $r_0 = \dfrac{1}{p-q}\left(\dfrac{1}{2}\log z + \log\dfrac{pA}{qB}\right)$.

The condition on the second derivative implies $p - q > 0$. The inequalities that must be satisfied are thus $p > q$ and $pA > qB$ (when $z = 1$). The equilibrium distance r_0 increases as the co ordinance number z.

(2) Regrouping terms in the expression for E_L we find

$$E_L(r_0) = \sqrt{z}\left(\frac{q}{p} - 1\right)Be^{-qr_0} \text{ or } E_L(r_0) = z\left(1 - \frac{p}{q}\right)Ae^{-pr_0}.$$

Taking the logarithm of one of these two expressions and using the expression for r_0, one finds

$$\log E_L(r_0) = \frac{p-2q}{2(p-q)}\log z + \log B\left(\frac{q-p}{p}\right) - \frac{q}{p-q}\log\frac{pA}{qB}$$

Thus E_L is of the form $E_L = Cz^{\alpha}$ with

$$\alpha = \frac{p-2q}{2(p-q)} \text{ and } C = B\left(\frac{q}{p} - 1\right)\left(\frac{pA}{qB}\right)^{\frac{q}{q-p}}$$

When $p = 2q$, $\alpha = 0$ and $E_L = C'$.
When $p = 4q$, $\alpha = 1/3$ and $E_L = C''z^{1/3}$.

(3) In the hypothesis where $p = 4q$, $\alpha = 1/3$ and the binding energy increases when the coordination increases. On the (100) face it will be maximal at the point C where $z = 4$.

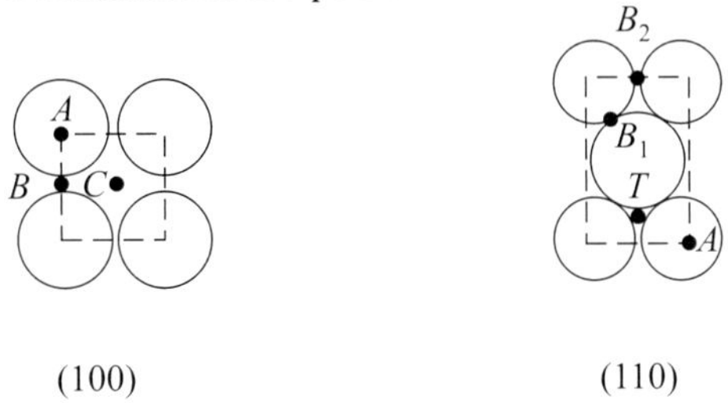

Figure 12

As shown in Fig. 12, the other possible sites are found at points such that B ($z = 2$) and A ($z = 1$). On the (110) face the most stable site is in T ($z = 3$). The points B_1 and B_2 are the equivalent of point B on the (100) face.

The binding energies vary as $z^{1/3}$ so that one easily obtains C″~ 4.4 eV; $E_L(C) = 7$ eV; $E_L(T) = 6.35$ eV; $E_L(B) = 5.55$ eV; $E_L(A) = 4.4$ eV.

If the binding energy increases with the coordination number z, this increase is relatively weak. The good part is, when $p = 2q$, one has $\alpha = 0$. This means that the position of the atom is indifferent since its binding energy is independent of z.

(4) $\omega = \sqrt{\dfrac{\beta}{m}}$ with $\beta = \left(\dfrac{d^2E}{dr^2}\right)_{r=r_0}$ or $\beta = (p-q) = \sqrt{z}qBe^{-qr_0}$,

taking into account $E_L(r_0)$, $\beta = -pqE_L(r_0)$. The vibration frequency of the adsorbed atom will vary as $\sqrt{E_L(r_0)}$ or as $z^{\alpha/2}$. This frequency depends therefore on the site chosen by this atom if $p > 2q$ (see Ex. 6).

β is, of course, positive because E_L is negative if one takes the origin of energies to be $r = \infty$.

$\beta = 18$ N/m; $\omega = 2.6 \times 10^{13}$ rad/sec; $\hbar\omega = 17$ meV.

Comments: Physi- and chemisorption; atom manipulation

When a polarized atom is located in a vacuum ε_0 near the surface its equivalent dipole is subject to the action of its dipolar image which will be attractive regardless of whether the surface is metallic or insulating. But in the latter case the attraction will be smaller.

Effectively, to a real charge q set in the vacuum in front of a semi-infinite medium, it corresponds an image charge Kq, where $K = (\varepsilon_0 - \varepsilon)/(\varepsilon_0 + \varepsilon)$, that is, symmetrically located into the medium of dielectric constant ε (Fig. 13a). For a metal, $\varepsilon = \infty$ and $K = -1$. For an insulator, $\varepsilon_0 < \varepsilon < \infty$, $K < 0$ with $|K| < 1$.

An adsorbed atom on a surface is subjected to physisorption (or physical adsorption) when it does not exchange electrons (e.g., if it does not form a chemical bond) with the substrate. Its binding energy, of van der Waals type, is weak and the distance r_0 of the atom–substrate is relatively large. An example is the condensation of rare gases on graphite surfaces. From a theoretical point of view, weak coupling between the atom and the substrate occurs when the electronic states of each partner (discrete level and band) will not be disturbed by one another. It is this hypothesis on which the present exercise is based and which concerns the absorption of metallic atoms on a metal.

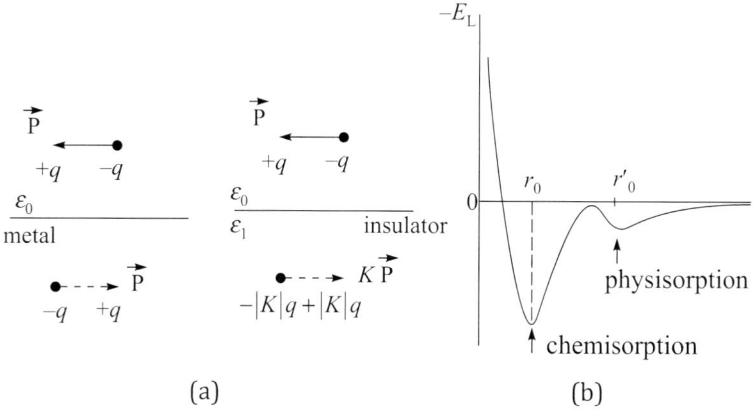

Figure 13

In chemisorption the electronic bonding atoms or molecules is changed and covalent or ionic bonds are formed from electron transfer of an electron (chemical bond. It results in a greater bond energy and a smaller distance r_0 (see Fig. 13b). When the formation of a superficial molecule with the appearance of bonding and anti-bonding states is considered the corresponding tight binding is more difficult to treat from a theoretical point of view.

From an experimental point of view, the occupied site of the ad-atoms may be visualized from the use of scanning probe microscopes (SPM) such as a scanning tunneling microscope (STM; see Chapter IV, Pb. 5, for details). Another powerful STM capability is the ability to move atoms and molecules. This is achieved by placing the tip close enough to the surface adsorbate so that the tip-adsorbate attraction is comparable to the surface corrugation barrier. In this regime, the molecule will follow the tip wherever it is moved along the surface, see Fig. 13c. One can then retract the tip, without causing the molecule to desorb from the surface. A fair example of such atom manipulation is shown in Fig. 13d and the interested readers are referred to C. Julian Chen [29].

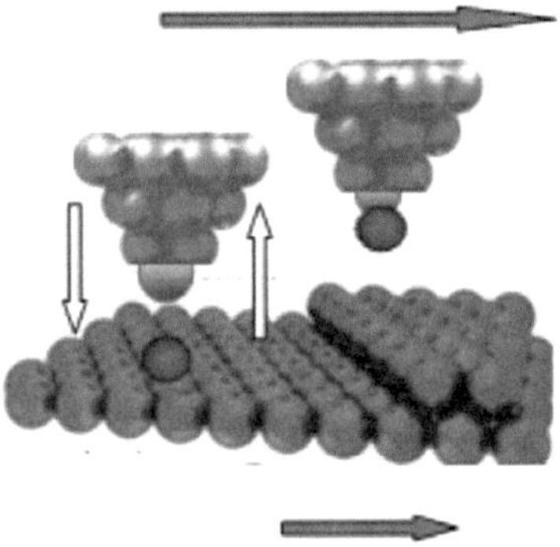

(c)

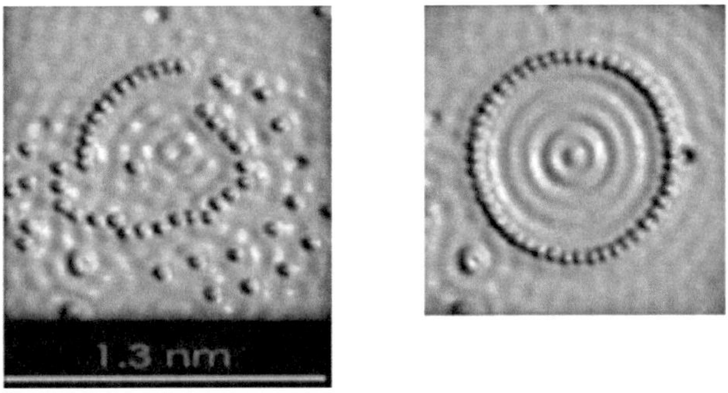

(d)

Figure 13

The examination of the diffraction diagram in LEED is also possible when the observed structure is periodic. The energy of quantum vibration of the ad-atom $\hbar\omega$ can be deduced from a measure of the energy gain or loss when an electron beam of slow and very mono-energetic electrons ($E_c \sim 5$ eV).

For further details, see M. C. Desjonqueres and D. Spanjaard [26] and *J. Phys. C., Solid State Physics* **15**, 1982, 4007, which provided the inspiration for this exercise as well as C. Noguera [18].

Exercise 12: Anisotropy of the thermal expansion of crystals

Aragonite is a calcium carbonate crystallized in an orthorhombic form ($a \neq b \neq c$ and $\alpha = \beta = \gamma = \pi/2$). Its thermal expansion coefficient in volume, α_v, is 62×10^{-6}/°C. When the temperature of the crystal changes from 0°C to 100°C the angle δ between the (100) and (110) planes decreases by 1′14 and the angle ε between the (001) and (011) planes increases by 2′24.

Knowing the ratio of the crystalline parameters at 0°C is such that a/b = 0.6224 and c/b = 0.7206, find the value of the 3 linear coefficients of thermal expansion (α_a, α_b, α_c) of the crystal.

Solution:

Figure 14a shows the aragonite crystal before (solid line) and after (dashed line) expansion. After expansion the volume of the crystal is $V+\Delta V = a(1+\alpha_a)\cdot b(1+\alpha_b)\cdot c(1+\alpha_c)$ or $\alpha_v = \dfrac{\Delta V}{V} = 62\times 10^{-6} = \alpha_a + \alpha_b + \alpha_c$

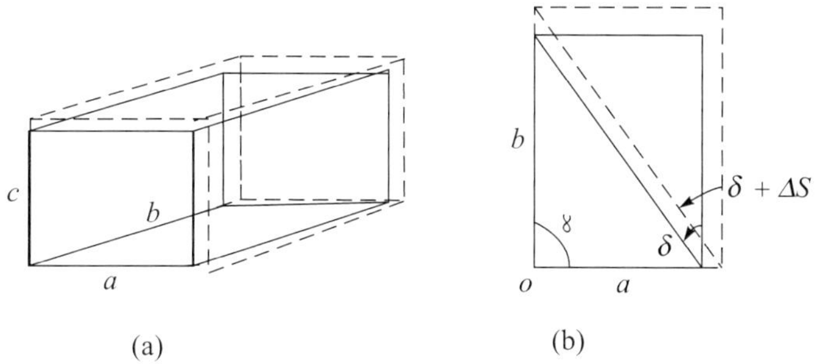

Figure 14

The angle δ between planes (100) and (110) shown in Fig. 14b is at 0°C such that $\tan\delta = a/b$. At 100°C one has

$$\tan(\delta+\Delta\delta) = \frac{a(1+100\alpha_a)}{b(1+100\alpha_b)}$$

which after simplifying becomes

$$\alpha_a - \alpha_b = \frac{\Delta\delta}{100\frac{a}{b}}\left(1+\frac{a^2}{b^2}\right) = -7.4\times10^{-6}$$

In the same way we find that the angle ε between planes (001) and (110) is such that tan = c/b from which

$$\alpha_c - \alpha_b = \frac{\Delta\varepsilon}{100\frac{c}{b}}\left(1+\frac{c^2}{b^2}\right) = 17.4\times10^{-6}$$

As a result we find

$$\alpha_a = 9.9\times10^{-6},\ \alpha_b = 17.3\times10^{-6},\ \alpha_c = 34.7\times10^{-6}$$

Exercise 13: Tension and compression in an isotropic medium: relations between S_{ij}, C_{ij}, E (Young's modulus) and σ (Poisson coefficient), λ and μ (Lamé coefficients)

A bar of length l and cross section S is shaped out of a homogenous and isotropic solid. When the bar is submitted to a longitudinal force F along l, its length variation Δl follows an algebraic variation of $\frac{\Delta a}{a} = \frac{\Delta b}{b}$ in each traversal dimension.

(a) Find the Young's modulus E $\left(E = \frac{F}{S}\cdot\frac{l}{\Delta l}\right)$ and the Poisson coefficient σ $\left(\sigma = -\frac{\Delta a/a}{\Delta l/l}\right)$ as a function of the elastic compliance coefficients S_{ij} and alternatively in terms of the elastic stiffness coefficients C_{ij}.

Starting from the expression:

$$\frac{F}{S} = \lambda\frac{\Delta V}{V} + 2\mu\frac{\Delta l}{l}$$

in which $\frac{\Delta V}{V}$ represents the relative variation of the volume of the bar when subject to the stress $\frac{F}{S}$, find the Lamé coefficients λ and μ.

(b) The bar is now subject to a hydrostatic pressure Δp which results in a volume variation of ΔV. What is the expression for the coefficient of compressibility $\beta = -\frac{1}{V}\left(\frac{\Delta V}{\Delta p}\right)$?

Express the results as a function of (α) σ and E, (β) S_{ij}, (γ) C_{ij}, and (δ) λ and μ.

(c) *Numerical application*: For aluminum $C_{11} = 1.07 \times 10^{11}$ N/m^2 and $C_{12} = 0.61 \times 10^{11}$ N/m^2. What are the corresponding numerical values of E and σ, λ and μ, and B?

Solution:

(a) In the simple situation of an isotropic solid, the useful relations between the stress and strain reduce to

$$e_{xx} = \frac{\Delta l}{l} = S_{11}\frac{F}{S} \quad (1)$$

$$e_{yy} = \frac{\Delta a}{a} = S_{12}\frac{F}{S} \quad (2) \Rightarrow \text{ or equivalently: } S_{11} = \frac{1}{E} \text{ and } S_{12} = -\frac{\sigma}{E}$$

$$e_{zz} = \frac{\Delta b}{b} = S_{12}\frac{F}{S} \quad (3)$$

In terms of the elastic stiffness constants one obtains:

$$(1'): \quad \frac{F}{S} = C_{11}e_{xx} + 2C_{12}e_{yy}$$

$$(2') = (3'): \quad 0 = C_{12}e_{xx} + (C_{11} + C_{12})e_{yy}$$

or equivalently,

$$E = C_{11} - 2C_{12}\sigma$$

$$0 = C_{12} - \sigma(C_{11} + C_{12})$$

The isotropy leads to $e_{yy} = e_{zz}$ or $C_{12} = C_{13} = C_{21} = C_{31}$. Then one obtains

$$E = \frac{1}{S_{11}} = \frac{(C_{11} - C_{12})(C_{11} + 2C_{12})}{C_{11} + C_{12}}, \sigma = \frac{-S_{12}}{S_{11}} = \frac{C_{12}}{C_{11} + C_{12}}$$

$$\text{and } S_{11} = \frac{C_{11} + C_{12}}{(C_{11} - C_{12})(C_{11} + 2C_{12})}, S_{12} = -\frac{C_{12}}{(C_{11} - C_{12})(C_{11} + 2C_{12})}.$$

These results may be obtained directly by identifying term by term the matrix of S_{ij}

$$\begin{bmatrix} S_{11} & S_{12} & S_{12} \\ S_{12} & S_{11} & S_{12} \\ S_{12} & S_{12} & S_{11} \end{bmatrix}$$

with the corresponding matrix

$$\frac{1}{E}\begin{bmatrix} 1 & -\sigma & -\sigma \\ -\sigma & 1 & -\sigma \\ -\sigma & -\sigma & 1 \end{bmatrix} \text{with } E \text{ and } \sigma.$$

The stress $\frac{F}{S}$ results in a volume variation of

$\frac{\Delta V}{V} = \frac{\Delta l}{l} + \frac{\Delta a}{a} + \frac{\Delta b}{b} = e_{xx} + 2e_{yy}$ from which may be written as

$$\frac{F}{S} = \lambda\frac{\Delta V}{V} + 2\mu\frac{\Delta l}{l} = (2\mu + \lambda)e_{xx} + 2\lambda e_{yy}$$

Comparing this with the equality 1′ above, one finds

$\lambda = C_{12}$ and $\mu = \frac{C_{11} - C_{12}}{2}$

$$\lambda = C_{12} = -\frac{S_{12}}{(S_{11} - S_{12})(S_{11} + 2S_{12})} = E\frac{\sigma}{(1+\sigma)(1-2\sigma)}$$

$$\mu = \frac{C_{11} - C_{12}}{2} = \frac{1}{2}\cdot\frac{1}{S_{11} - S_{12}} = \frac{1}{2}\cdot\frac{E}{1+\sigma}$$

(b) Under hydrostatic pressure, the relative volume variation of the sample is subject to normal stress in three main directions. The results are multiplied by three compared to that above:

$$\frac{\Delta V}{V} = -3\left(\frac{\Delta l}{l} + 2\frac{\Delta a}{a}\right) = -\frac{3(1-2\sigma)}{E}\Delta p$$

(the negative sign comes from $p = -\frac{F}{S}$). One also obtains

$$\beta = \frac{3(1-2\sigma)}{E} = 3(S_{11} + 2S_{12}) = \frac{3}{C_{11} + 2C_{12}} = \frac{3}{3\lambda + 2\mu}$$

The result $\beta = 3(S_{11} + 2S_{12})$ could be obtained directly by observing that the volumetric change caused by the pressure corresponds to the addition term by term of the equalities (1), (2), and (3) for the tension exerted in each of the three directions.

(c) The numerical values are

$$\sigma = 0.363,\ E = 0.627 \times 10^{11}\,\mathrm{N/m^2}\ (\mathrm{Pa})$$

$$\lambda = 0.609 \times 10^{11}\,\mathrm{N/m^2},\ \mu = 0.23 \times 10^{11}\,\mathrm{N/m^2}\ (\mathrm{Pa})$$

$$\beta = 1.31 \times 10^{-11}\,\mathrm{m^2/N}.$$

Exercise 14: Elastic anisotropy of hexagonal crystals

Using Cartesian co-ordinates Oxyz, one considers a homogenous hexagonal crystal, in the form of a rectangular parallelepiped, having its c-axis parallel to Oz as shown in Fig. 15. A tension T_z collinear to Oz is exerted normal to the face axb and parallel to xOy, the effect is a contraction of the lateral dimensions such that $\Delta a/a = \Delta b/b$. On the other hand, a tension T_y that is exerted normal to the face axc and parallel to xOz results in different relative contraction so that $\Delta a/a \neq \Delta c/c$.

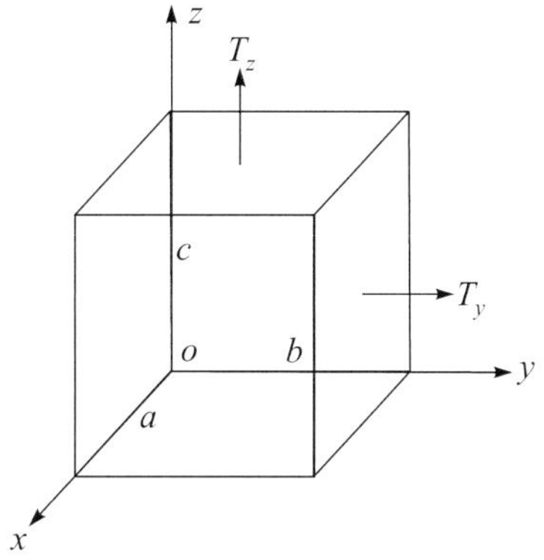

Figure 15

(a) In the Hooke's approximation and taking into account the symmetries of the problem, find the table of elastic compliance constants S_{ij} relating the stress components, e_{ij}, to the strain components, X_x, Y_y, Z_z, postulating that the S_{ij} tensor is symmetric ($S_{ij} = S_{ji}$).

(b) Find the linear compressibility $\beta_{\perp c}$ and $\beta_{||c}$ of the sample in terms of S_{ij} with a linear coefficient of compressibility of the form $\beta_l = -\frac{1}{l}\left(\frac{\Delta l}{\Delta p}\right)_T$ and with the volume coefficient of compressibility of the form: $\beta = -\frac{1}{V}\left(\frac{\Delta V}{\Delta p}\right)_T$.

Application: The compliance coefficients of zinc are S_{11} = 8.4; S_{12} = 1.1; S_{13} = −7.8; S_{33} = 28.7 (all units are in 10^{-12} m^2/N). Find the numerical values of Young's modulus $E_{\perp c}$ and $E_{||c}$ $\left(E = \frac{T}{S} \cdot \frac{l}{\Delta l}\right)$ of the different Poisson coefficients σ $\left(\sigma = -\frac{\Delta a/a}{\Delta c/c}\right)$ and of the linear and volumetric compressibilities.

Solution:

(a) The isotropy of the elastic properties in the xOy plane results in the Ox and Oy axes playing the same role so that $S_{11} = S_{12}$ and $S_{13} = S_{23}$. Thus under the single stress, $\frac{T_2}{ab} = Z_z$ one obtains

$$\frac{\Delta a}{a} = e_{xx} = S_{13} Z_z = \frac{\Delta b}{b} = e_{yy} = S_{23} Z_z.$$

When the same stress, $\frac{T_x}{b \cdot c} = \frac{T_y}{a \cdot c} = X_x$, is applied successively normal to the axc and next to bxc faces, the corresponding strain components are identical : $e_{xx} = S_{11} X_x = e_{yy} = S_{22} Y_y$. The resulting symmetric tensor reduces to

$$\begin{vmatrix} S_{11} & S_{12} & S_{13} \\ S_{12} & S_{11} & S_{13} \\ S_{13} & S_{13} & S_{33} \end{vmatrix}$$

It is completely determined from the knowledge of S_{11}, S_{12}, S_{13}, and S_{33}.

(b) When the sample is submitted to a hydrostatic pressure Δp, the relative length change is

$$\frac{\Delta a}{a} = -S_{11}\Delta p - S_{12}\Delta p - S_{13}\Delta p.$$

We thus find $\beta_{\perp c} = -\frac{1}{a}\frac{\Delta a}{\Delta p} = S_{11} + S_{12} + S_{13} = -\frac{1}{b}\frac{\Delta b}{\Delta p}$ and

$$\beta_{||c} = -\frac{1}{c}\frac{\Delta c}{\Delta p} = 2S_{13} + S_{33}.$$

In addition, the volume change is

$$\left(\frac{\Delta V}{V} = 2\frac{\Delta a}{a} + \frac{\Delta c}{c}\right)$$

and $\beta = 2\beta_{\perp c} + \beta_{||c} = 2S_{11} + 2S_{12} + 4S_{13} + S_{33}$

(c) $E_{||c} = \dfrac{1}{S_{33}} \equiv 0.35\times10^{11}\,\text{N/m}^2(\text{Pa}), E_{\perp c} = \dfrac{1}{S_{11}} = 1.2\times10^{11}\,\text{N/m}^2(\text{Pa})$

$$\sigma_{||c} = -\frac{\Delta a/a}{\Delta c/c} = -\frac{S_{13}}{S_{33}} = 0.27, \qquad \sigma_{\perp c} = -\frac{\Delta c/c}{\Delta a/a} = -\frac{S_{13}}{S_{11}} = 0.93$$

$$\sigma_{\perp c}(b) = -\frac{\Delta b/b}{\Delta a/a} = -\frac{S_{12}}{S_{11}} = -0.13$$

and

$$\beta_{\perp c} = 1.75\times10^{-12}\,\text{m}^2/\text{N}, \quad \beta_{||c} = 13.1\times10^{-12}\,\text{m}^2/\text{N},$$

$$\beta_{\text{v}} = 16.5\times10^{-12}\,\text{m}^2/\text{N}$$

Exercise 15: Shear modulus and anisotropy factor

A bar made from a homogenous cubic crystal in the form of a rectangular parallelepiped has the following dimensions: OA = a along Ox, OB = b along Oy, OC = c along Oz and the face OABD is fixed. A tangential force F_x is exerted on the opposite face CA′B′D′, resulting in a strain measured by the (small) angle α due to the tilt of the segments OC, BB′, AA′ and DD′, as shown in Fig. 16.

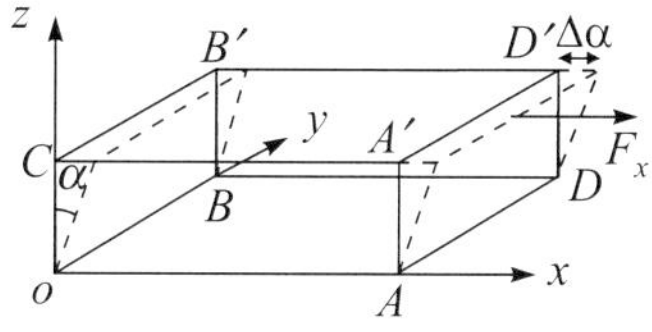

Figure 16

(1) The shear modulus μ is defined as: $\mu = \dfrac{1}{\alpha}\dfrac{F_x}{ab}$

After determining α as a function of the displacement Δa of the CB′D′A′ face, find the expression for μ as a function of the elastic compliance constants S_{44} and the elastic stiffness constants C_{44}.

(2) The bar used here has been cut from a cube with side l, of coordinate axes O$x'y'z'$ where the axis Oy' is parallel to Oy

and the base plane O′x′z′ coincides with Oxy following the arrangement shown in Fig. 17 where OC= c << a ≈ $l\sqrt{2}$.

When the cube is subject to the force F_0 exerted normal to on the side PMP′M′, the line O′M changes by length Δl and results in an algebraic variation Δe in each of the transverse directions.

(a) Find the angular variation α of the angle C$\hat{\text{O}}$A as a function of l, Δl, and Δe, noting that the angle C$\hat{\text{O}}$A is equal to the angle P$\hat{\text{I}}$M. Deduce the expression for the shear modulus μ of the initial bar.

(b) Find the elastic compliance constants S_{11} and S_{12} and the elastic stiffness constants C_{11} and C_{12}. Deduce the isotropy relation: $S_{44} = 2(S_{11} - S_{12})$.

(c) Find the expression for μ as a function of Young's modulus $E\left(E = \dfrac{F_0}{S}\dfrac{l}{\Delta l}\right)$ and the Poisson coefficient $\sigma\left(\sigma = -\dfrac{\Delta e/e}{\Delta l/l}\right)$.

Compare μ to the Lamé coefficient μ_L which obeys (see Ex. 13), the relation: $\mu_L = \dfrac{1}{2}\cdot\dfrac{E}{1+\sigma}$ and comment on the result.

(3) *Numerical application*: C_{11} = 1.07 × 10^{11} N/m^2, C_{44} = 0.28 × 10^{11} N/m^2 (or Pa) for Al. If the crystal was isotropic, what would the value of C_{44} be? What is the anisotropy factor A? $\left(A = \dfrac{2C_{44}}{C_{11} - C_{12}}\right)$

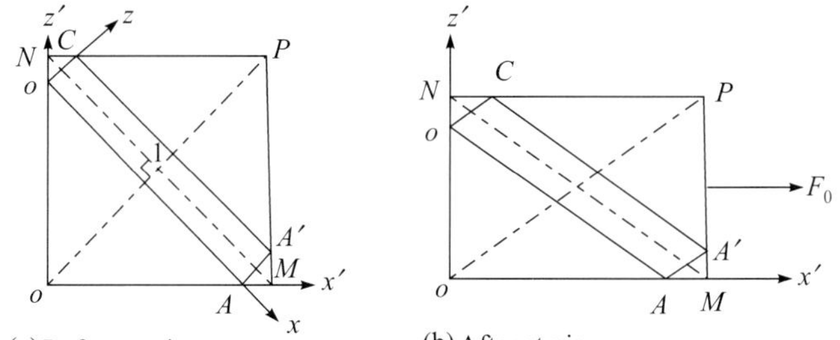

(a) Before strain (b) After strain

Figure 17

Solution:

(1) In this simple case, the useful relation between stress and strain reduces to

$$\frac{\Delta a}{c} = e_{xz} = \frac{\partial u}{\partial z} + \frac{\partial \omega}{\partial z} = S_{44} X_z \text{ with } X_z = \frac{F}{a \cdot b}$$

or the inverse relation: $X_z = C_{44}\, e_{xz}$.
Because α is small, one obtains $\alpha = \dfrac{\Delta a}{c}$ and

$$\mu = \frac{1}{\alpha} \cdot \frac{F}{a \cdot b} = \frac{1}{S_{44}} = C_{44}.$$

(2) We notice that we can always write $\hat{COA} = \hat{PIM} = \dfrac{\hat{POM}}{2}$.

(a) Before strain: $\tan(\hat{PO'M}) = \dfrac{l}{l} = 1$; $\hat{PO'M} = \dfrac{\pi}{4}$.

After strain:

$$\tan(\hat{PO'M}) = \frac{l + \Delta e}{l + \Delta l} = 1 + \frac{\Delta e}{l} - \frac{\Delta l}{l} = \tan\left(\frac{\pi}{4} - \frac{\alpha}{2}\right) = \frac{1 - \frac{\alpha}{2}}{1 + \frac{\alpha}{2}}$$

Then $\alpha = \dfrac{\Delta l}{l} - \dfrac{\Delta e}{l}$. For the face CA′B′D′, the shear stress is the relation between the projection of F_0 on this face, $F_0 \cos\varphi$, and the area of this face is $S' = \dfrac{l^2}{\sin\varphi}$.

The result thus is

$$\mu = \frac{1}{\alpha} \cdot \frac{F_0 \cos\varphi \cdot \sin\varphi}{l^2} = \frac{F_0}{l^2} \cdot \frac{1}{2\left(\frac{\Delta l}{l} - \frac{\Delta e}{l}\right)},$$

where the last relation holds because $\varphi = \dfrac{\pi}{2}$.

(b) When a single force is exerted normal to the side of the bar, the relations between the stress and the strain are reduced to (see Ex. 13):

$$e_{x'x'} = \frac{\Delta l}{l} = S_{11} \frac{F_0}{l^2} \text{ and } e_{y'y'} = \frac{\Delta e}{l} = S_{12} \frac{F_0}{l^2} \text{ so that}$$

$$\mu = \frac{1}{2(S_{11} - S_{12})}.$$

Directly or by matrix inversion we also obtain (see Ex. 13): $C_{11} - C_{12} = 1/(S_{11} - S_{12})$ which, on taking into account the initial results, gives

$$\mu = \frac{1}{2(S_{11} - S_{12})} = \frac{1}{S_{44}} = C_{44} = \frac{C_{11} - C_{12}}{2}$$

The isotropy relation can be written as: $2C_{44} = C_{11} - C_{12}$ or $S_{44} = 2(S_{11}-S_{12})$ with the indicated conventions in the Course Summary (Sections 3 and 4). An alternative convention would lead to $C_{44} = C_{11} - C_{12}$.

(c) It is sufficient to express the results of 2(a) as a function of E and σ, $\mu = \dfrac{E}{2(1+\sigma)}$.

This expression identifies the shear modulus as the Lamé coefficient, illustrating its physical meaning.

(3) If the relations established in 2(b) above were satisfied for Al, we would have $C_{44} = (C_{11} - C_{12})/2 = 0.23 \times 10^{11}$ N/m^2. Then the anisotropy factor is $A = 0.28/0.23 = 1.21$.

Exercise 16: Elastic waves in isotropic solids

(a) As a function of the elastic stiffness constants, C_{ij}, give the expressions for the velocity of longitudinal acoustic waves, V_L, along the [100], [110], and [111] directions of a cubic crystal. Same question for the velocity of the transverse waves, V_T [100] and V_T [110] along the [100] and [110] axes when the displacement of particles is only along the Oz [001] direction.

(b) How are the expressions simplified when the isotropy condition ($2C_{44} = C_{11} - C_{12}$) is satisfied?

How is the compression modulus B changed when in addition the Cauchy condition $C_{12} = C_{44}$ is satisfied?

What is the relation between the velocity of the longitudinal versus transversal elastic waves?

Solution:

(a) According to the relations demonstrated in Pb. 4 the following expressions are obtained:

$$V_L[100] = \left(\frac{C_{11}}{\rho}\right)^{1/2}; \; V_L[110] = \frac{(C_{11}+C_{12}+2C_{44})^{1/2}}{2\rho};$$

$$V_L[111] = \left(\frac{C_{11}+2C_{12}+4C_{44}}{3\rho}\right)^{1/2}; V_{T_{1,2}}[100] = \left(\frac{C_{44}}{\rho}\right)^{1/2} = V_{T_1}[110].$$

In addition, $B = 1/\beta = (C_{11} + 2C_{12})/3$ (see Ex. 13).

(b) When the isotropy condition is satisfied, all the longitudinal velocities are equal to $V_L = (C_{11}/\rho)^{1/2}$ and the transverse velocities become $V_T = (C_{44}/\rho)^{1/2}$. If in addition, the Cauchy condition, $C_{12} = C_{44}$, is satisfied, then $V_L/V_T = (C_{11}/C_{44})^{1/2} = \sqrt{3}$, with $C_{11} = 3 \cdot C_{44} = 3 \cdot C_{12}$.
Note that the transverse waves propagate less quickly than the longitudinal waves. This is a consequence of the fact that the interatomic spring constants β relative to a shear are less than those relative to a compression (see Chapter III, Ex. 2b, for instance), so that $C_{44} < C_{11}$.

Problems

Problem 1: Cohesion of sodium chloride

The crystal structure of NaCl is shown in Fig. 1e, Ex. 1, Chapter I. A small part of this structure is also shown in Fig. 18 in which the position of the Na^+ and Cl^- ions of charges of $+q$ and $-q$ are situated relative to Cartesian co-ordinates of Ox, Oy, and Oz axes parallel to the edges of the elementary cube. At the origin O, there is a Na^+ ion. The smaller distance between a Na^+ and a Cl^- ion is denoted by r. A kilomole of NaCl contains N molecules and $2N$ ions. For the following questions, use $q = 1.6 \times 10^{-19}$ C and $N = 1.602 \times 10^{26}$.

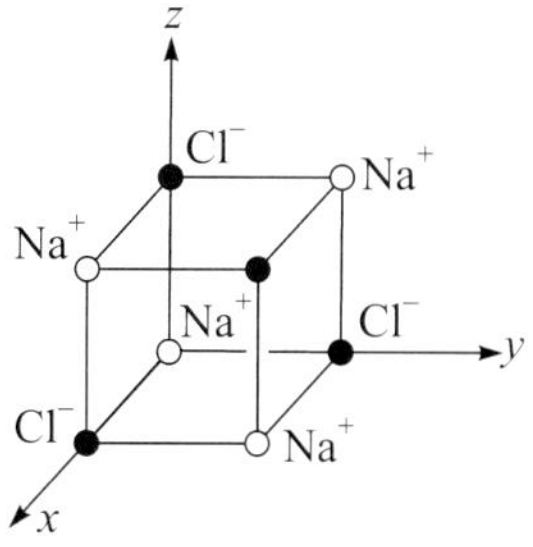

Figure 18 Structure of a NaCl crystal.

(1) The first step is to evaluate the electrostatic potential V created at O by all the other ions in the lattice, $V(O)$. For this step, the effect of ions in a sphere centered at O with radius μr is considered and the symbol ω is used, where ω is

$$\frac{4\pi\varepsilon_0 rV(O)}{q} = \omega$$

Evaluate ω as a function of μ^2 for $1 \le \mu^2 \le 12$. Show the results in the form of a table and show graphically ω as a function of μ^2.

(2) The above method does not allow a rapid evaluation of V(O) with the increase of μ^2. Another method, due to Evjen, may be used. This method takes into account only the effects of ions or fractions of ions contained in the interior of cubes centered at O with edges of length $2\mu r$ parallel to Ox, Oy, or Oz where μ is a whole integer and positive. The ions are assumed to be spherically charged and centered at their respective sites.

Evaluate ω for $\mu = 1$ and for $\mu = 2$. With these conventions, what is the total charge of ions and fractions of ion contained in the cube with edge $2\mu r$ for $\mu = 1$ and $\mu = 2$? For what physical reasons is this calculation more satisfying than that in para 1 or Pb. 1, Point (1)?

(3) A more complete calculation gives $\omega = -\alpha = -1.7467$, where α is the Madelung constant. Find the expression for the electrostatic energy U_1 of all the N ions per kilomole. Be careful not to count the mutual energy of two ions twice and neglect the finite dimensions of the crystal. Write U_1 with the form $U_1 = AN/\mathrm{r}$.

Numerical application: Calculate A.

(4) The energy U_1 corresponding to the electrostatic forces tends to move the ions closer to one another. These forces are opposite to repulsive forces which are necessary to maintain the stability of the crystal. Assume that the repulsive forces are only exerted between nearest neighbors of Na^+ and Cl^- and that the mutual repulsive energy of a couple of ions can be represented by the form $U_2 = \lambda e^{-r/\rho}$ where λ and ρ are constants.

Find the expression for the total repulsive energy U_2 of all the N ions (per kilomole) having the form $U_2 = \mathrm{B}Ne^{-r/\rho}$.

Express B as a function of λ.

(5) The cohesive energy of a crystal, U, is represented by the sum: $U = U_1 + U_2$. Knowing r_0, the distance between two neighboring ions Na^+ and Cl^- in equilibrium and β, the coefficient of compressibility of the crystal, one wishes to evaluate the constants λ and ρ, with the help of the following formulas:

$$P = -\frac{\partial U}{\partial V}, \quad \frac{1}{\beta} = -V\frac{\partial P}{\partial V}$$

P is the hydrostatic pressure applied to the crystal and V is the volume of a kilomole. Assume that $P = 0$ for $r = r_0$.

Find the explicit relation from which r_0/ρ can be expressed as a function of $1/\beta$ and of other data in the problem. Finally find the relation that allows the calculation of λ.

Numerical application: $r_0 = 2.814 \times 10^{-10}$ m; $\beta = 4.26 \times 10^{-11}$ $(N/m^2)^{-1}$. Calculate r_0/ρ, ρ and λ.

(6) Calculate the energies $U_1(r_0)$, $U_2(r_0)$, $U(r_0)$ in kilocalories per mole. Compare this last result to the experimentally value: $U_{exp}(r_0) = 184.7$ Kcal-mole^{-1}, 1 kilocalorie = 4180 J.

Solution:

(1) The potential created at the origin by point charges q_i distant from r_i is:

$$V(O) = \frac{1}{4\pi\varepsilon_0}\sum_i \frac{q_i}{r_i}.$$

For the case of NaCl the distance r_i between an ion M (mr, nr, pr) and the origin is $r_i = (m^2 + n^2 + p^2)^{1/2} r = \mu r$ where m, n, and p are integers. The Cl^- ions occupy the sites for which μ^2 is odd while Na^+ ions occupy the sites for which μ^2 is even.

The partial potential V_i created by n_i identical charges situated at the same distance μr from the origin is such that:

$$V_i = \frac{(\pm)|q| n_i}{4\pi\varepsilon_0 \mu r} \text{ or equivalently } 4\pi\varepsilon_0 r V_i / |q| = (-1)^{\mu^2} n_i/\mu.$$

Successive addition of different spheres of a given radius r_i allows to find the evolution of $\omega = 4\pi\varepsilon_0 r V(0)/q$ as a function of μ^2. Shown in Table 1 and Fig. 19, the corresponding results demonstrate that the convergence toward the exact value of α = 1.7476 is quite bad.

Table 1 Evolution of ω as a function of μ^2

Neighbor	Site (m, n, p)	Ion	μ^2	n_i	$(\pm)n_i/\mu$	ω
1st	$100, 00\bar{1}$ ...	Cl^-	1	6	−6	−6
2nd	$110, 0\bar{1}1$ ...	Na^+	2	12	+8.485	2.485
3rd	$111, 1\bar{1}1$ ...	Cl^-	3	8	−4.620	−2.135
4th	$200, 00\bar{2}$ ...	Na^+	4	6	+3	0.865
5th	$210, 1\bar{2}0$ ...	Cl^-	5	24	−10.733	−9.865
6th	$21\bar{1}, 1\bar{2}1$ ...	Na^+	6	24	+9.798	−0.070
7th	$220, 0\bar{2}\bar{2}$...	Na^+	7	12	+4.242	+4.172
8th	221, 300 ...	Cl^-	8	30	−10	−5.827
9th	310, ...	Na^+	9	24	−7.589	1.762
10th	311, ...	Cl^-	10	24	−7.236	−5.474
11th	222, ...	Na^+	11	8	+2.309	−3.165

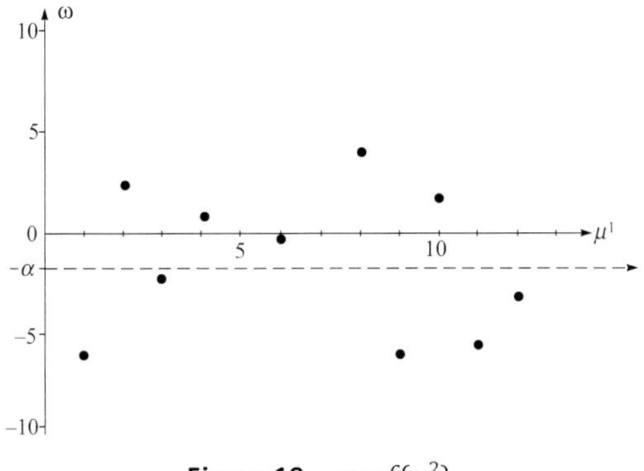

Figure 19 $\omega = f(\mu^2)$.

(2) The Evjen method consists in considering the fractions of ions inside successive cubes. The first cube ($\mu = 1$, side = $2r$) contains: 6 half ions of Cl^- at a distance r, 12 quarter ions Na^+ at a distance $\sqrt{2}\,r$, 8 eighths of ions Cl^- at a distance $\sqrt{3}\,r$.

The total charge of the first cube is thus $-q$ because it must be neutral (if the Na^+ ion at the origin is taken into account) and the corresponding value of ω is:

$\omega_1 = -3/1 + 3/\sqrt{2} - 1/\sqrt{3} = -1.456.$

The volume between the first and second cube ($\mu = 2$, side = $4r$) must have a total of zero charge. It contains:

6 × 1/2	Cl^-	at a distance of	r
12 × 3/4	Na^+	at a distance of	$\sqrt{2}\,r$

8 × 7/8	Cl^-	at a distance of	$\sqrt{3}\,r$
6 × 1/2	Na^+	at a distance of	$2\,r$
24 × 1/2	Cl^-	at a distance of	$\sqrt{5}\,r$
24 × 1/2	Na^+	at a distance of	$\sqrt{6}\,r$
12 × 1/4	Na^+	at a distance of	$\sqrt{8}\,r$
24 × 1/4	Cl^-	at a distance of	$3\,r$
8 × 1/8	Na^+	at a distance of	$\sqrt{12}\,r$

The contribution to ω of the second volume is:

$$\omega_2 - \omega_1 = -3 + \frac{9}{\sqrt{2}} - \frac{7}{\sqrt{3}} + \frac{3}{2} - \frac{12}{\sqrt{5}} + \cdots = -0.295$$

This brings the estimated value of the Madelung constant α to +1.75, which is very close to the exact value of 1.7476.

The Evjen method thus allows a faster convergence than the method in paragraph 1 because the charges contained in the 1st cube are equal to q and the other successive volumes are always neutral. The contribution of the first cube is thus the most important (polar contribution having a r^{-1} dependence). The additional contributions are small corrections (dipolar contribution having a r^{-2} dependence).

(3) The potential energy W_p of the ion Na^+ in O in the presence of $2N - 1$ other ions constituting a kilomole is $W_p = q\,V(O)$.

The electrostatic potential energy is equal to the sum of mutual energies of all pairs of ions. This energy corresponds to $U_1 = 2N \cdot W_p \dfrac{1}{2} = NqV(O)$. (The coefficient ½ accounts for the doubling of the interaction energy between two ions M and N).

From $\omega = -\alpha = \dfrac{4\pi\varepsilon_0 rV(O)}{q}$ one obtains: $U_1 = -Nq\dfrac{\alpha q}{4\pi\varepsilon_0}\cdot\dfrac{1}{r}$,

U_1 is of the form $U_1 = -AN/r$,

where $A = \alpha q^2/4\pi\varepsilon_0 = 40.26 \times 10^{-29}$ N/m² (Pa).

(4) To evaluate the total repulsive energy U_2, one must avoid the doubling of repulsive energy between two given ions, as was done for U_1, and take into account the six nearest neighbors of the Na^+ ion. This leads to

$U_2 = \dfrac{2}{2} Nzu_2 = Nz\lambda e^{-r/p} = NBe^{-r/\rho}$ from which one obtains

$B = z\lambda = 6\lambda$.

(5) In equilibrium ($r = r_0$) the total energy is minimum at $\left(\frac{\partial U}{\partial r}\right)_{r_0} = 0$ or $\frac{d}{dr}\left(Nz\lambda e^{-r/\rho} - \frac{\alpha q^2 N}{4\pi\varepsilon_0 r}\right) = 0$.

One obtains: $z\frac{\lambda}{\rho}e^{-r_0/\rho} = \frac{A}{r_0^2}$ or $zr_0^2 e^{-r_0/\rho} = \frac{\rho A}{\lambda}$ (1)

On the other hand, the volume occupied by an ion is equal to r^3 so that occupied by a kilomole is such that $V = 2\,Nr^3$.

Thus

$$P = -\frac{dU}{dV} = -\frac{dU}{dr}/\frac{dV}{dr} = \left(z\frac{\lambda^{-r/\rho}}{\rho} - \frac{A}{r^2}\right)\frac{1}{6r^2} \tag{2}$$

Insertion of Eq. 1 into Eq. 2 permits to verify that $P = 0$ at the equilibrium distance ($r = r_0$). The compressibility coefficient β is

$$\frac{1}{\beta} = -V\frac{dP}{dV} = -V\frac{dP/dr}{dV/dr} = \frac{A}{18r_0^4}\left(\frac{r_0}{\rho} - 2\right) \tag{3}$$

Knowing β, r_0, and A, it is easy to calculate r_0/ρ from (3) and then to deduce λ from relations (1) and (2). One obtains

$$\frac{r_0}{\rho} = \frac{18r_0^4}{A}\frac{1}{\beta} + 2 = 8.58, \quad \rho = 0.328 \times 10^{-10}\,\text{m}$$

$$\lambda = \frac{\rho A}{zr_0^2}e^{r_0/\rho} = 1480 \times 10^{-19}\,\text{J} = 925\text{ eV}$$

The small value of ρ as compared to r_0 implies that the repulsive forces are very short distance forces.

(6) In equilibrium we have

$U_1(r_0) = -\frac{AN}{r_0} = -86.13 \times 10^4$ J/mole = –206 kcal/mole

$U_2(r_0) = \lambda N e^{-r_0/\rho} = 10.04 \times 10^4$ J/mole = 24 kcal/mole

$U(r_0) = U_1 + U_2 = -76.08 \times 10^4$ J/mole = 182 kcal/mole

The last result corresponds to 7.9 eV/molecule, which is in excellent agreement with the experimental value of $U_{\text{exp}}(r_0)$ = 184.7 kcal/mole.

Problem 2: Cohesion and elastic constants of CsCl

The crystal structure CsCl is shown in Fig. 3a in Chapter I.

(a) Using the Evjen method find the Madelung constant α of this crystal considering the charges and fractions of charges contained in a cube of side $2a$ and centered on an ion Cs^+ where a is the dimension of the elementary cell.

Compare the result obtained with the exact value $\alpha = 1.7627$ and explain qualitatively the difference observed.

(b) Express the potential interaction energy U_p of the $2N$ ions of a kilomole as a function of α and r_0, the distance between nearest neighbors. Find the numerical value (in eV) of the energy for a "molecule" when $r_0 = 3.57$ Å.

(c) The repulsive energy u_r between two nearest neighbors can be represented by $u_r = \lambda e^{-r/\rho}$. Find the repulsive energy U_r for $2N$ ions and the total (cohesion) energy $U = U_p + U_r$.

(d) The Madelung constant can be calculated while the cohesive energy and r_0 can be obtained from experiments. Find the expression for ρ using the relation found in (c) and a relation that can be deduced from it at equilibrium. If $U(r_0) = -155.1$ kcal/mol, evaluate ρ.

(e) Find the expression and the numerical value for the compression modulus B in equilibrium with

$B = 1/\beta = -V\dfrac{dP}{dV}$, where P is the hydrostatic pressure ($P = -\dfrac{dU}{dV}$) and V is the volume of a kilomole.

(f) Compare the numerical results obtained for B to those found from the experimental elastic constants. Find the velocity of the longitudinal elastic waves propagating along the [100], [110], and [111] axes. The elastic constants in 10^{10} Pa units are $C_{11} = 6.64$, $C_{12} = 0.98$, $C_{44} = 0.80$.

Find the mass density d of CsCl using the periodic table (as a function of ε_0 and e).

Solution:

(a) Evjen Method:

One considers the cube centered on the Cs^+ ion 0 and side $2a$ (where a is the side of the elementary lattice). One obtains the relation $r_0 = \dfrac{a\sqrt{3}}{2}$.

The cube includes 8 ions of Cl^- at a distance r_0; $6/2$ ions of Cs^+ at a distance a; $12/4$ ions of Cs^+ at a distance $a\sqrt{2}$; $8/8$ ions Cs^+ at a distance $2r_0$ or $a\sqrt{3}$.

The total charge (including the central ion) of the cube is zero and its potential contribution at point O is

$$V(\mathrm{O}) = -\frac{q}{4\pi\varepsilon_0 r_0}\left(8 - \frac{3\sqrt{3}}{2} - \frac{3\sqrt{3}}{2\sqrt{2}} - \frac{1}{2}\right)$$

We thus find that α (Evjen) = 3.065. This value is rather far from the exact value (α = 1.7626). The Evjen method applied to structures of the CsCl type converges less quickly compared to those of the NaCl type (see previous problem) because the elementary dipoles constituting the charge fractions ($+q/n$ and $-q/n$) are distributed in NaCl in alternative planes perpendicular to Ox, Oy, Oz whereas in CsCl there are double dipolar layers between two planes. A better result can be obtained using the Ewald method.

(b) The potential energy of a Cs^+ ion in the presence of others is

$$\omega_\mathrm{p} = \frac{-q^2}{4\pi\varepsilon_0}\cdot\frac{\alpha}{r}$$

For $2N$ ions, $W_\mathrm{p} = ½\,(2N)\omega_\mathrm{p}$, which gives $U_\mathrm{p} = -\frac{Nq^2}{4\pi\varepsilon_0}\cdot\frac{\alpha}{r}$

For a molecule this corresponds to

$$\omega_\mathrm{p} = -\frac{q^2}{4\pi\varepsilon_0}\cdot\frac{\alpha}{r_0} = -11.4\times10^{-19}\,\mathrm{J} \quad \text{or} \quad -7.12\ \mathrm{eV}$$

(c) The mutual repulsive energy ω_r of the Cs^+ ion with eight nearest neighbors is $\omega_\mathrm{r} = 8u_\mathrm{r} = 8\lambda e^{-r/\rho}$ or $U_\mathrm{r} = ½(2N)\omega_\mathrm{r} = 8N\lambda e^{-r/\rho}$. The total energy for a kilomole will be $U = U_\mathrm{p} + U_\mathrm{r} = N\,[8\lambda e^{-r/\rho} - (\alpha\, q^2/4\pi\varepsilon_0 r\,)]$

(d) In equilibrium

$$\left(\frac{\partial U}{\partial r}\right)_{r_0} = 0 = -\frac{8\lambda}{\rho}e^{-r_0/\rho} + \frac{q^2}{4\pi\varepsilon_0}\frac{\alpha}{r_0^2} \qquad (1)$$

and $$\frac{U}{N} = 8\lambda e^{-r_0/\rho} - \frac{q^2}{4\pi\varepsilon_0}\frac{\alpha}{r_0} \qquad (2)$$

From (1), one obtains $8\lambda e^{-r_0/\rho} = \dfrac{q^2}{4\pi\varepsilon_0}\dfrac{\alpha}{r_0}\cdot\dfrac{\rho}{r_0}$

Inserting this into (2), one obtains $\dfrac{q^2}{4\pi\varepsilon_0}\dfrac{\alpha}{r_0}\left(\dfrac{\rho}{r_0}-1\right)=\dfrac{U}{N}$

Numerically this gives $\dfrac{\rho}{r_0}-1=-0.945$, $\dfrac{r_0}{\rho}\approx 18$ and $\rho = 0.198$ Å

(e) $B = V\dfrac{d^2U}{dV^2}$ with $\dfrac{dU}{dV}=\dfrac{dU}{dr}\Big/\dfrac{dV}{dr}$ and $V = Na^3 = \dfrac{8r^3}{3\sqrt{3}}N$ from which we obtain

$$B = V\frac{d^2U}{dr^2}\left(\frac{dr}{dV}\right)^2 + \frac{dU}{dr}\frac{d^2r}{dV^2} = V\frac{d^2U}{dr^2}\left(\frac{dr}{dV}\right)^2$$ because at equilibrium the second term of the summation is zero.

$$B = \frac{q^2\alpha}{8\sqrt{3}\cdot 4\pi\varepsilon_0 r_0^4}\left(\frac{r_0}{\rho}-2\right)$$, which gives

$B = 0.292\times 10^{11}\ \mathrm{N/m^2}$ (Pa).

(f) As a function of the elastic constants

$$B = \left(\frac{1}{3}\right)(C_{11}+2C_{12}) = 0.283\times 10^{11}\ \mathrm{N/m^2}\ \text{(see Ex. 13)}.$$

The agreement with the preceding result is satisfactory but it was obtained with a great precision in the previous numerical evaluations, most specifically in the values obtained for the ratio r_0/ρ.

The mass density of CsCl is $d = \dfrac{m_{\mathrm{Cl}}+m_{\mathrm{Cs}}}{a^3} = 4\times 10^3\ \mathrm{kg/m^3}$, which results in the following values for the speed of the elastic waves:

$$V_{\mathrm{L}}[100] = \left(\frac{C_{11}}{d}\right)^{1/2} = 4074\ \mathrm{m/s}$$

$$V_{\mathrm{L}}[110] = \left(\frac{C_{11}+C_{12}+2C_{44}}{2d}\right)^{1/2} = 3290\ \mathrm{m/s}$$

$$V_{\mathrm{L}}[111] = \left(\frac{C_{11}+2C_{12}+2C_{44}}{3d}\right)^{1/2} = 3120\ \mathrm{m/s}$$

(See Chapter III, Pb. 3)

Problem 3: Van der Waals–London interaction: cohesive energy of rare gas crystals

Classically one can account for the cohesive energy of crystals of rare gases by considering schematically that the atoms being polarizable their mutual attraction is analogous to that of electrostatic dipoles. In the first question we develop this idea to find the interaction energy.

(1) Calculate the interaction energy of two electrostatic dipoles.

(a) An electric dipole with dipolar moment $\vec{p}$ is placed at O:

- Recall without explanation the expression for the potential V and the electric field $\vec{E}$ created by such a dipole at a point M ($\overrightarrow{OM} = \vec{r}; \widehat{\vec{p}, \vec{r}} = \theta$).
- What is the potential energy of this dipole when it is placed in an external electric field, $\vec{E}_0$? What is the torque $\vec{C}$ is exerted by this field on the dipole?

(b) Two electric dipoles are placed at O and at A (OA = r), their dipolar moments are $\vec{p}$ and $\vec{p}'$, respectively, located in the same plane and we denote the angles made by $\vec{p}$ and $\vec{p}'$ with the OA axis as θ and θ', respectively (see Fig. 20).

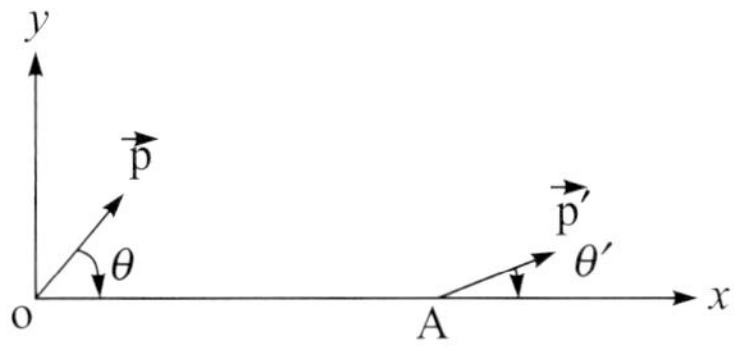

Figure 20

Show that the potential energy of two dipoles is

$$U_p = -\frac{p \cdot p'}{4\pi\varepsilon_0 r^3}[2\cos\theta \cdot \cos\theta' - \sin\theta \cdot \sin\theta'].$$

Find the torque C′ exerted by $\vec{p}$ on $\vec{p}'$ and then the torque C exerted by $\vec{p}'$ on $\vec{p}$. Find the stable equilibrium positions assuming that the dipoles can move around points O and A in the yOx plane.

(c) The dipole $\vec{p}'$ is in fact induced by the electric field $\vec{E}_A$ created at point A by the dipole $\vec{p}$:

$\overrightarrow{p'} = \varepsilon_0 \alpha \overrightarrow{E_A}$ where α is the polarizability of the dipole $\vec{p}'$. Show that the potential energy of the two dipoles can be written in the form $U_p = -D/r^6$ and find D when the dipoles are all parallel in the same direction as the Ox axis (θ, $\theta' = 0$). What is the value of D when $\alpha = 30.9 \times 10^{-30}$ m^{-3} (the polarizability of a krypton atom) and $\vec{E}$ corresponds to the dipolar moment of the torque formed by an ion and an electron at a distance of 1 Å?

(2) Cohesive energy of rare gas crystals.

Assume that the atoms in a rare gas crystal behave like the electric dipoles in (1). Thus the expressions and numerical values established in 1(c) are used below.

(a) Neglecting their kinetic energy, find an expression for the total energy of a pair of atoms by assuming that their mutual repulsion due to the overlap of electronic orbitals can be written as B/r^{12}.

(b) Knowing that rare gases crystallize in a face-centered cubic lattice and taking only into account the reciprocal interactions between a given atom and its nearest neighbors at a distance r_0, find an expression for the total energy (at 0 K and in the absence of pressure) of N atoms of a mole of rare gas as a function of B, D, and r_0.
Determine the value of B when r_0 = 4 Å (in equilibrium). Expressed in eV/atom, determine the numerical value of the cohesive energy U_T of the crystal.

(c) In fact the atoms vibrate Even at 0 K and the energy of this zero point vibration is equal to $U_c = 9/8\ k_B\theta_D$ per atom (see Chapter III, Ex. 20). What is the error on U_T when this vibrational energy is not taken into account. Choose $\theta_D \approx$ 70 K?

(d) Neglecting again the residual vibrations of atoms at 0 K, find the expression of the compression modulus $1/\beta$ in terms of D and r_0, and defined by $B = \dfrac{1}{\beta} = -V\dfrac{dP}{dV}$ (a variation of pressure dP results in a variation of volume dV). What is the corresponding numerical value?

(e, k_B, ε_0)

Solution

(1) (a) $V(r) = \frac{1}{4\pi\varepsilon_0} \cdot \frac{\vec{p} \cdot \vec{r}}{r^3}$, $\vec{E} = \overline{\nabla \cdot V}$

In polar coordinates (r and θ) this becomes:

$$E_r = \frac{2}{4\pi\varepsilon_0} \cdot \frac{p\cos\theta}{r^3} \quad \text{and} \quad E_\theta = \frac{1}{4\pi\varepsilon_0} \cdot \frac{p\sin\theta}{r^3}$$

This leads to $U_{\rm p} = -\vec{p} \cdot \vec{E}$ and $\vec{\rm C} = \vec{p} \times \vec{E}$.

(b) Explicitly calculating

$U_{\rm p} = -\overrightarrow{p'} \cdot \overrightarrow{E_{\rm A}} = -(p_r{}' \cdot E_r + p'_\theta \cdot E_\theta)$

One obtains:

$$U_{\rm p} = -\frac{pp'}{4\pi\varepsilon_0 r^3}[2\cos\theta \cdot \cos\theta' - \sin\theta.\sin\theta']$$

C and C′ are perpendicular to the plane containing the two dipoles:

$$C' = -\frac{\partial U_{\rm p}}{\partial \theta} = \frac{1}{4\pi\varepsilon_0} \cdot \frac{pp'}{r^3}(2\cos\theta \cdot \sin\theta' + \sin\cdot \cos\theta')$$

$$C = -\frac{\partial U_{\rm p}}{\partial \theta} = -\frac{1}{4\pi\varepsilon_0} \cdot \frac{pp'}{r^3}(2\sin\theta \cdot \cos\theta' + \cos\theta \cdot \sin\theta')$$

The equilibrium positions corresponding to $C = 0$ and $C' = 0$ are equivalent to $C + C' = 0$ and $C - C' = 0$, that is to say $\sin(\theta + \theta') = 0$ and $\sin(\theta - \theta') = 0$ simultaneously. The equilibrium will be stable (S) when $U_{\rm p}$ is negative. It will be unstable (U) when $U_{\rm p}$ is positive. The results are summarized in the following table:

$\frac{\theta}{\theta'}$	0	π	$-\frac{\pi}{2}$	$+\frac{\pi}{2}$
0	S	U		
π	U	S		
$-\frac{\pi}{2}$			U	S
$+\frac{\pi}{2}$			S	U

(c) When $\theta = \theta' = 0$ and with $\vec{p} = \varepsilon_0 \alpha \overrightarrow{E_{\rm A}}$:

$$E_A = \frac{2}{4\pi\varepsilon_0} \cdot \frac{p}{r^3} \text{ and}$$

$$U_p = -\frac{2pp'}{4\pi\varepsilon_0 r^3} = \frac{4\alpha p^2}{(4\pi)^2 \varepsilon_0 r^6} = -\frac{D}{r^6}, \text{ where } D = \frac{p^2\alpha}{4\pi^2\varepsilon_0}.$$

The numerical values are $p = 1.6 \times 10^{-29}$ C·m;
$D = 2.26 \times 10^{-77}$ J·m^{-6} = 142 eV·Å^{-6}.

(2) (a) U_T (pair) = $-D/r^6 + B/r^{12}$.

(b) In a fcc structure each atom has 12 first neighbors at equivalent positions. The evaluated energy until this point was a mutual energy, thus one must avoid double counting in the summation over the N atoms of a mole which leads to

$$U_T(N) = \frac{N}{2} \cdot 12(U \text{ pair}) = 6N\left(-\frac{D}{r^6} + \frac{B}{r^{12}}\right)$$

In equilibrium, $\left(\frac{\partial U_r}{\partial r}\right)_{r=r_0} = 0$ so that $B = \frac{D}{2} r_0^6 = 4.6 \times 10^{-14}$ J·Å^{12}

The cohesive energy U_T per atom is

$$U_T = 6\left(-\frac{D}{r_0^6} + \frac{B}{r_0^{12}}\right) = \frac{3D}{r_0^6} = -1.64 \times 10^{-20} \text{ J/at} \approx -0.1 \text{ eV/at}$$

(c) In fact the cohesive energy must also take into account the vibrational energy U_C of the atoms (with the half quantum at 0 K):

$$U_C = \frac{9}{8} k_B \theta_D = 6.7 \text{ x } 10^{-3} \text{ eV/at}$$

Then the corrected value of U_T is –0.093 eV/at. When not taken into account the error is of the order of 7%.

(d) At 0 K, the entropy is constant:

$$dU = -pdV \text{ and } \frac{1}{\beta} = \frac{Vd^2U}{dV^2}$$

Using the explicit volume occupied by an atom ($V = r_0^3/\sqrt{2}$) and taking into account $(\partial U/\partial r)_{r=r_0} = 0$ at equilibrium, the following result is obtained:

$$\frac{1}{\beta} = \frac{r_0^3}{\sqrt{3}} \cdot \frac{d^2U}{dr^2} \cdot \left(\frac{dr}{dV}\right)^2 = \frac{4 \cdot D\sqrt{2}}{r_0^9} = 4.8 \times 10^8 \text{ N/m}^2 \text{ (or Pa)}$$

Observe that for rare gas crystals B is two orders of magnitude smaller than for the alkali halides (Pb. 1) and three orders of magnitude smaller than for diamond (see Pb. 4).

Problem 4: Velocity of elastic waves in a cubic crystal: application to aluminum and diamond

In a homogenous cubic crystal of infinite dimension and of mass density, ρ, an elementary cube Δx, Δy, Δz is considered parallel to the Ox axis, stresses are applied to the $\Delta y{\cdot}\Delta z$ faces. There are two normal stresses $-X_x$ (for the face located at point x along the x-axis) and $X_{x+\Delta x}$ (located at point $x + \Delta x$). In addition, there are shearing forces (a) $-X_z$ and $X_{z+\Delta z}$ which are applied tangentially to the two faces $\Delta x{\cdot}\Delta y$ at z and $z + \Delta z$ respectively, and (b) X_y and $X_{y+\Delta y}$ which are applied to the two faces $\Delta x{\cdot}\Delta z$ respectively in y and $y + \Delta y$.

(a) What differential equation relates the displacement u_x of an elementary cube in the direction Ox to the stresses applied in the same direction?

By symmetry deduce the analogous differential equations relating the displacements u_y (and u_z) of an elementary cube in the y (and z) direction to the stresses exerted in these same directions.

(b) How do these equations change when taking into account the Hooke's law, relating stresses to strain via the elastic coefficients C_{ij}?

(c) For the elastic displacements, we search plane wave solutions of the form:

$\vec{u} = \overrightarrow{u_0} \exp.i(\omega t - \vec{K}\vec{r})$,

where $\vec{u} = (u_x, u_y, u_z), \vec{K} = (K_x, K_y, K_z)$ and $\vec{r} = (x, y, z)$.

Show that the equations of motion reduce to three linear homogenous relations in u_x, u_y, and u_z.

(d) Determine the equation for the velocity ($V = \omega/K$) for longitudinal waves which propagate along the [100], [110], and [111] axes, denoted as $V_L[100]$, $V_L[110]$, and $V_L[111]$. What is the ratio between these different velocities when the isotropy condition $C_{11} - C_{12} = 2C_{44}$ is satisfied?

(e) What are the velocity of $V_T[100]$ and $V_T[110]$ of the transverse waves that propagate along the [100] and [110] axes for a displacement of particles only parallel to the Oz [001]?

(f) Find the numerical values of the different velocities for single crystals of aluminum and of diamond (units in 10^{11} Pa or N/m^2).
For Al use: $C_{11} = 1.07$; $C_{12} = 0.61$; $C_{44} = 0.28$
For C (diamond) use: $C_{11} = 9.2$; $C_{12} = 3.9$; $C_{44} = 4.9$
$\rho(\text{Al}) = 2.73\ \text{g/cm}^3$; $\rho(\text{diamond}) = 3.5\ \text{g/cm}^3$. Comment on the difference between the two elements.

Solution

(a) In the Ox direction, the normal stress on the faces $\Delta y \cdot \Delta z$ results in $-X_x + [(X_x + (\partial X_x/\partial x)\Delta x)] = (\partial X_x/\partial x).\Delta x$ and the corresponding force is $(\partial X_x/\partial x)\Delta x \cdot \Delta y \cdot \Delta z$. Taking into account the tangential stresses limited to the first order and after simplification of the volume element $\Delta x \cdot \Delta y \cdot \Delta z$, the fundamental equation of dynamics in the Ox direction can be written as

$$\rho\frac{\partial^2 u_x}{\partial t^2} = \frac{\partial X_x}{\partial x} + \frac{\partial X_y}{\partial y} + \frac{\partial X_z}{\partial z}.$$

Similarly in the other directions, we find

$$\rho\frac{\partial^2 u_y}{\partial t^2} = \frac{\partial Y_x}{\partial x} + \frac{\partial Y_y}{\partial y} + \frac{\partial Y_z}{\partial z} \quad \text{and} \quad \rho\frac{\partial^2 u_z}{\partial t^2} = \frac{\partial Z_x}{\partial x} + \frac{\partial Z_y}{\partial y} + \frac{\partial Z_z}{\partial z}.$$

(b) The different components of the stress X_x, X_y, etc. are related to the strains $e_x = \frac{\partial u_x}{\partial x}$, $e_{xy} = \frac{\partial u_x}{\partial y} + \frac{\partial u_y}{\partial x}$ by a matrix of the C_{ij} coefficient which in a cubic crystal reduces to

	e_{xx}	e_{yy}	e_{zz}	e_{yz}	e_{zx}	e_{xy}
X_x	C_{11}	C_{12}	C_{12}	0	0	0
Y_y	C_{12}	C_{11}	C_{12}	0	0	0
Z_z	C_{12}	C_{12}	C_{11}	0	0	0
Y_x	0	0	0	C_{44}	0	0
Z_x	0	0	0	0	C_{44}	0
X_y	0	0	0	0	0	C_{44}

After substitution we find

$$\rho\frac{\partial^2 u_x}{\partial t^2} = C_{11}\frac{\partial^2 u_x}{\partial x^2} + C_{44}\left(\frac{\partial^2 u_x}{\partial y^2} + \frac{\partial^2 u_x}{\partial z^2}\right)$$

$$+(C_{12}+C_{44})\left(\frac{\partial^2 u_y}{\partial x\partial y} + \frac{\partial^2 u_z}{\partial x\partial z}\right)$$

The other two relations can be deduced by permutation of the coordinates x, y, and z.

(c) For u_x, for instance, the proposed solution takes the form $u_x = u_{x_0}e^{i[\omega t-(K_x x+K_y y+K_z z)]}$. Introducing this into the equations in (b), the following result is obtained:

$$\rho\omega^2 u_x = [C_{11}K_x^2 + C_{44}(K_y^2 + K_z^2)]u_x + (C_{12}+C_{44})K_x K_y u_y$$
$$+(C_{12}+C_{44})K_x K_z u_z$$

Permutating the coordinates of the two other relations one obtains by analogy

$$\begin{bmatrix} C_{11}K_x^2 + C_{44}\ (K_y^2+K_z^2)-\rho\omega^2 & (C_{12}+C_{44})K_xK_y & (C_{12}+C_{44})K_xK_z \\ (C_{12}+C_{44})K_yK_x & C_{11}K_y^2+C_{44}(K_x^2+K_z^2)-\rho\omega^2 & (C_{11}+C_{44})K_zK_y \\ (C_{12}+C_{44})K_xK_z & (C_{12}+C_{44})K_yK_z & C_{11}K_z^2+C_{44}(K_x^2+K_y^2)-\rho\omega^2 \end{bmatrix}$$

(d) Longitudinal waves:

- [100] direction: $u = u_x; u_y, u_z = 0, K = K_x, K_y = K_z = 0;$

$$V_L[100] = \frac{\omega}{K} = \sqrt{\frac{C_{11}}{\rho}}$$

- [110] direction: $u_x = u_y; u_z = 0; K_x = K_y = \frac{K}{\sqrt{2}}, K_z = 0;$

$$V_L[110] = \left[\frac{C_{11}+C_{12}+2C_{44}}{2\rho}\right]^{\frac{1}{2}}$$

- [111] direction: $u_x = u_y = u_z; K_x = K_y = K_z = \frac{K}{\sqrt{3}};$

$$V_L[111] = \left[\frac{C_{11}+2C_{12}+4C_{44}}{3\rho}\right]^{\frac{1}{2}}$$

When the isotropy condition is satisfied: $C_{11} - C_{12} = 2C_{44}$ (see Exs. 14–16), the three velocities are equal:

$$V_L[100] = V_L[110] = V_L[111] = \left(\frac{C_{11}}{\rho}\right)^{\frac{1}{2}}$$

(e) Transverse waves: If the displacement occurs only along u_z, we have $u_x = u_y = 0$.

- [100] direction: $K_x = K,\ K_y = K_z = 0.\ V_T[100] = (C_{44}/\rho)^{1/2}$
- [110] direction: $K_x = K_y = \dfrac{K}{\sqrt{2}}, K_z = 0.\ V_T[100] = (C_{44}/\rho)^{1/2}$

When, in addition to isotropy, the Cauchy condition ($C_{12} = C_{44}$) is satisfied, $V_L = \sqrt{3}\,V_T$ because $C_{11} = 3C_{44}$.

(f) *Numerical application*:

Al : V_L [100] = 6,300 m/s
V_L [110] = 6,400 m/s
V_L[111] = 6,450 m/s
V_T[100] = V_T[110] = 3,200 m/s

Diamond: V_L [100] = 16,200 m/s
V_L [110] = 17,600 m/s
V_L[111] = 18,000 m/s
V_T[100] = V_T[110] = 11,000 m/s ($\bar{u} \parallel [001]$).

None of the two crystals satisfies the isotropy condition: A(Al) = 1.21 (see Ex. 15) and A (diamond) = 1.62.

Comment: Numerical values for diamond

It should be noticed that the velocity of the elastic waves is exceptionally high in diamond. At the microscopic scale this is the result of a large covalent tight binding between the nearest neighbor atoms. At the macroscopic scale the consequence are the large numerical values of the C_{ij} and of compression moduli β^{-1} of the order of 5×10^{11} Pa. In comparison, the alkali metals have C_{ij} two orders of magnitude less than those of diamond and thus velocities of elastic waves are the lowest of the solids despite their very small densities. Another important consequence on the specific properties of diamond is its large thermal conductivity. This thermal conductivity

Kth is larger than that of metals because of the large velocity of the elastic waves (phonon transport) despite the lack of contribution of conduction electrons: diamond is an insulator from the point of view of electrical conductivity but it is the best for the thermal conductivity: a specific property exploited in microelectronic devices.

Additional Note: The calculations in this exercise postulate that the materials of interest are homogenous. This postulate is only valid when the wavelength of vibrations is large compared to the interatomic distances and the velocities thus determined provides the slope of the tangent to the origin of the phonon dispersion curves: a point investigated in detail in the exercises in Chapter III.

Problem 5: Strains in heteroepitaxy of semiconductors

The Bravais lattices of GaAs and AlAs are fcc of the zincblend type (see Chapter I, Ex. 1). The lattice parameter of GaAs, denoted as a_s is slightly smaller than that of AlAs, denoted a_0, but this difference is sufficiently small that it is possible to realize a single crystalline layer of AlAs on a substrate of GaAs (called heteroepitxy). The growth of AlAs is realized on the (001) face of GaAs and at the interface, the lattice of AlAs (001 face) fits that of the substrate so that it undergoes two deformations in the plane of the layer (e_1 and e_2) and a third e_3 perpendicular to the layer. The objective here is to determine the lattice parameter of AlAs in this last direction, $a^{\perp}$, assuming that there is a perfect fit between the two lattices at the (001) interface, that the growth of AlAs occurs in planes parallel to the interface and free from shear.

(1) Find the expressions of the strain e_1, e_2, e_3 as a function of a^0, a_s and $a^{\perp}$.

(2) The surface layer being free and the crystals being cubic, find the relation between e_1, e_2, e_3 and the elastic coefficients C_{11} and C_{12} of GaAs. Deduce the expression of $a^{\perp}$ as a function of a^0, a_s, C_{11} and C_{12}. Deduce also the relation giving the difference: $a^{\perp} - a_s$.

(3) *Numerical application*: a_s = 5.6528 Å; a^0 = 5.6612 Å; $C_{11} = 11.88 \times 10^{10}$ Pa; $C_{12} = 5.38 \times 10^{10}$ Pa. Evaluate $a^{\perp}$ and its difference with a_s.

Solution

(1) The fit of the AlAs lattice to that of GaAs at the interface implies:

$$e_1 = e_2 = \frac{(a_s - a_0)}{a^0}$$

In addition (by definition):

$$e_3 = \frac{(a^{\perp} - a^0)}{a^0}$$

(2) In cubic crystals the relations between stress and strain involve only the C_{11}, C_{12}, and C_{44} coefficients (see Pb. 4). The surface epitaxial layer being free one has $Z_z = 0$.
We thus obtain

$$2C_{12}e_1 + C_{11}e_3 = 0; e_3 = -\left(\frac{2C_{12}}{C_{11}}\right)e_1$$

Then $a^{\perp} - a_s = \left(1 + 2\frac{C_{12}}{C_{11}}\right)(a^0 - a_s)$ and $\frac{a^{\perp} - a^0}{a^0} = -2\frac{C_{12}}{C_{11}}\frac{(a_s - a^0)}{a^0}$

(3) $e_1 = e_2 = -15 \times 10^{-4}$; $e_3 \cong 14 \times 10^{-4}$

$$a^{\perp} - a_s = 0.016\ \text{Å}; \quad \frac{(a^{\perp} - a_s)}{a_s} \cong 28 \times 10^{-4}$$

The epitaxied lattice of AlAs is constrained along the (001) plane of the layer, it is expanded in the perpendicular direction. Strictly speaking, the epitaxial layer is not cubic but tetragonal.

Despite the constraints associated with the strain (easily calculable), the growth of thick single crystalline layers of AlAs on GaAs ($t \geq 1$ μm) is possible. Of course it is easier for tertiary layers such as $Al_xGa_{1-x}As$, where $0 \leq x \leq 1$, with strains vanishing when x approaches 0. On the other hand, when there are larger lattice differences the stresses relax in the form of dislocations rendering the epitaxy very difficult, if not impossible.

For additional details on surface reconstructions and relaxations, Chapter I, Pb. 4; Chapter V, Pb. 9; and comments therein.

Questions

Q.1: Why is NaCl lattice fcc when CsCl lattice is simple cubic?

Q.2: The parameters of the Lennard-Jones potential of a given rare gas crystal of the fcc structure are σ = 3.4 Å and ε = 10.4 meV. Give spontaneously the value of the cohesive energy and of the lattice parameter a.

Q.3: Why is the cohesive energy of MgO greater than four times that of NaCl?

Q.4: On keeping a crystal of NaCl is in water, why do its corners start dissolving first?

Q.5: What is physisorbtion? What is chemisorption?

Q.6: Consider an atom on the surface of a crystal set in vacuum. Why is an atom more attracted toward a conductive crystal than an insulating crystal?

Q.7: In an isotropic or cubic crystal, why is the C_{44} coefficient far smaller than the C_{11} coefficient? What is the consequence on the velocity of the transversal elastic waves along the [100] direction compared with longitudinal ones?

Q.8: Why should the Poisson coefficient, σ, be equal to 0.5 if there is no change in volume during traction?

Q.9: Give the order of magnitude of Young's moduli, E, of solids.

Q.10: If compressibility of NaCl is of the order 4×10^{11} Pa^{-1}, what is the order of magnitude of Young's moduli? Starting from atmospheric pressure, the crystal is introduced into an ultra-high vacuum chamber. What is its relative change in volume?

Q.11: In ionic crystals, why is the Madelung constant always greater than 1?

ANSWERS AT THE END OF THE BOOK

Chapter III

Atomic Vibrations and Lattice Specific Heat

Course Summary

1. Vibrations in a Row of Identical Atoms

Consider a row of identical atoms of mass M denoted by indices $n - 1$, n, and $n + 1$ that are equidistant by "a" at equilibrium and which are displaced from their equilibrium positions by u_{n-1}, u_n, and u_{n+1}. Using the Hooke's approximation (linearity between stress and strain), the restoring force exerted on the atom n by all the other atoms in the row takes the form $F = \sum_{j\neq 0}\beta_j(u_{n+j} - u_n)$. When limited to the force between the nearest neighbors ($j = \pm 1$), the equation of motion is $M\ddot{u}_n = \beta(u_{n+1} + u_{n-1} - 2u_n)$ with solutions in the form of waves of type $u_n = A \exp i(\omega t - kx)$ and the dispersion relation, $\omega = f(k)$, is of the form $\omega = 2(\beta/M)^{1/2}\sin(|k|a/2)$.

The propagation of the waves is along the row, $k||x$, but the spatial displacement of atoms, u, is referenced to the three coordinates with different force constant, β_L and β_T: (i) along the row, u_x, longitudinal, L or (ii) normal to the row, u_y and u_z, transversal T_1 and T_2 that are equivalent for a row because $\beta_{T1} = \beta_{T2}$. Thus as a function of the polarization, $\vec{u} \perp \vec{k}$ or $\vec{u} \,||\, \vec{k}$, the dispersion curves,

Understanding Solid State Physics: Problems and Solutions
Jacques Cazaux

ISBN 978-981-4267-89-2 (Hardcover), 978-981-4267-90-8 (eBook)
www.panstanford.com

$\omega = f(k)$, are generally different but tend toward zero with $\vec{k}$, the slope corresponding to the velocity of elastic waves of the same type as those in a homogenous medium explaining their name: acoustic modes (either transverse TA or longitudinal LA). In addition, this instantaneous atomic position can be represented by multiple values of the wave vector $\vec{k}$, which can be deduced from one to another by a translation vector or the reciprocal lattice $\vec{G}$. Thus the representation of the dispersion relation, ω versus k, may be limited to k vectors within the first Brillouin zone (BZ): $\frac{\pi}{a} \leq k \leq \frac{\pi}{a}$ in 1D or to $0 \leq k \leq \frac{\pi}{a}$ depending upon the boundary conditions being used (see Section 3).

In addition, the group velocity of the atomic vibrations, $v_g = \frac{\partial \omega}{\partial k}$, can be identified to the propagation of elastic waves along a continuous string when their wavelength is far larger than the inter-atomic distance (see Chapter II, Course Summary, Section 5a).

$$v_s = \sqrt{\frac{c}{\rho}} = \sqrt{\frac{\beta a^2}{M}} = v_{g(k \to 0)}.$$

The additional influence of neighboring atoms other than the nearest neighbors is studied in Exs. 3–7.

2. Lattices with More Than One Atom per Unit Cell

Consider a 1D crystal with two kinds of atoms of mass M and m where the distance between the nearest neighbors is "a" and the force constant is β in the Hooke's approximation. Limiting the interactions to the nearest neighbors only, the equations of motion for these two types of atoms can be written as

$$m\ddot{u}_{2n} = \beta(u_{2n+1} + u_{2n-1} - 2u_{2n}),\ M\ddot{u}_{2n+1} = \beta(u_{2n+2} + u_{2n} - 2u_{2n+1})$$

From solutions of the form

$$u_{2n} = A\exp i(\omega t - k \cdot 2na) \text{ and } u_{2n+1} = B\exp i[\omega t - k(2n+1)a]$$

the dispersion relation, ω versus k, is

$$\omega^2 = \beta\left(\frac{1}{M} + \frac{1}{m}\right) \pm \beta\left[\left(\frac{1}{M} + \frac{1}{m}\right)^2 - \frac{4}{Mm}\sin^2 ka\right]^{1/2}$$

The dispersion curves have two branches with a second branch called "optical branch," which extends to a domain of high frequencies compared to the acoustic branch. The optical branch corresponds to the vibrations of two consecutive atoms in opposite direction, A/B = −1, that can be excited by electromagnetic (EM) radiation in the infrared region for ionic crystals thus leading to selective absorption of this EM radiation (see Ex. 10).

This optical branch can exist even if the atoms of the basis are chemically identical but differ in their crystallographic environment and then on the inter-atomic force constant (see Ex. 1). Again in a 1D situation with two types of atoms at a distance of "a," the lattice parameter is $2a$ and the total length of the first BZ is π/a. As a result the number of modes for a given polarization (L or T) corresponds again to the number of atoms able to move. For linear crystal consisting of a basis of p atoms, one obtains three acoustic branches (1L and 2T) and $3(p-1)$ optical branches (see Ex. 2b and Pb. 8).

The main results are summarized in Fig. 1.

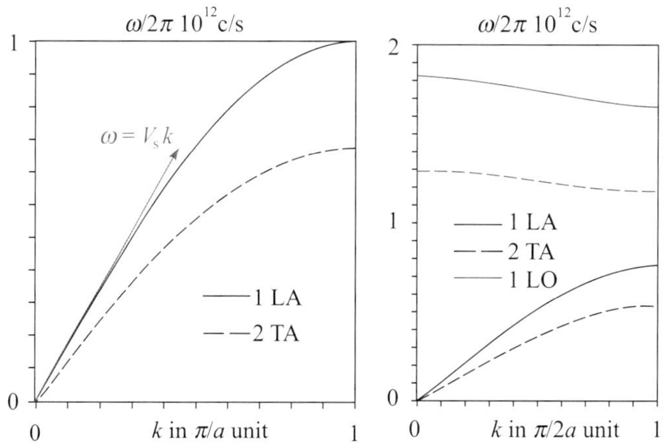

Figure 1 Frequency versus wave vector for a mono-atomic lattice, left with $\beta_L = 2\beta_T$, and for a diatomic lattice (right) with $M = 1.5$ m. The slope of the LA curve at the origin (the arrow) corresponds to the sound velocity, v_S.

3. Boundary Conditions

Two types of boundary conditions are used for the wave vector $\vec{k}$:

(a) Fixed boundary conditions: $u_0 = 0$, $u_N = 0$.

As for a homogenous string, the conditions $u = 0$ for $x = 0$ and $x = Na = L$ implies standing waves of the forms $u = 2A \sin(kx) \cdot \sin(\omega t)$. The wave vector k ($=2\pi/\lambda$) is positive with steps equal to $n\pi/L = n\pi/Na$ (n: positive integer) and the wave vector interval in the first BZ is $0 < k < \pi/a$. For a homogenous string the possible modes correspond to $\lambda = 2(L/n)$; for a row formed of discrete elements, there are $N - 1$ permitted values for k that are equidistant by π/Na and thus are nearly equal to the number of moving particles.

(b) Periodic (Born, von Karman) boundary conditions: $u_n = u_{N+n}$. They are applied to running waves, $u = A \exp i(\omega t - kx)$, with positive and negative values for k and step variations, $2\pi/L$, that are twice that of fixed boundary conditions. The total number of modes, corresponding to a wave vector included in the first BZ, $-\pi/a < k < \pi/a$, is equal to the number of atoms that can vibrate independently: $k = n2\pi/L$ and $n \leq 0$ or >0.

4. Generalization to 3D

In the general case we have already noted (see Chapter II, Course Summary, Section B: 5b and *Theory of Elasticity* by Landau and Lifshitz, p. 139) that for each given wave vector $\vec{k}$, there are three waves having orthogonal displacement vectors $\vec{u}$ being not necessarily transverse or longitudinal. To simplify the problem, however, the present considerations concern situations for which the spatial displacement of atoms can be described by two transverse waves, T_1 and T_2, such that $\vec{u} \perp \vec{k}$ and one longitudinal wave where $\vec{u} \parallel \vec{k}$. Two types of the boundary conditions may again be used: Fixed conditions of the form $u(0) = u(L_x) = 0$ on the faces of a parallelepiped $L_x L_y L_z$ or, for large samples, periodic Born-von Karman conditions of the form: $u(x, y, z) = u(x + L_x, y, z) = u(x, y + L_y, z) = \ldots$. In the first case, the waves are standing waves with $u = A \sin k_x x \cdot \sin k_y y \cdot \sin k_z z$. The components k_x, k_y, and k_z of the wave vector are all positive and vary discretely with steps such as

$$k_x = n_x(\pi/L_x), k_y = n_y(\pi/L_y), k_z = n_z(\pi/L_z),$$

where n_x, n_y, $n_z > 0$. For periodic BC, there are running waves: $u = A\exp i(k_x x + k_y y + k_z z)$ where k_x, k_y, and k_z can be positive or negative (and also 0 for one or two of them).

With the exception of objects that have one or several reduced dimensions such as thin films where L is of the order of a few lattice constants (see Chapter IV, where these phenomena are understood more easily), it is essential to point out that the two methods lead to the same results in particular for the (same) density of modes in the k space, denoted as $g(k)$, that is to say the same number of oscillators with a given wave vector of modulus $|k|$. In 3D for a given polarization, this density, $g(k)$, can be evaluated from the volume of a spherical shell defined by radiuses k and $k + dk$ with respect to the volume of an elementary cell $(2\pi/L_x \cdot 2\pi/L_y \cdot 2\pi/L_z)$ when the periodic BC are used. When fixed BC are used one considers the eighth of the preceding shell (k_x, k_y, $k_z > 0$) occupied by cells that are eight times smaller, see Fig. 2 below. For both cases this density in the k-space is $g(k) = \frac{k^2}{2\pi^2}$ V.

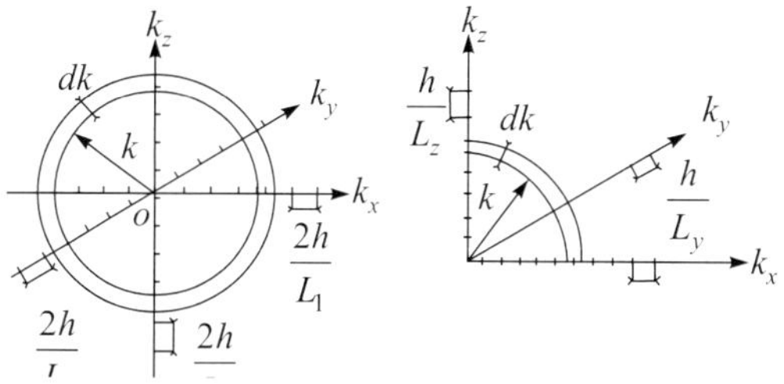

Figure 2 (a) Periodic boundary conditions (PBC) and (b) fixed boundary conditions (FBC).

5. Phonons

In the Hooke's approximation each atom is considered as a spatial harmonic oscillator or three linear harmonic oscillators (1L + 2T). The energy E of each linear oscillator is quantified by $E = (n + ½)$ $\hbar\omega$ with $n = 0, 1, 2, \ldots$. This implies that the action of an external stress (temperature, irradiation with photons or neutrons), each oscillator can win or can lose one (or several) quantum of energy, $\hbar\omega = h\nu$, called phonons and of momentum $\hbar\vec{k}$ (where the wave vector $\vec{k}$ is only defined up to a vector $\vec{G}$ of the reciprocal lattice). Thus an increase of heat to the crystal corresponds to an increase in the population of phonons (but not in the number of oscillators) or in the increase of the vibration energy (and therefore of the amplitude) of atoms in the lattice. In the same way, the interaction of a neutron or a photon (either IR absorption or Raman diffusion) with a lattice will conserve energy and momentum of the system formed by the photon and the phonon (see Pb. 7). At 0 K it remains a residual vibration energy ($1/2\hbar\omega$) due to Heisenberg's uncertainty principle: even 0 K the atoms vibrate (see Ex. 20a and b).

Note that this quantization of energy is independent of the quantization of the wave vector which is a result of the boundary conditions (see Final Remark in Ex. 11).

6. Internal Energy and Specific Heat

To evaluate the vibration energy U of a crystal at absolute temperature T and its specific heat C_v, it is sufficient to sum all of the possible frequencies produced by the density of linear oscillators $g_T(\nu)$ from which the frequency is between ν and $\nu + d\nu$ by the mean energy $\vec{E}$ of each oscillator of frequency ν:

$$U = \int_0^{\nu_M} g_T(\nu) \cdot \vec{E} \cdot d\nu \text{ and } C_v = \left(\frac{\partial U}{\partial T}\right)_V \text{ with } \int_0^{\nu_M} g_T(\nu) \cdot d\nu = 3N.$$

The mean energy of a linear oscillator is given by Bose–Einstein statistics:

$$\vec{E} = [\vec{n} + (1/2)]h\nu \text{ with } \vec{n} = \left(e^{\frac{h\nu}{k_B T}} - 1 \right)^{-1}$$

- The evaluation of $g_T(\nu)$ for each of the three polarizations can be extracted from the (experimental or theoretical) dispersion curves by applying the general formula (Ex. 19):

$$g(\omega)d\omega = \frac{V}{2\pi^3} \int\int_{s(\omega_0)} \frac{dS_\omega}{\left|\Delta_k \omega\right|} d\omega$$

  in which the integration extends over the surface of iso-frequency $S(\omega_0)$. Such a calculation, most often numerical, is of little interest for determining the specific heat of a lattice because it does not necessarily result in the correct value of C_v, within the precision of the calculations or experiments, but it does not allow one to determine the essential parameters for the evolution of $C_v(T)$.
- In the Debye model, the dispersion relations are reduced to a

linear form, $\omega = v_S k$. In 3D and using the results from Section 3, this model leads to the following results:

$$g_T(v)dv = 3g(v)dv = 3g(k)dk = \frac{4\pi V}{v_S^3} v^2 \times 3dv$$

The vibration energy is thus: $U = U_0 + U(T)$
where

$$U(T) = 9N\left(\frac{T}{\theta_D}\right)^3 \cdot k_B T \int_0^{x_D} \frac{x^3 \cdot dx}{e^x - 1}$$

In equation $U(T)$, v_D is the Debye frequency: $v_D = v_s\left(\frac{3N}{4\pi V}\right)^{1/3}$, which corresponds to the maximum vibration frequency of the oscillators, and x_D is $x_D = \frac{hv_D}{k_B T} = \frac{\theta_D}{T}$ with v_s (speed of sound).

Starting from $C_v = \left(\frac{\partial U}{\partial T}\right)_v$, the evolution of $C_v = f\left(\frac{T}{\theta_D}\right)$ may be established numerically and the result is shown in Fig. 3.

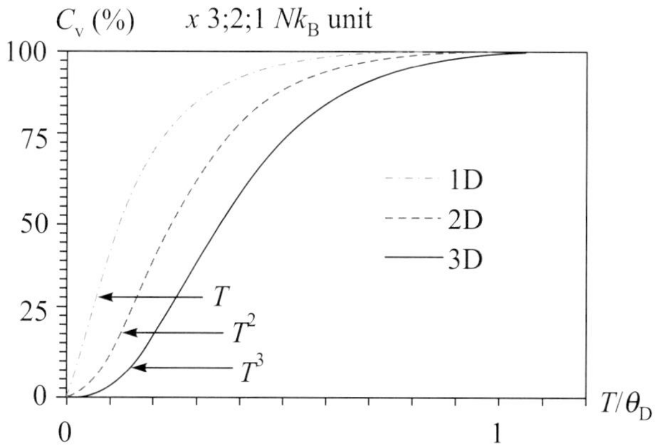

Figure 3 Heat capacity, C_v as a function of T/θ_D for 1D, 2D, and 3D solids. The vertical scale is in Nk_B unit to multiply by 1, 2, or 3 as a function of the degree of freedom for the atom vibrations. Note the initial evolution in T, T^2, or T^3 as a function of the dimensionality of the solid.

A physical discussion of the two limiting cases is more interesting:

(a) $T/\theta_D > 1 : C_v \approx 3Nk_B$ = constant. This is the classical limit corresponding to the law of Dulong and Petit, which can be directly obtained by considering $3N$ classical linear oscillators where each has a mean energy $\bar{E} = k_B T$.

(b) $T/\theta_D << 1$ (T being an order of magnitude less than the Debye temperature θ_D):

$$C_v \approx \frac{12\pi^4}{5} Nk_B \left(\frac{T}{\theta_D}\right)^3 = 1941\left(\frac{T}{\theta}\right)^3 \text{ J/mol}\cdot\text{K}$$

The Debye model describes successfully the variation of C_v as T^3 at low temperatures. Thus the Debye temperature of a given solid is an important parameter that allows one to state at which temperature range the lattice vibrations can be treated classically (when $T \geq \theta_D$) or when on the contrary it evolves as T^4 ($T < \theta_D/10$).

This model can be improved by distinguishing the velocity of the longitudinal waves from the velocity of the transverse waves (Pb. 4). The previous model of Einstein supposes that the 3N oscillators vibrated at the same frequency υ_E. It leads to $C_v \to 3Nk_B$ at high temperature and to $C_v \to 0$ as $T \to 0$, but the details of the latter evolution did not match the experiments. This model remains interesting for the optical frequencies or for certain specific cases (see Exs. 12 and 13).

In 1- or 2D systems, the density of vibration modes is different from that obtained in 3D. Thus the evolution of C_v as a function of T at low temperature is different as shown in Fig. 3. In addition, one may imagine that atoms are able to vibrate only in certain directions (e.g., normal to the plane or along a row) and such a constraint necessarily influences the determination of the vibration energy and the corresponding specific heat. The table in Ex. 18 summarizes a variety of possible arrangements. The conversion formulas presented at the end of this Course Summary are useful for the corresponding calculations.

7. Thermal Conductivity

In an isotropic medium the thermal conductivity is the parameter K_{th} in the Fourier expression for the heat flux, q: $\vec{q} = -K_{th}\vec{\nabla}T$ where $\vec{q}$ is the heat flux (amount of heat flowing per second and per unit area in W/m^2) and $\vec{\nabla}T$ the temperature gradient is in K/m. Thus,

K_{th} is expressed in W/m·K. The sign in the expression is chosen always $K_{th} > 0$ as heat always flows from a high temperature to a low temperature. This expression has the same form as the Ohm's law, $\vec{j} = -\sigma\vec{\nabla}V$ (j: current density; σ: electrical conductivity) when the electric field E is expressed as $\vec{\nabla}V$.

For metals, K_{th} is the sum of two contributions, conduction electrons and phonons. For poorly conductive materials, the contribution of phonons remains only and at the microscopic scale it is expressed as $K_{th} = (1/3)\ C_v\ v_s\ \Lambda_{ph}$, where Λ_{ph} is the phonon mean free path. Generally K_{th} is less for insulating materials than for conductive materials. An important exception for applications (evacuation of heat in microelectronic devices) is the case of diamond (a poor electrical conductor combined to an excellent thermal conductor). The reason is the large velocity of sound, v_s (18000 m/s instead of 3000–5000 m/s for most of the solids; see Chapter II, Pb. 5) in diamond and diamond-like materials. In turn, this large velocity is a consequence of the large force constant β between C atoms—covalent tight binding—and their light mass, m, via v_s that is proportional to $\sqrt{\beta/m}$. The thermal contribution to K_{th} is also significant for 2D materials (see Ex. 9b) such as graphene where Λ takes larger values than that in graphite.

Finally thermal expansion is a macroscopic phenomenon that cannot be explained microscopically by modeling the atoms as simple harmonic oscillators (that is to say with a potential energy in u^2 or a restoring force that obeys Hooke's law). The nonharmonic behavior is simply addressed in Pb. 6 and the distinction between specific heat at constant pressure and constant volume (resulting from thermal expansion) is the subject of Pb. 5.

Useful formulas:

$$F1: \int_0^\infty \frac{x\mathrm{d}x}{e^x - 1} = \frac{\pi^2}{6} = 1.64; F2: \int_0^\infty \frac{x^2\mathrm{d}x}{e^x - 1} = 2.4; F3: \int_0^\infty \frac{x^3\mathrm{d}x}{e^x - 1} = \frac{\pi^4}{15} = 6.5.$$

Exercises

Exercise 1: Dispersion of longitudinal phonons in a row of atoms of type C=C–C=C–C=

Consider a linear lattice with constant "a" having as a basis two identical atoms situated in a line and spaced from equilibrium by b ($b < a/2$), see Chapter I, Ex. 15. The instantaneous position of the atoms is x_1, x_2, ... x_{2n-1}, x_{2n}, x_{2n+1}, ..., and the distance along the line relative to their equilibrium position is u_1, u_2, ... u_{2n-1}, u_{2n}, u_{2n+1}.

We consider only interactions between nearest neighbors which are characterized by the force constants β_1 and β_2 as shown in Fig. 4. Find the equations of motion of the two species of atoms constituting the basis.

Starting from solutions of form:

$$u_{2n} = A\exp i(\omega t - k\,x_{2n})\,;\; u_{2n+1} = B\exp i(\omega t - k\,x_{2n+1}),$$

find the dispersion relations $\omega = f(k)$, the acoustic and optical longitudinal branches as a function of β_1, β_2, m (the mass of an atom), and a. Also determine the ratio of the amplitudes A/B for each of these branches in the center of the BZ ($k = 0$). Comment on the result.

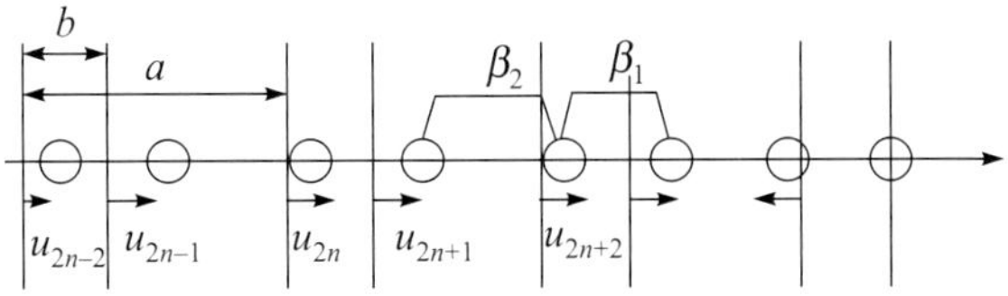

Figure 4

Knowing that the sound velocity along the row is $v_s = 5000$ m/s (experimental value) find the frequency of atomic oscillations in the center and at the limit of the BZ with $a = 5$ Å, $b = 1.25$ Å, $\beta_1/\beta_2 = b/(a-b)$. Show the corresponding dispersion curves and indicate the forbidden frequency interval.

Solution:

In equilibrium the sum of the forces exerted on each atom is zero. Limiting the result to interactions between nearest neighbors the movement of an atom in position u_{2n} and that of an atom in position u_{2n+1} obeys the following equations:

$$m\frac{d^2u_{2n}}{dt^2} = \beta_1(u_{2n+1} - u_{2n}) - \beta_2(u_{2n} - u_{2n-1})$$

$$m\frac{d^2u_{2n+1}}{dt^2} = \beta_2(u_{2n+2} - u_{2n+1}) - \beta_1(u_{2n+1} - u_{2n})$$

The instantaneous positions (x_{2n}) are related to the deviations of equilibrium positions (u_{2n}) by the following relations:

$$x_{2n} = na + u_{2n},\ x_{2n+1} = na + b + u_{2n+1},\ x_{2n+2} = (n+1)a + u_{2n+2}$$
$$x_{2n-1} = (n-1)a + b + u_{2n-1}$$

From solutions of the form:

$$u_{2n} = A\exp i(\omega t - kx_{2n}) \approx A\exp i(\omega t - k\,na)$$
$$u_{2n+1} = B\exp i(\omega t - kx_{2n+1}) \approx B\exp i(\omega t - kna - kb)\ \text{etc.,}$$

the following Kramers system in A and B is obtained:

$$-m\omega^2 A = \beta_1\{B\exp[-ikb] - A\} + \beta_2\{B\exp[ik(a-b)] - A\}$$
$$-m\omega^2 B = \beta_2\{A\exp[-ik(a-b)] - B\} + \beta_1\{A\exp[ikb] - B\}$$

The system of linear homogenous equations with two unknowns (A and B) has a non-zero solution when the corresponding determinant is zero:

$$\omega^4 - \frac{2(\beta_1+\beta_2)}{m}\omega^2 + \frac{2\beta_1\beta_2}{m^2}(1-\cos ka) = 0.$$

The dispersion relation for acoustic phonons and optical phonons respectively obeys the (–) and (+) solutions of

$$\omega^2 = \frac{\beta_1+\beta_2}{m} \pm \left[\left(\frac{\beta_1+\beta_2}{m}\right)^2 - \frac{4\beta_1\beta_2}{m^2}\sin^2\frac{ka}{2}\right]^{\frac{1}{2}}$$

When $k = 0$, $\omega = 0$ (acoustic branch) and $\omega^2 = (2/m)\,(\beta_1 + \beta_2)$ (optical branch). By introducing these values in the two linear equations we obtain the relation between the vibration amplitudes of adjacent atoms:

$A/B = 1$ (acoustic branch) and $A/B = -1$ (optical branch).

In the first case the atoms vibrate in phase and in the other they vibrate in phase opposition with a fixed center of gravity.

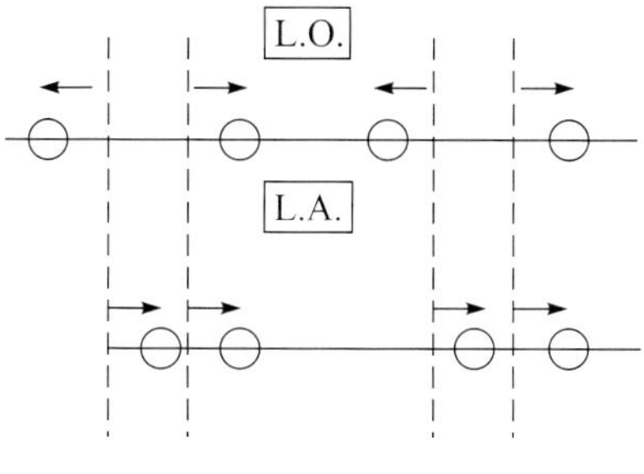

Figure 5

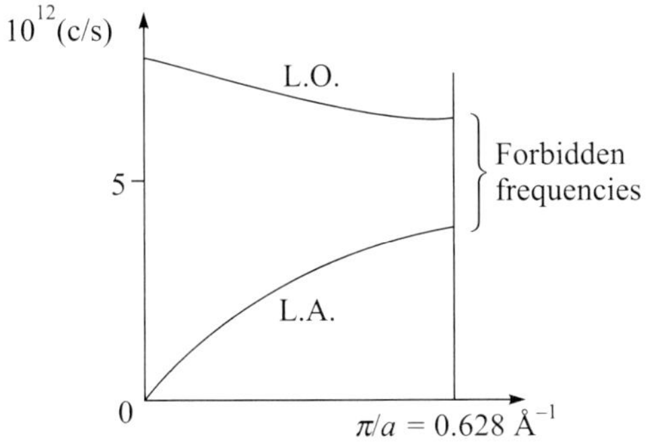

Figure 6

The velocity of sound corresponds to the slope of the tangent at the origin of the acoustic branch:

For small k one obtains $v_s = (\partial\omega/\partial k)_{k=0}$

$$\omega^2 \cong \frac{\beta_1+\beta_2}{m} - \left[\left(\frac{\beta_1+\beta_2}{m}\right)^2 - \frac{4\beta_1\beta_2}{m^2}\left(\frac{ka}{2}\right)^2\right]^{\frac{1}{2}}$$

which leads to $\omega = \frac{ka}{2}\sqrt{\frac{2\beta_1\beta_2}{\beta_1\beta_2}\cdot\frac{1}{m}}$ and $v_s = \frac{\partial\omega}{\partial k} = \frac{a}{2}\left[\frac{2\beta_1\beta_2}{m(\beta_1+\beta_2)}\right]^{\frac{1}{2}}$

From $\beta_1 b = \beta_2(a-b)$ and $\beta_1 = 3\beta_2$ one obtains $\beta_1/m = 8\, v_s^2/a^2 = 8\times10^{26}\ \mathrm{s}^{-2}$.

The other numerical values are

$$\text{L.O.}\,(k=0) \Rightarrow \nu = \frac{1}{2\pi}\left[\left(2\frac{\beta_1+\beta_2}{m}\right)\right]^{\frac{1}{2}} = 7.65\times10^{12}\,\text{c/s.}$$

$$\text{L.O.}\left(k=\frac{\pi}{a}\right) \Rightarrow \nu = \frac{1}{2\pi}\left[\frac{2\beta_1}{m}\right]^{\frac{1}{2}} = 6.36\times10^{12}\,\text{c/s.}$$

$$\text{L.A.}\left(k=\frac{\pi}{a}\right) \Rightarrow \nu = \frac{1}{2\pi}\left[\frac{2\beta_2}{m}\right]^{\frac{1}{2}} = 3.67\times10^{12}\,\text{c/s.}$$

The dispersion curves represented in Fig. 6 show that the forbidden interval of frequencies is between $3.67\times10^{12} < \nu < 6.36\times10^{12}$ c/s.

Note: The existence of the optical branch is related to the presence of a basis containing two atoms even if they are identical as in Si, Ge, diamond, or graphene (see Ex. 9b), where the masses of the atoms are identical but their environment and thus their force constants differ.

The rows studied here can be considered as a row of polymer chains C=C–C=C–C or resulting from a dimerization along the [100] direction of a 2×1 surface reconstruction (see Chapter I, Fig. 49 and comments in Pb. 4) or from a phase change (see comments in Exs. 2 and 3 of this chapter).

Exercise 2a: Vibrations of a 1D crystal with two types of atoms *m* and *M*

Consider a row of equidistant atoms having alternative masses m of species Z_1 and M of species Z_2 (see Chapter I, Ex. 3). Relative to the origin situated on one of these atoms, the average position of atoms from species Z_1 is $2na$ with n: integer. The Hooke's approximation is used and only is taken into account the action of nearest neighbors characterized by a force constant β.

(a) Starting from the equation of motion of these two types of atoms, find the dispersion relation for vibrations that propagate along the row with solutions of the form: $u_{2n} = A\exp i(\omega t - 2kna)$ and $u_{2n+1} = B\exp i[\omega t - (2n+1)ka]$.

(b) Find the expressions for ω and for the ratio $U = u_{2n+1}/u_n$ in the following cases:

(i) $k << \pi/a$

(ii) the extremity of k is in contact to the first BZ

(c) In each of the two situations above (i, ii), draw the elongation of several atoms as a function of their position: $u = f(x)$.

(d) What is the velocity of sound v_s along the row?

(e) Find the evolution of the dispersion curve, which has two branches toward the curve that has only 1 as the value for M is progressively changed toward that of m.

(f) *Numerical application to NaCl*:

Taking $a \approx 2.8$ Å and $\beta \approx 42$ N/m, find the numerical value of v_s and that of ω at characteristic points [m (Na) = 23, M (Cl) = 35.5]. (*N*)

Solution:

(a) See Course Summary, Section 4, for the equations of motion.

$$-m\omega^2 A = \beta B(e^{-ika} + e^{ika}) - 2\beta A \qquad (1)$$

$$-M\omega^2 B = \beta A(e^{-ika} + e^{ika}) - 2\beta B \qquad (2)$$

$$(\omega^2 - 2\beta/m)A + (2\beta/m)\cos ka \cdot B = 0 \qquad (1')$$

$$(2\beta/M)\cos ka . A + (\omega^2 - 2\beta/M)B = 0 \qquad (2')$$

Δ must be zero which leads to

$$\omega^4 - 2\beta\left(\frac{1}{M} + \frac{1}{m}\right)\omega^2 + \frac{4\beta^2}{Mm}\sin^2 ka = 0$$

The roots (giving ω^2) are mentioned in the Course Summary.

(b & c) The important points correspond to the value of k at the center (i) and the boundary of the BZ (ii).

(i) $k << \pi/a$

$$\omega_1' = 0; \qquad U = 1$$

$$\omega_2' = \left[2\beta\left(\frac{1}{M} + \frac{1}{m}\right)\right]^{1/2}; \qquad U = -1$$

The wavelength of the vibrations is very large compared to a. The atoms vibrate en phase for the acoustic branch and in phase opposition for the optical branch. They are running waves.

(ii) $k = \pi/2a$ (at not π/a, see Chapter I, Ex. 18)

$$\omega_1'' = (2\beta/M)^{1/2} \quad A = 0$$

$$\omega_2'' = (2\beta/m)^{1/2} \quad B = 0$$

The waves are stationary (λ = 4a). When the atoms "m" vibrate, the atoms "M" do not move and conversely. The two sub-lattices are decoupled (see Fig. 7).

(d) Near the center of the BZ ($k << \pi/a$) the dispersion relation of the acoustic branch reduces to:

$$\omega_1' = \sqrt{\frac{2\beta}{m+M}}ka \text{ from which one obtains:}$$

$$v_s = (\partial\omega/\partial k)_{k\to 0} = (2\beta a^2/[m+M])^{1/2}$$

When $m = M$, one obtains the velocity for a row of identical atoms: $v_s = (\beta a^2/m)^{1/2}$.

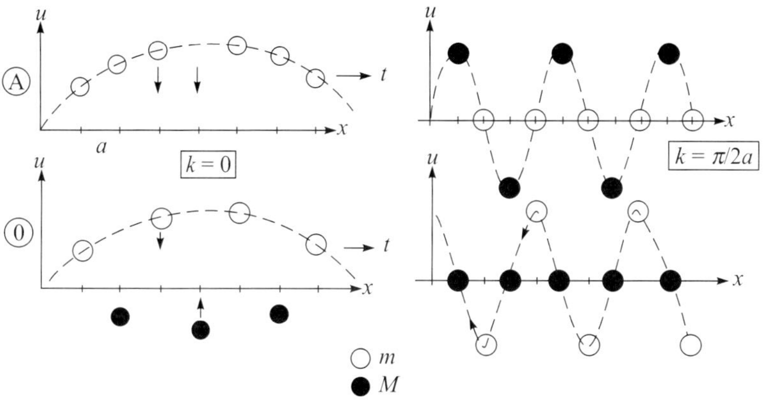

Figure 7

(e) When $m \neq M$, one obtains the curves shown in Fig. 8a. The dashed lines symbolize that the phonon wave vector $\vec{k}$ is only defined with $\pm\vec{G}$ (here $n|\vec{A}| = n\pi/a$). When $M \to m$, the interval for forbidden angular frequencies becomes narrower (ω_1'' and ω_2'' tend toward a common value) and the acoustic branch (in the first BZ) extends as an optical branch in the second BZ.

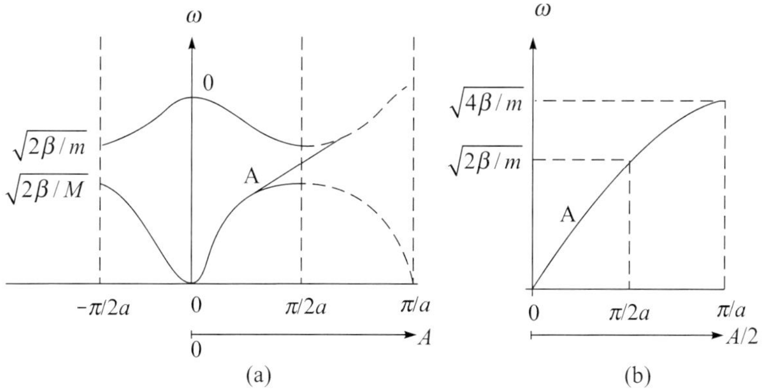

Figure 8

When $m = M$, this branch, initially optical, becomes acoustical and it is located in the first BZ because its lattice vector becomes a and not $2a$ (the zone boundary is at π/a and not at $\pi/2a$).

The atoms shown in Fig. 7 are now identical. If their instantaneous position is analogous to what is shown in the representation, one must note that the waves corresponding to $|\vec{k}| = \pi/2a$ are running waves now while the former running optical waves (at $k \approx 0$) is now a standing wave (at $k = \pi/a$).

(f) *Numerical application*:

$$v_s \approx 8.2 \times 10^3\, \text{m/s};\ \omega_0(k=0) = 6 \times 10^{13}\, \text{rad/s}$$

$$\omega_A(\pi/2a) \approx 3.77 \times 10^{13}\, \text{rad/s};\ \omega_0(\pi/2a) \cong 4.68 \times 10^{13}\, \text{rad/s}$$

Comments

Starting from a monoatomic basis, one can follow the opposite way which consists of studying the appearance of the forbidden frequency bands as a consequence of the diatomic basis. The influence of second nearest neighbors in 1D is the theme of Ex. 3. In the present exercise, taking into account the second nearest neighbors would change the dispersion curve, especially around $\left|\vec{k}\right| = \pi/2a$.

Exercise 2b: Vibrations of a 1D crystal with a tri-atomic basis

Consider a linear crystal with a lattice constant of $3a$ and a basis composed of an atom P, with mass M at the origin O and two atoms Q, with mass m located at 1/3 and –1/3 (or 2/3) see figure of Ex. 2b in Chapter II. We denote β_1 as the force constant between two atoms P and Q (first nearest neighbors) and β_2 as the force constant between two atoms of type Q (first nearest neighbors also).

(a) Limiting the problem of interactions between nearest neighbors, write the equations of motion of each atom constituting the pattern. These atoms are identified by their positions $3na$, $(3n + 1)a$, and $(3n - 1)a$.

(b) Starting from solutions of the form $u_{3n} = A \exp i(\omega t - 3kna)$ for atoms of type P and of the form $u_{3n+1} = B \exp i[\omega t - k(3n + 1)a]$ and $u_{3n-1} = C \exp i[\omega t - k(3n - 1)a]$ for atoms of type Q (at 1/3 and −1/3), show that the dispersion relation can be deduced from the determinant of type:

$$\Delta = \begin{bmatrix} a & b^x & b \\ b & c & d^x \\ b^x & d & c \end{bmatrix} = 0$$

Express the different terms of this determinant as a function of the given parameters and establish the dispersion relation in a form correlating a, c, k, β_1, and β_2.

(c) For $k = 0$ (center of the BZ), find the possible expressions for ω^2. For each of these expressions, find the corresponding relation between A, B, and C. Illustrate the atomic positions for a transversal and a longitudinal mode. What happens when $\beta_2 = -\beta_1/2$ with $\beta_1 > 0$? Why?

(d) What can be said about the expressions for ω^2 compared to the other characteristics of k (contact with the BZ and at mid-distance to this point)? Qualitatively describe the expected results without finding the exact expressions for ω^2.

Solution:

(a) $$M\frac{\mathrm{d}^2u}{\mathrm{d}t^2}3n = \beta_1(u_{3n-1} + u_{3n+1} - 2u_{3n})$$

$$m\frac{\mathrm{d}^2u_{3n+1}}{\mathrm{d}t^2} = \beta_1(u_{3n} - u_{3n+1}) + \beta_2(u_{3n+2} - u_{3n+1})$$

$$m\frac{\mathrm{d}^2u_{3n-1}}{\mathrm{d}t^2} = \beta_1(u_{3n} - u_{3n-1}) + \beta_2(u_{3n-2} - u_{3n-1})$$

Note that the two atoms of type Q have different left/right environments and therefore have two different equations (see Ex. 1).

(b) The proposed solutions reduce to:

$$u_{3n+1}/u_{3n} = (B/A)e^{-ika};\ \ u_{3n-1}/u_{3n} = (C/A)e^{ika};$$

$$u_{3n+2}/u_{3n+1} = (C/B)e^{-ika};\ \ u_{3n-2}/u_{3n-1} = (B/C)e^{ika}$$

Inserting these solutions into the equations of motion, we obtain, after elementary manipulations, a system of three linear and homogenous equations in A, B, and C which have a non-zero solution when the determinant Δ is zero. From Δ we have:

$a = M\omega^2 - 2\beta_1$; $c = m\omega^2 - \beta_1 - \beta_2$; $b = \beta_1 e^{ika}$; $b^x = \beta_1 e^{-ika}$; $d = \beta_2 e^{ika}$ and $d^x = \beta_2 e^{-ika}$

The dispersion relation results from $\Delta = 0$:

$a(c^2 - \beta_2^2) - 2\beta_1^2 c + 2\beta_1^2 \beta_2 \cos 3ka = 0$

(c) When $k = 0$, this relation takes the form: $[a(c+\beta_2) - 2\beta_1^2](c-\beta_2)$ or $\omega^2(m\omega^2 - \beta_1 - 2\beta_2)[Mm\omega^2 - \beta_1(M+2m)]$.

The three roots of ω^2 are obvious. By reinserting their value into the system of linear equations we obtain relations between A, B, and C.

Thus for $\omega^2 = 0$, the acoustic branch: $A = B = C$.

For $\omega^2 = \dfrac{(\beta_1 + 2\beta_2)}{m}$, the first optical branch: $A = 0$; $B = -C$.

For $\omega^2 = 2\beta_1\left(\dfrac{1}{M} + \dfrac{1}{2m}\right)$, the second optical branch: $A = -m/M\,(B + C)$ and $B = C$.

We find that the results thus obtained are coherent.

For the acoustical branch, $\omega = 0$, and it depends neither on β_1 nor on β_2 because the atoms vibrate in phase.

For the first optical branch ω does not depend on M because the corresponding P atom is effectively stationary because the two atoms of Q vibrate at a phase opposition of 180°. As a result the restoring forces on P cancel each other.

For the second optical branch, ω does not depend on β_2 since the two atoms of type Q are in phase with one another while still be in phase opposition with atom of type P: the center of mass of the molecule QPQ does not move.

Figure 9 summarizes these conclusions for the longitudinal and transverse vibrations.

In the case of ionic crystals of type $Ca^{2+}Cl_2^{-}$, the force constant β_2 between two ions of the same sign can be negative (repulsive). If $\beta_2 = -\beta_1/2$, the optical branch disappears and the cohesion of the crystal disintegrates to allow the formation of molecules of type QPQ: a result easy to predict from the evaluation of the Madelung constant (see Chapter II, Ex. 2b).

(d) At the boundary of the BZ, $k = \pi/3a$ and $\cos 3ka = -1$.

At mid-distance $k = \pi/6a$ and $\cos 3ka = 0$.

In the dispersion relation obtained in (b), the variation of k only affects the weight of the term independent of ω^2 which is a product of roots in a third-degree equation. On the other hand, the sum of the roots is unchanged because it is independent of k. As a consequence, when k increases the expected increase of the acoustic branch must correspond to decreasing frequencies of the optical branches since, regardless of k:

$$\omega_A^2 + \omega_{01}^2 + \omega_{02}^2 = 2\left(\frac{\beta_1}{M} + \frac{\beta_1}{m} + \frac{\beta_2}{m}\right)$$

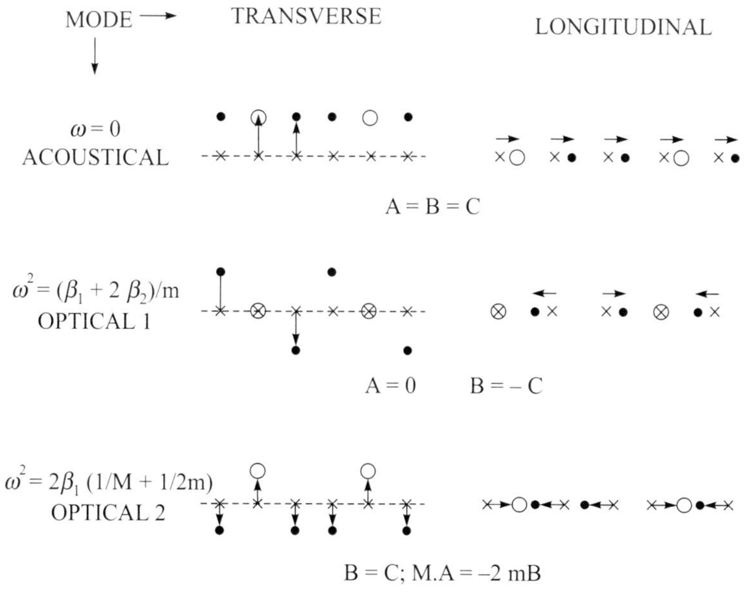

Figure 9

In particular one may verify that a similar result applies also to diatomic crystals at the boundary of the BZ (see Exs. 1 and 2) and that, here, the solution $\omega^2 = \beta_1/m$ is easily deduced from $(c+\beta_2)[a(c+\beta_2)-2\beta_1^2]=0$ when $k = \pi/3a$.

This exercise shows the fact that for a basis of p atoms we obtain a given polarization (L or T), an acoustic branch, and $(p - 1)$ optical branches. The linear and symmetric basis investigated here is relevant also for solid BeH_2 and CO_2.

Exercise 3: Vibrations of a row of identical atoms. Influence of second nearest neighbors

Consider a linear lattice with parameter a formed from identical atoms of mass m. Each atom is submitted to a force constant β_1 by its nearest neighbors and to β_2 from its second nearest neighbors.

(a) Find the equation governing the displacement of atom n, given by u_n (compared to its equilibrium position), as a function of u_{n+1}, u_{n-1}, u_{n+2}, u_{n-2}.

(b) Find the dispersion relation of longitudinal phonons, $\omega = f(k)$, starting from the solution in the form of a plane wave of type: $u_n = A\exp i(\omega t - kx_n) \approx A\exp i(\omega t - kna)$. In this relation highlight the corrective factor $S(k)$, related to the influence of second nearest neighbors.

(c) Find the expression for the velocity of sound. Indicate the characteristics of the dispersion curve by considering successively the hypothesis $\beta_2 > 0$ and $\beta_2 < 0$ (β_1 is necessarily >0). Also sketch the reference curve $\beta_2 = 0$. Show the displacement of atoms u_{n+j} as a function of their coordinates x_{n+j} for $k = \pi/2a$ and $k = \pi/a$. Explain what happens when $\beta_2 = -\beta_1/4$.

Solution:

(a) $m\ddot{u}_n = \beta_1(u_{n+1} + u_{n-1} - 2u_n) + \beta_2(u_{n+2} + u_{n-2} - 2u_n)$

(b) $u_{n+p} = u_n e^{-ikpa}$

$-m\omega^2 = \beta_1(2\cos ka - 2) + \beta_2(2\cos 2ka - 2)$, which gives

$$\omega^2 = \frac{4\beta_1}{m}\sin^2\frac{ka}{2} + \frac{4\beta_2}{m}\sin^2 ka = P^2(k) + S^2(k)$$

$$\omega = 2(\beta_1/m)^{1/2} \cdot \left|\sin\frac{ka}{2}\right| \left[1 + 4(\beta_2/\beta_1)\cos^2\left(\frac{ka}{2}\right)\right]^{1/2}$$

(c) $$v_s = \left(\frac{\partial\omega}{\partial k}\right)_{k\to 0} = [1 + 4(\beta_2/\beta_1)]^{1/2} \cdot (\beta_1 a^2/m)^{1/2}$$

When $\beta_2 = 0$, find the result in the Course Summary, Section 1.

When β_2 is positive, the second neighbors have an action that adds to the restoring forces exerted by the first neighbors. Accounting for them they increase the value of the sound velocity ($k \to 0$). The maximal influence of β_2 occurs when $k = \pi/2a$ because the second neighbors are in phase opposition to the reference atom. Their action cancels out at π/a because they are in phase with these same atoms (see Fig. 10).

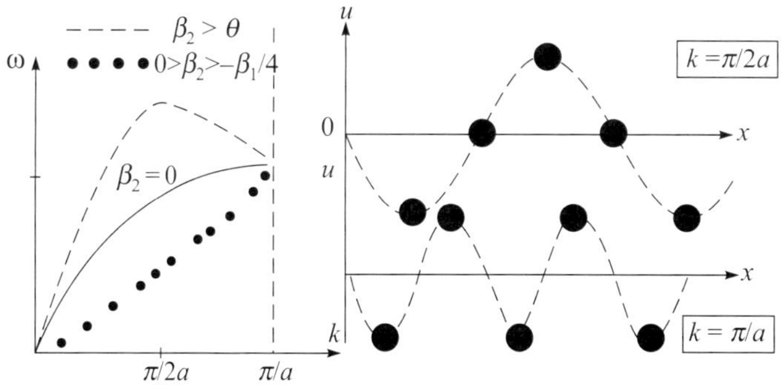

Figure 10

When β_2 is negative (the case for ions of opposite sign), the interactions of the second nearest neighbors result in a repulsive force. Their influence is opposite of that described above. The situation is only stable when the attractive forces are greater than the repulsive forces: $\beta_2 > -\beta_1/4$. When this is not the case, the crystal undergoes a phase change to attain a more stable state. This would be the case for example when the basis consists of a diatomic molecule (with two identical atoms and two force constants between different nearest neighbors. Such a dimerization, studied in Ex. 1, is also seen in the previous exercise. In the following exercise, we explore in more detail this instability and the appearance of soft modes that are precursors of this instability. The electronic aspect of dimerization is considered in Chapter V, Ex.4.

Exercise 4: Vibrations of a row of identical atoms: Influence of the *n*th nearest neighbor

Consider a row of identical atoms with mass m and spaced equidistantly by a. The force constant between the atom in position n and an atom in position $n + p$ (or $n - p$) is denoted by β_p.

(1) In the Hooke's approximation, find the equation of motion of the atom in position n (where the displacement from equilibrium is denoted by u_n) as a function of the displacement of other atoms in the row. Looking at solutions of the form $u_n = A \exp i\,(\omega t - kna)$, find the relation between ω^2 and k in the form of a sum including β_p.

(2) Explicit the relation above in the limit of long wavelengths and deduce the expression for the velocity of sound v_s.

In fact the restoring forces are alternatively positive and negative and decrease in $1/r^3$: $\beta_p = (-1)^{p+1}(\beta_1/p^3)$. What is the relation between the speed of sound thus obtained and that obtained when the action is limited to nearest neighbors?

(3) What happens to the expression for ω^2 as the extremity of the wave vector approaches the boundary of the first BZ? From physical arguments, explain why only odd atoms in the row are influenced by this result? Deduce the values of the $|k|$ for which the neighbor of order p has a maximal action and those for which this action is zero.

Using the law for β_p [given in Q. 2 above], find the relation between the vibration frequency obtained at the boundary of the BZ and that obtained when the action is limited to nearest neighbors.

Solution:

(1) $$m\ddot{u}_n = \sum_{p=1}^{\infty} \beta_p (u_{n+p} + u_{n+p} - 2u_n)$$

By inserting solutions of the type $u_{n+p} = A\exp i[\omega t - k(n+p)a]$, one obtains

$$\omega^2 = (4/m)\sum_{p>0} \beta_p \sin^2(kpa/2)$$

(2) In the limit of long wavelengths $\sin x = x$ so that

$$\omega^2 = (a^2/m)\sum_{p>0} p^2 \beta_p$$

$$v_s = (\partial\omega/\omega k)_{k\to 0} = (a^2/m)^{1/2}\left(\sum_{p>0} p^2 \beta_p\right)^{1/2}$$

When $\beta_p = (-1)^{p+1}(\beta_1/p^3)$, the term between parenthesis is just an alternating series:

$$1 - \frac{1}{2} + \frac{1}{3} - \frac{1}{4} + \cdots = \log 2 \text{ from which } v_s = (\beta_1 a^2/m)^{1/2}[\log 2]^{1/2}$$

The ratio between the velocities is $[\log 2]^{1/2} = 0.833$.

All the neighbors contribute to the velocity of sound, but here their actions are alternating and the final velocity is less than that resulting only from the action of nearest neighbors.

(3) When $k = \pi/a$, the expression from (1) becomes

$$\omega^2 = (4/m)\sum_{p>0} \beta_p \sin^2(p\pi/2)$$

The even values of p cancel the corresponding sine, and the odd values of p result in unitary values for the corresponding $\sin^2$:

$$\omega^2 = (4/m)\sum_{j=1}^{\infty} \beta_j \quad \text{where } j = 2p+1$$

When $k = \pi/a$, the vibration wavelength λ is $\lambda = 2a$, the odd atoms of the row (3rd, 5th, etc.) vibrate in phase opposition and the corresponding restoring force is maximum. On the other hand, even neighbors vibrate in phase and the restoring force they exerted is zero at any time (see Ex. 3, Fig. 10).

The action of a neighbor of order p will be maximal when $\sin^2(kpa/2) = 1$ or $k = p^{-1}(\pi/a)$. This situation corresponds to the opposite phase between atoms n and $n + p$ where β_p has the maximum effect and it concerns the same odd multiples (3, 5, ...) contained in the interval $0 < k \leq \pi/a$. Conversely, for values such as $k = 2p^{-1}\,\pi/a$ and even multiples contained in the same interval, the atoms n and $n + p$ always vibrate in phase and the corresponding constants β_p play no role.

The ratio can easily be obtained by summing the first four terms of the series in $(2n + 1)^{-3} \approx 1.048$ (For another way of doing this summation, see also the following exercise.)

Exercise 5: Soft modes

Consider a 1D lattice with parameter a composed of identical atoms of mass m. In equilibrium the abscissa of the atom n is na and its displacement relative to this position, u_n, is such that $|u_n| << a$.

Using the Hooke's approximation, take into account the forces which are exerted between a given atom and all the others without limitation to just nearest neighbors. Thus the force constant between the atom n and the atom $n + p$ (the same as that between n and $n - p$) is denoted as β_p.

(1) Find the dispersion relation $\omega(k)$ taking into account all of the possible β_p ($p > 1$).

(2) One specificity of the dispersion curves $\omega(k)$ is the possible existence of "soft modes." We say that a mode is soft when ω^2 tends toward zero for a value of $k \neq 0$. The appearance of a soft mode in a crystal corresponds to instability of the structure and to a possible phase change.
To show the possibility of ω^2 going to zero, consider a chain of ions of the same mass and alternating charges of $+e$ and $-e$ with $(-1)^n e$ for the position n.

(a) Show that the force constant is electrostatic in origin and for an order p is:

$$\beta_p = \frac{e^2}{4\pi\varepsilon_0}\frac{2}{a^3}\frac{(-1)p}{p^3}$$

(b) Taking into account another force constant β repulsive and limited to the nearest neighbors (in order to prevent the overlapping of atoms), find the expression for ω^2/ω_0^2 as a function of k.

Use $\omega_0^2 = \frac{4\beta}{m}$ and $\sigma = \frac{e^2}{4\pi\varepsilon_0}\frac{2}{\beta a^3}$

Show that σ must be smaller than $1/\log 2 = 1.443$ for $|k|$ near zero and ω^2/ω_0^2 positive.

(3) Evaluate ω^2/ω_0^2 for $|k| = \pi/a$. For this evaluation use the value of the Riemann function $\zeta(3) = \sum_{S=1}^{\infty} S^{-3} = 1.202$.

Deduce that if for some physical reason σ increases above a certain value σ_0, ω^2 goes to zero for a value of $|k|$ between 0 and π/a: a soft mode thus appears for $\sigma = \sigma_0$ and the lattice is unstable for $\sigma > \sigma_0$.

Solution:

(1) $m\ddot{u}_n = \sum_{p=1}^{\infty} \beta_p (u_{n+p} + u_{n-p} - 2u_n)$ (1)

(2) (a) The Coulomb force exerted between two charges separated by a distance r_{ij} is $F_{ij} = q_i q_j / 4\pi\varepsilon_0 r_{ij}^2$. Here, $q_n = (-1)^n e; q_{n+p} = (-1)^{n+p} e; r_{n,n+p} = pa + u_{n+p} - u_n$.

Using the limited expansion of the form $(1+\varepsilon)^{-2} \approx 1 - 2\varepsilon$, we find:

$$F_{n,n+p} = \frac{(-1)^p e^2}{4\pi\varepsilon_0 (pa)^2}\left[1 - 2\frac{u_{n+p} - u_n}{pa}\right]$$

The first term contributes to the crystal cohesion by its contribution to the Madelung constant (see Chapter II, Ex. 1), whereas the second term represents the restoring force that an atom $n + p$ exerts on another atom n. By identifying this second term to be $\beta(n_{n+p} - u_n)$, we find the proposed expression for β_p.

(b) Substituting in (1) β_p by their expressions and taking into account the constant β, one obtains:

$$\omega^2 = \frac{4}{m}(\beta - \beta_1)\sin^2\left(\frac{ka}{2}\right) + \frac{4}{m}\sum_{p>1} \beta_p \sin^2\left(p\frac{ka}{2}\right)$$

or

$$\left(\frac{\omega}{\omega_0}\right)^2 = (1-\sigma)\sin^2\left(\frac{ka}{2}\right) + \sum_{p>1}(-1)^p \frac{\sigma}{(2p)^3}\sin^2 p\frac{ka}{2}. \quad (2)$$

When $k << \pi/a$, that is to say for large wavelengths (center of the BZ), we can identify the sine functions and their angles.

$$\left(\frac{\omega}{\omega_0}\right)^2 = \left[1 - \sigma\left(1 - \frac{1}{2} + \frac{1}{3} - \frac{1}{4}\cdots\right)\right]\left(\frac{ka}{2}\right)^2$$

The alternated series is involved in the Madelung constant of the crystal $\left(\ln 2 = 1 - \frac{1}{2} + \frac{1}{3} - \frac{1}{4}\cdots\right)$, and $\left(\frac{\omega}{\omega_0}\right)$ will only be positive for $1 - \sigma\ln 2 > 0$; which is the case only when $\sigma < 1.443$.

(3) In the inverse hypothesis, $k = \pi/a$ (boundary of the BZ), expression in 2(b) takes the form:

$$\left(\frac{\omega}{\omega_0}\right)^2 = \left[1+\sigma\sum_{p>1}\frac{(-1)^p}{p^3}\sin^2\left(p\frac{\pi}{2}\right)\right]$$

The summation Σ is limited to the terms that are odd in p because the sines relative to even p are zero. This summation is thus

$$\sum = \zeta(3) - \frac{1}{2^3}\zeta(3) = 1.05\,.$$

For $\left(\frac{\omega}{\omega_0}\right) > 0$, we must have σ < 0.95.

In conclusion, when σ ($\propto \beta^{-1}a^{-3}$) has a value between 0.95 < σ < 1.443, ω^2 tends toward zero (and the lattice is unstable) for a wave vector such that $0 < k < \pi/a$.

To know more, see Solid State Phase Transformations in Metals and Alloys, Les editions de Physique (1987), in particular, the article by M. Gerl, pp. 459–521, and Chapter V, Ex. 4.

Exercise 6: Kohn anomaly

Consider a 1D lattice, row, of identical atoms equidistant by a. It is supposed to be metallic in such a manner that, via the conduction electrons, the force constant exerted on an atom in position $n + p$ by an atom at position n is of the form $\beta_p = A\frac{\sin pk_0 a}{pa}$ in which A and k_0 are constants. Find, in form of an unlimited series, the dispersion relation $\omega = f(k)$. Without trying to evaluate this summation, show that the derivative $\partial(\omega^2)/\partial k$ become infinite for a particular value of k that should be determined. Graph the dispersion curve.

Solution:

$$-m\omega^2 \ddot{u}_n = \sum \beta_p (u_{n+p} + u_{n-p} - 2u_n)$$

Substituting β_p by the given expression and taking into account that $u_{n+p} = u_n e^{-ikpa}$ one obtains

$$\omega^2 = \frac{2A}{m}\sum_{p>0}\frac{\sin(pk_0 a)}{pa}(1-\cos pka)$$

which leads to

$$\frac{\partial(\omega^2)}{\partial k}=\frac{2A}{m}\sum_{p>0}\sin pk_0a\cdot\sin pka=\frac{A}{m}\sum_{p>0}[\cos p(k_0-k)a-\cos p(k_0+k)a]$$

Then $\partial(\omega^2)/\partial k=\infty$ when $k = \pm k_0$ because one of the two cosines becomes equal to unity and then one must sum p from $p = 1$ to ∞.

The dispersion curve exhibits one point that goes to zero for $\vec{k}=\vec{k^\circ}\pm\vec{G}$. Its characteristics are shown in Fig. 11 (after Fig. 4 from the article by Gerl mentioned at the end of the previous exercise).

In reality, the constant k_0 is such that $\vec{k^\circ}=2\vec{k}_F$ (where $\vec{k}_F$ is the Fermi vector, see Chapter IV.)

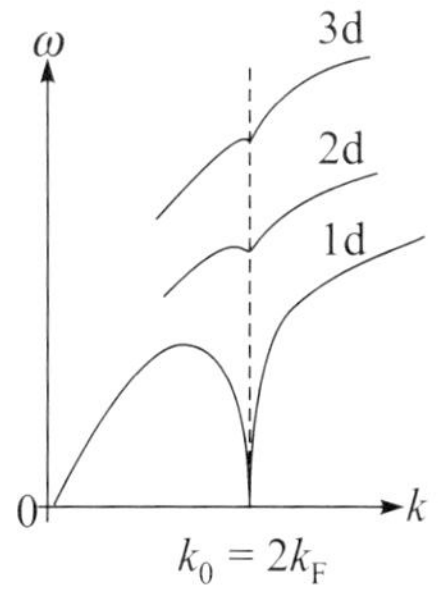

Figure 11

Exercise 7: Localized phonons on an impurity

Consider a linear chain of identical atoms of mass m each equidistant by a.

(a) Limiting the problem to interactions between nearest neighbors characterized by the force constant β, give the dispersion relation of longitudinal phonons and derive the expression of their group velocity. Express the results as a function of "a" and of the velocity of sound, v_S. What is maximum frequency of atomic vibration, ν_M, when $v_s = 5000$ m/s?

(b) An atom in the middle of the chain is replaced by the atom of an impurity with a mass m' (where $m' < m$). Write the displacement equations, u_0, for this impurity atom and the displacement "u_n" for an atom located at a distance "na." Assume that the force constant β between nearest neighbors is the same regardless of the atom considered.

(c) The impurity atom being lighter than the others it has an oscillation frequency ν'_m larger than ν_M and it drags the other neighbor atoms by its motion. We try to describe its effect with the expression of a damped wave of form:

$u_n = u_0 \exp - \alpha |x| \exp i(\omega t - kx)$

What is the value of the frequency ν'_M, which is compatible with the equations of motion when the wave vector is near the boundary of the first BZ? Verify that when $m = m'$, the solution matches that obtained in (a).

(d) After having numerically evaluated ν'_m and a when $m' = m/2$, sketch the instantaneous position of atoms as a function of x.

Solution:

(a) By limiting interactions between nearest neighbors and in the Hooke's approximation, the equation of motion for an atom in position n is:

$$m \frac{d^2 u_n}{dt^2} = \beta(u_{n-1} + u_{n+1} - 2u_n) \quad (1)$$

Introducing a solution sinusoidal of the form: $u_n = u_0 \exp i(\omega t - kx) \approx u_0 \exp i(\omega t - kna)$ we find (see Course Summary)

$$\omega = 2(\beta/m)^{1/2} \sin\left|\frac{ka}{2}\right|, \; v_g = \partial\omega/\partial k = (\beta a^2/m)^{1/2} \cdot \cos\left(\frac{ka}{2}\right),$$

$$v_s = v_{g(k\to 0)} = (\beta a^2/m)^{1/2}$$

Thus the following results are established:

$\omega = (2v_s/a) \cdot \sin(ka/2)$ and $v_g = v_s \cos(ka/2)$

Numerical results: $\nu_M = v_s/\pi a = 3.18 \times 10^{12}$ c/s

(b) To the equation of motion (1) for any atom in the chain, one must add the equation of motion of the impurity given by

$$m' \frac{d^2 u_0}{dt^2} = \beta(u_{-1} + u_1 - 2u_0) \quad (2)$$

(c) Inserting in (1) and (2) the proposed solution for $k = \pi/a$

$$u_n = u_0 \exp - \alpha|x| \cdot \exp i(\omega' t - kx)$$

$$\approx u_0 \exp - \alpha|x| \cdot \exp i\left(\omega' t - \frac{\pi}{a} na\right)$$

$\approx u_0 \cdot (-1)^n \cdot \exp -\alpha|n|a \cdot \exp i\omega' t$, we thus obtain

$(1'): \omega'^2 = (\beta/m)(2 + e^{-\alpha a} + e^{\alpha a}), (2'): \omega'^2 = (\beta/m')(2 + 2e^{-\alpha a})$,

These expressions are compatible with each other only for values of $e^{\alpha a}$, solutions of the second-order equation: $x^2/2m - [(1/m') - (1/m)x + (1/2m) - (1/m') = 0$, obtained with $x = e^{\alpha a}$.

Since one solution leads to $e^{\alpha a} = -1$, the only acceptable solution corresponds to $e^{\alpha a} = \frac{(2m - m')}{m}$, which gives $\omega'^2 = \frac{4\beta}{m'} \frac{m}{2m - m'}$.

When $m = m'$, this solution can be identified with that obtained in (a) for the undisturbed chain at the boundary of the BZ, that is to say $e^{\alpha a} = 1$ or $\alpha = 0, \omega' = \omega = (4\beta/m)^{1/2}$ and $u = u_0(-1)^n \exp i\omega t$.

Mathematically $e^{\alpha a}$ and ω'^2 must be positive, $m' < 2m$, but physically a damping term is needed: $\alpha > 0$ and $e^{\alpha a} > 1$. This can happen only for $m' < m$. In this case, $m' < m$, the ratio ω'/ω increases and the vibration of the impurity is more localized when the mass is smaller and smaller compared to the mass m of the other atoms of the lattice.

(d) When $m' = m/2$, $e^{\alpha a} = 3$, $\nu'_M = 2\,\nu_M/\sqrt{3} = 3.67 \times 10^{12}$ c/s.

The corresponding displacement u of atoms, $u = f(x)$, is shown in Fig. 12a. The wave being longitudinal, the position of the atoms in real space is that shown in Fig. 12b.

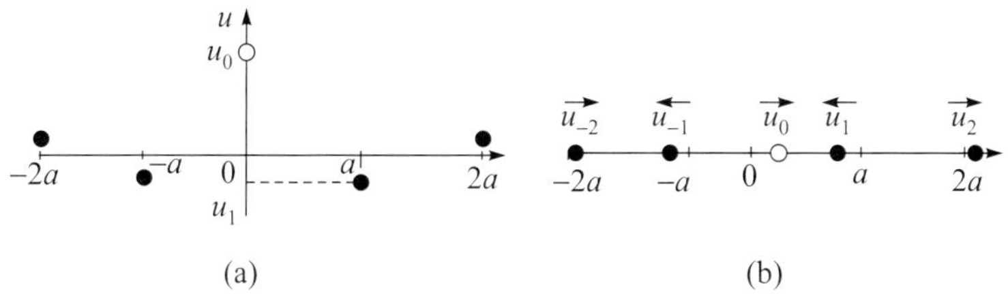

Figure 12

Exercise 8: Surface acoustic modes

Consider a semi-infinite row of atoms that are equidistant by "a" and numbered 0, 1, 2, ... n, ... The surface atom (n = 0) as a mass M'

and is submitted to the force constant β' with its nearest neighbor ($n = 1$). The other atoms in the row are identical with a mass M and are submitted only to the actions of their nearest neighbors with force constant β.

(a) Write the equations of motion of atoms: 0, 1, and n (where $n \geq 2$).

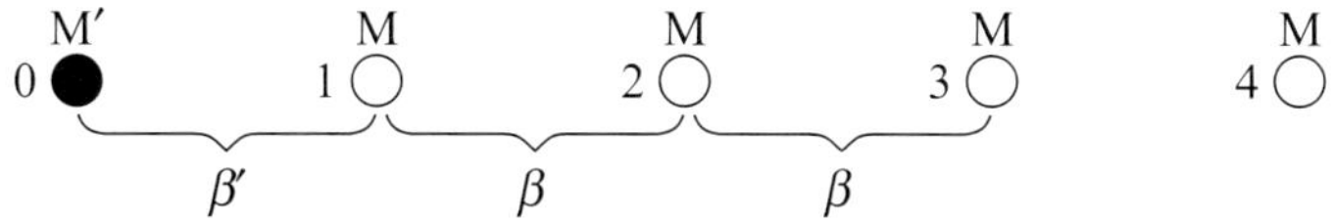

(b) For the semi-infinite chain, suppose that the displacements u relative to a localized excitation (surface mode) obeys the following relations:
$u_0 = Be^{i\omega t}$ and $u_n = C(-1)^n e^{-kna} e^{i\omega t}$,
where B and C represent constant amplitudes. How do the equations of motion change? What relation between ka, β'/β and M'/M must be satisfied for nontrivial solutions ($B, C \neq 0$) to exist?

(c) How can the relation between β'/β and M'/M be simplified when one observe that k must be positive for physical waves located near the surface? How are the frequencies of these surface modes situated with respect to the corresponding volume modes?

Solution:

(a) $$M'\ddot{u}_0 = \beta'(u_1 - u_0) \tag{1}$$

$$M\ddot{u}_1 = \beta(u_2 - u_1) + \beta'(u_0 - u_1) \tag{2}$$

$$M\ddot{u}_n = \beta(u_{n+1} + u_{n-1} - 2u_n) \tag{3}$$

(b) Inserting the proposed solution into Eq. (3), we do not obtain the usual result (see Ex. 3) because the wave is damped:

$$\omega^2 = (\beta/M)(e^{ka} + e^{-ka} + 2) \tag{3'}$$

(c) Inserting the proposed solutions into Eqs. (1) and (2) we find the following system:

$$(\beta' - M'\omega^2)B + \beta' e^{-ka} C = 0 \tag{1'}$$

$$\beta' B + e^{-ka}(\beta' + \beta - M\omega^2 + \beta e^{-ka})C = 0 \qquad (2')$$

In order for Eqs. (1′) and (2′) to have nontrivial solutions the determinant of the coefficients must be nonzero. Taking also into account Eq. (3′), we find

$$\left[\beta' - \frac{M'}{M}\beta(e^{-ka} + e^{ka} + 2)\right][\beta' - \beta - \beta e^{ka}] - \beta'^2 = 0.$$

Using $x = e^{ka}$, this condition becomes

$$x^2 + [2 - (\beta'/\beta) - (M/M') \cdot (\beta'/\beta)]x + 1 - (\beta'/\beta) = 0.$$

(d) For the wave to have a physical meaning, k must be positive, which leads to $x > 1$ and $-C > 1 + \beta$ resulting in

$$\frac{\beta'}{\beta} > \frac{4}{2 + (M/M')} \quad \text{or} \quad \frac{M}{M'} > 4\left(\frac{\beta}{\beta'}\right) - 2.$$

The corresponding frequencies (3′) are such that

$$\omega^2 = (2\beta/m)(ch\,ka + 1) \geq \omega_{\mathrm{M}}^2$$

where ω_{M} represents the maximum angular frequency for an infinite row of identical atoms [$\omega_{\mathrm{M}} = (4\beta/M)^{1/2}$].

Thus the acoustic surface modes can only exist with frequencies greater than those of the volume modes. This situation can only occur when the force constant β' (relative to the surface atom) is sufficiently larger than β (relative to the row atoms) or if the mass M' of this surface atom is small compared to those in the row of atoms.

One may point out similarities between the present exercise and that treating localized phonons (see Ex. 7). The results differ because of different geometries: $m' < m$ (Ex. 7); $m' < m/2$ when $\beta' = \beta$ and $k = \pi/a$, here. The key point is the existence of localized modes having larger frequencies than those of the indefinite crystal but only when some restrictive conditions are satisfied.

Exercise 9: Atomic vibrations in a 2D lattice

Consider a 2D square lattice of parameter "a" formed from identical atoms of mass m that are submitted to a force constant β between nearest neighbors and are forced to move perpendicular to the lattice plane (Z mode: associated with out-of-plane atomic motions).

(a) Find the equation relative to the displacement u_{lm} of an atom belonging to the lth column and the mth row.

(b) Starting from a solution of the form of running waves $u = A\exp i(\omega t - k_x x - k_y y)$, find the dispersion relation for the z-mode vibrations, $\omega = f(k)$.

(c) Sketch the corresponding curves in the [10] and [11] directions. What is the maximum vibration frequency ν_M ? Draw a few curves of isofrequencies within the first BZ, notably the curves where $\omega << \sqrt{\frac{4\beta}{m}}$ and those where $\omega = \sqrt{\frac{4\beta}{m}}$.

(d) Starting from periodic boundary conditions, with periodicity L along x and along y, find the density of vibration modes $g(k)$ in k space and deduce the density of modes $g(\nu)$ in frequency space (limiting the evaluation to the frequency domain corresponding to wavelengths which are large compared to the inter-atomic distances).

(e) After evaluating the components of the group velocity, indicate the method that would allow to obtain $g(\nu)$ over all allowed frequencies. Just describe the principles of this method, without the corresponding tedious calculations.

Solution:

(a) Limiting the problem to interactions between the four nearest neighbors and within the Hooke's approximation, the equation of motion of an atom in position l, m is

$$m\frac{\mathrm{d}^2 u_{l,m}}{\mathrm{d}t^2} = \beta(u_{l+1,m} + u_{l-1,m} - 2u_{l,m}) + \beta(u_{l,m+1} + u_{l,m-1} - 2u_{l,m})$$

(b) Inserting the running wave solution

$$u_{l,m} = A\exp i(\omega t - k_x x - k_y y) \approx A\exp i(\omega t - k_x la - k_y ma)$$

into this equation we find

$$-m\omega^2 u_{l,m} = \beta[\exp(-ik_x a) + \exp(+ik_x a) - 2]u_{l,m} + \beta[\exp(-ik_y a) + \exp(+ik_y a) - 2]u_{l,m}$$

from which we find the dispersion relation:

$$\omega^2 = \frac{2\beta}{m}(2 - \cos k_x a - \cos k_y a) \text{ or } \omega^2 = \frac{4\beta}{m}\left(\sin^2\frac{k_x a}{2} + \sin^2\frac{k_y a}{2}\right)$$

(c) Along the [10] axis where $k_y = 0$, we find $\omega^2 = \frac{4\beta}{m}\sin^2\frac{ka}{2}$. This result is the result obtained for a row of identical atoms within the same simplifying assumptions and values of k_x that have a physical meaning in the interval.

What was said for k_x, $-\pi/a \le k_x \le \pi/a$ applies also for k_y and the region in $\vec{k}$ space $\vec{k}\begin{vmatrix} k_x \\ k_y \end{vmatrix}$ for which independent solutions exist into the first BZ, see Fig. 13a.

In the [11] direction: $k_x = k_y = \frac{|k|}{\sqrt{2}}$ and the dispersion relation becomes $\omega^2 = (8\beta/m)\sin^2(|k|a/2\sqrt{2})$.

In this direction, the maximal frequency ν_M is $\nu_M = \frac{1}{\pi}(2\beta/m)^{1/2}$ and it is $\sqrt{2}$ times greater than that obtained in the [10] direction because the modulus of the wavevector is $\sqrt{2}$ times larger: $|k| = (k_x^2 + k_y^2)^{1/2} = \frac{\pi\sqrt{2}}{a}$

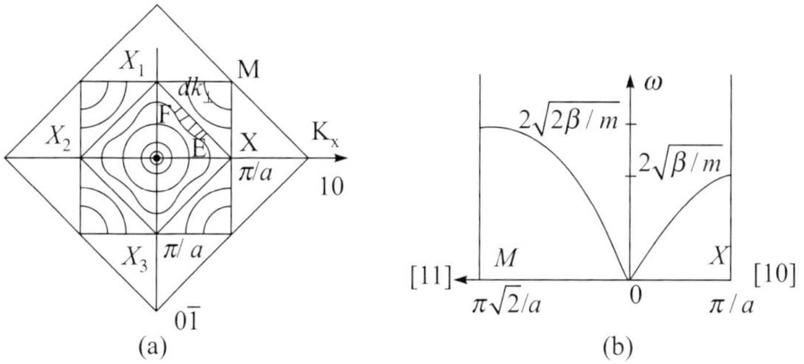

Figure 13 First BZ of a square lattice with isofrequency curves at left and dispersion curves at right.

For the wavelengths larger than inter-atomic distances ($ka << 1$), the dispersion relation reduces to

$$\omega^2 = \frac{4\beta}{m}\left[\left(\frac{k_x a}{2}\right)^2 + \left(\frac{k_y a}{2}\right)^2\right] = \left(\frac{\beta}{m}\right)k^2a^2$$

and the isofrequency lines around the origin are essentially circles.

This is no longer the case for the isofrequency lines far away from the origin. We observe, in particular in Fig. 13b, for the same frequency 2 $(\beta/m)^{1/2}$, the modulus of k is $\sqrt{2}$ smaller in the [11] direction than in the [10] direction. The isofrequency lines obey the general relation: $\sin^2(k_x a/2) + \sin^2(k_y a/2) =$ constant with $-\pi/a \le k_x \le \pi/a$ and $-\pi/a \le k_y \le \pi/a$. They are shown schematically in Fig. 13a. Specific curve of isofrequency $\omega^2 = 4\beta/m$ (constant = 1) is the square marked by X, X_1, X_2, X_3, the corresponding lines obeying to $k_x = \pi/a \pm k_y$. An analogous situation exists for the isoenergy curves of electrons in a square lattice (see Chapter V, Ex. 2, Fig. 2b).

(d) The periodic boundary conditions of the type Born, von Karman, can be written as $u(x, y) = u(x + L, y)$ and $u(x, y) = u(x, y + L)$. By inserting them into the expression of the running wave [see Q. (b)] one obtains a variation of k_x and k_y in steps of $2\pi/L$ because $\exp -(ik_xL) = 1$ and $\exp -(ik_yL) = 1$.

Inside of the first BZ, each vibration mode corresponds to a point at $2\pi n/L$ on the x-axis and at $2\pi m/L$ on the y-axis; n, m being integers such that $-L/a \le m, n \le L/a$.

The density of mode $g(k)\mathrm{d}k$ corresponds to the number of modes contained in a circular ring of inner radius k and outer radius $k + \mathrm{d}k$ (see Fig. 14a).

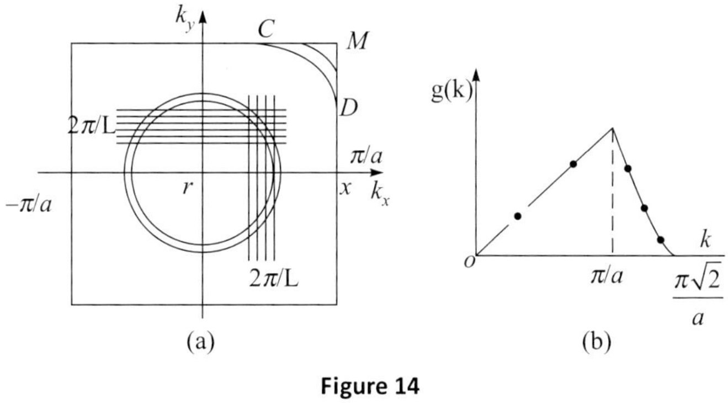

Figure 14

- For $|k| \le \pi/a, g(k) = \dfrac{L^2}{2\pi}|k|$ because

$$g(k)\mathrm{d}k = \frac{2\pi k \mathrm{d}k}{\left(\frac{2\pi}{L}\right)^2} = \frac{\text{Area of the ring}}{\text{Area occupied by 1 mode}}$$

- For $\pi/a \le |k| \le \pi\sqrt{2}\,/a$,

$$g(k) = \frac{L^2}{\pi^2}|k| \arccos \frac{2\pi[k^2 - (\pi^2/a^2)]^{1/2}}{ak^2}$$

Taking into account the length of the arc $\widehat{CD}$ such that

$$\widehat{CD} = k \arccos\left[\frac{2\pi}{ak^2}\left(k^2 - \frac{\pi^2}{a^2}\right)^{1/2}\right]$$

For large wavelengths, the isofrequency lines are circles and the dispersion relation reduces to:

$$\nu = \frac{1}{2\pi}\left(\frac{\beta a^2}{m}\right)^{1/2} \cdot |k|;\ g(\nu)\mathrm{d}\nu = g(k)|\mathrm{d}k| = \frac{L^2}{2\pi}|k||\mathrm{d}k| \text{ or}$$

$$g(\nu) = L^2 \frac{2\pi\nu}{(\beta a^2/m)}$$

The curve $g(k)$ is shown in Fig. 8b. The curve of $g(\nu)$ only follows the same linear evolution in ν when $k << \pi/a$.

(e) Differentiating the dispersion relation in (b), we find two components of the group velocity, $\vec{v}_g$:

$$\vec{v}_g = \overrightarrow{\nabla_k \omega} = \begin{cases} \partial\omega/\partial k_x = (\beta a/m\omega)\sin k_x a \\ \partial\omega/\partial k_y = (\beta a/m\omega)\sin k_y a \end{cases}$$

To evaluate the exact density of vibration modes $g(\nu)$, one must first evaluate the number of modes contained in the area $d\vec{s}/d\vec{k}_\perp$ (see Fig. 14a), in which $d\vec{s}$ represents the arc element, $\widehat{EF}$, measured along the isofrequency curve $\nu = \nu_0$ (see Ex. 19, Fig. 20).

Integrating along the length of the complete curve we find:

$$g(\nu_0)d\nu_0 = \left(\frac{I}{2\pi/L}\right)^2 \oint ds \cdot dk_\perp = \frac{L^2}{4\pi^2}\oint \frac{ds}{|v_g|} 2\pi \mathrm{d}\nu.$$

This formula is a 2D application of the general expression for the density of modes (Ex. 19). By writing the terms explicitly in rectangular coordinates:

$$g(\nu_0)d\nu_0 = \frac{L^2}{2\pi}d\nu_0 \int_{\nu=\nu_0} \frac{\sqrt{(dk_x)^2+(dk_y)^2}}{\left[\left(\frac{\partial\omega}{\partial k_x}\right)^2+\left(\frac{\partial\omega}{\partial k_y}\right)^2\right]^{1/2}}$$

By differentiating the dispersion relation
$2m\omega d\omega = 2\beta a\ (\sin k_x a\ dk_x + \sin k_y a\ dk_y) = 0$,
one obtains

$$g(\nu_0) = \frac{L^2 m\nu_0}{\beta a}\oint \frac{dk_x}{\sin k_y a}.$$

The resolution of the integral leads to tedious calculations. We can limit the resolution to large wavelengths, so that $\int = 2\pi/a$ in order to verify that the result obtained is identical to that in Question (d) (see Ex. 18; Chapter IV, Ex. 14; Chapter V, Ex. 2).

Exercise 10: Optical absorption of ionic crystals in the infrared

In ionic crystals the evaluation of the resonant frequency of ions (TO branch at $k = 0$) results in the following relation:
$\omega^2 = 2\beta/\mu$ where $\frac{1}{\mu} = \frac{1}{M} + \frac{1}{m}$ (see Course Summary) in which the constant force β may be identified to that of a Coulomb force between two neighbor ions separated by a distance r_0 at equilibrium.

Using this simplified scheme, find numerically the wavelength of absorption of Na^+Cl^- and $Zn^{2+}S^{2-}$ in the infrared with r_0(NaCl) = 2.815 Å and r_0(ZnS) = 2.34 Å [see Chapter I, Fig. 1 (c & e)]. Compare the results to experimental values: λ(NaCl) = 61 μm and λ(ZnS) = 35 μm.

Note: The constants involved in this problem include M(Cl); m(Na); M(Zn); m(S) and they are indicated in the periodic table in the beginning of this book. The other constants such as e, ε_0, c, and N are also given in Table I at the beginning of this book.

Solution:

The restoring force F between two ions is $F \approx \beta r_0$ and the Coulomb force $F = Z^2q^2/4\pi\varepsilon_0 r_0^2$. In the latter equation Zq corresponds to the charge of each ion. It is easy to deduce β numerically:

β(NaCl) = 10 N/m with $Z = 1$; β(ZnS) = 72 N/m with $Z = 2$. The wavelength of optical absorption corresponds to $\lambda = c/\nu$ in which the optical frequency of absorption coincides with the resonant frequency of ions $\upsilon = \omega/2\pi$.

λ(NaCl) = 64 μm; λ(ZnS) = 29.7 μm.

Remark: It is surprising that there is such an excellent agreement between this rather approximate model and the experimental values. A more rigorous calculation of the restoring force between ions would use the Madelung constant and the force of repulsion which partially compensates for the small differences.

A rigorous calculation would follow the procedure for CsCl (Pb. 3) using the preliminary results from Chapter II, Pb. 2.

We may keep in mind the main trends such as a resonant frequency of the ions proportional to $\sqrt{\beta/\mu}$ with a coefficient β that varies as r_0^{-3}. This frequency is larger when the ions are closer together and are light ions as it is for ions with low atomic number. This overall tendency is confirmed by the following experimental results (R. H. Bube, p. 43 [4]).

Crystal	Li H	Li F	Li Cl	Li Br	K Br	Cs Cl	Cs I	Tl Br
λ (μm)	17	33	53	63	88	100	158	233

The resonant frequency increases also with the increase of the charge carried by the ions. These points explains why the absorption wavelengths of MgO (λ = 25 μm) and of ZnS are shorter and of the order of that of LiF.

Exercise 11: Specific heat of a linear lattice

Consider a linear lattice consisting of N identical equidistant atoms separated by a.

(a) In such a lattice what is the density of vibrations in k-space, $g(k)$, for the single possible longitudinal mode.

(b) Suppose that the dispersion relation of phonons can be written in the Debye approximation as $\omega = v_s|k|$ where v_s is the sound velocity. Find the density of vibrations in ν-space, $g(\nu)$. State the maximum vibration frequency ν_D (the Debye frequency) of atoms in such a lattice.

(c) In the form of a definite integral, find the expression for the internal energy due to these lattice vibrations and deduce the behavior of the specific heat of a linear lattice at high ($k_B T >> h\nu_D$) and low ($k_B T << h\nu_D$) temperatures.

Application: Use a = 3 Å and v_s = 3000 m/s. Find numerically the Debye temperature θ_D (such that $h\nu_D = k_B\theta_D$) and the specific heat of atomic vibrations at 10 K.

Note: One can use either fixed or periodic boundary conditions but in the latter case, one must take into account that wave vector k can be either positive or negative. Use F1 (Course Summary, end of Section 6). (e, h, k_B)

Solution:

(a) The atomic waves are of the form:

$$u(x) = A\exp\cdot i(\omega t - kx)$$

- If we use periodic boundary conditions of the Born von Karman type, we must have $u(x) = u(x + L)$ or exp (ikL) = 1 and discrete values of $k = 2\ \pi/L$ in the interval from $-\pi/a \le k \le \pi/a$ (first BZ) and the corresponding waves are traveling waves.
- If we use the fixed boundary conditions $u(0) = 0$ and $u(L) = 0$, the first condition implies an incident wave of the type A exp $i(\omega t - kx)$ and a reflected wave B exp $i(\omega t + kx)$ such that $A = -B$. The resulting waves A exp $i(\omega t - kx) - A$ exp $i(\omega t + kx) = -2Ai\sin(kx)$ exp $i\omega t$ is a standing wave. This wave is the result of two traveling waves of the wave vector k with the same modulus but propagating in the opposite direction (see Final Remark). The possible values of k are therefore positive $0 \le k \le \pi/a$ and the steep variation is given by: $u(L) = 0$ (second condition), or sin kL = 0, which results in $k = n\pi/L$.

  In the second hypothesis, $g(k) = L/\pi$ with $0 \le k \le \pi/a$, whereas in the first hypothesis $g(k) = L/2\pi$ with $-\pi/a \le k \le \pi/a$ but in both cases the number of possible values of k is equal to $L/a = N$, which is the total number of atoms contained in L.

(b) As shown in Fig. 15, we have $g(\omega)$ d$\omega = g(k)$ or, in both cases

$$g(\omega) = \frac{1}{v_s}\cdot\frac{Na}{\pi}$$

From which we obtain

$$g(v) = \frac{2Na}{v_s}.$$

The Debye frequency of such a lattice is $\int_0^{v_D} g(v)\,dv = N$, which gives $v_D = \frac{v_s}{2a}$.

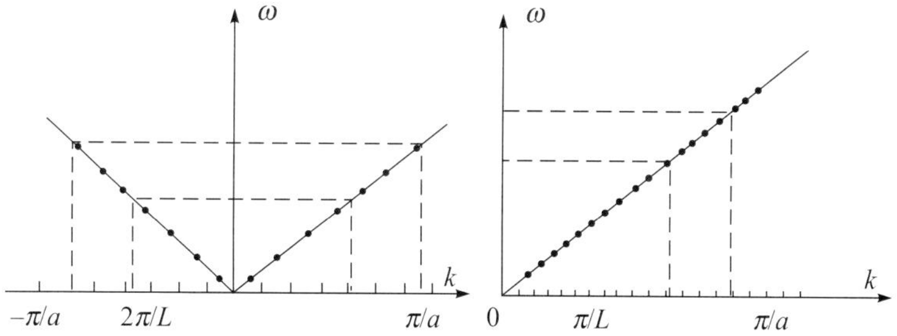

Figure 15 Periodic boundary conditions (left) and fixed boundary conditions (right).

(c) The internal energy U due to longitudinal vibrations of the lattice is

$$U = \int_0^{v_D} g(v)\left(\overline{n(v)} + \frac{1}{2}\right)\cdot hv\mathrm{d}v, \text{ where } \overline{n(v)} = \frac{1}{\exp(hv/k_BT) - 1}$$

Neglecting the half quantum which does not enter the determination of the specific heat we have

$$U = \frac{2Na}{v_s}\int_0^{v_D} \frac{hv\mathrm{d}v}{\exp(hv/k_BT) - 1}.$$

This integral may be evaluated numerically and the evolution of C_v may be deduced next (see Course Summary, Section 6, Fig. 3). More interesting is the following discussion.

- At high temperature $k_BT >> hv_D$, $e^{hv/k_BT} \approx 1 + \frac{hv}{k_BT} + \cdots$ and

$$U = \frac{2Na}{v_s} k_BT \int_0^{v_D} \mathrm{d}v = Nk_BT$$

We find the classical limit: k_BT with $k_BT/2$ for the kinetic energy and $k_BT/2$ for the potential energy per oscillator and per degree of freedom 1 and $C_v = Nk_B$.

- At low temperatures $k_B T << h\nu_D$, which implies $\int_0^{\nu_D} \approx \int_0^{\infty}$; substituting $\frac{h\nu}{k_B T} = x$, we find

  $$U = \frac{2Na}{v_s}\frac{(k_B T)^2}{h}\int_0^{\infty}\frac{x\mathrm{d}x}{e^x - 1} = \frac{Na}{v_s}\frac{\pi^2}{3}\frac{(k_B T)^2}{h}$$

  which results in

  $$C_v = \frac{2\pi^2}{3}\frac{Na}{v_s}\cdot\frac{k_B^2}{h}T$$

  If we use the Debye temperature θ_D, such that $k_B\theta_D = h\nu_D = h v_S/2a$ we find

  $$C_v = \frac{\pi^2}{3}Nk_B\frac{T}{\theta_D}\quad\left(U = \frac{\pi^2}{6}Nk_B\frac{T^2}{\theta_D}\right).$$

In the case of linear lattice, the variation of C_v at low temperature is in (T/θ_D), when the lattice is two-dimensional it is in $(T/\theta_D)^2$ and in 3D in $(T/\theta_D)^3$ (see the paragraph at the end of Ex. 18).

Numerically $\theta_D = \frac{h}{k_B}\cdot\frac{v_s}{2a} = 239\ \text{K}$

At 10 K, $C_v/N = 0.138\,k_B = 0.19\times10^{-23}$ J/deg $= 0.12\times10^{-4}$ eV/deg. The expressions C_v and U concern only one polarization and must be multiplied by three if taking into account the two transverse vibrations and the one longitudinal vibration.

***Final Remark*:** The fixed boundary conditions impose $k = n\pi/L$ or $\lambda = 2L/n$, which is none other than the resonant condition for vibrating cord pinched at O and L and, since we neglect the dispersion, $(\partial\omega/\partial k = v_s = Cste)$, the possible vibration frequencies obey $\nu_n = n\cdot v_S/2L$. If there is an absolute identity between the problem of a vibrating cord and the present exercise, it is the tight analogy with the discrete frequencies of an EM wave imprisoned between two parallel mirrors separated by L and the problem of a free electron trapped in a well of linear potential with infinite walls. In the latter case, the density $g(k)$ is identical to that established in (a) but the dispersion relation is different $(E = h^2k^2/2m)$ because $g(E)$

is different (see Chapter IV, Exs. 11 and 14). In addition, for electrons we must also include the factor of 2, related to their spin, $g(E)$.

The discrete variation of wave vectors of vibrating atoms is imposed by the limiting conditions of a crystal in 3D is similar to the problem of a free electron enclosed in a crystal, as well as the study of the resonant wavelengths of EM waves into a parallelepiped cavity with the discrete variations of the resonant frequencies. It should be emphasized that this discretization imposed by the classical boundary conditions is not what justifies the name "phonons" given to vibrations of a lattice (in the same way that the idea of "photon" is not associated with the discretization of frequencies of EM waves into parallelepiped cavities). It is the quantization of the energy of vibration such as it appears, for example, in the calculation of the specific heat in part (c) (see also Exs. 20 and 21 that justify this appellation).

Said in another way, the expression for an atomic wave (or electronic or light) is of the form $A \exp i(\omega t - kx)$, the notion of phonon (or photon) is associated with the fact that the energy of the wave AA* is proportional to $nxh\nu$ (plus ½ of a quantum for phonons).

Exercise 12a: Specific heat of a 1D ionic crystal

Consider the row formed by $2N$ charged ions of charge alternatively equal to $\pm q$ and equidistant by R_0 in equilibrium as in Ex. 2. Using the dispersion relation for longitudinal acoustic phonons in the Debye model, $\omega = v_s k$, and the dispersion relation for optical phonons in the Einstein model where $\nu_{\text{optical}} = \nu_E$ = constant, find the specific heat at low and high temperature for the longitudinal polarization only.

Numerical application: After finding first the numerical values of the Debye temperature θ_D and the Einstein temperature θ_E, calculate the specific heat of the crystal. Specify the respective contribution (by pair of ions) of optical and acoustical phonons at 20 K with $h\nu_D = k_B\theta_D$ and $h\nu_E = k_B\theta_E$ with $\nu_E = 9.5 \times 10^{12}$ c/s ; $R_0 = 3$ Å ; $v_s = 8800$ m/s. Use F1. (h, k_B, e)

Solution:

Using the results of Ex. 2, we sketch the dispersion of the acoustic and optical branches in Fig. 16. For each branch the density of states in k-space, using periodic boundary conditions, is: $g(k) = L/2\pi = 2NR_0/2\pi$, where $-\pi/2R_0 \leq k \leq \pi/2R_0$ (see preceding exercise). We verify that the number of oscillators per branch is

$$\frac{\pi/R_0}{2\pi/2NR_0} = N,$$

where N is the number of ion pairs contained in the chain.

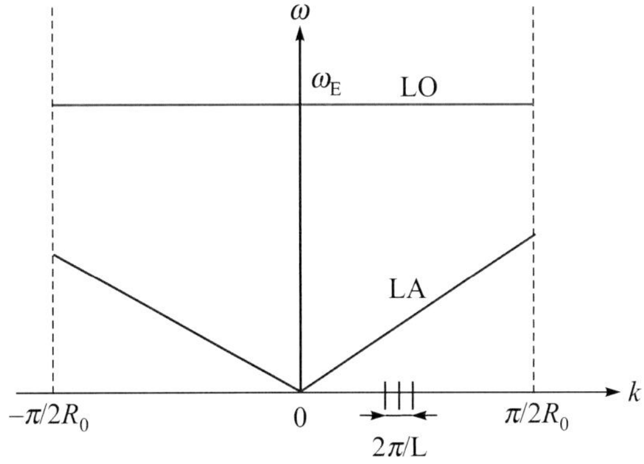

Figure 16

(a) Specific heat associated with the acoustical branch:
Following the demonstration step by step from the preceding exercise, we find: $C_{VA} = \frac{\pi^2}{3} Nk_B \frac{T}{\theta_D}$ at low temperature and $C_{VA} = Nk_BT$ at high temperature.

(b) Specific heat associated with the optical branch:
N oscillators have a frequency ν_E, the total energy of corresponding vibrations, neglecting the half quantum, is

$$U_0 = Nh\nu_E \cdot \frac{1}{\exp\left(\frac{h\nu_E}{k_BT}\right) - 1}.$$

The high temperature limit $\left(e^{(h\nu_E/k_BT)} \cong 1 + \frac{h\nu_E}{k_BT} + \cdots \right)$ is $E_0 = Nk_BT$.

For the low temperature case $(e^{(h\nu_E/k_BT)} >> 1)$, we have $E_0 = Nh\nu_E e^{(-\theta_E/T)}$, where $C_{vo} = Nk_B\left(\frac{\theta_E}{T}\right)^2 \cdot e^{(-\theta_E/T)}$.

(c) In total, we find

- At low temperature, $k_BT << h\nu_D, h\nu_E$,

$$C_v = Nk_B\left(\frac{T}{\theta_D} + \frac{\theta_E^{\ 2}}{T^2}e^{(-\theta_E/T)}\right)$$

- At high temperature, $C_v = 2Nk_B$, or $2k_B$ per pair of ions.

Numerical application:

$$\theta_E = \frac{h\nu_E}{k_B} = 456\,\text{K}, \theta_D = \frac{h\nu_D}{k_B} = \frac{h\nu_s}{4k_BR_0} = 352\,\text{K}$$

At 20 K, the respective contributions of acoustical and optical phonons to the specific heat for a pair of ions are 0.48×10^{-4} eV/K and 2.5×10^{-7} eV/K. As we can see the specific heat at low temperature is uniquely due to the excitation of acoustic phonons and at high temperatures, both types of phonons contribute with the same weight. We have limited our study here to only the longitudinal vibrations and the expressions for U and C_v should be multiplied by 3 to take into account the three polarizations (two transverse and one longitudinal).

Exercise 12b: Debye and Einstein temperatures of graphene, 2D, and diamond, 3D

The crystal basis of graphene and of diamond is composed of two carbon atoms in nonequivalent position (see Chapter I, Fig. 1d in Ex. 1 for diamond and Fig. 9a in Ex. 9b for graphene). Thus their dispersion curves are composed of acoustical branches and optical branches. Like in the above Ex. 12, Fig. 16, their acoustical branches are assumed to obey to the Debye approximation: $\omega = v_s|k|$ and their optical branches are assumed to obey to the Einstein model $\omega = \omega_E = Cst$. Deduce the numerical values of their Debye and Einstein temperatures from their crystal structure and their common sound velocity, $v_s = 18{,}000$ m/s with also ν_E (Einstein frequency) at about 4×10^{13} Hz.

From the C_v graph shown in Course Summary, Section 6, evaluate the specific heat of diamond at room temperature, 290 K. (h, k_B)

Solution:

The Einstein temperature is derived easily from $h\nu_E = k_B\theta_E$. Then θ_E = 1800 K.

For graphene, the Debye wave vector, k_D, may be derived from $\pi k_D^2 = N(2\pi/L_x)(2\pi/L_y)$, where N is the number of primitive cells contained in a surface of dimension $L_x \times L_y$.

For diamond, the Debye wave vector, k_D, may be derived from $(4\pi/3)k_D^3 = N(2\pi/L_x)(2\pi/L_y)(2\pi/L_z)$, where N is the number of primitive cells contained in a volume of dimension $L_x \times L_y \times L_z$.

The area of one hexagon of graphene is $6d^2\sqrt{3}/4$ with d = 1.42 Å. This area contains the equivalent of two C atoms or one primitive cell. $\pi k_D^2 = 8\pi^2/(3\sqrt{3}d^2)$; k_D = 1.175 Å^{-1}; $\nu_D = v_s k_D/2\pi = 33.7 \times 10^{12}$ Hz; $\theta_D \sim 1615$ K.

The volume of the cubic cell of diamond is a^3 with a = 3.56 Å. This volume contains the equivalent of eight C atoms or four primitive cells: $(4\pi/3)k_D^3 = 4(2\pi)^3/a^3$; $k_D = 2\pi(3/\pi)^{1/3}/a$ = 1.74 Å^{-1}; $\nu_D = v_s k_D/2\pi = 50 \times 10^{12}$ Hz; $\theta_D \sim 2340$ K.

At room temperature, T/θ_D is ~0.125 for diamond. This value is slightly outside the end of the C_v increase in $(T/\theta_D)^3$. From Fig. 3 shown in Section 6 of Course Summary, one obtains $C_v \sim 4$ J/mol·K. The present evaluation of θ_D may be improved in taking into account the difference between longitudinal and transverse modes in simplified dispersion relations (see Pb. 4) but the order of magnitude for θ_D remains: the currently admitted value is ~2230 K. Thus diamond has the highest Debye temperature of any material that is one of its many superlative features. As mentioned previously in Ex. 9b for graphene, the microscopic causes are the tight binding between neighbor atoms, β, combined to their light mass, m, explaining the value taken by the sound velocity (proportional to $\sqrt{\beta/m}$).

Exercise 13: Atomic vibrations in an alkaline metal: Einstein temperature of sodium

A model for the restoring forces exerted on alkaline metal atoms is to consider a lattice of ions (+e, M) immersed in a sea of conduction (free) electrons (with charge of −e)—the Fermi sea. These ions are in equilibrium when located on the lattice points. If an ion is displaced from a distance u, the restoring force that will bring it back to equilibrium is in large part due to the electronic charge contained in the sphere of radius u centered on the position of equilibrium where $u << a$; a is the lattice parameter of the alkaline crystal lattice, which is always body-centered cubic. Each atom gives one free electron uniformly distributed in the crystal.

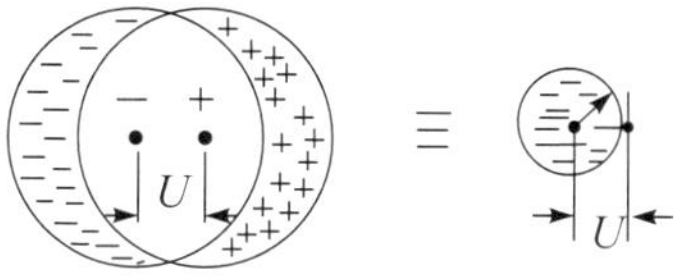

Figure 17

(1) Find the expression of the force constant β as a function of the parameters e and a.

(2) Find the frequency of oscillation of the ion

Numerical application: For sodium, use $M = 29$ and $a = 4.22$ Å. Compare the obtained result to the maximum frequency of acoustic waves $\nu = 3.8 \times 10^{12}$ Hz (experimental value).

(3) In the model described above, the atoms vibrate independently from each others; the density of modes is therefore described by the Einstein model. Find the expression and the numerical value for the Einstein temperature θ_E of sodium. Compare the result with the experimental value of the Debye temperature θ_D (Na) ≈ 160 K. (e, ε, N)

Solution:

(1) The atomic concentration (and therefore the electronic concentration) is $n = 2/a^3$–bcc. The electronic field created by the gas of conduction electrons at a distance u is (Gauss' Theorem):

$$\overrightarrow{E_r} = \frac{\rho}{3\varepsilon_0}\vec{u} \text{ with } \rho = -ne$$

By identifying the Coulomb force with the force constant we find

$$\frac{-ne^2}{3\varepsilon_0}u = -\beta u;\ \beta = \frac{2e^2}{3\varepsilon_0 a^3}$$

(2) $\omega = \sqrt{\beta/M}$ because we are considering independent (noncoupled) harmonic oscillator.
β = 26 N/m; ν_E = 3.7 × 10^{12} Hz. The agreement with the experiment is excellent.

(3) $h\nu_E = k_B\theta_E$; $\theta_E = \dfrac{h}{k_B}\sqrt{\dfrac{2e^3}{3\varepsilon_0 Ma^3}} = 178\ \text{K}$

For the temperatures, the agreement between the calculated value of θ_E and the experimental value of θ_D (160 K) is just fortuitous, given the underlying hypothesis for each of them leading to $\theta_E \approx \theta_D/\sqrt{3}$ (see Ex. 21).
The important point is that a rather rough approximation is sufficient to understand the basic physical phenomena.

Exercise 14: Wave vectors and Debye temperature of mono-atomic lattices in 1-, 2-, and 3D

Consider successively the following lattices (a) a linear lattice with parameter a; (b) a 2D centered rectangle ($a \times b$); (c) a body-centered cubic lattice with lattice parameter a. The basis in each case consists of a single atom.

Knowing that each cell in the k-space can have only one oscillator (for one mode of vibration L or T), we fill the cells starting by those corresponding to the smallest wave vectors up to a complete filling where the Debye wave, k_D, is attained. Find the expression for k_D for each of the three lattices. Use periodic boundary conditions.

(1) In the case of lattice (b), sketch the reciprocal lattice, the first BZ and the limit of the filling defined by k_D (see above). Take a = 3 Å and b = 4 Å. Answer the same question for the lattice in (a).
(2) The sound velocity in the three lattices v_s being known, find the expression for the frequency and the Debye temperature ν_D and θ_D.

Solution:

Using periodic boundary conditions, the dimension of the cells in k-space will be $2\pi/L_x$, $2\pi/L_y$, $2\pi/L_z$. We must distribute one oscillator per cell and the number of oscillator is L_x/a oscillators for (a), $2L_xL_y/ab$ oscillators for (b) and $2L_xL_yL_Z/a^3$ oscillators for (c).

The result is (see Fig. 18a)

$$\text{(a)}\ 2k_\text{D} = \frac{L_x}{a} \times \frac{2\pi}{L_x} = \frac{2\pi}{a}; \qquad k_\text{D} = \frac{\pi}{a}$$

$$\text{(b)}\ \pi k_\text{D}^2 = \frac{2L_xL_y}{ab} \times \frac{2\pi}{L_x} \times \frac{2\pi}{L_y};\ \ k_\text{D} = \left(\frac{8\pi}{ab}\right)^{1/2}$$

$$\text{(c)}\ \frac{4\pi}{3}k_\text{D}^3 = \frac{2L_xL_yL_z}{a^3} \cdot \frac{8\pi^3}{L_xL_yL_z}; k_\text{D} = \frac{(12\pi^2)^{1/3}}{a}$$

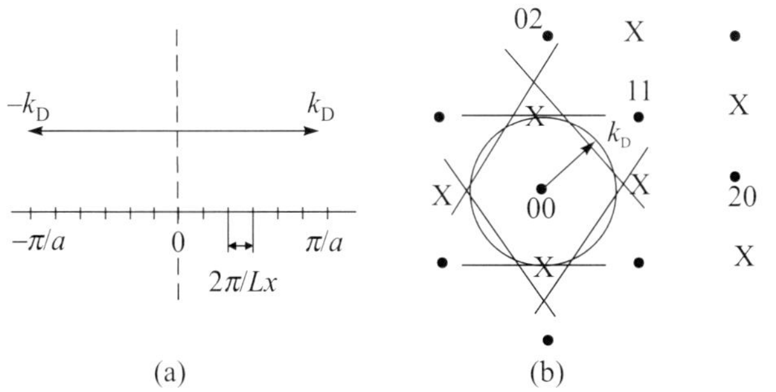

Figure 18

The result (a) is foreseen because the first BZ contains all of the oscillators of the crystal. The results for (b) and (c) include the fact that there are two atoms in each lattice considered.

The results could have been expressed as a function of the linear, surface, or volume atomic concentrations, respectively. See the table in Ex. 18.

The reciprocal lattice in hypothesis (b), has already been studied in Chapter I, Ex. 19; see also Fig. 18.

The area delimited by the circumference of Debye is identical to that of the first BZ but the form is different because the Debye circumference zone is not affected by the crystalline lattice (and it is uniquely a function of the atomic concentration). This is also the case in 3D.

The Debye model is based on a dispersion curve of the form $\omega = v_s k$, or $\nu_\text{D} = v_s k/2\pi$. In addition $h\nu_\text{D} = k\theta_\text{D}$. As a result:

(a) $\nu_D = \dfrac{v_S}{2a}$

(b) $\nu_D = v_S\sqrt{\dfrac{2}{\pi ab}}$

(c) $\nu_D = \dfrac{v_S(3/2\pi)^{1/3}}{a}$

We observe that these results are compatible with those given in the table of Ex. 18 (taking into account that the two atoms for the lattice $a \times b$ in (b) and for the volume a^3 in (c).

Note: (1) The method adopted here will be reconsidered point by point to determine the Fermi wave vector k_F of free electrons in Chapter IV. The only difference concerns the fact that each cell can accept two electrons (↑↓) instead of an oscillator as considered here (associated with the number of phonons function of T). The result is that the number of occupied cells of monovalent metal ($1e^-/at$) will be two times smaller than here.

(2) The method developed in this exercise allows the quick estimate for an order of magnitude of θ_D for a given solid and from this one can deduce its resulting thermal properties of a solid. It is sufficient to know the sound velocity (directly accessible from the macroscopic elastic constants or the microscopic force constant). An interesting application is given in the following exercise.

Exercise 15: Specific heat at two different temperatures

At 300 K the specific heat of a lattice of an element, C_v, is 20 J/mol/deg. Using the curve of $C_v = f(T)$, shown in the Course Summary, find the Debye temperature of this element and its specific heat in cm^3 at 150 K. Assume that its atomic density is $N = 5 \times 10^{22}$ cm^{-3}. (N)

Solution:

Graphically 20 J/mol/deg corresponds to $T/\theta_D \approx 0.45$, which gives θ_D = 666 K.

At 150 K, $C_v \approx 11.3$ J/mol/deg or (by multiplying by $N/\mathcal{N}$)

$C_v = 0.94$ J/cm^3/deg.

Exercise 16: Debye temperature of germanium

Find the Debye temperature of germanium starting from its specific heat measured at 3.23 K: $C_v = 12.5 \times 10^{-4}$ J/mol/deg.

Compare this result with that obtained from evaluating the average sound velocity $v_s = 3.75 \times 10^3$ m/s. The atomic density of Ge is $N \approx 4.42 \times 10^{22}$ cm^{-3}. (N, k_B, h)

Solution:

At 3.23 K, C_v varies as T^3:

$$C_v = \frac{12\pi^4}{5} N\, k_B \left(\frac{T}{\theta_D}\right)^3 = 1941 \left(\frac{T}{\theta_D}\right)^3 = 12.5 \times 10^{-4} \text{ or}$$

$$\left(\frac{T}{\theta_D}\right)^3 = 0.643 \times 10^{-6},\ \theta_D = 374 \text{ K}$$

To evaluate θ_D using the sound velocity, we write $k_D = (6\pi^2 N/V)^{1/3}$, followed by $v_D = v_s k_D / 2\pi$ and finally $\theta_D = h v_D / k_B$.

This gives

$$k_D = 1.378 \times 10^{10}\,\text{m}^{-1}, v_D = 8.22 \times 10^{12}\,\text{c/s};\ \theta_D = 394 \text{ K}.$$

For a discussion of the different results obtained by these two methods, see the comments in Pb. 4.

Exercise 17: Density of states and specific heat of a monoatomic 1D lattice from the dispersion relation

Consider a row of N identical atoms of mass m and equidistant by a.

(a) Limiting the problem to interactions between nearest neighbors with a force constant β, find the expression for the dispersion relation of acoustic longitudinal vibrations propagating along this row. What is the maximum frequency v_m of atomic vibrations, written in terms of v_s and a? Compare the value obtained with that of v_D deduced from the Debye model.

(b) Find the corresponding expression for the density of states $g(v)$.

(c) Write the expressions for internal energy U and for the heat capacity C_v. Find the limits as these functions approach high temperatures ($hv_m << k_B T$).

(d) What is the evolution of C_V at low temperature? Compare the result obtained here with that deduced from the Debye model (Ex. 11). Use F1 and eventually F3.

Solution:

(a) The equation of motion of the *n*th atom is

$$m\frac{d^2u_n}{dt^2} = \beta(u_{n+1} + u_{n-1} - 2u_n)$$

Starting from solutions of the form

$$u_n = A\exp i(\omega t - kx) \approx \exp i(\omega t - kna)$$

we find $\omega = 2\left(\frac{\beta}{m}\right)^{1/2} \sin\left|\frac{ka}{2}\right|$ and $\nu_m = \frac{1}{\pi}\left(\frac{\beta}{m}\right)^{1/2}$

The sound velocity v_s is such that $v_s = (\partial\omega/\partial k)_{k\to 0} = (\beta a^2/m)^{1/2}$ or $\nu_m = v_s/\pi a$ so that the Debye model ($\omega = v_s k$) leads to $\nu_D = v_s/2a$ (see Ex. 11).

(b) The density of states $g(\nu)$ is given by $g(\nu)d\nu = g(k)dk$ with $g(k) = L/\pi = Na/\pi$ (see Ex. 11) and $\nu = \nu_m \sin(|k|a/2)$:

$g(\nu) = 2N/\pi(\nu_m^2 - \nu^2)^{1/2}$.

The corresponding evolution is shown (solid line) in Fig. 19 with the density of states from the Debye model in dashed lines for comparison. We find in both cases that at the origin $g(0) = 2L/v_s$. Thus the areas below the curves $g(\nu)$ and $g_D(\nu)$ are the same because

$$\int_0^{\nu_D} g_D(\nu)\,d\nu = \int_0^{\nu_m} g(\nu)\,d\nu = N,$$

where N is the number of oscillators of the chain.

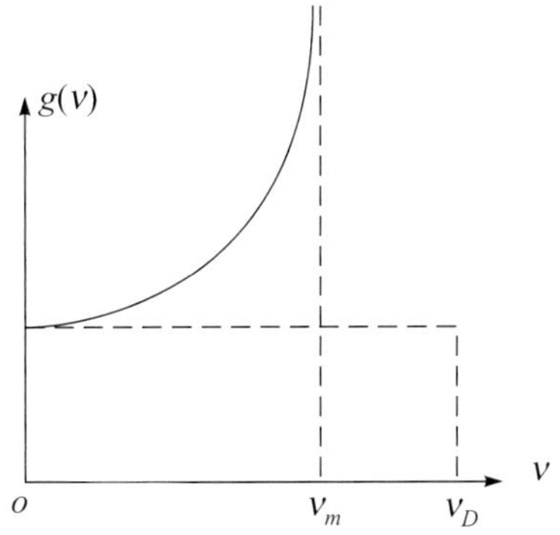

Figure 19

(c) The vibration energy U of a 1D solid corresponds to

$$U=\int_0^{\nu_m} g(\nu)\left(\bar{n}+\frac{1}{2}\right)h\nu d\nu,$$

where $\bar{n}=\dfrac{1}{e^{(h\nu/k_B T)}-1}$.

The specific heat C_v corresponds to $C_v=\left(\dfrac{\partial U}{\partial T}\right)_V$

At high temperatures: $h\nu \le h\nu_m << k_B T$, so that $\bar{n}\approx k_B T/h\nu$

$U=k_B T\int_0^{\nu_m} g(\nu)\,d\nu = Nk_B T$ (neglecting the half quantum).

Thus $C = Nk_B$.

These values correspond to the classical limit (law of Dulong and Petit for a 1D solid) and are independent of the form of the density of states adopted because the average of this density of states will always be N.

(d) At low temperatures: $k_B T << h\nu_m$, so that $\int_0^{\nu_m}\approx\int_0^{\infty}$
Substituting $x = h\nu/k_B T$ one obtains

$$U=\frac{2N}{h\nu_m\pi}\cdot k_B^2T^2\int_0^{\infty}\frac{xdx}{(e^x-1)(1-x^2/x_m^2)^{1/2}}$$

$$U=\frac{2N}{\pi h\nu_m}k_B^2T^2\int_0^{\infty}\frac{xdx}{e^x-1}\left(1+\frac{1}{2}\frac{x^2}{x_m^2}+\cdots\right)$$

$$U=\frac{\pi^2}{3}Nk_B\frac{T^2}{\theta_m}+\frac{\pi^3}{15}Nk_B\frac{T^4}{\theta_m^3}+\cdots \text{ with } h\nu_m=k_B\theta_m,$$

from which we obtain

$$C_v=\frac{2\pi^2}{3}Nk_B\frac{T}{\theta_m}+\frac{4\pi^3}{15}Nk_B\left(\frac{T}{\theta_m}\right)^3+\cdots=\alpha T+\beta T^3.$$

If we use the sound velocity in this result, we note that the term linear in T is identical to that obtained in the model of Debye

$$C_v=\frac{2\pi^2 aNk_B^2}{3hv_s}T \text{ (from Ex. 11c).}$$

The two results differ only in the βT^3 term. This result is easy to understand: The excited phonons at low temperatures correspond to a spectral region (low frequencies) in which the two dispersion curves and therefore, $g(\nu)$ (see Fig. 19) are practically merged.

Exercise 18: Specific heat of a 2D lattice plane

Consider a square lattice plane with parameter a, formed from identical atoms of mass m and submitted to a force constant β between nearest neighbors (as described in Ex. 9).

(a) Find the density of states of transverse modes $g(\nu)$ perpendicular to the plane and the corresponding Debye frequency ν_D. In addition to the isotropy of the dispersion relation: $\omega = v_s k$, the first BZ is limited to a circumference with radius k_D.

(b) In the form of the definite integral, what is the vibration energy of a unitary area? Deduce the behavior of the specific heat C_R at high and low temperatures ($T >> \theta_D$ and $T << \theta_D$ with $k_B\theta_D = h\nu_D$).

Numerical application: What is the numerical value of C_R (for one atom) at 20 K with a = 3 Å and v_s = 3,000 m/s?

(c) The atoms are forced to move in the plane of the lattice. What is the vibration energy of this lattice and the evolution of its specific heat at low and high temperature?

(d) Compare the relative results to $g(\nu)$, ν_D, and C_R obtained in the present exercise with those obtained in the study of a linear lattice (see Ex. 11) and a cubic lattice with constant a (see Ex. 20). Use F2. (h, k_B, e)

Solution:

(a) Using the periodic boundary conditions, we find

$$u_{n,l} = u_{(n+N),l} = u_{n,l+N} \text{ or } \exp\cdot i(\omega t - k_x na - k_y la) =$$

$$\exp\cdot i[\omega t - k_x(n+N)a - k_y\, la] = \exp\cdot i[\omega t - k_x na - k_y(l+N)a],$$

which leads to $k_x, k_y = \pm\dfrac{2\pi}{Na}s \quad (s = 0, 1, 2, 3, \ldots)$.

In the Ok_xk_y plane, the first BZ is a square of side $2\pi/a$ divided in N^2 elementary squares of dimension $(2\pi/Na)$, each of which corresponds to the extremity of a single wave vector composed of k_x and k_y (see Ex. 9, Fig. 14a).

The density of states in k-space can easily be deduced from the number of elementary cells contained in the circular shell of interior radius k and of exterior radius k + dk. We therefore find

$$g(k)\mathrm{d}k = 2\pi k\,\mathrm{d}k\left(\frac{2\pi}{Na}\right)^{-2}$$

This result has already been obtained for the solution of question (d) (see Ex. 9). But, in order to be able to evaluate the specific heat and following the implicit simplifications from the Debye theory for a lattice in 3D (see for instance Ref. [15], p. 212), we assume here

(i) The isotropy of the dispersion relation reduces to $\omega = v_s k$, which results in the fact that the iso-frequency curves are circular.

(ii) The fact that N^2 independent modes are contained in a circumference of radius k_D such that $(\pi k_D^2)\cdot(Na/2\pi^2) = N^2$ implies that the first square first BZ is replaced by a circle of the same surface. This simplifies the exact evaluation of (k) when $\pi/a \le |k| \le \pi\sqrt{2}/a$, as was described in the solution of Ex. 9.

The density of modes $g(\nu)$ obeys the relation $g(\nu)\mathrm{d}\nu = g(k)\mathrm{d}k = kN^2(a^2/2\pi)\mathrm{d}k$

or (taking into account $2\pi\nu = v_s k$): $g(\nu) = (2\pi N^2 a^2/v_s^2)\nu$ with $\nu \le \nu_D = v_s/a\sqrt{\pi}$.

Remark: To evaluate $g(k)$, we should also be able to use fixed boundary conditions: the step variation of k_x and k_y will be π/Na instead of $2\pi/Na$ as in periodic boundary conditions (see Course Summary), but the evaluation of $g(k)$ will be limited to the number of cells contained in the portion of the circular shell k, k + dk, limited to the quadrant k_x, $k_y > 0$ and the final result will be, of course, the same.

(b) The internal energy U due to vibrations of the lattice is $U = \int_0^{\nu_D} g(\nu)\left(\overline{n(\nu)} + \frac{1}{2}\right)h\nu\,\mathrm{d}\nu$ where $\overline{n(\nu)} = \frac{1}{e^{h\nu/k_B T} - 1}$ from which (neglecting the contribution of the half quantum which disappears in the specific heat calculation):

$$U = \frac{2\pi N^2 a^2}{v_s^2}\int_0^{\nu_D} \frac{h\nu^2\,\mathrm{d}\nu}{e^{h\nu/k_B T} - 1}$$

- At high temperatures $k_B T >> h\nu_D$: $e^{h\nu/k_B T} = 1 + \frac{h\nu}{k_B T} + \cdots$

  and $U = \frac{2\pi N^2 a^2}{v_s^2} k_B T \int_0^{\nu_D} \nu d\nu = N^2 k_B T$.

  We find the classical limit which corresponds to the energy of $k_B T$ per linear oscillator.

  By unitary area, the number of oscillators is $1/a^2$, the corresponding energy $k_B T/a^2$ and the specific heat is $C = k_B/a^2$.

- At low temperature $h\nu >> k_B T$, $U = \frac{2\pi N^2 a^2}{v_s^2} \int_0^{\infty} \frac{h\nu^2 d\nu}{e^{h\nu/k_B T} - 1}$,

  using the substitution $\frac{h\nu}{k_B T} = x$

  $$C_r = 14.4 N^2 k_B \left(\frac{T}{\theta_D}\right)^2$$

  we find:

  $$U = \frac{2\pi}{h^2 v_s^2} (k_B T)^3 \int_0^{\infty} \frac{x^2}{e^x - 1} dx = \frac{4.8\pi (k_B T)^3}{h^2 v_s^2} = 4.8 N^2 k_B \frac{T^3}{\theta_D^2}$$

  with $h\nu_D = k_B \theta_D = \frac{h v_s}{a\sqrt{\pi}}$

  Numerically one obtains (with v_s = 3000 m/s and a = 3 Å) θ_D = 270 K and

  $$C_r = 14.4 k_B \left(\frac{T}{\theta}\right)^2 \approx 0.08 k_B \approx 0.067 \times 10^{-4} \text{ eV/K}.$$

(c) If now the atoms of the lattice are forced to move in the plane, each of the oscillators has two degrees of freedom which correspond to the propagation of a longitudinal wave and of a transverse wave. If the force constant is different for these two types of waves (β_l and β_T), we find two velocity and two distinct Debye temperatures θ_{DL} and θ_{DT}. In addition, the density $g(k)$ is identical to that already evaluated in (a), the corresponding internal energy and the specific heat result in the addition of the two contributions for these two types of waves:

- At high temperatures $U = 2N^2 k_B T$ and $C = 2k_B N^2 = \frac{2k_B}{a^2}$

- At low temperatures $U = 4.8N^2 k_B T^3 \left(\frac{1}{\theta_{DL}^2} + \frac{1}{\theta_{DT}^2} \right)$ and

$$C = 14.4N^2 k_B T^3 \left(\frac{1}{\theta_{DL}^2} + \frac{1}{\theta_{DT}^2} \right).$$

The above results apply when the force constants are different: $\beta_l \neq \beta_T$. When the force constant β is isotropic, the result is obtained by taking $\theta_{DL} = \theta_{DT}$.

To facilitate the comparison, we assume isotropy and identity of the velocities for the different types of lattices, which result in slightly different frequencies ν_D and temperatures θ_D. The following table summarizes the results for 1-, 2-, and 3D in complement to Fig. 3 of Course Summary, Section 6. For a more general approach to the Debye model in n dimensions, see A. A. Valladeres, *Am. J. Phys.* **43**, 1975, 308.

Table 1 Density of states and specific heat due to vibrations in lattices of 1-, 2-, and 3D using the Debye model $\omega = v_s k$

		1D	2D	3D
$g(k)$		$\frac{L}{\pi}$(cste)	$\left(\frac{L^2}{2\pi}\right)k$	$\left(\frac{L^3}{6\pi^2}\right)k$
$g(\nu)$		$\frac{2L}{v_s}$(cste)	$\left(\frac{2\pi L^2}{v_s^2}\right)\nu$	$\left(\frac{4\pi}{v_s^3}\right)\nu^2$
ν_D*		$n_l \frac{v_s}{2}$	$v_s\sqrt{\frac{n_s}{\pi}}$	$v_s\left(\frac{3n_v}{4\pi}\right)^{1/3}$
$\Delta U(T)$ (1 at.)	$T << \theta_D$	$1.64k_B \frac{T^2}{\theta_D}$	$4.8k_B \frac{T^3}{\theta_D^2}$	$19.5k_B \frac{T^4}{\theta_D^3}$
	$T > \theta_D$	k_BT	k_BT	k_BT
C_r (1 at.)	$T << \theta_D$	$3.29k_B\left(\frac{T}{\theta_D}\right)$	$14.4k_B\left(\frac{T}{\theta_D}\right)^2$	$77.9k_B\left(\frac{T}{\theta_D}\right)^3$
	$T > \theta_D$	k_B	k_B	k_B

* $n_l = \frac{N}{L}$; $n_s = \frac{N}{L^2}$; $n_v = \frac{N}{L^3}$ are the atomic densities per unit length, unit surface, and unit volume, v_s is the velocity of the waves.

Important Note: Each oscillator is supposed to have a single degree of freedom. For atoms forced to move in the plane, the expressions of the density of states should be multiplied by 2, for spatial oscillators, multiply by 3 while ν_D and θ_D are unchanged.

That is to say, in the 1D of an atomic row the waves can propagate in one direction, parallel to the row but the atomic displacement can be spatial ($\vec{u} \parallel \vec{k}$ and two transverse vibrations $\vec{u} \perp \vec{k}$. In this case one must multiply the terms of the 1D column by 3, with the exception of ν_D. On the other hand, in a 3D object one might be only interested in the displacements along the longitudinal direction ($\vec{u} \parallel \vec{k}$), the terms of the 3D column should not be modified.

Final Remark: To end this discussion, the readers may have a look on the Exs. 11 and 14 of Chapter IV, where the Fermi energy and the electronic specific heat for free electron metals in 1-, 2-, and 3D are explored. It pointed out an analogy between the evaluation of the Fermi energy E_F and the present evaluation of the Debye frequency ν_D. In this way we leave it up to the reader to verify the possibility to obtain ν_D, without using the intermediary of k_D, but from the integral $\int_0^{\nu_D} g(\nu)\mathrm{d}\nu = N$, where N represents the number of basis per unit length, surface, or volume (see also Chapter IV, Ex. 14).

Exercise 19: Phonon density of states in 2- and 3D: evaluation from a general expression

(a) In a 3D solid and starting from the general formula:

$$g(\omega) = \frac{V}{(2\pi)^3} \int_{S\omega} \frac{\mathrm{d}S_\omega}{v_g},$$ find the expression for the density of states $g(\nu)$ of phonons when their dispersion relation is taken to be isotropic:

$$\text{(i)}\ \omega = v_s k \ \text{(Debye)}; \quad \text{(ii)}\ \omega = \omega_m \sin\left|\frac{ka}{2}\right|.$$

(b) Find the general expression corresponding to a 2D lattice and find $g(\nu)$ for the dispersion relations corresponding to (i) and (ii) above.

Solution:

(a) The group velocity $v_g = \dfrac{d\omega}{dk}$ is respectively equal to v_s (condition i) and to $\dfrac{\omega_m a}{2}\cos\left|\dfrac{ka}{2}\right|$ (condition ii). In addition, as the dispersion relations are supposed to be isotropic and the isofrequency surfaces spherical, the integral $n\int_{S_\omega}\dfrac{dS_\omega}{v_g}$ reduces to $\dfrac{S_\omega}{v_g}$. We thus obtain

(i) $g(\omega_0) = \dfrac{V}{(2\pi)^3}\cdot\dfrac{4\pi k_0^2}{v_s} = \dfrac{V}{(2\pi)^3}\cdot\dfrac{4\pi\omega_0^2}{v_s^3}$

or $g(\nu)d\nu = g(\omega)d\omega$, where $\omega = 2\pi\nu$, or $g(\nu_0) = \dfrac{4\pi V}{v_s^3}\nu_0^2$ (for a transverse or longitudinal mode).

(ii) $g(\omega_0) = \dfrac{V}{(2\pi)^3}\cdot\dfrac{4\pi k_0^2}{\omega_m \dfrac{a}{2}\cos(k_0 a/2)} = \dfrac{4V}{\pi^2 a^3 \omega_m}\cdot\dfrac{\left(\arcsin\dfrac{\omega}{\omega_m}\right)^2}{(1-\omega_0^2/\omega_m^2)^{1/2}}$

or $g(\nu_0) = \dfrac{V}{a^3}\cdot\dfrac{[(2/\pi)\arcsin(\nu_0/\nu_m)]^2}{(\nu_m^2-\nu_0^2)^{1/2}}$ (still for a single mode).

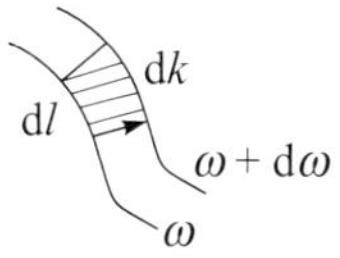

Figure 20

(b) To obtain the general expression for the density of states $g(\nu)$ in 2D, it is $\left(\dfrac{2\pi}{L}\right)^2$ sufficient to follow the same procedure as in 3D and evaluate in the plane of wavevectors (k_x, k_y) the number of cells between two iso-frequency lines $l(\omega)$ and $l(\omega + d\omega)$. An element of this surface is given by

$$\overrightarrow{\mathrm{dS}} = \overrightarrow{\mathrm{dl}_\omega} \wedge \overrightarrow{\mathrm{dk}} = \mathrm{d}l_\omega \cdot \mathrm{d}k_\perp = \frac{\mathrm{d}l_\omega \cdot \mathrm{d}\omega}{\left|\overrightarrow{\nabla_\mathrm{k} \acute{\mathrm{E}}}\right|} = \frac{\mathrm{d}l_\omega \cdot \mathrm{d}\omega}{\left|v_\mathrm{g}\right|}$$

$$\text{or } g(\omega) = \left(\frac{L}{2\pi}\right)^2 \int_{l(\omega=\mathrm{Cst})} \frac{dl_\omega}{v_\mathrm{g}}.$$

If the dispersion relation is isotropic, the isofrequency lines are circumferences and the integral reduces to $\frac{l(\omega)}{v_\mathrm{g}}$. We thus obtain

$$\text{(a)}\quad g(\omega_0) = \left(\frac{L}{2\pi}\right)^2 \cdot \frac{2\pi k_0}{v_\mathrm{s}} = \frac{L^2\omega_0}{2\pi v_\mathrm{s}^2}, \text{ which leads to } g(v_0) = \frac{2\pi L^2 v_0}{v_\mathrm{s}^2}.$$

$$\text{(b)}\quad g(\omega_0) = \left(\frac{L}{2\pi}\right)^2 \cdot \frac{2\pi k_0}{\frac{\omega_\mathrm{m} a}{2} \cos\left(\frac{k_0 a}{2}\right)} = \left(\frac{2L}{a}\right)^2 \cdot \frac{\arcsin(\omega_0/\omega_m)}{2\pi(\omega_m^2 - \omega_0^2)^{1/2}},$$

$$\text{which leads to } g(v) = \left(\frac{2\pi}{a}\right)^2 \cdot \frac{\arcsin(v_0/v_\mathrm{m})}{(v_\mathrm{m}^2 - v_0^2)^{1/2}}.$$

Remark: As the 1D limit is approached, the line reduces to two points (k_0 and $-k_0$) and the general expression for $g(\omega_0)$ becomes

$g(\omega_0) = 2 \cdot \frac{L}{2\pi}\frac{l}{v_g}$ which leads to $g(v) = \frac{2Na}{v_\mathrm{s}}$ in the Debye model and to $g(v) = \frac{2N}{\pi}(v_\mathrm{m}^2 - v_0^2)^{1/2}$ in the model (b).

These results agree with those established in Exs. 11 and 17 and represented in Fig. 19. In the same way the results obtained in the question are identical to those in the table in Ex. 18.

The density of states for free electrons, $g(E)$, can be obtained by adopting the same method (see Chapter IV, Ex. 11).

When the dispersion relation of phonons or electrons has a horizontal tangent (v_g = 0), as in $a\beta$ and $b\beta$, the density of states becomes infinite and the corresponding points (critical points known as Van Hove singularities) play an important role in the properties of solids (see Chapter V, Pb. 6) and the comments following it.

Exercise 20a: Zero point energy and evolution of the phonon population with temperature

Applying the Debye model to a lattice in 3D:

(a) Find the vibration energy at 0 K. Show that this energy is the same order of magnitude as the increase in thermal energy of the same lattice between 0 K and θ_D, respectively $U(0\text{ K})$ and $U(\theta_D) - U(0\text{ K})$.

(b) Find the number of phonons N_p at low and high temperatures ($T \leq \theta_D$ and $T \geq \theta_D$) as a function of T/θ_D.

(c) Sketch the curves $U(T/\theta_D)$ and $C_v(T/\theta_D)$. Explain their characteristics and show the evolution of the average energy, $\bar{E}$, of an oscillator of frequency ν as a function of the reduced variable $x = h\nu/k_BT$ and superpose the two curves representing the density of modes in the Debye model $g(\nu)$ for $T = 2\theta_D$ and $T = \theta_D/4$.

Using $\vec{E} = k_BT$ for x from $0 \leq x < 1$ and $\vec{E} = 0$ for $x \geq 1$, explain the behavior of C_v at high and low temperature and find the low $C_v \propto T^3$ at low temperatures. Use F2.

Solution:

(a) Taking into account the half quantum, the vibration energy of a 3D lattice is

$$U = \int_0^{\nu_m} \left(\frac{1}{2} + \frac{1}{e^{\frac{h\nu}{k_BT}} - 1} \right) h\nu\, g(\nu)\, d\nu$$

In the initial Debye model, the density of modes $g(\nu)$ obeys the relation (which takes into account the three polarizations (1L + 2T):

$$g(\nu) = 3 \times \frac{3N\nu^2}{\nu_D^3},$$

which results in the zero point energy.

$$U(0) = \int_0^{\nu_D} \frac{9Nh\nu^3 d\nu}{2\nu_D^3} = \frac{9}{8} Nh\nu_D = \frac{9}{8} N_{k_B} \theta_D.$$

When $T/\theta_D = 1$, the specific heat of the lattice C_v is slightly less than its asymptotic value $3Nk_B$; see Course Summary, Section 6, Fig. 21a.

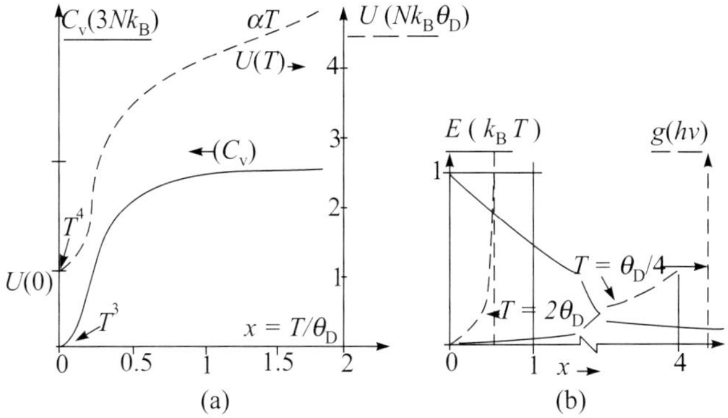

Figure 21

When going from T = 0 K to T= θ_D, the increase of thermal energy is $U(\theta_D)-U(0)\approx 3Nk_B\theta_D$ and it is not significantly far larger than the zero point energy.

(b) The average number of phonons $\bar{n}$ associated with an elastic wave of frequency v is given by:

$$\tilde{n}=\frac{1}{\exp(hv/k_BT)-1}.$$

The total number of phonons N_p included in the density of modes $g(v)$ is $N_p=\int_0^{v_m} g(v)\tilde{n}\,dv$.

In the Debye model and taking into account the two transverse waves and the one longitudinal wave, N_p can be expressed as:

$$N_p=\int_0^{v_D}\frac{9Nv^2}{v_D^3}\cdot\frac{dv}{\exp(hv/k_BT)-1}.$$

- At low temperatures ($T << \theta_D$) this expression becomes:

  $$N_p=9N\left(\frac{T}{\theta_D}\right)^3\int_0^{\infty}\frac{x^2dx}{e^x-1}=21.6N\left(\frac{T}{\theta_D}\right)^3.$$

  Comparing this result to that giving the corresponding specific heat, $C_v=234\,N\,k_B\left(\frac{T}{\theta_D}\right)^3$, we notice $C_v = 11\,k_B\,N_p$.

  That is to say the specific heat is proportional to the number of phonons in this temperature range.

- At high temperatures ($T > \theta_D$), N_p is such that

$$N_p = \frac{9Nk_BT}{h\nu_D^3}\int_0^{\nu_D} \nu d\nu = \frac{9}{2}N\left(\frac{T}{\theta_D}\right).$$

  By comparing this result to the classical expression for specific heat in the same temperature range, $C_v = 3Nk_B$, we note that the number of phonons increases with T while C_v remains constant.

(c) The behavior of $U(T)$ and $C_v(T)$ as a function of T/θ_D is shown in Fig. 21a. To understand these behaviors, one must take into account the evolution of $\bar{E} = \tilde{n}h\nu$ (solid lines in Fig. 21b) and especially the density of modes $g(\nu)$ (dashed line) for $T = 2\theta_D$ and $T = 0.25\theta_D$.

For a 3D structure $g(\nu)$ is parabolic (in ν^2 or in x^2) up to the maximum frequency ν_D (or $x_m = \theta_D/T$). If $\bar{E}$ is reduced to $\bar{E} = k_BT$ for $x \leq 1$, it is clear that the N linear oscillators have increased each by k_BT when going from 0 K to T when $T \geq \theta_D$. On the contrary at low temperatures, only the low-frequency oscillators (hatched area) are able to get such an increase.

Their number $n \propto \int_0^1 x^2 dx$ compared with $N \propto \int_0^{x_m} x^2 dx$ is N/x_m^3

It corresponds to an increase in energy $\Delta U = nk_BT = N(T/\theta)^3 k_BT$ and leads to C_v varying as T^3 ($T << \theta_D$).

The same analysis applies to 1D [$g(\nu)$ = constant] and a 2D [$g(\nu) \propto \nu$] lattices. It allows one to better understand the physical reason for the evolution of C_v in T and T^2 as established previously on a more rigorous basis (see Table 1 in Ex. 18).

Note: In fact the T^3 law is only valid when $T \leq \theta_D/10$. The choice here of $\theta_D/4$ is uniquely related to the convenience of representation.

In summary, when the temperature increases from a low temperature there is first a simultaneous increase in the number of excited oscillators and in the number of phonons. At high temperatures the number of oscillators is saturated ($3N$ linear oscillators) and the behavior is classical, resulting in a constant C_v. Nevertheless, at these elevated temperatures the population of phonons continues to increase because the amplitude (and energy) of oscillations increases with T.

Exercise 20b: Vibration energy at 0 K of 1-, 2-, and 3D lattices (variant of the previous exercise)

Knowing that in a 1-, 2-, or 3D lattices, the density of modes respectively obeys $g(\nu) = A_1$ (constant), $A_2\nu$, or $A_3\nu^2$, find A_1, A_2, and A_3 as a function of the number N of spatial oscillators of each lattice and their Debye frequency ν_D. Express the residual energy of vibration, $U(0)$ (at 0 K), as a function of N and the Debye temperature θ_D.

Compare $U(0)$ to the increase of energy $\Delta U = U(T) - U(0)$ when the temperature goes from 0 K to T K when $T > \theta_D$.

Solution:

$$\int_0^{\nu_D} g(\nu)\,d\nu = 3N, \text{where } A_1 = 3N/\nu_D;\ A_2 = 6N/\nu_D^2;\ A_3 = 9N/\nu_D^3.$$

$$U(0) = \int_0^{\nu_D} g(\nu)(h\nu/2)\,d\nu;\ U_1(0) = A_1 h\nu_D^2/4 = (3/4)Nk_B\theta_D;$$

$$U_2(0) = A_2 h\nu_D^3/6 = Nk_B\theta_D;\ \ U_3(0) = A_3 h\nu_D^4/8 = (9/8)Nk_B\theta_D$$

When $T \geq \theta_D$, $\Delta U = 3Nk_BT$ (law of Dulong and Petit) from which we find

$$U_1(0)/\Delta U = \theta_D/4T; U_2(0)/\Delta U = \theta_D/3T; U_3(0)/\Delta U = 3\theta_D/8T.$$

From an analogous problem concerning free electrons (see Chapter IV, Ex. 14), a direct evaluation of k_D (and therefore θ_D) is possible and a comparison between the residual vibration energy and the kinetic energy of a gas of free electrons (at 0 K) is suggested.

Exercise 21: Average quadratic displacement of atoms as a function of temperature

(a) A longitudinal elastic wave of frequency ν and amplitude A_ν propagates along a row of identical atoms of mass m. The displacement of an atom in this row obeys: $u_\nu = A_\nu \cos 2\pi\nu t$. Find the average quadratic value $\overline{u}_\nu^2$ as a function of A_ν, T, θ_E, and m when all the atoms vibrate thermally at a frequency ν_E ($h\nu_E = k_B\theta_E$) in the high temperature (classical) limit.

(b) In the Debye model we consider that the spatial displacement of N identical atoms in a 3D lattice is isotopic. Find the total average displacement $\overline{r}^2$ in integral form including the amplitude A_ν of each of the $3N$ vibrations and the density of modes $g(\nu)$. Find A_ν as a function of ν starting from the quantized energy of vibration (including the half quantum) and then find $g(\nu)$ as a function of the Debye temperature θ_D. Find in integral form the expression for the square of the average total displacement $\overline{r}^2$ and state its value at 0 K and evolution at high temperatures ($T > \theta_D$).

(c) *Numerical application*:
Find the values of $\overline{r}^2$ and $\sqrt{\overline{r}^2}/d_0$ (d_0 is the distance between nearest neighbors) for sodium and silicon when the two elements are near their melting temperature T_m:

Na :	$\theta_D = 158$ K	$T_m = 370$ K	$d_0 = 3.65$ Å	$M_{Na} = 23$
Si :	$\theta_D = 645$ K	$T_m = 1690$ K	$d_0 = 2.35$ Å	$M_{Si} = 28$

Solution:

(a) For a wave of frequency ν, the average of the quadratic displacement is: $\overline{u_\nu}^2 = \frac{A_\nu^2}{2}$.
At high temperatures the classical limit is reached:

$$\frac{1}{2}m(2\pi\nu_E)^2\overline{u_\nu}^2 = \frac{1}{2}k_B T,$$

which gives $\overline{u_\nu}^2 = \frac{\hbar^2}{4\pi^2 m k_B \theta_E}\left(\frac{T}{\theta_E}\right)$

(b) In the Debye model, all the waves are not of the same frequency and the average total displacement can be expressed by adding the effect of $3N$ linear vibrations: $r^2 = \int_0^{\nu_D} \frac{A_\nu^2}{2} g(\nu) d\nu$,

where $g(\nu) = \frac{3\nu^2 N}{\nu_D^3} \times 3$ (see, e.g., Exs. 18 and 20).

The expression for A_ν can be obtained from two different methods for the evaluation of the total energy E_ν of an elementary wave:

- The energy of a harmonic oscillator of frequency ν and amplitude A_ν is $2\pi^2\nu^2 m A_\nu^2$ and since the N atoms are subject to identical vibrations, the energy of the elementary wave is: $E_\nu = 2\pi^2 N \nu^2 m A_\nu^2$.
- This quantum energy of vibration can also be written for an atom:

$$E = h\nu\left(\frac{1}{2} + \frac{1}{\exp(h\nu/k_B T) - 1}\right)$$

Combining the two methods for A_ν^2 and using the result in the integral expression for $\bar{r}^2$, we find:

$$r^2 = \frac{9h}{4\pi^2 m \nu_D^3}\int_0^{\nu_D}\left(\frac{1}{2} + \frac{1}{\exp(h\nu/k_B T) - 1}\right)\nu d\nu.$$

The first term is independent of temperature and corresponds to the zero point energy (0 K):

$$r^2(0\,\mathrm{K}) = \frac{9h^2}{16\pi^2 m k_B \theta_D}.$$

The second term which results in a temperature evolution of the form: $e\int_0^{x_0}\frac{x dx}{e^x - 1}$. At high temperature $\left(e^{h\nu/k_B T} = 1 + \frac{h\nu}{k_B T} + \cdots\right)$ it goes

to $\bar{r}^2(T^{0\,\mathrm{K}}) = \frac{9h^2}{4\pi^2 m k_B \theta_D}\left(\frac{T}{\theta_D}\right)$ where $h\nu_D = k_B\theta_D$.

Taking into account the isotropy of the oscillations: $\bar{r}^2 = 3\bar{u}^2$, this last result coincides well with the average quadratic value of the linear displacement that was established classically in (a) using the condition $\nu_E = \frac{\nu_D}{\sqrt{3}}$.

(c) *Numerical application*:

For silicon at 1690 K: $\bar{r}^2 = 6.9 \times 10^{-2}\,\text{Å}^2$ and $\frac{\sqrt{\bar{r}^2}}{d_0} = 0.111 \approx \frac{1}{9}$.

For sodium at 370 K: $\bar{r}^2 = 30 \times 10^{-2}\,\text{Å}^2$ and $\frac{\sqrt{\bar{r}^2}}{d_0} = 0.152 \approx \frac{1}{7}$.

We thus have verified the law establishing that a crystal melts when the average displacement attains $d/8$.

This result can explain why helium is liquid, even at 0 K, because at this temperature the oscillation amplitude associated with the half quantum is of order $d/3$ due to its small atomic mass and θ_D (v_s = 240 m/s).

Remark: The calculation of $\bar{r}^2$ in the present exercise allows the explicit evaluation of the influence of temperature on the intensity I of the X-ray diffracted waves: $I = I_0 \exp\left[-\frac{1}{3}\bar{r}^2 \cdot G_{\text{hkl}}^2\right] = I_0 \exp. -2M$, where M is the Debye–Waller factor (see Ref. [15c], p. 85, or Ref. [10], p. 513).

Problems

Problem 1: Absorption in the infrared: Lyddane–Sachs–Teller relation

Consider a row of ions equidistant by a. The even ions, $x = 2na$, have a charge $+e$ and a mass m. The odd ions, $x = (2n + 1)a$, have a charge $-e$ and a mass M, as in Ex. 2.

This row is submitted to the transversal electric field E_z of a sinusoidal EM wave of angular frequency ω propagating along the x-direction of the row: $E_z = E_0 \exp i(\omega t - kx)$.

(1) The angular frequency ω of the EM wave is found in the infrared, that is to say, in the domain of the self-oscillations of the ions for their optical transverse mode ω_T corresponding to

$\omega_T = \left[2\beta\left(\frac{1}{m} + \frac{1}{M}\right)\right]^{1/2} \approx 10^{12} - 10^{13}$ rad/s where β is the force constant between nearest neighbors.

Show that the wavelength of the EM wave λ_e is very large compared to a. Sketch the action of $\vec{E}$ on the ions and show that this wave will excite the transverse vibrational modes of the small wave vectors. Find this result by reasoning in terms of phonons and photons.

(2) Assume that the propagation term of the field is negligible ($k = 0$) and establish the equations of motion of the ions under the action of E_Z with restoring forces limited to nearest neighbors. Deduce the expressions for the elongations: A $(m, +e)$ and B $(M, -e)$, of the ions as a function of ω_t, ω, M, m, and e. Also find the dipole moment along z formed by two consecutive atoms.

(3) The row belongs to a solid containing N pairs of ions per unit volume. Find the expression for the polarization $\vec{P}$ (per unit

volume) taken by the ionic crystal under the action of the electric field, $\vec{E}$, taking into account among other elements, the electronic polarizability α_1 and α_2 of each ion species.

Find the evolution of the relative dielectric constant, $\varepsilon_r(\omega)$, of the ionic solid as a function of ω. Determine the limiting values of ε_r when $\omega = 0$ and $\omega = \infty$, respectively ε_0 and ε_∞. Show that $\varepsilon_r(\omega)$ can be written in the form:

$$\varepsilon_r(\omega) = \varepsilon_\infty + (\varepsilon_s - \varepsilon_\infty)/[1-(\omega/\omega_T)^2].$$

(4) When the self-vibrations of the ions are longitudinal (optical vibrations of pulse ω_L and $k << \pi/a$) and in the absence of external forces, a longitudinal electric field is induced by the ionic motion. Starting from the Maxwell–Gauss equation, show that the dielectric constant must be zero at this frequency. Deduce the Lyddane–Sachs–Teller (LST) relation:

$$(\omega_T/\omega_L)^2 = \varepsilon_\infty/\varepsilon_s$$

(5) *Numerical application*: For NaCl, $\varepsilon_s = 5.62$; $\varepsilon_\infty = n^2 = 2.25$ (where n is the index of refraction in the visible spectrum); in addition $\varepsilon_r = 1$ for $\lambda_r = 31$ µm.

Find ω_L and ω_T and compare the results with the corresponding measurements: $\lambda_L = 37.7$ µm, and $\lambda_T = 61.1$ µm.

Sketch the graph of $\varepsilon_r(\omega)$ for NaCl. Comment on the sign of ε_r and the optical properties in the interval $\omega_T < \omega < \omega_L$.

Solution:

(1) $\lambda_e = \frac{c}{v} = 2\pi c/\omega$. When $\omega \approx 10^{12} - 10^{13}$ rad/s, $\lambda_e \approx 20$–200 µm, whereas $a \approx 3$–5 Å. The wavelength of the EM wave is very large compared to a; the wave vector of this wave is very small compared to the dimensions of the first BZ. This electric field will excite transversal ionic vibrations (forced oscillations) of small wave vectors k with respect to π/a.

In terms of particles, one may consider that the photons of the EM wave are absorbed and create optical phonons of the same energy and the same wave vector (conservation of energy and momentum):

$$h\nu \text{ (EM wave)} = h\nu \text{ (phonons)}$$
$$\vec{k} \text{ (EM wave)} = \vec{k} \text{ (phonons)}.$$

The optical phonons thus excited will be transversal and situated at the center of the BZ. Figure 22 shows the mechanism of this excitation where, under the action of the instantaneous electric field, the consecutive ions of opposite sign are subject to the opposite forces $F = \pm qE$, which naturally induces the excitation of transverse optical waves $\vec{u} \perp \vec{k}$, with long wavelength ($\lambda >> a$).

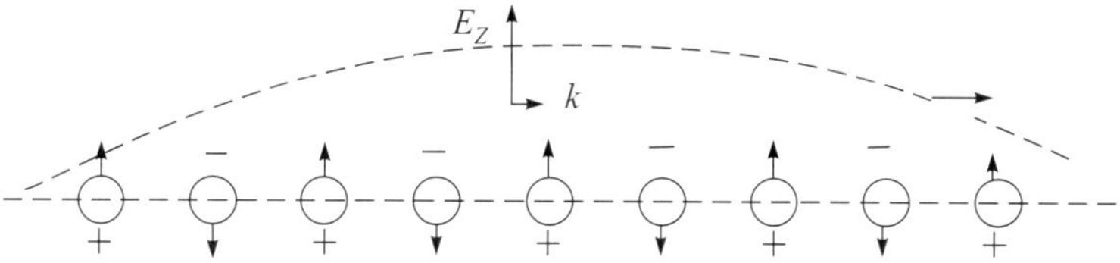

Figure 22

(2) The equations of motion of the ions are

$$m\ddot{u}2_n = \beta(u_{2n+1} + u_{2n-1} - 2u_{2n}) + \mathrm{e}E_0 e^{i\omega t}$$

$$M\ddot{u}2_{n+1} = \beta(u_{2n} + u_{2n+2} - 2u_{2n+1}) - eE_0 e^{i\omega t}$$

We look for solutions in the form of sine waves:

$u_{2n} = u_{2n+2} = Ae^{i\omega t}$; $u_{2n-1} = u_{2n+1} = Be^{i\omega t}$ in which the phase difference is negligible. Then

$$-m\omega^2 A - 2\beta(B - A) = +eE_0$$

$$-M\omega^2 B - 2\beta(A - B) = -eE_0$$

From which we find

$$A = \frac{eE_0/m}{\omega_\mathrm{T}^2 - \omega^2}; \quad B = \frac{-eE_0/M}{\omega_\mathrm{T}^2 - \omega^2}$$

$$P_\mathrm{z} = \frac{e^2}{\omega_\mathrm{T}^2 - \omega^2}\frac{E_0}{\mu}, \text{ where } \frac{1}{\mu} = \frac{1}{M} + \frac{1}{m}$$

(3) $$|\vec{\mathrm{P}}| = N\varepsilon_0\left(\alpha_1 + \alpha_2 + \frac{e^2/\xi_0}{\mu(\omega_\mathrm{T}^2 - \omega^2)}\right)E_0$$

$$\vec{\mathrm{D}} = \varepsilon_0\vec{\mathrm{E}}_0 + \vec{\mathrm{P}} = \varepsilon_0\varepsilon_\mathrm{r}\vec{\mathrm{E}}_0$$

$$\varepsilon_\mathrm{r} - 1 = N\left(\alpha_1 + \alpha_2 + \frac{e^2/\xi_0}{\mu(\omega_\mathrm{T}^2 - \omega^2)}\right)$$

When $\omega = 0$, we find $\varepsilon_\mathrm{s} - 1 = N\left(\alpha_1 + \alpha_2 + \dfrac{e^2/\xi_0}{\mu\omega_\mathrm{T}^2}\right)$

When $\omega = \infty$, we find $\varepsilon_\infty - 1 = N(\alpha_1 + \alpha_2)$

Combining these expressions the result is

$$\varepsilon_r(\omega) = \varepsilon_\infty + (\varepsilon_s - \varepsilon_\infty)/\left[1 - \frac{\omega^2}{\omega_T^2}\right]$$

(4) $\mathrm{div}\vec{D} = \rho = 0; \mathrm{i}\vec{k} \cdot \vec{D} = 0$ from which we have $\varepsilon_r(\omega_L) = 0$

Even though the displacement vector D is zero, the longitudinal electric field is not as well since $\varepsilon_0 \vec{E} + \vec{P} = 0$.

By substituting ω_L for ω and writing $\varepsilon_r(\omega_L) = 0$ in (3), we find the LST equation: $(\omega_T/\omega_L)^2 = \varepsilon_\infty / \varepsilon_s$

(5) The same procedure applied to ω_r with $\varepsilon(\omega_r) = 1$ leads to $\omega_r^2/\omega_T^2 = (\varepsilon_s - 1)/(\varepsilon_\infty - 1)$; $\lambda_T = 1.92\ \lambda_r = 59.6$ µm; $\lambda_L = 0.63\ \lambda_r = 37.5$ µm.

Note the excellent agreement with the experimental values: 61.1 for λ_T and 37.7 for λ_L.

Remarks: In the interval $\omega_T < \omega < \omega_L$, ε_r is negative and the optical index is purely imaginary. The amplitude of the reflection coefficient at normal incidence, $r = (1 - n)/(1 + n)$ has a unitary modulus. This corresponds to total reflection and an evanescent wave.

The angular frequency ω_r is easily obtained experimentally because it corresponds to a zero reflection coefficient $\varepsilon_r = n = 1$ which leads to $r(\omega_r) = 0$ (see Fig. 23).

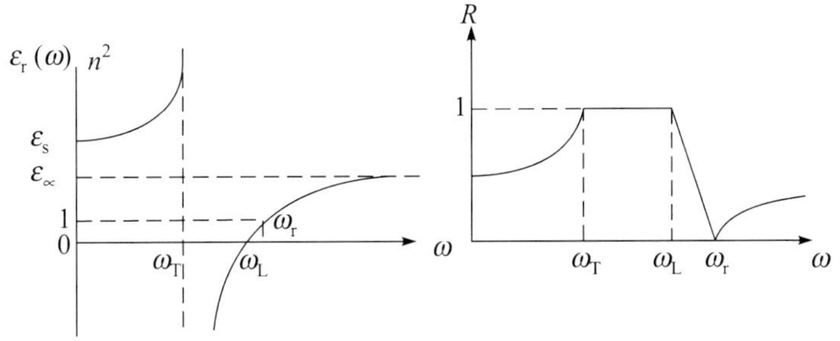

Figure 23

$\omega = 0$ (question 3) corresponds to the application of an electrostatic field where the elongation of ions is fairly independent of their mass ($\omega = 0$ in the equations of motion).

Below the Curie temperature, T_c of ferroelectric materials, such as $BaTiO_3$, they are paraelectric with a dielectric constant, ε_s, which tends toward infinity when cooling toward T_c. This phenomena is known as the dielectric catastrophe and can be explained by the fact that the force constant β_t is softened because the ions have the tendency to find another stable position of equilibrium which results in $\omega_r \to 0$ for $T \to T_c$ and therefore via the LST relation $\varepsilon_s \to \infty$.

When the frequency is very high, the inertia of the ions prevents them from following the rapid variations of the field. In reality ε_s is measured in the visible part of the EM spectrum, a region that is below to the frequencies in the UV range where electronic interband transitions may occur as well-collective (plasmon) excitations of valence electrons of the crystal ($\omega_\infty < \omega_{gap} < \omega_p$).

The behavior of these ionic crystals in UV range and more generally that of band gap materials, forbidden band E_g, is strongly dependent on the behavior of valence electrons being excited from the valence band to the conduction band with an average energy $\hbar\omega_T$ where $\hbar\overline{\omega_t} > E_g$ (see Chapter V, Ex. 25 and Pb. 6). The evolution of the dielectric constant in the corresponding spectral domain, far UV, shows the same characteristics as that encountered here by the motion of ions with a resonance around ω_T (in the UV and not in the IR). This evolution presents a 0 of $\varepsilon_l(\omega)$ to a pulsation ω_p' for which the longitudinal electric field is always susceptible to propagate.

The corresponding excitations are known as "plasmons" (here polaritons; see following problem) and the evolution of $\varepsilon_l(\omega)$ is also given by

$$\varepsilon_r(\omega) = 1 + \frac{\omega_p^2}{\overline{\omega_T}^2 - \omega^2}$$

in which ω_p, plasmon angular frequency, must take into account the density and mass of the valence electrons (in the place of the density and reduced mass of ions, which explains the displacement from the IR to the UV range: $m_0 << \mu$). We thus find another spectral domain of total reflection (between $\overline{\omega_t}$ and ω_p') characterized also by evanescent waves because ε_1 is negative.

If we take into account that the propagation term of the EM wave ($k \neq 0$), we obtain a dispersion relation $\omega = f(k)$ of the EM waves in ionic crystals (see following Pb. 2, Polaritons) that has characteristics analogous to those of the an EM wave in a plasma (plasmons; see Chapter IV, Pbs. 5 and 6).

Finally, note the identification of the local field to the applied field at such high frequencies (Q. 1 to Q. 3).

Problem 2: Polaritons

In an ionic crystal, the evolution of the relative dielectric constant, $\varepsilon_r(\omega)$, as a function of the angular frequency of the electric field of an EM wave, ω, can be described by the expression (see proceeding problem):

$$\varepsilon_r(\omega) = \varepsilon_\infty + (\varepsilon_s - \varepsilon_\infty)/1 - \left(\frac{\omega}{\omega_T}\right)^2$$

in which ε_s and ε_∞ are the limiting values taken by ε_r when respectively $\omega = 0$ and $\omega = \infty$; ω_T is the pulsation of transverse optical phonons relative to the very small wave vectors ($<<\pi/a$).

Consider the propagation along x of an EM sinusoidal plane wave where the transverse electric field is $E_z = E_0 \exp i(\omega t - kx)$.

(1) Starting from the propagation equation of such an EM wave in a medium characterized by $\varepsilon_0\varepsilon_r$ and μ_0, show that the relation between ω and k (dispersion relation of polaritons) is of the form: $\varepsilon_\infty \omega^4 - \omega^2(\varepsilon_\infty \omega_L^2 + k^2c^2) + k^2c^2\omega_T^2 = 0$ in which $c^2 = (\varepsilon_0\mu_0)^{-1}$ and ω_L^2 is defined by the LST relation: $(\omega_T/\omega_L)^2 = \varepsilon_\infty/\varepsilon_s$.

(2) Find the solutions in $\omega^2 = f(k)$ and discuss the limiting values for small and large k.

(3) Find the characteristics of the dispersion curves $\omega = f(k)$, knowing that $\varepsilon_\infty < \varepsilon_s$.

Solution:

(1) The propagation equation of E_z is

$$\frac{\partial^2 E_z}{\partial x^2} - \varepsilon_0\varepsilon_r\mu_0 \frac{\partial^2 E_z}{\partial t^2} = 0 \text{ or } \frac{\partial^2 E_z}{\partial x^2} = -k^2 E_z \text{ and } \frac{\partial^2 E_z}{\partial t^2} = -\omega^2 E_z$$

from which we find

$$-k^2 + \frac{\omega^2}{c^2}\left[\varepsilon_\infty + (\varepsilon_s - \varepsilon_\infty)/(1 - \omega^2/\omega_T^2)\right] = 0$$

Taking into account the LST relation and rearranging the terms we obtain

$$\varepsilon_\infty \omega^4 - \omega^2(\varepsilon_\infty \omega_L^2 + k^2c^2) + k^2c^2\omega_T^2 = 0 \quad (1)$$

(2) The solution above (Eq. 1) can be written as

$$2\varepsilon_\infty \omega_{1,2}^2 = \omega_T^2\varepsilon_s + c^2k^2 \pm \sqrt{(\omega_T^2\varepsilon_s + c^2k^2)^2 - 4\omega_T^2k^2c^2\varepsilon_\infty} \quad (2)$$

When k is small, these two solutions of (Eq. 2) become

$$\omega_2^2(k) = \omega_L^2 + \frac{c^2k^2}{\varepsilon_\infty} \text{ and } \omega_1^2(k) = \frac{c^2k^2}{\varepsilon_s}$$

When k is large such that $(k >> \frac{\omega}{c}\sqrt{\varepsilon_s})$ but nevertheless such as $ka << 1$, we have $\omega_2^2(k) = c^2k^2/\varepsilon_\infty$ and $\omega_1^2 = \omega_T^2$

There are thus two branches of elementary excitations:
$0 < \omega_1(k) < \omega_T$ and $\omega_L < \omega_2(k) < \infty$.

(3) The characteristics of the dispersion curves are shown in Fig. 24.

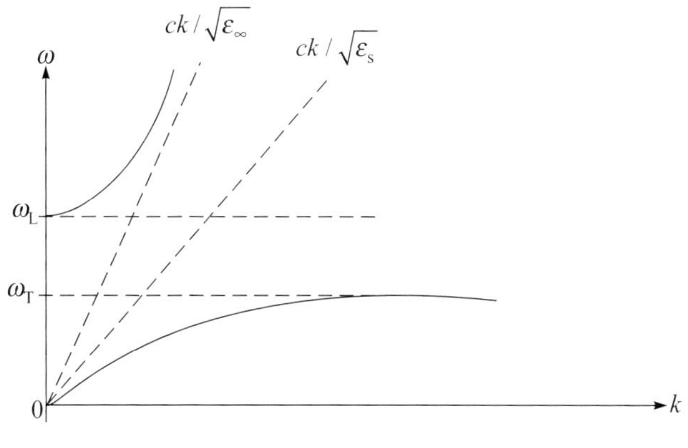

Figure 24

This exercise is an extension of the preceding one (we find the interval of forbidden frequencies between ω_T and ω_L). It explains the coupling between the photons of the EM wave and the vibration phonons of the lattice: elementary excitations known as “polaritons.”

On resonance, it is this coupling which modify the nature of the EM propagation when the two wave vectors of two different particles are approximately equal.

Problem 3: Longitudinal and transverse phonon dispersion in CsCl

The crystalline structure of cesium chloride is a simple cubic with side a and basis with a Cl^- at (1/2, 1/2, 1/2) and a Cs^+ ion at (0, 0, 0) (see Chapter I, Ex. 1, Fig. 3a). The repulsive energy between the two nearest neighbors is of the form $u_2 = \lambda e^{-r_0/\rho} = \frac{\rho}{8}\frac{q^2}{4\pi\varepsilon_0}\frac{\alpha}{r_0^2}$ where α is the Madelung constant and r_0 is the distance between two nearest neighbors in equilibrium (see Chapter II, Pb. 2).

(a) In the Hooke's approximation, find the resultant forces (both the attractive Coulomb force and the repulsive force) which is exerted on an atom in the (100) plane when, starting from their equilibrium position, all the atoms of this plane are subject to a translation u_n in the [100] direction, while the nearest neighbors (100) planes are subject to a translation of u_{n-1} and u_{n+1} in the same direction (see Fig. 25). Show that the main action result arises from the repulsive forces and can be written in the form: $F = \beta_L(u_{n+1} + u_{n-1} - 2u_n)$.

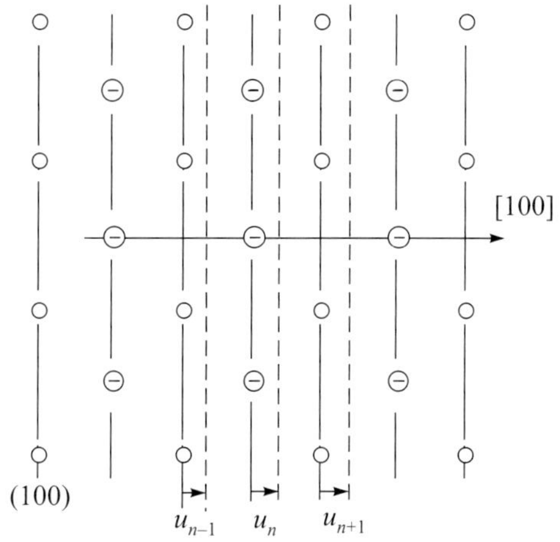

Figure 25

By limiting the problem to nearest neighbors, find the expression and the numerical value of the constant β_L (100) for the longitudinal waves.

(b) Starting from the equations of motion, find the dispersion relations of longitudinal phonons in the [100] direction. Specify the literal value and then give the numerical value of the optical longitudinal frequency ω_L in the center of the BZ (at $k = 0$). The numerical data for CsCl are

$r_0 = 3.57 \times 10^{-10}$ m; $r_0/\rho = 18$; $\alpha = 1.7627$ [for m(Cl) and m(Cs), see Table III at the beginning of the book.]

(c) Starting from the LST formula (given in Pbs. 1 and 2), find the expression and the numerical value of the transverse optical frequencies ω_T at $k = 0$ and compare the results to the experimental value $\omega_T = 1.85 \times 10^{13}$ rad/s (minimum of transmission of infrared light at normal incidence). Take $\varepsilon_r(0) = 7$ and $\varepsilon_r(\infty) = 2.62$.

(d) Find the expression and the numerical value of the constant β_T [100]. Compare the value of the velocity of transverse waves along the [100] axis as obtained using β_T [100] with that found using the elastic constant C_{44} ($C_{44} = 0.8 \times 10^{10}$ N/m^2).

(e) After having calculated the characteristic values for ω when k is in contact to the first BZ, sketch the dispersion curves of the longitudinal and transverse waves in the [100] direction. (These can be compared with those obtained by G. Mahler and P. Engelhardt, *Physica Status Solidi* (b) **45** (1973), 453—the article which motivated this exercise.)

Solution:

(a) When the atomic planes are displaced longitudinally, the coordinates of the central ion Cs$^+$ at M (and the notations adapted throughout this chapter) are $\frac{u_n}{a},0,0$ and those of nearest neighbors are: $\left(-\frac{1}{2}+\frac{u_n-1}{a},\pm\frac{1}{2},\pm\frac{1}{2}\right)$ and $\left(\pm\frac{1}{2}+\frac{u_n+1}{a},\pm\frac{1}{2},\pm\frac{1}{2}\right)$. The distance r_{MP} which separates the ion Cs$^+$ at M and the ion Cl$^-$ at P $\left(-\frac{1}{2}+\frac{u_n+1}{a},+\frac{1}{2},+\frac{1}{2}\right)$ is

$$r_{\mathrm{MP}} = \left[\left(-\frac{a}{2} + u_{n-1} - u_n \right)^2 + \frac{a^2}{4} + \frac{a^2}{4} \right]^{1/2}.$$

By denoting r_0 as the equilibrium distance between two nearest neighbors $\left(r_0 = \frac{a\sqrt{3}}{2} \right)$ and limiting the expansions do the terms in $\left(\frac{u_{n-1} - u_n}{a} \right)$ (Hooke's approximation), we find:

$$r_{\mathrm{MP}} \approx r_0 \left(1 + \frac{1}{\sqrt{3}} \frac{u_n - u_{n-1}}{r_0} \right).$$

The Coulomb force exerted on an ions at M and P is

$\vec{F}_e = \frac{q^2}{4\pi\varepsilon_0} \frac{\vec{u}_r}{r_{\mathrm{MP}}^2}$, which has a perpendicular component in the (100) plane

$$F_x \cong \frac{-q^2}{4\pi\varepsilon_0} \frac{a}{2 \cdot r_0^3} \left(1 + 0 \frac{u_n - u_{n-1}}{r_0} \right).$$

The resulting Coulomb forces that are exerted on the Cs^+ ion are zero (in terms up to u^3/r^5).

The repulsive forces originate from a smaller energy but for a much larger gradient and they contribute mostly to the restoring force. The contribution along O_x [100] of the repulsive force exerted by the ion at P on the ion M is

$F_{\mathrm{rx}} = \frac{\lambda}{\rho} e^{-r_{\mathrm{PM}}/\rho} \cdot \frac{x_{\mathrm{M}} - x_{\mathrm{P}}}{r_{\mathrm{PM}}}$, which gives

$$F_{\mathrm{rx}} \approx \frac{\lambda}{\rho} e^{-r_0/\rho} \cdot \left(1 - \frac{u_n - u_{n-1}}{\rho\sqrt{3}} \right) \left(u_n - u_{n-1} + \frac{a}{2} \right) \left(1 - \frac{1}{\sqrt{3}} \frac{u_n - u_{n-1}}{r_0} \right)$$

$$\approx \frac{\lambda}{\rho r_0} \cdot e^{-r_0/\rho} \cdot \left[\left(\frac{a}{2} + \frac{1}{3} \left(2 - \frac{r_0}{\rho} \right) (u_n - u_{n-1}) \right) \right].$$

The resulting repulsive forces exerted on the eight nearest neighbors will be

$$\sum F_{\mathrm{rx}} = \frac{4\lambda}{3\rho r_0} e^{-r_0/\rho} \cdot \left(\frac{r_0}{\rho} - 2 \right) (u_{n-1} + u_{n+1} - 2u_n)$$

They take the form $\sum F = \beta_L (u_{n+1} + u_{n-1} - 2u_n)$

with $\beta_L(100) = \dfrac{4\lambda}{3\rho r_0} e^{-r_0/\rho} \cdot \left(\dfrac{r_0}{\rho} - 2\right)$.

By replacing λ by its equivalent and substituting real values we find

$$\beta_L(100) = \frac{q^2}{4\pi\varepsilon_0} \frac{\alpha}{6r_0^3}\left(\frac{r_0}{\rho} - 2\right) = 23.9 \text{ N/m}.$$

(b) The equations of motion are

$$m_{Cs}\frac{d^2u_n}{dt^2} = \beta(u_{n+1} + u_{n-1} - 2u_n),\ m_{Cl}\frac{d^2u_{n+1}}{dt^2} = \beta(u_{n+1} + u_n - 2u_{n+1}).$$

The solutions are of the form $u_n = A\exp i(\omega t - \dfrac{kna}{2})$ and $u_{n+1} = B\exp i(\omega t - k\dfrac{n+1}{2}a)$.

We thus find a system of linear and homogenous equations with the two unknowns A and B (see Course Summary and preceding problems), which lead to non-zero solutions when the values of ω satisfy:

$$m_{Cl}m_{Cs}\omega^4 - 2\beta(m_{Cl} + m_{Cs})\omega^2 + 4\beta^2 \sin^2\frac{ka}{2} = 0.$$

For small wave vectors, $k << \pi/a$, the two roots of this equation are

$$\omega_{LO}^2 = 2\beta_L\left(\frac{1}{m_{Cl}} + \frac{1}{m_{Cs}}\right) \text{ (optical branch)}$$

$$\omega_{LA}^2 = \frac{\beta_L}{2(m_{Cl} + 2m_{Cs})} \cdot k^2a^2 \text{ (acoustical branch)}$$

$\omega_{LO} = 3.2 \times 10^{13}$ rad/s

(c) The LST formula is $\left(\dfrac{\omega_T}{\omega_L}\right)^2 = \dfrac{\varepsilon(\infty)}{\varepsilon(0)}$ which leads to the relation

$\omega_{TO}^2 = 2\beta_L \dfrac{\varepsilon(\infty)}{\varepsilon(0)}\left(\dfrac{1}{m_{Cl}} + \dfrac{1}{m_{Cs}}\right)$ and $\omega_T = 1.96 \times 10^{13}$ rad/s.

(d) The transverse vibrations obey to identical equations of motion, with β_T substituted for β_L, to those obtained in (b).

Then $\omega_{\mathrm{T}}^2 = 2\beta_{\mathrm{T}}\left(\dfrac{1}{m_{\mathrm{Cl}}} + \dfrac{1}{m_{\mathrm{Cs}}}\right)$ for $\omega_{\mathrm{T}}(k=0)$. Taking into account the LST relation it results $\beta_{\mathrm{T}} = \dfrac{\varepsilon(\infty)}{\varepsilon(0)} \cdot \beta_{\mathrm{L}} = 8.95\ \mathrm{N/m}$.

The numerical value of the corresponding elastic wave velocity will be

$$v_{\mathrm{T}}[100] = \left[\frac{\beta_{\mathrm{T}}}{2(m_{\mathrm{Cl}} + m_{\mathrm{Cs}})}\right]^{1/2} a = 1650\ \mathrm{m/s},$$

which is slightly larger than that deduced directly from the experimental value of C_{44}: $v_{\mathrm{T}}[100] = \left(\dfrac{C}{d}44\right)^{1/2} = 1414\ \mathrm{m/s}$ (see Chapter II, Pb. 2, Q. f).

Remark: We would have obtained v_{T} = 1560 ms if, when evaluating β_{T}, we had kept the experimental value of ω_{T} instead of the calculated value.

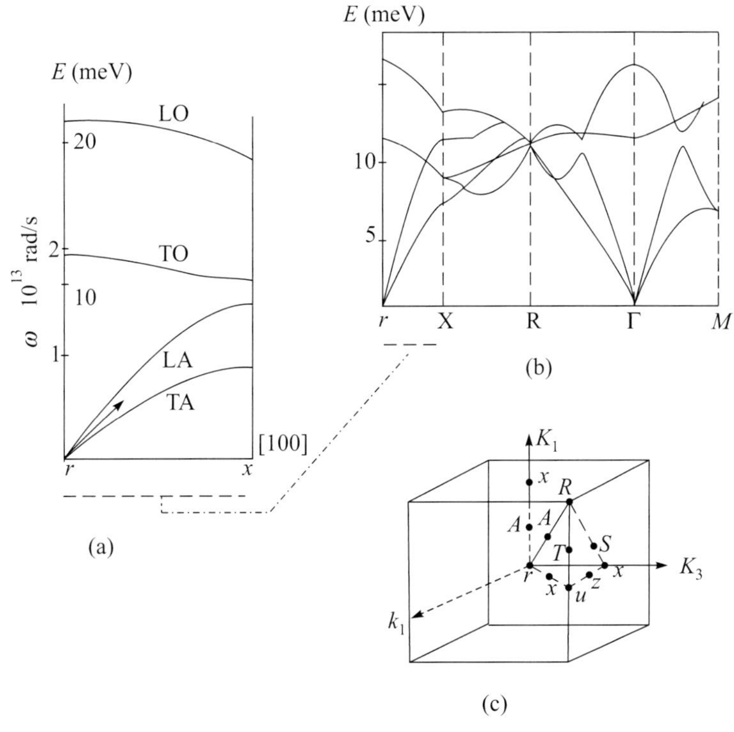

Figure 26

(e) The characteristic values of ω at the limit of the zone ($k = \pi/a$) are the following:

$$\omega_{\mathrm{LA}} = (2\beta_{\mathrm{L}}/m_{\mathrm{Cs}})^{1/2} = 1.47\times10^{13}\,\mathrm{rad/s};\quad \omega_{\mathrm{TA}} = (2\beta_{\mathrm{T}}/m_{\mathrm{Cs}})^{1/2} = 0.9\times10^{13}\,\mathrm{rad/s}$$

$$\omega_{\mathrm{LO}} = (2\beta_{\mathrm{L}}/m_{\mathrm{Cl}})^{1/2} = 2.85\times10^{13}\,\mathrm{rad/s};\quad \omega_{\mathrm{TO}} = (2\beta_{\mathrm{T}}/m_{\mathrm{Cl}})^{1/2} = 1.74\times10^{13}\,\mathrm{rad/s}$$

The dispersion curves in the [100] direction, such as deduced from the present exercise are shown in Fig. 26a and can be compared to those in the article of Mahler and Engelhardt (Fig. 26b) by limiting the comparison to the ΓX direction (the other curves are relative to the dispersion of phonons along other crystallographic directions: the points Γ, X, R, and M represent respectively the center of the first BZ and the intersections with the [100] (X), [111] (R), and [110] (M) axes (see Fig. 26c).

Qualitatively the only notable difference concerns the crossing of the LA and TO branches, which does not appear in Fig. 26a.

Comment

This simplified exercise illustrates the procedure followed in crystal dynamics to find the characteristics of the phonon dispersion curves in a partly or totally ionic crystal. To obtain a better precision of these curves, one can take into account the second and third nearest neighbors and substituted the model of a rigid ion considered here for the model of a rigid shell (in which the movement of the nucleus and inner electrons is disassociated with that of its outer electron to take into account the polarizability of the ions), or even better the model of a deformable shells which takes into account the movement of the nucleus + inner electrons and the shell, in which symmetry is not conserved. Figure 27 illustrates schematically these different models.

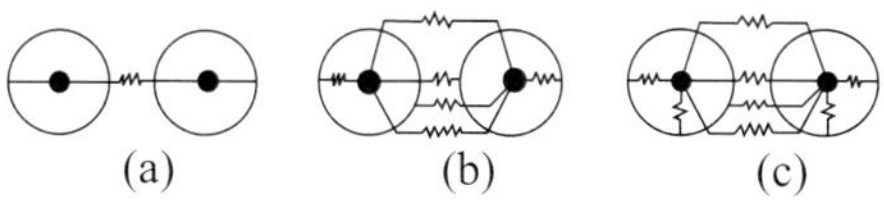

Figure 27 Illustration of various models: (a) rigid ions, (b) rigid shells, (c) deformable shells.

The coefficients of the dynamic matrix (the matrix of the coefficients of the equations of motion and therefore containing the values corresponding to the normal modes of vibration) are then calculated numerically from experimental data relative to U_0, r_0, C_{ij}, ω_T, ε_0, $\varepsilon(\infty)$, etc. See the article mentioned previously and more generally that of A. Maradudin, *Solid State Physics*, Sup 3 (1963).

Problem 4: Improvement of the Debye model: determination of θ_D from elastic constants application to lithium

One can improve the initial Debye theory for the specific heat by taking into account that the velocity of the transverse waves is distinct from that of longitudinal waves ($v_t \neq v_l$) in the relation $\omega = vk$, while still assuming that the crystal is an isotropic solid and assuming that the first BZ is a sphere.

(a) Find the maximal frequency of longitudinal, ν_l, and transverse, ν_t waves. Graph the frequency spectrum $g(\nu)$ when $\nu_l = 1.8\ \nu_t$.

(b) Find the expression for the Debye temperature θ_D as a function of θ_L and θ_T (where $h\nu_l = k_B\theta_L$ and $h\nu_t = k_B\theta_T$) from the relation, at low temperatures for the specific heat, $C_v(T)$.

(c) Find the numerical Debye temperature θ_D of lithium from the elastic coefficients along the [100] axis of the crystal.
For lithium at 300 K, use $C_{11} = 1.35 \times 10^{10}$ N/m^2, $C_{44} = 0.88 \times 10^{10}$ N/m^2, ρ (volumetric mass) $= 0.542 \times 10^3$ kg/m^3, and $N/V = 4.7 \times 10^{28}$ at m^{-3}. Use F3.

Solution:

(a) In the Debye approximation, the density of states in vector space, $g(k)$, corresponds, for each polarization, to

$$g(k)\mathrm{d}k = \frac{4\pi k^2\,\mathrm{d}k}{\left(\frac{2\pi}{L}\right)^3}$$

Taking into account $2\pi\nu = u_l k$ and $2\pi\nu = u_t k$, we find

$$g_l(\nu)\mathrm{d}\nu = \frac{V\cdot 4\pi\nu^2\,\mathrm{d}\nu}{u_r^3}$$ for the longitudinal polarization and

$$g_t(\nu)\mathrm{d}\nu = \frac{V\cdot 4\pi\nu^2\,\mathrm{d}\nu}{u_t^3}$$ for the two transverse polarization.

The frequency limits ν_l and ν_t can both be evaluated starting from

$$\int_0^{\nu_l} g_l(\nu)\,d\nu = N \text{ and } \int_0^{\nu_t} g_t(\nu)\,d\nu = N.$$

Or equivalently using $\frac{4\pi}{3}\frac{k_m^3}{\left(\frac{2\pi}{L}\right)^3} = N$, where $2\pi\nu_t = u_l k_m$

(or $2\pi\nu_t = u_l k_m$)

This results in $\nu_l = u_l\left(\frac{3N}{4\pi V}\right)^{\frac{1}{3}}, \nu_t = u_t\left(\frac{3N}{4\pi V}\right)^{\frac{1}{3}}$

The resulting density of states $g(\nu)$ is the sum

$$g(\nu) = 6N\frac{\nu^2}{\nu_t^3} + 3N\frac{\nu^2}{\nu_l^3} = 9N\frac{\nu^2}{\nu_D^3} \text{ for } 0 < \nu < \nu_g$$

and it reduces to $g(\nu) = 3N\frac{\nu^2}{\nu_l^3}$ for $\nu_t < \nu < \nu_l$.

(By anticipation, we have imposed $\frac{2}{\nu_t^3} + \frac{1}{\nu_l^3} = \frac{3}{\nu_D^3}$, see below).

Figure 28a shows this frequency spectrum when $\nu_l = 1.8\nu_t$ [see Remark (a) at the end of the problem].

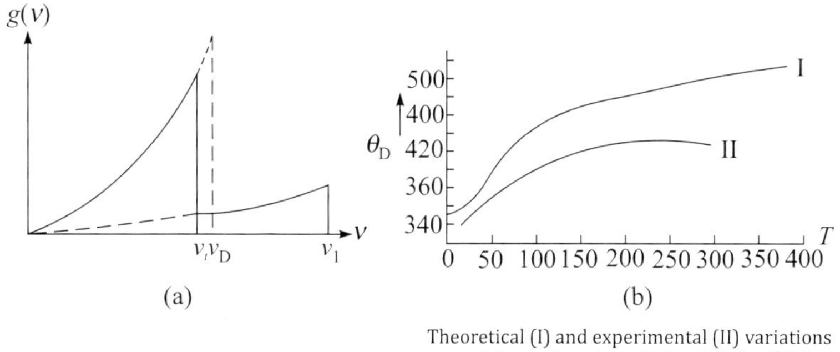

Theoretical (I) and experimental (II) variations of θ_D as a function of T for Li (after [25]).

Figure 28

(b) Neglecting the half quantum, the internal energy can be written

$$U(T) = \int_0^{\nu_l} g_l(\nu)\cdot\frac{h\nu}{\exp\left(\frac{h\nu}{k_B T}\right) - 1}d\nu + \int_0^{\nu_t} 2g_t(\nu)\frac{h\nu}{\exp\left(\frac{h\nu}{k_B T}\right) - 1}d\nu$$

Substituting $x = \dfrac{h\nu}{k_B T}$, $h\nu_l = k_B\theta_L$ and $h\nu_t = k_B\theta_T$, we find

$$U(T) = 3N\left(\frac{T}{\theta_L}\right)^3 k_B T\int_0^{x_l}\frac{x^3dx}{e^x - 1} + 6N\left(\frac{T}{\theta_T}\right)^3 k_B T\int_0^{x_t}\frac{x^3dx}{e^x - 1}.$$

At low temperatures $k_B T << h\nu_l$, t, which gives x_l, $x_t \to \infty$ and leads to

$$U(T) = \frac{\pi^4}{5}Nk_B T\left[\left(\frac{T}{\theta_L}\right)^3 + 2\left(\frac{T}{\theta_T}\right)^3\right] \text{ and}$$

$$C_v = \frac{4\pi^4 Nk_B}{5}\left[\left(\frac{T}{\theta_L}\right)^3 + 2\left(\frac{T}{\theta_T}\right)^3\right]$$

Identifying this result with that obtained starting from the simplified Debye theory where there is no difference between u_l and u_T (and between ν_l and ν_t, and θ_L and θ_T)

$$C_v = \frac{12\pi^4 Nk_B}{5}\left(\frac{T}{\theta_D}\right)^3 = 234Nk_B\left(\frac{T}{\theta_D}\right)^3$$

We find $\dfrac{3}{\theta_D^3} = \dfrac{1}{\theta_L^3} + \dfrac{2}{\theta_T^3}$, a result that can be deduced directly from $\dfrac{3}{\nu_D^3} = \dfrac{1}{\nu_l^3} + \dfrac{2}{\nu_t^3}$.

(c) As a function of elastic constants, the velocity of longitudinal and transverse waves is respectively

$$u_l = \left(\frac{C_{11}}{\rho}\right)^{\frac{1}{2}} = 5000 \text{ m/s and } u_t = \left(\frac{C_{44}}{\rho}\right)^{\frac{1}{2}} = 4030 \text{ m/s}.$$

$\nu_l = 11.2 \times 10^{12}$ c/s, $\theta_L = 537$ K, $\nu_t = 9.02 \times 10^{12}$ c/s, $\theta_T = 433$ K and which gives $\theta_D = 459$ K.

Remarks: (a) The advantage of the method developed here, due to M. Born, is that one can calculate the Debye temperature starting from the elastic coefficients C_{ij} or from the Lamé coefficients λ and μ: $C_{11} = \lambda + 2\mu + 8\nu$ ($\nu = 0$ for the isotropic case); $C_{12} = \lambda$, $C_{44} = \mu$; and even from the Young's modulus E and the Poisson coefficient σ:

$2\sigma = \frac{\lambda}{\lambda + \mu}$ and $E = \frac{\mu}{\lambda + \mu}(3\lambda + 2\mu)$ (see Chapter II, Ex. 13 and Pb. 4).

On the other hand, it would be better to evaluate the average longitudinal velocity and the average transverse velocity rather than the velocities corresponding to the propagation along specific directions because the Cauchy relation relative to the elastic isotropy: $C_{11} = 3C_{44}$ is far from being satisfied for lithium. When the isotropy condition is satisfied: $v_l = \sqrt{3}v_t$ and $v_l \approx 1.8v_t$, the choice made in question (a) is correct (see Chapter II, Pb. 4 and Ex. 15).

(b) The numerical value obtained for θ_D is rather good at ambient temperature but, for lithium, this value is far from being independent of T (see Fig. 28b) because the elastic constants are themselves dependent on temperature.

Problem 5: Specific heats at constant pressure C_p and constant volume C_v: $(C_p - C_v)$ correction

Consider $V = f(p, T)$, the equation of state for a solid where V represents the volume of one mole. In an infinitesimal reversible transformation, the quantity of heat δQ received can be written in the form $\delta Q = C_v dT + l\, dV = C_p dT + h\, dp$.

(a) Express the difference $C_p - C_v$ as a function of l and find the relation between l and h.

(b) We denote U as the internal energy, H as the enthalpy, and S as the entropy of a mole. Using thermodynamics show that

$$C_p - C_v = -T\frac{\left(\partial V/\partial T\right)_p^2}{\left(\partial V/\partial P\right)_T}$$

(c) Find the result for the preceding function as a function of the compressibility β and of the coefficient of linear thermal expansion α_l. Find as a function of temperature the evolution of the ratio C_p/C_v when the quantity γ, $\gamma = \frac{3\alpha_l \cdot V}{C_v \cdot \beta}$ is assumed to be independent of T (according to E. Grüneisen).

(d) *Numerical application*: Starting from the curve $C_v = f(T)$ shown in the Course Summary, find the ratio $\frac{C_p - C_v}{C_v}$ at an ambient temperature of 280 K for NaCl and Li. Use the following parameters:

$$\mathrm{NaCl}: \beta = 4.26\times10^{-11}\ \mathrm{m^2/N}, \quad \alpha_l = 40\times10^{-6}/\mathrm{K},$$

$$\theta_\mathrm{D} = 280\ \mathrm{K}, \quad V = 268\times10^{-7}\,\mathrm{m^3}.$$

$$\mathrm{Li}: \beta = 8.62\times10^{-11}\,\mathrm{m^2/N}, \quad \alpha_l = 45\times10^{-6}/\mathrm{K},$$

$$\theta_\mathrm{D} = 460\ \mathrm{K}, \quad V = 128\times10^{-7}\,\mathrm{m^3}.$$

Solution:

(a) According to the equation of state $V = f(p, T)$, the total differential volume is

$$dV = \left(\frac{\partial V}{\partial p}\right)_\mathrm{T} dp + \left(\frac{\partial V}{\partial T}\right)_\mathrm{p} dT$$

Substituting this for dV in the two expressions for δQ we find

$$\delta Q = C_\mathrm{v} dT + l dV = \left[C_\mathrm{v} + l\left(\frac{\partial V}{\partial T}\right)_\mathrm{p}\right] dT + l\left(\frac{\partial V}{\partial P}\right)_\mathrm{T} dp = C_\mathrm{p}\, dT + h\, dp$$

which gives $C_\mathrm{p} - C_\mathrm{v} = l\left(\frac{\partial V}{\partial T}\right)_\mathrm{p}$ and $h = l\left(\frac{\partial V}{\partial p}\right)_\mathrm{T}$

(b) According to the second principle of thermodynamics, the elementary variations of the internal energy U, the enthalpy H, and the entropy S are exact differentials. Taking T and p as independent variables, the differentials of H and S are ($H = U + pV$):

$$dH = \delta Q + V\, dp = C_\mathrm{p} dT + (h+V)dp \text{ and } dS = \frac{\delta Q}{T} = \frac{C_\mathrm{p}}{T} dT + \frac{h}{T} dV$$

In each of these expressions the coefficients of dT and dp are first-order derivatives of H (or of S):

$$\left(\frac{\partial H}{\partial T}\right)_p = C_\mathrm{p}, \left(\frac{\partial H}{\partial p}\right)_T = h + V, \left(\frac{\partial S}{\partial T}\right)_p = \frac{C_\mathrm{p}}{T}, \left(\frac{\partial S}{\partial V}\right)_T = \frac{h}{T},$$

from which we can then find

$$\frac{\partial^2 U}{\partial T \partial p} = \frac{\partial C_\mathrm{p}}{\partial p} = \frac{\partial (h+V)}{\partial T} \quad \frac{\partial^2 S}{\partial T \partial p} = \frac{\partial}{\partial p}\left(\frac{C_\mathrm{p}}{T}\right) = \frac{\partial}{\partial T}\left(\frac{h}{T}\right).$$

After multiplying this last quantity by T, we find

$$\frac{\partial C_\mathrm{p}}{\partial p}=\frac{\partial h}{\partial T}-\frac{h}{T}$$

Since the independent variables are T and p, by setting the two expressions for $\frac{\partial C_\mathrm{p}}{\partial p}$ equal to each other, we find

$$h=-T\left(\frac{\partial V}{\partial T}\right)_\mathrm{p} \text{ and } l=-\frac{T\left(\frac{\partial V}{\partial T}\right)_\mathrm{p}}{\left(\frac{\partial V}{\partial p}\right)_\mathrm{T}}, \text{ which gives}$$

$$C_\mathrm{p}-C_\mathrm{v}=-T\frac{\left(\frac{\partial V}{\partial T}\right)_\mathrm{p}^2}{\left(\frac{\partial V}{\partial p}\right)_\mathrm{T}}.$$

(c) Taking into account the definitions of α_l and β,

$\alpha_\mathrm{v}=\frac{1}{V}\left(\frac{\partial V}{\partial T}\right)_\mathrm{p}=3\alpha_\mathrm{l}$ and $\beta=-\frac{1}{V}\left(\frac{\partial V}{\partial p}\right)_\mathrm{T}$, the difference $C_\mathrm{p}-C_\mathrm{v}$ becomes

$C_\mathrm{p}-C_\mathrm{v}=TV\frac{9\alpha_\mathrm{l}^2}{\beta}$, which leads to the ratio $\frac{C_\mathrm{p}}{C_\mathrm{v}}=1+3\gamma\alpha_\mathrm{l}T$.

The molar heat that is measured usually is C_p but the theoretical calculations (see previous exercises) concern C_v.

Thus to compare experiment and theory one must take into account the correction $TV\frac{9\alpha_\mathrm{l}^2}{\beta}$.

At ambient temperature, such a comparison does not present any difficulty because one can measure the different terms of the correction but at high temperature the determination of β can be tricky because one must use a linear extrapolation of the values measured at ambient conditions. Such a methodology can be avoided if one uses, for a given body, the Grüneisen parameter γ, which is a constant as a function of temperature. Unfortunately, the Grüneisen relation assumes that interatomic forces are central, which is far from being the case for metals and covalent materials (see Ref. [25], p. 153).

(d) *Numerical application*:

NaCl at 280 K

$$\frac{T}{\theta_D} \approx 1; \quad C_v \approx 24 \text{ J/mol deg};$$

$$\gamma = 3.17; \frac{C_p - C_v}{C_p} \approx 3\gamma\alpha_1 T \approx 10\%.$$

Li at 280 K

$$\frac{T}{\theta_D} \approx 0.6; \quad C_v \approx 22 \text{ J/mol deg}$$

$$\gamma = 0.9; \quad \frac{C_p - C_v}{C_v} \approx 3.4\%.$$

Problem 6: Anharmonic oscillations: thermal expansion and specific heat for a row of atoms

As shown in Fig. 29, the potential felt by an atom in a crystal can be approximately described by

$$V(x) = \frac{1}{2}ax^2 - bx^3 - cx^4$$

in which all the coefficients are positive and such that bx^3, $cx^4 << ax^2$ and x measures the distance relative to the equilibrium position.

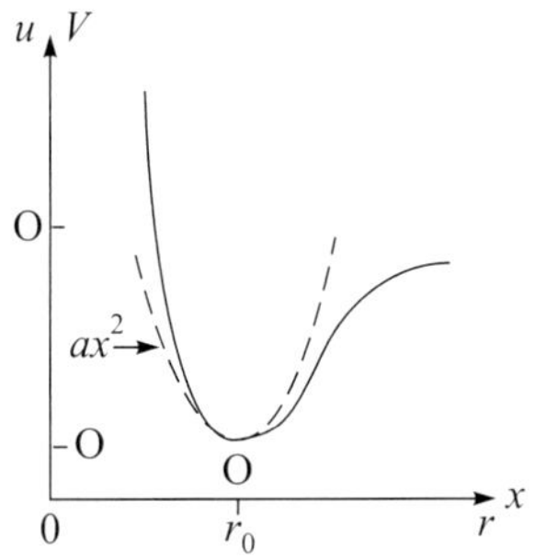

Figure 29

(a) Find the equation of motion for an atom and solve it by successive approximations. Show that the presence of an anharmonic term b results in oscillations with pulsation $2\omega_0$ [ω_0 is the fundamental pulsation of the harmonic oscillator] and that the average position of the atom is proportional to the square of the amplitude A of the harmonic oscillator.

(b) Using the Boltzmann distribution function, find (in the classical approximation), the position of the atom as a function of temperature T. Find this result as a function of the partition function $Z(\beta)$ where $\beta = 1/k_B T$. Deduce the coefficient of linear thermal expansion α_l for a row of identical atoms equidistant of r_0.

(c) Starting from the partition function, find the approximate expression of the specific heat C of this oscillator.

(d) Find the numerical anharmonic contribution to the specific heat for $T = 100$ K and $T = 1000$ K using the following values:

$\frac{b}{a} = \frac{c}{b}$, $a = 10$ N/m, $\alpha_l = 5\times10^{-5}$/K, and $r_0 = 3$ Å

Recall that

$$\int_{-\infty}^{+\infty} \exp-\alpha x^2 dx = \left(\frac{\pi}{\alpha}\right)^{1/2} \text{ and that}$$

$$\int_{-\infty}^{+\infty} x^{2n}.\exp-\alpha x^2 dx = \frac{1\cdot3\cdot5...(2n-1)}{2^{n+1}\alpha^n}\left(\frac{\pi}{\alpha}\right)^{1/2}$$

Neglect the term in cx^4 for parts a and b to simplify the result.

Solution:

(a) The force $\vec{F}$ exerted on an atom is such that $\vec{F} = \overrightarrow{-\text{grad}V}$

Using the expression for V, we find:

$$F_x = -dV/dx = -ax + 3bx^2 + 4cx^3$$

The equation of motion of oscillator of mass m will therefore be $md^2x/dt^2 = -ax + 3bx^2 + 4cx^3$.

In the 0th order approximation, we neglect b and c and obtain the harmonic solution of form:

$$x_0 = A\cos\omega_0 t \text{ where } \omega_0 = (a/m)^{1/2}$$

Next, we seek the first-order approximation; and thus assume a solution of the form $x_1 = A\cos\omega_0 t + \varepsilon$. Substituting this into the equation of motion we find

$$m\frac{d^2\varepsilon}{dt^2} + a\varepsilon - \frac{3b}{2}A^2 - \frac{3bA^2}{2}\cos2\omega_0 t = 0,$$

which results in the trivial solution $\varepsilon_0 = \frac{3bA^2}{2a}$ and more generally $\varepsilon = \varepsilon_0 + \varepsilon_1 \cos 2\omega_0 t$, where $\varepsilon_1 = -\frac{bA^2}{2a}$. The final result is thus $x_1 = A\cos\omega_0 t + \frac{3bA^2}{2a} - \frac{bA^2}{2a}\cos 2\omega_0 t$.

We observe that, in addition to the fundamental pulsation ω_0, there is an additional harmonic of $2\omega_0$, whose amplitude is proportional to the square of the fundamental oscillation (A^2) and is therefore proportional to the energy of the oscillator. In addition, the average position of the atom is no longer at the origin but $\frac{3bA^2}{2a}$ shifted of. This shift of the atom is also proportional to the vibrational energy (see below).

(b) The Boltzmann distribution balances possible values of x by their thermodynamic probability:

$$\bar{x} = \frac{\int_{-\infty}^{+\infty} dx\, x\, e^{-\beta V(x)}}{\int_{-\infty}^{+\infty} dx\, e^{-\beta V(x)}}, \text{ where } \beta = \frac{1}{k_B T}$$

By expanding

$$\exp -\beta\left(\frac{a}{2}x^2 - bx^3 - cx^4\right) \approx (1 + \beta bx^3 + \beta cx^4)\exp -\left(\beta a\frac{x^2}{2}\right)$$

and limiting to the first non-zero term, we find for the numerator

$$\int_{-\infty}^{+\infty} xdx(1+\beta bx^3)e^{-\frac{\beta ax^2}{2}} = \frac{3b\sqrt{2\pi}}{\beta^{3/2}\cdot a^{5/2}}$$

and for the denominator

$$\int_{-\infty}^{+\infty} dxe^{-\frac{\beta ax^2}{2}} = \left(\frac{2\pi}{\beta a}\right)^{1/2},$$

which results in $\bar{x} = \frac{3b}{a^2\beta} = \frac{3b}{a^2}k_B T$

The coefficient of linear thermal expansion α_l for a row of atoms of length l corresponds to $\alpha_l = \frac{1}{l}\frac{\partial l}{\partial T}$. For a distance r

between two atoms: $\alpha_l = \frac{1}{r_0}\frac{\partial(r_0+x)}{\partial T} = \frac{3b}{r_0 a^2}\cdot k_{\mathrm{B}}$

Remark: In this question, one can obtain the initial expression for $\bar{x}$ using the partition function:

$Z = \int\int dp dx e^{-\beta E}$, where $E = \frac{p^2}{2m} + V(x)$.

For N oscillators $\bar{x}$ is such that $\bar{x} = \frac{N}{Z}\int\int \mathrm{dp}\cdot \mathrm{dx}\cdot x e^{-\beta E}$,

which gives $\bar{x} = \dfrac{N\int_{-\infty}^{+\infty} dx x\cdot e^{-\beta V}}{\int_{-\infty}^{+\infty} dx\cdot e^{-\beta V}} \cdot \dfrac{\int_{-\infty}^{+\infty}\exp -\left(\frac{\beta p^2}{2m}\right)dp}{\int_{-\infty}^{+\infty}\exp -\left(\frac{\beta p^2}{2m}\right)dp}$.

(c) The average energy per oscillator is

$$\bar{E} = \frac{\int\int dp dx E e^{-\beta E}}{\int\int dp dx e^{-\beta E}} = -\frac{\partial}{\partial\beta}(\log Z).$$

To evaluate $\bar{E}$, one can therefore either follow a method parallel to that used above to evaluate $\bar{x}$ or use the partition function Z. In the latter case we have

$$Z(\beta) = \int_{-\infty}^{+\infty} dp \exp -\left(\frac{\beta p^2}{2m}\right)\int_{-\infty}^{+\infty} dx e^{-\beta V}$$

$$= \left(\frac{2\pi m}{\beta}\right)^{1/2}\int_{-\infty}^{+\infty} dx\cdot e^{\frac{-\beta a x^2}{2}}\left(1+\beta b x^3+\frac{1}{2}\beta^2 b^2 x^6+\beta c x^4\right).$$

As the terms above in $\int_{-\infty}^{+\infty} x^{2n+1}\cdot e^{-\alpha x^2}dx$ cancel each other by symmetry, the result is

$$Z(\beta) = \frac{2\pi}{\beta}\left(\frac{m}{a}\right)^{\frac{1}{2}}\left(1+\frac{3c}{\beta a^2}+\frac{15b^2}{2a^3\beta}\right)$$

and thus

$$\bar{\mathrm{E}} = -\frac{\partial(\log Z)}{\partial\beta} = k_{\mathrm{B}}T + \left(\frac{3c}{a^2}+\frac{15b^2}{2a^3}\right)k_{\mathrm{B}}^2 T^2$$

where we have taken into account that $\log(1+\varepsilon) = \varepsilon - \frac{\varepsilon^2}{2}$

The heat capacity of an anharmonic oscillator is

$$C = \frac{\partial \bar{E}}{\partial T} = k_B \left[1 + \left(\frac{6c}{a^2} + \frac{15b^2}{a^3} \right) k_B T \right].$$

The first term represents the specific heat of a linear harmonic oscillator $\left(\frac{1}{2} k_B T = E_C \text{ and } \frac{1}{2} k_B T = E_p \right)$, while the last two correcting terms are of comparable weight when b/a and c/b are of the same order of magnitude.

(d) Starting from the literal expression for linear thermal expansion, we find

$$b = \frac{\alpha_l r_0 a^2}{3k_B} = 3.6 \times 10^{10}\,\text{N/m}^2,\; c = \frac{b^2}{a} = 13 \times 10^{19}\,\text{N/m}^3.$$

The anharmonic contribution to the total specific heat is

$$\frac{C(\text{anh})}{C} \approx \frac{21c}{a^2} k_B T = 3.75 \times 10^{-4}\,\text{T},$$

which corresponds to 4% at 100 K and 27% at 1000 K.

Finally, we observe that when the amplitude of oscillations approaches the melting point the harmonic term is $ax \approx 3 \times 10^{-10}$ N, whereas the anharmonic terms $3bx^2$ and $4cx^3$ are respectively equal to 10^{-10} N and 0.14×10^{-10} N.

Problem 7: Phonons in germanium and neutron diffusion

Figure 30 shows the dispersion curve of phonons along the [100] axis of germanium.

(a) Determine graphically the velocity of longitudinal acoustic v_L and transverse v_T waves. Justify the limiting value of the abscise for points D and F, using $k_{max} = 0.177$ Å $\times 2\pi$, and the existence of the branches HF and HG.

(b) Deduce the numerical value of the macroscopic elastic constants C_{11} and C_{44} as well as the frequencies and Debye temperatures corresponding to each polarization (ν_L and θ_L as well as ν_T and θ_T).

What is the weighted Debye temperature θ_D such that

$$\frac{3}{\theta_D^3} = \left(\frac{1}{\theta_L^3} \right) + \left(\frac{2}{\theta_T^3} \right)?$$

Compare the result with the accepted value: $\theta_D = 375$ K.

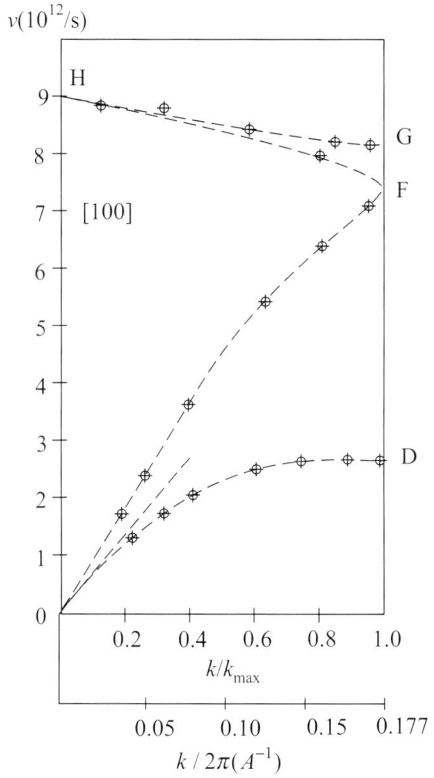

Figure 30

(c) We expect that this material will be absorbing in the infrared. Why? State the numerical value of the wavelength λ at which this absorption will take place.

(d) The curve shown was obtained by inelastic scattering of neutrons with associate initial wavelength of λ = 1.52 Å. Supposing that the wave vector kon of incident neutrons contained in the plane $\vec{A}$, $\vec{B}$ in reciprocal space, show that the extremity of the diffuse wave vectors which leads to the determination of the point H is situated on a sphere and find its radius. The method to follow can be inspired by the Ewald construction, but including the inelastic effects to adapt it to this problem. It will be useful to compare the initial kinetic energy of the neutrons and the maximum energy of excited phonons. Generalize this reasoning to determine a point on the dispersion curve.

Data for these calculations: Ge is fcc with lattice parameter a = 5.66 Å with basis of two atoms at 000 and at ¼ ¼ ¼. A(Ge) = 72.5; M(neutron) = 1.67×10^{-27} kg. (h, e, k_B, and N).

Solution:

(a) The longitudinal acoustic branch is the OF branch and the two other acoustic branches represented by OD are the transverse degenerate branches. The constant force between atoms is larger for the longitudinal vibrations, which propagate by successive compression and elongation, than for transverse vibrations, which correspond to shear forces (it is easier to slide the (100) planes toward one another than to reduce the distance separating them). The degeneracy of the transverse vibrations can be understood because the two branches correspond to atomic displaces $\bar{u} \parallel [010]$ and to the [001] which are equivalent: the propagation being such that $\bar{k} \parallel [100]$. The requested velocities can be deduced using $\omega = vk$, from the slopes of the tangents at the origin for the two acoustic dispersion curves (tangents being suggested in Fig. 30). We find $v_L[100] \approx 4800$ m/s; $v_T[100] \approx 3400$ m/s.

The reciprocal lattice of Ge is cubic centered and the indices of the points of this lattice have the same parity. In the [100] direction, the median plane between the origin and the point at 200—the point at 100 being forbidden—gives for the abscissa $2\pi/a$ with $1/a = 0.177$ Å^{-1}, where a = 5.66 Å. (It is the point labeled X in Chapter I, Ex. 14, Fig. 12).

The existence of a two-atom basis in non-equivalent positions is sufficient to induce the existence of the optical branches (even if these atoms are chemically identical): HF:LO and HG:2TO (degenerate).

(b) $v_L[100] = \sqrt{C_{11}/\rho}$ and $v_T[100] = \sqrt{C_{44}/\rho}$

There are eight atoms of Ge in a volume of a^3 from which we find a mass density $\rho = 8A/N a^3$. $\rho = 5.31 \times 10^3$; $C_{11} = 1.2 \times 10^{11}$ Pa and $C_{44} = 6.1 \times 10^{10}$ Pa.

In addition, $v_{L,T} = v_{L,T}(3n/4\pi)^{1/3}$ (see Pb. 4) where $n = 8/a^3$.

$\nu_L \approx 10.5 \times 10^{12}\ s^{-1}; \nu_T \approx 7.45 \times 10^{12}\ s^{-1}$

$\theta_L \approx 500\ K; \theta_T \approx 357\ K; \theta_D \approx 386\ K.$

The agreement with the accepted value is surprising. It is due to the overestimation of the weighted acoustic frequencies compensating for the optical frequencies, which are not taken into account in the Debye model. (Compare ν_L to the experimental value ν_{max}.)

(c) The IR absorption leads to the excitation of transverse optical phonons with very short wave vectors ($k \sim 0$). It is therefore related to the existence of the optical branches (see the above answer in a):

$\lambda(IR) = c/\nu$, where $\nu_0 = 9 \times 10^{12}\ s^{-1}$ and $\lambda(IR) \approx 33\ \mu m$.

The order of magnitude of β may be deduced from the expression for ν_0 relative to a basis consisting of two atoms of different masses m and M

$$(2\pi\nu_0)^2 = 2\beta\left(\frac{1}{M} + \frac{1}{m}\right)$$

Here there is the identity of the two masses. Then, $\beta = m(2\pi\nu_0)^2/4$, where $m = A/N = 1.2 \times 10^{-25}$ kg and $\beta \approx 96$ N/m.

In fact the reality is much more complex. The resolution of the dynamical matrix leads to the introduction of two force constants from which the sum corresponds sensitively on the evaluation of β determined below. But this is not sufficient to correctly describe all of the experimental results. The complete study must also take into account the polarizability of the atoms and the interaction of atoms which is not just limited to nearest neighbor interactions. (For further details see Ref. [2], pp. 92–98.)

(d) Resulting from the inelastic interactions of neutrons with Ge, the conservation of energy is $E_{on} = E'_n \pm h\nu$, where the $\pm$ corresponds to the creation (annihilation) of a phonon with frequency ν. The energies E_{on} and E'_n are related to the wave vectors k_{on} and k'_n via $E_{on} = h^2k^2_{on}/2M$ and $E'_n = h^2k'^2_n/2M$. The modulus of the scattered wave vector is therefore: $|k'_n| = [k_0^2 \mp (4M\pi\nu/\hbar)]^{1/2}$.

Starting from the origin of the wave vector $\overrightarrow{\mathrm{k}_{\mathrm{on}}}$, of the incident neutrons, one can therefore construct the two spheres (circumference in the plane) of the radii $k'_n(+)$ and $k'_n(-)$. The procedure is analogous to that of Ewald apart from the inequality, here, of k_{on} and k'_n.

In addition, the momentum conservation, $\hbar\vec{\mathrm{k}}_{\mathrm{on}} = \hbar\vec{\mathrm{k}}'_n + \hbar(\vec{\mathrm{k}} + \vec{\mathrm{G}})$, must also be satisfied with possible translation of a vector $\vec{\mathrm{G}}$ of the lattice, for the phonon vector $\vec{\mathrm{k}}$ to be included in the first BZ.

For the point H, $\vec{\mathrm{k}}(H) = 0$ and the vector equality implies that the two spheres are connected by a point on the reciprocal lattice, that is $\vec{\mathrm{k}}_{\mathrm{on}} - \vec{\mathrm{k}}'_n = \vec{\mathrm{G}}$.

Starting from the formula $\lambda = \dfrac{h}{\sqrt{2ME}}$ (see Chapter I, Course Summary and Pb. 7), the energy of incident neutrons is $E_{\mathrm{on}} \approx 40$ meV, whereas the phonon energy at point H is $h\nu = 37$ meV. This leads to the illustration in Fig. 31 where the representation in reciprocal space is limited to the allowed points in the plane AB; the circumference of radius k' (absorption of a phonon) and k'_+ (creation of a phonon) encircles the Ewald sphere of elastic diffusion. Regardless of the incidence, the conservation conditions will not be simultaneously satisfied and it will be necessary to orient the crystal correctly.

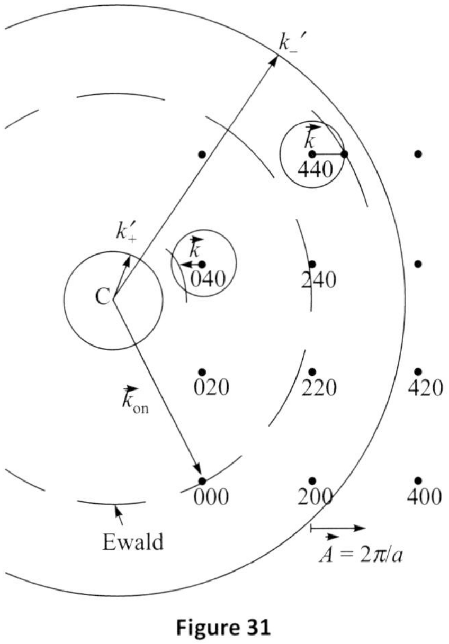

Figure 31

For the other points of the dispersion curve, where the exchange of energy $h\nu$ is weaker, the corresponding spheres are necessarily situated between the extreme spheres k'_+ and k'_- mentioned above. The directions of the diffuse neutrons will be determined graphically by the intersection of these spheres with the spheres of radius k (phonon) centered on the different points of the reciprocal lattice.

For phonons propagating along the [100] axis, only the $\vec{k}$ vectors parallel to this axis will be relevant. To these constraints, which are shown for the points at 040 and 440 in Fig. 31, must be added those relative to the position of the neutron detector which imposes that the vectors $\vec{k'}$ be orthogonal to $\vec{k}_{on}$. Finally, we observe that at 0 K, the neutrons cannot acquire an energy $h\nu$ because the atoms only have a half quantum of energy.

Comment: Nobel Prize in physics in 1994

This problem therefore illustrates particularly the experimental method used to reveal point by point the dispersion curve of phonons starting from the measurement of kinetic energy and from the direction of inelastic thermally diffused neutrons that are initially monochromatic. For further details the reader can consult the article by Brockhouse and Iyengar, *Phys Rev III*, 1958, 747. In addition to their elastic interactions (i.e., λ is not changed and the process is coherent (see comment in Chapter I, Pb. 7), there are also collisions that involve a change in energy with phonons and magnons. B. N. Brockhouse shared with C. G. Shull the Nobel Prize in physics, 1994, for this type of investigation.

Problem 8: Phonon dispersion in a film of CuO_2

Consider a 2D crystalline structure defined by its square lattice with parameter $2a$ and its basis consisting of one copper atom at 0,0 and two oxygen atoms at ½,0 and 0,½.

(1) Draw the crystal structure.

(2) The vibrations perpendicular to the surface are investigated by limiting the analysis to the actions of nearest neighbors with force constant β.

Find the equations of motion of the atom Cu with mass M which is at a position $2l$ and $2n$ and which is thus at a distance

$2la$ along x and lna along y, from the origin. Also find the equations for the oxygen atoms of mass m whose positions correspond respectively to $2l + 1$, $2n$, $2l$, and $2n + 1$. Starting from solutions of the form $\exp i(\omega t - \vec{k} r)$ where the amplitudes will be successively A, B, and C, show that the dispersion relation of phonons is deduced from a determinant equal zero and express its coefficients.

(3) Find the solutions of this relation at the center and extremities of the BZ ($k = 0$) in the [11] and [10] directions. It is useful to express the results as a function of $\omega_m^2 = 2\beta/m$ and $\omega_M^2 = 4\beta/M$. Show the dispersion curves in the two directions mentioned above and the density of states $g(\omega)$. Take into account that $M \approx 4m$ [A(Cu) = 63.6 and A(O) = 16]. Comment on the results.

(4) Now consider the effect of second nearest neighbors limiting the analysis to the interactions between oxygen atoms characterized by a force constant $\beta_0 < 0$. Use $\omega_r^2 = -4\beta_0/m$.

Solution:

(1) See Fig. 32.

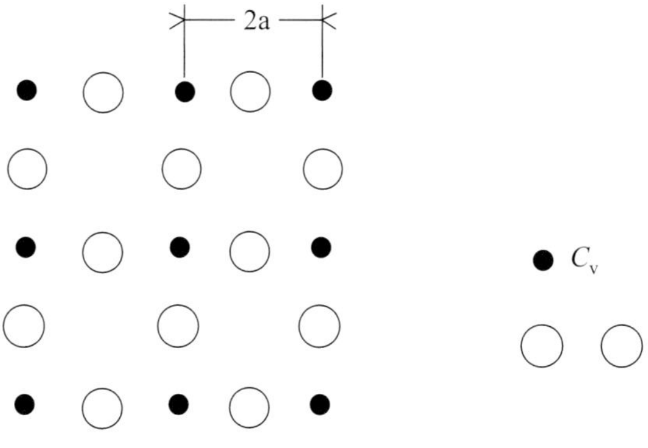

Figure 32

(2)
$$M\ddot{u}_{2l,2n} = \beta(u_{2l+1,2n} + u_{2l-1,2n} - 2u_{2l,2n}) + \beta(u_{2l,2n+1} + u_{2l,2n-1} - 2u_{2l,2n})$$
$$m\ddot{u}_{2l+1,2n} = \beta(u_{2l,2n} + u_{2l+2,2n} - 2u_{2l+1,2n})$$
$$m\ddot{u}_{2l,2n+1} = \beta(u_{2l,2n} + u_{2l,2n+2} - 2u_{2l,2n+1})$$

The solutions are of the type

$$u_{2l,2n} = A\exp i(\omega t - 2lk_x a - 2nk_y a)$$

$$u_{2l+1,2n} = B\exp i(\omega t - (2l+1)k_x a - 2nk_y a)$$

$$u_{2l,2n+1} = C\exp i[\omega t - 2lk_x a - (2n+1)k_y a]$$

We obtain

$$\begin{vmatrix} M\omega^2 - 4\beta & 2\beta\cos k_x a & 2\beta\cos k_y a \\ 2\beta\cos k_x a & m\omega^2 - 2\beta & 0 \\ 2\beta\cos k_y a & 0 & m\omega^2 - 2\beta \end{vmatrix} = 0$$

(3) $$(\omega^2 - \omega_m^2)\left[\omega^4 - (\omega_m{}^2 - \omega_M{}^2)\omega^2 + \frac{\omega_m^2\omega_M^2}{2}(\sin^2 k_x a + \sin^2 k_y a)\right] = 0$$

In the center, point O or Γ of the BZ, $k_x = k_y = 0$ so we have:

$$\omega''' = 0(\mathrm{TA});\ \omega' = \omega_m(\mathrm{TO}_1);\ \omega'' = (\omega_m^2 + \omega_M^2)^{1/2}(\mathrm{TO}_2)$$

At $k_x = k_y = \pi/2a$, point M of the BZ:

$\omega' = \omega'' = \omega_m$ (TO_1 and TO_2); $\omega''' = \omega_M = (\mathrm{TA})$

At point X: $k_x = \pi/2a$, $k_y = 0$. In addition to the square root $\omega''' = \omega_M$, we find:

$$2(\omega')^2, 2(\omega'')^2 = \omega_m^2 + \omega_M^2 \pm \sqrt{\omega_m^4 + \omega_M^4}$$

The most remarkable result concerns the TO_1 branch which corresponds to a vibration of two atoms of oxygen in phase opposition (B = –C and A = 0) regardless of the wavelength of vibration. This leads to a large density of states at $\omega = \omega_m$ because the integrated area under the curves $g(\omega)$ is the same for the three branches.

(4) Taking into account the second interactions between atoms of oxygen, to the number 4, modifies neither the coefficients of the first row nor those of the first column of the determinant (D_{11}, D_{12}, D_{13}). The diagonal coefficients become $D_{22} = m\omega^2 - 2\beta - 4\beta_0$ and those that are initially zero, such as D_{23} now become $D_{23} = 4\beta_0 \cos k_x a \times \cos k_y a$. As expected, the effect of the 2^d neighbors concerns mostly the oxygen atoms of the TO_1 branch with angular frequencies obeying to $\omega^2 = (\omega_m^2 - 2\omega_r^2)$, when $k = 0$, and $\omega^2 = (\omega_m^2 - \omega_r^2)$, when

$k_x = k_y = \pi/2a$. At the center of the BZ the other branches are not affected because the O atoms vibrate in phase. Nevertheless, if the repulsive forces are very strong, we can foresee a decrease in the sound velocity and the appearance of soft modes. (See Exs. 2b, 3, and 5 for more details.)

The dashed curves on the Fig. 33 show the main changes due to the second nearest neighbors compared to the solid lines, which deal with the nearest neighbors only.

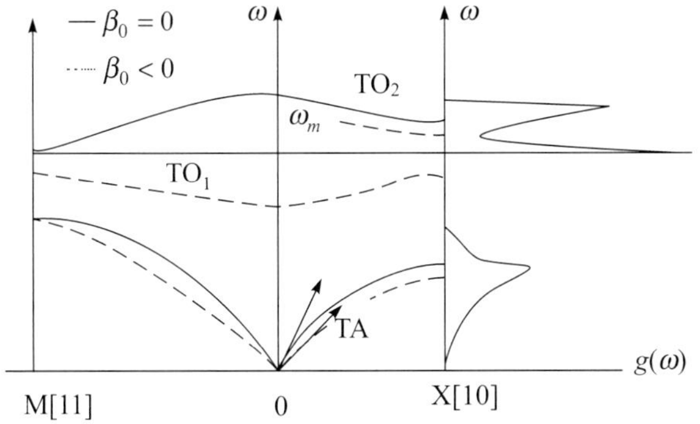

Figure 33

Comment: See also Chapter IV, Pb. 7, and Chapter V, Ex. 2b.

In metallic superconductors, the John Bardeen, Leon Neil Cooper, and John Robert Schrieffer (BCS) theory shows that the critical temperature can be written as: $T_c = 1.14\theta_D \exp{-\alpha^{-1}}$, where $\alpha = U \cdot g(E_F) << 1$ with $g(E_F)$ is the density of states of electrons at the Fermi level, θ_D is the Debye temperature and U is the interaction energy of electrons with phonons. The present problem concerns ceramics of YBaCuO type that have a high transition temperature between normal and superconducting states. In this context and to explain the superconductivity of ceramics based on copper oxides, it is interesting to study the atomic vibrations of layers of CuO_2, which make up the basis of these new superconductors (see Chapter V, Ex. 2b).

Problem 9: Phonons dispersion in graphene

Graphene is a 2D single layer of carbon atoms, mass m, arranged in a honeycomb structure as shown in Fig. 14a. The distance between two neighboring atoms is d = 1.42 Å. These atoms are submitted to a force constant β_z limited to each nearest neighbors and are forced to move perpendicular to the lattice plane (Z modes associated with out-of-plane atomic motions).

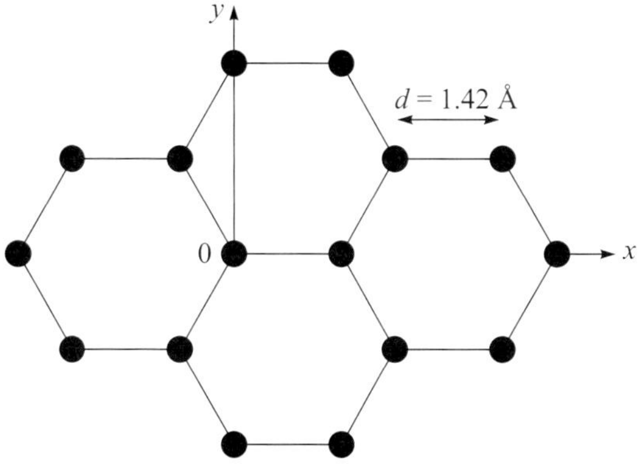

Figure 34 Graphene.

(1) Describe the crystal structure of graphene: Bravais lattice and basis. Following an approach similar to that used in Ex. 9, above, find the equation of motion relative to the displacement u_0 of the atom at the origin 0 and those relative to the displacement u_i of atoms of index i.

(2) Starting from solutions of the form of running waves propagating only in the y-direction, exp $i(\omega t - k_y y)$, find the dispersion relation between ω and α with $\alpha = k_y d\sqrt{3}/2$. Show the corresponding curves using a vertical scale in $\sqrt{\beta_z/m}$ unit and give the characteristic values taken by ω at selected points Γ and K of the first BZ (point Γ is in the BZ center and point K corresponds to the point where the wave vector k is in contact with the BZ in the y-direction).

(3) The phonon dispersion for all polarizations and crystallographic directions in graphene lattice has been extensively investigated as it may be seen in the literature. Inspired from Nika et al. *Phys. Rev. B.* **79**, 2009, 155413, an example is shown in Fig. 35 (where calculations are based on the valence-force field method including the effects of the 2^d nearest neighbors).

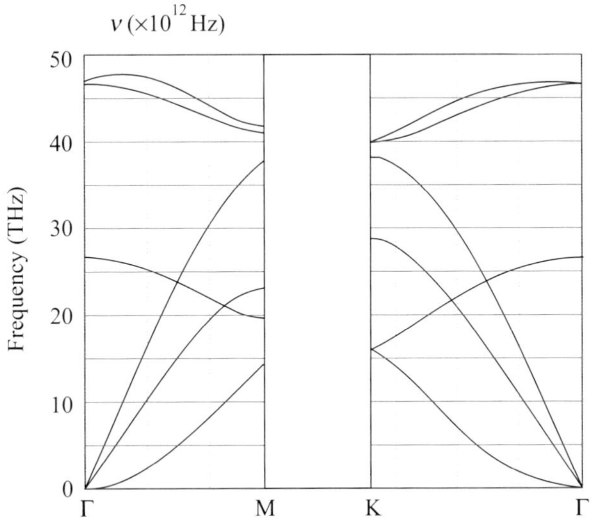

Figure 35

From simple physical arguments, identify the out-of-plane modes (Z) the in-plane longitudinal modes (L), and the in-plane transverse modes (T) and their direction in the BZ. Indicate which of them correspond to the above calculations and suggest a simple method to draw the missing branches of the curves asked in (2). From the numerical values of ν (vertical scale in Fig. 14b), evaluate the sound velocity along a graphene layer.

Solution:

(1) See Chapter I, Ex. 17, and Fig. 36.

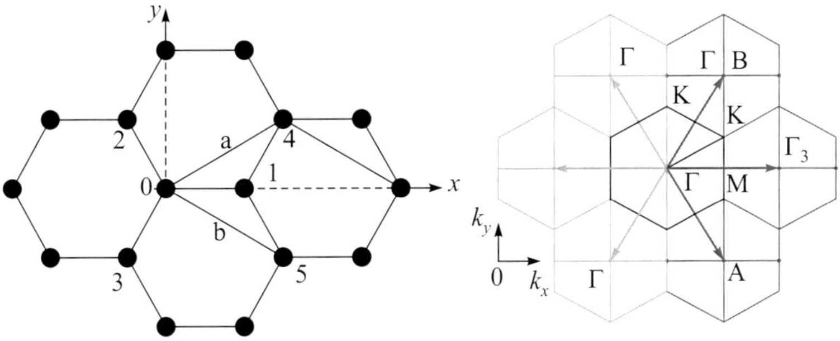

Figure 36 Graphene: Direct space and reciprocal space with $\Gamma M = 2\pi/3d$ and $\Gamma K = 4\pi/3\sqrt{3}d$.

The rhombic unit cell can be defined by two basis vectors $a = (3; \sqrt{3}/2)d/2$ and $b = (3; -\sqrt{3}/2)d/2$. The basis is composed of one C atom in (00) and another one in (1/3; 1/3). Thus there are two atoms in non equivalent position, thus one must expect two types of branch: acoustical and optical (see Ex. 1). Each of the two atoms has three nearest neighbors

$m\ddot{u}_0 = \beta_z\,[(u_1 - u_0) + (u_2 - u_0) + (u_3 - u_0)]$

$m\ddot{u}_1 = \beta_z\,[(u_0 - u_1) + (u_4 - u_1) + (u_5 - u_1)]$

(2) We look for solutions of the form exp $i[\omega t - (k_x x + k_y y)]$, where x and y are the co-ordinates of the atoms of interest: $1(d, 0)$; $2(-d/2, \sqrt{3}d/2)$; $3(-d/2, -\sqrt{3}d/2)$; $4(1.5d, \sqrt{3}d/2)$; $5(1.5d, -\sqrt{3}d/2)$.

For running waves in the y-direction, the displacements are $u_0 = A\ e^{i\omega t}$; $u_1 = B\ e^{i\omega t}$; $u_2 = B\ e^{i\omega t}\ e^{-i\alpha}$; $u_3 = B\ e^{i\omega t}\ e^{i\alpha}$; $u_4 = B\,e^{i\omega t}\,e^{-i\alpha}$; $u_5 = B\,e^{iwt}\,e^{i\alpha}$, where $\alpha = k_y d\sqrt{3}/2$.

On substitution in the equations of motion, one obtains

$-\omega^2 A = (\beta_z/m)[B + 2B\cos\alpha - 3A]$

$-\omega^2 B = (\beta_z/m)[A + 2B\cos\alpha - 3B]$

This set of homogeneous equations has a non-trivial solution only if the determinant, Δ, of the coefficients A and B vanishes. $\Delta = 0$ when

$$\omega^4 - 2(\beta_z/m)(3 - \cos\alpha)\,\omega^2 + 8(\beta_z/m)^2(1 - \cos\alpha) = 0 \qquad (1)$$

The solutions are $\omega' = [2(\beta_z/m)(1 - \cos\alpha)]^{1/2}$ for the z-acoustical mode; $\omega' = 0$ for $k_y = 0$; and $A = B$.

$\omega'' = 2(\beta_z/m)^{1/2}$ for the optical mode. $A = -B$ for $k_y = 0$.

(3) Under the same simplified assumptions, the dispersion curves for the other modes obey to the same equations.

The unique changes are the respective values of the force constants, β_t or β_l. The stronger force constant is β_l: compression effect. The weaker is β_z because of the ease of the atoms to move normal to the graphene layer. The results are shown in Fig. 37 (right), where the ratios between them are chosen for an approximate fit with the Mika's calculations: $\beta_t/\beta_z \sim 3.3$ and $\beta_l/\beta_z \sim 5.5$.

The sound velocity corresponds to $v_s = 2\pi\,\partial\nu/\partial k$ for the LA branch, compression, around the point Γ. From the slopes of the dashed arrows in Fig. 37 (left), one obtains $v_s = 18000$ m/s.

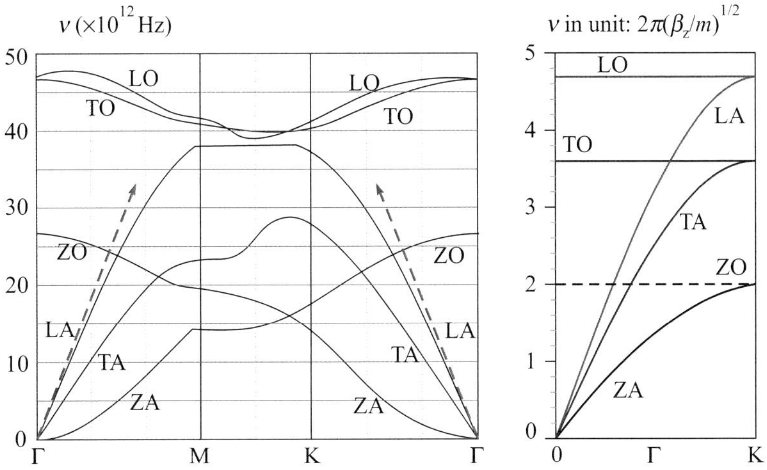

Figure 37 Phonon dispersion in graphene: (Left) from Mika et al. (Right) present calculations.

Remarks: (1) The value of the sound velocity is consistent to the value being measured in diamond (see Chapter II, Pb. 4). Like for diamond-like materials, the reason is the large value of the constant force β_1 (C–C covalent binding) combined to the light mass of carbon atoms: vs is proportional to $\sqrt{\beta_1/m}$. Thus graphene layers have a good thermal conductivity, *K*th (see Course Summary, Section 7), but also a good electrical conductivity along the layers (see Chapter V).

(2) The present problem is an oversimplified approach for the investigation of the phonon dispersion in graphene. In general, atomic vibrations are partially screened by filled electronic states. This screening may lead to Kohn anomaly (see Ex. 6) that may occur to only some specific points of the Fermi surface of graphene (Maultzsch et al., *Phys. Rev. Lett.* **92**, 2004, 075501).

In addition, the heat transport in basal planes of bulk graphite slightly differs from that in a single-layer graphene. The thermal conductivity, *K*th, is ~2000 W/mK for graphite and ~7000 W/mK for graphene. The reason is the presence of a phonon dispersion curve parallel to the *c*-axis in graphite, making possible a strong coupling with the plane phonon modes and propagation of heat in all directions; phenomena that cannot exist in graphene. This lack explains that a phonon mean free path Λ_{ph} of ~775 nm has been measured in graphene near room temperature (D. L. Nika et al. *Appl. Phys. Lett.* **94**, 2009, 203103; S. Ghosh et al. *Appl. Phys. Lett.* **92**, 2008, 151911). Such specific properties open opportunities for tuning the properties of phonons in a way similar to electrons, as it may be deduced from the recent works in particular those of Balandin group. Such an approach for controlling the phonon spectrum of materials for specific applications was termed phonon engineering or nanophononics. (Chapter V, Pb. 11, deals with the electrons in graphene.)

Questions

Q.1: In what medium does sound propagate faster: lead or diamond? Why?

Q.2: The Debye temperature of lead is $\theta_D = 100$ K.

What physical arguments justify such a small value?

What is the vibration energy of an atom of lead at 300 K?

Q.3: The Debye temperature of diamond is $\theta_D = 2230$ K.

What physical arguments justify such a large value?

Q.4: Find the ratio of the specific heat at 50 K and 5 K for diamond. Find the thermal conductivity at these two temperatures for a very small sample.

Q.5: Why is the velocity of a transverse acoustic wave in a cubic crystal smaller than that of the longitudinal wave?

Q.6: At 300 K, the thermal conductivity of germanium is $K = 80$ W/m·K, its specific heat is $C \approx 17 \times 10^5$ J/m^{3}K and the average sound velocity is $v_s = 4500$ m/s. What is the mean free path of phonons at this temperature?

Q.7: What is the asymptotic value ($T > \theta_D$) of the specific heat of a row of identical atoms when considered as a row of spatial oscillators? At low temperatures ($T << \theta_D$) is the dependence T, T^2, or T^3?

Q.8: What is the asymptotic value of the specific heat of a 3D solid from which the atoms will be constrained to vibrate in only one direction? At low temperatures will the evolution of C_v have a T, T^2, or T^3 dependence?

Q.9: Is the concept of "phonon" related to a boundary condition?

Q.10: What is the order of magnitude of the amplitude of atomic vibrations for solids when their temperature is approaching their melting point?

Q.11: Why is helium a liquid even at 0 K (at atmospheric pressure)?

Q.12: What is a soft mode?

Q.13: What are the microscopic causes of the thermal expansion of a solid?

Q.14: What is the Kohn anomaly?

Q.15: Why does silicon absorb in the infrared?

Q.16: Classify the following alkali halides by their infrared absorption wavelength in increasing magnitude: LiF, NaCl, KBr, RbI, and CsI.

Q.17: The static dielectric constant ε_s and the optical index n (in the visible) of several alkali halides are as follows:

LiF ($\varepsilon_s = 8.9$; $n = 1.38$); NaCl ($\varepsilon_s = 5.9$; $n = 1.5$); AgCl ($\varepsilon_s = 12.3$; $n = 2$).

Find the weight of ionic vibrations relative to the static dielectric susceptibility. What is the relation between the longitudinal and transverse resonant frequencies of ions, ω_L/ω_T, for $k = 0$?

Q.18: What are the microscopic causes of a dielectric constant smaller than unity, eventually negative, in the infrared? (Specify the reasons in the case of ionic crystals). Experimentally, what are the optical consequences of this phenomenon?

Q.19: Why diamond is a good thermal conductor despite its very low electrical conductivity?

Q.20: What is the specificity of the thermal conductivity of graphene with respect to that of graphite?

ANSWERS AT THE END OF THE BOOK

Chapter IV

Free Electrons Theory: Simple Metals

Course Summary

1. Hypothesis

We assume that electrons can propagate freely in certain solids without 'seeing' the ions of the lattice. This model is satisfactory for valence electrons in certain metals, in particular the alkaline metals, and can explain many of their physical properties.

2. Dispersion Relation and the Quantization of the Wave Vector

Free electrons satisfy $\lambda = h/m\gamma = h/p$ or $\hbar\vec{k} = m\vec{v}$. Their energy is uniquely kinetic: $E = \frac{1}{2}mv^2 = \hbar^2k^2/2m$, which corresponds to the dispersion relation.

(a) 1D case:

With V (potential energy) = 0, the Schrödinger equation reduces to $\frac{d^2\psi}{dx^2} + \frac{2mE\psi}{\hbar^2} = 0$.

Substituting $E = \frac{\hbar^2k^2}{2m}$, the solutions are

Understanding Solid State Physics: Problems and Solutions
Jacques Cazaux

ISBN 978-981-4267-89-2 (Hardcover), 978-981-4267-90-8 (eBook)
www.panstanford.com

- Running waves of the form $\Psi = Ae^{ikx}$ with periodic boundary conditions:
  $\Psi(x) = \Psi(x + L)$, implying $k = 2\pi n/L$ where n is a nonzero integer, > or <0 (see Fig. 3 in Ex. 2).
- Standing waves of the form $\Psi = C \sin kx$, with fixed boundary conditions $\Psi(0) = \Psi(L) = 0$, implying $k = \pi n/L$, where n is a positive integer (see Fig. 2 in Ex. 1).

(b) 3D case:

We generalize the above results: $(-\hbar^2/2m)\nabla^2\psi = E\psi$.

- Solutions in the form of running waves: $\psi = \frac{1}{\sqrt{V}} \exp i(k_x x + k_y y + k_z z)$ with periodic boundary conditions (PBC): $k_x = 2\pi n_x/L_x$, $k_y = 2\pi n_y/L_y$, $k_z = 2\pi n_z/L_z$, where n_x, n_y, n_z are integers where one or two of them can be zero.
- Standing wave solutions of form: $\psi = \left(\frac{8}{V}\right)^{1/2} \sin k_x x \cdot \sin k_y y \cdot \sin k_z z$ with fixed boundary conditions, FPC: $k_x = \pi n_x/L_x$, $k_y = \pi n_y/L_y$, $k_z = \pi n_z/L_z$ where n_x, n_y, and n_z are positive integers (see Ex. 3).

For infinite objects in 3D with a large number of states to be filled, the choice of periodic or fixed boundary conditions is free because it leads to the same evaluation for the density of states $g(k)$ and for $g(E)$. It is not the same for the first possible states of objects limited along one or several directions, for instance surfaces of a metal bounded by vacuum, a thin layer with parallel sides and of thickness of a few lattice distances (Ex. 7) or aggregates (clusters) of atoms (Exs. 5 and 6).

The discrete values of n_x, n_y, n_z correspond to the quantum numbers that a free electron (excluding spin) can take, as n, l, m, represent the quantum numbers of an electron subject to a central potential of an ion in atomic physics.

3. Electron Distribution and Density of States at 0 K: Fermi Energy and Fermi Surface in 3D

As for an atom, the Schrödinger equation only allows the determination of all the possible states of electrons. The states that are occupied will be obtained by filling each state $(2\pi/L_x)\cdot(2\pi/L_y)\cdot(2\pi/L_z)$ in k-space with two electrons of anti-parallel spin (↑↓). The electronic states thus differ from one another by at least different quantum numbers n_x, n_y, n_z or spin s (from the Pauli exclusion principle) starting with the lowest energy levels. When N free electrons of the volume L_x, L_y, L_z are thus distributed, we obtain a sphere in k-space limiting the occupied and the empty states at 0 K. The radius k_F of this Fermi sphere is $\frac{4\pi}{3}k_F^3/\frac{(2\pi)^3}{L_xL_yL_z}=\frac{N}{2}$, where $k_F=(3n\pi^2)^{1/3}$ and $n = N/V$. The electron energy is uniquely kinetic and the velocity $\vec{v}$ is proportional to $\vec{k}$ $(m\vec{v}=\hbar\vec{k})$. The Fermi sphere visualizes the velocity vectors of the free electrons in a metal. The maximum kinetic energy is given by $E_F=\frac{\hbar^2k_F^2}{2m}=\frac{\hbar^2}{2m}\cdot(3n\pi^2)^{2/3}$.

For alkali metals, the orders of magnitude are $n = 5\times10^{22}$ e. cm^{-3}, $k_F = 1.2\times10^8$ cm^{-1}, $v_F \approx 1.3\times10^8$ cm/sec, $E_F = 5$ eV.

- Density of states: In k-space, the density of states $g(\mathrm{k})$ between k and $k + dk$ is the same as that evaluated previously for lattice vibrations (Chapter III). However, $g(E)$ must take into account the particular dispersion of free electrons with two electrons of opposite spin per state. We thus obtain $g(E)\cdot dE = 2g(k)\cdot dk$. In 3D, this results in $g(E) = (V/2\pi^2)\cdot(2m/\hbar^2)^{3/2}E^{1/2}$.
  The Fermi energy at 0 K can also be deduced from $\int_0^{E_F} g(E)\,dE = N$.

4. Influence of Temperature on the Electron Distribution: Electron-Specific Heat

Electrons obey Fermi–Dirac statistics: These are fermions because of their non-integer spin leading the Pauli exclusion principle (in contrast to phonons, Chapter III that are bosons from the Bose–Einstein statistics). As a function of temperature, the occupation probability of an electronic state is given by $f(E) = [e^{(E-E_F)/k_BT}+1]^{-1}$. The distribution obtained for $T \neq 0$ differs from that at 0 K (where $f(E) = 1$ for $E < E_F$) only for energies that are very close (several k_BT) to the Fermi energy.

By definition $E_F(T)$ is deduced from $\int_0^{\infty} f(E)\cdot g(E)\cdot dE = N$. Thus, the total energy of an electron gas U_e obeys $U_e = \int_0^{\infty} f(E)\cdot g(E)\cdot E\cdot dE$, which in 3D becomes $U_e = \frac{3}{5}NE_F(0) + (\pi^2/4)\,k_B^2T^2/E_F(0)$ (see Exs. 8 and 9).

The electronic specific heat $C_e = (\partial U/\partial T)$ can be easily deduced: $C_v(\text{electronic}) = [\pi^2/2E_F(0)]\cdot N\cdot k^2_B T$, where N is the number of free electrons in a volume V.

As opposed to the vibration energy of atoms, the kinetic energy of electrons in a metal is already large at 0 K, but it varies relatively little with temperature (see Ex. 14a). The electronic specific heat is always linear in T (in 1D, 2D, and 3D), but the coefficient of proportionality depends on the dimension (see the table in Ex. 14a).

5. Electronic Conductivity

- $\vec{J} = -ne\overrightarrow{v_e}; \overrightarrow{v_e} = -(e\tau/m)\vec{E}$ from which one obtains the Drude Law, $\sigma = ne^2\tau/m$, where $n = N/V$.
- The time of flight of electrons is inversely proportional to the probability of collisions between electrons with phonons, impurities and deformations with the crystal lattice $1/\tau = P_p + P_i + P_d$, from which we obtain the Matthiessen's rule $\rho = 1/\sigma = \rho_p + \rho_i + \rho_D$. The electric resistivity of a metal increases with temperature due to phonons and with the concentration of impurities and with deformations of the lattice. Note that the influence of the first two parameters is strictly the opposite of what is observed in semiconductors (see Ex. 16, and Chapter V, Pb. 4).

Note that the action of an electric field has the effect of adding a small unidirectional velocity $\overrightarrow{v_e}$ to the large but isotropic one $(\overrightarrow{v_F})$ of electrons. The mean free path Λ between two collisions is given by $\Lambda = v_F\tau$.

In the case of thin films, the mean free path can also be limited by the thickness of the film t, which leads to an increase in the resistivity of such films for $t < \Lambda$ (Ex. 20).

In addition, the presence of a magnetic field leads to an electrical conductivity in the steady regime characterized by a tensor (Pb. 2), which describes the Hall effect and magneto-resistance. Even if the Hall effect is more prominent in semiconductors as compared with metals, one can formally treat it by considering a gas of free electrons and find evidence of quantum effects (Ex. 26).

In the sinusoidal regime, when the skin depth is smaller than the mean-free path, an anomalous skin effect appears (Ex. 21).

6. Wiedemann–Franz Law

The thermal conductivity K of a gas of particles obeys $K = 1/3\ \mathrm{Cv}\Lambda$. For electrons in a metal, we have $K = (\pi^2/3)\,(nk_B^2T/\mathrm{m})\,\tau$, where $C = C_v$ (electronic) and $v = v_F$ from which we obtain the Wiedeman–Franz law: $K/\sigma = (\pi^2/3)(k_B^2/e^2)T = LT$ (see Ex. 12).

7. Other Successful Models Obtained from the Free Electron Formalism

With additional ingredients, the model of free electrons can be used to evaluate the cohesive energy of free electron metals (Ex. 27) and the formation of a surface barrier at the metal/vacuum interface (Ex. 28a), while the correlated work function effect plays an important role in the electron emission into vacuum (Ex. 29) with applications in particular in the thermo-electronic emission (Exs. 30–32), the tunnel effect and in scanning tunneling microscopy (Pb. 3), X-ray photoelectron Auger electron emission spectroscopy. The same model explains the paramagnetism of simple metals (Ex. 22) and the reflecting power of alkali metals in the ultraviolet (UV) (Ex. 32) or in the infrared (IR) (Ex. 34) as well as the index of refraction of X-rays (Ex. 33).

Superconductors form a class of very particular materials. They are considered in this chapter (Pb. 7–9) for convenience even though their specific magnetic properties require the aid of perfect conductor (Pb. 7). Their specific heat and band structure are considered in Chapter V (see Exs. 2b and 23).

Useful formulas for using the Fermi–Dirac function, f(E):

$$F(1): \int_0^\infty f(E)\varphi(E)dE = \int_0^{E_F} \varphi(E)dE + \frac{(\pi k_B T)^2}{6}\varphi'(E)_{E_F} + \varepsilon T^4$$

$$F(2):\int_0^\infty \frac{E^p dE}{e^{\gamma\left(\frac{E}{\chi}-1\right)}+1} = \chi^{p+1}\left(\frac{1}{p+1}+\frac{\pi^2}{6}\frac{p}{\gamma^2}\right) \text{ when } \gamma \gg 1.$$

Exercises

Exercise 1: Free electrons in a 1D system: going from an atom to a molecule and to a crystal

Consider a segment with length L along which electrons are able to move freely ($V = 0$). Outside this segment, their potential energy V is infinite ($V = \infty$ for $x \geq L$ and $x \leq 0$).

(1) What is the general form of the solutions to Schrödinger's equation? Give the solutions in the case of fixed boundary conditions. Sketch the graph for the first three wave functions.

(2) Deduce the allowed quantized energy levels. What is the expression for the first three distinct energy levels denoted E_1, E_2, and E_3.

(3) Application to an atom: L = 3 Å. What are the numerical values (in eV) for the E_1, E_2, and E_3? The atom has two electrons assumed to be free, what is the minimal energy that these electrons must have to go from the ground state to the first excited state?

(4) Application to a molecule: L = 15 Å. What are the numerical values (in eV) for the E_1, E_2, and E_3? The formula for the molecule could be $H_2C{=}CH{-}CH{=}CH_2$ in which the symbol-- represents the existence of a π electron that is susceptible to propagate freely along the molecule. What is the minimal energy that one of these electrons must have in order to be excited from the ground to the first excited state?

(5) Application to a metal: L = 3 mm. What are the numerical values (in eV) for the E_1, E_2, and E_3? The row consists of identical divalent atoms (2 e^- free/atom) equidistant by a distance a = 3 Å.

How many energy levels are occupied in the ground state? What is the energy E_F of the last level occupied?

Show the dispersion curve of free electrons using fixed boundary conditions. Give the numerical values of k_F and E_F and thus the minimal energy δE for an electron to go from the lowest level to the first unoccupied level. Comment on the transition from an atom to a crystal. $\left[(h, m, e), \dfrac{\hbar^2}{2m} = 3.8\ \text{eV} \cdot \text{Å}^2\right]$

Solution:

(1) The Schrödinger equation reduces to $-\dfrac{\hbar^2}{2m}\dfrac{d^2\psi}{dx^2} = E\psi$ or

$$\frac{d^2\psi}{dx^2} + k^2\psi = 0, \tag{1}$$

where $k^2 = \dfrac{2mE}{\hbar^2}$. The solutions are of the form $\psi = A\sin kx + B\cos kx$.

The fixed boundary conditions imply $\psi(x = 0) = \psi(x = L) = 0$.

We thus find $\psi = A \sin kx$ with $k = n\pi / L$, where n is a positive integer. (2)

The graphs of the first three wave functions (n = 1, 2, 3) are shown in Fig. 1.

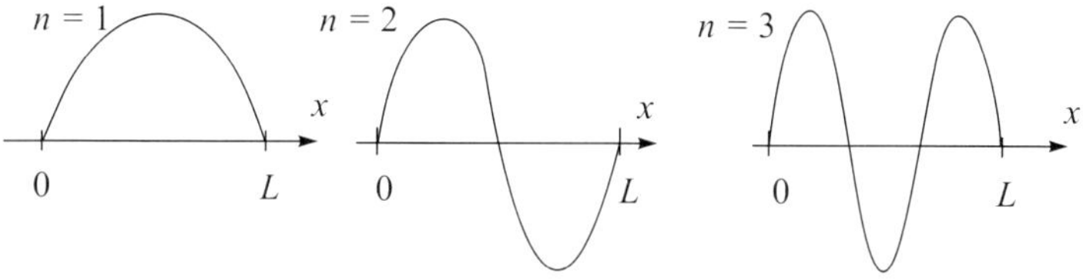

Figure 1

(2) Taking into account the relations (1) and (2), we obtain:

$$E = \frac{\hbar^2}{2m}\left(\frac{\pi}{L}\right)^2 n^2, \text{ which gives}$$

$$E_1 = \frac{\hbar^2}{2m}\left(\frac{\pi}{L}\right)^2; E_2 = \frac{\hbar^2}{2m}4\left(\frac{\pi}{L}\right)^2; E_3 = \frac{\hbar^2}{2m}9\left(\frac{\pi}{L}\right)^2.$$

(3) In the case of an atom, we find: E_1 = 4.2 eV; E_2 = 16.7 eV; E_3 = 38 eV. Among the possible states (deduced from the Schrödinger equation), only the lowest energy states will be occupied. We

fill these states by electrons that differ from each other by either quantum number (n) or spin ($s = \pm 1/2$).

(4) In the case of a molecule, we find $E_1 = 0.167$ eV; $E_2 = 0.67$ eV; $E_3 = 1.5$ eV. When L increases, the allowed levels move closer in energy. Here also, only one level will be occupied (the grounds state) by 2 e^- π ($\uparrow$ and $\downarrow$). $E_m = E_1 - E_2 = 0.5$ eV.

(5) Compared with an atom, in the case of a metal considered here, L is 10^7 times larger and the energies will therefore be 10^{-14} times smaller. $E_1 = 4.2 \times 10^{-14}$ eV; $E_2 = 16.7 \times 10^{-14}$ eV; $E_3 = 38 \times 10^{-14}$ eV. However, we must fill these energies starting at the lowest level with 2×10^7 free electrons, that is to say that 10^7 states will be occupied in the ground state (from $n = 1$ to $n = 10^7$).

The energy of the last occupied level is $E_F = \frac{\hbar^2}{2m} N^2 \left(\frac{\pi}{L}\right)^2$. The corresponding wavevector is $k_F = N\frac{\pi}{L}$. We have $N = \frac{L}{a} = 10^7$, which leads to $k_F = \frac{\pi}{a}$ and $E_F = \frac{\hbar^2}{2m}\left(\frac{\pi}{a}\right)^2 = 4.2$ eV.

The minimal energy between the last occupied state and the first empty state is

$$\delta E = E(N+1) - E(N) \approx E_1 \times 2N = 2N\frac{\hbar^2 \pi^2}{2mL^2} \approx 8.4 \times 10^{-7}\ \text{eV} \cdot$$

We can also find this result starting from

$$E = \frac{\hbar^2 k^2}{2m}; \quad \delta E = \frac{\hbar^2 k_F}{m}\delta k, \text{ where } \delta k = \frac{\pi}{L}.$$

Figure 2 (top) shows schematically the transition from an atom (a) to a crystal (b). The electronic densities have been chosen to be identical in the two cases by taking E_F of the crystal to be identical to the energy E_1 of the atom. Apart from this coincidence (related to a linear identical electronic density), the essential result is that the distance between allowed energy levels is reduced, leading to those in a metal being quasi-continuous. As in an atom, the different electrons must differ by at least a quantum number (here either n or s).

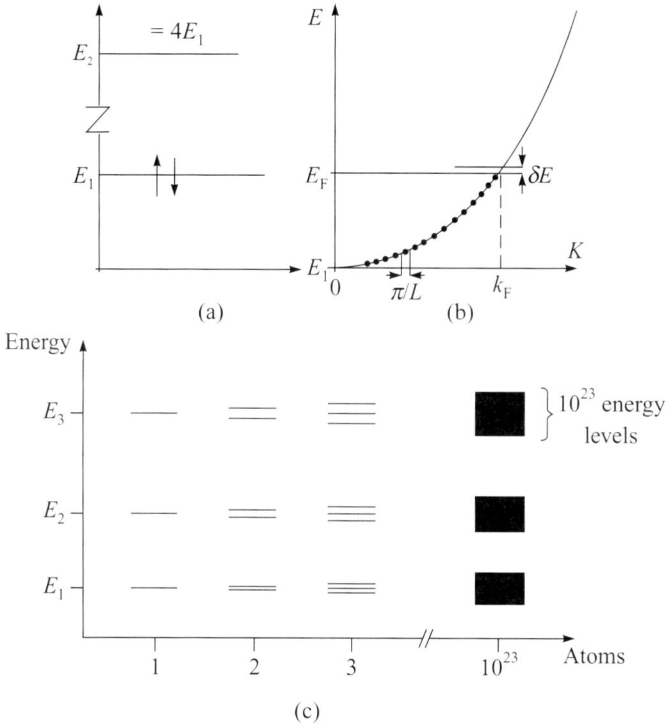

Figure 2

In fact, if there are multiple atoms side by side and they are interdependent, the discrete energy levels are fanned out as shown in Fig. 2c.

Exercise 2: 1D metal with periodic boundary conditions

Consider an infinite row of identical atoms that are equidistant by a.

(1) Find the general expression of the electron wavefunction that can move along the row.

(2) This wavefunction satisfies periodic boundary conditions, PLC, with a periodicity L: $\psi(x) = \psi(x + L)$. Find the wavevector quantization and the possible quantized energy levels for the electrons.

(3) Find the expression and then the numerical value for the first three distinct energy levels in the case of a monovalent

element ($1e^-$/at.) in which the atoms are equidistant by a = 3 Å (take L = 3 mm). Find the expression and the numerical value characterizing the last occupied level at 0 K in terms of k_F and E_F (the wave vector is in unit of Å^{-1} and the Fermi energy in eV).

(4) Find the answer for Question (3) using instead fixed boundary conditions, FBC, $\psi(0) = \psi(L) = 0$; see previous exercise. Illustrate the similarities and differences using a comparative graph. $(\hbar, m, e)$

Solution:

(1) As in the previous exercise, the Schrödinger equation reduces to $\dfrac{d^2\psi}{dx^2} + k^2\psi = 0$, where $k^2 = \dfrac{2mE}{\hbar^2}$

When using periodic boundary conditions, it is best to use solutions in the form of $\psi(x) = C_e{}^{ikx}$, which represents a running wave (where k can take negative or positive values) compared with a standing wave (written in terms of sine wave with k > 0), resulting from two running waves propagating in opposite directions.

(2) $\psi(x) = \psi(x + L)$ implies $e^{\text{ikL}} = 1$, where $k = \dfrac{n2\pi}{L}$ (with n as a non-zero integer): $E = \dfrac{\hbar^2}{2m}k^2 = \dfrac{\hbar^2}{2m}\left(\dfrac{2\pi}{L}\right)^2 n^2$.

(3) The first three energy levels correspond to n = 1, 2, 3 are $E_1 = 16.7 \times 10^{-14}\,\text{eV}$; $E_2 = 67 \times 10^{-14}\,\text{eV}$; $E_3 = 1.5 \times 10^{-12}\,\text{eV}$.
On must place N electrons in cells with dimensions $2\pi/L$ because of the two electrons per state (↓↑) and taking into account that n (and k) are nonzero integers: $N = L/a = 10^7$.

$$k_F = \frac{1}{4} \cdot N \cdot \frac{2\pi}{L} = \frac{\pi}{2a} = 0.52\ \text{Å}^{-1};\ E_F \approx 1\ eV.$$

(4) This is the same as Question (5) in the previous exercise but with a monovalent element.

$$E_1 = 4.2 \cdot 10^{-14}\,\text{eV};\ E_2 = 16.7 \cdot 10^{-14}\,\text{eV};\ E_3 = 38 \cdot 10^{-14}\,\text{eV}.$$

$$k_F = \frac{1}{2} \cdot N \cdot \frac{\pi}{L} = \frac{\pi}{2a} = 0.52\ \text{Å}^{-1};\ E_F \approx 1\ \text{eV}.$$

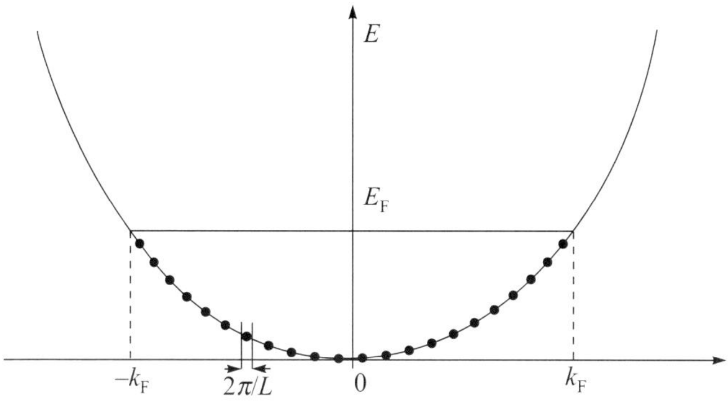

Figure 3

Comparing the two boundary conditions shown in Fig. 2 (FBC) and Fig. 3 (PBC), the differences concern essentially the first occupied energy levels. The energies E_1, E_2, and E_3 are four times greater for periodic boundary conditions than fixed boundary conditions, but the corresponding states can be filled by twice as many electrons due to the negative values taken into account by $n = -1, -2, -3$, and by k.

In total, the number and the size of the cells should be multiplied by 2 in the case of periodic boundary conditions compared with fixed boundary conditions leading to identical values for wave vectors and the Fermi energies, for given equal electronic densities. This result will be the same for the density of states (number of electronic states contained in an energy range between E and $E + dE$) when the number of electrons is large.

In general, the choice of the boundary conditions thus leads to the same result. Exercise 7 and Pb. 1 show specific exceptions to this rule.

Exercise 3: Free electrons in a rectangular box (FBC)

Consider an electron of mass m subjected to zero potential energy inside of a rectangular box with sides a, b, and c along Ox, Oy, and Oz axes (Fig. 4). The potential function is infinite outside the rectangular box.

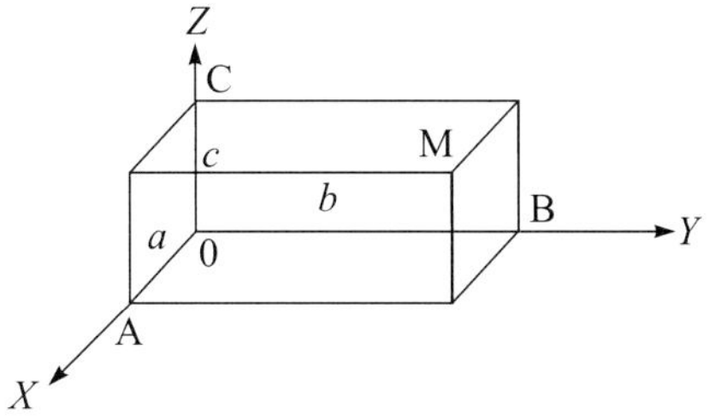

Figure 4

(1) What is the differential equation describing the wave function of the electron?

(2) Find a solution to this equation using separation of variables of the type: $\phi(x,y,z)=\phi_x(x)\cdot\phi_y(y)\cdot\phi_z(z)$.

Show that the equation of (1) can be written in the form $\frac{1}{\phi_x}\frac{\partial^2\phi_x}{\partial^2 x}+\frac{1}{\phi_y}\frac{\partial^2\phi_y}{\partial^2 y}+\frac{1}{\phi_z}\frac{\partial^2\phi_z}{\partial^2 z}=-\frac{2m}{\hbar^2}E$,

and that it is sufficient to resolve the equations $\frac{d^2\phi_x}{dx^2}+k_x^2\phi_x=0, \frac{d^2\phi_y}{\partial y^2}+k_y^2\phi_y=0, \frac{d^2\phi_z}{dz^2}+k_z^2\phi_z=0,$

where $k_x^2+k_y^2+k_z^2=E\frac{2m}{\hbar^2}$.

(3) Integrate the preceding differential equations.

(4) Show that the limiting conditions at surfaces of the cavity impose solutions of the type

$$\phi_x=A\sin\left(\frac{n_x\pi}{a}x\right),\ \phi_y=B\sin\left(\frac{n_y\pi}{b}y\right),\ \phi_z=C\sin\left(\frac{n_z\pi}{c}z\right)$$

in which the quantum numbers n_x, n_y, n_z are integers greater than zero. Find the complete expression of the resulting wave function.

(5) Express the energy quantization as a function of n_x, n_y, and n_z and the dimensions of the cavity.

(6) If the cavity is now assumed to be cubic ($a = b = c = L$):

(a) Find the different wave functions for the first three distinct energy levels.

(b) Find the energy of each level.

(c) Determine the degeneracy of each level, that is, the number of independent wave functions having the same energy; neglect spins.

Solution:

(1) The electron obeys Schrödinger's equation $-\frac{\hbar^2}{2m}\nabla^2\psi + V\psi = E\psi$, where $V = \infty$ outside the box (implying that $\psi = 0$ and $V = 0$ inside the box. This results in $-\frac{\hbar^2}{2m}\left(\frac{\partial^2\psi}{\partial x^2} + \frac{\partial^2\psi}{\partial y^2} + \frac{\partial^2\psi}{\partial x^2}\right) = E\psi.$

(2) If we seek solutions using separation of variables for the spatial part of the wave function, after substitution we obtain $\phi_y\phi_z\frac{d^2\phi_x}{dx^2} + \phi_x\phi_z\frac{d^2\phi_y}{dy^2} + \phi_x\phi_y\frac{d^2\phi_z}{dz^2} = -\frac{2mE}{\hbar^2}\phi_x\phi_y\phi_z$ or $\frac{1}{\phi_x}\frac{d^2\phi_x}{dx^2} + \frac{1}{\phi_y}\frac{d^2\phi_y}{dy^2} + \frac{1}{\phi_z}\frac{d^2\phi_z}{dz^2} = -\frac{2mE}{\hbar^2}$, which is of the form $f(x) + g(y) + h(z) = Cste < 0.$

The variables x, y, z being independent, such an equation may be solved only when the functions $f(x)$, $g(y)$, and $h(z)$ are separately equal to three constants k_x^2, k_y^2, and k_z^2.

The initial equation thus reduces to the following three relations:

$\frac{d^2\phi_x}{dx^2} + k_x^2\phi_x = 0, \frac{d^2\phi_y}{dy^2} + k_y^2\phi_y = 0, \frac{d^2\phi_z}{dz^2} + k_z^2\phi_z = 0$, where

$\frac{\hbar^2}{2m}(k_x^2 + k_y^2 + k_z^2) = E = Cste.$

(3) The solutions are of the type $\phi_x = A\sin k_x x + D\cos k_x x$ (1)
or equivalently $\phi_x = A'\exp i\,k_x x + D'\exp -i\,k_x x$ (2)

(4) The boundary conditions suggested by the question are the FBC and the wave function is zero at the borders of the cavity:

$\phi_x(0) = \phi_y(0) = \phi_z(0) = 0;\ \phi_x(a) = \phi_y(b) = \phi_z(c) = 0.$

The first boundary conditions (fixed) impose that the constants of integration such as D are zero.

The second boundary conditions results in $k_x a = n_x \pi$, $k_y b = n_y \pi$, $k_z c = n_z \pi$, where n_x, n_y, and n_z are positive integers.

The resulting wave function is $\phi(x, y, z) = ABC \sin \frac{n_x \pi}{a} x \cdot \sin \frac{n_y \pi}{b} y \cdot \sin \frac{n_z \pi}{c} z$, where ABC $= (8/abc)^{1/2}$ is obtained after normalization $\int_0^a \int_0^b \int_0^c \phi^2(x,y,z)\, dxdydz = 1$.

(5) Starting from $E = \frac{\hbar^2}{2m}(k_x^2 + k_y^2 + k_z^2)$ and after substitution, we note that the energy depends on the quantum numbers n_x, n_y, and n_z because $E = \frac{\hbar^2 \pi^2}{2m}\left(\frac{n_x^2}{a^2} + \frac{n_y^2}{b^2} + \frac{n_z^2}{c^2}\right)$.

(6) Under the hypothesis of a cubic cavity, the general expression reduces to $\phi(x,y,z) = \left(\frac{8}{L^3}\right)^{1/2} \sin \frac{n_x \pi}{L} x \cdot \sin \frac{n_y \pi}{L} y \cdot \sin \frac{n_z \pi}{L} z$ and the energy is $E = \frac{\hbar^2 \pi^2}{2mL^2}(n_x^2 + n_y^2 + n_z^2)$. The first distinct energy levels correspond to $n_x^2 + n_y^2 + n_z^2 = 3,\ 6,\ 9$. We exclude zero values of n_x, n_y, and n_z because the resulting ϕ is zero.

The energy levels, their degeneracy (neglecting spin), and their corresponding wave functions are shown in the following table.

Levels	Energy	Degeneracy	Wave functions
1	$3\left(\frac{\hbar^2\pi^2}{2mL^2}\right)$	1: $n_x = n_y = n_z = 1$	$\phi(111) = \sqrt{\frac{8}{L^3}} \sin\left(\frac{\pi x}{L}\right) \sin\left(\frac{\pi y}{L}\right) \sin\left(\frac{\pi z}{L}\right)$
2	$6\left(\frac{\hbar^2\pi^2}{2mL^2}\right)$	3: $n_x = 2, n_y = n_z = 1$	$\phi(211) = \sqrt{\frac{8}{L^3}} \sin 2\left(\frac{\pi x}{L}\right) \sin 1\left(\frac{\pi y}{L}\right) \sin 1\left(\frac{\pi z}{L}\right)$
		$n_y = 2, n_x = n_z = 1$	$\phi(121) = \sqrt{\frac{8}{L^3}} \sin 1\left(\frac{\pi x}{L}\right) \sin 2\left(\frac{\pi y}{L}\right) \sin 1\left(\frac{\pi z}{L}\right)$
		$n_z = 2, n_x = n_y = 1$	$\phi(112) = \sqrt{\frac{8}{L^3}} \sin 1\left(\frac{\pi x}{L}\right) \sin 1\left(\frac{\pi y}{L}\right) \sin 2\left(\frac{\pi z}{L}\right)$
3	$9\left(\frac{\hbar^2\pi^2}{2mL^2}\right)$	3: $n_x = n_y = 2, n_z = 1$	$\phi(221) = \sqrt{\frac{8}{L^3}} \sin 2\left(\frac{\pi x}{L}\right) \sin 2\left(\frac{\pi y}{L}\right) \sin 1\left(\frac{\pi z}{L}\right)$
		$n_x = n_z = 2, n_y = 1$	$\phi(212) = \sqrt{\frac{8}{L^3}} \sin 2\left(\frac{\pi x}{L}\right) \sin 1\left(\frac{\pi y}{L}\right) \sin 2\left(\frac{\pi z}{L}\right)$
		$n_y = n_z = 2, n_x = 1$	$\phi(122) = \sqrt{\frac{8}{L^3}} \sin 1\left(\frac{\pi x}{L}\right) \sin 2\left(\frac{\pi y}{L}\right) \sin 2\left(\frac{\pi z}{L}\right)$

Exercise 4: Periodic boundary conditions, PBC, in a 3D metal

An infinite metal is considered in which the electrons can propagate freely.

(1 & 2) Using periodic boundary conditions (with a periodicity of a along Ox, b along Oy, and c along Oz), answer to Questions (1) and (2) of Ex. 3 above.

(3) Starting from the solutions to Schrödinger's equation, show that PBC leads to a quantization of the energy of free electrons. Express this energy as a function of the quantum numbers n_x, n_y, and n_z and of a, b, and c.
Find the resulting expression for the wave functions.

(4) Assume $a = b = c = L$
   (a) Give the wave function for the first three distinct energy levels.
   (b) Find the energy for each level.
   (c) Neglecting spin, determine the degeneracy of each level.

(5) For clarity, limit now the situation to 2D in order to show, in the wave vector (phase) k space, the distribution of states imposed by the periodic boundary condition. Compare this distribution with that resulting from fixed boundary conditions.

Solution:

This problem is a generalization to 3D of that in Ex. 2 using periodic boundary conditions, as the preceding exercise was a generalization to 3D of Ex. 1 using fixed boundary conditions.

(1 & 2) See Ex. 3 for the solutions to the first two questions.

(3) We choose to represent the running waves by solutions of the form $\phi_x = A' \, exp \; i \; k_x x$; $\phi_y = B' \, exp \; i \; k_y y$; $\phi_z = C' \, exp \; i \; k_z z$; where k_x, k_y, and k_z are >0 or <0 or eventually =0, except at least one.

Periodic boundary conditions applied to these solutions result in $k_x a = 2\pi n_x$; $k_y b = 2\pi n_y$; $k_z c = 2\pi n_z$, (1)
where n_x, n_y, and n_z are whole numbers with at least one different from zero.

$$E = \frac{h^2}{2m}(k_x^2 + k_y^2 + k_z^2) = \frac{\hbar^2}{2m} 4\pi^2 \left(\frac{n_x^2}{a^2} + \frac{n_y^2}{b^2} + \frac{n_z^2}{c^2} \right).$$

The resulting wave function will have the form

$\phi(x,y,z) = D' \exp i\,(k_x x + k_y y + k_z z) = D' \exp i\,(\vec{k}\cdot\vec{r})$,
where $D' = A'B'C'$.
The components of $\vec{k}$ are given by relation (1).

(4) **Level 1:** $E_1 = \left(\frac{\hbar^2}{2m}\right)\frac{4\pi^2}{L^2}$, where $n_x = \pm 1; n_y, n_z = 0$ or any other result obtained by circular permutation of n_x, n_y, and n_z.

For example, $\phi(-1,0,0) = D' \exp -i\left(\frac{2\pi}{L}\right)x$

- The degeneracy is of 6: (100); (−100), (010), (0−10); (001) and (00−1)

Level 2: $E_2 = 2E_1$, where $n_x = \pm 1$, $n_y = \pm 1$, $n_z = 0$, and any other result obtained by circular permutation.

For example, $\phi(0,-1,1) = D' \exp +i\left(\frac{2\pi}{L}\right)(z-y)$.

- The degeneracy is of 12: (110); (−110); (1−10); (−1−10); (101); (10−1); (−101); (−10−1) and (011) (0−11) (01−1) (0−1−1).

Level 3: $E_3 = 3E_1$; $n_x = \pm 1$, $n_y = \pm 1$, $n_z = \pm 1$.
For example, $\phi(1-11) = D' \exp +i\,(2\pi/L)(x+z-y)$.

- The degeneracy is of 8: (111); (−1,1,1); (1−11); (11−1); (−1−11); (−11−1); (1−1−1) and (−1−1−1).

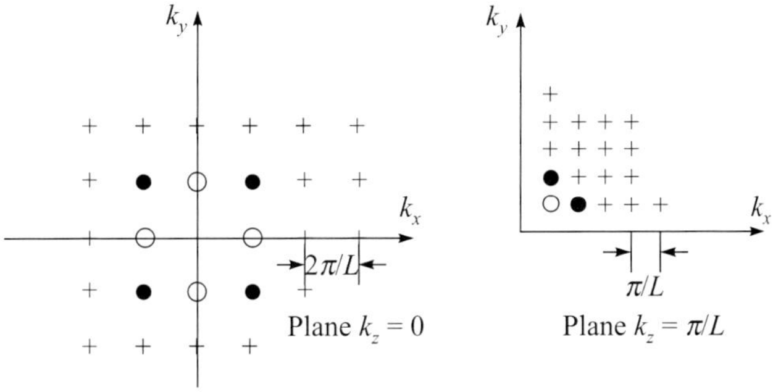

(a) PBC (b) FBC

Figure 5

(5) Fig. 5a shows the end points of the $\vec{k}$ vectors relative to periodic boundary conditions: the white circles correspond

to level 1, and the black circles to level 2 (in the $k_z = 0$ plane).

Figure 5b uses the same symbols to show the end points of the $\vec{k}$ vectors relative to fixed boundary conditions in the $k_z = \pi/L$ plane.

Exercise 5a: Electronic states in a metallic cluster: influence of the cluster size

Consider three metallic clusters in the form of a cube constituting, respectively, 8, 27, and 64 identical monovalent atoms. The center of each atom is located on the point of a cubic simple lattice with parameter a and we consider that the electron liberated by each atom can propagate freely in the interior of cubes, represented by dashes in Fig. 6, with edges successively equal to 2a, 3a, and 4a. We note that the choice of outside volume of these aggregates takes into account the steric hindrance of the atoms while still maintaining the electronic density n in the three clusters.

(1) Using fixed limits, find the expression of the quantized kinetic energies of free electrons enclosed by such cubic boxes.

(2) At $T = 0$ K, find in each case the numerical value of the quantum numbers n_x, n_y, and n_z relative to the last occupied state. Compare the corresponding energies, $E_M(1)$, $E_M(2)$, and $E_M(3)$ at the Fermi energy E_F of an infinite sample of the same nature and therefore has the same electronic density n.

(3) Compare the same average electronic energies, $\overline{E_1}$, $\overline{E_2}$ and $\overline{E_3}$, with the average kinetic energy $\overline{E}$ of a free electron propagating in an infinite metal.

(4) For each cluster, find the energy difference $\Delta E = E_e - E_M$ corresponding to an electronic transition between the last occupied level and the first unoccupied and non-equivalent level.

(5) Each cluster receives an additional free electron (e.g., by substituting a divalent atom into a monovalent atom). Characterize the electronic state (n_x^0, n_y^0, n_z^0) and the energy E^0 taken by this electron.

(6) *Numerical application*: What are the values of the energies E_M, $\overline{E}$, ΔE, and E^0 for $E_F = 5$ eV?

(7) Comment on the physical results of this exercise. In particular, define the size limit of an aggregate for which it will acquire a metallic character (referring to the Kubo criteria: $\Delta E = k_B T = 25$ meV at ambient temperature) and find the order of magnitude of the number of atoms N_C of the corresponding aggregate.

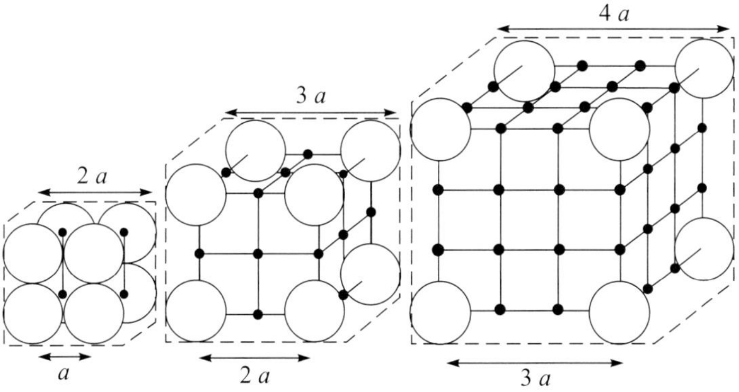

Figure 6

Solution:

(1) We use directly a consequence of Ex. 3:

$\psi(x, y, z) = A \sin k_x x \cdot \sin k_y y \cdot \sin k_z z$, where

$k_x = \dfrac{n_x \pi}{L_x}$; $k_y = \dfrac{n_y \pi}{L_y}$; $k_z = \dfrac{n_z \pi}{L_z}$ and n_x, n_y, and n_z are positive whole numbers.

When $L_x = L_y = L_z = pa$, where $p = 2, 3$, or 4, we find

$E(p) = (\hbar^2/2m)(\pi^2/p^2a^2)(n_x^2 + n_y^2 + n_z^2)$

(2) The *first cluster* contains eight electrons that, when taking into account spin (±1/2), will be distributed on four characteristic energy levels by a combination of different n_x, n_y, and n_z.
First level: 111 (2e)
Second level: 211, 121,112 (6e)
The energy of this degenerate level is
$E_M(1) = (3\pi^2/2)(\hbar^2/2ma^2)$.
The Fermi energy of a free electron in an infinite metal is
$E_F = (\hbar^{2}/2m)(3\pi^2/n)^{2/3}$, where $n = a^{-3}$. Thus, we find
$E_M(1)\ E_F = (3\pi^2)^{1/3}/2 = 1.54$.

The *second cluster* contains 27 free electrons. The 27th must occupy the 14th level obtained by classifying the results from the sum $n_x^2 + n_y^2 + n_z^2$ by increasing order and taking into account the degeneracy of certain levels: 1 when $n_x = n_y = n_z$; 3 when $n_x = n_y \neq n_z$; and 6 when $n_x \neq n_y \neq n_z$. The desired sequence is

111(1); 211(3); (221)(3); 311(3); (222)(1); (321)(6).

The last five electrons occupy the levels of the 321 type, which may contain 12 electrons. The corresponding energy is

$E_M(2) = (14\pi^2/9)\,(\hbar^2/2ma^2)$.

Thus, we have $E_M\,(2)\,/E_F = 1.6$.

The third cluster contains 64 free electrons and 32 distinct levels. By prolonging the sequence, we find 322(3); 411(3); 331(3); 421(6), and from which we find the result in $E_M(3) = (21\pi^2/16)\,(\hbar^2/2ma^2)$ which leads to $E_M(3)/E_F = 1.35$. Note that the levels of type 421 are occupied by the maximum number of electrons they may accept. The corresponding cluster would be an insulator at least at 0 K.

(3) $\overline{E(1)} = [2E\,(111) + 6(112)]/8 = (7/8)E_M\,(1)$ and the average energy of free electrons in a metal is $\overline{E} = \dfrac{3}{5}E_F$ which leads to $\overline{E(1)}\,/\,\overline{E} \cong 2.25$.

Taking into account the quantity $\hbar^2\pi^2/18ma^2$ as a unit of energy, the half sum of electronic energies of the 2nd cluster corresponds to $3 + (6\times3) + (9\times3) + (11\times3) + 12 + (14\times2.5) = 128$,

which leads to $\overline{E}(2)/\overline{E_M} = 0.677$ and to $\overline{E}(2)/\overline{E} = 1.8$.

For the third cluster, the above sum filled up to the 421 level leads to $\overline{E}(3) = 474/\,3\overline{2}$ in units of $\hbar^2\pi^2/32ma^2$.

$\overline{E}(3)/E_M(3) \approx 0.705$ and $\overline{E}(3)/\overline{E} \cong 1.59$.

(4) $\Delta E(1)/E_M(1) = (9-6)/6 = 0.5$

$\Delta E(2)/E_M(2) = (17-13)/13 = 0.3$

$\Delta E(3)/E_M(3) = (22-21)/21 = 0.048$

(5) For cluster 1, the levels of type 211 will all bc occupied and an additional electron will be found in a level of type 221 from which the energy $E^\circ(1)$ will be
$E^\circ(1) = E_M(1) + \Delta E_1 = 1.5E_M(1)$.

Similar result holds for cluster 3, the level 421 being filled and the next level to be filled is 332: $E^\circ(3) = (22/21)E_M(3)$.
On the contrary, for cluster 2, we find $E^\circ(2) = E_M(2)$ because the corresponding levels, of type 221, are incompletely filled.

(6) $E_M(1) = 7.7$ eV ; $E_M(2) = 8$ eV ; $E_M(3) = 6.7$ eV ; $E_F \sim 5$ eV

$\overline{E}(1) = 6.7$ eV; $\overline{E}(2) = 5.4$ eV; $\overline{E}(3) = 4.7$ eV; $\overline{E} = 3$ eV

$\Delta E(1) = 3.85$ eV; $\Delta E(2) = 2.4$ eV; $\Delta E(3) = 0.32$ eV

$E^\circ(1) = 11.55$ eV; $E^\circ(2) = 8$ eV; $E^\circ(3) = 7$ eV; $E^\circ = 5$ eV

(7) Starting from a small number of atoms, the increase of the size of the aggregates leads therefore to a reduction in the spacing between subsequent energy levels. By extrapolating the numerical values, we obtain an estimate of the limiting size for a transition from a non-metal to a metal in the order of several thousand atoms. For an order of magnitude, we can follow the method of Ex. 1: $E = \dfrac{\hbar^2 k^2}{2m}\, \delta E = \dfrac{\hbar^2 k_F}{m} \delta k$ (where $\delta k = \pi/L$), which leads to $\delta E/E_F = \delta k/2k_F = \pi/2.3L$. $k_F = 1.15$ Å^{-1} so that L = 270 Å and $N_C = nL^3 \approx 7 \times 10^5 - 10^6$ atoms. Thus, the cluster 3 is an insulator at $T = 0$ K but not at room temperature where $k_B T = 25$ meV.

For more details, see J. Friedel, *Helvetica Physica Acta* **56**, 1983, 507 and W. P. Halperin, *Rev. Mod. Phys.* **58**, 1983, 533.

Comments: Metallic clusters and nanoparticles

There is no sharp discrimination between the terms "cluster" and "nanoparticle" even if clusters are mostly considered as species, exactly defined in chemical composition and structure, and nanoparticles often linked with their dimension in the nanometer range. Thus clusters are groups of bound atoms intermediate in size between molecules and bulk solids and exhibiting specific physical and chemical properties, between molecules and solid state, related to their size dependence or "size quantization."

The size quantization occurs in many nano-objects that have been the subject of a great attention in solid state physics. The present exercise, Ex. 5a, and the following, Ex. 5b, are limited to metallic clusters while other exercises and problems related to semiconducting quantum wells or to carbon-based objects (graphene; single and multi-walls C-nanotubes and fullerenes in the form of soccer balls) are considered in Chapter V. The Jahn–Teller effect occurring in alkali halides is considered in the present section, Ex. 6, because it is easily deduced from the present developments of quantization in rectangular or cubic boxes.

Despite its large simplifications, the present exercises, Ex. 5a and 5b, illustrate well certain aspects of small metallic clusters that have electronic states intermediate between an isolated atom (with very large separations between the levels) and an infinite metal (where the levels are quasi-continuous). The cohesive energy being decreasing when the kinetic energy of the electrons is increased, the present exercise shows simply that the maximum and average kinetic energies of an electron in a cluster are all greater than the energies corresponding to an infinite sample. In addition, it shows that this difference decrease as the size of the clusters increases because successive levels move closer together more quickly than the increase in electronic density (the dependence is essentially the p^{-2} factor in the expression for E(p) in 1). This result is physically satisfying because the cohesion energy of an infinite metal is surely larger than that of clusters (otherwise the metal would break into clusters) and the increase in kinetic energy of the constituents of a solid would tend to reduce its cohesive energy. The fact that $E_M(2) > E_M(1)$ is accidentally related to the discrete and nonlinear character of the progression because some clusters are more stable than others. The relatively more stable metallic clusters are composed with a "magic" number of atoms (2, 8, 20, 34, 40, 58, ...) where the last occupied states are fully occupied and where the addition of one more electron induces a jump in their mean kinetic energy.

From the theoretical point of view, a more precise evaluation of the magic numbers is detailed in M. K. Harbola, *Proc. Nat. Acad. Sci. U S A.* **89,** 1992, 1036 and in E. Engel and J. P. Perdew, *Phys. Rev. B* **43**, 1991, 1331 for the asymptotic size dependence of electronic properties within a Thomas Fermi approximation for jellium spheres. Also mentioned in Ex. 27 for the cohesive energy of free electron metals there is the more general and more sophisticated use of density functional theory (DFT) to describe complex electronic structures, to accurately treat large systems and to predict physical and chemical properties (Density functional theory (DFT) of molecules, clusters, and solids, W. Kohn et al., *J. Phys. Chem.* **100**, 1996, 12974.)

From the point of view of applications of metallic clusters and besides a wide range of specific physical properties such as luminescence and molecular magnetism, one must mention their decisive role in catalysis. Because catalysis by metals is a surface phenomenon, many technological catalysts contain small (typically nanometer-sized) supported metal particles with a large fraction of the atoms exposed. Metal catalysts are used on a large scale for refining of petroleum, conversion of automobile exhaust, hydrogenation of carbon monoxide, hydrogenation of fats, and many other processes. The smaller the metal particles, the larger the fraction of the metal atoms that are exposed at surfaces, where they are accessible to reactant molecules and available for catalysis. Such a considerable industrial impact justifies the intense research on the influence of the size and of composition of mono or bimetallic catalysts that are in a finely dispersed form as particles on a high-area porous metal oxide support.

Exercise 5b: Electronic states in metallic clusters: influence of the shape

Consider three aggregates of metal atoms (clusters) with the same volume $V = L^3$ but, as shown in Fig. 7, each having a different form: aggregate $A = L \times L \times L$; aggregate $B = 2L \times \frac{L}{\sqrt{2}} \times \frac{L}{\sqrt{2}}$; aggregate $C = L \times L\sqrt{2} \times L/\sqrt{2}$.

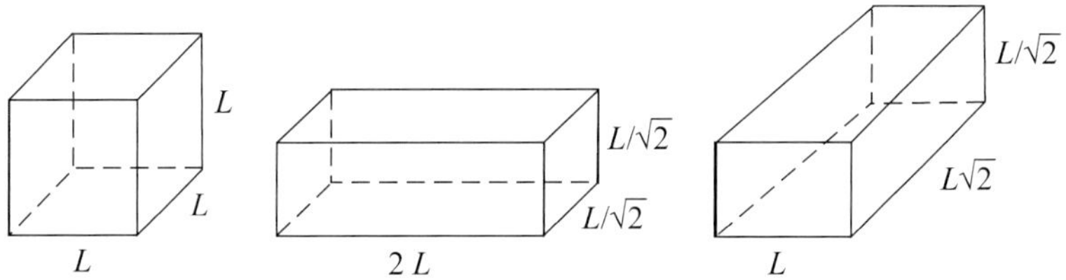

Figure 7

For each one, find the increasing sequence of energy levels specifying the degeneracy of each level (but neglecting spin). Limit the discussion to the first four distinct levels, take the quantity $\hbar^2/2mL^2$ as unity, and use fixed boundary conditions to take into account the reduced size of the clusters.

What is the relation between L_x, L_y, and L_z to obtain non-degenerate electronic levels (except the spin)?

Solution:

$E = (\hbar^2\pi^2/2m)/(n_x^2/L_x^2 + n_y^2/L_y^2 + n_z^2/L_z^2)$ (see Ex. 5a above).
For aggregates A, B, and C, we find, respectively, (in $\hbar^2/2mL^2$ units):

$E_A = n_x^2 + n_y^2 + n_z^2$

$E_B = n_x^2/4 + 2n_y^2 + n_z^2$

$E_C = n_x^2 + n_y^2/2 + n_z^2$

The table below is easily obtained (D: means degeneracy). It shows the influence of the geometry of the clusters on their electronic states.

Aggregate		First level	Second level	Third level	Fourth level
A	n_x, n_y, n_z	111	211, 121, 112	221, 122, 212	311, 131, 113
	E_A	3	6	9	11
	D	1	3	3	3
B	n_x, n_y, n_z	111	211	311	411
	E_B	4.25	5	6.25	8
	D	1	1	1	1
C	n_x, n_y, n_z	111	121	211	131
	E_C	3.5	5	6.5	7.5
	D	1	1	1	1

The energy of the initial level 111 in clusters B and C is larger than in cubic cluster A but the subsequent levels are closer together. Consequently to determine which form is the most stable, one must take into account the degeneracy of the levels in a cubic cluster. Thus, if we suppose that the minimal kinetic energy of an ensemble of electrons is the only stability criteria, the clusters A and C will be more stable than the cluster B when they are filled with six free electrons and cluster A will be more stable than C when filled with eight electrons.

In order to not obtain the same energies in the general expression, we should have $L_x/L_y \neq n_x/n_y$ and $L_x/L_z \neq n_x/n_z$, that is, to say that the ratio of dimensions has to be irrational.

Exercise 6: F center in alkali halide crystals and Jahn–Teller effect (variation of Ex. 5a and 5b)

An F-center in an alkali halide crystal occurs when there is the lack of a negative ion, a vacancy, in the lattice, creating a cubic cavity in which an electron can be trapped. The side of this cavity is nearly equal to the lattice parameter a. It is αa, where $\alpha \approx 1.13$. Find the expression of the energy E^0 for the different states of this electron.

Find the energy $h\nu$ and the wavelength λ of photons, which, when absorbed, can induce an electronic transition from the ground state E_1^0 toward the first excited state E_2^0 in LiF (a = 4 Å) and in RbI (a = 7.34 Å).

When the electron is in its excited state, suppose that the cavity spontaneously deforms at a constant volume to take the form of a parallelepiped in which one of its dimensions is double than the two others. What is the new expression for the energy E' of the new electronic states?

Find the energy $h\nu'$ and the wavelength λ' of fluorescent photons that are emitted from the transition of an electron from its excited state E'_2 to its ground state, E'_1. ($\hbar^2/2m$ = 3.8 eV·Å^2)

Solution:

The solution to the present exercise is formally identical to that in Exs. 5a and 5b with both cases highlighting the implicit presence of positive ions (those in the lattice of the atomic clusters and here the positive charge carried by the vacancies to assure the neutrality of the crystal.)

For a cubic cavity, $E^0 = \dfrac{\hbar^2}{2m} \cdot \dfrac{\pi^2}{(\alpha a)^2}(n_x^2 + n_y^2 + n_z^2)$.

The only quantitative difference here is the reduced dimensionality of this cavity. The ground and first excited states are $E_1^0 = \dfrac{\hbar^2}{2m} \cdot \dfrac{3\pi^2}{(\alpha a)^2}; E_2^0 = 2E_1^0$.

We find $h\nu$ = 5.5 eV, λ = 2250 Å (in UV range) for LiF; $h\nu$ = 1.53 eV, λ = 7580 Å (red in the visible range) for RbI.

The effective dimension of the side is of the cavity results from the distortion of the lattice around the vacancy (via the Madelung constant effect) and the attraction of the trapped electron by the positive ions surrounding it. The nature of an alkaline halide only comes from the value of the lattice parameter that increases with the size of the ions. The name "color centers" given to such defects is a result of the fact that when lit by natural light, the spectrum of light re-diffused by the crystal no longer contains the color that has been absorbed.

Optical absorption measurement thus represents an elegant method to probe electronic levels. The experimental verification of the law $h\nu = C(\alpha a)^{-2}$, called the Mollwo–Yvey law, permits the evaluation of α.

The deformation of the cavity at a constant volume allows a partial lifting of the degeneracy between the levels 211, 121, and 112. This deformation is spontaneous because one of the levels takes a lower energy than the degenerate state E_2^0 (see table in Ex. 5b). This lower energy stabilizes the F center: it is the Jahn–Teller effect, which is also observed in semiconductors.

In the case, the statement we find $E' = \dfrac{\hbar^2}{2m} \dfrac{\pi^2\, 2^{2/3}}{(\alpha a)^2} \left[\dfrac{n_x^2}{4} + n_y^2 + n_z^2 \right]$,

where $L_x = 2L_y = 2L_z$ and $L_xL_yL_z = (\alpha a)^3$.

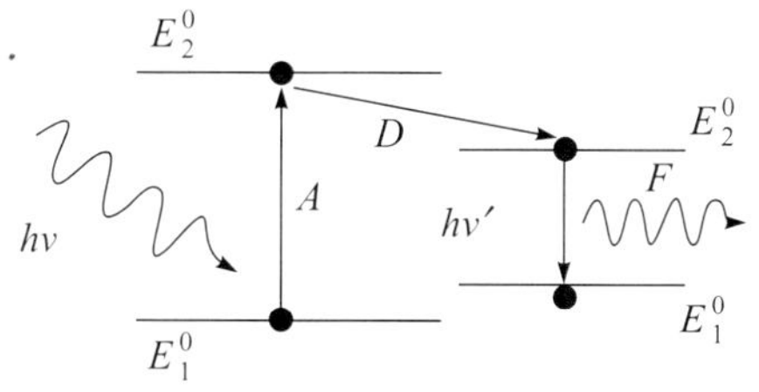

Figure 8

The state E'_2, is the 211 state. Its energy is less (by 79.3%) than that of E_2^0 because $3(2)^{2/3} < 6$, $E'_2(\text{LiF}) = 8.73$ eV; E'_2 (RbI) = 2.56 eV.

On the contrary, the ground state E'_1 is larger (by 20%) than E_1^0, as shown in Fig. 8. E_1^0 (LiF) = 6.55 eV; E_1^0 (RbI) = 1.94 eV, $h\nu'$(LiF) = 2.18 eV; $\lambda' = 5690$ Å (yellow) and $h\nu'$(RbI) = 0.62 eV; $\lambda = 2$ μm (IR).

As an additional exercise, the reader can verify that the ratio $L_x/L_y = 2$ is a minimum minimorum for the energy of the 211 state. To know more, see Y. Queré, [21], p. 150 and 353), which was the inspiration of this exercise.

Exercise 7: Fermi energy and Debye temperature from F and P boundary conditions for objects of reduced dimensions

Consider a gas of electrons enclosed in a cubic cavity (C) of sufficiently small dimension L to impose the use of fixed boundary conditions (FBC). Assume also that the number of electrons, N, is sufficient to allow the evaluation of k_F and E_F by the construction of a type of Fermi sphere without the evaluation of a discrete summation of too many states (see Ex. 5a).

The boundaries extend successively toward infinity in 1-, then 2-, and finally 3D. The initial cube thus changes into a rod with a square cross-section (T) next to a plate with parallel faces (L) and finally to an infinite crystal (I) by substituting fixed boundary conditions, FBC, by periodic boundary conditions, PBC, over a length conserved L as the dimensions are elongated. The electron density n ($= N/L^3$) remains a constant in these different situations. We thus propose to establish the evolution of the Fermi energy as a function of L when the FBC are changed into PBC.

(1) For each situation, state the general expression of the wave functions and of the energy levels as a function of the quantum numbers n_x, n_y, and n_z. Indicate the constraints imposed by the FBC and next by the PBC for these quantum numbers.

(2) In wave vector space (k_x, k_y, k_z), show the Fermi surface specifying the position of the excluding planes for the ends of the $\vec{k}$ vectors.

(3) Find the expressions relating the Fermi wave vectors $k_F(T)$, $k_F(L)$, and $k_F(I)$ to $k_F(C)$. Same question for the relations between the Fermi energies $E_F(T)$, $E_F(L)$, and $E_F(I)$ to the Fermi energy $E_F(C)$.

(4) *Numerical application*: $k_F(C) = 1$ Å^{-1}, so that $E_F = 3.8$ eV with $L = 100$ Å and next $L = 1000$ Å. Find the relative variation of energy, $\nabla E_F/E_F$, for the different systems specifying the sign of the change relative to E_F (C), taken as a reference. Comment on the result when L tends to infinity.

(5) Consider a plate bounded with parallel and infinite planes. Its initial thickness L_z is very large, but it decreases progressively keeping the electronic density constant. How will $E_F(L_z)$ vary as a function of L_z? Continuing this decrease of L_z until the

thickness a of a single monolayer, one can hope to go from a 3D electron gas to a 2D electron gas. Is it the case? Why?

(6) We are interested now in the lattice vibrations in objects of type C, T, L, and I. Briefly reconsider the preceding questions to find the relative variation of the Debye temperature θ_D of the different objects as a function of the characteristic dimension L of these objects when fixed boundary conditions are used. The approach consists of pointing out the similarities of the conditions imposed on k for electrons and for phonons to use next the phonon relation $\omega = v_S k$ (where v_S is the speed of sound).

Find the qualitative evolution of the lattice specific heat (for a single atom as unit) as a function of the size reduction of the sample when fixed boundary conditions are used.

Solution:

(1) The general wave function is the product of three wave functions $\Phi_x \cdot \Phi_y \cdot \Phi_z$. For the directions (x,y,z), when fixed boundary conditions are imposed, the corresponding wave functions are of the form: $\Phi_x = A \sin k_x x$, where $k_x = n_x \left(\frac{\pi}{L_x}\right)$ and n_x is a necessarily a positive whole number because $n_x = 0$ results in a null value for Φ_x (see Ex. 3 for details).

For directions in which we consider periodic boundary conditions, the corresponding wave function is of the form $\Phi_x = A' e^{ik_x x}$, where $k_x = n_x \left(\frac{2\pi}{L_x}\right)$ and n_x is a whole number that can be positive, negative, or zero. When $n_x = 0$, $\Phi_x = A'$ and the only forbidden situation is when n_x, n_y, and n_z are zero simultaneously.

The general expressions for the energies will thus be

$$E(C) = \frac{\hbar^2}{2m}\left(\frac{\pi^2}{L^2}\right)(n_x^2 + n_y^2 + n_z^2), \text{ where } n_x, n_y, n_z = N^x$$

$$E(T) = \frac{\hbar^2}{2m}\left(\frac{\pi^2}{L^2}\right)(4n_x^2 + n_y^2 + n_z^2), \text{ where } n_y, n_z = N^x;\ n_x = Z$$

$$E(L)=\frac{\hbar^2}{2m}\left(\frac{4\pi^2}{L^2}\right)\left(n_x^2+n_y^2+\frac{n_z^2}{4}\right), \text{ where } n_z = N^{x};\ n_x, n_y = Z$$

$$E(I)=\frac{\hbar^2}{2m}\left(\frac{4\pi^2}{L^2}\right)(n_x^2+n_y^2+n_z^2), \text{ where } n_x, n_y, n_z = Z$$

(2) See Fig. 9 that shows the exclusion planes of the fixed boundary conditions, planes defined by $k_x = 0$ and $k_y = 0$; $k_x = 0$ and $k_z = 0$; $k_y = 0$ and $k_z = 0$. These are the planes that introduce a difference between fixed and periodic boundary conditions when evaluating the corresponding k_F wave vectors. Effectively, with the exception of these planes, the transition from fixed to periodic conditions in one given direction results in a doubling of the step variation of k in this direction ($\pi/L \rightarrow 2\pi/L$) and also in the covered domain ($n_x > 0$ goes to $|n_x| > 0$ or <0).

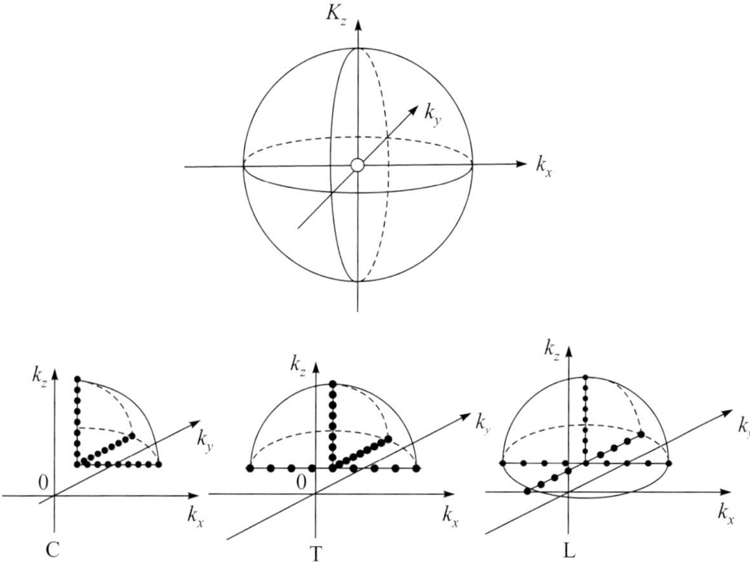

Figure 9

(3) For the cubic cavity, taking into account fixed boundary conditions as a reference, we find

$$\frac{N}{2}=\frac{4\pi}{3}\left(\frac{k_F}{\pi}\right)^3\frac{V}{8} \quad \text{with } V = L^3 \text{ and } K_F(C) = (3\pi^2 n)^{1/3}.$$

In the case of a long rod, a part of the N electrons occupy the states in the $k_y k_z$ quadrant of the plane:

$$\frac{N}{2}=\frac{4\pi}{3}\left(\frac{k_F}{\pi}\right)^3\frac{V}{2\cdot 4}+\frac{\pi k_F^2}{4\pi^2}\cdot L^2 \text{, which gives}$$

$$k_F^3(C)=k_F^3(T)+\frac{3\pi}{2L}k_F^2(T) \text{ or}$$

$$k_F(C)=k_F(T)\left[1+\frac{3\pi}{2k_F L}\right]^{1/3}.$$

In the case of a plate, the quadrant becomes a half-circle (plane k_y, k_z). In these planes, the surface of the states will double $\left(\frac{\pi^2}{L^2}\rightarrow\frac{2\pi^2}{L^2}\right)$, which leads to a doubling of the surface states compared to case T and the expected result to which one must add the states situated along the k_z axis (>0):

$$k_F(\mathrm{C})=k_F(\mathrm{L})\left[1+\frac{3\pi}{k_F L}+\frac{6\pi}{k_F^2 L^2}\right]^{1/3}$$

Finally, for an infinite sample, the states located in the planes $k_x = 0$, $k_y = 0$, and $k_z = 0$ are logically tripled (compared to case T) and those located on the axis are tripled also compared with case L.

We have $k_F(C)=k(I)\left[1+\dfrac{9\pi}{2k_F L}+\dfrac{18\pi}{k_F^2 L^2}\right]^{1/3}$.

Taking the square of this and having neglected the second order terms in $(k_F L)^{-2/3}$, we find

$$E_F(C)\approx E_F(T)\left(1+\frac{\pi}{k_F L}\right)\approx E_F(L)\left(1+\frac{2\pi}{k_F L}\right)=E_F(I)\left(1+\frac{3\pi}{k_F L}\right).$$

Note that in these expressions, the wave vectors k_F are related to the Fermi energy of the corresponding system, for example, $E_F(\mathrm{L})=\dfrac{\hbar^2 k_F^2(\mathrm{L})}{2m}$

(4) *Numerical application*: There is a reduction in the Fermi energy when starting from a cube, we successively increase the 3D. This decrease is of the order $\dfrac{\pi}{k_F L},\dfrac{2\pi}{k_F L},\dfrac{3\pi}{k_F L}$ with respect to the

systems T, L, I so that $\frac{|\Delta E_\mathrm{F}|}{E_\mathrm{F}} = 3\%,\ 6\%,\ 9\%$ for L = 100 Å and for L = 1000 Å.

As L goes toward infinity, the four Fermi energies tend to a common limiting value $E_\mathrm{F}(\infty)$, which is the same. The choice of the boundary conditions is regardless for macroscopic objects in the three directions (typically for $L > 1$ μm).

(5) Taking as a reference the infinite object (periodic boundary conditions) and denoting by δk_F the increase of the Fermi wave vector $k_\mathrm{F}(\mathrm{I})$ related to the introduction of fixed boundary conditions on the two surfaces of the infinite plate we find from (4) that

$$(k_\mathrm{F}+\delta k_\mathrm{F})^3 + \frac{3\pi}{L}(k_\mathrm{F}+\delta k_\mathrm{F})^2 + \frac{6\pi}{L^2}(k_\mathrm{F}+\delta k_\mathrm{F}) = k_\mathrm{F}^3 + \frac{9\pi}{L}k_\mathrm{F}^2 + \frac{18\pi}{L^2}k_\mathrm{F}$$

Limiting this to the first order (in δk_F), this becomes $\left(3k_F + \frac{6\pi}{L^2}\right)\delta k_\mathrm{F} = \frac{3\pi}{2L}k_\mathrm{F} + \frac{12\pi}{L^2}$. When the terms in L^{-2} related to the contribution of the axes are negligible, we have

$$\delta k_\mathrm{F} = \pi/2L_z \text{ and } E_\mathrm{F}(L_z) = E_\mathrm{F}(\infty)\left(1 + \frac{\pi}{k_\mathrm{F}L_z}\right).$$

When L_Z decreases, the Fermi energy increases as expected. The asymptotic value for $L_Z = a$ is of the order of $2.5E_\mathrm{F}(\infty)$ taking $k_\mathrm{F} = (3\pi^2 n)^{1/3}$ and $n = a^{-3}$ because the first energy state is $(\hbar^2/2m)(\pi^2/a^2)$.

But for this method, to go from 3D to 2D is not correct because one of the first consequences of using fixed boundary conditions is that $n_x, n_y, n_z \geq 1$ or equivalently $k_x, k_y, k_z \neq 0$ (see solution to Ex. 1), thus forbidding the propagation of electronic or acoustic waves parallel to the planes or edges where the fixed boundary conditions are applied. In the case of an infinite plane and with fixed boundary conditions, we cannot make a continuous transition from 3D to 2D by reducing the thickness L_z toward zero, even if we change the solutions by adopting cosines, which will cancel at $\pm L_z/2$. On the contrary, by using periodic boundary conditions, we can reduce L_z to zero because $e^{ik_z z}$ goes to 1 (whereas $\sin k_z z = 0$) and we thus obtain the correct wave functions of form $Ae^{ik_x x}\, e^{ik_y y}$.

(6) For atomic displacements $\vec{u}$, we seek sinusoidal solutions with form (see Course Summary, Chapter III): $\vec{u} = \vec{A}\sin k_x x \cdot \sin k_y y \cdot \sin k_z z \cdot e^{i\omega t}$ for fixed boundary conditions (C) and $\vec{u} = \vec{A}\exp i\,(k_x x + k_y y + k_z z)e^{i\omega t}$ for periodic boundary conditions (I).

In the intermediary situations (T,L), we use the product of sine and exponential functions, resulting in strictly the same conditions on k for phonons and for free or bound electrons.

The differences appear when are the different dispersion relations are used.

The Debye wave vector k_D for a cube will be $k_D = (6\pi^2 n_a)^{1/3}$, where n_a is the atomic density. The relations established in (3) for k_F apply to k_D as well as to ν_D,$(2\pi\nu_D = v_s k_D)$, and to θ_D $(h\nu_D = k_B\theta_D)$, because all the quantities are proportional to each other.

We thus find

$$\theta_D(C) = \theta_D(T)\left[1 + \frac{3\pi}{2k_D(T)\cdot L}\right]^{\frac{1}{3}}$$

$$\theta_D(C) = \theta_D(L)\left[1 + \frac{3\pi}{k_D(L)\cdot L} + \frac{6\pi}{k_D^2(L)\cdot L^2}\right]^{\frac{1}{3}}$$

$$\theta_D(C) = \theta_D(I)\left[1 + \frac{9\pi}{2k_D(I)\cdot L} + \frac{18\pi}{k_D^2(I)\cdot L^2}\right]^{\frac{1}{3}}$$

As for electrons, starting from very large dimensions $\theta_D(\infty)$, we expect an increase of the Debye temperature when one or several of its dimensions are reduced progressively.

At constant T, the smaller the object, the more the specific heat (brought to an atom) will decrease, since T/θ_D will decrease. The weight of this decrease, however, will depend on the value of T. When $T >> \theta_D$, the decrease will be negligible because the law of Dulong and Petit applies. When $T << \theta_D$, in the region where C_v increases in T^3, this decrease will be important, see figure in Chapter III, Course Summary.

In fact when studying atomic vibrations, one may ask the question of the physical meaning of fixed boundary conditions because they impose to atoms located on external planes to remain fixed. For electrons, on the contrary, this point is not important because it is obvious that their wave functions must go to zero on the surface or at a point close to it. See Ex. 26, for instance.

The present exercise is related to academic objects of the type *L* and *T*, but it opens the investigation of important devices and materials such as semiconducting quantum wells or layers of graphene and carbon nanotubes in Chapter V.

Exercise 8: Fermion gas

In this problem, we study several aspects of the theory of a fermion gas in which the particles obey Fermi–Dirac statistics and where the Pauli principle limits the occupation of their quantum states.

(1) (a) In the momentum space, we consider the number of fermions for which the end of momentum vector $\vec{p}$ is situated in an elementary volume $dp_x\, dp_y\, dp_z$. This number dN is given by $dN = \dfrac{gV}{h^3} f(E) dp_x dp_y dp_z$, where

$$f(E) = \frac{1}{1 + \exp\left(\dfrac{E - \mu}{k_B T}\right)}.$$

V is the volume of the space considered; *g* is the maximal number of fermions per quantum state; $\hbar$ is the Planck constant; *E* is the kinetic energy of the fermion; μ, known as the chemical potential or the Fermi level, varies weakly with temperature but will be considered in Question (1c) as a positive constant, k_B is the Boltzmann constant, *T* is the thermodynamic temperature of the gas. The fermions are assumed to be non-interacting with each other or with their environment so that their potential energy is zero. Sketch the representative curves for *f*(*E*) qualitatively and specify its asymptotic value when $T \to 0$ K and when $T \to \infty$.

(b) The chemical potential μ is a function of *T* and it is determined from the number of fermions *N* enclosed in the volume *V*. We denote $n = N/V$ and we assume that the energies involved are much smaller than the speed of light. Find the distribution of dN/dE as a function of *E*. Calculate

μ_0, the value of μ at absolute zero, as a function of the fermion mass. Find the average energy $\overline{E_0}$ at $T = 0$ K and express it as a function of μ_0.

(c) Now consider variable temperatures but always such that $k_B T << \mu_0$. Calculate μ and the average energy $\bar{E}$ as a function of T using a limited expansion in powers of $k_B T/\mu_0$. (Use F1 in the Course Summary.)

(2) Electron gas

(a) Assume that in the interior of a metal electrons are subject to a constant potential that does not influence the above calculations but which prevents the electrons from escaping into vacuum (Ex. 28a).

Consider silver (atomic mass = 107, mass density, ρ = 10,500 kg/m^3) and assume that there is one free electron per atom. Take $g = 2$ and justify this choice. Calculate μ_0 in eV. Find μ at $T = 300$ K and compare the result with $k_B T$. Calculate the speed of the most energetic electrons at T = 0 K.

The relation $\mu_0 = k_B T_F$ defines the Fermi temperature T_F. Find T_F. For what value of E does the function $f(E)$ equal to 0.99 and next to 0.01? Remark on the results.

(b) Calculate the heat capacity of this electron gas, C_e, assuming again that it is without interactions and therefore that the changes in energy with temperature are due only to the change of kinetic energy as calculated previously.

Compare the numerical value of C_e thus obtained for N (Avogadro's number) of metal atoms at $T = 1$ K and next at $T = 300$ K with C, the lattice heat capacity at a constant volume. In the first case, one can use $C = \alpha T^3$, where $\alpha = \dfrac{2.4\pi^4 k_B \mathcal{N}}{\theta_D^3}$ and $\theta_D = 225$ K for Ag. The second case C corresponds to the classical limit. Use F(1). (e, h, k_B, m, N)

Solution:

(1) (a) Using $\dfrac{E-\mu}{k_B T} = x$, the function $f(E)$ takes the form $\dfrac{1}{e^x + 1}$.

When $T = \infty$, $x = 0$ and $f(E) = ½$.

When $T = 0$, it becomes a step function with $f(E) = 1$ for

$E - \mu < 0$ ($x = -\infty$) and $f(E) = 0$ for $E - \mu > 0$ ($x = +\infty$). Outside of these two limiting values ($T = 0$ and $T = \infty$), the calculation of its first derivative shows that the function f(E) is constantly decreasing and the 2nd derivative shows the presence of an inflection point at $x = 0$, where $f(E) = ½$.

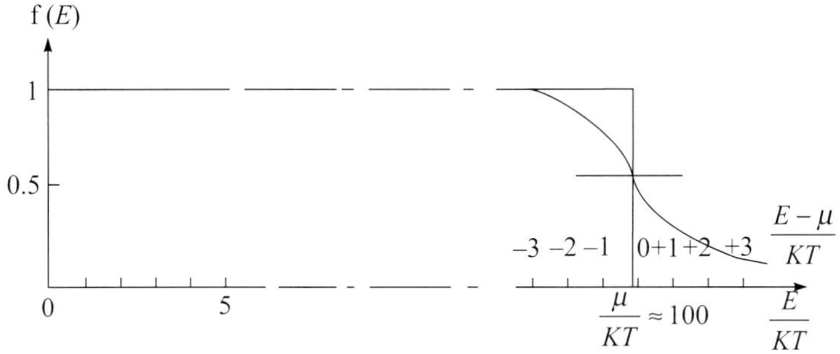

Figure 10

In a solid, the temperature T of a gas of electrons is limited by its melting point even though μ, of the order of several eV, corresponds to Fermi temperatures ($\mu = kT_F$) of the order of 50,000 K. The quantity μ/k_BT is thus always large compared with 1 (Question (1c)) and $f(E)$ varies essentially in the energy range near μ. To specify this variation, we consider the following values:

$E - \mu = k_BT \Rightarrow F(E) = 0.27$;

$E\text{-}\mu = 3k_BT \Rightarrow F(E) = 0.047$

$E - \mu = -k_BT \Rightarrow F(E) = 0.73$;

$E\text{-}\mu = -3k_BT \Rightarrow F(E) = 0.952$

(b) When fermions are free and non-relativistic, their energy, which is uniquely kinetic, is given by $E = p^2/2m$ $(= \hbar^2k^2/2m)$. In momentum space (proportional to wave vector space by $\vec{p} = \hbar\vec{k}$), the number dN of fermions in which the energy is between E and $E + dE$ have a vector $\vec{p}$ of which the extremity is between the spheres of radius p and $p + dp$ where $dE = p.dp/m$.

If in the formula from the statement, the elementary volume $dp_xdp_ydp_z$ is replaced by the volume in momentum space $4\pi p^2dp$, we find

$$dN = \frac{gV}{h^3} f(E)\cdot 4\pi p^2 dp = \frac{4\pi\sqrt{2}gV}{h^3}\cdot m^{3/2} f(E)\cdot E^{1/2}\cdot dE,$$

which leads to $\dfrac{dN}{dE} = \dfrac{4\pi\sqrt{2}gV}{h^3} m^{3/2}\cdot f(E)\cdot E^{1/2}$.

In the general case, we deduce $\mu(T)$ from $N = \int_0^\infty dN$.

Here $T = 0$ and $f(E) = 0$ for $E > \mu_0$ and we, therefore, obtain μ_0 starting from the integration:

$N = \int_0^{u_0} dN = \dfrac{8\pi\sqrt{2}}{3}\cdot\dfrac{gV}{h^3}\cdot m^{3/2}\mu_0{}^{3/2}$, which gives

$\mu_0 = \left(\dfrac{3}{8\pi\sqrt{2}g}\right)^{2/3}\cdot\dfrac{h^2}{m}\cdot n^{2/3}$, where $n = N/V$.

The average energy of a fermion will thus be given by the integral

$$\overline{E} = \frac{E_{\text{Total}}}{N} = \frac{\int_0^{u_0} E\cdot dN}{\int_0^{u_0} dN} = \frac{A\cdot\int_0^{u_0} E^{3/2}\cdot dE}{A\cdot\int_0^{u_0} E^{1/2}\cdot dE} = \frac{3}{5}\mu_0,$$

where $A = \dfrac{4\cdot\sqrt{2}\cdot\pi gV}{h^3}\cdot m^{3/2}$.

(c) If $T \neq 0$, we can deduce $\mu(T)$ from

$\mu(T) de\ N = A\int_0^\infty f(E)\cdot E^{1/2}\cdot dE$.

Taking into account F1, which is valid when $k_B T << \mu_0$ (general case in solids), we find

$$N = A\left[\frac{2}{3}\mu^{3/2} + \frac{(\pi k_B T)^2}{6}\cdot\frac{1}{2}\mu^{-1/2} + \cdots\right] = \frac{2}{3}A\mu_0^{3/2}.$$

Replacing μ by μ_0 in the expansion, we note that the error is in $\left(\dfrac{k_B T}{\mu_0}\right)^4$ from which we have $\mu = \mu_0\left[1 - \dfrac{\pi^2}{12}\left(\dfrac{k_B T}{\mu_0}\right)^2 + \cdots\right]$.

Following the same procedure for $\overline{E}(T)$, we find $\overline{E} = \dfrac{A}{N}\int_0^\infty f(E)\cdot E^{3/2}\cdot dE = \dfrac{2}{5}\dfrac{A\mu^{5/2}}{N}\left[1 + \dfrac{5}{8}\pi^2\left(\dfrac{k_B T}{\mu_0}\right)^2 + \cdots\right]$ or equivalently $\overline{E} = \frac{3}{5}\mu_0\left[1 + \frac{5\pi^2}{12}\left(\frac{k_B T}{\mu_0}\right)^2 + \cdots\right]$.

Remarks:

(a) The Fermi level decreases weakly with temperature ($\sim 10^{-4}$ for $k_B T/\mu_0 \sim 10^{-2}$). We can compare this evolution, obtained in 3D, with that deduced in 1- and 2D (see Ex. 14a).

(b) The energy E of a gas of fermions is the sum of a very large term independent on T and of a residual term in T^2 thus the influence of the temperature will affect this energy only a little.

The electronic specific heat $C_e = \frac{\partial E}{\partial T}$ will be linear in T and zero at 0 K, as in 1- and 2D (see Ex. 14a).

(2) (a) The spin of an electron, ±1/2, implies that one can only place two electrons per quantum state, thus $g = 2$. The concentration of free electrons in silver is such as 6×10^{23} electrons occupy 1.02×10^{-5} m^3, from which we find $n = 5.9 \times 10^{28}$ e/m^3; $\mu_0 = 8.86 \times 10^{-19}$; J = 5.52 eV.
At 300 K, $k_B T$ is of the order of 26 meV, therefore

$k_B T/\mu_0 \approx 4.7 \times 10^{-3}$, which results in a correction of the Fermi level of $\frac{\pi^2}{12}\left(\frac{k_B T}{\mu_0}\right)^2 \approx 1.8 \times 10^{-5}$.

The most energetic electrons are at the Fermi energy. Their speed v_F corresponds to $\frac{1}{2}mv_F^2 = \mu_0$, or $v_F = 1.4 \times 10^6$ m/s (approximately 1/200 the speed of light) and the corresponding Fermi temperature is T_F = 64,200 K.

- $f(E) = 0.99$ when $\exp\frac{E-\mu}{k_B T} = 0.01$, or

$$E = E_1 = \mu - k_B T \log 100 = \mu - 4.6 k_B T$$

- $f(E) = 0.01$ when $\exp\frac{E-\mu}{k_B T} \approx 100$, or

$$E = E_2 = \mu + k_B T \log 100 = \mu + 4.6 k_B T.$$

The important variation zone is a few k_BT around μ (as already discussed in 1(a).

(b) $C_e = \frac{\partial E}{\partial T}\text{total} = \mathcal{N}\frac{\partial \bar{E}}{\partial T} = \mathcal{N}\frac{\pi^2}{2}\frac{k_B{}^2 T}{\mu_o} = \mathcal{N}\frac{\pi^2}{2}k_B\left(\frac{T}{T_F}\right)$ from which we find $C_e = 6.4 \times 10^{-4}$ J/K at 1 K and $C_e = 0.19$ J/K at 300 K. The specific heat of the lattice at these same temperature is $C = 1.7 \times 10^{-4}$ J/K (at 1 K) and $C = 3R = 25$ J/K (at 300 K).

The total specific heat ($C_e + C$) is of the form $\gamma T + \alpha T^3$, the electronic term dominates at 1 K but it is comparable to the lattice term from 2 K, and it becomes negligible at higher temperatures. It is thus necessary to operate at very low temperature to determine C_e from the sum. In practice, one built a staight line $\sum C/T = f(T^2)$ from which the intercept at the origin gives γ (see Ex. 10).

Exercise 9: Fermi energy and thermal expansion

Consider a sample of volume V containing N free electrons.

(a) Find the expression for the Fermi energy at temperature T, E_F (T), as a function of the Fermi energy at 0 K, $E_F(0)$, taking into account both (i) the conservation of the total number of electrons N and (ii) the volumetric thermal expansion coefficient α_V.

(b) Compare numerically the weight of each of these corrections for $\alpha_V \approx 15 \times 10^{-5}$ K^{-1} and $T_F = 50{,}000$ K.

(c) Find the average energy of an electron and the corresponding electronic specific heat. Remark on the influence of the thermal expansion on C_e.

Solution:

(a) The definition of $E_F(T)$ is based on the conservation of the total number of electrons N: $N = \int_0^\infty g(E)f(E)dE$, where in 3D $g(E) = \Lambda.V.\sqrt{E}$ and $A = \frac{1}{2\pi^2}\left(\frac{2m}{\hbar^2}\right)^{3/2}$. Taking $V = V_o(1 + \alpha_v T)$ and using F1, we have

$$N/V_o = A(1+\alpha_v T)\left[\frac{2}{3}E_F^{3/2}(T)+\frac{(\pi k_B T)^2}{12}E_F^{-1/2}(T)\right].$$

The same formula applied at $T = 0$ K gives $N/V_o = \frac{2A}{3}E_F^{3/2}(0)$.

Identifying the two expressions for N/V_o and noting that $E_F^{-1/2}(T) \sim E_F^{-1/2}(0)$ in the correction term, we have $E_F(T) \approx E_F(0)[1+\alpha_v T]^{-2/3}\left[1-\frac{(\pi k_B T)^2}{12\cdot E_F^2(0)}\right]$.

The first term (between the []) takes into account the thermal expansion. Its variation, essentially in $-\frac{2}{3}\alpha_v T$, could be foreseen because $E_F = (\hbar^2/2m)(3\pi^2 n)^{2/3}$ where $n = N/V$. Thus, an increase of volume $\alpha_v T$ leads to a proportional decrease in the electronic density, n, which in turn, leads to a relative decrease in E_F, by, taking into account the exponent 2/3 in the expression for E_F.

The second term (between the []) is the normal contribution to the decrease in the Fermi level at constant volume (see Ex. 8).

(b) When T increases, the two contributions both tend to reduce the Fermi energy, but they have significantly different weights. For an increase of 1 K, the weight of (i) will be of the order -10^{-4}, whereas of (ii) will be of the order -4×10^{-10}.

(c) The total energy of an electron gas is

$$U_e = AV\int_0^\infty E^{3/2} f(E).dE = AV\left[\frac{2}{5}E_K^{\frac{5}{2}}(T)+\frac{(\pi k_B T)}{4}E_F^{\frac{1}{2}}\right]$$

By expressing $E_F(T)$ as a function of $E_F(0)$, and then dividing by $N = \frac{2AV_0}{3}E_F^{3/2}(0)$, we obtain the average energy of an electron: $\overline{E_e} = \frac{3}{5}E_F(0)(1+\alpha_v T)^{-2/3}\left[1+\frac{5}{12}\frac{(\pi k_B T)^2}{E_F^2}\right]$.

We find correctly that the average energy of a free electron in 3D at $T = 0$ is $(3/5)E_F$.

The conventional term for the specific heat at constant volume for a free electron results from the derivative with respect to T

of the third term, that is, $C_e = \frac{\partial \overline{E_e}}{\partial T} = \frac{\pi^2}{2} k_B \left(\frac{T}{T_F}\right)$, which leads to the law of form $C_e = \gamma T$.

As in (a), the term α_V describes the decrease of E_F due to the thermal expansion. Its derivative with respect to T leads to the introduction of the terms $-\frac{2}{3}\alpha_V + \frac{10}{9}\alpha_V^2 T$.

Reduced to one electron, the specific heat at constant pressure becomes

$$C_e = -\frac{2}{5}\alpha_V k_B T_F + \frac{2}{3}\alpha_V^2 k_B T_F T + \frac{\pi^2}{2} k_B T / T_F.$$

The numerical application allows us to note that the contribution of the thermal expansion to the term linear in T is 7.6 times greater than the coefficient γ.

Exercise 10: Electronic specific heat of copper

Measurements of the specific heat of copper at constant pressure, C_P, and low temperature give the following results:

$C_P = 3.53 \times 10^{-4}$ J mole^{-1} deg^{-1} at 0.5 K

$C_P = 7.41 \times 10^{-4}$ J mole^{-1} deg^{-1} at 1 K

$C_P = 17.66 \times 10^{-4}$ J mole^{-1} deg^{-1} at 2 K

By assimilating C_p to C_V, extract the electronic contribution and compare the result with the theory for free electrons. In addition, find the Debye temperature θ_D, the Fermi temperature T_F, as well as the temperature T_0 from which the two expected contributions will have the same weight. The atomic density of Cu (which is monovalent) is $N = 8.45 \times 10^{28}$ m^{-3}. ($\mathcal{N}$, k_B)

Solution:

At low temperature, the specific heat of metals is described by C = $\gamma T + \alpha T^3$ in which the first term represents the electronic contribution (where theoretically $\gamma = (\pi^2/2T_F0)k_B$ per free electron) and the second term is the contribution from atomic vibrations (where $\alpha = 12\pi^4 k_B/5\theta_D^3$ per oscillator).

As seen in Fig. 11, the experimental determination of γ can be done by determining the value of the ordinate at origin of $C/T = f(T^2)$.

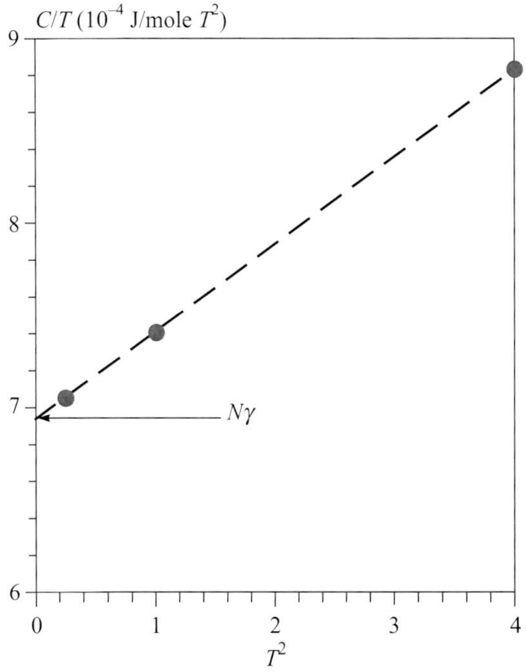

Figure 11

Another possibility is to solve the two equations for the last two values of C for γ and α and next verify that the results are compatible with the first experimental data. We thus obtain $\gamma = 1.153 \times 10^{-27}$ J at^{-1} K^{-2} and $\alpha = 7.86 \times 10^{-29}$ J at^{-1} K^{-4}.

With E_F = 7 eV, T_F = 81,200 K, the theoretical value of γ is 0.838×10^{-27} J at^{-1} K^{-2}. The difference with the experimental value can be justified by the use of an effective thermal mass for the electrons $m_e^x = 1.38\, m_0$.

By identifying the experimental values of α with the theoretical one, we find $\theta_D = 345$ K.

The two contributions are equal at a temperature T_0 such that $T_0 = \sqrt{\gamma/\alpha}$ = 3.83 K.

Exercise 11: Density of electronic states in 1, 2, and 3D from a general formula

(a) In a 3D solid and starting from the general formula, $g(E) = \dfrac{2V}{(2\pi)^3}\iint \dfrac{dS_E}{\left|\overrightarrow{\nabla_k E}\right|}$, find the expression for the electronic density of states: (i) for free electrons, (ii) for electrons that obey the isotropic dispersion relation: $E = -E_o - 2\gamma\cos\left(\frac{ka}{2}\right)$, where E_0, $\gamma > 0$.

(b) After having established the corresponding expression for 2D, find g(E) for the hypothesis (i) and (ii) above.

(c) Same question for 1D.

Solution:

(a) The general expression giving the density of electronic states $g(E)$ is formally similar to that giving the density of vibrations for a lattice $g(\nu)$ except a coefficient of 2 for the spin of electrons. It is thus sufficient to follow the method previously applied to phonon density in 2D and 3D (see Chapter III, Ex. 19) by evaluating $\left|\overrightarrow{\nabla_k E}\right| = \hbar\vec{v}g$ using the following hypotheses:

(i) Free electrons: $E = \hbar^2 k^2/2m$ leading to $\left|\overrightarrow{\nabla_k E}\right| = \hbar^2 k/m$.

(ii) Bound electrons: $E = -E_o - 2\gamma\cos(ka/2)$ leading to $\left|\overrightarrow{\nabla_k E}\right| = \gamma a \cdot \sin\frac{ka}{2}$.

Taking into account the isotropy, we can replace the integral by the quantity $S_E/\left|\overrightarrow{\nabla_k E}\right|$, we find

- $$g(E) = \frac{2V}{(2\pi)^3}\cdot\frac{4\pi k^2}{\hbar^2 k/m} = \frac{V}{\pi^2}\cdot\frac{m^{3/2}}{\hbar}\cdot(2E)^{1/2}$$

- $$g(E) = \frac{2V}{(2\pi)^3}\cdot\frac{4\pi k^2}{\left|\gamma a\sin\frac{ka}{2}\right|} = \frac{4V}{\pi^2 a^3}\cdot\frac{\left(\text{arc cos}-\frac{E_0+E}{2\gamma}\right)^2}{\gamma\left[1-\left(\frac{E_0+E}{2\gamma}\right)^2\right]^{1/2}}$$

where $-E_0 - 2\gamma \le E \le -E_0 + 2\gamma$.

(b) As in Chapter III, Ex. 19, we note that in 2D, the isoenergy surfaces become lines in the wave vector planes so that $g(E)$ obcys: $g(E_0) = 2\left(\frac{L}{2\pi}\right)^2 \int_{l(E_0=E_0)} \frac{dl_E}{\left|\overrightarrow{\nabla_k E}\right|}$.

In an isotropic case, this leads to $g(E)=2\left(\frac{L}{2\pi}\right)^2\cdot\frac{2\pi k}{\left|\overrightarrow{\nabla_k E}\right|}$

Thus

(i) $g(E)=\frac{mL^2}{\pi\hbar^2}=Cste.$

(ii) $$g(E)=\frac{2L}{\pi^2a^2}\frac{\arccos-\frac{E_0+E}{2\gamma}}{\gamma\left[1-\left(\frac{E_0+E}{2\gamma}\right)^2\right]^{1/2}},$$

where $-E_0-2\gamma\leq E\leq E_0+2\gamma$.

(c) Finally, in 1D, the isoenergetic line reduces to two points (k and –k) and the density of states obeys the general relation $g(E)=2\cdot\frac{2L}{2\pi}\frac{1}{\left|\nabla_k E\right|}$ form which we find

(i) $g(E)=\frac{2L}{h\pi}\left(\frac{m}{2E}\right)^{1/2}$

(ii) $$g(E)=\frac{2L}{\pi a}\frac{1}{\gamma\left[1-\left(\frac{E_0+E}{2\gamma}\right)^2\right]^{1/2}}$$

Remarks:

(a) The results obtained in Questions a(i), b(i), and c(i) are identical to the evaluations in Exs. 13 and 14.

(b) When $\left|\overrightarrow{\nabla_k E}\right|=0.$, we find $g(E)=\infty$, the corresponding critical points coincide with large electronic concentrations known as Van Hove singularities, which occur when $E=-E_0-2\gamma(k=0)$ for the particular case of bonded electrons [case (ii) in the exercise]. The schematic evolution of the curves $E=f(k)$ and $g(E)$ nearly correspond to 1D system and are shown in Fig. 1 of Chapter V, Ex. 1.

Exercise 12: Some properties of lithium

(1) Knowing that lithium crystallizes in a cubic system with lattice

[simple cubic, body-centered cubic (bcc), or face-centered cubic (fcc)] taking into account its atomic mass (7) and its volumetric mass (546 $kg \cdot m^{-3}$).

(2) Knowing that the valence electrons of this metal (1 per atom) behave as free electrons, find the shape of the Fermi surface and its expression and then calculate its characteristic dimension k_F.

(3) Compare k_F obtained in (2) to the distance d_m, which in reciprocal space separates the origin from the first boundary of the first Brillouin zone nearest the origin. (Evaluate d_m using simple geometric considerations without having to sketch the first Brillouin zone.)

(4) Find the Fermi energy of lithium E_F, the Fermi temperature T_F, and the speed of v_F of the fastest free electrons.

(5) Knowing that the resistivity ρ of lithium is of the order 10^{-5} $\Omega \cdot cm$ at ambient temperature, find the time of flight, τ, and the mean free path Λ of conduction electrons.

(6) Find the drift velocity v_d of conduction electrons subject to a electric field of 1 V/m and compare it with the Fermi velocity v_F.

(7) Starting from the relation $k_e = \frac{1}{3} C_e v_F \Lambda$ (or with the help of the Wiedemann–Franz expression), find the thermal conductivity due to electrons K_e of lithium at ambient temperature $T = 300$ K.

(8) From the point of view of the photoelectric effect, free electrons are considered to be enclosed in a potential well with height E_0 equal to 6.7 eV, measured by taking as the origin the energy of electrons at zero speed. What is the work function ϕ of lithium and the wavelength λ_0 relative to the photoelectric threshold? $(\hbar, m, k_B, e, \mathcal{N})$

Solution:

(1) The mass of an atom of lithium $\frac{M \cdot A}{\mathcal{N}}$ = is 1.166×10^{-26} kg and the mass contained in a cube of edge a is $546 \times (3.48)^{-3} \times 10^{-30}$ $\approx 2.3 \times 10^{-26}$ kg. There are therefore two atoms per cube and lithium must be bcc.

(2) For free electrons, the Fermi surface is spherical and k_F obeys the relation $(4\pi/3)k_F^3;(2\pi/L)^3 = N/2$, which gives $k_F = (3\pi^2 n)^{1/3}$, where $n = N/V = 4.7 \times 10^{22}$ e/cm^3, $k_F = 1.11 \times 10^8$ cm^{-1}.

(3) The reciprocal lattice of the bcc is fcc and the closest point to the origin in reciprocal space is 110, at a distance from 0 of $2\pi\sqrt{2}/a$, so that we have $d_m = \pi\sqrt{2}/a = 1.27 \times 10^8$ cm^{-1}.

- $d_m > k_F$, the Fermi sphere is entirely enclosed in the first Brillouin zone without any contact with it. We can find this result by considering that the first authorized reflection at normal incidence that electrons can undergo is the reflection 110 ($h + k + l$ = even); the wavelength will thus be $\lambda = 2d_{110} = a/\sqrt{2}$ and the corresponding wavevector reaching the first Brillouin zone will have a modulus $d_m = \pi\sqrt{2}/a$.

(4) $$E_F = \frac{\hbar^2 k_F^2}{2m} = \frac{\hbar^2}{2m}(3\pi^2 n)^{2/3} \approx 4.72\ eV$$

$$T_F = \frac{E_F}{k_B} = 64{,}800\ \text{K},\ v_F = \frac{\hbar k_F}{m} = 1300\ \text{km/s}$$

(5) $$\sigma = \frac{ne^2\tau}{m} \text{ so we have } \tau = \frac{m}{ne^2}\cdot\frac{1}{\rho} = 0.76\times 10^{-14}\,\text{s}$$

$$\Lambda = v_F\tau \approx 100\ \text{Å}$$

$$\vec{j} = \sigma\vec{E} = -ne\vec{v}_d$$

(6) The drift velocity is $v_d = 0.133$ cm/s. Its value is very small compared with the isotropic velocity v_F.

(7) The electronic specific heat of a metal is (see Ex. 8):

$$C_e = \frac{\pi^2}{2} n k_B^2 \frac{T}{E_F}, \text{ which leads to}$$

$$\hat{K}_e = \frac{1}{3}C_e v_F \Lambda = \frac{1}{3}C_e v_F^2 \tau = \frac{\pi^2 n k_B^2}{3m} T\tau.$$

Using the Wiedemann–Franz law, we have

$$K_e \cdot \rho = \frac{K_e}{\sigma} = \frac{\pi^2}{3}\left(\frac{k_B}{e}\right)^2 T.$$

$$\hat{K}_e = \frac{L_T}{\rho} = 0.735\ \text{W/cm-degree}.$$

(8) See figure and Ex. 28a for details on the origin of ϕ:

$$E_o = E_F + \Phi_o, \Phi_o \approx 3eV, \lambda_o = \frac{hc}{\Phi_o} = 4133\ \text{Å}.$$

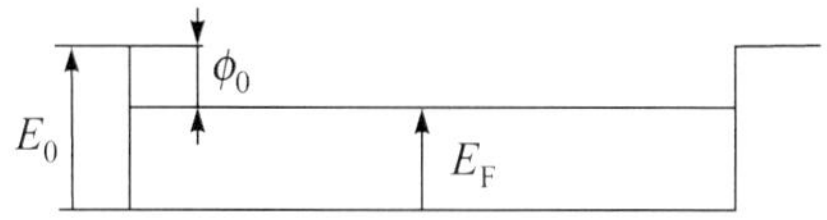

Exercise 13: Fermi energy, electronic specific heat, and conductivity of a 1D conductor

Consider a linear lattice of N identical atoms distant by 'a' from one to each of its nearest neighbors (see Chapter III, Ex. 11). One electron per atom can propagate freely along the line.

(a) Recall the dispersion relation $E = f(k)$ of these free electrons and deduce the energy $E_F(0)$ and the Fermi temperature $T_F(0)$ at 0 K. Numerical application for E_F in eV when a = 3 Å.

(b) What is the density of electronic states $g(E)$ in energy space. Express the result as a function of $E_F(0)$ and show the corresponding characteristic curve. Establish the expression for $E_F(0)$ at 0 K starting from $g(E)$ and deduce the average energy $\bar{E}$ of an electron in this metal.

(c) Specify the evolution of E_F as a function of temperature T where $T << T_F$, and find the internal energy U_e of the electron gas at temperature T.

(d) Deduce the evolution of the electronic specific heat $C_e(T)$ at constant volume as a function of T.

Compare this result with that obtained approximately by considering only neighboring electrons within one k_BT energy near from the Fermi energy will gain k_BT in energy when the metal goes from 0 K to T K.

(e) Compare the exact expressions of $C_e(T)$ and $C_R(T)$, where C_R is the specific heat of a lattice at low temperature: $C_R = (\pi^3/3) \cdot Nk_B \dfrac{T}{\theta_D}$ (see Chapter III, Ex. 11), and determine the numerical value of the ratio C_e/C_R where θ_D = 240 K.

(f) Consider a metal made of identical parallel rows along the axis Ox, analogous to those studied above. In the yOz plane, these rows form a square lattice with side b. Find the electric resistivity ρ along the axis Ox, starting from the numerical values given in the preceding steps and assuming that the mean free path of electrons along this axis is such that $\Lambda = 10a$ and $b = 5$ Å.
Use F2 or F1 from the Course Summary. (h, m, e, k_B)

Solution:

(a) The dispersion relation for free electrons is $E = \dfrac{\hbar^2 k^2}{2m}$. As the density of states in k-space, $g(k) = L/\pi$, is constant (Chapter III, Ex. 11), the first Brillouin zone contains N states with two possible electrons (↓↑) per state. The wave vector k_F and the Fermi energy E_F are thus:

$$k_F = \frac{1\pi}{2a};\; E_F(0) = \frac{\hbar^2}{2m}\frac{\pi^2}{4a^2};\; E_F = 1.04 \text{ eV when } a = 3 \text{ Å};$$

$$T_F = \frac{E_F}{k_B} = 12{,}000 \text{ K}.$$

(b) For electrons, the density of cells in the k-space is the same as for phonons in the same space. The only difference is the different dispersion as a function of frequencies or energies:

$g(k) = \dfrac{Na}{2\pi}$, where $-\dfrac{\pi}{a} \le k \le \dfrac{\pi}{a}$, which leads to $g(|k|) = 2g(k)$

$g(E)\,dE = 2g(|k|)\cdot dk$, $(2 \Rightarrow \uparrow\downarrow)$, which gives

$$g(E) = 4\cdot\frac{Na}{2\pi}\cdot\frac{dk}{dE} = 2\frac{Na}{\hbar\pi}\left(\frac{m}{2E}\right)^{1/2} = \frac{N}{2(E_F(0)\cdot E)^{1/2}}$$

(Note that this is the result of Ex. 11c)

The Fermi energy at 0 K is $\int_0^{E_F(0)} g(E)dE = N$ or $E_F(0) = \dfrac{\hbar^2\pi^2}{8ma^2}$.

The average energy $\bar{E}$ of an electron in a linear metal is

$$\overline{E} = \frac{\int_0^{E_F(0)} g(E)\cdot E\cdot dE}{\int_0^{E_F(0)} g(E)\cdot dE} = \frac{E_F(0)}{3}.$$

(c) At the temperature T, E_F is given by $\int_0^{\infty} g(E)\,f(E)\,dE = N$, where $f(E) = \dfrac{1}{e^{\frac{E-E_F}{k_B T}}+1}$.

This gives $\dfrac{N}{2\left[E_F(0)\right]^{1/2}}\cdot\displaystyle\int_0^{\infty}\frac{E^P dE}{e^{\gamma[(E/\chi)-1]}+1} = N$, where $p = -1/2$,

$\chi = E_F(T)$, $\gamma = \dfrac{E_F(T)}{k_B T}$, and using F2.

Since $I_{1/2} \approx [E_F(T)]^{1/2}\left(2 - \dfrac{\pi^2}{2\cdot 6}\dfrac{k_B^2 T^2}{E_F^2(T)} + \cdots\right)$, we deduce

$\left[\dfrac{E_F(T)}{E_F(0)}\right]^{1/2}\left(1 - \dfrac{\pi^2}{24}\dfrac{k_B^2 T^2}{E_F^2(T)}\right) \approx 1$ and

$E_F(T) = E_F(0)\left[1 + \dfrac{\pi^2}{12}\cdot\dfrac{T^2}{T_F^2(0)}\right]$ by substituting the correction term $E_F^2(T)$ by $E_F^2(0) = [k_B T_F(0)]^2$. This result can be obtained directly using F1.

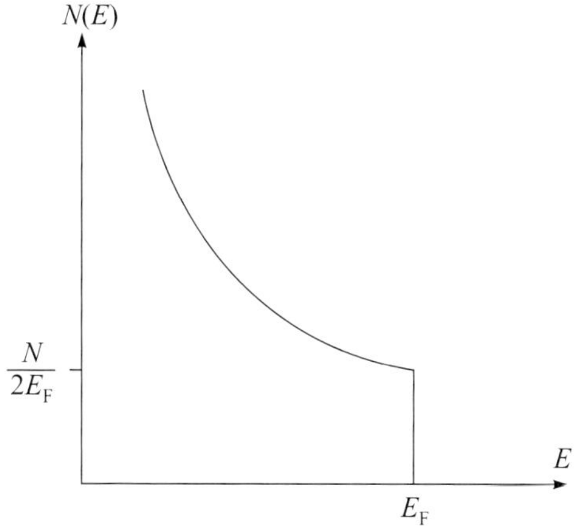

Figure 12 Evolution of g(E) for a 1D free electron gas.

In the same way, the expression for the energy of an electron gas corresponds to

$U_e = \int_0^\infty g(E)\cdot f(E)\cdot E\cdot dE$. Using F2,

$$U_e = \frac{N}{2[E_F(0)]^{1/2}}\cdot\int_0^\infty \frac{E^{p+1}dE}{E^{\gamma[(E/\chi)-1]}+1}, \text{ one may deduce}$$

$$U_e = \frac{N}{2[E_F(0)]^{1/2}}\cdot E_F^{3/2}(T)\left[\frac{2}{3}+\frac{\pi^2}{12}\cdot\frac{(k_BT)^2}{E_F^2(T)}\right] \text{ or equivalently}$$

$$U_e = \frac{NE_F(0)}{3}\left[1+\frac{\pi^2}{8}\left(\frac{T}{T_F(0)}\right)^2\right],$$ taking into account the previously calculated evolution of $E_F(T)$. The first term $\left(\frac{N}{3}E_F(0)\right)$ is the internal energy of an electron gas at 0 K.

(d) $$C_e = \left(\frac{\partial U_e}{\partial T}\right)_V = N\frac{\pi^2}{2}k_B\frac{T}{T_F(0)}$$

We could have obtained more rapidly an expression close to this result by considering that only the n neighboring electrons near the Fermi energy E_F within energy k_BT ($n \approx g(E_F)\cdot k_BT$) can win an energy k_BT when the metal temperature goes from 0 K to T K. The corresponding increase in electronic energy is $\Delta U_e = \frac{N}{2E_F(0)}\cdot k_B^2T^2$ and $C_e \approx \frac{\partial(\Delta U_e)}{\partial T} = Nk_B\frac{T}{T_F(0)}$.

These two results correspond to the same temperature dependence excepting the exact coefficient $\frac{\pi^2}{2} \approx 5$.

(e) In 1D, the lattice specific heat is also linear in T at low temperatures (see Chapter III, Ex. 11). The ratio $\frac{C_e}{C_R} = \frac{1}{2}\frac{\theta_D}{T_F}$ is of order 10^{-2} and is independent of T.

We cannot experimentally separate the two contributions by varying T, which is different from the 3D case (see Exs. 8 and 10), where we can obtain the contributions from a plot of $C/T = (\gamma + \alpha/T^2)$.

(f) Measured along the axis Ox, the electric conductivity obeys the relation $\sigma \approx \frac{ne^2\tau}{m}$, where $\tau = \frac{\Lambda}{v_F} = \Lambda\left(\frac{m}{2E_F}\right)^{1/2}$. Numerically we find $n = \frac{1}{75 \cdot A^3}$ = 1.33×10^{28} e/m^3, $\tau = 5 \times 10^{-15}$ s,

$\sigma = 1.8 \times 10^6\ \Omega^{-1}\text{m}^{-1}$ or $\rho = 55 \times 10^{-6}\ \Omega\cdot\text{cm}$.

Note: 1D Conductors

Certain compounds of platinum such as $K_2Pt(CN)_4Br_{0.3}\cdot 2.3H_2O$ are good examples of quasi-1D conductors because the overlapping of D_{Z^2} orbitals (see Fig. 13) leads to good electrical conductivity along the chains of Pt–Pt while they are insulators perpendicular to the chains.

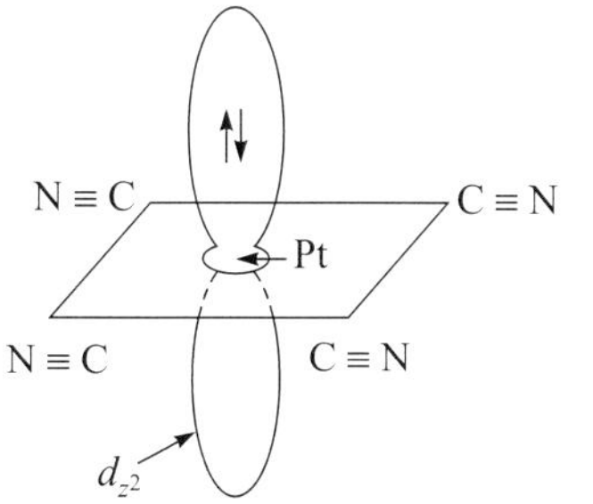

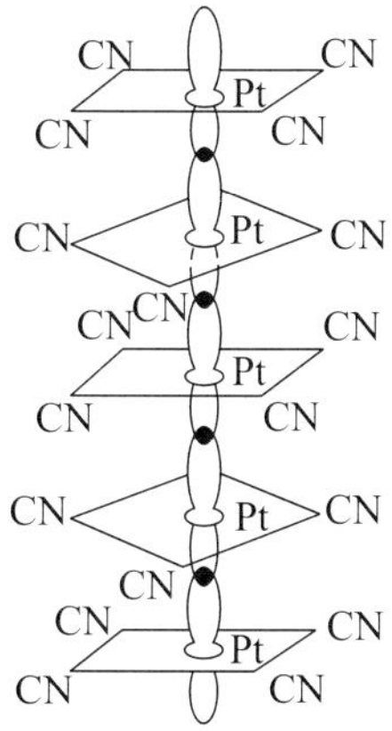

Figure 13

The discovery of this phenomenon by Gando et al. (*Solid State Comm* **12,** 1973, 1125), in combined molecules of tetracyanoquinodimethane (TCNQ) and tetrathiofulvalene (TTF) resulted in numerous theoretical and experimental researches and the mechanisms of these phenomena stimulated investigations of specific effects related to 1D and 2D that contributed to the development of molecular electronics. We note that this exercise neglects electron–phonon interactions related to the Kohn effect and to Peierls distortions (presence of a forbidden band in the neighborhood of the Fermi energy; see Chapter V, Ex. 4). The present purpose is to focus the readers' attention on these fascinating materials (to which we can also add chains of type S—N—S—N--) without going into a more realistic but more complicated description. A detailed analysis of the specific properties of another class of 1D objects, carbon nanotubes, is developed in Section V.

Exercise 14a: Fermi energy and electronic specific heat of a 2D conductor

In the xOy plane, consider identical atoms distributed at the points of a square lattice with side a (see Chapter III, Exs. 9 and 18).

(a) Show the first Brillouin zone in reciprocal space. Recall the dispersion relation $E = f(k)$ for free electrons and specify the energy (E_X and E_M) of electrons for which the wave vectors ($\overrightarrow{OX}$ and then $\overrightarrow{OM}$) of origin O (or Γ) are in contact to the 1st Brillouin zone, X and M, in the 10 direction and the 11 direction. What are the numerical values of E_X and E_M when a = 3 Å?

(b) Starting from cyclic Born-von Karman conditions taken for the length L = Na along x and y in direct space, find in the reciprocal lattice, the density of states $g(k)dk$ contained in the circumference of radii k and $k + dk$. Deduce the energy density of states $g(E)$ of a 2D gas of free electrons.

(c) Find the expression and the numerical value of the Fermi wave vector, $\overrightarrow{k_F}$, the speed, v_F, the energy, E_F, and the Fermi temperature, T_F, of the metal assuming that a single electron per atom can move freely in the plane (a = 3 Å). What is the average kinetic energy $\bar{E}$ of such an electron? Sketch the Fermi curve on the scheme of Question (a).

(d) Establish that, in a 2D lattice, the Fermi energy does not vary with temperature T and deduce the expression for energy $U(T)$ and electronic specific heat $C_e(T)$. Was it possible to obtain directly an order of magnitude for the latter result?

(e) Knowing that the specific heat at low temperatures associated with atomic vibrations in the plane is of the form $C_R = 28.8N^2k_B\left(\frac{T}{\theta_D}\right)^2$ (see Chapter III, Ex. 18), find at what temperature, T_c, the electronic contribution to the total specific heat is equal to that of lattice vibrations. Find its value for θ_D = 270 K.

(f) Compare the results of $g(E)$, $\bar{E}$, and $C_e(T)$ obtained here with those obtained for a linear chain in the previous exercise as well as for a 3D metal. Use F1 or F2. (e, m, h, k_B)

Solution:

(a) If $\vec{a}$ and $\vec{b}$ are the basis vectors in the direct space ($\vec{b} \perp \vec{a}$ and $|\vec{a}| = |\vec{b}| = a$), the basis vectors $\vec{A}$ and $\vec{B}$ of the reciprocal lattice are ($\vec{A} \cdot \vec{a} = \vec{B} \cdot \vec{b} = 2\pi$ and $\vec{A} \cdot \vec{b} = \vec{B} \cdot \vec{a} = 0$) orthogonal to each other and have the same modulus $\dfrac{2\pi}{a}$. In reciprocal space, the median of segment 00-10 and thus the medians of equivalent segments 00–01, 00–0$\bar{1}$, 00–$\bar{1}$0 delimit a square of side $\dfrac{2\pi}{a}$, which limits the first Brillouin zone (see Chapter III, Fig. 14 and Ex. 9).

The dispersion relation for free electrons is $E = \dfrac{\hbar^2 k^2}{2m} = \dfrac{\hbar^2}{2m}(k_x^2 + k_y^2)$, which leads to the following values:

$$E_X = \frac{\hbar^2}{2m} \cdot \frac{\pi^2}{a^2}, \quad E_M = \frac{\hbar^2}{2m} \cdot \frac{2\pi^2}{a^2}.$$

Numerically we have $E_x = 4.23$ eV and $E_M = 8.45$ eV.

(b) The density of states in wave vector space is the same as that of phonons in the same lattice (see Chapter III, Exs. 9 and 18):

$$g(k) = \frac{2\pi k}{\left(\dfrac{2\pi}{Na}\right)^2} = \frac{N^2 a^2 k}{2\pi} = \frac{Sk}{2\pi}.$$

Taking into account the dispersion relation and $g(k)dk = g(E)dE$, we find $g(E) = \dfrac{N^2 a^2}{\pi} \dfrac{m}{\hbar^2} = Cste$. This result takes into account two electrons per state.

(c) The system has one free electron per atom: N^2 electrons are distributed in $N^2/2$ states and are contained in the circle of radius k_F such that $\dfrac{\pi k_F^2}{\left(\dfrac{2\pi}{Na}\right)^2} = \dfrac{N^2}{2}$ leading to

$$k_{\text{F}} = \frac{(2\pi)^{1/2}}{a} = 0.8\pi / a = 0.835 \ \text{Å}^{-1}.$$

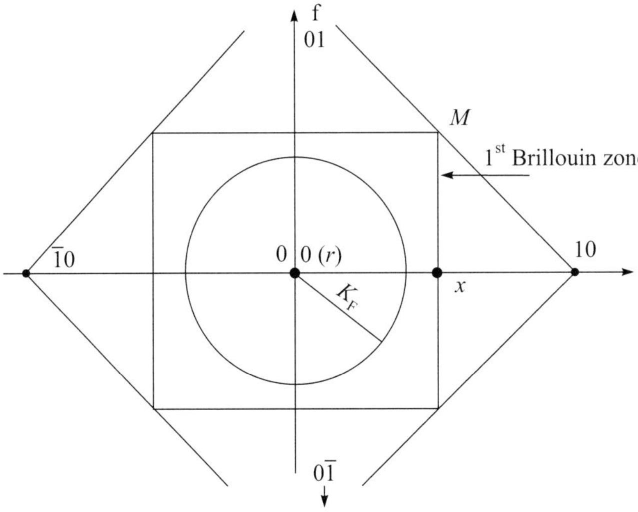

Figure 14

We can also find this result by considering that the area (volume in 3D) of the Fermi surface is, for a monovalent metal, equal to ½ the area of the first Brillouin zone (volume in 3D) because the first Brillouin zone contains as many cells as atoms in the crystal (N^2) and we can put two electrons (↑↓) into each cell inside the Fermi surface so that $\pi k_{\text{F}}^2 = \left(\frac{2\pi}{a}\right)^2 \cdot \frac{1}{2}$.

The Fermi velocity $(\hbar \vec{k}_{\text{F}} = m\vec{v}_{\text{F}})$, the Fermi energy E_{F} and the Fermi temperature T_{F} can be deduced simultaneously:

$$v_{\text{F}} = \frac{\hbar}{m} \cdot \frac{(2\pi)^{1/2}}{a}, T_{\text{F}} = \frac{E_{\text{F}}}{k_{\text{B}}}, \text{ where } E_{\text{F}} = \frac{\hbar^2 \pi}{ma^2}$$

Taking into account this result, we find $g(E) = \frac{N^2}{E_{\text{F}}}$, which could also be deduced from $\int_0^{E_{\text{F}}} g(E) \cdot dE = N^2$.

At 0 K, the average energy is $\overline{E} = \frac{\int_0^{E_{\text{F}}} g(E)\, E dE}{\int_0^{E_{\text{F}}} g(E)\, dE} = \frac{E_{\text{F}}}{2}$.

The numerical values are $V_F = 967 \times 10^3$ m/s $\approx 10^3$ km/s, $E_F = 2.7$ eV, $T_F = 31{,}300$ K.

The Fermi circumference is shown in Fig. 14.

(d) The Fermi energy $E_F(T)$ can be deduced from $\int_0^\infty g(E)\cdot f(E)\,dE = N^2$, where $f(E) = \frac{1}{e^{(E-E_F)/k_BT}+1}$.

In 2D, $g(E) = \frac{N^2}{E_F(0)} = cste$, and can be removed from the integral so that we find

$$\frac{N^2}{E_F(0)}\int_0^\infty \frac{dE}{e^{\gamma(E/\chi)-1}+1} = N^2, \text{ where } \chi = E_F(T) \text{ and } \gamma = \frac{E_F(T)}{k_BT} \approx \frac{T_F}{T}.$$

The integral can be solved using F2 and making $p = 0$ we find

$$\frac{E_F(T)}{E_F(0)} = 1 + 0\cdot\left(\frac{T}{T_F}\right)^2.$$

In a 2D lattice, the Fermi energy does not vary with temperature (see note at end of exercise).

The energy $U_e(T)$ of an electron gas obeys the relation

$$U_e = \int_0^\infty E\cdot g(E)\cdot f(E)\cdot dE = \frac{N^2}{E_F(0)}\int_0^\infty \frac{EdE}{e^{\gamma[(E/\chi)-1]}+1},$$

from which we find $U_e = N^2E_F(0)\left[\frac{1}{2} + \frac{\pi^2}{6}\left(\frac{T}{T_F}\right)^2\right]$.

In this expression, the term that is independent of T corresponds to the energy of N^2 electrons at 0 K is $N^2\overline{E} = \frac{N^2E_F}{2}$.

The electronic specific heat C_e is given by

$$C_e = \left(\frac{\partial U_e}{\partial T}\right)_V = \frac{N^2\pi^2}{3}k_B\frac{T}{T_F}.$$

If we consider that the only electrons having an energy E between $E_F - k_BT$ and E_F gain energy k_BT when an electron gas goes from 0 K to T K, the corresponding increase in energy of the electron gas is such that $\Delta U_e \approx g(E_F)k_BT\cdot k_BT = \frac{N^2k_B^2T^2}{E_F}$. The order of magnitude of C_e will thus be $C_e = \left(\frac{\partial(\Delta U_e)}{\partial T}\right)_V = 2N^2 k_B \frac{T}{T_F}$.

We find that C_e is linear as a function of T and that only the numerical coefficient (2 instead of 3.3) is different.

(e) At low temperatures, the total specific heat C_T is

$$C_T = C_R + C_e = N^2 k_B \left[28.8\left(\frac{T}{\theta_D}\right)^2 + 3.3\left(\frac{T}{T_F}\right) \right] = \alpha T^2 + \gamma T.$$

The electronic contribution will equal the contribution of the lattice to the temperature $T_c = \gamma/\alpha = 0.27$ K.

Above this temperature, the contribution of phonons is greater than that of electrons. We can experimentally determine the coefficients α and γ by considering the curve $C_T/T = f(T)$, which is theoretically a line in 2D and therefore the intercept at the origin gives γ.

(f) To facilitate the comparison, we assume that the parameter a is the same for all the 1-, 2-, and 3D lattices and we assume that the objects all have z free electrons per elementary cell of the lattice.

The results obtained in the present exercise and in Exs. 8 and 13 are summarized in Table 1 (see also the table in Ex. 22 relative to the paramagnetism of free electrons in 1-, 2-, and 3D).

n_l, n_s, n_v are respectively the density of electrons in 1-, 2-, and 3D; N and V represent the total number of free electrons and the total volume. The density of states $g(E)$ is proportional to $E^{-1/2}$, constant and proportional to $E^{1/2}$ for lattices in respectively 1-, 2-, and 3D but contrarily to the corresponding result for lattice vibrations (see Chapter III, table of Ex. 18), this does not affect the linearity of the electronic specific heat as a function of temperature.

Note: The expression $\frac{N^2}{E_F(0)}\int_0^\infty \frac{dE}{e^{(E-E_F)/k_B T}+1} = N^2$ can be integrated directly to obtain the exact result:

$E_F(0) = E_F(T) + k_B T \log(e^{-E_F(T)/k_B T}+1)$. This confirms that E_F is not sensitive to temperature for a 2D electron gas.

Table 1 Essential parameters for free electrons in 1-, 2-, and 3D objects

	1D	2D	3D
k_F	$\frac{\pi}{2}n_1 = \frac{\pi z}{2a}$	$\sqrt{2\pi n_s} = \frac{\sqrt{2\pi z}}{a}$	$(3\pi^2 n_v)^{1/3} = \frac{(3\pi^2 z)^{1/3}}{a}$
E_F	$\frac{\hbar^2}{2m}\frac{\pi^2}{4}n_1^2$	$\frac{\hbar^2}{2m}2\pi n_s$	$\frac{\hbar^2}{2m}(3\pi^2 n_v)^{2/3}$
$\overline{E}$	$\frac{E_F}{3}$	$\frac{E_F}{2}$	$\frac{3E_F}{5}$
$g(E)$	$\left(\frac{L}{\pi}\right)\left(\frac{2m}{\hbar^2}\right)^{\frac{1}{2}}(E)^{\frac{-1}{2}}$ $\frac{N}{2\sqrt{E_F}}\cdot(E)^{-1/2}$	$\left.\left(\frac{S}{2\pi}\right)\left(\frac{2m}{\hbar^2}\right) \quad \frac{N}{E_F}\right\vert_{cste}$	$\left(\frac{V}{2\pi}\right)\left(\frac{2m}{\hbar^2}\right)^{3/2}\sqrt{E}$ $\left(\frac{3N}{2E_F^{3/2}}\right)\sqrt{E}$
C_e 1 free e^-	$\frac{\pi^2}{12}k_B\left(\frac{T}{T_F}\right)$	$\frac{\pi^2}{3}k_B\left(\frac{T}{T_F}\right)$	$\frac{3\pi^2}{4}k_B\left(\frac{T}{T_F}\right)$

Exercise 14b: π-electrons in graphite (variation of Ex. 14a and simplified approach for graphene)

Graphite is a lamellar crystal in which the atoms of carbon in a given graphene layer are distributed in regular hexagons (of side d = 1.42 Å) to form a honeycomb structure (see Chapter I, Ex. 17). The π-electrons of graphite, one electron per atom, are restricted to move in these layers to form a 2D (assumed here to be free) electron gas.

(a) What is the density of states $g(E)$? Find the Fermi energy $E_F(0)$, the Fermi temperature of graphite $T_F(0)$, and the average energy $\overline{E}$ of these electrons at 0 K as a function of the surface electronic density n_s. What are the numerical values of $T_F(0)$ and $E_F(0)$?

(b) In reciprocal space, draw the Fermi circumference and the first Brillouin zone. Are there intersections between the two?

(c) After having shown that the Fermi level does not change with temperature, calculate the total energy of an electron gas at temperature T ($T << T_F$) and deduce the electronic specific heat C_e per mole. Find the numerical value of C_e at ambient temperature. (h, m, e, k_B)

Solution:

(a) $g(k) = \dfrac{LxLy}{2\pi}k;\ g(E) = \dfrac{LxLy}{\pi}\dfrac{m}{\hbar^2};\ \int_0^{E_F(0)} g(E)dE = N;$

$E_F(0) = \dfrac{\hbar^2}{m}\cdot\pi\cdot\dfrac{N}{L_x L_y}$ in which $\dfrac{N}{L_x L_y} = n_s$ (electronic density per unit area).

Here $n_s = \dfrac{2}{\text{Hex. area}} = \dfrac{4}{3\sqrt{3}d^2} = 38\times10^{18}\ \text{e/m}^2;\ E_F(0) \approx 9eV;$

$T_F(0) \approx 10^5(°K), \bar{E} = E_F/2$ (see Ex. 14a).

(b) Graphite can be considered as monovalent ($1\pi e^-$ per atom) but there are two atoms in non-equivalent position in the unit cell of order 1. One must therefore distribute these two electrons ↑↓ per cell. The surface occupied by the Fermi circumference is therefore equal to that of the first Brillouin zone (and not equal to half of it as seen in Ex. 14a above). The Fermi circumference thus intercepts the first Brillouin zone and this can be verified numerically by comparing $k_F = 1.545\ \text{Å}^{-1};\ |\vec{A}|/2 = 1.474\ \text{Å}^{-1}$ Effectively $|\vec{a}| = 2d\cos 30° = 2.46$ Å and $\vec{a}\cdot\vec{A} = 2\pi = |\vec{A}|\cdot|\vec{a}|\cos 30°$, see Chapter I, Ex.17.

Note that the Fermi circumference leaves unoccupied cells near the summits of the hexagon in the 1st Brillouin zone and it leads to occupied cells in 2*d* Brillouin zone along the sides of the hexagon. The density of holes (unoccupied states) is therefore equal to the density of electrons.

In fact the crystal potential modifies the parabolic dispersion relation of electrons in particular at intersections between Brillouin zones and Fermi surfaces. This is of specific importance for graphene layers either from the theoretical point of view or for the practical applications in spintronics: see the more detailed investigation in Chapter V.

(c) See Ex. 14a.

$$E_F(0) = E_F(T) + k_B T\log(e^{-E_F(T)/k_B T} + 1)$$

$$U_e = \mathcal{N}\cdot E_F(0)\cdot\left[\frac{1}{2} + \frac{\pi^2}{6}(T/T_F)^2\right]$$

$$C_e = \frac{N\pi^2}{3}k_B(T/T_F) = 82\times10^{-2}\ \text{J/deg.}$$

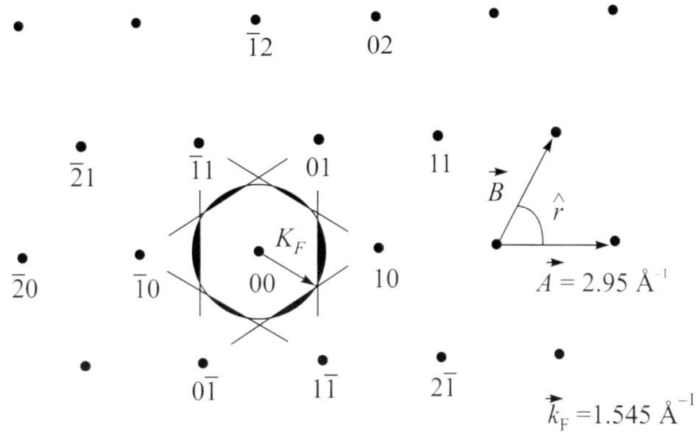

Figure 15

Exercise 14c: Fermi vector and Fermi energy (at 0 K) of an electron gas in 1-, 2-, and 3D—comparison with the residual vibration energy of atoms

Consider N atoms of a monovalent metal that are successively used to build a 1D structure of length L, a 2D structure of surface L^2 and a 3D structure of volume L^3.

(1) Find the expression for the Fermi wave vector k_F in each case. Compare k_F to k_D, the Debye wave vector relative to the atomic vibrations. Compare the dimension of the Fermi length, 1D, circle, 2D, radius, 3D, to the corresponding dimensions of the 1st Brillouin zone.

(2) Knowing that the density of states of free electrons $g(E)$ in 1-, 2-, and 3D take respectively the forms: $g(E) = B_1/\sqrt{E}$ (1D); $g(E) = B_2$(constant)(2D); $g(E) = B_3\sqrt{E}$ (3D), find the constants B_1, B_2, and B_3 as a function of N and of the Fermi energy E_F. Express the average energy of a free electron $\bar{E}$ as a function of E_F and the total energy $U_e(0)$ of N electrons at 0 K.

(3) Compare this energy $U_e(0)$ to the residual atomic vibration energy $U_e(0)$ at the same temperature by expressing the ratio $U_e(0)/U_P(0)$ as a function of Fermi temperature T_F and of the Debye temperature θ_D. In the latter take into account that $g(\nu)$ for the vibrations has the form $g(\nu) = A_1$, $A_2\,\nu$, $A_3\,\nu^2$ in 1-, 2-,

and 3D, respectively (Chapter III, Ex. 20b). Find the ratio of the increased energies $\Delta U_e(T)/\Delta U_P(T)$ when the temperature changes from 0 K to T K where $T > \theta_D$. For $\Delta U_e(T)$ an estimate is sufficient instead of the exact value.

Solution:

(1) Because there are two electrons (↑↓) per cell of dimension(s)

$$\frac{2\pi}{L}\left(\frac{2\pi}{L}\cdot\frac{2\pi}{L}\right), \text{ we find}$$

$$1\text{D}:\ 2k_F = \frac{N}{2}\cdot\frac{2\pi}{L} \rightarrow k_F = \pi N/2L$$

$$2\text{D}:\ \pi k_F^2 = \frac{N}{2}\left(\frac{2\pi}{L}\right)^2 \rightarrow k_F = \frac{\sqrt{N2\pi}}{L}$$

$$3\text{D}: \frac{4}{3}\pi k_F^3 = \frac{N}{2}\left(\frac{2\pi}{L}\right)^3 \rightarrow k_F = (3N\pi^2)^{1/3}/L$$

k_D is respectively 2, $(2)^{1/2}$, and $(2)^{1/3}$ times larger than the corresponding k_F values because there is one oscillator per cell (see Chapter III, Ex. 20b).

For the same reason, the length, surface, and volume occupied by electrons are always two times smaller than the corresponding values for the oscillators (Debye).

(2) $\int_0^{E_F} g(E)\,dE = N$, where

$$B_1 = N/2E_F^{1/2};\quad B_2 = N/E_F;\quad B_3 = 3N/(2E_F^{3/2})$$

$$\bar{E} = \int_0^{E_F} Eg(E)dE \Big/ \int_0^{E_F} g(E)dE;$$

$$\bar{E}(1\text{D}) = E_F/3;\quad \bar{E}(2\text{D}) = E_F/2;\quad \bar{E}(3\text{D}) = 3E_F/5.$$

(3) $U_e(0) = N\bar{E}$ that gives $U_e(0)/U_p(0) = (4/9)(T_F/\theta_D)$ for 1D, $(T_F/3\theta_D)$ for 2D, and $(8/15)(T_F/\theta_D)$ for 3D.

The expressions for $U_p(0)$ can be deduced from k_D (evaluated in Question (1)). We find $v_D = uk_D/2\pi$ and $\theta_D(hv_D/k_B)$, taking into account $g(\nu)$ (see Chapter III, Ex. 20a, which used the same method).

Since $T_F \approx$ several 10^4 K and $\theta_D \approx$ several 10^2 K, $U_e(0)$ is always larger than $U_p(0)$.

Going from 0 to T K, ΔU_e is of order of $\Delta U_e \approx g(E_F) \cdot k_B T \cdot k_B T$ (see reasoning in Ex. 13d) and $\Delta U_D = 3Nk_B T$ when $T > \theta_D$.

$\Delta U_e / \Delta U_D \approx T/6T_F(1D); \approx T/3T_F(2D); \approx T/2T_F$ (3D)

This ratio corresponds to the ratio of specific heats. It shows that if the initial energy of electrons at 0 K is larger than that of the lattice vibrations, its increase as a function of T is far smaller.

Exercise 15: Surface stress of metals

Consider a monovalent (1 electron/atom) fcc metal with lattice parameter a.

(1) Assume that the atom in the center of a face (100) is in contact with its four neighbors and sketch the atomic distribution on the (100), (110), and (111) faces.

(2) Express the atomic density n_v (at. number/volume) and the surface density n_s along the following surfaces: (100), (110), and (111).

(3) Write the general expressions for the density of state, $g_v(E)$, and the Fermi energy, E_{FV}, for a 3D electron gas as a function of the electronic density n_v. Find the average energy of an electron $\overline{E_V}$ as a function of E_{FV}. Specify E_{FV} as a function of a.

(4) Assume that, the electrons of surface atoms behave as a free electron gas in 2D over the thickness of a single atomic layer. For each surface, find the general expression of the density of state $g_s(E)$, the Fermi energy E_{FS} of such an electron gas as a function of n_s. Express the average energy $\overline{E}$ as a function of E_{FS}.

(5) Evaluate the surface Fermi energy for the (100), (110), and (111) faces, denoted as $E_{FS}(100)$, $E_{FS}(110)$, and $E_{FS}(111)$. Class E_{VS} and the different E_{FS} in order of increasing energy (in units of $\hbar^2\pi/ma^2$). Sketch the different curves, $g(E)$, and indicate the position of E_F and $\overline{E}$.

(6) For obvious thermodynamic reasons, the maximal kinetic energy, E_{FS}, of surface electrons is the same as E_{FV} of volume electrons (alignment of the Fermi level). Taking this common value as a new origin, sketch the new corresponding graph. State (in units of $\hbar^2\pi/ma^2$) the energetic distance E_0 between the minimum of the surface energy (relative to each of the three faces considered) and the minimum volume energy.

(7) Deduce the difference $\Delta\bar{E}$ between the average energy of a surface electron and that of a volume electron:
$\Delta\bar{E} = \bar{E}_s - \bar{E}_v$. Express this difference as a function of a for the faces considered: $\Delta\bar{E}(100)$, $\Delta\bar{E}$ (110) and $\Delta\bar{E}$ (111).
Numerical application: Find the different values of $\Delta\bar{E}$ for $\hbar^2/2m = 3.8$ eV Å^2 and $a = 3.5$ Å.

(8) Estimate the surface stress σ (energy per surface: $\sigma = \Delta\bar{E}/S$) of the different surfaces: $\sigma(100)$, $\sigma(110)$ and $\sigma(111)$.
Numerical application: Express σ in J/m^2. Comment on the results.

Solution:

(1) See Fig. 16.

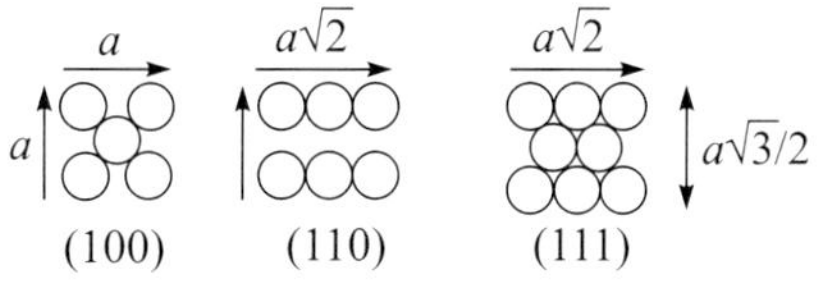

Figure 16

(2) $n_v = 4/a^3$; $n_s(100) = 2/a^2$; $n_s(110) = \sqrt{2}/a^2$; $n_s(111) = 4/a^2\sqrt{3}$.

(3) $g_v(E) = \dfrac{L_x L_y L_z}{\pi^2}\dfrac{\sqrt{2m^3}}{\hbar^3}\sqrt{E}$; $E_{FV} = \dfrac{\hbar^2}{2m}(3\pi^2 n_e)^{2/3}$;

$$\bar{E} = \frac{3}{5}E_{FV};\ E_{FV} = \frac{\hbar^2}{2m}\left(\frac{12\pi^2}{a^2}\right)^{2/3} \text{ with } n_v = n_e.$$

(4) $g_s(E)dE = \dfrac{L_X L_Y}{\pi}\dfrac{m}{\hbar^2}$; $E_{FS} = \dfrac{\hbar^2}{2m}2\pi n_s$; $\bar{E} = E_{FS}/2$.

(5) $E_{FS}(100) = \dfrac{\hbar^2}{m}\dfrac{2\pi}{a^2}$; $E_{FS}(110) = \dfrac{\hbar^2}{m}\dfrac{\pi\sqrt{2}}{a^2}$; $E_{FS}(111) = \dfrac{\hbar^2}{m}\dfrac{4\pi}{a^2\sqrt{3}}$.

$E_{FS}(110) < E_{FS}(100) < E_{FS}(111) < E_{FV}$

since $\sqrt{2} < 2 < \dfrac{4}{\sqrt{3}} \approx 2.31 < 3.84$

The graph is shown in Fig. 17.

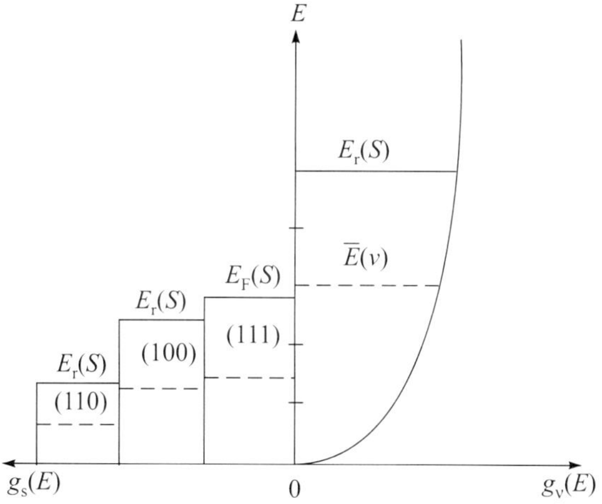

Figure 17

(6) See Fig. 18.
$E_0 = -E_{FS} + E_{FV}$; $E_0 = 1.53, 1.84, 2.43$ (in units of $\hbar^2\pi/ma^2$) for $E_{FS}(111)$, $E_{FS}(100)$, and $E_{FS}(110)$, respectively.

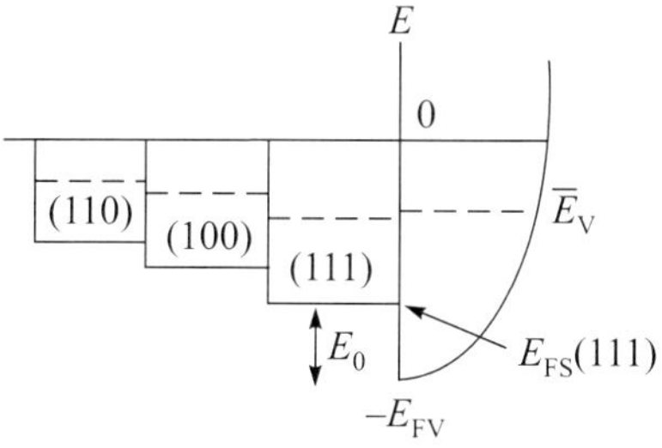

Figure 18

(7) $\overline{\Delta E} = -(1/2)E_{FS} + (2/5)E_{FV}$, which gives $\overline{\Delta E}(111) = -0.114$, $\overline{\Delta E}(100) = +0.536$ and $\overline{\Delta E}(110) = 0.83$, all in unities of $\hbar^2\pi/ma^2$ that equals 1.95 eV for $a = 3.5$ Å.

Thus we have $\overline{\Delta E}(111) = -3.23$ eV; $\Delta E(100) = 1.05$ eV;
$\overline{\Delta E}(110) = 1.62$ eV

(8) $\sigma = n_s \overline{\Delta E}$; $\sigma(111) = -2.43\ \text{J/m}^2$; $\sigma(100) = 2.7\ \text{J/m}^2$; $\sigma(110) \approx 3\ \text{J/m}^2$.

We thus have a contraction when $\sigma > 0$ and lateral expansion when $\sigma < 0$.

Note: Units

Thus, although joule has the same dimensions as the Newton-meter ($1\ \text{J} = 1\ \text{N}\cdot\text{m} = 1\ \text{kg}\cdot\text{m}^2\cdot\text{s}^{-2}$), these units are *not* interchangeable. Joule is the unit of energy while Newton-meter (N·m) is the unit of torque.

Comment

The effect of surface stress is to reconstruct the surface in order to minimize its average energy $\Delta\overline{E}$. The atomic position of reconstructed surface atoms is different from that of an ideal surface, as imagined in this problem. More sophisticated approaches involving the change of the work function with crystalline orientation of surfaces are detailed in Ex. 26.

Experimentally, this reconstruction can be studied by diffraction of slow electrons at normal incidence (see Chapter I, Pb. 3) or rapid electrons at grazing incidence (see Chapter I, Pb. 4) or with the widespread use of a scanning tunneling microscope, Pb. 3.

Exercise 16: Effect of impurities and temperature on the electrical resistivity of metals: Matthiessen's rule

(a) A beam of electrons with uniform density n and velocity $\vec{v}$ strikes different spheres of radius r that are homogenously diluted in a target with a density per unit volume of n_s.

After evaluating the number/second of electrons that strike the target, find the probability P_i of electron collision with the scattering centers and deduce when this probability is equal to 1 for an electron mean free path of Λ where $\Lambda = \dfrac{1}{\pi r^2 \cdot n_s}$. (To find this result, assume that the electron is a point charge and strikes hard spheres.)

(b) Now consider a monovalent metal, sodium, in which there are impurities with atomic concentration c (c = atomic density of impurities/atomic density of sodium) and conduction electrons with velocity v_F that interact only with these impurities following the rule in a).

Find the electrical resistivity ρ_i due to the impurities. What is the numerical value of ρ_i when the concentration of impurities is 1% and the radius of an impurity is ≈ 1 Å?

(c) Due to thermal vibrations, the atoms of sodium also act like scattering spheres having an apparent radius of r_p. Starting from elementary considerations concerning the collision probability, show that the total resistivity ρ_t of a metal with impurities follows the Matthiessen's rule: $\rho_t = \rho_i + \rho_p$, where ρ_p is the electrical resistivity due to thermal vibrations.

At ambient temperature (300 K), the resistivity of pure sodium is 4.3×10^{-8} Ω-m, find the value of the resistivity for sodium containing 1% impurities.

(d) At high temperature ($T > \theta_D$), one may admit that the apparent radius r_p follows the relation $r_p^2 \approx \overline{u}^2$, where $\overline{u}^2$ represents the average quadratic displacement of a metal atom. Find the expression of ρ_p as a function of T by considering that each Na atom behaves classically as a harmonic oscillator vibrating at the Debye frequency $\nu_D(h\nu_D = k_B\theta_D)$. Compare the numerical result to the experimental value. Find the thermal coefficient of sodium (which obeys $\alpha = \frac{1}{\rho}\frac{\partial \rho}{\partial T}$) and compare it with the experimental value of $\alpha \approx 4 \times 10^{-3}$.

For sodium, use the following values: $M(\mathcal{N}) = 23$ g, $\theta_d = 160$ K, $v_F = 1.07 \times 10^6$ m/sec. (e, m, k_B, h, $\mathcal{N}$)

Solution:

(a) The number of electrons per second that strike a scattering center is contained in a cylinder of length v and section area πr^2 is $n \cdot v\pi r^2$, as shown in Fig. 19. The total number of collisions per second between electrons of the beam and the diffuse centers will be: $n \cdot v \cdot \pi r^2 \times S \cdot v \cdot n_s$ (where S is the cross-section area of the electron beam). The probability of collisions will thus be $P_i = \pi r^2 \cdot v \cdot n_s$.

The time of flight between two collisions is $\tau = 1/P_i = 1/\pi r^2 \cdot v \cdot n_s$ and the mean-free path $\Lambda = v \cdot \tau = 1/\pi r^2 n_s$ (in which πr^2 is the scattering cross-section).

(b) $\rho_i = \frac{1}{\sigma} = \frac{m}{ne^2\tau} = \frac{m}{ne^2}(\pi r^2 \cdot v_F \cdot n_s)$, $\rho_i = \frac{\pi r^2 \cdot m v_F}{e^2} \cdot c$, where $c = n_s/n$. Since sodium is monovalent, n is both the density of conduction electrons and the density of sodium atoms.
For an impurity concentration of 1%, $\rho_i = 1.2 \times 10^{-8}$ Ω-m.

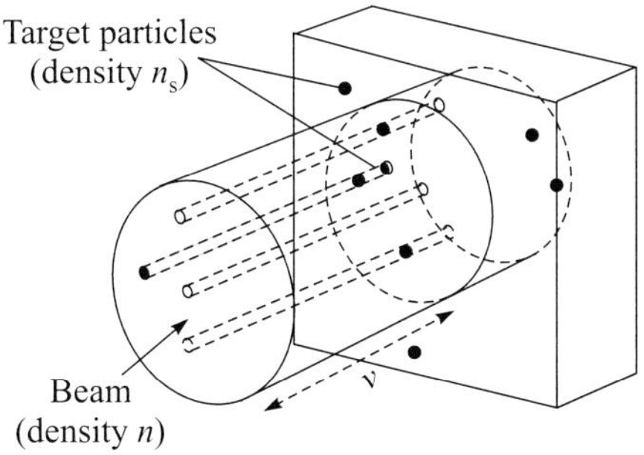

Figure 19 Scattering probability (or cross-section).

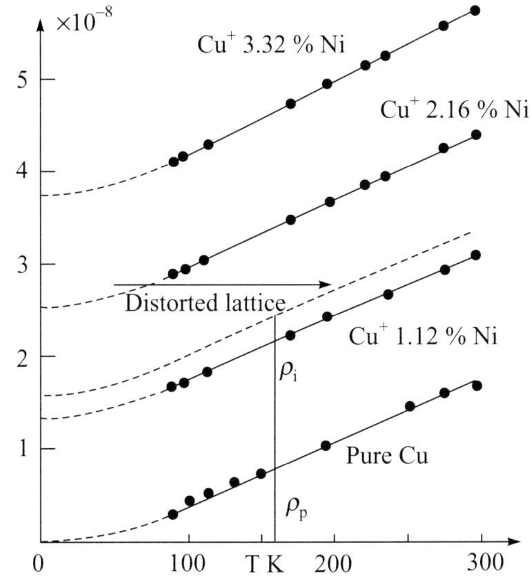

Figure 20 Addition of the different components of resistivity: Temperature ρ_p and impurities ρ_i.

(c) When there are two types of scattering centers with cross-sections σ_1 and σ_2 and densities n_1 and n_2, the electron-scattering probability is equal to the sum of probabilities:

$P = P_1 + P_2 = (\sigma_1 n_1 + \sigma_2 n_2)v = \frac{1}{\tau}$. By writing $\frac{1}{\tau_1} = \sigma_1 n_1 v$ and $\frac{1}{\tau_2} = \sigma_2 n_2 v$, the time of flight is $1/\tau = 1/\tau_1 + 1/\tau_2$. Thus, taking into account the general expression of the resistivity, $\rho = m/ne^2\tau$, we find the Matthiessen's rule: $\rho = \rho_1 + \rho_2$.

Numerically, we note that the resistivity due to an impurity concentration of 1% is of the same order as the intrinsic resistivity: $\rho_t = \rho_i + \rho_p = 5.5 \times 10^{-8}$ Ωm.

(d) Each atom of Na acts as a classical oscillator of mass M and of frequency ν_D. Therefore, its energy of vibration is of $3k_BT$. The average quadratic displacement thus obeys (see also Chapter III, Ex. 21b) $3k_BT = M\,(2\pi\nu_D)^2 \cdot \bar{u}^2$, which is equivalent to

$$\bar{u}^2 = \frac{3\hbar^2 T}{M \cdot k_B \theta_D^2} \text{ and } \rho = \frac{m}{ne^2\tau} = \frac{m}{e^2}\pi\bar{u}^2 \cdot v_F = \frac{3\pi m v_F \cdot \hbar^2 T}{e^2 M \cdot k_B \theta_D^2}.$$

At high temperature, the electrical resistivity is linear in T and the temperature coefficient α is $1/T$.

We have $\rho = AT$, $\Delta\rho = A\Delta T$ and $\alpha = \frac{1}{\rho}\frac{\partial\rho}{\partial T} = T^{-1}$.

Numerically, we find $\rho_p = 7.9 \times 10^{-8}$ Ω-m and $\alpha = 3.33 \times 10^{-3}$, which taking into account the implicit approximations made, is in good agreement with the experimental values.

Comment

Electronic waves propagate nearly without attenuation in an ideal crystal: Λ can attain several centimeters in very pure metals such as Ag at very low temperatures. Deviations from an ideal crystal are due to impurity atoms or crystalline defects, and they also result in the thermal vibration of atoms that is non-zero even at 0 K and are responsible for the electrical resistivity of real metals.

The rigorous calculation of the electrical conductivity of a metal outside the context of the present exercise can explain the T^{-5} evolution of σ at low temperature by taking into account the electron waves scattered by phonons (from which the average quadratic displacement was considered in Chapter III, Ex. 21a).

Despite the simplifications of this exercise in metals, it explains:

- The additivity of partial resistivities and the influence of impurities.
- The linear increase (as a function of T) of the resistivity at high temperatures ($T > \theta_D$).
- The thermal coefficient that is equal to $1/T$ and is independent of the metal considered ($\alpha \approx 3$–5×10^{-3} for most metals excluding specific alloys such as manganin (86% Cu, 12% Mn, and 2% Ni) where $\alpha \approx 10^{-5}$.

These results, illustrated in Fig. 20 (see also Exs. 17 and 18), can be completed by the size effect studied in Ex. 20. They are characteristics of metals and radically different from the electric behavior of semiconductors (see Chapter V).

Exercise 17: Effect of the vacancy concentration on the resistivity of metals

According to Peter Matthiessen, the resistivity ρ of a metal obeys the relation $\rho = \rho_T + \rho_i$ where ρ_T is uniquely a function of temperature and ρ_i depends on the concentration of impurities or atomic vacancies. The concentration of vacancies, c_1, follows a low of the form $c_1(T) = e^{-\frac{E_F}{k_B T}}$, where E_f is the formation energy of vacancies.

The resistivity of two pure gold wires is measured at 78 K after having heated one, denoted A_1, to a temperature T_1 = 920 K, and the other, denoted A_2, to a temperature T_2 = 1220 K, and then quenched immediately to 78 K, in order to stabilize the concentration of vacancies. The relative change of resistivity is $\Delta\rho/\rho$ = 0.4% for A_1 and 9% for A_2 using as a reference value $\rho = \rho_T$ of a sample that was not subject to any thermal treatment. Find the energy of formation of vacancies, in eV, and their concentrations.

Solution:

$$\rho(78) = \rho_T;\ \rho(78/T/78) = \rho_T + B\exp-\left(\frac{E_f}{k_B T}\right)$$

$$\Delta\rho(78/T/78)/\rho = B\exp-\left(\frac{E_f}{k_B T}\right)/\rho T,$$ from which we find

$$\ln \frac{\Delta\rho(A_1)/\rho}{\Delta\rho(A_2)/\rho} = \ln\left(\frac{9}{0.41}\right) = -\frac{E_f}{k_B}\left(\frac{1}{T_1} - \frac{1}{T_2}\right)$$, resulting in

E_f = 0.996 eV; c_1 (78/920/78) = 3/5 × 10^{-6}; c_2 (78/1220/78) = 7.8 × 10^{-5}.

This exercise illustrates the influence of thermal treatments on the resistivity of metals with the significant increase of ρ for a density of vacancies that is relatively small.

In metallurgy, this simple method, easy to operate, allows one to find E_f and c_1. It gives satisfactory results when the measurements are done at low temperatures and when the concentration of vacancies is not too large (otherwise $\Delta\rho$ is not proportional to c_1). Another experimental approach simultaneously combines the measurements of the linear expansion ($\Delta l/l$) and of the change in the lattice parameters, $\Delta a/a$ (see Ref. [20] for further details).

Exercise 18: Effect of impurity concentration on the resistivity

At low temperatures, the addition of impurities to a pure metal of resistivity ρ leads to an increase $\Delta\rho$ of this resistivity that is proportional to the concentration of impurities, c_i, weighted by a coefficient $(\Delta z)^2$, where Δz is the difference between the valence of the impurity and that of the host metal.

Find the resistivity ρ of a copper sample containing, respectively, (a) 2 × 10^{-3} at/at of In^{+3}; (b) 5 × 10^{-4} at/at of Sn^{4+}; (c) 10^{-3} at/at of Sb^{5+}; (d) a 10^{-3} concentration of vacancies.

The initial resistivity of pure Cu is $\rho(Cu^+)$ = 10^{-10} Ω-m and becomes 5 × 10^{-10} Ω-m when it contains 10^{-3} at/at of cadmium Cd^{2+}.

Solution:

$\Delta\rho = (\Delta z)^2 c_i A$, where A = 4 × 10^{-10} Ω-m

(a) ρ (Cu + 2 × 10^{-3} In) = 33 × 10^{-10} Ω-m

(b) ρ (Cu + 5 × 10^{-4} Sn) = 19 × 10^{-10} Ω-m

(c) ρ (Cu + 10^{-3} Sb) = 65 × 10^{-10} Ω-m

(d) ρ (Cu + 10^{-3} vacancies) = 5 × 10^{-10} Ω-m

The increase of the resistivity related to the addition of impurities is due in part to the distortion of the lattice from the change in the size of the impurities atoms and in part to the electrostatic interaction of metallic ions (Rutherford scattering cross-section in Z^2) or vacancies that act as missing ions⁺. When the metal contains magnetic impurities (Fe or Ni for instance), the resistance exhibits a low-temperature minimum resulting from the scattering of electrons by magnetic moments, an effect known as Kondo effect not taken into account by the Matthiessen law.

One application of the present exercise is the control of the degree of purity of a metal from the measurement of its residual electrical resistivity at a very low temperature (T < 20 K).

This exercise also permits to qualitatively understand why the resistivity of certain alloys, for instance CuZn, can be greater than that of the constituents mainly when the alloy lattices are disordered [20]. This exercise was inspired by Ref. [22].

Exercise 19: Another expression for the conductivity σ

When an electron gas is subject to an applied electric field, $\vec{E}$, each electron feels a drift velocity $\vec{v}_e = -\frac{e\vec{E}\tau}{m}$, which adds to its initial velocity resulting from Fermi Dirac statistic. Geometric considerations show that the electric conductivity σ can take the form $\sigma = \frac{1}{3} e^2 \cdot v_F^2 \cdot \tau \cdot g(E_F)$, where v_F is the Fermi velocity and $g(E_F)$ is the density of states per unit volume.

(a) Verify that the proposed expression gives the classical result for σ in the case of a free electron gas. What physical arguments argue for the use of this expression instead of the classical one?

(b) Can you establish simply this expression, up to a coefficient of 1/3, from the use of a representation of the displacement of the Fermi sphere in k-vector space?

Solution:

(a) We have $g(E) = A\sqrt{E}$, where $A(1/2\pi^2)(2m/\hbar^2)^{3/2}$ for a unit volume.

Therefore, we can write $g(E_F) = A\sqrt{E_F}$ and $\int_0^{E_F} g(E)\,dE = n = (2/3)\,AE_F^{3/2}$.

Consequently $g(E_F)$ =(3/2)n/E_F and the proposed expression thus gives $\sigma = ne^2\ \tau/m$ by taking into account that $(1/2)mv_F^2 = E_F$.

The proposed expression is physically more acceptable because under the action of the electric field, the Fermi volume is translated by a momentum of $\hbar\overline{\delta k}$ (or equivalently $m\overline{v_e}$), which is equal to $e\tau\overline{E}$. Figure 21a amplifies this displacement 00' that is small compared with $\hbar k_F$. Thus, as shown in Fig. 21b, the only electrons in the neighborhood of the Fermi energy—and Fermi density $g(E_F)$—are in fact implied in the current transport.

If the two expressions lead to the same result when the electrons are perfectly free, the proposed expression can also explain experimental results observed in materials in which the electrons do not strictly follow the free electron theory. When conduction electrons are sensitive to the crystalline potential, their dispersion relation is no longer parabolic, the Fermi surface is no longer spherical and the density of states in the neighborhood of E_F no longer follows a $\sqrt{E}$ law. This is practically the case for all materials except the alkali metals. In particular, the proposed expression takes into account the conductivity of divalent metals for which the measurement of the Hall constant leads to a density of carriers significantly different from the expected 2e per atom. Also, a large density of Fermi states can explain the superconducting properties of certain copper oxides (see Chapter V, Ex. 2b) and $g(E_F)$ may be involved into a more elaborate expression for the electronic specific heat (see Chapter V, Ex. 8).

(b) We can find the proposed expression up to a 1/3 coefficient by considering the current density $\vec{j}$, corresponding to the displace of δn electrons crossing from the white to the hatched part in Fig. 21a; $\vec{j} \cong \delta n \cdot (-e) \cdot v_F$, where $\delta n = g(k_F) \cdot \delta k$, where δk represents the average thickness of the depleted zone $(= eE\tau/\hbar)$ and v_F is the average speed imparted to δn electrons. This results in $g(k_F) = g(E_F) \cdot (\partial E/\partial k)_{k=k_F}$ and $J \cong g(E_F) \cdot (\hbar^2 k_F/m) \cdot (eE\tau/\hbar) \cdot ev_F$, leading thus to $\sigma \approx g(E_F) e^2 v_F^2\ \tau$.

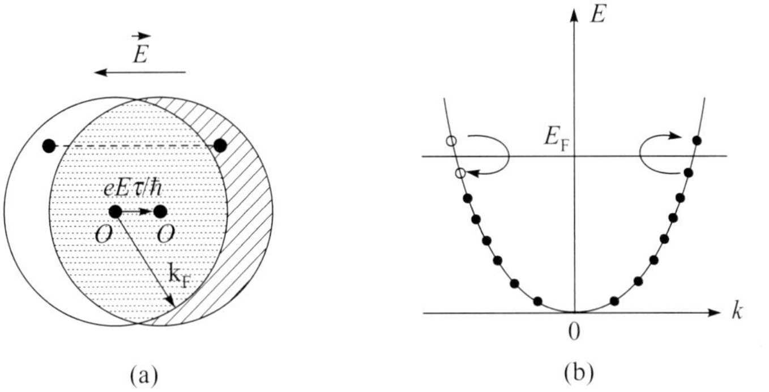

Figure 21

Exercise 20: Size effects on the electrical conductivity of metallic films

Consider a metallic thin film between the planes defined by $z = 0$ and $z = a$ where the thickness, a, is smaller than the mean free path λ_0 of conduction electrons in the bulk metal $(a/\lambda_0 \leq 1)$.

(a) Find the mean free path $\lambda(z)$ of an electron scattered at a point P at z assuming that the scattering probability in a solid angle is isotropic. Next, assume that the scattering probability is the same for all points between $0 \leq z \leq$ a and show that the mean free path of electrons $\bar{\lambda}$ in the metal film obeys $\bar{\lambda}/\lambda_0 = \frac{1}{2}k[\log(1/k)+(3/2)]$ for $k = a/\lambda_0 \leq 1$.

(b) Show the sketch of the curve $\sigma/\sigma_0 = f(\log k)$ where σ is the electrical conductivity of the thin film and σ_0 is the electrical conductivity of the bulk material.

Numerically specify the resistivity and the resistance, R, along the length l of a ribbon of gold with the following two thicknesses: $a = \lambda_0$ and $a = \lambda_0/10$, where the length l and the width b are equal such that $l = b >> \lambda_0$.

Use σ_0 (Au) $= 0.44 \times 10^6\ \Omega^{-1}\text{-cm}^{-1}$, λ_0 (Au) ≈ 420 Å.

Recall that $\int x \log x\, dx = \frac{x^2}{2}\log x - \frac{x^4}{2}$.

Solution:

(a) From Fig. 22 , one may observe that the electron mean free path in thin films is equal to λ_0 only for angles θ such that $\theta_1 < \theta < \theta_2$.

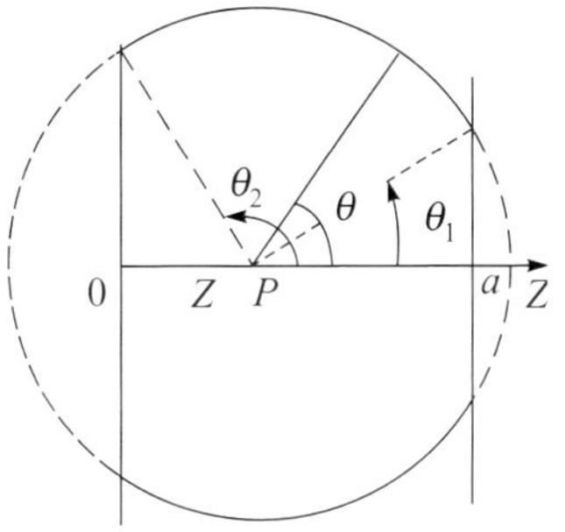

Figure 22

On the contrary, for the other angles, one has

$$\lambda(\theta) = \frac{a-z}{\cos\theta} \text{ for } 0 \le \theta \le \theta_1, \text{ where } \cos\theta_1 = \frac{a-z}{\lambda_0}, \text{ and}$$

$$\lambda(\theta) = \frac{-z}{\cos\theta} \text{ for } \theta_2 \le \theta \le \pi, \text{ where } \cos\theta_2 = \frac{-z}{\lambda_o}$$

The scattering probability over 4π steradian is equal to 1 so that the isotropic scattering probability in a solid angle dΩ is $P = \dfrac{d\Omega}{4\pi}$.

Taking into account $d\Omega = 2\pi \sin\theta\, d\theta$, we have $P = \dfrac{1}{2}\sin\theta.d\theta$, which results in

$$\bar{\lambda}(z) = \frac{1}{2}\left[\int_0^{\theta_1}(a-z).t\,g(\theta)d\theta + \lambda_o\int_{\theta_1}^{\theta_2}\sin\theta d\theta - \int_{\theta_2}^{\pi} ztg(\theta)d\theta\right]$$

$$\bar{\lambda}(z) = \frac{1}{2}\,[(z-a)\log\frac{a-z}{\lambda_o} + a - z\log\,\frac{z}{\lambda_o}].$$

Postulating that the scattering probability obeys to the same law for all the points between 0 and a, one obtains $\lambda = \dfrac{1}{a}\displaystyle\int_0^a \bar{\lambda}$

$$(z)dz = \frac{1}{2a}\left[\frac{a^2}{2} - a^2\log\frac{a}{\lambda_o} + a^2\right] \text{ or } \frac{\lambda}{\lambda_o} = \frac{3a}{4\lambda_o} - \left(\frac{a}{2\lambda_o}\right)\log\left(\frac{a}{\lambda_o}\right),$$

which corresponds to the proposed expression with $k = a/\lambda_o$.

(b) The electrical conductivity σ of the ribbon is such that $\sigma = \dfrac{ne^2\tau}{m} = \dfrac{ne^2\lambda}{mv_F}$.

It differs from that of the bulk material σ_0 only by the mean free path λ that is limited by the boundaries of the ribbon. Thus, $\dfrac{\sigma}{\sigma_0} = \dfrac{\lambda}{\lambda_0} = \dfrac{k}{2}\left(\log\dfrac{1}{k} + \dfrac{3}{2}\right)$.

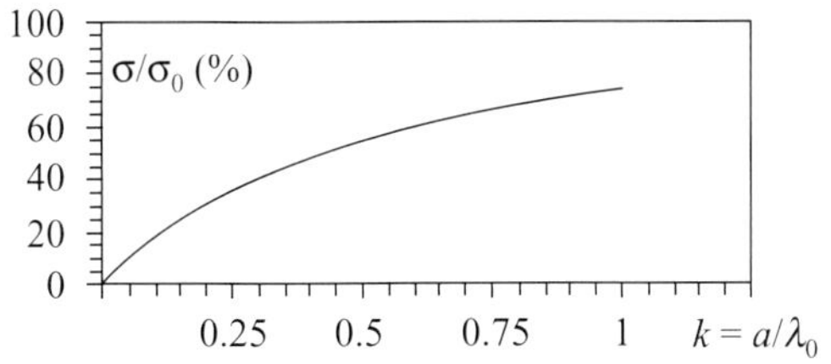

Figure 23

The result is shown in Fig. 23 wherein the lower limit of k has to correspond to a realistic lower limit for the minimal thickness of a, a few atomic layers, corresponding to uniform thicknesses.

Numerical application:

- $a = \lambda_0$; $k = 1$; $\dfrac{\sigma}{\sigma_0} = 0.75$; $1/\sigma$ (Au, 420 Å) = 3×10^{-6} Ω.cm; R(Au, 420 Å) = $(1/\sigma)$ (l/ba) = 0.71 Ω.
- $a = \dfrac{\lambda_0}{10}$; $k = 0.1$; $\sigma/\sigma_0 = 0.19$; $1/\sigma$ (Au, 42 Å) = 12×10^{-6} Ω.cm; $R(Au$, 42 Å) = 28.6 Ω.

Comment

This problem was inspired by the article of J. J. Thomson, published in 1901, to explain an experimental phenomenon known since 1898: the electrical resistivity of thin films is larger than that of the corresponding bulk material. The present classical calculation can be improved by using the Boltzmann transport equation and supposing that electrons are subject to a partial specular reflection on the surfaces. These classical size effects should be added to the quantum effects related to the discrete variation of the electron wave vector of the material, k_z ($k_z = n_z \dfrac{2\pi}{a}$) (see also Ex. 26). One can find a more detailed description of this phenomenon and the corresponding experimental techniques in the article of D.C. Larsen *Physics of Thin Films* **6**, 1971, 81.

Exercise 21: Anomalous skin effect

An electromagnetic plane wave with its electric field, $E_x(z)$, polarized along x, propagates along Oz with an angular frequency ω in free space of permittivity ε_0 and permeability μ_0. At $z = 0$, it encounters the surface Oxy of a metal, which occupies the half space for z positive.

(a) Starting from the Maxwell's equations and neglecting the displacement current, find the evolution of the electric field $E_x(z)$ in the metal characterized by its magnetic permeability μ_0 and its electrical conductivity σ_0. Express as a function of μ_0, σ_0, and ω, the skin depth δ_0, the distance at which the amplitude of the incident wave has decreased to $1/e$: $E(\delta_0) = E(0)e^{-1}$. Also determine the surface impedance $Z = R + iX$ [defined by the ratio between the electric and magnetic field at $z = 0 : E(0)/H(0)$]. Finally state the classical expression for σ_0 as a function of the electronic density n and the relaxation time τ and then as a function of n, v_F, and λ_0 (where v_F is the Fermi speed and λ_0 is the electron mean free path).

(b) When the wave frequency increases up to $f_0 = \omega_0/2\pi$, the skin depth decreases down to δ_0 and is of the order of λ_0. For frequencies greater than f_0, it is no longer possible to admit that the n electrons move in the constant electric field between collisions, as implied in the classical expression above. The mean free path of electrons is thus limited by this new skin depth δ'. Find the new expression for the conductivity σ ($f \geq f_0$) as a function of σ_0, λ_0, and δ'. Next deduce the expression of thickness of the anomalous skin depth δ' as a function of μ_0, σ, and ω and next as a function of μ_0, σ_0, λ_0, and ω. What is the corresponding expression for the surface impedance?

(c) By choosing the appropriate scale, find schematically the evolution of $\delta = F(f)$ and $R = F(f)$ around the frequency f_0.

Specify the numerical values of R and λ (wavelength of the incident electromagnetic wave of frequency f_0) for copper at 300 K with $\sigma = 6 \times 10^5\ \Omega^{-1}\text{-cm}^{-1}$, $\lambda = 3 \times 10^{-6}$ cm.

Solution:

(a) This first question, relative to the normal skin effect, is treated by the majority of texts on electromagnetism using the following differential equations:

$$\vec{\nabla}\times H=\vec{j}+\frac{\partial \vec{D}}{\partial t}\cong\sigma_0\vec{E}$$

$$\vec{\nabla}\times \mathrm{E}=-\partial\vec{\mathrm{B}}/\partial t=-i\omega\mu_0\vec{\mathrm{H}}\ \text{(in the sinusoidal regime)}$$

$\vec{\nabla}\times(\vec{\nabla}\times\mathrm{E}) = -i\omega\mu_0\sigma_0\vec{E}=-\Delta E+\nabla(\vec{\nabla}\cdot\mathrm{E})$, where the Laplacian symbol is $\Delta = \nabla^2$.

In cartesian coordinates and taking into account the polarization of $\vec{E}$ along O_x, we find

$$\partial^2E_x/\partial z^2-i\omega\mu_0\sigma_0E_x=\partial^2E_x/\partial z^2-(2i/\delta_0^2)E_x=0\,,$$

where $\delta_0=\left(\dfrac{2}{\omega\sigma_0\mu_0}\right)^{\frac{1}{2}}$.

Using $\sqrt{i}=\dfrac{1+i}{\sqrt{2}}$ and keeping only the solution that introduces the damping, we find

$$E_x(z,t)=E_x(0)\exp(-z/\delta_0)\exp i(\omega t-z/\delta_0)\ \text{and}$$

$$H_y=\frac{1}{i\omega\mu_0}\frac{\partial E_x}{\partial z}=\frac{1-i}{\omega\mu_0\delta_0}E_x^{\circ}\exp(-z/\delta_0)\exp i(\omega t-z/\delta_0)$$

so that we have

$$Z=\frac{E_x(0)}{H_y(0)}=\frac{\omega\mu_0\delta_0}{1-i}=\frac{(1+i)\omega\mu_0\delta_0}{2}=(1+i)\left(\frac{\omega\mu_0}{2\sigma_0}\right)^{\frac{1}{2}}$$

Finally, we write $\sigma_0=\dfrac{ne^2\tau}{m}=\dfrac{ne^2\lambda_0}{mv_{\mathrm{F}}}$

(b) When the mean free path is limited to the depth of δ', we have

$$\sigma=\frac{ne^2\delta'}{mv_{\mathrm{F}}}=\sigma_0\cdot\frac{\delta'}{\lambda_0}$$

By substituting σ_0 for σ, the skin depth becomes

$$\delta'=\left(\frac{2}{\omega\mu_0\sigma}\right)^{1/2}=\left(\frac{2\lambda_0}{\omega\mu_0\sigma_0\delta'}\right)^{1/2}$$

or equivalently $\delta'=\left(\dfrac{2\lambda_0}{\omega\mu_0\sigma_0}\right)^{1/3}$.

By the same substitutions, we find

$$Z = (1+i)\left(\frac{\omega\mu_0\lambda_0}{2\sigma_0\delta'}\right)^{1/2} = (1+i)\left(\frac{\omega^2\mu_0^2\lambda_0}{4\sigma_0}\right)^{1/3}$$

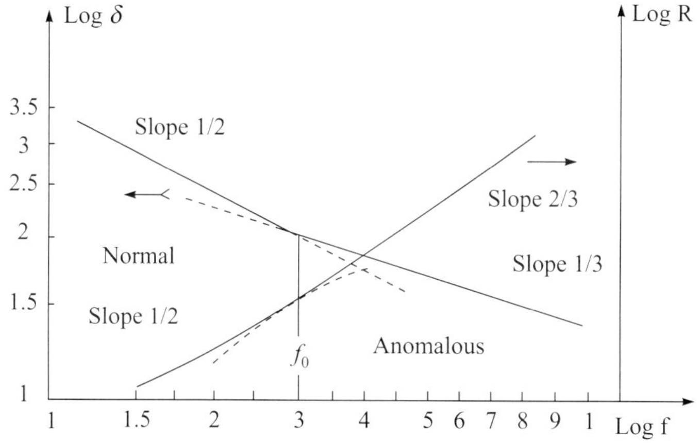

Figure 24

(c) The evolution of δ and R as a function of frequency f is of the form $\delta = \mathrm{A}f^{-1/2}(f < f_0)$ and $\delta = \mathrm{B}f^{-1/3}(f > f_0)$; $R = Cf^{1/2}(f < f_0)$ and $R = Df^{2/3}(f > f_0)$.

This is shown schematically in Fig. 24 where we have used logarithmic axes.

Numerical application:
At 300 K, $\lambda_0 = \delta = 300$ Å, $\lambda = \pi\lambda_0{}^2\sigma_0(\mu_0/\varepsilon_0)^{1/2} = 64\ \mu\mathrm{m}$.

$$R = \frac{1}{\sigma_0\lambda_0} = 0.55\ \Omega$$

Remark: This exercise was inspired by the method followed by Pippard to take into account qualitatively the anomalous skin effect with some slightly different physical hypotheses (see Sondheimer, E. H. in *Advances in Physics* 1-1952-1).

This method can correctly explain the evolution of the essential parameters, R and δ, as a function of frequency and shows that the anomalous skin effect appears in the IR for ordinary metals at ambiant temperature and down to the microwave region for pure metals at low temperature.

Physically, the measurement of Z that is obtained from the measurement of the reflection coefficient can be used to obtain the quantity σ_0/λ_0 and therefore to evaluate the density of electrons "n" participating in the conduction. One can also use this to evidence anisotropic effects in single crystals.

Exercise 22: Pauli paramagnetism of free electrons in 1-, 2-, and 3D

Consider N free electrons per unit volume of a metal set in a magnetic induction $\vec{B}$ applied along the Oz axis. The magnetic moment of each electron is $\vec{\mu}$ proportional to its spin and its projection on the Oz axis can take only two values: $+\mu_B$ and $-\mu_B$ (where $\mu_B = \dfrac{e\hbar}{2m}$ is the Bohr magnetron).

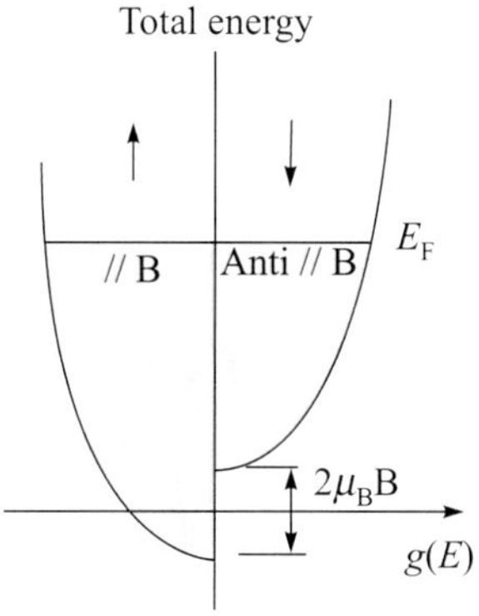

Figure 25

(a) Find the total energy E of electrons, which have a kinetic energy E_c and a magnetic moment that can be parallel or anti-parallel to $\vec{B}$. Find the magnetization per unit volume of the sample $\vec{M}$ as a function of the number of electrons $N_\uparrow$ and $N_\downarrow$ that have a magnetic moment $\vec{\mu}$, respectively, parallel and anti-parallel to $\vec{B}$. Also deduce the paramagnetic susceptibility χ of the sample.

(b) 3D at 0 K. From the expression of the electron density of states in 3D and with the help of Fig. 25, establish the expression of $N_\uparrow$ and $N_\downarrow$ in the form of integrals. Deduce the susceptibility χ as a function of the electronic density N, μ_B, μ_0 (magnetic permeability of vacuum), and T_F (Fermi temperature). Use the hypothesis that $\mu_B B << E_F$ and check initially that E_F is practically independent of $\vec{B}$.

(c) 2D at 0 K: Following the same method, find the susceptibility χ of a layered metal in which the free electrons move only in the planes.

(d) 1D at 0 K: Same question as in (b) but assuming that the electrons move only along lines.

(e) 3D at T K: Taking into account the Fermi distribution $f(E)$ in the integral expressions $N_\uparrow$ and $N_\downarrow$. Find the expression of $\chi(T)$ as a function of $\chi(0)$, T, and $T_F(0)$ after the evaluation of E_F as a function of T K.

(f) 2D at T K: Same question as in (e) but for a 2D layered metal.

(g) 1D at T K: Same question as in (e) but for a 1D metal.

(h) *Numerical application*: Summarize the previous results in a table. Next, evaluate numerically $\chi(0)$ in 3D where $N = 10^{22}$ cm^{-3}, $\mu_B = 0.927 \times 10^{-23}$ A·m^2. (use F1 in Questions (e), (f), and (g)) $(\mu_o, \hbar, m, e)$

Solution:

Note that the first two questions of this problem are treated in detail in the many textbooks (see for instance Ref. 15b). Note also that the magnetic moment of an electron is opposite its spin because $\vec{\mu} = -g\mu_B \vec{S}$, where $g = 2$.

(a) The energy of the magnetic moment of an electron $\vec{\mu}$ placed in the magnetic induction $\vec{B}$ is $W_m = -\vec{\mu}\,\vec{B}$. The total energy E of a free electron is therefore

$$E = E_c - \mu_B B\,(\vec{\mu}\uparrow\,\vec{B}\uparrow)$$

$$E = E_c + \mu_B B\,(\vec{\mu}\downarrow\,\vec{B}\uparrow)$$

The magnetization per unit volume of the sample is: $M = N\uparrow(\mu_B) + N\downarrow(-\mu_B) = (N\uparrow - N\downarrow)\mu_B$

The paramagnetic susceptibility χ corresponds to: $\vec{M} = \chi\vec{H} = (\chi B)/\mu_0$ so that $\chi = \dfrac{M\mu_0}{B} = (N_\uparrow - N_\downarrow)\mu_0\mu_B/B$.

The remaining parts of the problem thus reduce to the evaluation of the population difference $N_\uparrow - N_\downarrow$.

(b) As shown in Fig. 25, the magnetic induction increases the population of electrons having $\vec{\mu}$ parallel to $\vec{B}$ while decreasing the anti-parallel population and maintaining the position of the Fermi level.

At 0 K, the two populations are thus $N_{\uparrow} = \int_{-\mu_B B}^{E_F} g(E + \mu_B B)\, dE$ and $N_{\downarrow} = \int_{\mu_B B}^{E_F} g(E - \mu_B B)\, dE$, while in 3D we have (see Exs. 8 and 11): $g(E) = \alpha E^{1/2}$ where $\alpha = \frac{2\pi V}{h^3}(2m)^{3/2}$.

(Note the elimination of coefficient in Ex. 8 because it was including the 2 spin orientations.) After integration, we find $N_{\uparrow} = \frac{2}{3}\alpha(E_F + \mu_B B)^{3/2}$ and $N_{\downarrow} = \frac{2}{3}\alpha(E_F - \mu_B B)^{3/2}$

The Fermi energy E_F is derived from the sum $N = N_{\uparrow} + N_{\downarrow}$, thus

$$N \approx \frac{4}{3}\alpha E_F^{3/2}\left(1 + \frac{3}{8}\frac{\mu_B^2 B^2}{E_F^2}\right) \approx \frac{4}{3}\alpha E_F^{3/2}$$

and the population difference is

$N_{\uparrow} - N_{\downarrow} = 2\alpha E_F^{3/2} \cdot \frac{\mu_B B}{E_F} = \frac{3N\mu_B B}{2E_F}$, which gives the 3D susceptibility at 0 K: $\chi(3d, 0°\text{K}) = \frac{3}{2}\frac{N\mu_0\mu_B^2}{E_F}$

(c) In 2D: $g(E)$ is a constant (see Ex. 14a) and we have $N_{\uparrow} = C(E_F + \mu_B B), N_{\downarrow} = C(E_F - \mu_B B)$, which gives $N = 2CE_F$, $N_{\uparrow} - N_{\downarrow} = 2C\mu_B B$, and a susceptibility of $\chi(2d, 0°\text{K}) = \frac{N\mu_0\mu_B^2}{E_F}$.

(d) In 1D: $g(E) = \beta E^{-1/2}$ (see Ex. 14a) and we have $N_{\uparrow} = 2\beta(E_F + \mu_B B)^{1/2}, N_{\downarrow} = 2\beta(E_F - \mu_B B)$

$$N = N_{\uparrow} + N_{\downarrow} = 4\beta E_F^{1/2}\left(1 - \frac{3}{8}\frac{\mu_B^2 B^2}{E_F^2}\right) \approx 4\beta E_F^{1/2}$$

$N_{\uparrow} - N_{\downarrow} = 2\beta\frac{\mu_B B}{E_F^{1/2}} = \frac{N}{2}\frac{\mu_B B}{E_F}$ and a susceptibility of:

$$\chi(1d, 0°\text{K}) = \frac{N}{2}\frac{\mu_0\mu_B^2}{E_F}$$

(e) When the temperature is taken into account, $N_\uparrow$ and $N_\downarrow$ must be evaluated using $N_\uparrow = \int_{-\mu_B B}^{\infty} f(E+\mu_B B) g(E+\mu_B B)\,dE$ and $N_\downarrow = \int_{\mu_B B}^{\infty} f(E-\mu_B B) g(E-\mu_B B)\,dE$.

As previously, we deduce $E_F(T)$ from $N = N_\uparrow + N_\downarrow$ and then use the result for the population difference. Taking into account F1, we have:

$$N_\uparrow = \frac{2}{3}\alpha(E_F+\mu_B B)^{3/2} + \frac{(\pi k_B T)^2}{6}\frac{1}{2}\alpha(E_F+\mu_B B)^{-1/2},$$

$$N_\downarrow = \frac{2}{3}\alpha(E_F-\mu_B B)^{3/2} + \frac{(\pi k_B T)^2}{6}\frac{1}{2}\alpha(E_F-\mu_B B)^{-1/2}$$

and thus we find $E_F(T)$ (see Ex. 8) as:

$E_F(T) \approx E_F(0)\left[1-\frac{(\pi k_B T)^2}{12E_F^2(0)}+\cdots\right]$ from which we find

$$N = N_\uparrow + N_\downarrow = \frac{4}{3}\alpha E_F^{3/2}(T) + \frac{(\pi k_B T)^2}{6}\alpha E_F^{-1/2} = \frac{4}{3}\alpha E_F^{3/2}(0)$$

The population difference is:

$$N_\uparrow - N_\downarrow = 2\alpha\mu_B B E_F^{1/2}(T) - \frac{(\pi k_B T)^2}{12}\alpha E_F^{-3/2}\mu_B B$$

$$= 2\alpha\mu_B B E_F^{1/2}(0)\left[1-\frac{\pi^2}{12}\left(\frac{T}{T_F}\right)^2+\cdots\right]$$

The 3D susceptibility at temperature T is:

$$\chi(3d,T^\circ) = \chi(3d,0^\circ\mathrm{K})\left[1-\frac{\pi^2}{12}\left(\frac{T}{T_F}\right)^2\right]$$

(f) In 2D, the density of states is a constant, thus $N_\uparrow$ and $N_\downarrow$ reduce to $N_\uparrow = C[E_F(T)+\mu_B B]$ and $N_\downarrow = C\,[E_F(T)-\mu_B B]$.
We thus find $N = 2CE_F(T) = 2CE_F(0)$ and $\chi(2d, T^\circ\mathrm{K}) = \chi(2d, 0^\circ\mathrm{K})$.
In 2D, neither the Fermi energy nor the magnetic susceptibility of free electrons is influenced by temperature up to $(T/T_F)^4$.

(g) Taking into account the density of states in 1D, $g(E) = \beta E^{-1/2}$, the expressions for $N_\uparrow$ and $N_\downarrow$ are:

$$N_{\uparrow} = 2\beta[E_{\rm F}(T) + \mu_{\rm B}B]^{1/2} - \frac{1}{2}\beta[E_{\rm F} + \mu_{\rm B}B]^{-3/2}\frac{(\pi k_{\rm B}T)^2}{6}$$

$$N_{\downarrow} = 2\beta[E_{\rm F}(T) - \mu_{\rm B}B]^{1/2} - \frac{1}{2}\beta[E_{\rm F} - \mu_{\rm B}B]^{-3/2}\frac{(\pi k_{\rm B}T)^2}{6}$$

which leads to $N \approx 4\beta E_{\rm F}^{1/2}(T) - \frac{(\pi k_{\rm B}T)^2}{6}\beta E_{\rm F}^{-3/2}(0) = \beta E_{\rm F}^{1/2}(0)$

and $E_{\rm F}(T) = E_{\rm F}(0)\left[1 + \frac{(\pi k_{\rm B}T)^2}{12E_{\rm F}^2(0)} + \cdots\right]$.

The population difference is thus

$$N_{\uparrow} - N_{\downarrow} = 2\beta\frac{\mu_{\rm B}B}{E_{\rm F}^{1/2}(T)} + \frac{3}{12}\frac{(\pi k_{\rm B}T)^2}{E_{\rm F}^{5/2}(0)}\cdot\mu_{\rm B}B$$

$$= \frac{2\beta\mu_{\rm B}B}{E_{\rm F}^{1/2}(0)}\left[1 - \frac{\pi^2}{24}\left(\frac{T}{T_{\rm F}}\right)^2 + \frac{3\pi^2}{24}\left(\frac{T}{T_{\rm F}}\right)^2 + \ldots\right]$$

The magnetic susceptibility in 1D at arbitrary temperature T is as follows: $\chi(\text{1D}, T) = \chi(\text{1D}, 0^\circ\text{K})\left[1 + \frac{\pi^2}{12}\left(\frac{T}{T_{\rm F}}\right)^2 + \cdots\right]$

Numerical application: $N = 10^{28}$ m^{-3}, $E_{\rm F} = 2.44 \times 10^{-19}$ J = 1.525 eV, $\chi(0^\circ\text{K}) = 6.6 \times 10^{-6}$.

The following table summarizes the main results for the paramagnetism of free electrons (also known as Pauli paramagnetism) in 1-, 2-, and 3D objects. Compared with atomic paramagnetism (Brillouin and Langevin) where $\chi = C/T$, note that here the effects of temperature on χ are comparatively small because $E_{\rm F}$ introduces a corrective term $(T/T_{\rm F})^2$ which at ambient is of order 10^{-4} smaller than the unity for 1- and 3D systems.

	1D	**2D**	**3D**
$g(E)$	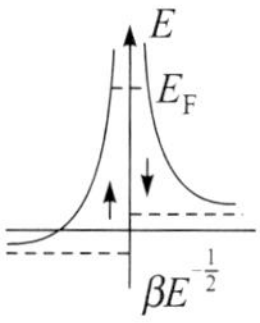	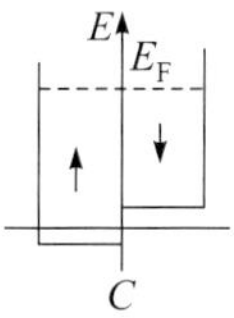	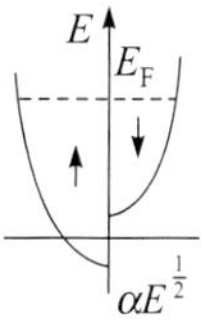

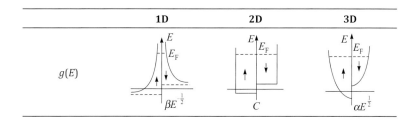

$\chi(0°\ \mathrm{K})$	$\frac{N}{2}\frac{\mu_0\mu_B^2}{E_F}$	$\frac{N\mu_0\mu_B^2}{E_F}$	$\frac{3}{2}\frac{N\mu_0\mu_B^2}{E_F}$
$\chi(T°\ \mathrm{K})$	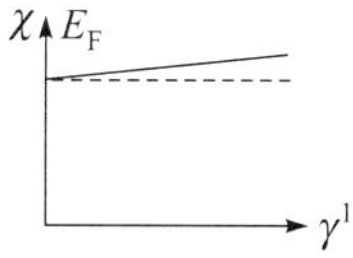 $\chi(0)\left[1+\frac{\pi^2}{12}\left(\frac{T}{T_F}\right)^2+\cdots\right]$	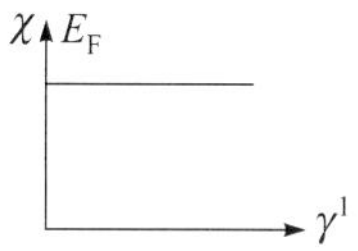 $\chi(0)$	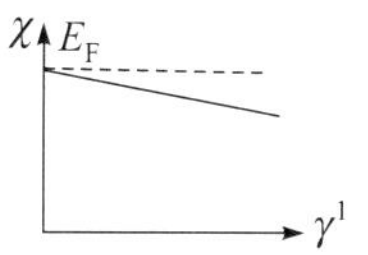 $\chi(0)\left[1-\frac{\pi^2}{12}\left(\frac{T}{T_F}\right)^2+\cdots\right]$
$E_F(T)$	$E_F(0)\left[1+\frac{\pi^2}{12}\left(\frac{T}{T_F}\right)^2+\cdots\right]$	$E_F(0)$	$E_F(0)\left[1-\frac{\pi^2}{12}\left(\frac{T}{T_F}\right)^2+\cdots\right]$

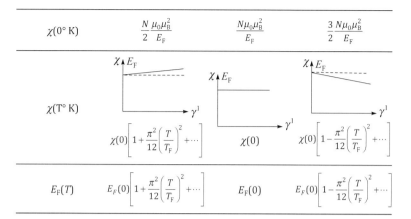

Exercise 23: Quantum Hall effect

Consider a parallelepiped sample (L_x, L_y, c) subject to a magnetic induction B along the z-axis (// to c) and in which a current I_x flows when the Hall voltage V_y is measured.

(1) Express the ratio V_y/I_x as a function of the density per unit volume n_v of free electrons, V and the thickness c. Suppose that c approaches zero and reconsider the results as a function of n_s (the surface density of conduction electrons in the x–y plane).

(2) In an appropriate arrangement (see comment), n_s is increased by increasing the Fermi level of a 2D electron gas. Express the density of 2D electronic states as a function of E_F to deduce next the ratio $V_y/I_x = f(E_F)$.

(3) Show the curve of $V_y/I_x = f(E_F)$.

(4) In fact under the action of B the density of states (evaluated in 2) is not continuous but it is formed with discrete (Landau) levels localized at energies $E_L = \left(\frac{\hbar eB}{m}\right)\left(j+\frac{1}{2}\right)$, where $j = 0, 1, 2, 3$. The $E_L(j)$ level contains as many electronic states as initially distributed in the interval $\Delta E_L = E_L(j+1) - E_L(j)$ in the absence of B. Find the number of electronic states ΔN contained in one Landau level and the corresponding quantity Δn_s. Starting from the result established in 1), find the evolution of the Hall resistance V_y/I_x when E_F crosses a Landau level (E_F varying from $E_L(j)$ to $E_L(j+1)$). Sketch the curve $V_y/I_x = f(E_F)$. Compare the result with that found in 3). Numerically evaluate $V_y/I_x(j) - V_y/I_x(j-1)$ for $j = 1$. Comment on the results. (h, e)

Solution:

(1) In a classical Hall device (see Pb. 2), the trajectory of conduction electrons is rectilinear when $-e\vec{v}\Lambda\vec{B}-e\vec{E}_{\mathrm{H}}=0$ or equivalently $\vec{E}_{\mathrm{H}}=-\vec{v}\Lambda\vec{B}$. The current density is $\vec{J}_x=-n_{\mathrm{v}}e\vec{v}$ from which the Hall field can be written as: $\left|E_{\mathrm{H}}\right|=\dfrac{j_x\cdot B}{n_{\mathrm{v}}e}$. We thus have $V_y=L_y\cdot\left|E_{\mathrm{H}}\right|$; $I_x=j_x\cdot L_y\cdot c$; and $n_s=n_v\cdot c$. This leads to $V_y/I_x = B/en_{\mathrm{v}}c = \mathrm{B}/en_s$.

(2) In 2D the density of states is a constant (see, e.g., Ex. 14a): $g(E)=\dfrac{L_xL_y}{\pi}\dfrac{m}{\hbar^2}$ and $n_{\mathrm{s}} = N/L_xL_y$ so that $n_{\mathrm{s}}=\dfrac{1}{L_xL_y}\int_0^{E_{\mathrm{F}}} g(E)dE=\dfrac{m}{\hbar^2\pi}E_{\mathrm{F}}$, which results in the ratio $\dfrac{V_y}{I_x}=\dfrac{B\pi\hbar^2}{emE_{\mathrm{F}}}$.

(3) See Fig. 26 (left). The Hall resistance decreases when the density of carriers (and thus E_{F}) increases.

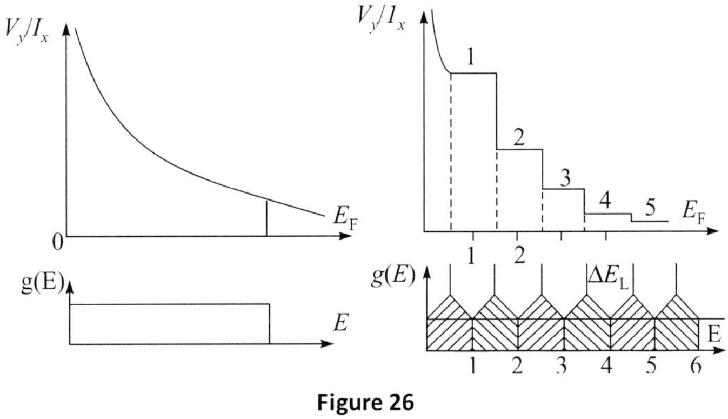

Figure 26

(4) $\Delta N=g(E)\Delta E_{\mathrm{L}}=\dfrac{L_xL_ye}{\pi\hbar}B$ which gives $\Delta n_{\mathrm{s}}=\dfrac{2eB}{h}$.

When E_{F} crosses the jth Landau level, the Hall resistance is discontinuous and changes from $\mathrm{B}/\mathrm{e}(\mathrm{j}-1)\,\Delta n_{\mathrm{s}} = \mathrm{h}/2e^2(\mathrm{j}-1)$ to $\mathrm{B}/\mathrm{ej}\,\Delta n_{\mathrm{s}} = h/2e^2 j$.

The corresponding evolution is shown in Fig. 26, right. Observe that the frequency associated with the Landau levels is the cyclotron frequency, which implies a large magnetic induction or small effective masses so that the different steps can be easily observed. The difference between two consecutive steps corresponds to $\frac{h}{2e^2}\left(\frac{1}{j}-\frac{1}{j+1}\right)$, which is 6.45 kΩ for $j = 1$.

Comment: Quantum Hall effect; Nobel Prizes in physics in 1985 and 1998

The quantum Hall effect was observed by von Klitzing et al. (*Phys. Rev. Lett.* **45**, 1980, 494) in a field-effect transistor (MOSFET). At the semiconductor/oxide interface of such a transistor, one can consider that the conduction electrons move freely with effective mass m^* in the plane of the interface. One can change the surface density by changing the Fermi level via a gate voltage. The experiment was performed at a very low temperature, $T = 1.5$ K, and with very large magnetic induction ($B = 15$ T). It can be more easily observed in epitaxial hetero-structures (see Chapter V, Pb. 9, for properties of such heterostructures), for instance at $T = 4$ K and $B = 8$ T (Tsui and Gossard; *Appl. Phys. Lett.* **38**, 1991, 550).

Von Klitzing's experiment, for which he won the Nobel Prize in 1985, contains a great much more physics than is presented in this exercise. In particular, one of the interests of the quantum Hall effect is that, for each step, the Hall resistance is a unique function of $h/2e^2$ = 25.812 kΩ (see the result in Solution (4) above), which allows the establishment of a resistance standard based only on h and e. Conversely a precise measurement of the Hall resistance (to 10^{-8}) can be used to determine the fine structure constant α via a new and independent method.

Finally, we have to point out also the existence of the fractional quantum Hall effect (FQHE) that has been reported for the first time by Tsui et al. (*Phys. Rev. Lett.* **48,** 1982, 1559). This FQHE effect) is a physical phenomenon in which the Hall conductance of 2D electrons shows precisely quantized plateaus at fractional values of e^2/h. It is a property of a collective state in which electrons bind magnetic flux lines to make new quasi-particles, having a fractional elementary charge of magnitude $e^* = e/q$ (where q is the integer) that are neither bosons nor fermions and exhibit anyonic statistics. When the Hall resistance is a constant the longitudinal resistance, $V_x/I_x = 0$.

The Nobel Prize in physics in 1998 was awarded to R. Laughlin, H. Störmer and D. Tsui for the discovery and explanation of the FQHE

(H.L. Störmer *Rev. of Modern Physics* **71**, 1999, 875; R.G. Haug, *Surf. Sci.* **196**, 1988, 242).

Exercise 24: Simplified evaluation of the interatomic distance, compression modulus, B, and cohesive energy of alkali metals

The cohesive energy of a solid is equal to the difference between the total energy of free atoms and the total energy of the same atoms in a crystal lattice. For an alkali metal, this energy (for a single atom), approximately obeys the relation: $E_c = E_i - E_e - \overline{E_F}$ where E_i is the ionization energy of an isolated atom, $\overline{E_F}$ is the average kinetic energy of a conduction electron and E_e is its electrostatic energy in a field of ions and other electrons.

(a) Evaluation of E_e: Assuming that the conduction electron is uniformly distributed with a density ρ inside a sphere of radius r_0, find the potential energy of this electronic distribution submitted to a central ion, $E(-e, i)$, next its interaction energy $E(-e, -e)$ when limiting this evaluation to the sphere of radius r_0. Deduce the resulting expression of the energy, E_e, as a function of a single variable r_0.

(b) Find the average kinetic energy, $\overline{E_F}$, as a function of r_0 for a conduction electron assumed to be free.

(c) Deduce an expression for the cohesive energy (where E_i is a constant) and thus the expression and numerical value of r_0 at equilibrium. Knowing that the volume of the sphere of radius r_0 is equal to the volume occupied by an atom in the crystal (Wigner–Seitz method) and that the alkali metals crystallize in a cubic-centered structure, find the numerical value of the side a.

(d) Find the expression for the stiffness modulus B at 0 K: $B = -V\left(\frac{\partial P}{\partial V}\right)_T = V\frac{\partial^2(E_e + \overline{E_F})}{\partial V^2}$.

What are the numerical values taken by the Fermi energy E_F and by B? Compare with those of lithium:

E_F = 4.7 eV, $B = 1.16 \times 10^{10}$ Pa; a = 3.49 Å. $(\varepsilon_0, \hbar, m, e)$

Solution:

(a) The electric field created by a uniform distribution of electronic charges is given by Gauss's theorem and it corresponds to $E_r = \frac{4\pi}{3}\frac{r^3\rho}{\varepsilon_0}\frac{1}{4\pi r^2} = \frac{\rho r}{3\varepsilon_0}(r \le r_0)$; $E_r = \frac{4\pi r_0^3\rho}{3\varepsilon_0 4\pi r^2} = \frac{\rho r_0^3}{3\varepsilon_0 r^2}(r \ge r_0)$.

Taking into account the limiting conditions V(∞) = 0 and V at $r = r_0$, the corresponding expressions for the potential created by isolated electrons are $V(r) = \dfrac{\rho r_0^3}{3\varepsilon_0 r}(r \geq r_0)$ and

$V(r) = -\dfrac{\rho r_0^2}{6\varepsilon_0} + \dfrac{\rho r_0^2}{2\varepsilon_0}(r \leq r_0)$, where $\rho \cdot \dfrac{4\pi}{3} r_0^3 = -e$.

The potential energy of the electronic distribution in presence of a single central ion is equal to the potential energy of the central ion in the presence of $\rho(r)$: $E_\mathrm{p} = (i, -e) = -3e^2/(8\pi\varepsilon_0 r_0)$. This result can also be obtained directly using

$$E_\mathrm{p}(-e, i) = \iiint \rho \cdot V_i(r) dv = \rho \int_0^{r_0} \frac{e}{4\pi\varepsilon_0 r} \cdot 4\pi r^2 dr.$$

The electrostatic interaction of electrons obeys $E(-e,-e) = \dfrac{1}{2}\iiint \rho \cdot V_\mathrm{e}(r) dv = \dfrac{3e^2}{4\pi\varepsilon_0 \cdot 5r_0}$ and the total electrostatic energy E_e is thus $E_\mathrm{e} = -\dfrac{9e^2}{40\pi\varepsilon_0 r_0}$.

(b) In 3D, the average kinetic energy of a free electron is equal to 3/5 of the Fermi energy: $\overline{E}_\mathrm{F} = \dfrac{3E_\mathrm{F}}{5} = \left(\dfrac{3}{5}\right)\dfrac{\hbar^2}{2m}(3\pi^2 n)^{3/3}$ (see Ex. 8). The electronic concentration is $n = \dfrac{1}{4}\left(\dfrac{\pi}{3} r_0^3\right)$. We thus find $\overline{E}_\mathrm{F} = \dfrac{3}{10} \cdot \dfrac{\hbar^2}{m}\left(\dfrac{9\pi}{4}\right)^{2/3} \cdot \dfrac{1}{r_0^2}$.

(c) The cohesive energy $E_\mathrm{c} = -(E_\mathrm{e} + \overline{E}_\mathrm{F} - E_\mathrm{i})$ is of the form $E_\mathrm{C} = +(A/r_0) - (C/r_0^2) + E_\mathrm{i}$. It reaches a maximum for r_0 corresponding to $(\partial E_\mathrm{c}/\partial r)_{r=r_0} = 0$:

$r_0 = 2C/A = (h^2\varepsilon_0/me^2)(3/2\pi)^{1/3}$

$r_0 = 1.3 \times 10^{-10}$ m and $a_0 = (8\pi/3)^{1/3} \cdot r_0 = 2.65$ Å

We note that for this value of r_0, the electrostatic energy term is, in absolute value, two times larger than the kinetic energy term.

(d) $B = V \cdot \dfrac{\partial^2 E}{\partial V^2} = V \cdot \left(\dfrac{\partial r}{\partial V}\right)^2 \cdot \dfrac{\partial^2 E}{\partial r^2} = \left[\dfrac{1}{12\pi r}\left(\dfrac{\partial^2 E}{\partial r^2}\right)\right]_{r_0}$

$$B = \frac{1}{6\pi r_0^4} \cdot \frac{A}{2} = \frac{3e^2}{160\pi^2 \varepsilon_0 r_0^4} = 1.93 \times 10^{10} \text{ Pa}, E_F = 7.4 \text{ eV}$$

Although very simplified, the present model gives an acceptable value for r_0 and thus for E_F and B. Note that because the ionic radius has been ignored, it leads to the same value for all the alkali metals. A better precision can be obtained by assuming a value of r_0 (or electron density n) specific to each alkali metal. (See following exercise.)

Exercise 25: Pressure and compression modulus of an electron gas: application to sodium

(a) After having calculated the total energy of a free electron gas of density n, find the pressure P at 0 K that these electrons exert on the walls of the sample containing them. Find the result starting from the kinetic theory of gases $\left(P = \frac{1}{3} n m v_{\text{eff}}^2\right)$.

(b) Deduce the modulus of compressibility B of this gas: $B = -V \frac{\partial P}{\partial V}$.

(c) Numerically find P and B for sodium, where $n = 2.54 \times 10^{22}$ cm^{-3} and compare it to the experimental value of $B = 0.68 \times 10^{10}$ Pa. ($\hbar, m$)

Solution:

This exercise is analogous to the preceding exercise but here the kinetic energy of electrons is only taken into account and it is combined to the experimental value of r_0 (and thus for n).

(a) We can calculate the pressure P starting from the Helmholtz free energy F: $P = -\left(\frac{\partial F}{\partial V}\right)_T$, which at T = 0°K becomes

$$P = -\left(\frac{\partial U}{\partial V}\right)_{T=0} = -\left[\frac{\partial (N\bar{E})}{\partial V}\right]_{T=0}.$$

The average energy $\bar{E}$ of a free electron is such that $\bar{E} = \frac{3}{5} E_F$ or $U = \frac{3}{5} N E_F$ for N electrons. We have

$$U = \frac{3}{5}\frac{\hbar^2}{2m}(3\pi^2)^{2/3} \cdot \frac{N^{5/3}}{V^{2/3}} \text{ (see Ex. 8)},$$

$$P = \frac{1}{5}\frac{\hbar^2}{m}(3\pi^2)^{2/3} n^{5/3}, \text{ where } n = N/V.$$

The same result may be obtained directly from the kinetic theory of gases:

$$P = \frac{1}{3} nmv_{\text{eff}}^2, \text{ where } \frac{1}{2} mv_{\text{eff}}^2 = \frac{3}{5}\bar{E}_{\text{F}}.$$

(b) At 0° K, $B = -V\dfrac{\partial P}{\partial V} = \dfrac{1}{3}\dfrac{\hbar^2}{m}(3\pi^2)^{2/3} n^{5/3}$

(c) For sodium where $n = 2.65 \times 10^{22}$ cm^{-3}, we obtain $P = 49.7 \times 10^8$ Pa = 49,000 atm, $B = 0.83 \times 10^{10}$ Pa (which is comparable to the experimental value of 0.68×10^{10} Pa).

In the previous exercise, the cohesive energy E_c includes an electrostatic term E_e that is twice the kinetic term $\bar{E}_F$ but when calculating B, the kinetic term was 1.5 times larger than the electrostatic term. This explains why the estimates of B in the two different exercises remain comparable. A more sophisticated approach is given in Ex. 27.

Exercise 26: Screening effect

In a metal or a plasma or an electrically neutral electrolyte, the potential created by positive ions ($+e$) attracts the electrons ($-e$) and the electronic concentration $n(r)$ in the vicinity of a given ion is greater than the average electronic concentration n_0: this electron cloud screening the potential created by a single ion. In this problem we study two methods that allow the evaluation of this screening effect and which, in a spherically symmetric system centered on a positive ion, consists of finding two equations that relate the potential Φ (r) at a point r and the concentration of electronic charges $n(r)$ at this same point.

(1) *The Thomas–Fermi Model*: It assumes that the total energy of an electron at point r is constant and equal to the Fermi energy of a gas of free electrons with an average concentration of n_0. The kinetic energy of this electron is equal to the Fermi energy of a gas of free electrons with concentration $n(r)$:

(a) What is the relation between $n(r)$ and $\Phi(r)$ that results from this hypothesis? How can this result be simplified if we assume that the difference of the densities around equilibrium, $n(r) - n_0$, is weak compared to the average electronic concentration, n_0?

(b) Compare this result with that given by the Poisson equation relative to the same difference in density. Show that the potential function $\Phi(r)$ obeys:

$$\Delta\Phi = \frac{d^2\Phi}{dr^2} + \frac{2}{r}\frac{d\Phi}{dr} = k_S^2\Phi$$

Find k_s as a function of n_0 and integrate the equation using the change of variables $U = r\,\Phi$; find the solution which satisfies the boundary conditions: $\Phi = 0(r \to \infty)$ and $\Phi = \frac{1}{4\pi\varepsilon_0}\frac{e}{r}(r \to 0)$

(c) Find the screening length $r_{TF} = 1/k_s$ for silver where $n_0(\mathrm{Ag}) = 5.85 \times 10^{28}\,\mathrm{m}^{-3}$.

(2) *Debye Potential*: Suppose that the electronic concentration (n_e) and the ionic concentration (n_i) obey the relations: $n_e(r) = n_0 \exp -\frac{W_e(r)}{k_B T}$ and $n_i(r) = n_0 \exp -\frac{W_i(r)}{k_B T}$ in which W_e and W_i represent the potential energy of electrons and ions at a point r. Write the Poisson equation and then simplify it taking into account the fact that $e\Phi(r)/k_B T$ is much smaller than 1. Show that the potential function obeys an analogous equation to that established in 1b. What is the expression for the screening length r_D? Compare r_{TF} and r_D assuming that $T = T_F$. $(e, m, \hbar, \varepsilon_0, k_B)$

Solution:

(1) (a) $\frac{p_i^2(r)}{2m} - e\Phi(r) = E_F(n_0) = \frac{\hbar^2}{2m}[3\pi^2 n_0]^{2/3}$ where

$$\frac{p_i^2(r)}{2m} = \frac{\hbar^2}{2m}[3\pi^2 n(r)]^{2/3}$$

We thus deduce: $e\Phi\,(r) = \frac{\hbar^2}{2m}\,(3\pi^2)^{2/3}\,[n^{2/3}\,(r) - n_0^{2/3}]$ or

$$\text{equivalently } e\Phi\,(r) = \frac{\hbar^2}{2m}(3\pi^2 n_0)^{2/3}\cdot\frac{2}{3}\frac{n(r)-n_0}{n_0}$$

$$= \frac{2}{3}E_F(n_0)\cdot\frac{n(r)-n_0}{n_0}.$$

Taking into account the development:

$$n^{2/3}(r) = n_0^{2/3}\left(1 + \frac{2}{3}\frac{n(r) - n_0}{n_0} + \cdots\right)$$

(b) If we combine this result with Poisson equation relative to the density of electronic fluctuations, where the average electronic density is neutralized by the average ionic density, we have

$\Delta\Phi = -\dfrac{\rho}{\varepsilon_0} = \dfrac{e}{\varepsilon_0}[n(r) - n_0]$, and we find

$\Delta\Phi = \dfrac{2}{3\varepsilon_0}\dfrac{n_0 e^2 \Phi}{E_F(n_0)} = k_S^2\Phi$, where $k_S^2 = \left(\dfrac{3}{\pi^4}\right)^{1/3}\dfrac{m}{\varepsilon_0}\left(\dfrac{e}{\hbar}\right)^2 n_0^{1/3}$.

From a solution of the form $U = r\Phi$, we find

$\dfrac{d^2U}{dr^2} - k_S^2 U = 0$ and $U = A\exp(-k_S r) + B\exp(k_S r)$.

The stated boundary conditions allow the determination of A and B ($r \to \infty : \Phi = 0$ from which $B = 0$; $r \to 0 : U \to A$) and the complete solution takes the form:

$\Phi(r) = \dfrac{e}{4\pi\varepsilon_0} \cdot \dfrac{1}{r}\exp - k_S r$

(c) *Numerical application*: $k_S = 1.78 \times 10^{10}\ \mathrm{m}^{-1}$, $r_{TF} = 0.56$ Å. The corresponding influence of the screening effect is illustrated in Fig. 27.

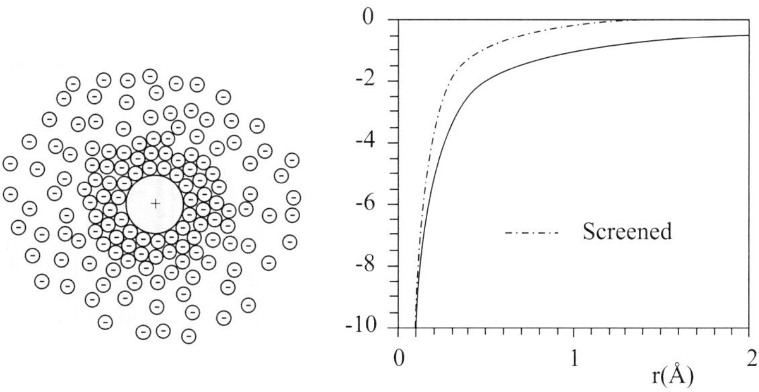

Figure 27 (Left) Screening effect. (Right) Electron potential energy attracted by a positive ion with and without screening effect for Ag with $r_{TF} = 0.56$ Å.

(2) $W_e = -e\Phi(r), W_i = e\Phi(r)$ from which we have

$$n_e(r) = n_0 \exp\frac{e\Phi(r)}{k_B T} \text{ and } n_i(r) = n_0 \exp-\frac{e\Phi(r)}{k_B T}.$$

Inserting these expressions in Poisson's equation, we find

$$\Delta\Phi = -\frac{\rho}{\varepsilon_0} = -\frac{n_0 e}{\varepsilon_0}\left[\exp\left(-\frac{e\Phi}{k_B T}\right) + \exp\left(\frac{e\Phi}{k_B T}\right)\right] \text{ or equivalently}$$

$$\frac{d^2\Phi}{dr^2} + \frac{2}{r}\frac{d\Phi}{dr} = \frac{2n_0 e}{\varepsilon_0} sh \frac{e\Phi}{k_B T}$$

When $k_B T >> e\Phi$, this equation reduces to the form $\Delta\Phi = k_D^2\Phi$,

where $k_D^2 = \dfrac{2n_0 e^2}{\varepsilon_0 k_B T}$

We note that at a temperature of an electron gas $T = T_F$, the two screening lengths r_{TF} and r_D differ only by a numerical coefficient of order of unity (because $k_B T = E_F$): $k_D/k_s = r_{TF}/r_D = 2/\sqrt{3} = 1.15$.

Note: Debye length and Fermi length

The Debye method applies especially to electrolytes, plasma and semiconductors. In doped Si semiconductors the Debye length is $L_D = [\varepsilon_{Si} k_B T/e^2 N_d]^{1/2}$ where ε_{Si} is the Si dielectric constant and N_d is the density of dopants (either donors or acceptors). It is the distance, screening radius, over which local electric field affects distribution of free charge carriers. It decreases when increasing the free carrier density.

The Thomas–Fermi method allows the calculation of the surrounding potential, in dilute alloys, of impurity atoms of metal A dissolved in metal B. In this latter case, the asymptotic value of the potential $Ze^2/4\pi\varepsilon_0 r$ must take into account the difference between the valence of the impurity and that of the solvent, for instance, Z = 3 – 1 = 2 for Al dissolved in Cu or Z = 1 for atomic vacancies in copper. More precise calculations show that screening in a metal results in an oscillatory behavior in the electron density. These "Friedel" oscillations provide the mechanism for long-range interactions and hence dictate many physical phenomena briefly considered in Pbs. 2a and 2b.

Exercise 27: Thermionic emission: the Richardson–Dushman equation

A metallic plane surface (cathode) perpendicular to the axis Oz is held at a temperature T and the extraction potential of free electrons in the conductor is Φ (work function $e\Phi$).

(a) In vector space, k_x, k_y, and k_z, one considers a volume element of dimension dk_x, dk_y, and dk_z. What is the condition on k_z, expressed in the form of an inequality, for electrons to escape from the cathode? Find the corresponding expression for the elemental current density dJ_z as a function of T.

(b) Find the total current density emitted from the cathode. Show that the final result can be put in the form of the Richardson–Dushman equation: $J_z = AT^2 \exp\left(-\frac{e\Phi}{k_B T}\right)$ and find the expression for A.

Recall that $\int_{-\infty}^{\infty} e^{-\alpha x^2} dx = \left(\frac{\pi}{\alpha}\right)^{1/2}$ and assume $k_B T << e\Phi$.

(c) In fact, the experimental values of A are less than the theoretical values which neglect internal reflection on the metal/vacuum interface. Starting from the experimental values of A and Φ, find the current density emitted from the cathodes indicated below, each one given at a working temperature T_f slightly less than the melting temperature. Comment on the results. (k_B, h, m, e)

Cathodes	W	BaO + $e\Phi$ on Ni	LaB_6
A (10^4 A/m^2·°K^2)	75	0.05	40
Φ (volts)	4.5	1	2.4
T (in° K)	2700	1100	1800

Solution:

(a) As shown in Fig. 28, the electrons that can escape the cathode must have a velocity component in z such that: $\frac{1}{2}mv_z^2 \geq E_F + e\Phi$ (where the origin is taken to be the bottom of the conduction band. From $mv_z = \hbar k_z$, we have $\frac{\hbar^2 k_z^2}{2m} \geq \frac{\hbar^2 k_{zo}^2}{2m} = E_F + e\Phi$.

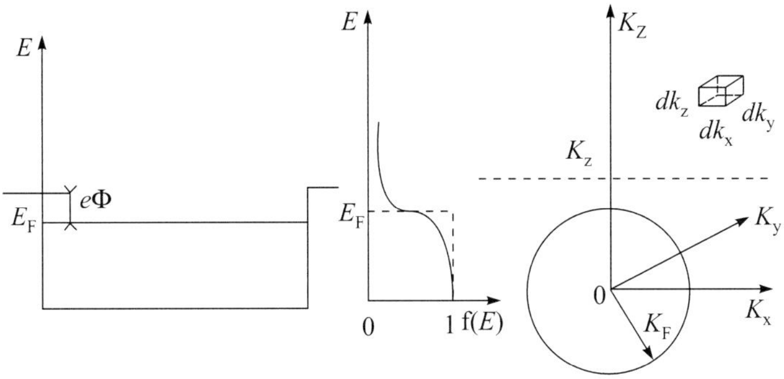

Figure 28

The infinitesimal current density dJ_z corresponding to the dn electrons contained in the element $dk_x\, dk_y\, dk_z$ is $dJ_z = -dn\, e\, v_z$, where $dn = \dfrac{2dk_x dk_y dk_z}{\frac{(2\pi)^3}{V}} \cdot f(E)$.

Here the volume is unitary and $f(E) = \dfrac{1}{e^{\left(\frac{E-E_F}{k_B T}\right)}+1}$ reduces to $e^{\frac{E_F}{k_B T}} \cdot e^{-\frac{E}{k_B T}}$ since $E \geq E_F + e\Phi$ and $e\Phi/k_B T >> 1$. We thus have

$$dJ_z = -e\frac{dk_x dk_y dk_z}{4\pi^3} \cdot e^{\frac{E_F}{k_B T}} \cdot e^{-\frac{E}{k_B T}} \cdot \frac{\hbar k_z}{m}.$$

(b) Writing E explicitly as a function of the wave vector: $E = (\hbar^2/2m)(k_x^2 + k_y^2 + k_z^2)$, the integral expression of the current density becomes (omitting the minus sign only indicates that the current flows positively, in the opposite direction of the electron motion):

$$|J_z| = \frac{e\hbar}{4\pi^3 m} e^{\frac{E_F}{k_B T}} \int_{-\infty}^{+\infty} e^{-\alpha k_x^2}\, dk_x \int_{-\infty}^{+\infty} e^{-\alpha k_y^2}\, dk_y \int_{k_{z0}}^{+\infty} e^{-\alpha k_z^2}\, k_z dk_z,$$

where $\alpha = \dfrac{\hbar^2}{2mk_B T}$. The product of the first two integrals is $\dfrac{\pi}{a}$. The result of the 3rd is $\left(\dfrac{1}{2\alpha}\right)e^{-\alpha k_{z0}^2}$.

We thus obtain the complete Dushman formula where $A = 4\pi m e k_B^2/h^3 = 1.6 \times 10^6\ \mathrm{A/m^2K^2}$.

(c) The numerical values are as follows:

Tungsten (at 2700 K) $\rightarrow J = 2.2\ \mathrm{A/cm^2}$, where $\dfrac{e\Phi}{k_B T} \approx 19.3$

Alkaline–earth oxide on nickel (at 1100 K)$\rightarrow J \approx 1.6\ \mathrm{A/cm^2}$, where $\dfrac{e\Phi}{k_B T} \approx 10.5$

Lanthanum hexaboride (1800 K) $\rightarrow J \approx 25\ \mathrm{A/cm^2}$, where $\dfrac{e\Phi}{k_B T} \approx 15.4$.

The temperatures mentioned in this problem are just an indication: one can increase J by increasing T, but to the detriment of the lifetime of the filament.

Notes: Production of electron beams

(a) The thermo-electronic power of tungsten is comparable or slightly better than that of oxide cathodes. The latter require a much smaller heating power so that the energetic width of the emitted beam is much narrower as seen in the following exercise. The oxide cathodes however are more prone to contamination and do not support well contact with air. They are almost always exclusively used in sealed tubes (such as oscilloscopes). In electron microscopy, filaments of tungsten were competed with cathodes of LaB_6, which deliver current densities of at least 10 times greater and constitute an emissive source with a smaller size.

Since the 2000s, modern electron microscopes are based all on the use of a different principle: the field electron emission. A field emission gun (FEG) is a type of electron gun in which a sharply pointed Müller-type emitter is held at several kilovolts negative potential relative to a nearby electrode, so that there is sufficient potential gradient at the emitter surface to cause field electron emission. A comparison between the principle of thermal electron emitters and of field electron emitters is illustrated in Fig. 29.

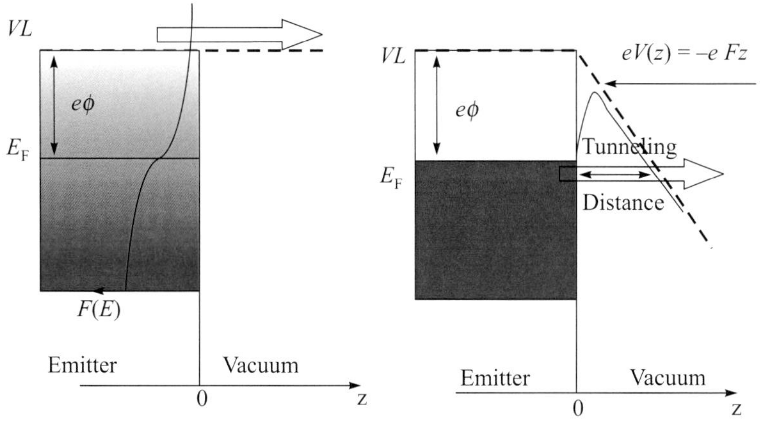

Figure 29 Principle of thermo-electronic emission (left) and of field-electron emission (right).

The open arrows indicate the energetic position of most of the emitted electrons.

In the FEG, a cold cathode is usually made of single crystal tungsten sharpened to a tip radius of about 100 nm and this electron source is de-magnified by a set of electron lenses in order to obtain a small electron spot on the sample to be imaged. The result in both scanning and transmission electron microscopy is a significant improvement of the signal-to-noise ratio and of the spatial resolution compared with thermionic devices. The same principle of field electron emission is used in scanning tunneling microscopy where the end of the emitting tip is a single atom very close to the sample surface, a few angstorms, and thus without any electron optical devices (see Pb. 3 for details).

The devices based on a cathode field emission are governed by the Fowler–Nordheim law of the form:

$$j = \frac{e^3}{8\pi\hbar t^2(y)} \cdot \frac{F}{\phi} \exp\left\{ -\frac{8\pi\sqrt{2m}\phi^{\frac{3}{2}}}{3\hbar eF} \theta(y) \right\},$$

where F is the electric field strength; ϕ the work function of the metal, and $t(y)$ and $\theta(y)$ tabulated functions of the variable $y = e\sqrt{e} \cdot \frac{F}{\phi}$. The approximate relation is

$$j \approx 1.4 \times 10^{-6} \frac{F^2}{\phi} \times 10^{4.39/\sqrt{\phi}} \times 10^{-2.82 \times 10^7 \phi \sqrt{\phi^{3/2}}/E}$$

(b) To experimentally verify the Dushman law, it is sufficient to measure the saturation current of a vacuum diode as a function of the temperature of its cathode and trace a straight line:

$\log\left(\frac{J}{T^2}\right) = f\left(\frac{1}{T}\right) = (\log A) - \frac{e\Phi}{k_B T}$. The slope $\left(\frac{e\Phi}{k_B}\right)$ can be used to evaluate ϕ and intercept can be used to determine A.

Exercise 28a: Thermal field emission: the energy width of the emitted beam

A metallic plane surface (the cathode) is raised to a temperature T and the work function of the (free) conduction electrons is ϕ (see preceding exercise).

(a) In vector space, find the position corresponding to the vacuum level of electrons where electrons having energy between E and $E + dE$ and contained in the emission angle between θ and $d\theta$, measured relative to the normal $z' - z$ at the surface. Take $E = 0$ to be at the bottom of the conduction band.

(b) Find the number $n(E, \theta)\ dEd\theta$ of electrons occupied at temperature T assuming that $k_B T << e\Phi$. Independently deduce the number of electrons emitted $N(E)dE$ between E and $E + d(E)$ and the energy distribution $j_z(E)$ of the current emission density. Starting from the latter result deduce the total current density J_z, known as the Dushman law.

(c) Sketch $j_z(E)$. Indicate the approximate numerical value of the energy width (measured a half-width) when the temperature of the cathode is changed from 1100 K to 2700 K.

Solution:

(a) In k-space, electrons that can leave the metal must have wave vector $\vec{k}$ of which k_z (normal to the surface) such that

$\frac{\hbar^2 k_z^2}{2m} \geq E_F + e\Phi = \frac{\hbar^2 k_{z0}^2}{2m}$, where the zero energy is measured at the bottom of the conduction band.

Figure 30 shows the conditions for electron emission in which the energy is between E and $E + dE$, the angle is between θ and $d\theta$ are situated in the half space $k_z \geq k_{z0}$ and the hatched part between the two spherical shells with radius k and $d + dk$ are such that $E = \hbar^2 k^2/2m$.

(b) The corresponding volume of the spherical shell is $V_{c.s.} = kd\theta \cdot dk \cdot 2\pi k \cdot \sin\theta$. Taking into account the volume of a state is $2\pi/L^3$ ($L^3 = 1$ is the unitary volume of the cathode) and that the occupation probability of the states is $2f(E) \approx 2e^{E_F/k_BT} \cdot e^{-E/k_BT}$, the number $n(E,\theta)dEd\theta$ of electrons occupying these states is such that

$n(k,\theta)\cdot dk \cdot d\theta = [2k \cdot d\theta \cdot dk \cdot 2\pi k \sin\theta / (2\pi)^3] \cdot e^{(E_F - E)/k_BT}$, where $k_z \geq k_{z0}$ and $0 < \text{Arc}\cos\theta < K_{z0}/k$ (see Fig. 30a).

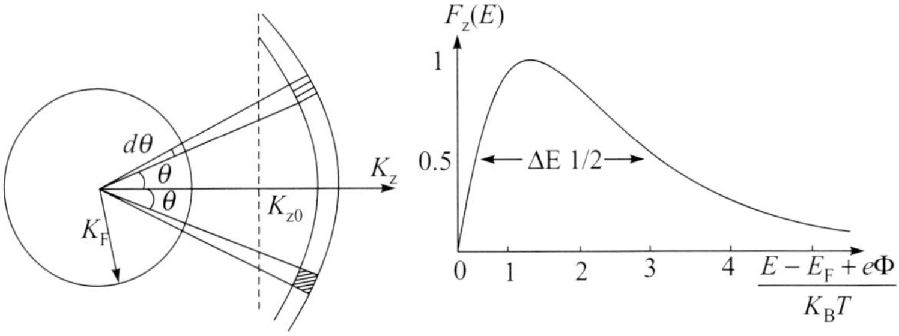

(a) (b)

Figure 30

Taking into account $E = \hbar^2 k^2/2m$ for free electrons, we find

$$n(E,\theta) = \frac{4\pi\sqrt{2}}{h^3} m^{3/2} \cdot E^{1/2} \sin\theta\, e^{(E_F - E)/k_BT}$$

and writing $\cos\theta_0 = k_{z0}/k_z = \left(\frac{E_F + e\Phi}{E}\right)^{1/2}$, we find

$$N(E) = \int_0^{\theta_0} n(E,\theta)d\theta = \frac{4\pi\sqrt{2}}{h^3} m^{3/2} E^{1/2} \cdot e^{(E_F - E)/k_BT} \cdot \left[1 - \left(\frac{E_F + e\Phi}{E}\right)^{1/2}\right].$$

The partial current density along z due to the motion $n(k,\theta)$ of electrons is such that

$$\frac{dj_z(k)}{d\theta} = n(k,\theta)\cdot(-e)\cdot v_z = n(k,\theta)(-e)\frac{\hbar k}{m}\cos\theta$$

or equivalently

$$\frac{dj_z(k)}{d\theta} = -\frac{eh}{4\pi^3 m} \cdot k^3 dk \sin\theta \cdot \cos\theta \, e^{(E_F - E)/k_B T}$$

We thus find: $\left|j_z(E)\right| \cdot dE = \frac{8\pi em}{h^3} E \cdot dE e^{(E_F - E)/k_B T} \int_0^{\theta_0} \sin\theta \cos\theta d\theta$

$$= \frac{4\pi meEdE}{h^3} e^{\frac{E_F - E}{k_B T}} \left(1 - \frac{E_F + e\Phi}{E}\right).$$

Completing the integration, $J_z = \int_{E_F + e\Phi}^{\infty} j_z(E) \cdot dE$, we find the Dushman expression (see previous exercise): $J_z = AT^2 e^{\frac{-e\Phi}{k_B T}}$, where $A = \frac{4\pi me}{h^3} k_B^2$.

The functions $N(E)$ and $J_z(E)$ are of the form

$$N(E) \propto e^{\frac{-E}{k_B T}} \left[E^{\frac{1}{2}} - (E_F + e\Phi)^{\frac{1}{2}}\right] \propto e^{-x}(\sqrt{x} - \sqrt{a})$$

and $J(E) \propto e^{\frac{-E}{k_B T}} [E - (E_F + e\Phi) \propto e^{-x}(x - a)]$,

where $x = \frac{E}{k_B T}$ and $a = \frac{E_F + e\Phi}{k_B T}$.

(c) These two functions are zero when $E = E_F + e\phi$. Their evolution as a function of E are comparable at every point because $J(E)$ is obtained by multiplying $N(E)$ by $(\sqrt{x} + \sqrt{a})$, which is essentially constant $(2\sqrt{a})$ when E and $E_F + e\phi$ differ by only few $k_B T$. They both go through a maximum at $E = E_F + e\Phi + k_B T$ and their full width at half max corresponds to $x_2 - x_1 \approx 2.5$ (where $x_1 = a + 0.23$ and $x_2 = a + 2.7$). This is depicted in Fig. 30b.

Numerically, we obtain the following:

$T = 1100$ K, $\Delta E = 0.24$ eV; $T = 2700$ K, $\Delta E = 0.6$ eV.

The energetic width of the emitted beam is thus highly dependent on the temperature of the emitting filament.

Exercise 28b: Thermionic emission in 2D

A metal occupies the negative half space of z in 2D along xOz. This metal is characterized by its Fermi energy E_F and its extraction potential ϕ and is held at temperature T.

(a) What condition is imposed on the component k_z of the wave vector $\vec{k}$ for an emitted electron ($z \geq 0$)?

(b) After evaluating the number of electrons $nd(s)$ contained in the surface element dk_x, dk_z in $\vec{k}$-space, find the corresponding current density $dj_z(s)$.

(c) Find the areal current density j(s) emitted by the cathode (where $k_\mathrm{B}T << e\phi$).

Show that $j_s = BT^n \exp\left(-\dfrac{e\Phi}{k_\mathrm{B}T}\right)$ and find B and n explicitly.

Solution:

Note that this exercise is a variation in 2D of Ex. 27.

(a) Emitted electrons must have $k_z \geq k_{z0}$, where $\hbar^2 k_{z0}^2/2m = E_\mathrm{F} + e\phi$ (see Fig. 28).

(b) $dn(s) = 2\dfrac{dk_x dk_z}{(2\pi)^2} \cdot S \cdot f(E)$. Here S is unitary and $f(E)$ reduces to $e^{\frac{E_F - E}{k_\mathrm{B}T}}$. We also have $dj_z(s) = dn(s) \cdot (-e) \cdot v_z$.

(c) $v_z = \dfrac{\hbar k_z}{m}$; $E = \dfrac{\hbar^2}{2m}(k_x^2 + k_z^2)$. The integration (limited to k_x and k_z) is analogous to that developed in Ex. 30b. We find

$$|j(s)| = B \cdot T^n \cdot e^{\frac{-e\phi}{k_\mathrm{B}T}}, \text{ where } B = \frac{e\sqrt{m}}{\hbar^2}(2k_\mathrm{B})^{3/2} \text{ and } n = 3/2.$$

We can compare this result leads to the Dushman formula in 3D (Ex. 30):

$$j(s) \times \frac{(2\pi m k_\mathrm{B} T)^{1/2}}{\hbar} \times \frac{1}{2\pi} = AT^2 e^{\frac{-e\phi}{k_\mathrm{B}T}}.$$

Exercise 29: UV reflectivity of alkali metals (simplified variation of Pb. 6)

The behavior of alkali metals in UV light is correctly described by the complex dielectric of the form:

$\tilde{\varepsilon}_r(\omega) = 1 - \dfrac{\omega_p^2}{\omega^2 - i\gamma\omega}$, where $\omega_p^2 = \dfrac{Ne^2}{m\varepsilon_0}$, γ is the damping, which is equal to τ^{-1}; N is the density of free electrons, which is equal to atomic density for monovalent metals.

Starting from this complex dielectric constant $\varepsilon_1 - i\varepsilon_2$, we can define the complex propagation speed for the electromagnetic wave $\tilde{v}_\varphi = (\varepsilon_0 \tilde{\varepsilon}_r \mu_0)^{-1/2}$, and the complex index of reflection $\tilde{N} = n - ik (= c/\tilde{v}_\varphi)$. We can thus generalize the wave propagation in a vacuum $\overline{E}_x = \overline{E}_0 \exp i\,\omega\left(t - \dfrac{z}{c}\right)$ to that in a medium characterized by $\varepsilon_0 \tilde{\varepsilon}_r$ and μ_0.

(1) Find the expression for a medium characterized by $\tilde{\varepsilon}_r$. Find the existing relations between n and k and ε_1 and ε_2.

(2) Neglecting the damping term, γ, find the pulse interval $0 << \omega < \omega_p$. Sketch the evolution of $\varepsilon_1(\omega)$ and specify it sign. What is the value of value of ε_2. Discuss the nature of the wave which propagates in the plasma (formed at the interface between two planar medium). Also discuss the amplitude $\vec{E}_r$ of the reflected wave that is associated with the incident wave $\vec{E}_i$, which originally propagated in the vacuum and encountered the plasma at normal incidence.

(3) Explore the interval $\omega_p < \omega < \infty$ when γ is small but non negligible compared with ω. What is the form and nature of the wave propagating in the plasma? What is the amplitude of the wave reflected into the vacuum by the plasma at normal incidence?

Recall that: $|r| = \left|\dfrac{\vec{E}_r}{\vec{E}_i}\right| = \dfrac{|1 - \tilde{n}|}{|1 + \tilde{n}|}$, where r is the amplitude of the coefficient of reflection.

(4) Evaluate the plasmon energy, $\hbar\omega_p$, for Na with one free electron per atom and $N(\text{at}) = 2.5 \times 10^{22}$ at/cm^3. Describe its optical properties in the UV range, $\lambda < 4000$ Å and indicate the numerical value of the critical wavelength of the electromagnetic waves, λ_p, where these optical properties change suddenly. On the microscopic level, what is the physical significance of the dielectric constant smaller than 1 or even negative?

Solution:

(1) The wave propagating in a medium characterized by $\tilde{v}_\varphi$ will have the form: $\vec{E}_x = \bar{E}_0 \exp i\omega\left(t - \frac{z}{\tilde{v}_\varphi}\right)$, where $\tilde{v}_\varphi^{-1}$ can be replaced by $\tilde{N}/c$: $\vec{E}_x = \vec{E}_0 \exp -\frac{k\omega}{c} z \exp i\omega\left(t - \frac{nz}{c}\right)$.

The attenuation α of the wave amplitude corresponds to $\frac{k\omega}{c}$ and, in the phase term, the quantity nz represents is the optical path (in the general case where $n > 1$):

$\tilde{N} = \sqrt{\tilde{\varepsilon}_r}$ or $\varepsilon_1 = n^2 - k^2$ and $\varepsilon_2 = 2nk$.

(2) The evolution of ε_1 (ω) is negative in the interval $0 << \omega < \omega_p$. $\varepsilon_2 = 0$ because $\gamma = 0$ so we find $n = \sqrt{\varepsilon_1} = \pm i\sqrt{|\varepsilon_1|}$.

The dielectric constant is real but negative, which induces a purely imaginary optical index. The wave attenuation is not related to the value of k (or ε_2 or γ) different from zero (usual situation when $n > 0$) but to the fact that n is purely imaginary.

The wave takes the form $\bar{E}_x = \bar{E}_0 \exp -\left(\frac{\omega\sqrt{|\varepsilon_1|}z}{c}\right) \exp i\omega t$.

This type of wave is called evanescent wave.

At normal incidence, the reflection coefficient of amplitude $|r|$ corresponds to the ratio of two complex conjugates: $\frac{\left|1 - i\sqrt{|\varepsilon_1|}\right|}{\left|1 + i\sqrt{|\varepsilon_1|}\right|}$ and its modulus is equal to one. There is total reflection.

(3) When ω is such that $\omega >> \omega_p$, ε_1 tends to 1 and ε_2 tends to zero as $\frac{\gamma\omega_p^2}{\omega^3}$. n thus goes to 1 and $k \approx \frac{\varepsilon_2}{2} \approx \frac{\gamma\omega_p^2}{2\omega^3}$. The transmitted wave propagates normally with weak attenuation: $\alpha \approx \frac{\gamma\omega_p^2}{2\omega^2 c}$; the amplitude of the coefficient of reflection is essentially zero. For more details see the solution to Pb. 5.

(4) For Na, $\hbar\omega_p = 5.94$ eV and λ_p=2.088 Å.

The position of the plasmon energy and plasmon wavelength is shown in Fig. 31. It is situated in the UV range and it is the limit between the total reflection region and the total transmission region. The total reflectance in the visible region explains the mirror effect for freshly cut alkali metals and also for Ag and Ag. For other metals, electronic transition in this visible region explains their colors.

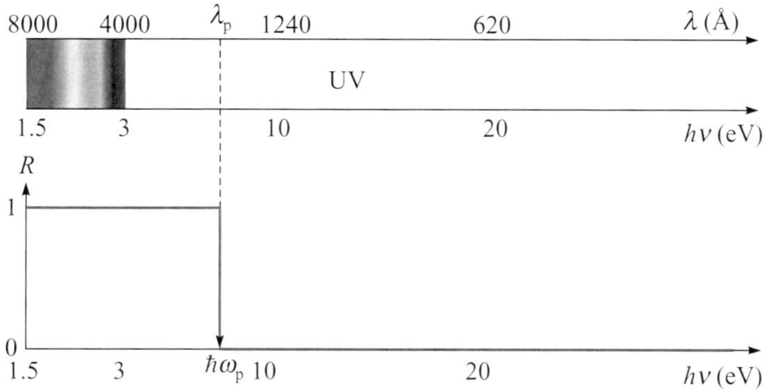

Figure 31 Evolution of the reflectivity of sodium as a function of the wavelength, λ (and photon energy $h\nu$) of the incident electromagnetic waves at normal incidence.

In the spectral domain considered here, the incident wave frequency is so elevated that the electrons cannot follow the rapid changes in field. The dipoles formed by each electron and one fixed ion is oriented opposite the electric field due to the delay of these electrons. (This situation is opposite to that encountered in electrostatic of dielectrics.) We always have $\vec{D} = \varepsilon_0 \vec{E} + \vec{P}$ where $\vec{P} = N\vec{p}$ but $\vec{p}$(and $\vec{P}$) are $\uparrow\downarrow \vec{E}$. The dielectric susceptibility χ is negative.

When $\omega < \omega_p$, the oscillation amplitude of electrons is so large that $\vec{P}$ dominates $\vec{E}$ so that $\chi < -1$.

When $\omega > \omega_p$, the electrons do not always follow the inversions of the electric field $\vec{E}$ and their elongation remains in phase opposition but with a much weaker amplitude. Thus $-1 < \chi < 0$.

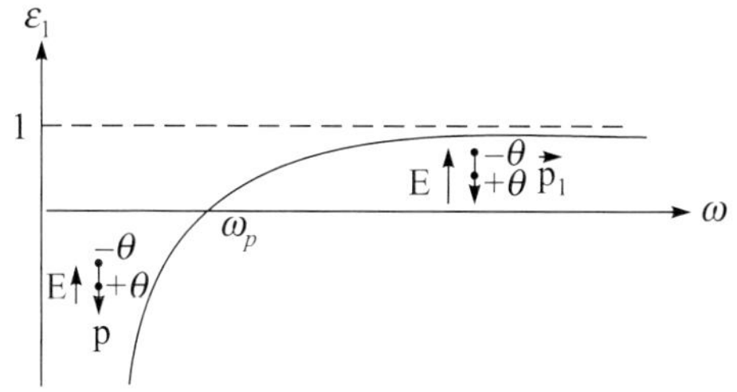

Figure 32

These two situations are shown in Fig. 32.

A similar explanation applies to the situation of ions in the IR (see Chapter III, Pb. 1) between ω_T and ω_L (or ω_r) outside the resonance of ions in the lattice. Here (UV), we identify always the local and applied fields and the behavior described here can be extended to atomic electrons in the X-ray region. (See following exercise.)

In the photon energy of interest here, 5 eV< $h\nu$ < 30 eV, the present approach applies also to Mg (with two free e^-/atom) and to Al (three free e^-/atom) and even to Si (four free e^-/atom), but on substituting $(h\,\omega_p)^2$ by $[(h\,\omega_p)^2 + E_T{}^2]$, where E_T is the mean transition energy of electrons between the valence and the conduction band. This approach does not work for metals such as Au or Cu because of electronic transitions influencing the evolution of $\varepsilon_2(\omega)$ (see Chapter V for details).

Technical point: $\varepsilon_1 + i\varepsilon_2$ or $\varepsilon_1 - i\varepsilon_2$?

In the literature, one often finds two forms for $\tilde{\varepsilon}(\omega)$. We write ε_1 – iε_2 if an electric field of form $E_0 e^{+i\omega t}$ is used, as in the present exercise. We write $\varepsilon_1 + i\varepsilon_2$ if an electric field of form $E_0 e^{-i\omega t}$ is used.

In both cases, ε_2 must be positive because it describes the dissipation of energy associated with the damping term, γ, which is always positive. The same remark applies to the energy loss function which will always be positive and equal to $\dfrac{\varepsilon_2}{(\varepsilon_1^2 + \varepsilon_2^2)}$, but can also be written $(\pm)\mathrm{Im}\left(\dfrac{1}{\tilde{\varepsilon}}\right)$, depending on the convention implicitly adopted for E_0. These remarks also concern $n + ik$ (or $n - ik$) where k is necessarily positive (and leads to an attenuation).

Exercise 30: Refractive index for X-rays and total reflection at grazing incidence

The complex dielectric constant of a free electron gas is of the form $\tilde{\varepsilon}_r = 1 - \omega_p^2/(\omega^2 - i\omega/\tau)$, where $\omega_p^2 = Ne^2/m\varepsilon_0$ in which N represents the density of free electrons (see Ex. 29).

(a) Find the expression for the index of refraction n for X-rays (with energy $h\nu$) that propagate in a solid with average atomic number Z and atomic density N_{at}. Use $n = 1 - \delta$ and find δ.

(b) Assume that the energy of these photons, $h\nu$, is such that all electrons of the atoms vibrate freely in such a way that $\omega >> 1/\tau$, ω_p. Find the value of the critical angle α_c (measured relative to the surface of the material) for which total reflection of X-rays occurs.

(c) Application to sodium: $Z = 11$; $N(\text{at}) = 2.5 \times 10^{22}$ at/cm^3 where X-photons have $h\nu = 10$ keV. What are the numerical values of δ and α_c. $(\hbar, m, e, \varepsilon_0)$

Solution:

(a) $n = \sqrt{\tilde{\varepsilon}_r} \approx \sqrt{\varepsilon_1}$ because ε_2 is negligible $(1/\tau << \omega)$

$$\varepsilon_1 = 1 - \left(\frac{\omega_p^2}{\omega^2}\right), \text{ where } \omega_p^2 = \frac{Ne^2}{m\varepsilon_0} = \frac{N(at)Ze^2}{m\varepsilon_0}$$

or $\omega >> \omega_p$;

$$n \approx \sqrt{\varepsilon_1} \approx 1 - \left(\frac{\omega_p^2}{2\omega^2}\right) = 1 - \delta$$

The optical index of all materials for X-rays is smaller but very close to unity, which here justifies the assimilation of the optical trajectory to the geometric trajectory in the Bragg's law (see Chapter I).

(b) $\sin i = \cos\alpha = n \sin r$.

$$\cos\alpha_c = 1 - \delta = 1 - \left(\frac{\alpha_c^2}{2}\right), \text{ where } \alpha_c = \sqrt{2\delta} = \frac{\omega_p}{\omega}.$$

(c) $\hbar\omega_p = 19.7$ eV, $\delta = 1 - n = 2 \times 10^{-6}$; $\alpha_c = 2 \times 10^{-3}$ rad.

Contrary to Ex. 32 where only one electron per atom was taken into account for the evaluation of the plasma frequency in the UV range, the use, here, of 11 free electrons per atom of Na is justified by the fact that the energy (or frequency) of X-rays, 10 keV, is far larger than the binding energy of all the atomic electrons of Na including the 1s electrons where E_B is ~1.072 keV (see also Chapter V, Ex. 27). Thus all the 11 electrons are considered to vibrate nearly freely.

Comment: Focusing of X-rays

This exercise explains the impossibility of realizing X-ray lenses based on the laws of refraction of Snell–Descartes because $n \approx 1$. This physical impossibility is regrettable given the promising medical applications such efficient focusing of X-rays would represent.

For physics experiments this difficulty is overcome from the use of mirrors (plane and elliptical) that function at grazing incidence in the total reflection domain. In addition to focusing with the aid of curved crystals (using the Bragg's law), other recent possibilities include using multi-layer mirrors (W/C/W/C . . .), already described in Pb. 10 of Chapter I and Fresnel lenses. These possibilities are often limited to the soft-X-ray regime.

Exercise 31: Metal reflectivity in the IR: the Hagen–Rubens relation

The complex dielectric constant of a free electron gas is of the form $\tilde{\varepsilon}_r = 1 - \omega_p^2/(\omega^2 - i\omega/\tau) = \varepsilon_1 - i\varepsilon_2$, where $\omega_p^2 = Ne^2/m\varepsilon_o$ in which N represents the density of free electrons of the metal and is of ~5 × 10^{22} e/cm^3.

(a) Show that in the IR $(\omega >> \omega_p, 1/\tau)$, $\varepsilon_2 >> |\varepsilon_1|$.
Find the simplified expression (as a function of ε_2 only) for the refection coefficient of amplitude r at normal incidence.

(b) Find the relation between ε_2 and the static electrical conductivity σ_0. Deduce that in the IR, the reflection coefficient of metals R can be put in the form (Hagen–Rubens relation):
$R = 1 - A(\omega/\sigma_0)^{1/2}$ or equivalently, $R = 1 - \alpha\lambda^{-\frac{1}{2}}$. Find A explicitly. Find the numerical value α for the case of sodium where $\sigma_0 = 2.1 \times 10^5\ \Omega^{-1}\text{cm}^{-1}$.

Solution:

(a) Far below the plasma frequency, in the IR, $\omega = 10^{12}$ rad/s; $\tau^{-1} \approx 10^{14}\ \text{s}^{-1}$; $\omega_p \approx 10^{16}$ rad/s (for $\hbar\omega_p \sim 6eV$).
Thus $\tilde{\varepsilon} \approx -i\omega_p^2/\omega\tau \approx -i\varepsilon_2$.
At normal incidence, $r = (1-n)/(1+n)$, where $n = \sqrt{\tilde{\varepsilon}_r} \approx \sqrt{i\varepsilon_2}$.

(b) $R = \dfrac{1+\varepsilon_2 - \sqrt{2\varepsilon_2}}{1+\varepsilon_2 + \sqrt{2\varepsilon_2}} \approx 1 - 2\sqrt{\dfrac{2}{\varepsilon_2}}$, with $\varepsilon_2 \approx Ne^2\tau/m\varepsilon_0\omega = \dfrac{\sigma_0}{\varepsilon_0\omega}$. This result can be obtained by writing ω_p explicitly and comparing to $\sigma = Ne^2\tau/m$ or, more generally, by $\tilde{\sigma} = \sigma_1 - i\sigma_2 = i\omega\varepsilon_0\,(\varepsilon_1 - i\varepsilon_2)\sigma_1$ (see Pb. 5). We thus find $R = 1 - A(\omega/\sigma_0)^{1/2}$, where $A = 2(2\varepsilon_0)^{1/2} = 8 \times 10^{-6}$. This result may be written in the form $R = 1 - \alpha\lambda^{-1/2}$, where $\alpha = 8 \times 10^{-5}$ in the case of sodium.

Remark: This exercise and the preceding one treat the optical properties of metals in the two extreme limits: (a) X-rays: $n = 1 - \delta$ (quasi-total transparence) and (b) IR: $R \approx 1$ (quasi-total reflection). For more details see R.W. Christy, *Am. J. Phys*, **40**, 1972, 1403.

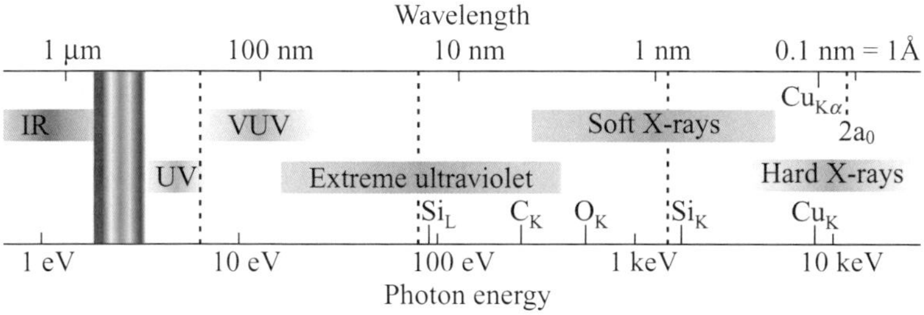

Figure 33

Problems

Problem 1: Cohesive energy of free electron metals

In the jellium model of free electron metals, the discrete nature of the ionic lattice is replaced with a smeared out uniform positive background exactly equal to that of the valence electron gas. Each element is completely specified by just the electron density of free electrons, $n = N/V$, where N is the number of conduction electrons in the crystal and V is its volume. Usually the electron density is given in terms of the so-called Wigner–Seitz radius, r_s, corresponding to the spherical volume available to one valence electron. In the literature, the mean binding energy of such a valence electron is given by

$$E = (2.21/r_s^2) - (0.916/r_s) - 0.1156 + 0.0313 \ln r_s \qquad (1)$$

In Eq. 1, r_s is expressed, as usual, in Bohr atomic units, $r_B = 0.53$ Å, and E in Rydberg (1 Ry = 13.6 eV) and this expression apply only between $2 < r_s < 7$.

(1) Express r_s in Å and k_F as a function of the valence electron density, n. Express Eq. 1 in eV for the energies and in angstorm for r_s. Show the evolution of E (in eV) when r_s changes from 2 to 7 in Bohr radius unit.

(2) Evaluate numerically the r_s value and the cohesive energy for the elements mentioned in the table using the indicated

atomic density, N, and taking into account the number of valence electrons, z. Evaluate also directly their Fermi energy, E_F, from their free electron density, n. Show the results in the form of a table and remark.

Symbol	Na	Mg	Al	Cs	Ba	Li
N ($\times 10^{22}$ at. cm^{-3})	2.652	4.3	6.02	0.905	1.6	4.7

(3) Could you explain the first term of Eq. 1 eventually from a comparison of its numerical value to the corresponding Fermi energy?

Solution:

(1) $4\pi r_s^3/3 = Nz/V = n$; $r_s = (3/4\pi n)^{1/3}$; $k_F = (3\pi^2 n)^{1/3}$;
$k_F x r_s = (9\pi/4)^{1/3} \sim 1.92$
$E\ (\text{eV}) = [8.44/r_s(\text{Å})]^2 - [6.6/r_s(\text{Å})] - 1.293 + [0.426 \ln r_s(\text{Å})]$
$r_s = 2$ corresponds to r_s (Å)= 1.06 Å; r_s= 7 corresponds to r_s (Å)= 3.71 Å

(2) See Fig. 34.

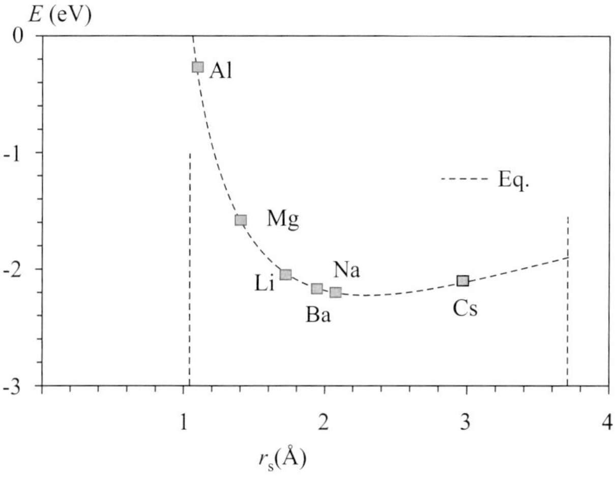

Figure 34

(3) The first term of Eq. 1 represents the mean kinetic energy of the conduction electrons. It is positive and it is a repulsive

term. For free electron metals in 3D, it corresponds to $3E_F/5$. The sum of the three other terms, the contributions from electron exchange and correlation, is attractive and thus negative. Therefore the equilibrium-bound state of minimum total energy results from a balance between the kinetic energy of the valence electrons, which tries to push the atoms apart, and the exchange–correlation energy, which tries to pull them together. The minimum in this binding energy curve for jellium occurs at $r_s/r_B = 4.2$ with a binding energy 2.2 eV/electron. This is quite close to the cohesive energy, E_{coh}, of real sp metals which fall at around 1–2 eV/atom and quite good agreement considering the simplicity of Eq. 1. The exceptions are Al where $r_s/r_B \sim 2$ and more Be where $r_s/r_B < 2$ gives a positive value for E_{coh}. The electron exchange and correlation effects are the glue that hold metals together and arises from the formation of the so-called exchange-correlation hole. This is a region of charge depletion around each electron due to the fact that electrons of like spin keep apart because of the anti-symmetry condition and the motion of electrons of unlike spin is correlated. The main consequence of this region of charge depletion around each electron is that each electron feels an attractive potential from the surrounding positive jellium background. The nature of such a binding explains the plasticity of metals such as Al or Ag when they are in a single crystal form free of dislocations and impurities.

Comments

The three last terms of Eq. 1 result from a parametrization of complicated quantum-mechanical many-body effects deduced from the use of the density functional theory (DFT) for the electronic structure problem. Briefly electron exchange arises because a many-body wave function must by antisymmetric under exchange of any two electrons since electrons are fermions. This antisymmetry of the wave function, which is simply a general expression of the Pauli exclusion-principle, reduces the Coulomb energy of the electronic system by increasing the spatial separation between electrons of like spin. Likewise electron correlation further reduces the Coulomb energy between electrons of unlike spin because the motion of each individual electron is correlated with the motion of all others, helping also to keep electrons of unlike spin spatially separated. The sum of these two quantum mechanical effects is incredibly difficult to describe. The interested readers are referred to the excellent article from which this problem is inspired, Ref. (30): A. Michaelides., & M. Scheffler, 2012, in *An Introduction to the Theory of Crystalline Elemental Solids and their Surfaces.*

Problem 2: Dipole layer and work function at surfaces of free electron metals

In the jellium model of free electron metals the discrete nature of the ionic lattice is replaced with a uniform positive background exactly equal to that of the valence electron gas (see previous exercise). At the surface of metals this positive uniform background is assumed to terminate abruptly along a plane at $z = 0$, filling the half-space $z < 0$ with the form: $n^+(z) = n; z < 0$ and $n^+ = 0; z \geq 0$, where n is the mean density of the positive charges in the ionic lattice.

For the electron density, $n^-(z)$, the result of sophisticated calculations (out of the present purpose) is displayed in the left side of Fig. 35, valid for Fermi energies less than ~10 eV (or for Wigner–Seitz radius, r_s, between 2 and 6). This figure shows that (i) the electron density spills into the vacuum and (ii) this density within the boundary oscillates (Friedel oscillations) with an amplitude that decreases asymptotically. For the sake of the simplicity, here, this electron density, $n^-(z)$, is supposed to decrease linearly across a transition region of thickness $D \sim \pi/k_F$, where k_F is the Fermi wave vector (see Fig. 30).

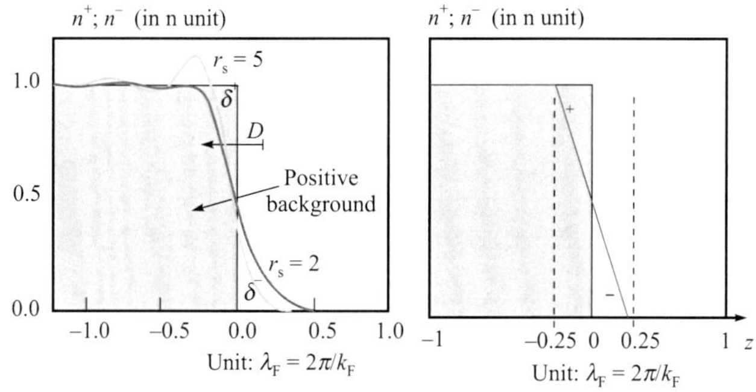

Figure 35 (Left) Calculated electron density distribution at the jellium surface (from N. D. Lang and W. Kohn, *Phys. Rev. B* **1**,1970, 4555 and **3**, 1971, 12158). (Right) Simplified model.

(1) Evaluate numerically (in Å) the thickness of the transition region, D, for E_F = 3.2 eV (Na) and 7.1 eV (Mg).

(2) Deduce the evolution of the electrostatic potential function, $V(z)$, across the interface by solving the Laplace–Poisson equation. For this, take into account the continuity of $V(z)$ at $z = 0$, the $V(z)$ value at $z = +\infty$, $V(+\infty) = 0$, and the zero electric field value at the two boundaries of the dipolar layer. Express $\Delta V = V(-\infty)$ and evaluate it numerically for Na and Mg.

(3) For the evaluation of the maximum total energy of free electrons into the metal with respect to the vacuum level one must add two contributions to the electrostatic potential energy, eV^{es}, with $V^{es} = V(-\infty)$ as evaluated above. The first contribution is the exchange/correlation energy, eV^{ex}, as derived from the sum of the three last terms of Eq. 1 in the statement of the previous exercise. The 2[d] contribution is the kinetic energy of the conduction electron.

From a simple diagram of energies, represent the position of the various energies in a metal and deduce the corresponding work function value, ϕ, that is to say the energy needed to excite an inner electron from the Fermi level (at 0 K) to become an outer electron with a zero kinetic energy into the vacuum.

Numerical application: Evaluate ϕ for Na and Mg with eV^{ex} = −4.14 eV for Na and −5.83 eV for Mg.

(4) The two semi-infinite metals are set parallel and in front of each other with a distance between them in the vacuum gap of L = 1 mm. The two are electrically connected to the ground without bias: Is there an electric field in the gap? If yes, give its direction and strength. ($\hbar^2/2m$, e, ε_0)

Solution:

(1) k_F(Na) = 0.92 Å^{-1}; k_F(Mg) = 1.37 Å^{-1}; D(Na) ~ 3.41 Å; D(Mg) ~ 2.3 Å

(2) The electrostatic problem is solved by integrating the 1d Laplace–Poisson equation: $d^2V/dz^2 = -\rho/\varepsilon_0$, applied to the dipolar distribution of electrical charges shown in Fig. 36.

The density of charge varies as $\rho = (n|e|/2)\ [1+(2z/D)]$ for $-D/2 < z < 0$ and as $\rho = (n|e|/2)\ [(-1+2z/D)]$ for $0 < z < D/2$.

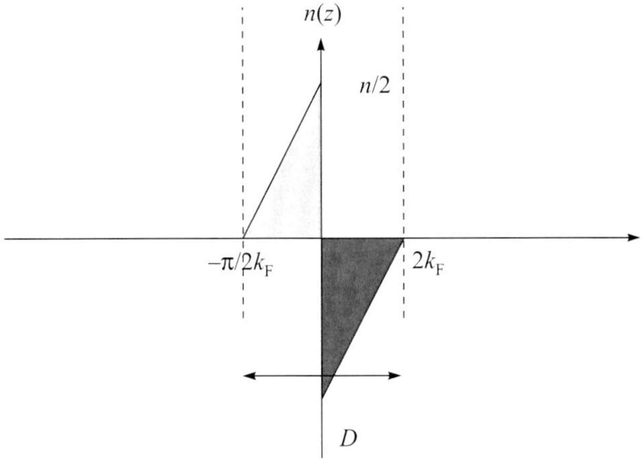

Figure 36

For $0 < z < D/2$, the first integration leads to the electric field expression with a change of sign:

$dV/dz = -(n|e|/2\varepsilon_0)[-z + (z^2/D) + C_1]$

$C_1 = D/4$ for obtaining a zero field at $z = D/2$.

A 2^d integration leads to

$V(z) = -(n|e|/2\varepsilon_0)[(-z^2/2) + (z^3/3D) + (Dz/4)]$ when the origin of the potential is taken at $z = 0$. When the origin of $V(z)$ is taken at $z = D/2$, a constant $C_2 = -(D^2/24)$ has to be added into the square brackets.

The same procedure applied between $-D/2 < z < 0$ leads to

$V(z) = -(n|e|/2\varepsilon_0)[(z^2/2) + (z^3/3D) + (Dz/4) - (D^2/24)]$

$\Delta V = V(-\infty) = 2V(0) = -(n|e|/2\varepsilon_0) \times 2C_2 = (n|e|D^2/24\varepsilon_0)$

The full evolution of $V(z)$ is shown in Fig. 37. $V(-\infty) = 2.33$ v. for Na and 3.4 v. for Mg.

(3) The two potential energies are negative, $e = -1.6 \times 10^{-19}$ C, the kinetic energy E_F is positive. Then the energy needed to excite an electron from the Fermi energy to the vacuum level is (see right side figure):

$\phi = -(eV^{es} + eV^{ex}) - E_F$

For Na: $\phi = 2.33 + 4.14 - 3.2 \sim 3.3$ eV.

For Mg: $\phi = 3.4 + 5.83 - 7.1 \sim 2.1$ eV.

(4) When two metals are set in contact, there are conduction electron transfers from M_2 to M_1 (open arrow), inducing the formation of a dipolar layer at the metal interfaces leading thus to the alignment of the Fermi levels and to the shift of the respective vacuum levels, VL. An electron at VL_1 would be attracted toward VL_2, whatever would be the connection: direct between the two metals via various intermediate connections. In a plane capacitor arrangement, an electric field is established from M_2 to M_1. Its strength would be $(\phi_1 - \phi_2)/|e|$. In the present situation with the obtained numerical values, the electric field is 1.2 kV/m and it is directed from Mg toward Na.

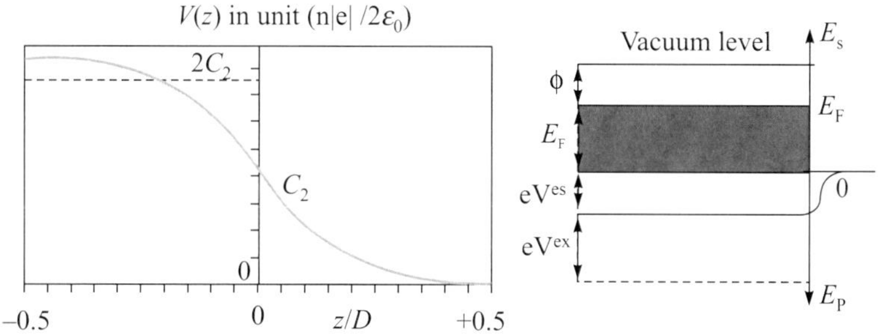

Figure 37 (Left) Calculated evolution of the electrostatic potential across the dipolar layer. (Right) Diagram of the conduction electron energies. *Note*: The origin of the kinetic energies of electrons into the metal, E_S, is at the bottom of the conduction band.

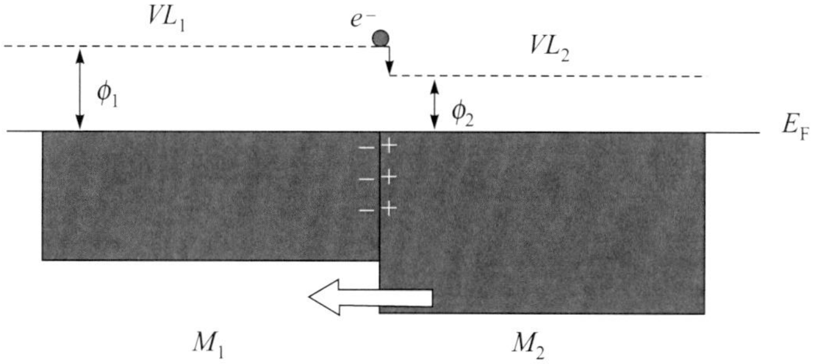

Figure 38 Voltage contact between two metals.

Comments

(1) ***Work function and surface barrier:*** Despite the obvious simplifications, the present problem provides a rather good value for the thickness of the transition layer D = one half of the Fermi wavelength, λ_F. Thus this thickness decreases with the increase of the free electron density or with r_s. It is a consequence of the overspill of free electrons into the vacuum. Here the main feature of the potentials sketched in the figure of answer (2) is as follows: V^{es} arises because the spread of electrons beyond the edge of the positive background renders the electrostatic potential in the vacuum, $V^{es}(\infty)$, higher than that in the metal interior, $V^{es}(-\infty)$. Although the electrostatic potential, V^{es} is a relatively small component of the barrier with respect to V^{ex}, it is of the utmost importance since it is closely related to the work function, ϕ. Thus an electron trying to leave the metal encounters an electrostatic surface dipole layer, *D*, with a electrostatic potential energy difference of height: a fact of prime importance in the emission of electrons into the vacuum.

Moreover the absolute value of ϕ obtained here fall correctly in the 2–4 eV range but the experimental value of ϕ(Mg) is larger than that of ϕ(Na) (see Table 2 and Da Silva et al. *Surf. Sci.* **600**, 2006, 703). One reason of the present discrepancy is the oversimplified model used in the statement of the present exercise: a linear decrease instead of the more precise description of the calculated electron density distribution at the jellium surface.

From the theoretical aspect there is a large dependence of the work function on the crystal structure and surface orientation (see Ex. 15) and also upon adsorption as a consequence of the change of the dipole distribution. Ashcroft and Mermin's classic book (1) has a long discussion on the work function in Chapter 18.

From the experimental aspect there is also the role of the surface contamination or oxidation. A spectacular example is given for Al where the work function changes from ~4 eV for samples evaporated in situ to ~2 eV for sample elaborated ex situ. This change is attributed to the spontaneous formation of a native layer when the exposure to air exceeds 50 L (1 Langmuir = exposure of 10^{-6} torr during 1 s). Another famous case is the adsorption of alkali metals which drastically lower the work function [e.g., from 5.7 eV down to

2.5 eV upon the adsorption of K on W(110)]. Also the experimental results obtained from different (thermo-ionic; photoelectric; contact potential difference) methods give slightly different results.

Table 2 Compilation of work function values (top) and Fermi energy values (bottom) of various elements (from Cazaux, *J. Electron. Spectrosc. Rel. Phen.* **187**, 2013, 23)

Metal	eV	Metal	eV	Metal	eV	Metal	eV	Metal	eV
	4.52–4.74		4.06–4.26		5.1–5.47		2.52–2.7		4.98
Ag:	**5.48**	Al:	**11.2**	Au:	**9.0**	Ba:	**3.65**	Be:	**14.14**
C:	4.7		2.87		5.0		4.5		2.14
Graphite:	**22.0**	Ca:	**4.68**	Co:	10.0	Cr:	7.8	Cs:	**1.58**
	4.53–5.10		4.67–4.81		4.32		4.09		2.93
Cu:	**7.0**	Fe:	8.9	Ga:	**10.35**	In:	**8.6**	Li:	**4.72**
	3.66		4.36–4.95		2.75		5.93		4.25
Mg:	**7.13**	Mo:	6.5	Na:	**3.24**	Os:	11.4	Pb:	**9.37**
	5.22–5.6		5.12–5.93		2.3		4.42		~2.6
Pd:	6.2	Pt:	10.6	Rb :	1.85	Sn:	**10.0**	Sr:	**3.95**
	4.00–4.80		4.33		4.3		4.32–5.22		3.6–4.9
Ta:	8.4	Ti:	6.0	V:	6.4	W:	10.1	Zn:	**9.39**

(2) ***Friedel oscillations:*** The shape of the electron density distribution at a jellium surface (Fig. 35) results from tedious numerical calculations but an approximate analytical expression is (Pb. 2b, below):

$$n(z)=\left(\frac{k_{\mathrm{F}}^{3}}{3\pi^{2}}\right)\cdot 1-3\left\{\frac{\sin(2k_{\mathrm{F}}z)-2k_{\mathrm{F}}z\cos(2k_{\mathrm{F}}z)}{(2k_{\mathrm{F}}z)^{3}}\right\}\quad \mathrm{u}$$

Via the reduced variable $k_F z$ the advantage of this expression is to show that the amplitude of the oscillations and the abruptness of the transition region increase with the free electron density, or with the k_F value, of the free electron metal of interest. The key result is that any interface between a metal and the vacuum would induce oscillations of the electron density at around the interface. This is also the case around an atomic vacancy which may be approximated to a spherical cavity. The asymptotic form of the electron density is similar to that of a plane interface except dependence in r^{-3} instead dependence in z^{-2}. The corresponding oscillations are named 'Friedel' oscillations (J. Friedel, *Philos. Mag.* **43,** 1952, 153; *Adv. Phys*. **3**, 1953, 446 *Nuovo Cimento* **7,** 1958, 287. These oscillations induce atomic relaxations around the vacancies. Some atoms may be attracted toward the center of the vacancy, Au, or pushed away, Al, as a function of the position of the maximum electric field acting on them and associated to these oscillations via Gauss's Equation. An atomic impurity in proximity of such a vacancy may also be attracted toward it more than the atoms of the host metal (Cu in Al). The result is an accumulation of impurities into Guinier–Preston zones increasing considerably the hardness of the corresponding Al alloys with its obvious applications in metallurgy and among others in the aircraft industry [21].

This remark illustrates how academic calculations may lead to considerable practical applications.

Problem 2b: Electronic density and Energy of metal surfaces: Breger–Zukovitski model

Consider an infinite metallic sheet in the xOy plane with thickness L along the Oz axis.

(a) Apply periodic boundary conditions (PBC) of period L along x and y and fixed boundary conditions along Oz between the two surfaces of the film. Find the wavefunction $\Phi(x,y,z)$ of the free electrons contained in the film.

(b) Compare the different components of the wave vector k obtained using PBC in 3D to that imposed here using fixed boundary conditions along the z-axis. Show that the latter results in the exclusion of states at $k_z = 0$, corresponding to electrons at the Fermi surface. Find the energy E^0 needed to obtain a surface of unit area.

Numerical application: Evaluate E^0 using $k_F = 1$ Å or $E_F = 3.8$ eV.

(c) Starting from $\Phi(x,y,z)$ and summing over all possible wave vectors, determine the evolution of the electronic density

$n(z)$ in the neighborhood of the surface area and sketch the corresponding curve.

Numerically find the minimal distance z_0 corresponding to $n(z) = n_0$, which corresponds to the electronic density of the bulk material. $\left(\dfrac{\hbar^2}{2m}\right) = 3.8 \text{ eV} \cdot \text{Å}^2$

Solution:

(a) Recall that fixed boundary conditions for a cube with side L result in a free electron wave function of the form (see Ex. 7):

$$\Phi(x,y,z) = \left(\frac{8}{L^3}\right)^{1/2} \sin k_x x \cdot \sin k_y y \cdot \sin k_z z, \text{ where}$$

$k_x = n_x \dfrac{\pi}{L}$, $k_y = n_y \dfrac{\pi}{L}$, $k_z = n_z \dfrac{\pi}{L}$, and $k_{x,y,z} > 0$; n_x, n_y, and n_z are positive integers.

When the cyclic Born-von Karman boundary conditions are used we find:

$$\Phi(x,y,z) = \frac{1}{L^{3/2}} e^{i(k_x x + k_y y + k_z z)},$$

where $k_x = n_x \dfrac{2\pi}{L}$, $k_y = n_y \dfrac{2\pi}{L}$, $k_z = n_z \dfrac{2\pi}{L}$ and n_x, n_y, $n_z = Z$.

Applying these results to the present problem, the wave function will have a $\sin k_z z$ form along Oz and $[\Phi(x,y,0) = \Phi(x,y,L) = 0] e^{i(k_x x + k_y y)}$ along Ox and Oy. The final result is thus $\Phi(x,y,z) = \left(\dfrac{2}{L^3}\right)^{1/2} \sin k_z z \cdot \exp i(k_x x + k_y y)$.

(b) The use of periodic conditions leads to divide the wave vector space into parallelepiped cells with dimensions $(2\pi/L) \times (2\pi/L) \times (2\pi/L)$ (see Fig. 39a). The use of fixed conditions along Oz, however, leads to cells with dimensions $(2\pi/L) \times (2\pi/L) \times (\pi/L)$ which are limited in the half space $k_z > 0$ and in which the values of $k_z = 0$ are excluded (see Fig. 39b).

The increase in energy of the electron gas is the surface energy $2E_s^0 L^2$ of the two new surfaces. To evaluate this, we may observe that it is as if the electrons contained initially in the equatorial plane have attained their Fermi energy.

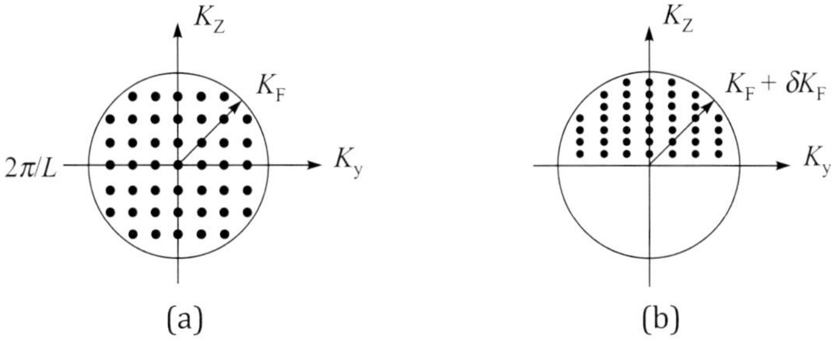

Figure 39

The initial kinetic energy is, on average, $\bar{E} = \frac{E_F}{2}$ for an electron gas in 2D and thus with a constant density of states (see Exs. 11 and 14). The number dn is equal to $2 \cdot \frac{\pi k_F^2}{(2\pi/L)^2}$. Their final energy is E_F and the change is thus $\Delta E = \frac{k_F^2 L^2}{2\pi} \cdot \frac{E_F}{2}$.

This difference corresponds to the formation of interfaces (from the fixed boundary conditions) with area L^2. We thus have $E_0 = \frac{\Delta E}{2L^2} = \frac{k_F^2 E_F}{8\pi}$.

Note that the increase of k_F and of E_F related to a decrease in the thickness L_z of the sheet has been previously evaluated in Ex. 7. Here we neglect this increase (E_F remains unchanged) and we evaluate only the increase in energy of all N electrons resulting from the conditions at $k_z = 0$. This last result only depends on the fixed boundary conditions and is independent of L_Z. It can therefore only be applied to macroscopic thicknesses.

Numerical application: $k_F = 1\ \text{Å}^{-1}$; $E_F = 3.8$ eV;

$$E_0 = 0.15\ \text{eV} \cdot \text{Å}^{-2}.$$

The estimated value is of the same order of magnitude as that accessible in experiments (~1–2 J·m^2) but the present problem has not introduced the anisotropy of the surface density as a function of the orientation of the relevant crystallographic faces. Accounting for this does not result in significant additional difficulties. First, it is sufficient to include the surface electron density n_s of each face and evaluate the initial energy of an electron in the $k_z = 0$ plane:

$$\overline{E} = \frac{E_F(2d)}{2} = \left(\frac{1}{2}\right)\left(\frac{\hbar^2}{2m}\right)(2\pi n_s).$$

Next, the evaluation of the electron fraction n_s, has to start from E_F(2d) and extends to E_F(3d). Exercise 15 illustrates this procedure.

(c) In direct space, the electronic density $n(z)$ can be obtained by summing the product $\Phi\Phi^x$ over all possible wave vectors taking into account the two electrons $(\uparrow\downarrow)$ per state. We obtain

$$n(z) = \sum_{|\vec{k}|\le k_F} \Phi\Phi^x = \frac{2}{L^3}\int_{|\vec{k}|\le k_F} \sin^2 k_z z \cdot dk_x dk_y dk_z \cdot \frac{L^3}{2\pi^3}$$

In cylindrical coordinates, this becomes (see Fig. 40):

$$n(z) = \frac{1}{\pi^3}\int_0^{k_F} dk_z \int_0^{(k_F^2-k_z^2)^{1/2}} 2\pi k_r \cdot dk_r \cdot \sin k_z z.$$

Performing the calculations, we obtain

$$n(z) = \frac{k_F^2}{3\pi^2}\left(1 - 3\frac{\sin 2k_F z - 2k_F z \cdot \cos 2k_F z}{(2k_F z)^3}\right).$$

When $k_F z$ is very large, the density $n(z)$ tends toward $k_F^2/3\pi^2$, which is just the electronic density n_0 in an infinite crystal. Substituting $2k_F z = x$, we note that the evolution of $n(z)$ obeys a relation of the form: $n(z) = n_0\left(1 - \frac{3\sin x}{x^3} + \frac{3\sin x}{x^2}\right)$, which is sketched in Fig. 40.

Starting from zero for $z = 0$, the electronic density attains the value n_0 for the first time when

$$tg(x) = x \text{ at } x_0 \approx 4.5 \text{ and } z_0 = \frac{4.5}{2k_F} = 2.25\ \text{Å}$$

Note that if we impose that the electron probability cancels at the surfaces, the resulting compression of the gas is very small because the thickness of the partially depleted region is of the order of the lattice parameter. In fact, if one defines its position as starting from the plane at which the volumetric density of positive charges (associated with the presence of ions) goes abruptly from

$n_0|e|$ to zero, the electronic density is not strictly zero at the surface. Regardless of this arbitrary choice concerning the position of the surface, these considerations result in the presence of a dipolar layer at the surface of a metal, see Fig. 39b and Fig. 35(left). The present model of Breger–Zukovitski is based on very different arguments from that of Kohn (Pb. 2) and it is surprising that the two lead to similar oscillations.

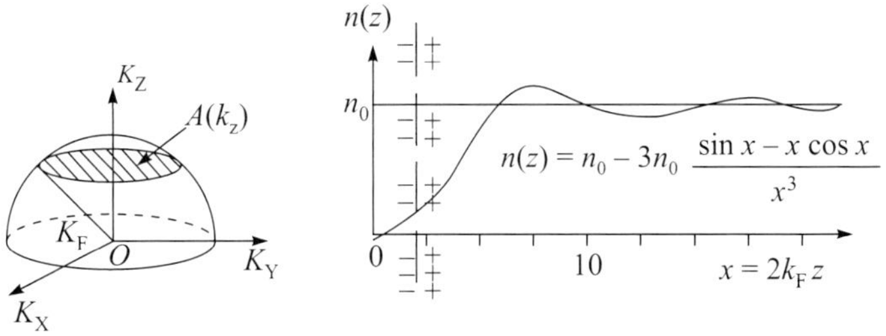

Figure 40

Remarks: As developed in Pb. 2, the continuation of this problem would consist of finding the evolution of the potential $V(z)$ by integrating the Poisson equation in 1D: $\frac{d^2V(z)}{dz^2} = -\frac{\rho(z)}{\varepsilon}$, where $\rho(z)$ includes the ion charges and the oscillations of the electronic density $n(z)$. Such a process would determine the potential barrier that keeps the electrons enclosed in the sheet.

Problem 3: X-ray photoelectron spectroscopy, X-ray absorption spectroscopy, extended X-ray absorption fine structure, Auger electron and X-ray photon emissions

When N isolated atoms condense to form a solid metal, electrons from each atomic outer shell (valence electrons) are distributed into an energy band, whose top is known as the Fermi energy, E_F. The inner-shell electrons (1s, 2s, 2p, ...) of the atoms are little affected by this condensation. Each binding energy $E_L(1s)$, $E_L(2s)$, keeps its discreteness and significantly its initial atomic value, and it is convenient to choose their origin at the Fermi level.

(1) X-ray photoelectron spectroscopy (XPS)

(i) A solid is irradiated with monochromatic X-ray photons (energy $h\nu$). Taking into account the work function potential ϕ (energy $e\phi$), find the inequality that the binding energy E an electron must satisfy (a) to absorb a photon and (b) to be ejected outside the solid. What is its kinetic energy E_c in the latter case?

(ii) *Numerical application*: Measured with respect to the Fermi energy, E_F, the binding energies, E_L, of atomic electrons of magnesium (Z = 12) and aluminum (Z = 13) are, respectively, Mg(1s) = 1305 eV; Al(1s) = 1560 eV; Mg(2s) = 90 eV; Al(2s) = 118 eV; Mg(2p) = 51 eV; Al(2p) = 73 eV.

An Al/Mg alloy is successively irradiated with Al K_α radiation and next with Mg K_α radiation (see Chapter I, Course Summary, Part B, Section 2 for the symbols). Sketch the spectrum of ejected X-ray photoelectrons as a function of their kinetic energy into vacuum. Take $e\phi$ = 3 eV and disregard the photoelectrons issued from the conduction band.

In fact, such photoelectrons are not all emitted into the vacuum because they undergo inelastic collisions (plasmon excitations) during their travel toward the surface. Thus the flux of the generated photoelectrons experiences an attenuation often simply described with an exponential expression of form $I = I_0 e^{-z/\lambda}$, where $\lambda(\text{Å}) = 1.8\dfrac{E_c^{3/4}}{E_p}$ (E_c and E_p are plasmon energy in eV). Specify the numerical value of λ for the Mg(1s) and Mg(2s) photoelectrons excited by Al K_α radiation (take E_p = 12.5 eV). Why does this technique allow the chemical analysis of the first atomic layers of a surface?

(2) X-ray absorption spectroscopy (XAS)

The same solid is now irradiated with continuous X-ray radiation: 50 eV < $h\nu$ < 2.5 keV, having a constant intensity over its spectral domain. The spectrum of transmitted X-rays through the sample, in a thin film form, is measured in order to infer the evolution with $h\nu$ of the absorption coefficient. Knowing that the main mechanism of X-ray absorption is due to the photoemission (photoelectrons either leaving the sample or not), find the position of threshold absorption relative to the Al/Mg alloy levels. Sketch the absorption spectrum $\mu(h\nu)$ knowing that beyond a threshold, the absorption decreases as

ΔE^{-3} (ΔE: difference between the energy of the photon, $h\nu$, and the minimum energy required to excite an atomic electron).

(3) Extended X-ray absorption fine structure (EXAFS)

Experimentally after each threshold, small oscillations of the absorption coefficient are observed being superimposed on the spectrum studied in (2). These oscillations are interpreted as interference of photoelectron waves associated with the ejected photoelectron of an atom being reflected by neighboring atoms back to their initial site.

(a) What is the kinetic energy of the photoelectrons of initial binding energy E_L and resulting from the radiation $h\nu$? What is their associated wave vector k? Knowing that the nearest neighbors are located at a distance r_0, find the relationship between k (next $h\nu$) and r_0 for which there is interference between outgoing photoelectrons and those reflected back to their initial position. Specify the energetic position of the first three oscillations beyond the threshold 1s(K) of aluminum, choosing r_0 = 3 Å.

(b) To establish the variation (beyond the threshold) of the absorption coefficient $\mu(E)$ around its mean value in μ_0, a more detailed theoretical analysis leads to the relationship:

$$\frac{\mu(E)-\mu_0}{\mu_0} \propto \sum_j \frac{N_j f_j \cdot e^{-2r_j/k}}{kr_j^2} \sin(2\vec{k}\vec{r}_j + \delta)$$

where r_j represents the distance between the emitting atom and one of its nearest 'j' neighbors. State the meaning of the various terms appearing on the right hand side, particularly, f_j, the exponential and the sine term?

(4) Auger electron and X-ray photon emissions

Following the ejection of corresponding photoelectrons, the atoms irradiated with X-rays present electron vacancies in their core electronic levels. An electronic vacancy is filled by an electron from an outer electron level and the released energy causes either the emission of an X-ray photon or the ejection of a third electron: the Auger emission process.

(a) Consider the initial ionization of the 1s Mg level. Find the energy of the subsequent X-ray radiation and the kinetic energy of the emitted Auger electrons. What is the order of magnitude of the escape depth of the Auger electron? Comment on their existence in a spectrum of photoelectrons (see Ex. 1).

(b) Replace the X-ray irradiation by irradiation with electrons of a few keV energy. The sample is a Al/Mg alloy. Find the X-ray photon energy and the kinetic energy of Auger electrons thus created. Comment on the identification of the components of a solid. ($\hbar^2/2m = 3.8$ eV $\cdot$ Å^2).

Solution:

(1) X-ray photoelectron spectroscopy

(a) $\alpha : h\nu > E_L$; $\beta : h\nu \geq E_L + e\phi$

From energy conservation: $h\nu = E_L + e\phi + E_e$.

(b) $h\nu(\mathrm{AlK}_\alpha) = E_L(1s) - E_L(2p)$, taking into account the selection rule $l' = l \pm 1$; $h\nu(\mathrm{Al}\ K_\alpha) \approx 1487$ eV;

$h\nu(\mathrm{Mg}\ K_\alpha) \approx 1254$ eV.

The spectrum of emitted photoelectrons reflects the spectrum of their initial binding energies with a shift of $h\nu - e\ \phi$, provided that the energy of the radiation is sufficient (see Fig. 41).

With Mg (K_α) radiation, the position of the photoelectron lines are Mg(2p) = 1200 eV; Al(2s) = 1178 eV;

Mg(2s) = 1161 eV; Al(2s) = 1133 eV.

With Al(K_a) radiation, these lines will have an additional kinetic energy of 233 eV. In addition, the Mg (1s) line will appear at 179 eV.

(c) $\lambda[\mathrm{Mg1s}] \approx 7$Å; $\lambda[\mathrm{Mg2s}] \approx 33$ Å

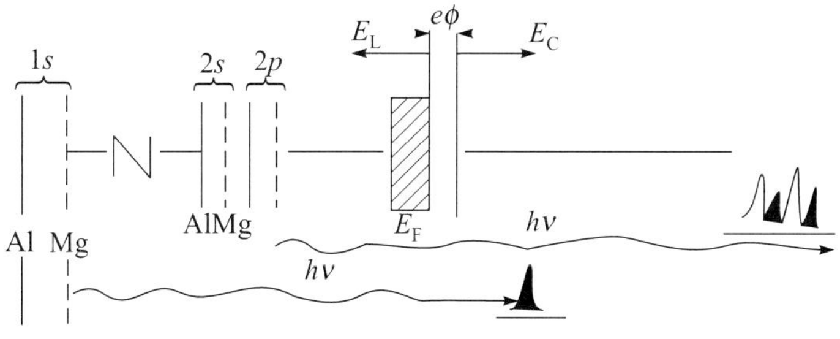

Figure 41

Comment: Nobel Prize in physics in 1981

The beauty of X-ray photoelectron spectroscopy lies in its simplicity. The spectrum of the emitted photoelectrons directly reflects the binding energy of the electrons prior to excitation. These binding energies have been tabulated (see *Review of Modern Physics*, **39**, 1967, 78). Thus the measurement of the photoelectrons kinetic energies permits to identify the elements composing the surface of the irradiated sample; see differences between Mg (Z = 12) and Al (Z = 13) in the problem. In fact, the binding energies and therefore the kinetic energies of photoelectrons also slightly depend (second-order effects of a few eV only) on the chemical environment of the excited atom. The corresponding chemical shift can be used to determine the atomic number of atoms of this environment. The choice of the Fermi level as origin for the measurement of the binding energies is driven by the alignment of the Fermi level when two metals are set in contact (here that of the sample and that of the electron spectrometer). Finally the photoelectrons have a very low escape depth [λ = a few atomic monolayers; see (c) above] and thus the resulting information is related to the chemistry of the surface. Popularized by K. Siegbahn (Nobel Prize in physics, 1981) and known as ESCA (electron spectroscopy for chemical analysis), this technique is widely used for analyzing the surfaces of semiconductors, polymers, catalysts, etc. Figure 41 shows a sketch of the instrument operated for surface characterization with XPS where incident X-rays are used and photoelectrons issued from the core electronic levels are analyzed instead of the UV–soft X-ray irradiation and the detection of photoelectrons from the valence band being concerned in the exploration of the surface electron states of Pb. 4.

(2) X-ray absorption spectroscopy

Absorption of a photon occurs when a photoelectron is excited to an allowed state (and therefore above the level of Fermi in metals). The thresholds for absorption are therefore exactly located at the values of $h\nu$ corresponding to the binding energy (statement of Pb. 1b). The characteristics of an X-ray absorption spectrum is given in Fig. 42.

(3) Extended X-ray absorption fine structure (EXAFS) oscillations

(a) $E_C = h\nu - E_L; k = \sqrt{2m(h\nu - E_L)}/\hbar$

$2r_0 = n\lambda = n2\pi/k$ (where n is an integer) or $2kr_0 = 2\pi n$

$\sqrt{2m(h\nu - E_L)} \cdot r_0 = nh/2$ so that $h\nu - E_L = (h^2/2m)(n^2/4r^2_0)$

$\Delta E = h\nu - E_L = 4.2$ eV ($n = 1$); 16.7 eV ($n = 2$);

37.4 eV (n = 3).

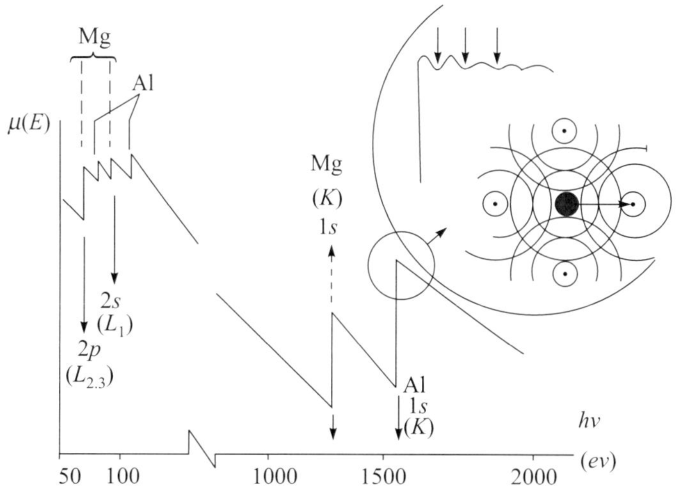

Figure 42

(b) N_j is the number of nearest neighbor atoms and f_j is the atomic form factor (see Chapter I, Ex. 22) of the photoelectrons scattered from these atoms (for $\theta = \pi$). The exponential describes the attenuation of the photoelectron wave (already mentioned in 1°C). The sine term $2kr_j$ is related to the interference evaluated above in (3a) to which a phase shift term, δ, is added and is the sum phase shifts to the emission and to the reflection. (The denominator is due to the spherical nature of the incident and the scattered waves.) An exponential term associated with the thermal agitation of atoms (the Debye–Waller factor) should also be added.

Comments: EXAFS oscillations

Oscillations (EXAFS or Kronig) have been observed for a long time but their extensive use has only been implemented since the development of spectrally white and intense X-ray sources such as those at synchrotrons. Unlike slow electron diffraction, this spectroscopic technique applies to disperse systems, amorphous solids, or liquids, and can locally evaluate the distance between first neighbors of an atom (characterized chemically by the position of its absorption threshold) and the number N_j of nearest neighbors. From the experimental acquisition of these oscillations up to 300 eV or 400 eV (but it is difficult to see the 1st) beyond the threshold one can determine r_j with precision of ±0.05 Å by a Fourier transform.

One can make this technique sensitive to the surface (S. EXAFS) by analyzing not the X-rays transmitted but the intensity of the photoelectrons or the Auger electrons as a function of $h\nu$. These intensities also reflect the photoabsorption mechanism but the information can only comes from the first atomic layers (see Woodruff, *Surf. Interf. Analysis* **11**, 1988, 25).

(4) Auger electron and X-ray photon emissions

(a) The emission process of a photoelectron leaves an electronic vacancy on the initial orbital of this photoelectron. This vacancy is quite spontaneously, ~10^{-15} s, filled with an outer shell electron via one of the two complementary processes: X-ray emission or Auger emission. The two processes always coexist but the probability a_{ijk} of Auger emission is greater than that for X-ray emission ω_{ij} when $E_L \leq 5$–10 keV.

For Mg (1*s*): ω_{KL} = 3% and $a_{KLL} \approx 97\%$. Figure 43 illustrates the competition between these two processes.

In the case of magnesium, both types of radiation are Mg $(K_\alpha) = E_K - E_{L2,3} = 1254$ eV and Mg $(K_\alpha) = E_K - E_{BC} \approx 1300$ eV.

Auger electrons are designated by the symbols relating to the three electronic levels involved in the process: vacancy level, electron filling this vacancy, and ejected electron. Multiple combinations are possible (different combinations of $KL_j L_k$ and KL_jM where j and k = 1 or 2, 3) because transitions leading to an Auger electron emission are not affected by selection rules, unlike X-emission which must obey $l' = l \pm 1$.

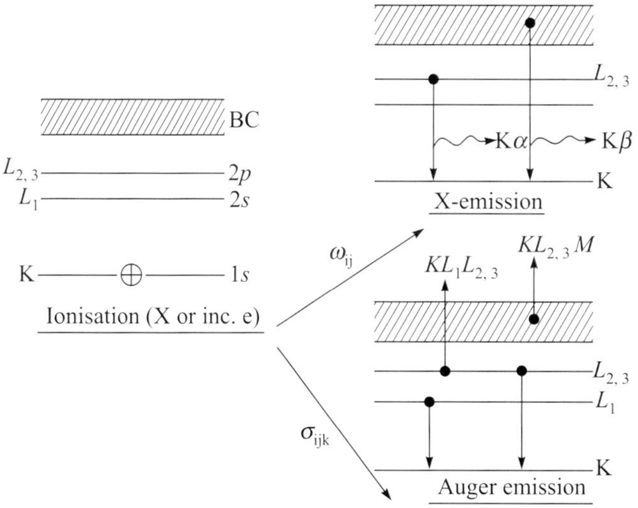

Figure 43

It will be for example (see Fig. 43):

Mg $(KL_1L_{2,3}) \approx E_L(K) - E_L(L_1) - E^x(L_{2,3}) - e\phi$ and

Mg $(KL_{2,3}M)$ or $(KL_{2,3}V)$,

where V signifies valence such that

Mg $(KL_{2,3}V) = E_L(K) - E_L(L_{2,3}) - E_L{}^x(V) - e\phi$,

$\mathrm{Mg}(KL_1L_{2,3}) \approx 1160$ eV, and Mg $(KL_{2,3}V) \approx 1245$ eV. (In fact, their energy is less than ≈ 20 eV because the ejected electron, e.g., $L_{2,3}$, has a binding energy closer to Al ($2p$) than Mg ($2p$) due to the influence of the hole created in the lower electronic shell which alters the $2p$ orbital, explaining the symbol x indicated in the above expressions.)

Most of the Auger electrons analyzed outside the sample have energies of the order of keV or less, thus their elastic escape depth is of a few atomic surface layers (see 1c). They appear also as lines in the spectra of photoelectrons (X-ray-induced Auger electrons). Auger electrons may be distinguished from photoelectrons because their kinetic energy is independent of the incident photon energy, $h\nu$. Thus when $h\nu$ is changed, their position will not be affected.

(b) If one replaces the beam of X-rays by an electron beam, the energy spectrum of X-rays emitted by the magnesium will not be affected because, as with the process of Auger electron emission, atomic decay is independent of the nature and the energy of incoming particles that created the initial electronic vacancy. The spectra of emitted X-rays from an Al–Mg alloy, is essentially composed of the Al(K_α) and Mg (K_α) radiations. The spectrum of the Auger electrons emitted from a Al–Mg alloy is composed of the series Al (KLL) between 1320 eV and 1400 eV, the Mg (KLL) series between 1100 eV and 1190 eV, the series Al (LVV) ≈ 50–70 eV and the series Mg (LVV) = 30–35 eV.

In all cases (XPS, X-ray emission, Auger emission), the identification of elements composing an alloy is obtained by comparing the position of the characteristic lines emitted by this alloy to those (tabulated) for all the elements of the classification. The estimate the concentration of these elements is obtained by the measurement of the corresponding intensities relative to those of pure elemental standards.

Comment: X-ray and Auger emission

The Auger effect was discovered in 1925 (P. Auger, *J. Phys. Radium* **6**, 1925, 205). As in X-ray emission or photoelectron spectroscopy, the energy position of Auger lines allows the determination of the chemical elements that have been excited either by a beam of X-rays, or an incident electron beam. As with XPS, since the 1960s, the Auger electron spectroscopy, AES, is one of the preferred methods for surface analysis of, especially metallurgical (corrosion) and semiconductors, because of the depth of analysis, of the order of nanometer, is given by the expression of λ in 1°C, where the kinetic energy is found in the range 50 eV < E_c < 2.5 keV.

The advantage of AES over XPS lies in the possibility of having a higher lateral resolution using a very fine electron probe (<0.1 μm). Its disadvantage is the difficulty in interpreting the chemical shift, in which 3 electronic levels are in competition instead of one.

Microanalysis of solids with X-ray emission spectroscopy (electron probe microanalysis) was pioneered by R. Castaing in the 1950s (*Advances in Electronics and Electron Physics*, **13**, 1960, 317), who developed an electron microprobe to analyze the concentration of metallurgical alloys and geological objects from their emitted X-ray spectra. The lateral resolution and depth analysis is of about ~1 μm because, unlike Auger electron, X-rays generated into a target suffer less attenuation in their path toward the surface being limited by the penetration depth of the incident electrons in microprobe, typically one micron for $E_0 \approx 20$–30 keV.

Finally the absorption of an incident X-ray photon, $h\nu$, may be followed by the emission of another X-ray photon, $h\nu'$ ($h\nu' < h\nu$) according to the de-excitation process shown in the right of Fig. 43. The corresponding analytical technique is X-ray fluorescence spectroscopy (XRF), which also allows the identification of the elements constituting a solid. This method has a high sensitivity (ppm) and it may be operated in the ambient atmosphere (instead of in the vacuum required for the other techniques) and on liquids but it suffers from poor lateral resolution, related to the difficulty to focus incident X-rays (see Ex. 30).

Problem 4: Refraction of electrons at metal/vacuum interface and angle-resolved photoemission spectroscopy (ARPES)

An often-used way to look on photoemission is the so-called three-step model:

(i) A photon of energy $h\nu$ is absorbed in the solid by an electron either bounded (atomic electron) or free (conduction electron). The electron is excited to an unoccupied final state and it is considered here as free with an energy E_S (measured with respect to the bottom of the conduction band).

(ii) The excited electron is brought to the surface with a wave vector $\boldsymbol{k}$ of components $k_{\parallel}$(in) parallel to the surface, and $k_{\perp}$(in) perpendicular to the surface.

(iii) The photo (excited) electron escapes into the vacuum after a refraction effect at the sample/vacuum interface. Its inner incident angle is β and its emission angle into vacuum is α (both angles with respect to the normal). In vacuum, the energy of the photoelectron, E_k, is measured with respect to the vacuum level, VL, and the components of its wave vector are $\boldsymbol{k}_{\parallel}$ (out) parallel to the surface, and $k_{\perp}$(out) perpendicular to the surface (see Fig. 44).

(1) From the typical values of the work function values of metals (Table 2 in the previous exercise), give an order of magnitude for the minimum energy, $h\nu$, and wave vector, $2\pi/\lambda$, of the incident photon needed for such an electron emission process into vacuum. Compare $2\pi/\lambda$ with the typical value of wave vector of an inner electron at Fermi level $\boldsymbol{k}_{\mathrm{F}}$ and remark.

(2) When the photoelectron crosses the sample/vacuum interface, the parallel component of k is preserved: $k_{\parallel}$(out) $= k_{\parallel}$(in) $= k_{\parallel}$. From this fact, express the refraction effect in the form of Snell's law relating E_{S} and β to E_{k} and α. Show the change of α as a function of β, $0 \le \beta \le 90°$ with a graph for the electrons at the bottom of the conduction band and at the Fermi level issued from Be ($E_{\mathrm{F}} \sim 14$ eV; $\phi \sim 5$ eV) when irradiated with a helium radiation (He $I_{\alpha} \sim 21.2$ eV) and next with a synchrotron radiation at $h\nu = 100$ eV.

(3) An electron spectrometer is set in front of the sample and its electrical connection to the sample leads to the Fermi level alignment of the two (Fig. 44). This spectrometer accepts just a small solid angle and analyses the electrons emitted in that solid angle. Then it is possible to explore the angular distribution of the emitted photoelectrons from a simple sample tilt changing α from 0° to 90°.

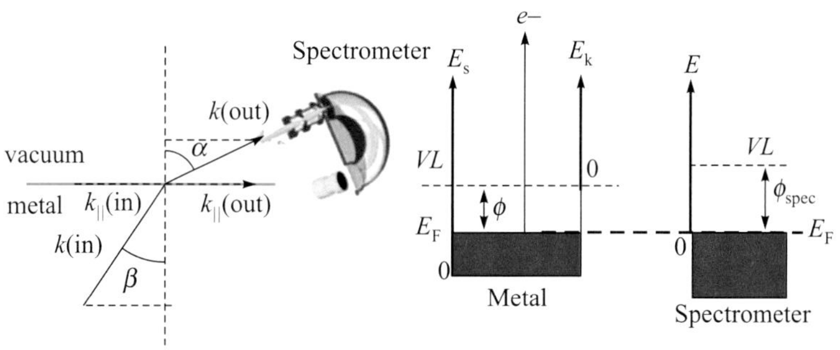

Figure 44 Geometry of the refraction effect (left). Energy diagram of the photoemission process (right). Note the different origins for E_{s} and E_{k} and also the alignment of the Fermi levels.

(a) Show the expected spectrum obtained from measurements at a normal emission angle, $\alpha = 0$, for Be measurements.

(b) The sample is tilted from 0° to 90°. Show how an exploration of the Fermi sphere is possible from an azimuth rotation combined to the tilts. Indicate the angular limits of such an exploration in the k-space for Be and $h\nu = 21.2$ eV. What are the corresponding limits for constant energy surfaces such as $E = E_F/2$ and $E = E_F/4$. From the results obtained at a nearly grazing incidence, $\alpha \sim 90°$, deduce that the spectra-obtained non-normal emission angles correspond to non-colinear wave vectors, k.

(c) The surface barrier influences the emitted intensities through the transmission probability, $T(\alpha)$, of the photoelectrons crossing the sample/vacuum interface. In elementary textbooks of quantum mechanics, this transmission probability is given by

$$T(\alpha) = \frac{4G^{1/2}}{\left[1 + G^{1/2}\right]^2} \text{ with } G = 1 + \frac{E_S - E_k}{E_k \cos^2\alpha} \tag{1}$$

Represent in polar coordinates this transmission probability for electrons issued from the Fermi level and from the bottom of the conduction band.

(4) From the conservation of the parallel component of k, establish the expressions correlating $k_{\parallel}$(in) and E_S inside the solid (prior to electron excitation) to the experimental values α and E_k. Evaluate numerically the maximum value of $k_{\parallel}$(in) being reached for Be and $h\nu = 21.2$ eV. Valid only for a parabolic dispersion of the final states, establish also the expressions correlating $k_{\perp}$(in) and E_S inside the solid (prior to electron excitation) to the experimental values α and E_k combined to the tilt. (h, e, c)

Solution:

(1) $h\nu$ has to be larger than ϕ (basic principle of the photoelectric effect). When $h\nu \sim 5$ eV, $2\pi/\lambda$ is $\sim 2.5\ 10^{-3}$ Å^{-1}. When $E_F \sim 5$ eV, $k_F \sim 1.1$ Å^{-1}. The momentum of the photon is small compared to the electron momentum and can therefore be neglected.

(2) The condition $k_{\parallel}(\text{out}) = k_{\parallel}(\text{in}) = k_{\parallel}$ leads to $\hbar\boldsymbol{k}(\text{out})\sin\alpha = \hbar\boldsymbol{k}(\text{in})\sin\beta$ and to $E_k \sin^2\alpha = E_S \sin^2\beta$ because $E_S = (\hbar^2/2m)\boldsymbol{k}^2(\text{in})$ and $E_k = (\hbar^2/2m)\,\boldsymbol{k}^2(\text{out})$.

$$\sqrt{E_k}\sin\alpha = \sqrt{E_S}\sin\beta \qquad (2)$$

Eq. 1 looks like the Snell's law of classical optics,

where $E_S = E_k + E_F + \phi$ (3)

The reference of the kinetic energies E_S is the bottom of the conduction band in the sample while it is the vacuum for the kinetic energies E_k of electrons in the vacuum level (see Fig. 37).

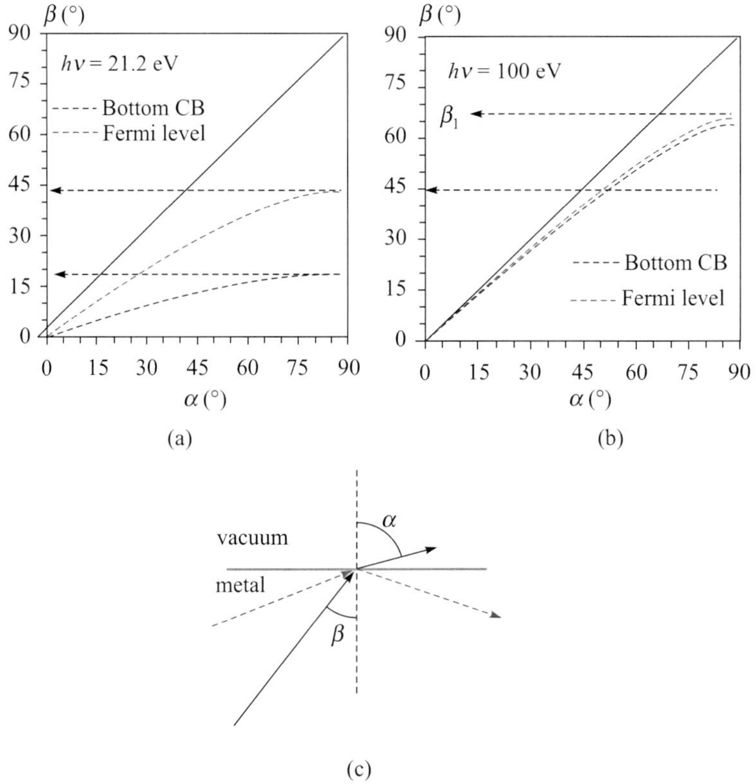

Figure 45 Correlation between α and β for Be irradiated with a UV lamp (a) and with a radiation $h\nu$ = 100 eV (b). The diagonal line corresponds to $\alpha = \beta$, no refraction effect. (c) Refraction effects (full arrows) and internal reflection for inner angles larger the critical angle β_l where $\sin\beta_l = \sqrt{E_k}/\sqrt{E_S}$.

For electrons excited from the Fermi level $E_S = E_F + h\nu = 35.2$ eV; $E_k = h\nu - \phi = 16.2$ eV with $h\nu = 21.2$ eV; $E_S = 114$ eV and $E_k = 95$ eV with $h\nu = 100$ eV. For electrons excited from the bottom of the conduction band $E_S = h\nu = 21.2$ eV; $E_k = h\nu - \phi - E_F = 2.2$ eV with $h\nu = 21.2$ eV; $E_S = 100$ eV and $E_k = 81$ eV with $h\nu = 100$ eV. As shown in Fig. 45, the refraction effects are very large in the present example of a UV radiation on Be. From the use of a synchrotron radiation, when the photon energy is increased up to $h\nu \sim 100$ eV, these effects decrease and lead to larger values for the critical angle β_1 for external emission.

(3a) The photoemission process is illustrated in Fig. 46 for $\alpha = 0°$. The density of states for the conduction electrons in Be is parabolic: $g(E_S)$ is proportional to $\sqrt{E_S}$. The reference energy is the bottom of the conduction band and their energy gain is $h\nu$ into the solid with a change in the reference level for the bulk conduction electrons when they escape into vacuum. At the solid/vacuum interface, a part of these conduction electrons are reflected back to the solid and this effect leads to a slight distortion of their measured spectral distribution, full line, with respect to that expected from a total transmission probability (dashed line).

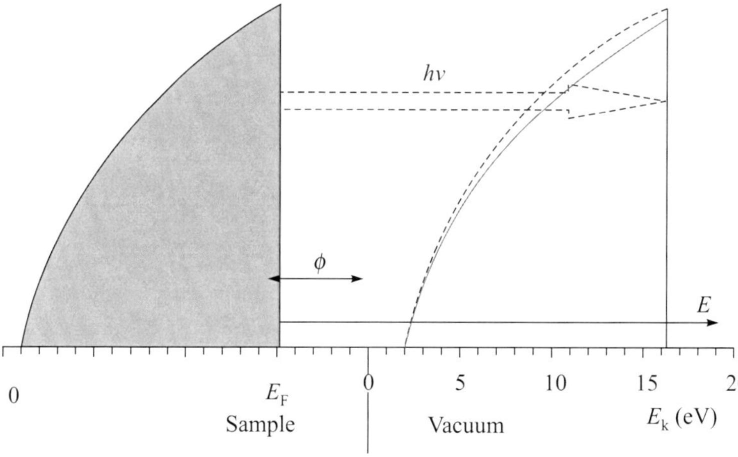

Figure 46 Photoemission process in Be irradiated with photons of energy $h\nu = 21.2$ eV.

There are other sources of distortion: the difference in the attenuation of electrons during their transport toward the surface; the change in the effective solid angle of collection with respect to the fixed instrumental angle, $\Delta\Omega$ (for details on this point see Cazaux, *J. Electron. Spectrosc. Rel. Phen.* **187**, 2013, 23). One key result is that the value of E_F may be easily deduced from experiments and that of ϕ is derived easily from the knowledge of $h\nu$.

(3b) From tilts, $0° < \alpha < 90°$ and rotations, the exploration of the Fermi surface is limited to the part of the sphere defined by the solid angle of the semi-apex angle equal the critical angle for external emission β_l as seen in Fig. 47; $\beta_l = 42°7$ for $k = k_F$. For a surface of constant energy $E_F/2$ or $k = k_F/\sqrt{2}$, $\beta_l = 34°3$. For a surface of constant energy, $E_F/4$ or $k = k_F/2$, $\beta_l = 28°7$. Then the spectral distribution of photoelectrons at grazing emergence is composed of electrons having non-collinear wave vectors. This fact may be interpolated for other angles except $\alpha = 0°$. Thus the need is to operate at a normal emission angle, as in 3α, for obtaining the correct density of occupied states of non-free electron materials.

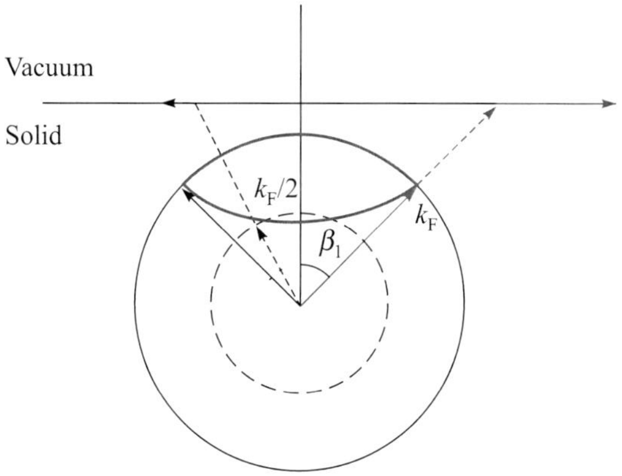

Figure 47 Limited exploration of the Fermi surface. The critical angle for external emission, β_l, decreases with the energy of the other constant energy surfaces, making the obtained density of states composed of electrons having non-collinear wave vectors.

(3c) As it was expected from Fig. 45, the influence of the refraction effects on the transmission probability is shown in Fig. 48 with an increase in α and a decrease in E_S.

(4) In vacuum, the parallel component of the wave vector is given by $k_{\|} = \sin\alpha(2m/\hbar^2)^{1/2}\sqrt{E_k}$. $k_{\|}$ being preserved, one obtains
$k_{\|} = \sin\alpha\,(2m/\hbar^2)^{1/2}\sqrt{E_k} = \sin\beta\,(2m/\hbar^2)^{1/2}\,(E_k + E_F + \phi)^{1/2}$ (4)
This equation is the key equation for converting the measured parameters, E_k and α, into the electron momentum and energy inside the solid prior to the electron excitation. As established above, E_F may be evaluated from the difference between the maximum and the minimum kinetic energies; E_k can be measured at a normal emission angle, α=0; and ϕ may be derived from the knowledge of $h\nu$.

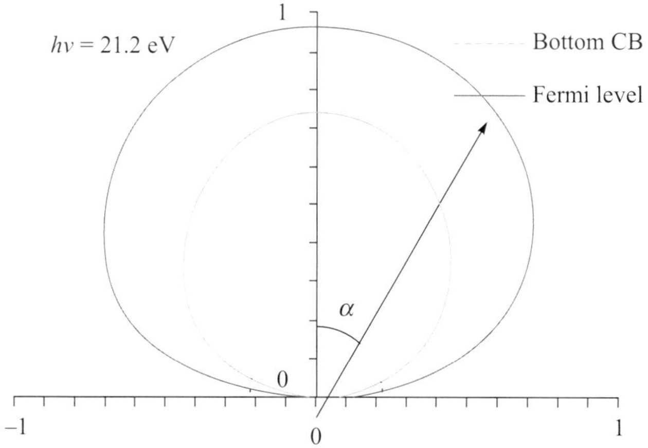

Figure 48 Transmission probability, $T(\alpha)$, for electrons issued from the Fermi level and from the bottom of the conduction band of Be for $h\nu$ = 21.2 eV polar co-ordinates.

From Fig. 45, the maximum value of $k_{\|}$ is $k_F \sin\beta_l$.
For Be k_F = 1.92 Å^{-1} and $k_{\|}$(max) = 1.3 Å^{-1}
$k_{\perp}(\text{in}) = (2m/\hbar^2)^{1/2}\cos\beta\sqrt{E_S}$ or (from Eqs. 1 and 2):
$k_{\perp}(\text{in}) = (2m/\hbar^2)^{1/2}(E_S - E_k\sin^2\alpha)^{1/2}$ or
$k_{\perp}(\text{in}) = (2m/\hbar^2)^{1/2}(E_F + \phi + E_k\cos^2\alpha)^{1/2}$.

Comments: ARPES and on surface states

ARPES: Angle-resolved photoemission spectroscopy, also called ARUPS with U for UV, is the technique for the experimental determination of bulk and surface electron band structures of crystals (as investigated in detail in Chapter V). By measuring the kinetic energy and angular distribution of the electrons photoemitted from a sample illuminated with sufficiently high-energy radiation, one can gain information on both the energy and momentum of the electrons propagating inside a material or at its surface. Equation 4 based on the conservation of the parallel component of momentum is the key equation for such a determination. For surface states, the normal component is null and the method is perfect for the experimental investigation of these states. For bulk states, the determination of the normal component of bulk states is more complicated than the present evaluation. First the periodicity of the crystal structure has to be taken into account. This periodicity implies that the transitions are allowed only by considering the extended zone scheme and employing a reciprocal lattice vector G or by limiting the *k*-values contained in the first Brillouin zone in a reduced energy band scheme (see Fig. 40 inspired from Damascelli, *Physica Scripta* **109**, 2004, 61). This last choice implies vertical transition between the occupied and the unoccupied states involved in the photoemission process (Fig. 49).

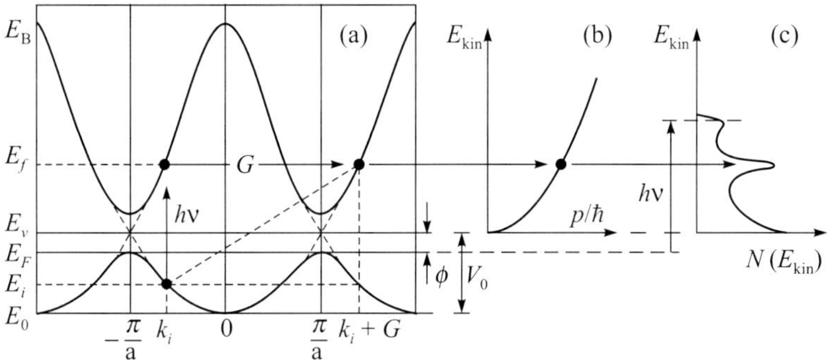

Figure 49 Photoemission process (a) electron transition in the solid vertical transition in the reduced band scheme (solid arrow) or oblique transition in the extended band scheme (dashed arrow); (b) free-electron final state in vacuum; (c) corresponding photoelectron spectrum.

Safely ignored in the case of a 2D sample and of surface states, the problem with $k_\perp$ may be circumvented by postulating a free electron–like behavior for the final state branches. To distinguish between surface states and bulk states, it is convenient to take a spectrum with two different photon energies but for the same $\boldsymbol{k}_\parallel$, bulk-related peaks will in general show a dispersion in the spectrum while surface state peaks stay at a fixed binding energy. For this strategy, the use of a synchrotron radiation as a light source (see Chapter I, Course Summary, Fig. 2) is obviously better than that of a gas-discharge lamp because of its tunability.

Only suggested in the present exercise, the use of ARPES permits to measure the density of states at a normal emission angle as well as to measure electronic band dispersions and to map the Fermi surfaces of bulk and surface states.

It is trivial but worth mentioning that photoelectron spectroscopy can only be used to study the occupied electronic states of the sample up to the Fermi level. The unoccupied states can be measured by a technique called inverse photoemission or bremsstrahlung isochromat spectroscopy (BIS), which is based on shooting electrons at a surface and detecting the emitted photons.

Surface States: Surface states are electronic states at the surface of materials. They result from the sharp transition from solid material that ends with a surface and extended only at the atom layers closest to the surface. When solving the Schrödinger equation one obtains two types of solutions. One type corresponds to bulk states which terminate in an exponentially decaying tail reaching into the vacuum illustrated in Fig. 35 of Pb. 2. The other type corresponds to surface states characterized by an exponential decay both into the vacuum and into the bulk crystal making the wave functions localized close to the crystal surface.

Historically the surface states are in turn classified into two types: The Shockley states and the Tamm states, depending upon the approximation being used for solving the Schrödinger equation at around the surface (see Chapter V for details on these approximations), which are the nearly free electron approximation, or Shockley states, and the tight-binding approximation, or Tamm states.

The first type, Shockley states, can be obtained for clean and ideal surfaces of both metals and some narrow-gap semiconductors. Indeed the hexagonal close-packed surfaces of Be, Mg, Al, Cu, Ag, and Au all possess occupied Shockley states. The Tamm states are characteristic of more tightly bound systems such as those in which the valence electrons are *d* states. Tamm states are suitable to describe transition metals and also wide-gap semiconductors. In fact, there is no real physical distinction between the two terms, only the mathematical approach in describing the surface states is different.

Another classification distinguishes the intrinsic surface states from the extrinsic ones. The intrinsic surface states originate from clean and well-ordered surfaces, including reconstructed surfaces, where the 2D translational symmetry gives rise to the band structure in the k space of the surface. Extrinsic surface states originate from surfaces with adsorbates or defects and from interfaces between two materials.

Thus, besides their interest from the fundamental aspect, the investigation of the properties of surface states is of a key importance for practical applications of the physical and chemical properties of surface.

Fair examples concern the interfaces between two materials such as a semiconductor–oxide or semiconductor–metal in the semiconductor technology where the presence of large density of surface states at these interfaces modifies the nature of the electrical contact, rectifying (Schottky) or non-rectifying (ohmic), via the pinning of the Fermi levels.

Surface states are routinely observed in experiment, notably with angle-resolved photoemission spectroscopy (ARPES) or also with scanning tunneling microscopy (STM). ARPES is the subject of very abundant literature among which one may select arbitrarily. A. Damascelli et al. *Reviews of Modern Physics* **75**, 2003, 473–541; P. Hofmann, *Surface Physics: An Introduction*, 2013; eBook in pdf format, ISBN 978-87-996090-0-0, F. Reinert and S. Hüfner, *New J. Phys.* **7**, 2005, 97; E.W. Plummer and W. Eberhardt, *Adv. Chem. Phys.* **49**, 1982, 533.

Problem 5: Scanning Tunneling Microscope (STM)

Consider two parallel metallic plates of the same composition and separated from each other by a vacuum of thickness s. The metal constituting the two plates is characterized by its Fermi energy E_F and its work function ϕ. A potential difference is applied between the two plates so that the electric field is uniform with one held at zero potential and the other at potential $V > 0$.

(1) Sketch the position of the Fermi levels as well as the variation of the vacuum level when one moves away from one plate to the other along the z-axis (where $eV < E_F$).

(2) Evaluate the number of electrons n (per unit volume), located in one of the plates, having enough energy and the satisfactory

direction to be able to occupy the available states in the other plate. Make the following simplifying assumptions: T = 0 K; eV << E_F. One must therefore evaluate the electronic density contained in a fraction of the spherical shell limited by a solid angle $\Delta\Omega$ and comprised between the two spheres of radii k_{F1} and k_{F2}. Choose the arbitrary value $\Delta\Omega$ = 1/6 sterad.

(3) In fact, in quantum mechanics, the probability P(E) that such electrons will traverse the potential barrier situated in the interval $0 \leq z \leq s$ is given by the Wentzel–Kramers–Brillouin (WKB) approximation:

$$\mathrm{P}(E) = e^{-a}, \text{where } a = 2\int_0^s |k(z)|\,dz$$

in which $|k(z)|$ represents the modulus of the (imaginary) wave vector of the electron at a point z situated in the barrier. Consider that the electrons that pass from electrode 1 to electrode 2 have a component k_z such that $k_z \approx k_F$ and show that P(E) takes the form P(E) = $e^{-2k_0 s}$. Find k_0.

(4) Deduce the expression for the current density j_z between the two electrodes and show that it may take the form: $j_z = AVe^{-2k_0 s}$. Find A.

(5) *Numerical application*: ϕ = 0.5 eV; E_F = 5 eV; V = 0.05 V; s = 5 Å What are the numerical values taken by $2k_0$ (in Å^{-1}), A (in Ω^{-1}m^{-2}) and j (in A/cm^2)?

(6) In fact one of the electrodes consists of a very sharp tip (ending in a single atom) without this form and significantly changes the results above. Knowing that it is possible to detect relative variations of current of the order of 10%, find the variation of s, Δs, that can be deduced (the depth resolution of a scanning tunnel microscope) for s = 5 Å.

(7) Evaluate the lateral resolution L of the microscope knowing that the majority of current flows in the interior of the solid angle $\Delta\Omega$ ($\approx \pi L^2/s^2$). $(e, \hbar, m)$

Solution:

(1) See Fig. 50. If electrode 2 is held at a positive potential V with respect to electrode 1, the Fermi level E_{F2} will be shifted below eV compared to E_{F1}, because of the electron energy diagram.

(2) The electrons flowing from 1 to 2 (associated with a current >0 in the opposite direction) must not only have an energy between E_{F1} – eV and E_{F1} but also the correct direction k_z.
At 0 K, we therefore consider that they are contained in the hatched volume v shown in Fig. 51.
The number of electrons contained in v is

$n = 2(\uparrow\downarrow)v/8\pi^3$, where $v = \frac{1}{6}4\pi k_F^2 \delta k_F$ and

$\delta k_F = k_{F1} - k_{F2} \approx k_F eV/2E_{F1}$, taking into account

$\hbar^2 k_{F2}^2/2m = E_{F1} - eV$. From this we find $n = \frac{2m}{3h^2}k_F eV$

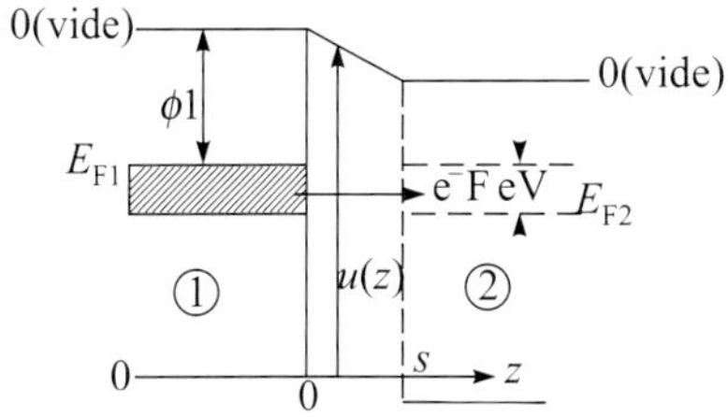

Figure 50

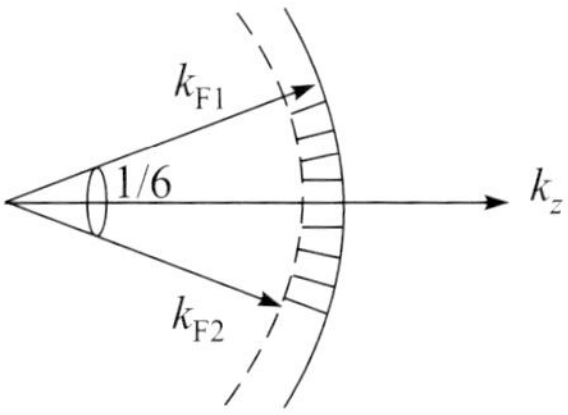

Figure 51

(3) The n particles above have kinetic energy E_F in the electrodes. In the trapezoidal potential barrier of height $E_F + \phi$ and $E_F + \phi$ – eV (measured from the bottom of the conduction band), they have a pure imaginary k wave vector such that:

$$\frac{\hbar^2 k^2}{2m} = \frac{\hbar^2 k_F^2}{2m} - u(z) = -\phi(z) = -\phi + eEz$$

from which $|k| = \frac{\sqrt{2m}}{\hbar}[\phi - eEz]^{1/2}$. The proposed integration leads to the expression for a:

$$a = \frac{4}{3}\frac{\sqrt{2m}}{eE\hbar}[(\phi)^{3/2} - (\phi - eV)^{3/2}], \text{ where } E.s = V.$$

Since $eV << \phi$, we find $P = \exp(-k_0 s)$, where $k_0 = \frac{\sqrt{2m\phi}}{\hbar}$.

(4) The current density in the electrodes corresponds to $|j_z| = n\,P\,e\,v_z = n\,P\,e\,v_F$ from which we find $j = A.V\,e^{-2k_0 s}$, where $A = \frac{8\pi m e^2}{3h^3}E_F$.

In the context of these approximations (eV << ϕ), the relation $j = f(v)$ is linear (in V). This is the ohmic regime.

(5) $2k_0 = 0.72\ \text{Å}^{-1}$; $A = 5.4\times 10^{14}\,\Omega^{-1}m^{-2}$; $j = 7.5\times 10^{7}\ \text{A/cm}^2$. Note that j corresponds to 7.5 nA/Å^2.

(6) $-2k_0 s = \text{Log } j - \text{Log } AV$; $|\Delta s| = \dfrac{1}{2k_0}\left|\dfrac{\Delta j}{j}\right|$, we find Δs = 0.14 Å.

(7) $tg\theta = L/s$; $\theta \approx 13°$; $L = s\sqrt{\Delta\Omega/\pi} = 1.15\text{Å}$.

In a scanning tunneling microscope, the resolution in depth is better than the lateral resolution.

Comment: The tunnel effect

The first questions of the exercise are relative to a simple tunnel effect. They could foreshadow further studies on the solid state physics of metal–insulator–metal junctions (MIM) or tunnel diodes (see [16], p. 185: L. Esaki, Nobel Prize in physics 1973). The calculation developed here is valid only under the hypothesis that $eV << \phi$, the ohmic region (see J.G. Simmons, *Journal of Applied Physics* **34**, 1963, 1793), which establishes more rigorously the law $j = AV \exp(-2k_0 s)$ but with $A = (e^2/\hbar)(k_0/4\pi^2 s)$.

At T ≠ 0, the electron density will be given by $n\,\alpha \int g_1(E) f_1(E) P(E) g_2(E) f_2(E)\,dE$, where $g_1(E)$ and $g_2(E)$ represent the density of states of electrodes 1 and 2; $f_1(E)$ and $f_2(E)$ represent the (F.D.) probability of occupation of these states (f_2 = 1 – F.D., because in 2 we are concerned with empty electron states).

WKB approximation: This approximation is used to integrate differential equations of the type $\psi''(z) + f(z)\,\psi(z) = 0$ (1)

We write $t = \dfrac{\psi'}{\psi}$ or $\psi' = t\psi$ and $\psi'' = t\psi' + t'\psi$.

Expression (1) becomes $t' + t^2 + f(z) = 0$ (2)

The approximation of order 1 consists of writing

$t' = 0$. $t^2 + f(z) = 0$ and leads to $t = \pm i\sqrt{f(z)} = \dfrac{\psi'}{\psi}$ or $\dfrac{d\psi}{\psi} = \pm i\sqrt{f(z)}dz$

from which $\psi(z) = \exp \pm i\int\sqrt{f(z)}dz$.

The approximation of order 2 consists of writing $t' = \pm u'$, where $u' = i\sqrt{f(z)}$. The expression (2) becomes $\pm u' + t^2 - u^2 = 0$ or

$$t = \pm\sqrt{u^2 \pm u'} \approx u\left(1 \pm \frac{u'}{2u^2}\right) \text{ and } \ln\psi = \pm\int u\,dz + \int \frac{u'}{2u}dz.$$

In total, $\psi(z) = \dfrac{1}{\sqrt{u}}\exp \pm \int u\,dz$.

Here we limit the approximation to the first order with the Schrödinger equation (1):

$$f(z)=\frac{2m}{\hbar^2}(E-u(z))=k_z^2<0.$$

Therefore $\psi(z)=\exp\pm\int_0^s|k_z|dz$ and the probability of transferring an electron via tunneling across a non-planar barrier will be $P(E)=\psi\psi^x=\exp-2\int_0^s|k_z|dz=\exp-a.$

Comment: Scanning tunneling microscopy; Nobel Prize in physics in 1986

The last part of the problem concerns the scanning tunneling microscope, invented by G. Binnig and H. Rohrer (*Helvetica Phys. Acta* **55**, 1982, 726) and for which they were awarded the Nobel Prize in physics in 1986. Figure 52 shows a version of this microscope as marketed in the 1990s. Image of the surfaces of graphite and Pt(111) are shown in Fig. 53a and 53b, respectively (see also Fig. 13c and 13d in Chapter II, Ex. 11 on atom manipulation).

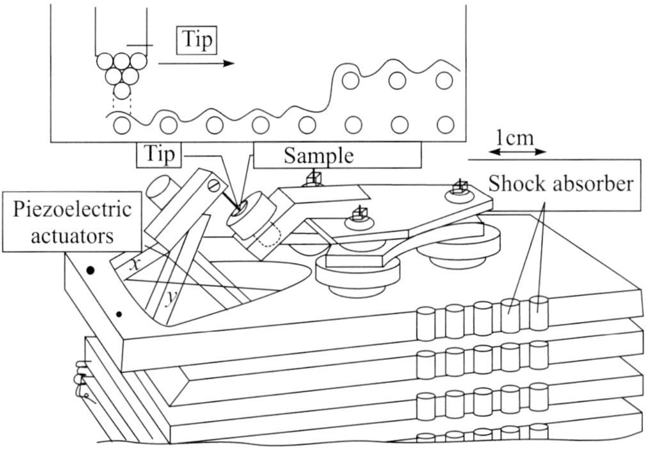

Figure 52 STM microscope in the 1990s.

For viewing surfaces at the atomic scale, one may simply (!) move a very fine tip (terminated by an atom) in the vicinity of the surface of the object (xOy). During the scan of the tip (provided by piezoelectric actuators), the tunnel current is monitored and is usually displayed in image form. In the invention of Binning and Rohrer, one does not know what to admire most: the simplicity of the idea, the optimization of its implementation, or the delicacy of its execution. Indeed, if the resolution in z is 0.1 Å, it is necessary that the amplitude of mechanical vibrations be smaller than this resolution. Scanning tunneling microscopy (STM) is a challenging technique, as it requires excellent vibration control combined with sharp tips and sophisticated electronics.

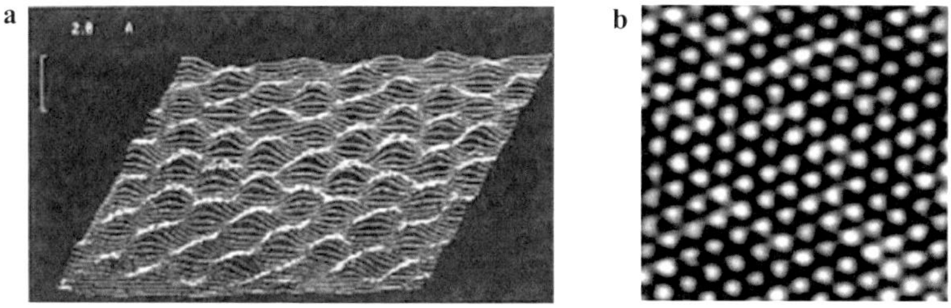

Figure 53 STM image of graphite in the 1990s (a) and of Pt(111) in the 2010s (b).

In reality there are two modes of operation. One uses a constant current procedure (such that s = constant) and determines the displacement in z of the tip, by measuring the feedback current. Variations in height (topography) are therefore detectable to ±0.1 Å, whereas the lateral resolution is of the order of 1 Å. The other mode is the constant height mode where the probe is not moved in the z-axis during the raster scan. Instead the value of the tunneling current is recorded and the image is usually referred to as a constant-height image.

The numerical results of the problem are perfectly realistic: $I \approx nA$; $V = 100$ mV, which allows the visualization of surface atoms (Si 7 × 7). However, the tunnel microscope measures only the local density of states (see the above expression) and while we can do spectroscopy of these, they can also lead to difficulties in the interpretation of the images. Therefore, on the surface of graphite represented in Fig. 53a, only one atom of the two per unit cell is discernible (compare the image with the graphite studied in Chapter I, Ex. 17).

Based upon STM, many other microscopy techniques (general labeling: scanning probe microscopy) have been developed. Among them there is photon scanning microscopy (PSTM), which uses an optical tip to tunnel photons; spin-polarized scanning tunneling microscopy (SPSTM), which uses a ferromagnetic tip to tunnel spin-polarized electrons into a magnetic sample; and atomic force microscopy (AFM), in which the force caused by interaction between the tip and the sample is measured. AFM and STM are the most commonly used for roughness measurements but, as indicated previously in Chapter II, Ex. 11, the same moving tip principle permits atomic scale manipulations such as atomic deposition of metals (Au, Ag, W, etc.) with any desired (pre-programmed) pattern, which can be used as contacts in nanodevices or as nanodevices themselves (see Chapter I, Pb. 4, and Chapter II, Pb. 11).

Problem 6: DC electrical conductivity: influence of a magnetic field

In a "normal" metal (such as copper) the charge carriers responsible for the electric current are "free" electrons with electric charge q, of mass m and n for their number per unit volume.

(1) A piece of copper, is set in a constant magnetic induction $\vec{B}$, that is uniform in the z-direction such that: $\vec{B} = (0, 0, B_z \equiv B)$ or $\vec{B} \equiv B\vec{e}_z$, where $\left|\vec{e}_z\right| \equiv 1$.

(a) Using the SI system show that if the electrons were free and in the absence of any electric field, their instantaneous velocity $\vec{v}$ would be given by

$v_x = -v_\perp \sin\omega t,\;\; v_y = v_\perp \cos\omega t,\;\; v_z = Cte$, where

$v_\perp \equiv$ | projection of $\vec{v}$ on xOy |
provided that ω has the special value, $\omega = \omega_c$, known as the cyclotron frequency. Express ω_c as a function of q, m, and B. Calculate numerically $|\omega_c|$ for electrons when $B = 1$ T.

(b) $\vec{B}$ is superimposed on a uniform constant electric field, $\vec{E} = (E_x, E_y, E_z)$, of components E_x, E_y, E_z relative to the axes $Oxyz$.

(i) Find the fundamental equation of motion for an electron with average drift velocity $\vec{u}$ in fields $\vec{E}$ and $\vec{B}$, which is subject to a damping force of form $-m\vec{u}/\tau$, where τ is the time of flight.

(ii) In the steady state regime, find the vector relation between the current density $\vec{j}$, the electric field $\vec{E}$ and the unit vector, $\vec{e}_z$ thus establishing a generalized form of Ohm's law $\vec{j} = \sigma\vec{E}$. Write the expression using the dimensionless number $(\omega_c\tau)$.

(iii) Project this relationship on the axes (*x, y, and z*) to solve the equations obtained from the components of $\vec{j}$. Show that

$$\begin{bmatrix} j_x \\ j_y \\ j_z \end{bmatrix} = \begin{bmatrix} \sigma_{xx} & \sigma_{xy} & \sigma_{xz} \\ \sigma_{yx} & \sigma_{yy} & \sigma_{yz} \\ \sigma_{zx} & \sigma_{zy} & \sigma_{zz} \end{bmatrix} \cdot \begin{bmatrix} E_x \\ E_y \\ E_z \end{bmatrix}, \quad (1)$$

where the components σ_{uv} of the matrix represent "conductivity tensor" and are functions σ and of $(\omega_c\tau)$, which are known values with *u* and *v* = *x* or *y* or *z*.

(iv) The tensor relation above means that the current density $\vec{j}$ is not generally parallel to the electric field $\vec{E}$ in the presence of a magnetic field $\vec{B}$. However, for a metal with relatively high values of *B*, the corresponding anisotropy is very low so that $\vec{j} = \sigma\vec{E}$. To clarify the degree of this approximation for copper, evaluate the three diagonal terms σ_{uv} and show that they are almost equal (how close?) and the non-zero non-diagonal terms σ_{uv} $(u \neq v)$ are comparatively very small (how much?) using a numerical calculation, with $B = 1$ T and $\tau = 2.4 \times 10^{-14}$ s.

(2) The phenomena related to the above anisotropy are especially important for semiconductors and depend on the geometry of the system studied.

Hall Geometry (Fig. 54): A uniform DC current flows through a rectangular bar of almost infinite length. This geometry implies a uniform current density *j* of the form $\vec{j} = j_x (\equiv j,0,0)$ along the axis O*x*.

(a) Using the equations in (1) show that

(i) There is a permanent regime with a transverse electric field, E_y in the conductor, called the Hall field, that is proportional to E_x.

(ii) The resistance of a piece of bar of length *l*, is unchanged from its value in the absence of magnetic field (no "magnetoresistance").

(b) Usually the Hall electric field is written in the form $E = R_H j_x B$, where R_H is the "Hall constant". Find the expression of R_H. Show that measurements of R_H and the conductivity σ allow the determination of the mobility μ of the charge carriers; μ being their velocity per unit electric field.

(c) Using a figure analogous to that in Fig. 54, show the direction of $\vec{E}_y$ for a "normal" metal such as copper, and give a physical interpretation of this electric field. (Hint: In the steady state the transverse component of the resulting force on the free electrons is zero.)

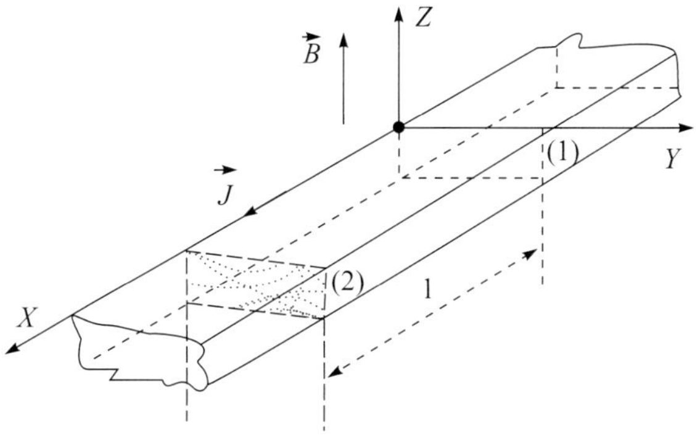

Figure 54

(d) The bar in Fig. 54, assumed to be made of copper and to be immobile, is subject to the Laplace "magnetic" force. Find the expression of this force per unit volume, and show its direction in a figure similar to that in Fig. 54. Since the transverse component of the resulting force on the conduction electrons is zero [see Question (c) above], there is a force exerted on the rest of the immobile conductor. State the direction and modulus of this electric force per unit volume exerted by the Hall field E_y, on the Cu^+ ions in an electrically neutral bar. What can conclusions can be made?

(e) For B = 1 T and $|j_x|$= 1 Å/mm^2, calculate the order of magnitude of the numerical value of $|E_y|$
   - in a metal ($n \approx 10^{22}$ cm^{-3}), expressing $|E_y|$ in μV/cm
   - in a semiconductor ($n \approx 10^{16}$ cm^{-3}, expressing $|E_y|$ in V/cm

A measurement of the Hall effect in metallic sodium leads to $|E_y| = 25\ \mu V/cm$ for $j = 1000\ A/cm^2$ and $B = 1$ T. Calculate the number of conduction electrons per cm^3 involved in this observation, and compare it to the number of atoms in 1 cm^3 of sodium, or $n(Na) = 2.53 \times 10^{22}\ cm^{-3}$. (q, m)

Solution:

(1) (a) Electrons having a velocity $\vec{v}$ are subject to the Lorentz force:

$$\vec{F}_L = q\vec{v} \wedge \vec{B} = m\frac{d\vec{v}}{dt}$$

In Cartesian coordinates the equations of motion become:

$$\frac{dv_x}{dt} = v_y\omega_c \quad (1)$$

$$\frac{dv_y}{dt} = -v_x\omega_c \quad (2)$$

$$\frac{dv_z}{dt} = 0 \quad (3)$$

where $\omega_c = \frac{qB}{m}$ have been used ($\omega_c < 0$ because $q = -q_e < 0$). Combining (1) and (2), we obtain $\frac{d^2v_x}{dt^2} = -\omega_c^2 v_x$, which has a solution of form $v_x = A\sin(\omega_c t + \varphi)$ leading to $v_y = A\cos(\omega_c t + \varphi)$.

The initial conditions $v_x(t = 0) = A \sin \varphi$ and $v_y(t = 0) = A \cos \varphi$ allow the evaluation of A:

$A = [v_x^2(t=0) + v_y^2(t=0)]^{1/2} = v_\perp$ and it results in

$$v_x = v_\perp \sin(\omega t + \varphi), \quad v_y = v_\perp \cos(\omega t + \varphi), \quad v_z = \text{cste}$$

Substitute $\omega = -\omega_c \cdot |\omega_c| = 0.175 \times 10^{12}$ rad/s.

(b) (i) $m d\vec{u}/dt = q\vec{E} + q\vec{u} \wedge \vec{B} - m\vec{u}/\tau$.

(ii) The steady state $\left(\frac{d\vec{u}}{dt} = 0\right)$ results in the relation

$$\vec{u} = \frac{q\tau}{m}\vec{E} + \frac{q\tau B}{m}(\vec{u} \wedge \vec{e}_z), \quad \text{where} \quad \vec{u} = \mu\vec{E} + \omega_c\tau(\vec{u} \wedge \vec{e}_z).$$

From which we find $\vec{j} = Nq\vec{u} = Nq\mu\vec{E} + (\omega_c\tau)(\vec{j} \wedge \vec{e}_z)$.

The projection on the axis gives

$$j_x = Nq\mu E_x + \omega_c \tau j_y;\ \ j_y = Nq\mu E_y - \omega_c \tau j_x;\ j_z = Nq\mu E_z,$$

where $\mu = q\tau/m$.

(iii) After rearranging terms we find

$$\begin{vmatrix} j_x \\ j_y \\ j_z \end{vmatrix} = \frac{\sigma}{1+\omega_c^2\tau^2} \begin{vmatrix} 1 & \omega_c\tau & 0 \\ -\omega_c\tau & 1 & 0 \\ 0 & 0 & 1+\omega_c^2\tau^2 \end{vmatrix} \begin{vmatrix} E_x \\ E_y \\ E_z \end{vmatrix}$$

where $\sigma = Nq^2\tau/m$ (obtained from $B = 0$).

(iv) We note that only σ_{zz} is strictly equal to the DC conductivity in the absence of B. As $\omega_c\tau$ is equal to 4×10^{-3}, the two other diagonal terms, σ_{xx} and σ_{yy}, are only 16×10^{-6} less than σ. The relative value of the non-diagonal terms σ_{xy} and σ_{yx} are only 4/1000 of σ.

(2) (a) In the Hall geometry, $j_y = 0$ so that

(i) $\sigma_{yx}E_x + \sigma_{yy}E_y = 0 \Rightarrow E_y = -\dfrac{\sigma_{yx}}{\sigma_{yy}}E_x = \omega_c\tau E_x$.

(ii) Using the value of E_y in the current density along x we have $j_x = \dfrac{\sigma}{1+\omega_c^2\tau^2}(E_x + \omega_c\tau E_y)$ and thus we obtain $j_x = \sigma E_x$, which shows the DC resistivity of the bar is not changed by the presence of the magnetic field.

(b) Starting from the two relations above ($E_y = \omega_c\tau E_x$ and $j_x = \sigma E_x$) and using the explicit values for ω_c and σ, we find $E_y = \dfrac{1}{nq}j_x B$ and $R_H = \dfrac{1}{nq}$.

Taking $\sigma = nq\mu = \dfrac{\mu}{R_H}$, the measurement of the Hall constant and of the conductivity permits effectively the determination of the mobility of charge carriers $\mu = R_H \cdot \sigma$, provided that there is only one carrier type. (For two carrier types, see Chapter V, Exs. 18 and 19).

If the DC current is due to electrons, E_y is negative when E_x is positive ($\omega_c < 0$). We find this result by considering the steady state regime when the transverse force: $\vec{F} = q\vec{u} \wedge \vec{B} + q\vec{E}_y$ is zero, or $\vec{E}_y = \vec{B} \wedge \vec{u}$.

(c) The Hall field for Cu is shown in the figure below. If $j_x > 0, E_y < 0$ then the current is created by a displacement of electrons ($\vec{j} = nq\vec{u}$ when $u_x < 0$ because $q = -|q_e|$).

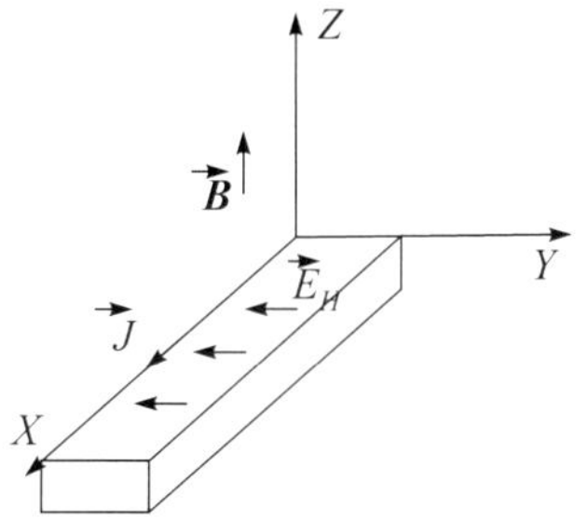

Figure 55

If $j_x < 0$ and $E_y > 0$, then the current is created by positive charges ($\vec{j} = n_+ q_+ \vec{u}$ when $u_x > 0$).

As a result, the direction of the Hall electric field (sign of the Hall voltage) can be used to determine the nature of majority carriers (holes or electrons).

(d) A unitary volume traversed by a uniform current density $\vec{j}$ is subject to the Laplace force $\vec{F} = (\vec{j} \wedge \vec{B})$.

For $\vec{j} = j_x \vec{e}_x$, where $j_x > 0$ and $\vec{B} = B \cdot \vec{e}_z$ for $B > 0$, we have $\vec{F} = F \cdot \vec{e}_y$, where $F = -j.B$.

We can find this force (both modulus and direction) by considering that it is created by a Hall electric field acting on the ions: $\vec{F}_e = nq_i \vec{E}_y = -nq_i \vec{u} \wedge \vec{B} = \vec{j} \wedge \vec{B}$.

(e) *Numerical application*: Since the Hall constant is inversely proportional to the volumetric density of charge carriers, the electric field is much weaker in a metal than in a semiconductor:

$E_y = 6.25\ \mu\text{V/cm}$ ($n = 10^{22}\text{cm}^{-3}$),

$E_y = 6.25\ \text{V/cm}$ ($n = 10^{16}\text{cm}^{-3}$)

The Hall electric field in metallic sodium will be $2.5\ \mu\text{V/cm}$ for a current density of $j = 1\ \text{A/mm}^2$, and the corresponding electronic concentration is therefore $n = 2.5 \times 10^{22}\ \text{cm}^{-3}$.

For other complements on magneto-resistance in semiconductors and on giant magneto-resistance, see Ex. 19.

Problem 7: Drude model applied to the reflectance of alkali metals in the ultraviolet and to characteristic electron energy losses

We consider an alkali metal to be a set of N fixed ions (per unit volume) and N free electrons (one free electron per atom) of mass m and immersed in a vacuum (ε_0).

(1) This electron gas is submitted to an alternating electric field of frequency ω with form $E = E_0 e^{i\omega t}$. Find the equation of motion for each electron knowing that there also exists a damping force that is proportional to their velocity v and of form $-\frac{m}{\tau}\vec{v}$, τ is the relaxation time. Find the steady state solution.

(2) Express the conduction current density corresponding to the motion of N electrons per unit volume N? Taking into account the displacement current density (which is that of the vacuum), find the expression for the total current density.

(3) Show that alkali metals can be characterized electrically either by the complex relative dielectric constant $\tilde{\varepsilon}_r(\omega) = \varepsilon_1(\omega) - i\varepsilon_2(\omega)$ or by a complex electrical conductivity $\tilde{\sigma}(\omega) = \sigma_1(\omega) + i\sigma_2(\omega)$. Simplify the expressions obtained using $\omega_2^p = Ne^2/m\varepsilon_0$.

(4) *Numerical application for potassium*: Specify the value of the plasma frequency ω_p as well as the corresponding energy (in eV). mass density (K) = 870 kg/m^3, atomic mass = 39.1, and τ = 2.64×10^{-14} s.

(5) Sketch the evolution of $\varepsilon_1(\omega)$ and $\varepsilon_2(\omega)$ specifying values when $\omega = 0$ and $\omega = \infty$. Give the value of ω at which $\varepsilon_1 = 0$.

(6) When initially mono-energetic electrons ($E^\circ \approx$ 20–50 keV) are transmitted through a thin metal film, they excite the collective plasma oscillations of the free electrons in the film via the longitudinal electric field they induce. Thus their energy loss is proportional to $\varepsilon_2 / (\varepsilon_1^2 + \varepsilon_2^2)$. Indicate the characteristics of this energy loss by specifying the position of its maximum ΔE and its full width at half maximum $\Delta E(1/2)$.

What are the numerical values of ΔE, $\Delta E(1/2)$, and E_F for potassium? (To simplify the calculations take into account that $\omega_p \tau >> 1$).

(7) The electric field is now a transverse field of an electromagnetic plane sine wave linearly polarized in x and propagating from the origin O in the positive z-direction.

Determine the equation of the electromagnetic wave propagation in the plasma. Solve it using $\overline{E}_{\mathrm{T}}$ its complex amplitude in z = 0 (where $\mu = \mu_0$). Show graphically the dispersion relation $\omega = f(k)$ for $\omega > \omega_{\mathrm{p}}$ assuming that the relaxation time is infinite. Find the expression for the phase velocity of the wave as a function of ω and ω_{p}. What is the nature of the wave when $\omega \leq \omega_{\mathrm{p}}$?

(8) Continuing in the hypothesis that $\tau = \infty$, find the expression for the magnetic excitation associated with the previous electric field as function of ε_{r} and next of ω_{p} and ω. What is the wave impedance Z_{p} in the plasma?

(9) In fact the metal only occupies the half-space of positive z and the wave calculated above is only the part of the incident monochromatic linearly polarized electromagnetic wave that propagates in a vacuum $(\varepsilon_0, \mu_0, z < 0)$ and is partially reflected at the z = 0 plane.

After recalling the expressions of incident waves ($\overline{E}_i$ and $\overline{H}_i$: complex amplitudes) and the reflected waves $(\overline{E}_{\mathrm{r}}, \overline{H}_{\mathrm{r}})$, determine, from boundary conditions on the z = 0 plane, the expression for the amplitude of the reflection coefficient $r = \overline{E}_{\mathrm{r}}/\overline{E}_i$ as a function of Z_0 and Z_{p} and next as a function of $\tilde{\varepsilon}_{\mathrm{r}}$ and ε_0.

(10) Show the variation of the ratio R between the reflected and incident intensity $(R = r \cdot r^x)$ as a function of ω after having first calculated the ratio $\omega/\omega_{\mathrm{p}}$ for which R = 1/m. Carry out a numerical calculation using m = 16.

Indicate the frequency domain where a total reflection is obtained and the frequency domain where the wave can be fully transmitted.

Determine for potassium the numerical value of the wavelength λ_0 of the incident electromagnetic wave in a vacuum that marks the boundary between these two regions.

(11) Show that the results obtained in (9) and (10) could be deduced immediately from the expression for the coefficient of optical reflection (in magnitude) at normal incidence: $r = \dfrac{1-\tilde{N}}{1+\tilde{N}}$,

where the plasma may be described by a complex optical index $\tilde{N} = n - ik$ that is a function of ε_1 and ε_2. $(e, m, \hbar, \varepsilon_0)$

Solution:

(1) $m\dfrac{d\vec{v}}{dt} = -e\vec{E} - \dfrac{m}{\tau}\vec{v}$;

If $\vec{E} = \vec{E}_0 e^{i\omega t} \Rightarrow \vec{v} = \vec{v}_0 e^{i\omega t}$ or after substitution

$$\vec{v}_0 = \frac{-e\vec{E}_0}{m\left(\dfrac{1}{\tau} + i\omega\right)}.$$

(2) $\vec{J}_C = -Ne\vec{v} = \dfrac{Ne^2}{m} \cdot \dfrac{\vec{E}}{\left(\dfrac{1}{\tau} + i\omega\right)}$

The density of total current is $\vec{J}_C + \dfrac{\partial \vec{D}}{\partial t} = \left[\dfrac{Ne^2}{m\left(\dfrac{1}{\tau} + i\omega\right)} + i\omega\varepsilon_0\right]\vec{E}$.

(3) One can electrically characterize the plasma by identifying the total current with the displacement current of the form:

$$\frac{\partial(\varepsilon_0 \tilde{\varepsilon}_r \vec{E})}{\partial t} = i\omega\varepsilon_0\tilde{\varepsilon}_r\vec{E} \quad \text{in which} \quad \tilde{\varepsilon}_r = 1 - \frac{\omega_p^2}{\omega^2 - i\dfrac{\omega}{\tau}},$$

where $\omega_p^2 = \dfrac{Ne^2}{m\varepsilon_0}$.

Separating the real and imaginary parts we obtain

$$\varepsilon_1 = 1 - \frac{\omega_p^2}{\omega^2 + \dfrac{1}{\tau^2}}, \quad \varepsilon_2 = \frac{\omega_p^2}{\omega\tau\left(\omega^2 + \dfrac{1}{\tau^2}\right)}.$$

One can also electrically characterize the medium by identifying the total current to $\vec{J}_T = \tilde{\sigma}\vec{E}$, where $\tilde{\sigma} = \sigma_1 + i\sigma_2$

with $\sigma_1 = \omega\varepsilon_0\varepsilon_2 = \varepsilon_0 \dfrac{\omega_p^2}{\tau\left(\omega^2 + \dfrac{1}{\tau^2}\right)}$ and

$$\sigma_2 = \omega\varepsilon_0\varepsilon_1 = \omega\varepsilon_0\left(1-\frac{\omega_p^2}{\omega^2+\dfrac{1}{\tau^2}}\right).$$

The choice of one or the other of the two (equivalent) quantities is based on practical considerations:

- $\tilde{\sigma}(\omega)$ in the experiments of electrical conductivity as a function of frequency (see Ex. 21).
- $\tilde{\varepsilon}_r(\omega)$ in optical experiments (see below).

Note: (a) We can also obtain the same evolution of $\tilde{\varepsilon}(\omega)$ by finding the stationary solution of the equation of motion of electrons based on their elongation x (and not as a function of their velocity v). The polarization per unit volume is then expressed as $\vec{P} = N\vec{p} = -Ne\vec{x}$. Assimilating the local field acting on an electron to the applied field, the result has to be included in the expression $\vec{D} = \varepsilon_0\tilde{\varepsilon}_r\vec{E} = \varepsilon_0\vec{E} + \vec{P}$. Such an approach was followed in the study of the movement of ions (see Chapter III, Pb. 1).

(b)
- ε_2 (like σ_1) describes the dissipative nature of the medium (damping , Joule effect) and ε_2 is always positive (see technical note of Ex. 29).
- σ_2 (like ε_1) describes the inductive nature of the medium: it can be >0 or <0.

(4) For potassium N = 1.34 × 10^{28} e/m^3 , where ω_p = 6.5 × 10^{15} rad/sec and $\hbar\omega_p = 4.3$ eV.

When $\omega_p \to 0$, $\varepsilon_1 \to -\infty$, $\varepsilon_2 \to \infty$. When $\omega \to \infty$, $\varepsilon_1 \to 1$, $\varepsilon_2 \to 0$.

In addition, $\varepsilon_1 = 0$ when $\omega = \left(\omega_p^2 - \dfrac{1}{\tau^2}\right)^{1/2} \approx \omega_p$ because $\hbar/\tau = 0.25$ eV and $\hbar\omega_p \approx 4.3$ eV.

(5) The corresponding curves are shown in Fig. 56.

(6) Writing the complex dielectric constant as a function of the energy losses, we find:

$$\mathrm{Im}\left(\frac{1}{\tilde{\varepsilon}_r}\right) = (\omega/\tau)\omega_p^2 / [(\omega^2-\omega_p^2)^2 + \omega^2/\tau^2]$$

It is maximal when ω approaches ω_p, a result that can be deduced from $\text{Im}(1/\tilde{\varepsilon}) = \varepsilon_2/(\varepsilon_1^2 + \varepsilon_2^2)$, which is maximum when $\varepsilon_1 \approx 0$ or $\omega = \omega_p$.

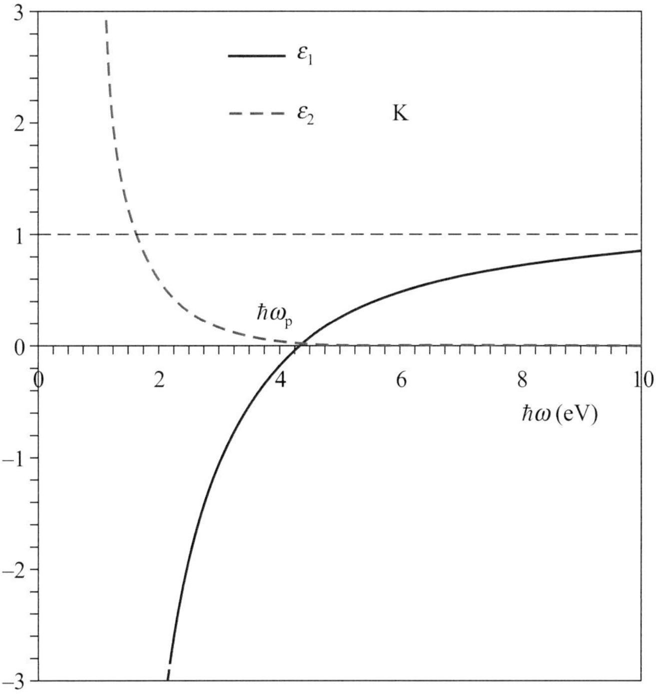

Figure 56

This maximum is equal to $\omega_p\tau$ and the full width at half-maximum ($2\Delta\omega$) can be obtained from

$$\frac{(\omega_p + \Delta\omega)\dfrac{\omega_p^2}{\tau}}{[(\omega_p + \Delta\omega)^2 + \omega_p^2]^2 + \left[\dfrac{\omega_p + \Delta\omega}{\tau}\right]^2} = \frac{\omega_p \tau}{2}$$

which gives $\Delta\omega = \pm\dfrac{1}{2\tau}$.

We note that the most probable energy loss corresponds to $\Delta E = \hbar\omega_p$ and its full width at half-maximum is at $\Delta E\left(\dfrac{1}{2}\right) = \dfrac{\hbar}{\tau}$.

The energy loss function is shown in Fig. 57 for potassium.

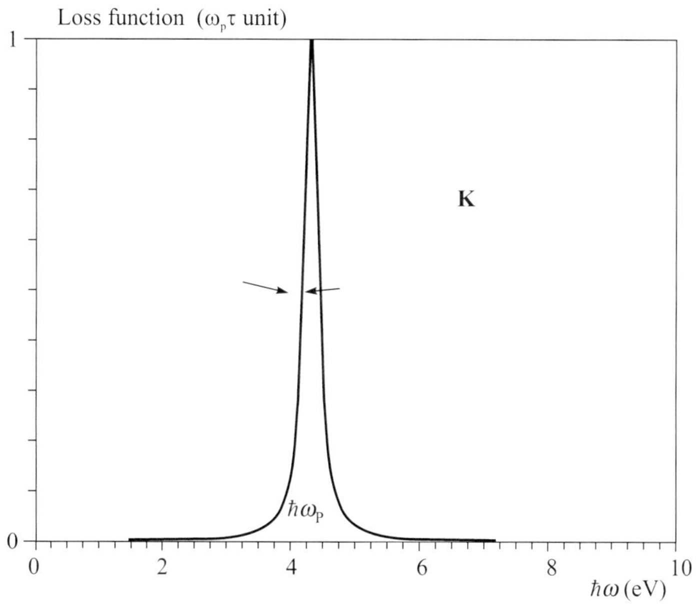

Figure 57 Energy loss function of free electrons.

Obtained by transmission of 28 keV incident electrons through an Al foil, an experimental spectrum is shown in Fig. 58 (from J. Cazaux, *Thesis*, Paris 1970). One may observe that incident electrons may experience successive plasmon losses (each equals to $\Delta E(0°) \sim 15$ eV for Al) when the film is relatively thick. The parabolic dependence as a function of the scattering angle, θ, obeys

$[\Delta E(\theta°) - \Delta E(0°)]/2E° \sim \alpha\theta^2$

Note also that there are elastically transmitted electrons (0 loss line) and the corresponding lines have the energetic width of the electron beams emitted by the source of incident electrons (see Ex. 28a).

Numerical application:

$\Delta E = 4.3$ eV, $\Delta E(1/2) = 0.25$ eV, $E_F = 2.1$ eV (for potassium)

(7) Starting from Maxwell's equations:

$$\begin{cases} \overrightarrow{curl H} = i\omega\varepsilon_0\tilde{\varepsilon}_r\vec{E} \\ \overrightarrow{curl E} = -i\omega\mu_0\vec{H} \end{cases}$$, we obtain $\nabla^2\vec{E} + \omega^2\varepsilon_0\mu_0\tilde{\varepsilon}_r\vec{E} = 0$.

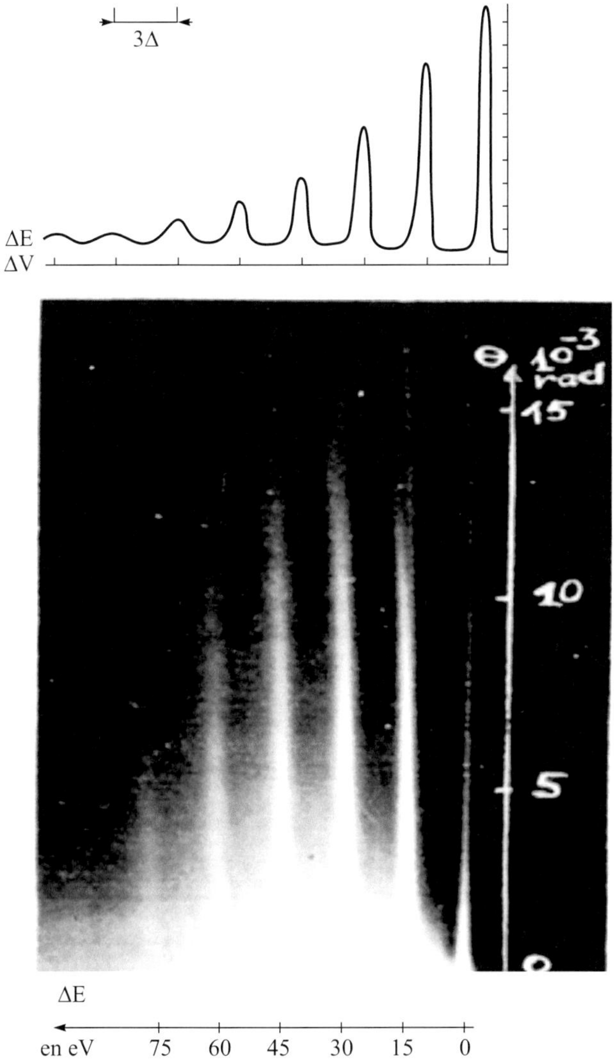

Figure 58 Experimental energy loss dispersion Al film, ~1000 Å thick; E° ~28 keV.

For the transverse excitation of a wave linearly polarized along x and propagating along z:

$$E_x = E_T \exp i(\omega t - \tilde{k}z), \quad \text{where} \quad \tilde{k} = \omega(\varepsilon_0\mu_0\tilde{\varepsilon}_r)^{1/2} \approx \frac{(\omega^2 - \omega_p^2)^{1/2}}{c}$$

because $\tau = \infty$ gives $\varepsilon_2 = 0$.

The graphical evolution of the dispersion relation is shown in Fig. 59 in the interval $\omega_n \le \omega \le \infty$.

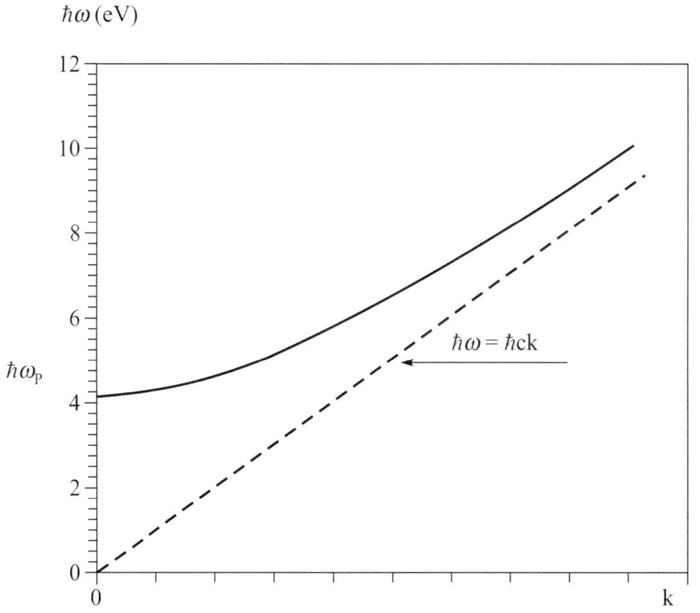

Figure 59

The phase velocity of the wave is

$$v_p = \frac{1}{(\varepsilon_0 \varepsilon_r \mu_0)^{1/2}} = \frac{c}{\left(1 - \frac{\omega_p^2}{\omega^2}\right)^{1/2}}.$$

It is greater than the speed of light when $\omega \ge \omega_p$.

In the interval $0 \le \omega \le \omega_p$, the wave vector is purely imaginary, $k = i|k|$, and the corresponding wave $E_x = E_T \exp - |k| z \exp i\omega t$ is an evanescent wave that does not penetrate into the plasma (see further on Ex. 29).

(8) By substitution into one of the two Maxwell equations (see preceding question), we obtain

$$E_x/H_y = (\mu_0 / \varepsilon_0 \tilde{\varepsilon}_r)^{1/2} = Z_p = 120\pi/(1 - \omega_p^2/\omega^2)^{1/2} \text{ (in } \Omega).$$

This wave impedance is purely imaginary when $\omega < \omega_p$. It is real and greater than the vacuum wave impedance

$$Z_0 = (\mu_0/\varepsilon_0)^{1/2} = 120\pi \ (\Omega), \text{ when } \omega > \omega_p$$

$$H_y = \left(\frac{\varepsilon_0 \tilde{\varepsilon}_r}{\mu_0}\right)^{1/2} \bar{E}_T \exp i(\omega t - \tilde{k}z)$$

(9) In the vacuum, the incident wave is of the form:

$E_x = \bar{E}_i \exp i(\omega t - k_0 z)$ $H_y = \dfrac{\bar{E}_i}{Z_0} \exp i(\omega t - k_0 z)$, where

$k_0 = \omega(\varepsilon_0 \mu_0)^{1/2}$

The reflected wave is

$$E_x = \bar{E}_r \exp i(\omega t + k_0 z) \qquad H_y = -\frac{E_r}{Z_0} \exp i(\omega t + k_0 z)$$

The continuity of the tangential components of the electric and magnetic field result in the following two relations (in $z = 0$):

$E_i + E_r = \bar{E}_t$ and $\bar{H}_i + \bar{H}_r = \bar{H}_t$

Then one obtains: $\dfrac{1}{Z_0}(\bar{E}_i + \bar{E}_r) = \dfrac{\bar{E}_t}{Z_p}$ and

$$r = \frac{\bar{E}_r}{\bar{E}_i} = \frac{Z_p - Z_0}{Z_p + Z_0} = \frac{1 - (\tilde{\varepsilon}_r)^{1/2}}{1 + (\tilde{\varepsilon}_r)^{1/2}}$$

(10) $R = r \cdot r^x = 1$ for $\sqrt{\tilde{\varepsilon}_r}$ purely imaginary, that is to say for $\omega \le \omega_p$.

$$R = \left[\frac{1 - (\tilde{\varepsilon}_r)^{1/2}}{1 + (\tilde{\varepsilon}_r)^{1/2}}\right]^2 \text{ for } \omega \ge \omega_p$$

When $R = \dfrac{1}{m}$, $r = \dfrac{1}{\sqrt{m}}$ and $\sqrt{\varepsilon_r} = \dfrac{(m)^{1/2} - 1}{(m)^{1/2} + 1} = \left(1 - \dfrac{\omega_p^2}{\omega^2}\right)^{1/2}$

from which $\dfrac{\omega}{\omega_p} = \dfrac{(m)^{1/2} + 1}{2m^{1/4}}$; for $m = 16$: $\dfrac{\omega}{\omega_p} = 1.25$, which leads to an abrupt fall when ω exceeds ω_p.

The evolution of R is shown in Fig. 60.

The plasma frequency corresponds to the limit between the region of total reflection and total transparency to electromagnetic waves. In the alkali metals, the electron density N is such that the corresponding wavelength is typically in the UV. This explains the transparency of these metals in the far UV and the total reflection to visible light, resulting in metallic aspects of these metals, when they are not oxidized. Here $\lambda_0 = \dfrac{2\pi c}{\omega_p} = 2900$ Å.

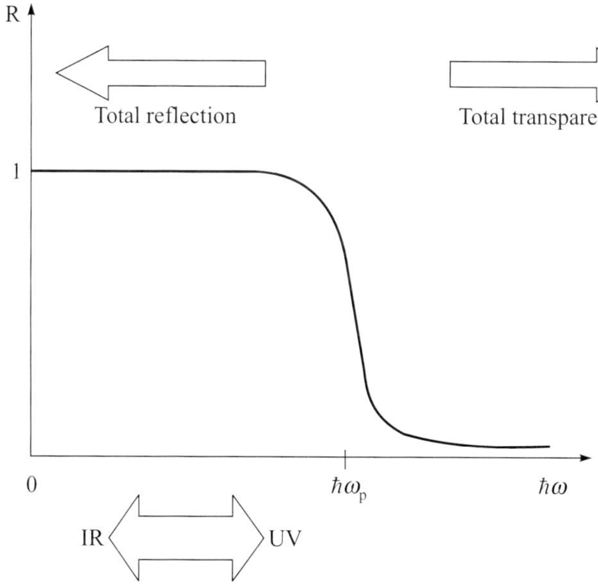

Figure 60

(11) The results in (9) and (10) could be deduced immediately from the optical reflection coefficient at normal incidence: $R = \dfrac{1-\tilde{N}}{1+\tilde{N}}$ because the complex index $\tilde{N}$ of a medium can be considered as the ratio between the phase velocity of the wave in the vacuum and phase velocity of the wave in the medium of interest (see also Ex. 29).

$$\tilde{N} = \frac{c}{v_p} = (\tilde{\varepsilon}_r)^{1/2} \text{ or } r = \frac{1-(\tilde{\varepsilon}_r)^{1/2}}{1+(\tilde{\varepsilon}_r)^{1/2}} \text{ and } \varepsilon_1 = n^2 - k^2,\ \varepsilon_2 = 2nk.$$

Comments

(a) On $\tilde{\varepsilon}(\omega)$ and on the Kramers–Kronig relations:

Very generally, the complex dielectric constant of a given body, regardless of the nature of this body (solid or liquid, metal or insulator, elemental or a compound), is such that its real part and imaginary part are not independent but obey the dispersion relation known as the Kramers and Kronig relations:

$$\varepsilon_1(\omega)-1=\frac{2}{\pi}\int_0^\infty \frac{\omega'\varepsilon_2(\omega')d\omega'}{\omega'^2-\omega^2} \text{ and } \varepsilon_2(\omega)=-\frac{2\omega}{\pi}\int_0^\infty \frac{\varepsilon_1(\omega')d\omega'}{\omega'^2-\omega^2}.$$

Their derivation is based only causality and linearity, here between $\vec{D}$ and $\vec{E}$, that is to say that D (or P) is the consequence of E and $\vec{D}=\varepsilon_0\tilde{\varepsilon}\vec{E}$ (or $\vec{P}=\tilde{\chi}\vec{E}$). They allow the calculation of the dispersion, $\varepsilon_1(\omega)$, at any frequency ω from the evolution of the absorption ε_2 for the entire spectrum of frequencies (theoretically zero to infinity) and reciprocally. We can deduce, in particular, that a medium which would be non-absorbent everywhere would also not be dispersive: If $\varepsilon_2 = 0$ $(0 \le \Omega \le \infty) \rightarrow \varepsilon_1 = 1$ $(0 \le \omega \le \infty)$, this medium can only be a vacuum. Any other medium, characterized by a dielectric function, will have a response that will depend on the frequency, and its optical index will therefore depend on the wavelength of the electromagnetic waves. The decomposition of the white light by transparent materials in the visible thus finds its justification (see also Chapter V, Ex. 26) in the name "dispersion" relations.

These relations have a direct impact on both real and imaginary complex electrical conductivity (see Pb. 3) but also on those with a complex index $\tilde{N}=n-ik$.

Also, note an evidence: The usual name "dielectric constant" is incorrect, ε changes with frequency and other strains, so that "dielectric function" would be better.

(b) On reflectivity and dielectric constant:

To obtain the experimental evolution of $\tilde{\varepsilon}(\omega)$ in the UV for a given sample, one measures the reflection coefficient (typically at normal incidence) in the widest possible spectral interval. Using certain extrapolations, one can then estimate numerically the phase shift of the reflected wave θ starting from $\theta(\omega)=-\frac{\omega}{\pi}\int_0^\infty \frac{\mathrm{Log}R(\omega')d\omega'}{\omega'^2-\omega^2}$, an expression resulting from the Kramers–Kronig relations. One can then deduce the values of ε_1 and ε_2 in the spectral range considered. This method, which allows the identification of the electron interband transitions in semiconductors (see Chapter V, Pb. 6) has been extensively developed since the use of intense and continuous electromagnetic polarized beams issued from synchrotron radiation sources. It later led to the construction all around the world of synchrotrons for the sole purpose of using such UV and X-sources for carrying out experiments in atomic, molecular physics, and condensed matter such as in crystallography (see Chapter I, Section 2).

(c) On plasmons:

If we consider an electron gas of density N surrounded by a lattice of immobile ions (to ensure the electrical neutrality of the system), and if we move a block of these electrons by a distance x, they will return to equilibrium via oscillations that must satisfy the local electromagnetic equations, in particular

$$\nabla \times H = j + \frac{\partial \vec{D}}{\partial t} = i\omega\tilde{\varepsilon}(\omega)\bar{E}\,.$$

As there is no magnetic excitation in this problem, we can deduce that the free frequency of oscillations corresponds to the zero(s) of the dielectric constant, $\tilde{\varepsilon}(\omega)=0$. If they concern free electrons whose behavior can be described by $\tilde{\varepsilon}(\omega)=1-(\omega_p^2/\omega^2)$, where the resonant frequency corresponds to $\omega=\omega_p=(Ne^2/m\varepsilon_0)^{1/2}$, the plasma frequency.

To obtain experimentally these zero(s), one can use electromagnetic waves that induce a transverse excitation with the response function given by $\tilde{\varepsilon}(\omega)$, with dissipative part $\mathrm{Im}\,\tilde{\varepsilon}$ or $\varepsilon_2(\omega)$. One can also use incident charged particles (electrons, for example) that induce longitudinal excitations. The response function is then represented by $1/\tilde{\varepsilon}(\omega)$, with dissipative part given by $\mathrm{Im}\,\tilde{\varepsilon}^{-1}$, which is maximal when $\varepsilon_1=0$.

For the alkali and simple metals of the Al, Mg, and Be type, the evolution of $\tilde{\varepsilon}(\omega)$ is well described by the Drude model as it is used in the present problem. In particular, the value of ω for which $\varepsilon_1=0$ (and therefore for which energy loss is maximum) corresponds to $\omega_p=(Ne^2/m\varepsilon_0)^{1/2}$.

For other metals with more complicated band structure (Ag, Au, W, etc.) as well as for semiconductors and insulators, the evolution of $\varepsilon_2(\omega)$ [and therefore that of $\varepsilon_1(\omega)$ obtained from the Kramers and Kronig relations, see above] does not follow this model and the resonant frequency to which ε_1 (ω) = 0 does not necessarily match $(Ne^2/m\varepsilon_0)^{1/2}$. Thus the corresponding loss function does not present a sharp "plasmon" peak and it would be better to speak of "collective" excitation region for the 3–40 eV interval even if the word "plasmon" is kept here. The case of graphite (Chapter V, Pb. 9) and some of the semiconductors and insulators (Chapter V, Ex. 25) are developed in Chapter V. It should be noted that the resonance condition $\varepsilon_1 = 0$, is also that of polaritons in a different spectral domain (Chapter III, Pb. 1).

From the experimental point of view ε_1 (ω) and ε_2 (ω) may also be deduced from the spectrum of energy losses induced by monochromatic incident electrons. The electron energy loss spectrum allows, after eliminating losses due to multiple processes with the excitation of surface plasmons, to obtain $\mathrm{Im}[\tilde{\varepsilon}^{-1}]$ in the interval between 3 eV and 40 eV. The Kramers and Kronig relations, also apply to $\tilde{\varepsilon}^{-1}$ so it is easy to deduce $\mathrm{Re}[\tilde{\varepsilon}^{-1}]$ from the imaginary part of the inverse dielectric constant, i.e., the energy loss function, to obtain $\varepsilon_1(\omega)$ and $\varepsilon_2(\omega)$ in the same energetic interval.

The two methods (reflectivity and energy loss) generally give comparable results. If the energy loss technique is potentially less accurate, because of the energy width of the incident electronic beams (see Ex. 28a), it is advantageous in that it allows the study of plasmon dispersion: $\omega_\mathrm{p}(\vec{q}) = \omega_\mathrm{p}(0) + \alpha q^2 + \cdots$ by simple measurement of the position of the energy losses as a function of diffusion angle θ, as seen in Fig. 58.

Further information can be found in the following:

- on optical methods in *Optical properties and band structure of semi-conductors,* DL Greenaway, and G. Harbeke. Pergamon (1970);
- on energy losses, in *Springer Tracts in Modern Physics* **38** (1965) p. 84. H. Reather, and **54** (1970) p. 78 Daniels J et al.
- see also solutions and comments of Chapter V, Ex. 25 and Pb. 5; for technical note on $\varepsilon_1 + i\varepsilon_2$ or $\varepsilon_1 - i\varepsilon_2$, see Ex. 29.

Problem 8: Dispersion of surface plasmons

Consider two media separated by the xOy plane: (1) $z > 0$ is the vacuum(ε_r = 1) and (2) $z < 0$ is characterized by a complex relative dielectric constant $\tilde{\varepsilon}_r$ and $\mu = \mu_0$ everywhere.

(a) Starting from the propagation equation in the sine regime (ω), find the expression for the modulus of the wave vector of an electromagnetic wave in medium (1), $\vec{k}(1)$, and in medium (2), $\vec{k}(2)$, as a function of ε and ω.

(b) The electric field and the wave vector are, in both medium, fully contained in the plane xOz.

What are the two relations imposed by the boundary conditions on the xOy plane on $\vec{E}_0$ and $\vec{H}$ between, $k_z(1)$, k (1), $k_z(2)$ and between k_x (1) and k (2)?

(c) Deduce expressions of k_x^2, $k_z^2(1)$, and $k_z^2(2)$ as a function of ω, $\tilde{\varepsilon}_r$, and c (speed of the light in the vacuum), having established that $k_x(1) = k_x(2)$.

(d) The medium (2) consists of a plasma of free electrons. Its relative dielectric constant is of the form $\varepsilon_r = 1 - \frac{\omega_p^2}{\omega_2}$ (see previous problem) and is therefore likely to vary between $-\infty$ and 1. Study the variations of the different components of $\vec{k}(i)$ in this interval. Deduce that a wave is likely to propagate to the plasma–vacuum interface with a dispersion relation of the form

$$\omega^2 = \frac{\omega_p^2}{2} + k_x^2 c^2 - \left(k_x^4 c^4 + \frac{\omega_p^2}{4} \right)^{1/2}$$

Comment on this result.

Find the expression of the wave when $\varepsilon_r = -2$.

(e) Plot the dispersion curve of the surface wave by specifying the asymptotic values $k_x \to 0$ and $k_x \to \infty$.

Numerical application: What is the electrostatic limit ($c = \infty$) of the plasmon energy $\hbar\omega_s$ propagating at the aluminum/vacuum interface with $\hbar\omega_p(\text{Al}) \approx 15$ eV ?

What is the electrostatic energy $\hbar\omega_s$ of the plasmon propagating at the aluminum/magnesium interface $\hbar\omega_p(\text{Mg}) \approx 10.5$ eV?

To answer this last question, it is necessary to first find the expression for the energy of a plasmon propagating at the interface between two plasmas, starting from the results obtained above.

Solution:

(a) Starting from Maxwell's equations: $\overrightarrow{curl\vec{H}} = \dfrac{\partial(\varepsilon_0\tilde{\varepsilon}_r\vec{E})}{\partial t}$ (see Pb. 5) and $\overrightarrow{curl\vec{E}} = \dfrac{\partial\vec{B}}{\partial t}$, one obtains the wave equation $\nabla^2\vec{E} - \varepsilon_0\mu_0\tilde{\varepsilon}_r\dfrac{\partial^2\vec{E}}{\partial t^2} = 0$ which yields a wave of form $\vec{E} = \vec{E}_0 \exp i(\vec{k}\cdot\vec{r} - \omega t)$ when $-k^2 + \omega^2\varepsilon_0\mu_0\tilde{\varepsilon}_r = 0$.

If the wave vector $\vec{k}$ is contained in the plane xOz, we deduce that:

(i) in vacuum (1): $(\tilde{\varepsilon}_r = 1)$: $k_x^2(1) + k_z^2(1) = k^2(1) = \omega^2\varepsilon_0\mu_0 = \dfrac{\omega^2}{c^2}$

(ii) in a medium (2) : $k_x^2(2) + k_z^2(2) = k^2(2) = \omega^2\varepsilon_0\mu_0\tilde{\varepsilon}_r = \dfrac{\omega^2}{c^2}\tilde{\varepsilon}_r$

(b) The magnetic excitations have only a component in y and are related to the corresponding electric field by

$$\vec{k}(1)\times\vec{E}(1) = \omega\mu\vec{H}(1) \qquad\qquad \vec{k}(2)\times\vec{E}(2) = \omega\mu\vec{H}(2)$$

(classic results that can be easily deduced from the Maxwell–Faraday equation taking into account the form of $\vec{E}$)

The boundary conditions in the plane $z = 0$ imply:

(i) The continuity of the tangential component of the electric field at the origin: $[E_x(1)]_0 = [E_x(2)]_0$ or $E(1)\cdot\dfrac{k_z(1)}{k(1)} = E(2)\cdot\dfrac{k_z(2)}{k(2)}$ (see Fig. 61)

(ii) The continuity of the tangential component of the magnetic excitation at the origin gives $[H_y(1)]_0 = [H_y(2)]_0$ from which we find $k(1)\cdot E(1) = k(2)\cdot E(2)$.

By dividing these two equalities term by term, we find $\dfrac{k_z(1)}{k^2(1)} = \dfrac{k_z(2)}{k^2(2)}$.

(iii) Finally, the components E_x or H_y must be continuous at every point in the xOy plane for every x. Thus for instance, one obtains $H_y(1)\exp i[k_x(1)x-\omega t]=H_y(2)\exp i[k_x(2)x-\omega t]$ or $k_x(1) = k_x(2)$.

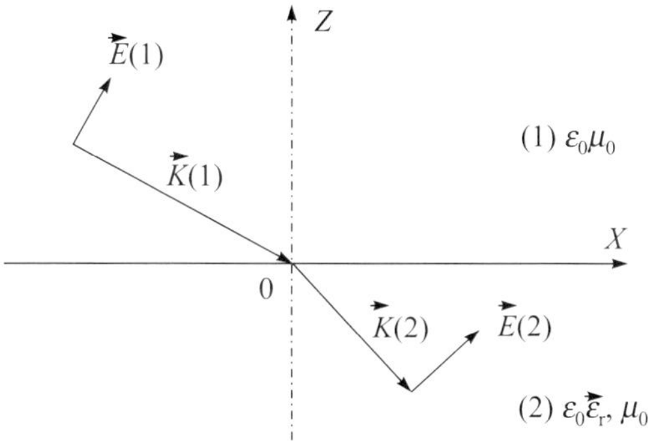

Figure 61

(c) Combining the above equalities with those established in (a), we deduce the following expressions:

- $k_x^2 = k_x^2(1) = k_x^2(2) = \dfrac{\omega^2}{c^2}\dfrac{\tilde{\varepsilon}_r}{\tilde{\varepsilon}_r + 1}$
- $k_z^2(1) = \dfrac{\omega^2}{c^2}\dfrac{1}{\tilde{\varepsilon}_r + 1}$
- $k_z^2(2) = \dfrac{\omega^2}{c^2}\dfrac{\tilde{\varepsilon}_r}{\tilde{\varepsilon}_r + 1}.$

(d) Neglecting the damping term, the evolution of the dielectric constant of a plasma of free electrons, shown in Fig. 56 (Pb. 7), obeys the relation $\tilde{\varepsilon}_r = 1 - \dfrac{\omega_p^2}{\omega^2}$ from which the following table with the variations of the different components of $\vec{k}$ can be constructed. From the table, three frequency domains can be distinguished:

ω	0		$\frac{\omega_p}{\sqrt{3}}$		$\frac{\omega_p}{\sqrt{2}}$			ω_p		$+\infty$
ε_r	$-\infty$	$\nearrow$	-2	$\nearrow$	-1		$\nearrow$	0	$\nearrow$	1
k_x^2	0	≥ 0 $\nearrow$	$\frac{2\omega^2}{c^2}$	$\nearrow$	$+\infty$	$-\infty$	≤ 0 $\nearrow$	0	$\nearrow$ ≥ 0	$\frac{\omega^2}{2c^2}$
$k_z^2(1)$	0	$\searrow$ ≤ 0	$-\frac{\omega^2}{c^2}$	$\searrow$	$-\infty$	$+\infty$	$\searrow$ ≥ 0	$\frac{\omega^2}{c^2}$	≥ 0 $\searrow$	0
$k_z^2(2)$	$-\infty$	$\nearrow$ <0	$-\frac{4\omega^2}{c^2}$	$\searrow$ <0	$-\infty$	$+\infty$	$\searrow$ >0	0	$\nearrow$ >0	$\frac{\omega^2}{2c^2}$

(i) $0<\omega<\frac{\omega_p}{\sqrt{2}}$: k_x is real and $k_z(1)$ and $k_z(2)$ are purely imaginary; the wave is guided along the surface from which it propagates without attenuation (if $\tau = \varepsilon_2 = 0$) even though it is exponentially damped when it deviates in the xOy plane.

The resulting wave has the form:

$$\vec{E}(1)=\vec{E}_0(1)\exp-|k_z(1)|z\exp i(k_x x-\omega t)$$

$$\vec{E}(2)=\vec{E}_0(2)\exp-|k_z(2)|z\exp i(k_x x-\omega t)$$

For the particular case when $\varepsilon_r=-2$ (that is to say $\omega=\frac{\omega_p}{\sqrt{3}}$), one obtains

$$z\geq 0:\vec{E}(1)=\vec{E}_0(1)\exp-\frac{\omega z}{c}\exp i\omega\left(\frac{\sqrt{2}x}{c}-t\right)$$

$$z\leq 0:\vec{E}(2)=\vec{E}_0(2)\exp\frac{2\omega z}{c}\exp i\omega\left(\frac{\sqrt{2}x}{c}-t\right)$$

(ii) $\frac{\omega_p}{\sqrt{2}}\leq\omega\leq\omega_p$: $k_z(1)$ and $k_z(2)$ are real, k_x is purely imaginary; it is the inverse of the preceding situation, the waves propagate without attenuation along the Oz axis but are exponentially damped when they move transversally along the Oz axis.

$$\vec{E}(1)=\vec{E}_0(1)\exp\pm(|k_x|x)\cdot\exp i(k_z z-\omega t)\begin{pmatrix}+\text{si } x<0\\ -\text{si } x<0\end{pmatrix}$$

(iii) $\omega > \omega_p$: All the wave vector components are real. It is the transparent region of the plasma (see Pb. 7).

The dispersion relation of the surface wave $\left(\omega < \frac{\omega_p}{\sqrt{2}}\right)$ is obtained by substituting ε_r into the expression $1 - \frac{\omega_p^2}{\omega^2}$ in the relation giving k_x^2. We find the proposed expression $\omega^2 = \frac{\omega_p^2}{2} + k_x^2 c^2 - \left(k_x^4 c^4 + \frac{\omega_p^4}{4}\right)^{1/2}$, in which the + sign that appears mathematically before the radical is omitted because it places ω outside the interval studied ($\omega > \omega_p$).

(e) When $k_x c << \frac{\omega_p}{\sqrt{2}}$, the dispersion relation is identified with that of an electromagnetic wave in a vacuum: $\omega = ck_x$ (straight "light" line).
When $k_x c \to \infty$ (electrostatic approximation $c \to \infty$ or $\lambda \to 0$), ω tends to $\omega_p/\sqrt{2}$. The shape of the corresponding curve is shown in Fig. 62.

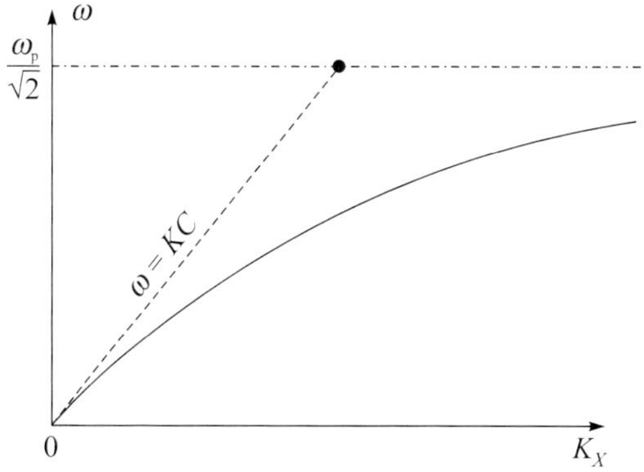

Figure 62

(f) The plasmon energy that propagates at the interface between aluminum and the vacuum is $\hbar\omega_s = \hbar\omega_p/\sqrt{2} = 10.6$ eV.

To obtain (without having to redo all the calculations) the plasmon energy that propagates at the interface between two metals, it is sufficient to replace the ratio of the relative dielectric constants of the plasma to the vacuum by the relative dielectric constants of one medium to another in the expression for k_x:

$$k_x^2 = \frac{\omega^2}{c^2} \frac{\varepsilon(2)/\varepsilon(1)}{[\varepsilon(2)/\varepsilon(1)]+1}.$$

The electrostatic limit corresponds to $\varepsilon(2)/\varepsilon(1) = -1$, which gives $\hbar\omega_s(1,2) = \hbar\left[\dfrac{\omega_p^2(1)+\omega_p^2(2)}{2}\right]^{1/2}$; $\hbar\omega_s(1,2) = 13\,\text{eV}$ for the plasmon at the Al/Mg interface.

Comments: Surface plasmons

One can excite surface plasmons by the following methods:

(a) Optically, for instance, by placing a thin film of silver ($\hbar\omega_s \approx 3.6\,\text{eV}$) very close and parallel to the hypotenuse of a prism with total reflection (see Fig. 63 and A. Otto, *Zeitschrift für Physik* **216**, 1968, 398) who coined this method, frustrated total reflection (FTR).

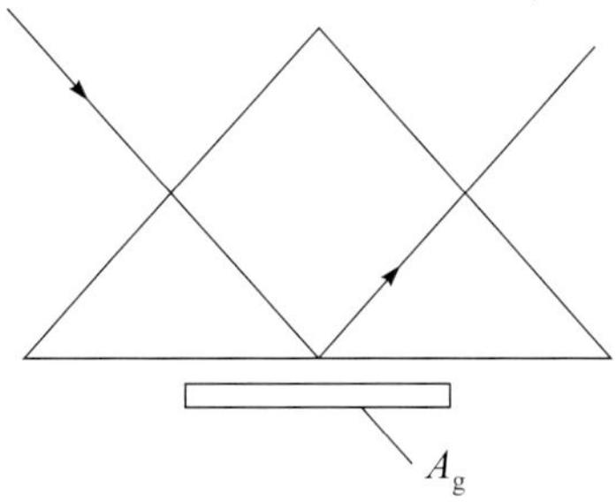

Figure 63

(b) Using electron beams by reflection on metals or by transmission through thin films. The energy losses of the incident electrons correspond to multiple integers of $\hbar\omega_s$ (surface plasmons) and $\hbar\omega_p$ (volume plasmons), and their intensity decreases as a function of the diffusion angle θ, following a θ^{-3} law for surface losses and θ^{-2} for volumetric losses. Using angular measurements, one can determine the experimental dispersion of these plasmons. The asymptotic value is very sensitive to topography and cleanliness of the interface. The experimental value of surface plasmons of aluminum goes from $\hbar\omega_s = 10.6\,\text{eV}$ to $\hbar\omega_s = \hbar\omega_p(1+\varepsilon_{\text{Al}_2\text{O}_3})^{1/2} = 6.5$ to 7 eV when the surface is coated with a thin film of aluminum. For non-planar interfaces, the results will also be modified.

Problem 9: Metallic superconductors, London equations, and the Meissner effect

In the simplest approximation superconducting properties can be explained by considering that N conduction electrons propagate without friction (perfect conductor: $\tau = \infty$) under the action of an electric field. Although incorrect, this approach predicts some of the unusual properties of superconductors.

(1) Write the equation of motion for one of these electrons ($-e$, m). Infer the relationship between $\partial\vec{j}/\partial t$ and $\vec{E}$ (where $\vec{j}$ is the current density) for a perfect conducting medium.

(2) Starting from Maxwell's equations (established in vacuum μ_0 and neglecting the displacement current density $\frac{\partial D}{\partial t}$), show that the variation of $\vec{B}$ as a function of time obeys the relationship:

$$\nabla^2\left(\frac{\partial\vec{B}}{\partial t}\right) - \frac{1}{\lambda^2}\frac{\partial\vec{B}}{\partial t} = 0 \qquad (\nabla^2: \text{Laplacian}) \quad (1)$$

Find λ and evaluate it numerically for the case of lead ($N = 6.72 \times 10^{28}\ \text{e/m}^3$).

(3) A perfect conducting material occupies the half-space where $x > 0$ and is subject to an external magnetic field, H_{ext}, uniform with a single component H_z that grows slowly over time from zero. When H_{ext} reaches a critical value H_c, the material becomes normal.

Starting from equation (1), determine the evolution of $\partial B_z/\partial t$ as a function of x into the perfect conducting material.

Deduce that at a point M inside the perfect conductor, where $\text{OM} = x_M \gg \lambda$, the magnetic induction $B(x_M)$ remains zero if it was initially zero. Show the graph of $B = f(H)$ of the first magnetization. Show that if H subsequently decreases from H_1 to $-H_1$ (where $|H_1| > H_c$), the material is characterized by a hysteresis curve. Sketch this hysteresis curve $B = f(H)$.

(4) In fact, superconductors do not have a hysteretic magnetic behavior but reversibly following the first magnetization curve established above, known as the Meissner effect. To account for this phenomenon, London postulated that the superconducting medium is characterized by the relationship

$$curl\,\vec{j} = -(1/\mu_0\lambda_L^2)\vec{B} \tag{2}$$

From Maxwell's equations [same assumptions as in (2)] show that B follows a differential equation similar to (1), which was established previously for $\partial B/\partial t$. Deduce the evolution of B_z as a function of x.

(5) Derive the London equation with respect to time and show that the result is compatible with that established in (1) for a perfectly conducting material. Deduce the expression of λ_L.

In fact, superconductivity of lead is explained by associating electrons in pairs, known as Cooper pairs with mass m^x. In this case give a more realistic expression for λ_L.

(6) Verify that, when subject to an external magnetic field similar to that of the Question (3), the superconducting material describes the characteristic cycle of the Meissner effect [mentioned in (4)]. What is the value of the magnetic susceptibility χ, that characterizes this behavior?

(7) From the evolution of B [established in (4)], deduce the corresponding evolution of the vector density of current, $j(x)$. Show that it exerts pressure on the surface of the superconductor. What is its direction?

The critical magnetic induction of lead is B_c = 0.08 T at 0 K. What is the numerical value of the critical current density, j_c? Give also the order of magnitude of the maximal pressure [to simplify, keep the numerical value of λ found in (2)]. ($\mu 0$, e, m)

Solution:

(1) $m\dfrac{\partial\vec{v}}{\partial t} = -e\vec{E}; \vec{j} = -Ne\vec{v}$ from which $\dfrac{\partial\vec{j}}{\partial t} = -Ne\dfrac{\partial\vec{v}}{\partial t} = \dfrac{Ne^2}{m}\vec{E}$

(2) $\overrightarrow{curl\ (\mu_0 \vec{H})} = \mu_0 \vec{j}, \quad \overrightarrow{curl\ \frac{\partial \vec{B}}{\partial t}} = \mu_0 \frac{\partial \vec{j}}{\partial t} = \frac{\mu_0 N e^2}{m} \vec{E} = \frac{1}{\lambda^2} \vec{E}$

and $\overrightarrow{curl \vec{E}} = -\frac{\partial \vec{B}}{\partial t}$, from which

$\overrightarrow{curl\, curl \left(\frac{\partial \vec{B}}{\partial t} \right)} = \frac{1}{\lambda^2} \overrightarrow{curl \vec{E}} = -\frac{1}{\lambda^2} \left(\frac{\partial \vec{B}}{\partial t} \right)$, which leads to Eq. (1)

$div\, \vec{B} = 0$ because $\lambda = \left(\frac{m}{N e^2 \mu_0} \right)^{1/2} = 205$ Å.

(3) The only physical solution for $\partial B_z / \partial t$ is

$$\frac{\partial B_z(x)}{\partial t} = \frac{\partial B_z(0)}{\partial t} e^{-x/\lambda}$$

For a point M such that $x_M >> \lambda$, in the interior of a perfect conductor, the variations of B_z as a function of time are zero: if at the initial instant $B_z(x_M)$ is zero, it will remain zero. On the other hand, if $B_z(x_M) = B_1 \neq 0$, this value will not change and the flux will be trapped. The curve of the first magnetization and the deduced hysteresis are shown in the left side of Fig. 64.

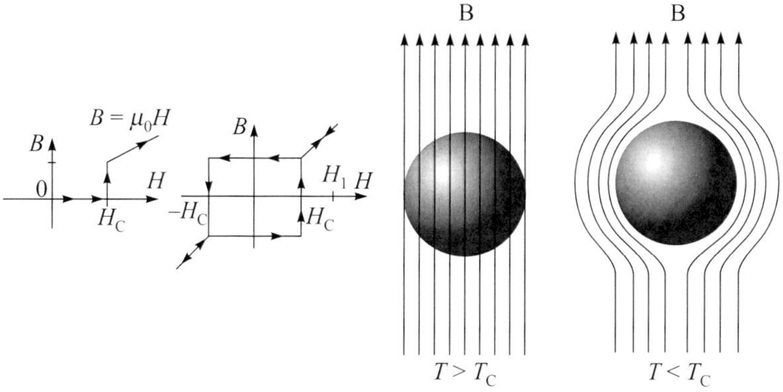

Figure 64

(4) $\overrightarrow{curl \vec{j}} = (1/\mu_0 \lambda_L^2) \vec{B}$ and $\overrightarrow{curl\ \mu_0 \vec{H}} = \mu_0 \vec{j}$

from which we find $\overrightarrow{curl\, curl \vec{B}} = \mu_0 \overrightarrow{curl \vec{j}} = \frac{-\vec{B}}{\lambda_L^2}$.

This result leads to $\nabla^2 \vec{B} - (1/\lambda_L^2)\vec{B} = 0$.
Here, the solution has the form $B_z(x) = B_z(0)e^{-x/\lambda_L}$.

(5) The magnetic induction B does not penetrate into the superconductor and is expelled. Deriving the London equation with respect to time we find

$curl \frac{\partial \vec{j}}{\partial t} = -\frac{1}{\mu_0 \lambda_L^2}\frac{\partial \vec{B}}{\partial t} = \frac{1}{\mu_0 \lambda_L^2} curl\, \vec{E}$, which includes the equality established in (1) where $\lambda_L = \lambda = (m/N_s e^2 \mu_0)^{1/2}$.

If we take into account Cooper pairs with mass m^*, charge $-2e$ and density $N_s/2$, we find

$\lambda_L = (m^x/2Ne^2\mu_0)^{1/2}$

(6) The left curve in Fig. 64 is now reversible, the flux is expelled as soon as the material becomes superconducting.

$\vec{B} = \mu_0(\vec{H} + \vec{M}) = \mu_0(1+\chi)\vec{H}$

In the superconducting state $B = 0$ and $\chi = -1$; it is a perfect diamagnetic material.

(7) $curl\, \vec{B} = \mu_0 \vec{j}$ or $B_z = B_z^0 \exp-(x/\lambda)$ from which

$$j_y^{(x)} = \frac{1}{\mu_0}\left(-\frac{\partial B_z}{\partial x}\right) = \frac{B_z^0}{\mu_0 \lambda}\exp-(x/\lambda).$$

The exponential damping of the current is correlated to that of B in the superconducting material.

In the volume dv, situated near the surface of the superconductor, the force exerted on the Laplacian is $d\vec{F} = \vec{j}dv \times \vec{B}$.

Per unit of orthogonal area at xx': $F\int_0^\infty \vec{j}(x) \times \vec{B}(x)\,dx$, and the critical current density J_c (that is to say, that created by the critical field at the surface of the superconductor) corresponds to $j_c = B_c/\mu_0\lambda$. The corresponding pressure in the same direction as Ox is $B_0^2/2\mu_0$.

Numerical application: $J_c = 3.1 \times 10^{12}\ \text{A/m}^2$; $P_c = 2.55 \times 10^3$ Pa.

Comment: Superconductors; Nobel Prize in physics in 1913 and 1987

The purpose of this problem (and the two that follow) is to familiarize the reader with the remarkable physical properties of superconductors (see also Chapter V, Exs. 2b and 23). Superconductivity of Hg was discovered by H. Kamerlingh Onnes

(Nobel Prize in physics, 1913) after the liquefaction of Helium but these properties concern not only the disappearance of electrical resistance but very specific magnetic properties that are referred to the Meissner–Ochsenfeld effect (discovered by Walther Meissner and Robert Ochsenfeld in 1933 and followed by the phenomenological theory of superconductivity by Fritz and Heinz London in 1935) and critical field and current density.

We know the enthusiasm sparked the discovery of high-temperature superconductors, which have a critical temperature higher than liquid nitrogen (J. G. Bednorz and K. A. Müller, Nobel Prize in physics in 1987, following their article in *Zeit Phys* **64**, 1986, 189 on superconductivity of copper oxides of type Y, Ba Cu, O). However, it should be noted that if the effects specified in this exercise (at least in their generality) concern all superconductors, the themes developed in the next two problems relate only to superconductors based on metals and metal alloys. The microscopic mechanisms of high-temperature superconductivity seem related to 2D behavior of the charge carriers in the CuO planes (see Chapter V, Ex. 2a).

Problem 10: Density of Cooper pairs in a metallic superconductor

In a simple model, superconductivity is modeled as made up of two paired electrons obeying Bose–Einstein statistics.

The superconductor therefore consists of two fluids, one formed by normal free electrons with density n_e and the other formed by such pairs of density n_p.

(1) In the normal state, but at very low temperatures (we neglect the dependence of the Fermi energy E_F and assume $T = 0$ K), find the expression for the density of free electrons n_e as a function of $E_F(N)$.

(2) In the superconducting state, the Fermi energy is slightly lower $E_F(S) = E_F(N)(1-\delta)$, where $0 < \delta << 1$.
What becomes of the expression n_0 for n_e? Deduce the expression for the density of pairs as a function of n_0 and δ.

(3) In an independent evaluation (based on Bose statistics), we find that the density of pairs is $n_p = 2.6\left(\frac{mk_B T_c}{\pi\hbar^2}\right)^{3/2}$ where m = mass of electrons and T_c = the critical temperature.

Numerically find n_e and n_p for the case of lead where $T_c = 7.2$ K and $E_F(N) = 6$ eV. Deduce the numerical value of δ. (h, k_B, m, e)

Solution:

(1) $E_F = (\hbar^2/2m)(3\pi^2 n)^{2/3}$ (see Course Summary) from which

$$n_0 = \frac{1}{3\pi^2}\left(\frac{2mE_F(N)}{\hbar^2}\right)^{3/2}$$

(2) When $E_F(N)$ becomes $E_F(N)(1-\delta)$, we have $n_e = n_0(1 - \delta)^{3/2} \approx n_0[1-(3/2\delta)]$, from which (by conservation of electrical charge): $n_p = \frac{1}{2}(n_0 - n_e) = (3/4)\delta \cdot n_0$.

(3) $n_0 = 6.72\times10^{28}\,\text{e/m}^3$; $n_p = 3.45\times10^{23}\,\text{e/m}^3$; $\delta = 6.8\times10^{-6}$.

If the superconductivity of the mixed copper oxides is not yet fully understood, the superconductivity of metal alloys is explained by the BCS theory (J. Bardeen, L. N. Cooper and J. R. Schrieffer, Nobel Prize in physics in 1972). This theory (*Physical Review B* **108**, 1957, 162) is based on the attraction of pairs of electrons through a phonon: an electron in the crystal attracts ions of the network which, in turn, attract a second electron and a Cooper pair is formed. This new state has a lower energy than the normal state and is separated from it by a forbidden band gap (0.3–3 meV). In the present problem, this band gap is $\delta^* E_F \approx 0.04$ meV, an order of magnitude lower than typical BCS superconductors (see Chapter V, Ex. 23).

Problem 11: Dispersion relation of electromagnetic waves in a two-fluid metallic superconductor

In a two-fluid superconductor, the total current density, J, is the sum of the normal current density, J_N, and the superconductor current density J_s: $\vec{J} = \vec{J}_{N+}\vec{J}_s$. $\vec{J}_N = \sigma_0 \vec{E}_0 \cdot \sigma_0$ describes the conductivity of a normal metal with n_e normal electrons and j_s obeys the London equation (in which λ_L is determined starting from the n_s, superconducting electrons per unit volume: $\sqrt{m/n_s e^2 \mu_0}$.

(1) Starting from Maxwell's equations and taking into account the displacement current, show that the dispersion relation of electromagnetic waves in such a superconductor is $k^2 = -i\mu_0\sigma_0\omega - (1/\lambda_L^2) + \omega^2\mu_0\varepsilon_0$.

(2) Compare the contribution of the normal current with that of the superconducting current and determine the frequency domain for which the super-current short-circuits the normal current. Express the result as a function of n_e, n_s, and τ (relaxation time of normal electrons) after having compared λ_L to the skin depth $\delta = \sqrt{2/\mu_0\sigma_0\omega}$.

In fact, the relationship is valid for photons with energy less than the forbidden band width $E_g = k_B T_c$. The critical temperature of lead is $T_c \approx 7$ K. Find the order of magnitude of the frequency limits imposed by the latter and deduce that in any case the displacement current density term is negligible.

Recall that the London equation is $\overrightarrow{curl\ j} = -\vec{B}/\mu_0\lambda_L^2$.

Solution:

(1) $$curl\ (\mu_0 \vec{H}) = \mu_0 j_n + \mu_0 j_s + \mu_0 \frac{\partial D}{\partial t},$$

where $j_n = \sigma_0 E$ such that $curl\ \vec{j}_s = -(1/\lambda_L^2)\vec{B}$ (London).

Taking the curl of the two terms of the equality and taking into account that $curl\ \vec{E} = -\partial\vec{B}/\partial t$, we find

$$-\nabla^2 \vec{B} = -\mu_0\sigma_0 \frac{\partial \vec{B}}{\partial t} - \frac{1}{\lambda_L^2}\vec{B} - \varepsilon_0\mu_0 \frac{\partial^2 \vec{B}}{\partial t^2}$$

By inserting a solution of the form $\vec{B} = \vec{B}_0 \exp i(\omega t - \vec{k}\vec{r})$, we find:

$$+k^2\vec{B} = -i\omega\mu_0\sigma_0 \vec{B} - \frac{1}{\lambda_L^2}\vec{B} + \frac{\omega^2}{c^2}\vec{B}.$$

(2) The right hand side of the equality links respectively the term related to the normal conductivity, the London equation, and the displacement current density.

Taking into account that $\sigma_0 = (n_e e^2 \tau)/m$ and also that $\lambda_L = (m/n_s e^2 \mu_0)^{1/2}$, the normal conduction current will be short-circuited by the superconducting current when $(1/\lambda_L^2) >> \omega\mu_0\sigma_0$, or equivalently $\omega\tau << n_s/n_n$.

In other terms, the superconductor becomes normal when the London length λ_L is smaller than the normal skin depth: $\delta = \sqrt{2/u_0\omega\sigma_0} = \sqrt{2m/\mu_0 n_e e^2 \tau}$.

$k_B T_c = 0.583$ meV, $v = k_B T_c/h = 14.10^{14}$ c/s or $\lambda = 2.1$ mm (where λ is the wavelength of the electromagnetic wave). The displacement current density is always negligible in this spectral region (this justifies the approximations made in Pb. 7) because $\frac{\omega^2}{c^2} = \frac{(2\pi)^2}{\lambda^2(\text{OEM})} << 1/\lambda_L^2$.

Questions

Q.1: Why does freshly prepared sodium shine so brightly?

Q.2: From atmospheric pressure, a sample of sodium is introduced into a ultrahigh vacuum chamber. Does this operation influence the average speed of conduction electrons? If yes, specify the sign and magnitude of the variation of the Fermi level knowing that the compressibility of the metal is of the order 1.6×10^{-10} m^2/N.

Q.3: Comparing the measured value of $\boldsymbol{\sigma}$ for a metal to the expression $\sigma_0 = ne^2\tau/m$, one obtains the mean time of flight τ for free electrons of order 10^{-14} s at ambient temperature. When such a metal is subjected to an electric field of order 100 V/m, the drift velocity of these electrons is such that $v_e = cE\tau/m = 0.1$ m/s and the distance traveled by an election between two collisions will be $\Lambda \approx v_e\tau = 10^{-15}$m or 10^{-5} Å or ~10^{-5} atom size! Where is the error? What is the correct order of magnitude for Λ?

Q.4: $\varepsilon_1 - i\varepsilon_2$ or $\varepsilon_1 + i\varepsilon_2$?

Q.5: In a range of temperature around room temperature, we can write $\frac{\Delta\rho}{\rho} = \alpha\Delta T$. What is the order of magnitude and the sign of α, the temperature coefficient of resistivity ρ of metals?

Q6: To focus X-rays why is not possible to build lenses based on the law of refraction of geometrical optics?

Why do we identify the optical path to the geometric path in Bragg's law?

Q.7: Why do we say superconductors are perfect diamagnetic materials?

Q.8: What is the order of magnitude of the resistivity of a metal such as copper at room temperature? Without recourse to superconductors, how can it be minimized during its elaboration and use?

Q.9: Why do we also call the Kramers–Kronig relationships: dispersion relations?

Q.10: EXAFS allows us to determine the distance between two neighboring atoms but the diffraction experiments (including slow electrons) allow access to the same information. What is the advantage of EXAFS?

Q.11: How is the quantum Hall effect useful for developing standards of resistances?

Q.12: How many atoms are needed for an aggregate so that it acquires a metallic character?

Q.13: What difficulties must be overcome when one is seeking a good electrical conductor and a poor heat conductor?

Q.14: What minimum thickness must a metallic film have to use the value of the bulk resistivity of the solid metal when calculating the resistance?

Q.15: What simple method can be used to evaluate the density of vacancies in a metal?

Q.16: Which problems and exercises in this chapter require knowledge of crystallography?

Q.17: What experimental methods should we implement to obtain, in a metal or a strongly doped semiconductor, the sign and density of carriers, as well as their mobility?

Q.18: An atomic electron is ejected from its initial level by irradiation of electrons or X-rays. Is the emission of a photon the only possible process to restore equilibrium?

Q.19: What is the best resolution which may be achieved in STM: the vertical one or the horizontal one?

Q.20: What is the experimental method to obtain the valence band density of states? Same question for the dispersion curve of the surface states.

Q.21: What is the physical origin of the work function in metals?

ANSWERS AT THE END OF THE BOOK

Chapter V

Band Theory: Other Metals, Semiconductors, and Insulators

Course Summary

1. Introduction

The free electron model can be successfully applied to certain metals. However, it does not account for the electrical properties of semimetals, semiconductors, and insulators because it neglects the potential energy of valence electrons in the periodic potential of the ions of crystal lattice.

2. Band Theory

The general solutions of the Schrödinger equation $\left[\frac{\hbar^2}{2m}\nabla^2 + V\right]\psi = E\psi$, where V takes into account the periodic nature of the lattice, are Bloch waves of form $\psi(\vec{r}) = u(\vec{r})e^{i\vec{k}.\vec{r}}$ in which u(ŕ) like V takes into account the periodicity of the lattice. In 1D, $u(x) = u(x + a)$ because $V(x) = V(x + a)$.

The wave vector $\vec{k}$, therefore, always refers to a spatial orientation, and the components of k are, as for free electrons, quantized (cyclic boundary conditions $k_x = n_x 2\pi/L_x$, $k_y = n_y\ 2i\pi/L_y$, $k_z = n_z\ 2\pi/L_z$) and the integers n_x, n_y, and n_z are good quantum numbers of electrons.

Understanding Solid State Physics: Problems and Solutions
Jacques Cazaux

ISBN 978-981-4267-89-2 (Hardcover), 978-981-4267-90-8 (eBook)
www.panstanford.com

Beyond the general form indicated above, rigorous analytical solutions are impossible to obtain. Depending on the nature of the crystal, various simplifications can be introduced to obtain the dispersion relation $E = f(k)$.

For metals, simplifications result in the nearly free-electron approximation. For a periodic crystalline potential, the perturbation on the propagation of the valence electrons is a small correction. The resulting dispersion relation deviates from the parabolic free electron relation only when the wave vector k approaches the limits of the Brillouin zones where the Bragg conditions are nearly satisfied. When the Bragg conditions are satisfied strictly, the corresponding electronic waves cannot propagate: the amplitude of reflected waves is equal to that of incident waves and results in standing waves. This causes the appearance of a forbidden energy band of width that is proportional to (the Fourier component of) the corresponding potential energy.

In contrast, for insulators and semiconductors, the tight-binding approximationis typically used. It starts from the initial electronic states of isolated atoms and considers the effect of condensing them into the solid state via a perturbation, which is responsible for the appearance of a chemical bond between atoms and their neighbors. Using wave functions satisfying the Bloch theorem, this approximation is also known as the linear combination of atomic orbitals (LCAO). It leads to a dispersion relation of the form:

$$E(k) = E_0 - \alpha - \gamma \sum_j e^{-i\vec{k}\vec{\rho}_j} \tag{1}$$

in which E_0 represents the initial electron energy, α is the energy of the orbital, and γ is the overlap energy between neighboring atoms (which are separated by a distance ρ_j and are generally limited to the nearest neighbors j). α and γ are positive so that α determines the cohesion of the crystal even though the width of the energy band will be proportional to γ: the weaker the coupling between neighbors, the narrower the band.

The determination of the detailed band structure is a complex task that is beyond the scope of this book. Assuming the validity of expressions similar to Equation (1), for example, the goal here is to study the influence of the crystal potential on the Fermi surface and the density of states in order to understand the electronic and optical properties of some important crystals.

Regardless of the calculation method, valence electron dispersion curves, $E = f(k)$ can be represented in either the extended or reduced schemes. Figures 1a and 1b illustrate these two types of representations for electrons obeying the nearly free-electron theory. In particular note that depending on the amplitude of the discontinuities in contact to the Brillouin zones, there can be an overlapping or no overlapping between allowed energy bands.

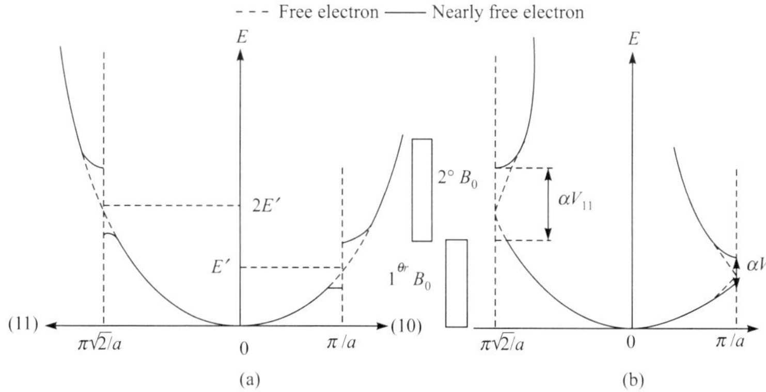

Figure 1

3. Filling of Available States: The Fermi Surface

- Each allowed band can only accept two valence electrons per basis.
- If there is a single electron per basis (monovalent alkali metals, for instance), the filling leaves the first band half-empty: the occupied states and the Fermi level are often very close to those obtained in the theory of free electrons, which justifies *a posteriori* the success of this theory (see Fig. 2).
- If there are two valence electrons in the initial lattice, we can have, in 2D, either a full first band or partially filled first band and a partially filled second band (depending on the amplitude of the discontinuities of the forbidden band; see Fig. 2 and Ex. 5a). Generalizing this reasoning and taking into account that a full band cannot conduct electricity, we can explain the existence of good (metals) and bad (semiconductors, insulators) electric conductors.

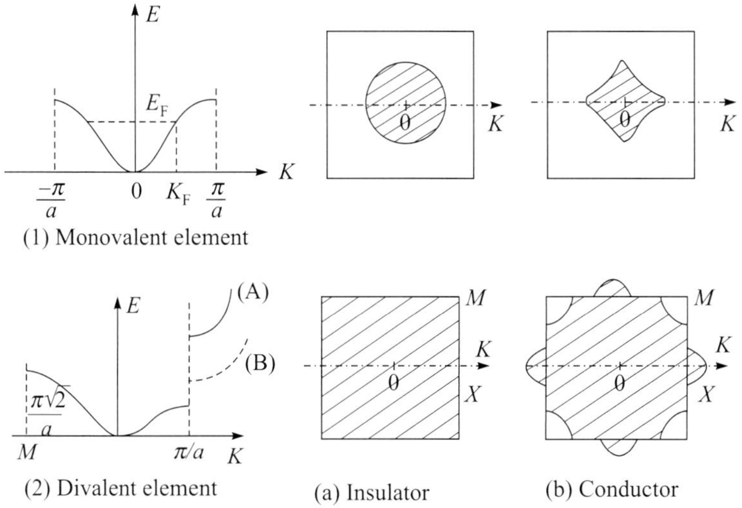

Figure 2

4. Density of states, effective mass, electrons, and holes

- The curvature of the band near the Brillouin zone results in an increase in the density of states at the corresponding energies:

$$g(E) = g(k)\frac{\mathrm{d}k}{\mathrm{d}E} = \frac{g(k)}{\hbar v_g}.$$

To take into account the effects of the periodic potential on the electron dynamics when electrons are subject to an external force $\vec{F}_e$, we assign these electrons an effective mass m^* such that $\frac{1}{m_{ij}^x} = \frac{1}{\hbar^2}\left(\frac{\partial^2 E}{\partial k_i \partial k_j}\right)$.

If $m^* < 0$ (at the Brillouin zone edge, as shown in Fig. 3), we consider the particles to be holes such that $m^* > 0$, $q = e$, instead of considering the nonphysical picture of electrons with a negative mass and a negative charge $(-e)$.

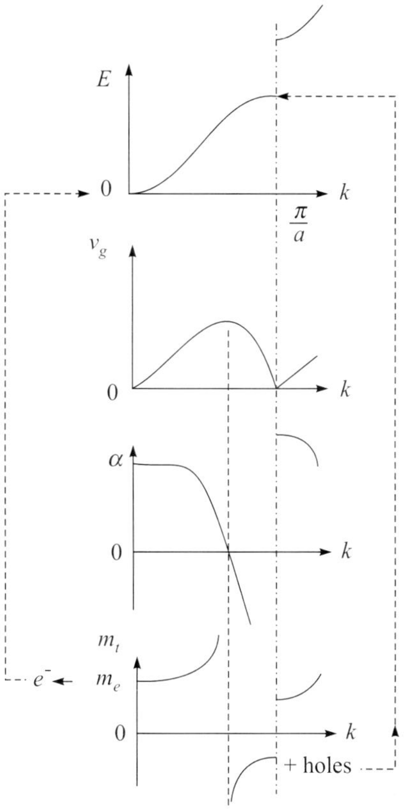

$-e, m$ subject to $\vec{F}_e$ + lattice =
e, m^* (holes) or / $-e, m^*$ (electron) subject to $\vec{F}_e$ only (where $m^* > 0$)

Figure 3

5. Success of Band Theory

- Describe the differences between metals, semimetals, semiconductors, and insulators (see Exs. 5 and 9) and the cohesion of transition metals (Ex. 8)
- Predicts the correct sign of hole conduction in the Hall effect (Ex. 18)
- Accounts for the color of insulators and semiconductors (Exs. 25 and 28)

- Predicts why Ni is ferromagnetic and not Cu (Ex. 7)
- Accounts for the phase changes in alloys (see Ex. 6), etc.

6. Semiconductors (Generalities)

- Semiconductors are materials (most often elements of column IV or binary alloys of the III–V columns) that have room temperature resistivities σ^{-1} intermediate between metals and insulators (10^{-2} Ω-cm < ρ < 10^{9} Ω-cm) and have a forbidden energy band E_g (0.1 eV < E_g< 2 eV).
- Their electric conductivity, σ, increases quickly with temperature and with the addition of impurities n_d: extrinsic semiconductors doped with *n*-type impurities such as As and Sb or *p*-type impurities such as Al, In, and B.
- With two types of carriers: $\vec{j} = \rho_1 \vec{v}_1 + \rho_2 \vec{v}_2$. When one deals with electrons and holes of concentrations n_e and n_h per unit volume, one obtains $\sigma = e(n_e\mu_e + n_h\mu_h)$ from $\vec{v}_e = -\mu_e \vec{E}$ and $\vec{v}_h = \mu_h \vec{E}$
- If n_d is negligible, the semiconductor is intrinsic $n_e = n_h = n_i$.

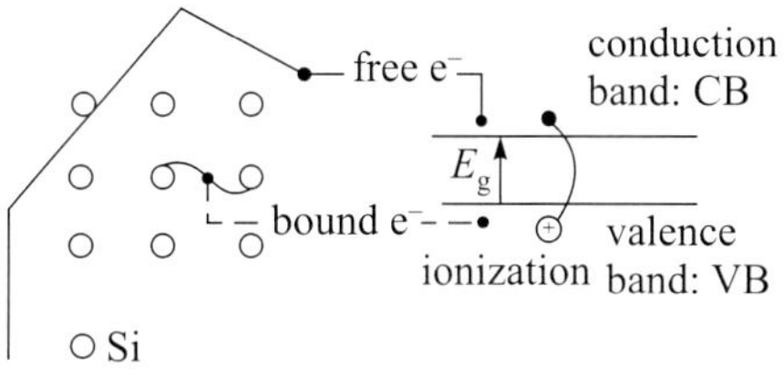

Figure 4

In real space and in a band model, the breaking of an electronic bond results in the transfer of an electron from the valence band, VB, to the conduction band, CB, and thus the creation of an electron–hole pair.

- The additional electron from a pentavalent impurity will be in the conduction band if it escapes itself from the impurity atom: its energy level will be in the forbidden energy gap if it rests bound to the initial impurity but its bonding energy will be weak.

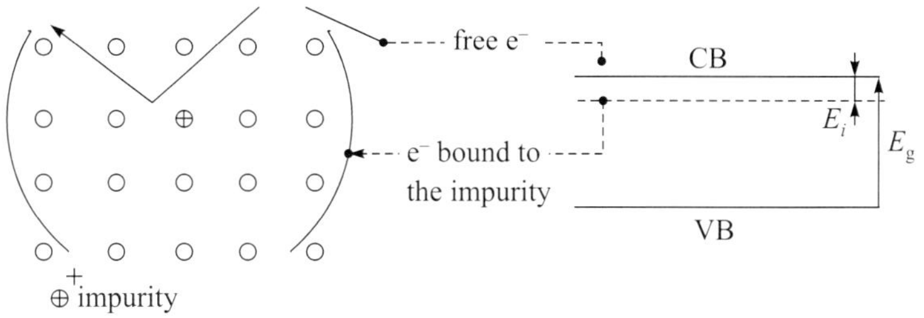

Figure 5

7. Law of Mass Action

$$n_e = \int_{E_g}^{\infty} g_e(E)\cdot f(E)\mathrm{d}E,\ n_h = \int_{-\infty}^{0} g_h(E)\cdot f_h(E)\mathrm{d}E$$

where the density of states $g(E)$ is referenced to unitary volume.

- $5k_BT < E_F < E_g - 5k_BT$: nondegenerated semiconductor.

$$n_e = N_C \exp\frac{E_F - E_g}{k_BT} \quad \text{where} \quad N_C = 2\left(\frac{2\pi m^*_e \cdot k_BT}{h^2}\right)^{3/2}$$

$$n_h = N_v \exp-\frac{E_F}{k_BT} \quad \text{where} \quad N_v = 2\left(\frac{2\pi m^*_h \cdot k_BT}{h^2}\right)^{3/2}$$

Useful formula (F1): $N_e = 2.5\left(\frac{m_e^x}{m_0}\right)^{3/2}\left(\frac{T}{300}\right)^{3/2} 10^{19}\,\mathrm{cm}^{-3}$

Law of mass action: $n_e n_h = N_C N_V \exp-\frac{E_g}{k_BT} = n_i^2$

From which for an intrinsic semiconductor we find

$$n_i = (N_C N_V)^{1/2} \exp-\frac{E_g}{2k_BT}$$

and $\sigma = n_i e(\mu_e + \mu_h) : \sigma \propto \exp-\frac{E_g}{2k_BT}$.

For a doped semiconductor:

If n_d is small or T is high: $n_e \approx n_d$

If n_d is large or T is low: $n_e \propto \exp-\frac{E_i}{2k_BT}$

The addition of n-type impurities results in a decrease in holes (known as compensation) and this increase in the electronic concentration results in an increase in the Fermi level (Ex. 10).

- If $E_g - 5k_BT < E_F < E_g + 5k_BT$: the semiconductor is partially degenerated.
- If $E_F > E_g + 5k_BT$: the semiconductor is degenerated (see Pb. 5).

8. Different Types of Semiconductors

Silicon remains the most widely used semiconducting material in electronic devices (see Pbs. 7 and 8); however, it has an indirect bandgap (e.g., the minimum of the conduction band (E_c) and the maximum of the valence band (E_v) are not located at the same place in the Brillouin zone).

For some applications direct bandgap III–V compounds are more efficient. Some of their properties can be deduced simply from the band structure (see Ex. 21). In addition, alloys of type $III_xIII'_{1-x}V$ can be grown with controlled concentration x of an element of the III column (or the fifth column) with respect to the concentration $(1 - x)$ of another element of the same column (III′) in order to adjust the bandgap energy E_g (see Ex. 21). If the material is homogenous such as GaAs, one can realize Gunn diodes (see Ex. 30). One can also grow heterostructures such as quantum wells and superlattices with specific properties (see Pb. 9a).

Table 1 and Fig. 6 provide specific values for different semiconductors.

Table 1 Bandgap energy E_g (i = indirect; d = direct) and mobility of carriers for several semiconductors

	E_g (eV)	Mobility (cm^2/V·s)	
Crystal	**300 K**	**Electrons**	**Holes**
Diamond	5.4 (i)	1800	1200
Si	1.14 (i)	1300	500
Ge	0.67 (i)	4500	3500
InSb	0.18 (d)	77000	750
InAs	0.35 (d)	33000	460
InP	1.35 (d)	4600	150
GaAs	1.43 (d)	8800	400

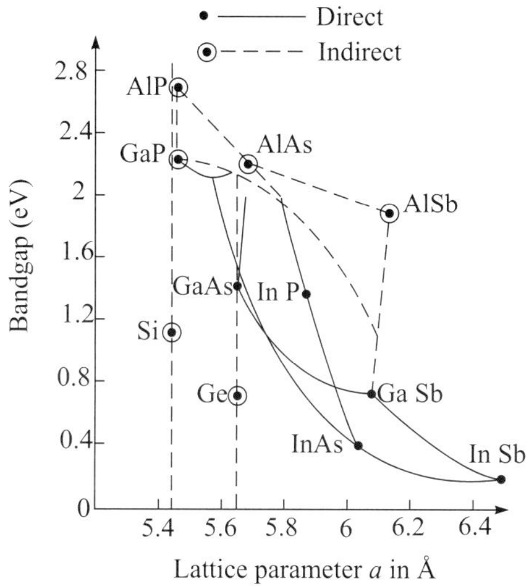

Figure 6 Bandgap energies of different semiconductors as a function of lattice parameter.

9. Allotropes of Carbon: Graphene, C-nanotubes, and Buckyballs

One may consider that the end of the last century was the silicon era because of the fantastic applications of semiconductors in, among many others, microelectronics. The beginning of the present century is the beginning of the era of spintronics based on graphene and others allotropes of carbon such as C-nanotubes and buckyballs. Graphene has quickly emerged as a promising electronic material and it could be a leading candidate to replace silicon in applications ranging from high-speed computer chips to biochemical sensors. Pb. 10 is an introduction to the 2D band structure of graphite with some of its optical properties. Pbs. 11 and 12 are a simple overview of the specific properties of graphene and of C-nanotubes also derived partly from the tight-binding approximation.

Exercises

Exercise 1: s-Electrons bonded in a row of identical atoms: 1D

In the tight-binding approximation, the energy E of s-electrons obeys the relationship:

$$E = -a - \gamma \sum_m e^{-i\vec{k}\cdot\vec{\rho}_m}$$

in which α and γ are positive energies that can be calculated, $\vec{k}$ is the wave vector of electrons, and $\vec{\rho}_m$ represents the vectors between the origin of each atom and its nearest neighbor.

(a) Find dispersion relation $E = f(k)$ for a row of equally spaced identical atoms equidistant by a.

(b) Deduce the expression of the density of states $g(E)$ and that of the effective mass m^*: how this last expression simplified when $ka << 1$?

The energy γ is related to the s-orbital overlap between nearest neighbors and its value decreases very quickly when the distance separating them increases. How does the bandgap change as atoms are made further apart from one another? What is the concomitant evolution of effective mass of electrons at the bottom of the band?

(c) At 0 K, what is the value of Fermi energy when the element is monovalent?

(d) $\alpha = 1$ eV, $\gamma = 0.5$ eV: show the curves $E = f(k)$ and $g(E)$.

With $a = 3$ Å, specify the numerical values of m^* for electrons found at the bottom of the band. What about the mass of electrons located at the top of band?

(e) The energy α represents the difference between an s-electron in an isolated atom and the average corresponding orbital energy in the crystal. Without calculation, find the order of magnitude of the cohesion energy (approximate expression and numerical value). $(e, m, \hbar)$

Solution:

(a) $\rho m = \pm a$ from which $E = -\alpha - 2\gamma \cos ka$

(b) We can evaluate $g(E)$ starting from the general expression:

$$g(E) = \frac{2V}{(2\pi)^3} \iint \frac{dS_E}{\left|\vec{\nabla}_k E\right|} \quad \text{(see Chapter IV, Ex. 11)}$$

or more simply here by using the equality:

$$g(E) \cdot dE = 4\, g(k) \cdot dk = 2g(|k|) \cdot dk \quad \text{where} \quad g(k) = \frac{Na}{2\pi}$$

(see Chapter IV, Ex. 13) from which

$$g(E) = 4 \cdot g(k) \cdot \frac{\mathrm{d}k}{\mathrm{d}E} = \frac{N}{\pi} \frac{1}{\gamma \sin ka} = \frac{N}{\pi\gamma} \frac{1}{\left[1 - \frac{E+\alpha}{2\gamma}^2\right]^{1/2}}$$

[compare to the result in Chapter IV, Ex. 11c (β)].

$$\frac{1}{m^*} = \frac{1}{\hbar^2} \frac{\partial^2 E}{\partial k^2} = \frac{2\gamma a^2}{\hbar^2} \cos ka = -\frac{E+\alpha}{\hbar^2} a^2 : m^*$$

$$\Rightarrow \frac{\hbar^2}{2\gamma a^2} \text{ when } k \to 0.$$

When the atoms are further away $\gamma \approx 0$; the electron level is discrete and without dispersion. The width 4γ of the band increases when the atoms become closer together. The bottom of the band has more curvature as the effective mass decreases (the electrons become lighter). In an isolated atom $m^* = \infty$, corresponding to the impossibility of electron propagation, which can be seen because $v_{\mathrm{g}} = 0$ when $\frac{\partial E}{\partial k} = 0$.

Note that for hydrogen $\gamma = 2E_{\mathrm{H}} \left(1 + \frac{\rho}{r_{\mathrm{B}}}\right) e^{-\frac{\rho}{r_{\mathrm{B}}}}$,

where E_{H} = 13.6 eV and r_{B} = 0.53 Å.

(c) $N = \int_{-\alpha-2\gamma}^{E_{\mathrm{F}}} g(E)\mathrm{d}E = \int_0^{k_{\mathrm{F}}} 2 \cdot g|k|\mathrm{d}k = \frac{2Na}{\pi}|k_{\mathrm{F}}|$ or $k_{\mathrm{F}} = \frac{\pi}{2a}$.

This is a logical result since when the band is full $\left(-\frac{\pi}{a} \leq k \leq \frac{\pi}{a}\right)$, it can accommodate two electrons per atom and the density $g(k)$ (in 1D) is constant.

From this, $E_{\mathrm{F}} = -\alpha$, which can alsobe obtained directly using

$$N = \int_{-\alpha-2\gamma}^{E_{\mathrm{F}}} g(E) \cdot \mathrm{d}E = \frac{N}{\pi\gamma} \int_{-\alpha-2\gamma}^{E_{\mathrm{F}}} \frac{\mathrm{d}E}{\left[1 - \left(\frac{E+\alpha}{2\gamma}\right)^2\right]^{1/2}}$$

$$\Rightarrow \text{Arcsin}\left[\frac{E_{\mathrm{F}}+\alpha}{2\gamma}\right] = 0 \text{ and } E_F = -\alpha.$$

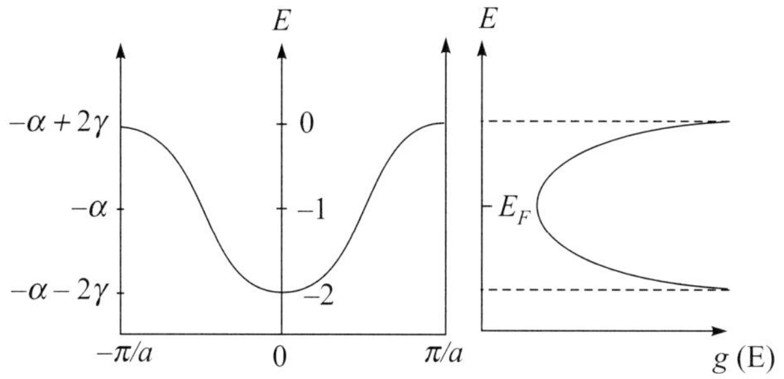

Figure 7

(d) $m^x = \dfrac{\hbar^2}{2\gamma a^2} = 0.4\, m_0$. Given the symmetry of the curve E = $f(k)$, the effective mass is the same in absolute terms at the top and at the bottom of the band. Note, however, that it has the opposite sign at the top of the band, which reflects the behavior of holes.

(e) Taking into account the choice of origin (0 eV here corresponds to the energy of the s-electrons in the isolated atom), it is sufficient to calculate the reduced average energy of $1e^-$ $\overline{E} = \dfrac{E_\mathrm{T}}{N}$ in the half-filled band with $E_\mathrm{T} = \int_{-\alpha-2\gamma}^{E_\mathrm{F}} E \cdot g(E)\mathrm{d}E$. For a free electron gas in 1D $\overline{E} = \dfrac{E_\mathrm{F}}{3}$, due to the high density of states at the bottom of the band.

As we find here the same form of $g(E)$ for $k \ll \pi/a$, we can adopt the same value: $E_\mathrm{c} \approx \alpha + \left(\dfrac{4}{3}\right)\gamma \approx 1.66\,\mathrm{eV}$ (per e^- our atom). See also the evaluation done in Ex. 4.

Exercise 2a: Electrons bounded in a 2D lattice

In the tight-binding approximation, energy E of the s-electrons obeys the relationship

$$E = -\alpha - \gamma \sum_m e^{-i\vec{k}\cdot\vec{\rho}_m}$$

in which α and γ have positive energies that can be calculated and $\vec{\rho}_m$ represents vectors that connect the atom located at the origin at each of its m nearest neighbors.

(a) In the xOy plane, identical atoms are distributed in a square lattice with lattice parameter a (see Chapter III, Exs. 9 and 18, and Chapter IV, Ex. 14). After having explained the valence electron dispersion relation $E = f(k)$ for such a lattice, specify the energy E_Γ of electrons with zero wave vector, then with theenergies E_X (and E_M) for which the end of the wave vector in X (then M) coincides, in reciprocal space, with the middle point situated between origin Γ and point 10 (then the point 11); $\alpha = 1$ eV, $\gamma = 0.5$ eV. What are the numerical values of E_Γ, E_X, and E_M?

(b) Find the relationship $E = f(k)$ relating to the directions $\overrightarrow{\Gamma X}$, $\overrightarrow{\Gamma M}$, and $\overrightarrow{XM}$. Find the corresponding curves and compare them with the dispersion curves of free electrons in the same relative directions.For clarity shift the representation $E' = f(k)$ relative to the free electron model and then evaluate numerically the corresponding energies E'_X and E'_M taking a common origin E_Γ with $a = 3$ Å.

(c) Find the general equation of the constant energy curves and specify their form around the points Γ and M.

Show the corresponding curves inside the first Brillouin as well as the line corresponding to $E_0 = -\alpha$. What are the characteristics of the Fermi line when the element under consideration is monovalent?

Without calculating it explicitly but relying on the above results, sketch the curve of the density of states g(E) and indicate the width of the energy bandgap.

(d) Find the tensor expression of the inverse effective mass tensor $\dfrac{1}{m^x}$. What is the numerical value of m^x near the Γ and M points, as well as around the point X when $a = 3$ Å. ($\hbar, e, m$)

Solution:

(a) In direct space, the atom at 0 has four neighbors: $\vec{\rho}_m = \begin{vmatrix} \pm a \\ 0 \end{vmatrix}$ and $\rho_m = \begin{vmatrix} 0 \\ \pm a \end{vmatrix}$. Then one obtains (the components of $\vec{k}$ being k_x and k_y):

$$E = -\alpha - 2\gamma(\cos k_x a + \cos k_y a)$$
$$(\Gamma): k_x = k_y = 0 : E = -\alpha - 4\gamma = -3\,\text{eV}.$$
$$(X): k_x = \pi/a, k_y = 0 : E_x = -\alpha = -1\,\text{eV}.$$
$$(M): k_x = k_y = \pi/a : E_\text{M} = -\alpha + 4\gamma = 1\,\text{eV}.$$

(b) The relations established in (a) reduce to

- $E = -(\alpha + 2\gamma) - 2\gamma \cos k_x a$ along $\overrightarrow{\Gamma X}(k_y = 0)$
- $E = -\alpha - 4\gamma \cos \frac{|k|}{\sqrt{2}} a$ along $\overrightarrow{\Gamma M}\left(k_x = k_y = \frac{|k|}{\sqrt{2}}\right)$.
- $E = -\alpha + 2\gamma - 2\gamma \cos k_y a$ along $\overrightarrow{XM}(k_x = \pi/a)$

while for free electrons we obtained (see Chapter IV, Ex.14):

$$E' = \frac{\hbar^2}{2m} k_x^2 \text{ along } (\overrightarrow{\Gamma X}),$$
$$= \frac{\hbar^2}{2m} k^2 \text{ along } (\overrightarrow{\Gamma M}), \text{ and}$$
$$= \frac{\hbar^2}{2m}\left(\frac{\pi^2}{a^2} + k_y^2\right) \text{ along} (\overrightarrow{XM}).$$

$$\left(\frac{\hbar^2}{2m}\right)\left(\frac{\pi}{a}\right)^2 \approx 4.2\,\text{eV}$$

or $E'_\text{X} \approx 1.2$ eV and $E'_\text{M} \approx 5.4$ eV, taking into account the 3 eV shift from the origin.

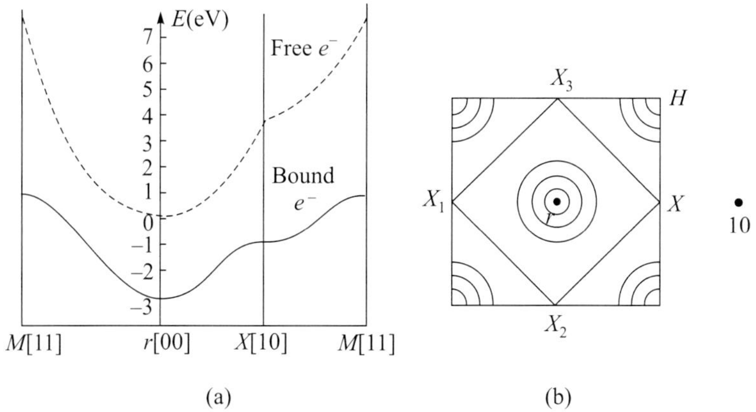

Figure 8

(c) $E = -\alpha - 2\gamma(\cos k_x a + \cos k_y a) = \text{Constant}$

Around point Γ:

$$\cos k_x a = 1 - \frac{(\Delta k_x \cdot a)^2}{2}, \cos_y a = 1 - \frac{(\Delta k_y \cdot a)^2}{2}$$

Hence, $\dfrac{E + \alpha + 2\gamma}{\gamma a^2} = \Delta k_x^2 + \Delta k_y^2$: the lines of constant energy are circles.

Around point M:

$$\cos k_x a = \cos(\pi - \Delta k_x \cdot a) = \frac{(\Delta k_x \cdot a)^2}{2} - 1$$

$$\cos k_y a = \cos(\pi - \Delta k_y \cdot a) = \frac{(\Delta k_y \cdot a)^2}{2} - 1$$

The lines of constant energy are also circles.

When $E = -\alpha$, $\cos k_x a = -\cos k_y a$

so that $k_y = \pm k_x \mp \dfrac{\pi}{a} \left(-\dfrac{\pi}{2} \le k_x, k_y \le \dfrac{\pi}{2} \right)$.

Inside the first Brillouin zone, the corresponding lines of constant energy delimit a square of length $\dfrac{\pi\sqrt{2}}{a}$ with the point X and its equivalent $X_1\left(-\dfrac{\pi}{a}, 0\right), X_2\left(0, -\dfrac{\pi}{a}\right)$, and $X_3\left(0, \dfrac{\pi}{a}\right)$ at the corners.

The area of this square is equal to half of the first Brillouin zone, it contains $-N/2$ states that can accept N electrons, the perimeter of this square is the Fermi surface of the element considered when it is monovalent (see analogies with phonons: Chapter III, Ex. 9).

We can compare this result with that obtained for free electrons occupying a square lattice (Chapter IV, Ex. 14).

At the bottom of the band, the surfaces of constant energy are circles, and the dispersion relation is so closely related to that of a 2D free electron gas and the density of states is a constant.

When we approach energy $E = -\alpha$, the lines of constant energy turn into a square while the slope of the curve $E = f(k)$ decreases.

Considering the expression $g(E) \propto \int \frac{\mathrm{d}I_{\mathrm{E}}}{\overrightarrow{\nabla_{\mathrm{k}} E}}$, we note that these two evolutions both tend to increase $g(E)$, as compared to the constant density of 2D free electrons. Finally and most importantly, the evolution of $E = f(k)$ around point X marks the presence of a horizontal tangent plane in X and this point X is a point of inflection for curves of type $\Gamma X \rightarrow XM$. The inflection is confirmed by the different sign attributed to effective masses depending on the direction of propagation, k. The evolution of $E = f(k)$ around the critical point X leads to a logarithmic singularity in the density of states curve (for the detailed calculation see Y. Quéré [21], p. 306).

Beyond $E = -\alpha$, we expect a symmetric evolution, as previously, and to be convinced it is sufficient to show the constant energy curves around the point M in the adjacent Brillouin zones to the first (such as the lines XX_3, which are symmetry axes for the constant energy curves.)

The characteristics of the curve $g(E)$, with a total width of 8γ, are shown in Fig. 9.

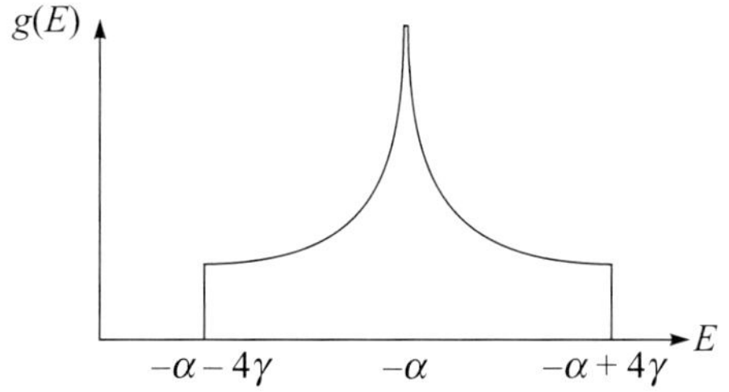

Figure 9

(d) $$\frac{1}{m^*} = \frac{1}{\hbar^2}\begin{bmatrix} \frac{\partial^2 E}{\partial k_x^2} & \frac{\partial^2 E}{\partial k_x \partial k_y} \\ \frac{\partial^2 E}{\partial k_x \partial k_y} & \frac{\partial^2 E}{\partial k_y^2} \end{bmatrix} = \frac{2\gamma a^2}{\hbar^2}\begin{bmatrix} \cos k_x a & 0 \\ 0 & \cos k_y a \end{bmatrix}.$$

Around the Γ point, $\cos k_x a = \cos k_y a \approx 1$, the electron effective mass is a positive scalar equals to $\frac{\hbar^2}{2\gamma a^2} = 0.4\, m_0$.

Around point M, $\cos k_x a = \cos k_y a = -1$, effective mass is negative/it is a hole region.

It should be noted, generally, that the sign of the effective mass around the extrema of the curves $E = f(k)$ is such that $m^* > 0$ for the minimum and $m^* < 0$ for the maxima.

Around the X point: $m^{*-1} = \dfrac{2\gamma a^2}{\hbar^2}\begin{bmatrix} 1 & 0 \\ 1 & -1 \end{bmatrix}$; we obtain a hole behavior when the force is along Ox and an electron behavior when it is along Oy.

Exercise 2b: Band structure of high-T_c superconductors: influence of 2D nearest neighbors (variation of Ex. 2a)

High-temperature superconductors contain quasi-planes of CuO_2 layers that are perpendicular to the c-axis of the crystal. The goal is to establish the structure of one of these n-bands, the most important, by considering a direct square lattice 2a, which consists of an atom of copper at (0, 0) and two atoms of oxygen in (½, 0) and (0, ½), as shown in the Fig. 10.

To achieve this goal, we consider the reciprocal space of a 2D lattice and use the tight-binding approximation, including the action of the second neighbors.

The initial expression if of the form:

$$E(k) = E_0 - \gamma_1 \sum_m e^{-i\vec{k}\vec{\rho}_m} - \gamma_2 \sum_j e^{-i\vec{k}\vec{\rho}_j}$$

where the sums over m concerns the first neighboring oxygen atoms with distance $\vec{\rho}_m$, and the sums over j is for second neighboring oxygen atoms with distance $\vec{\rho}_j$. The energy γ_1 is positive, while γ_2 restricted to the orbital anti-bonding between oxygen atoms (second neighbors) is negative.

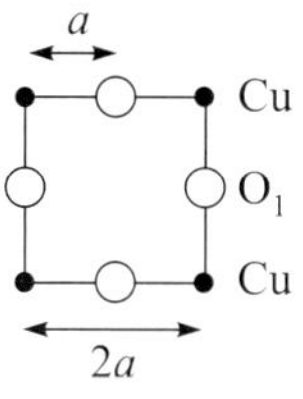

Figure 10

(1) Find the general form taken by the dispersion relation $E = f(k)$.

(2) Auxiliary calculations give $|\gamma_2|/\gamma_1 = 0.45$ and we wish to evaluate the corresponding energies for various vectors k.

For convenience use the interval $-\pi/a \leq k_x, k_y \leq +\pi/a$ even though the first Brillouin zone is limited to $-\pi/2a, +\pi/2a$.

Find the energies taken at the Γ point: $k_x = 0$, $k_y = 0$; point X: $k_x = \pi/a$, $k_y = 0$; at the intermediate point A: $k_x = \pi/2a$, $k_y = 0$; at point M: $k_x = \pi/a$, $k_y = \pi/a$, as well as the points corresponding to $\overrightarrow{\Gamma M}/2$ and to $\overrightarrow{XM}/2$.

Show the dispersion curves in the [10] direction, that is at $\overrightarrow{\Gamma X}$, in the [11] direction along $\overrightarrow{\Gamma M}$ as well as in the $\overrightarrow{XM}$ direction. Compare them with those obtained when only first neighbors are taken into account

(3) The Fermi energy is such that $E_F = E_0$. Plot this position energy on the dispersion curves. Next show the corresponding line in the 2D reciprocal space.

To ensure the continuity of this Fermi line, this space must be centered at point M by placing the points of type Γ at the corners of a square. Compare this with the lines of constant energy deduced when only first neighbors are considered.

(4) The addition of foreign atoms in the crystalline planes adjacent to the CuO_2 reference planes reduces the Fermi energy E_F via p-doping. Referring to the curves established in 2, find the decrease, ΔE_F, which should lead to a significant increase of the density of states at the Fermi level $g(E_F)$ and therefore to a significant increase in the conductivity σ (see the expression given in the Chapter IV, Ex.19).

(5) At right angles to the plane layers, the dispersion relation is of the form $E(k_z) = -\gamma_c|k_z|c$ (where $\gamma_c \ll \gamma_1$). Why is this relationship consistent with 2D behavior of charge carriers?

Solution:

(1) Taking the origin at an oxygen atom, there are four first neighbor Cu atoms at (±a, 0) and (0, ±a). In addition the four second O neighbors are characterized by vectors $\vec{\rho}_j$ such that $\vec{\rho}_j(\pm a, \pm a)$.

$$E(k) = E_0 - 2\gamma_1(\cos k_x a + \cos k_y a) + 4|\gamma_2|\cos k_x a \cdot \cos k_y a$$

(2) $E(\Gamma) = E_0 - 4\gamma_1 + 4|\gamma_2| \approx E_0 - 2.2\gamma_1$

$E(X) = E_0 - 4|\gamma_2| = E_0 - 1.8\gamma_1$

$E(M) = E_0 + 4\gamma_1 + 4|\gamma_2| \approx E_0 + 5.8\gamma_1$

$E(\Delta) = E_0 - 2\gamma_1; E(\overrightarrow{\Gamma M}/2) = E_0; E(\overrightarrow{XM}/2) = E_0 + 2\gamma_1$

These last three energies are not influenced by the second neighbors. The corresponding curves are shown in solid lines in Fig. 11, with dashed lines corresponding to $\gamma_2 = 0$.

It should be noted that there is a significant distortion resulting from the action of the second neighbors. Technically, it should be noted that the choice of interval $-\pi/a < k_s$, $k_y < \pi/a$ allows the exploration of the angular domain covered by $\cos(k_{x,y}a)$ between −1 and +1.

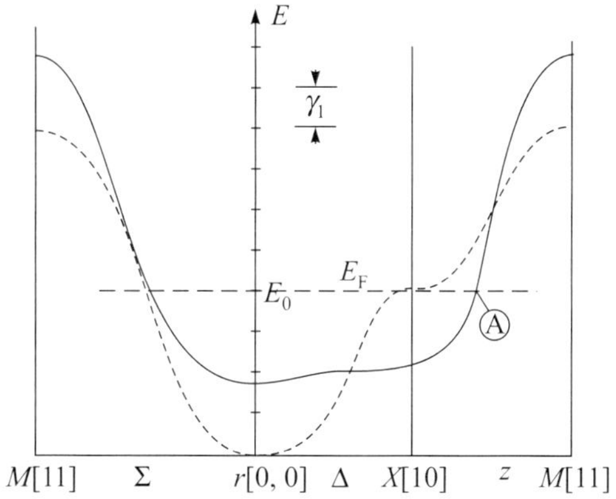

Figure 11

(3) See Fig. 12, in which the dashed curves show the square constant energy surfaces resulting from the action of only nearest neighbors (see previous exercise).

The action of the second neighbors has the effect of removing the intersection of E_F with $E = f(k)$ at the X point (see previous exercise) to reposition it between X and M, at point A in Fig. 11 and shown again in Fig. 12, taking symmetries into account. The topology of the Fermi surface is therefore deeply modified by the action of the second neighbors.

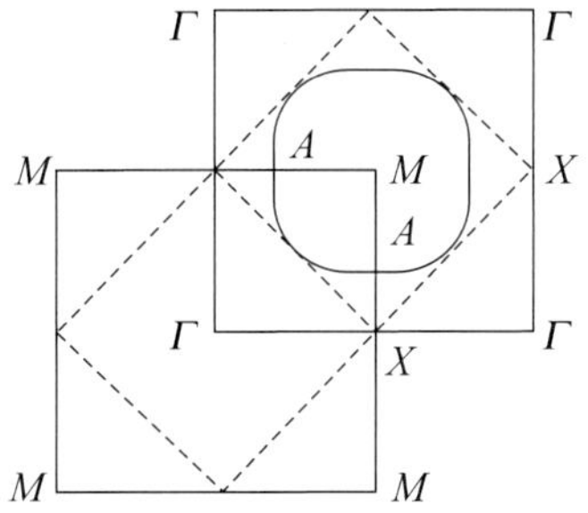

Figure 12

(4) We observe that the band between Γ and M is almost flat, which causes the existence of an area with a high density of states since $\overrightarrow{\nabla_k E} \approx 0$ and $g(E)$ is such that $g(E) \propto \int \frac{\mathrm{d}I(E = Cste)}{|\nabla_k E|}$, hence a large value of Γ(see Chapter IV, Ex. 19).
For the Fermi level to reach this region of high density of states, it must decrease of about $2\gamma_1$.

(5) $\partial^2 E/\partial k^2_z = 0$; the effective mass of particles measured along the *c*-axis is infinite and these particles are therefore bound to the plans of CuO_2.

Comment: High-T_c superconductors

The discovery of high-temperature superconductors stimulated considerably the imagination and research of solid state physicists (see Chapter IV, Pb. 7). The zoology of the materials of this type increases every day: see the review paper of T. Tohyama *Japanese Journal of Applied Physics* **51**, 2012, 010004. Other high-T_c superconductors such as iron-based superconductors have been recently discovered (Bianconi et al. *Nature Physics*, **9**, 2013, 536).

This exercise is inspired by the work of Yu and Freeman (*J Electron Spectros. Rel. Phenom.* **66**, 1994, 281 and Refs. 27 and 38, therein). According to these authors, the detailed mechanisms are not related to the existence of the observed singular point X in the density of states curve when first neighbors are only taken into account—see preceding exercise—but must be correlated to the actions of the second neighbor that induce distortions of the surface of Fermi and the dispersion curves that have just been studied. Figure 13a shows the unit cell of another high-T_c material: $HgBa_2CuO_4$, which has a critical temperature $T_c = 95$ K. Figure 13(b) completes their analysis, according to which the addition of 0.38 hole per unit cell is sufficient to lower the Fermi level (from 0.4 eV) at the height of the maximum of g(E), as suggested in the Question (4). Except for the notations used, the results obtained here are in perfect agreement with those of this reference article.

For the parallels between the study of the dispersion of electrons and that of phonons, we refer the reader to Chapter III, Exs. 3–5, which highlight the significant influence of the second neighbor on the shape of the curves $\omega = f(k)$.

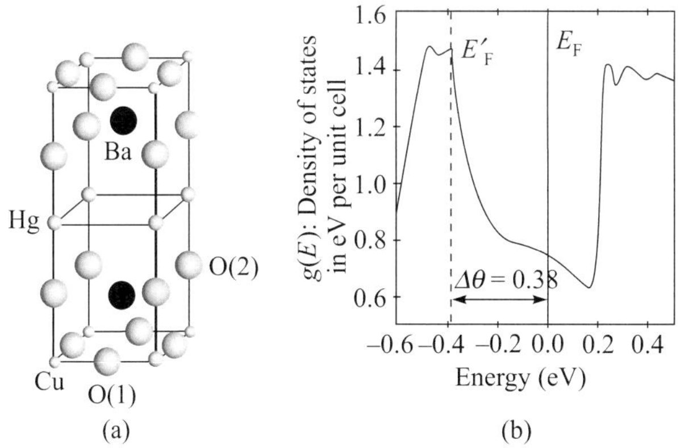

Figure 13

Exercise 3a: Tight binding in a simple cubic lattice (3D)

In the tight-binding approximation, the dispersion relation of the s-valence electrons is given by:

$$E = E_0 - \alpha - \gamma \sum_m e^{-\vec{k}\vec{\rho}_m}$$

where E_0, α, and γ have positive energies, $\vec{k}$ is the wave vector, and $\vec{\rho}$ is the vector connecting the position of the initial atom to its m nearest neighbors.

(1) We consider identical atoms distributed in a simple cubic lattice with parameter a.

Find the dispersion relation of s-electrons. Specify the form it takes along the following directions: [100] or $\overline{\Gamma X}$, [110] or $\overline{\Gamma M}$, and [111]or $\overline{\Gamma W}$ and express the energy at the X, M, and

W points on the first Brillouin zone as well as at the origin Γ in the zone center (see Fig. 14).

Compare the evolutions along $\overline{\Gamma X}$ and then $\overline{\Gamma M}$ with the corresponding evolution of a 2D square lattice (Ex. 2a).

Show the dispersion curves along $\overline{\Gamma X}, \overline{\Gamma M}, \overline{\Gamma W}$ as well as $\overline{XW}$.

(2) Find the expression of the effective mass, m^x, for particles located in the vicinity of Γ. Indicate the characteristics of the constant energy curves in this region (ka << 1) as well as the energy curve $E' = E_0 = -\alpha - 2\gamma$ obtained in the plane ΓXM (k_z = 0). Relying on the symmetry of the object,describe the constant energy surface $E' = E_0 - \alpha - 2\gamma$. Also show that corresponding to $E'' = E_0 - \alpha$ by considering the shape of the constant energy lines for different values of k_z and also possibly the curves obtained in 2D (see Ex. 2a).

Which of these two surfaces is the Fermi surface of a monovalent element?

Without detailed calculations, find the shape of the curve for the density of states $g(E)$.

(3) If we consider p-states, the atomic wave functions no longer have a spherical symmetry and the dispersion relation given by the tight-binding approximation becomes

$$E = E_1 - \alpha - 2\gamma_1 \cos k_x a - 2\gamma'_1 (\cos k_y a + \cos k_z a)$$

in which E_1, α_1, and γ'_1 are positive while γ_1 is negative such that $|\gamma_1| > \gamma'_1$.

For k_z = 0, show the shape of the s- and p-bands first in the k_x direction and then in the k_y direction when $E_1 - \alpha_1 > E_0 - \alpha + 2\gamma$. Also show the constant energy curves of the p-band in the k_z = 0 plane.

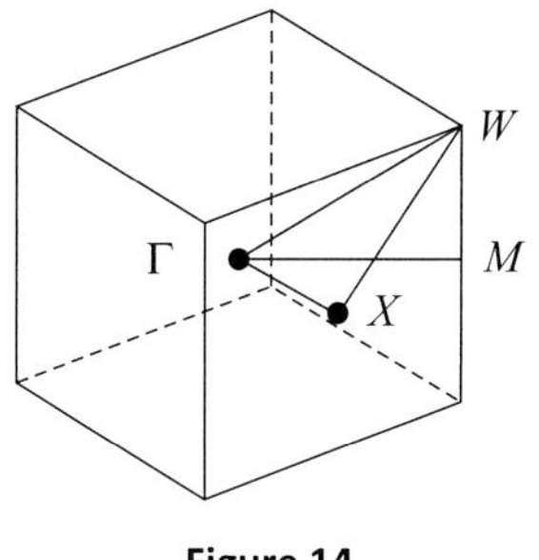

Figure 14

Solution:

(1) Each atom has six nearest neighbors: ρ = ($\pm a$,0,0); (0,$\pm a$,0); and (0,0, $\pm a$)

$$E = E_0 - \alpha - 2\gamma(\cos k_x a + \cos k_y a + \cos k_z a)$$

At the Γ point: $k_x = k_y = k_z = 0$; $E(\Gamma) = E_0 - \alpha - 6\gamma$

Along [100], $\overrightarrow{\Gamma X}$; $k_x = k,\ k_y = k_z = 0$;

$E = E_0 - \alpha - 4\gamma - 2\gamma \cos ka$

and $E(X) = E_0 - \alpha - 2\gamma$.

Along [110], $\overrightarrow{\Gamma M}: k_x = k_y = k/\sqrt{2};\ k_z = 0$

$E = E_0 - \alpha - 2\gamma - 4\gamma\cos(ka/\sqrt{2})$ and $E_M = E_0 - \alpha + 2\gamma$

Note that up to a translation in energy, the same results are obtained for a 2D square lattice with $E(\Gamma) = E_0 - \alpha - 6\gamma$

Along [111], $\overrightarrow{\Gamma W}: k_x = k_y = k_z = k/\sqrt{3}$;

$E = E_0 - \alpha - 6\gamma\cos(ka/\sqrt{3})$ and $E(W) = E_0 - \alpha + 6\gamma$

Figure 15 shows three of the four evolutions. For the fourth ($\overrightarrow{\Gamma M}$), see curve 2b obtained in Ex. 2a.

(2) Near the zone center, Γ: $\cos k_x a \approx 1 - \frac{(k_x a)^2}{2}$,

$E = E_0 - \alpha - 6\gamma + \gamma a^2 k^2$.

The constant energy surfaces are spherical (isotropic) and are occupied with electrons of mass $m^x = \dfrac{\hbar^2}{2\gamma a^2}$.

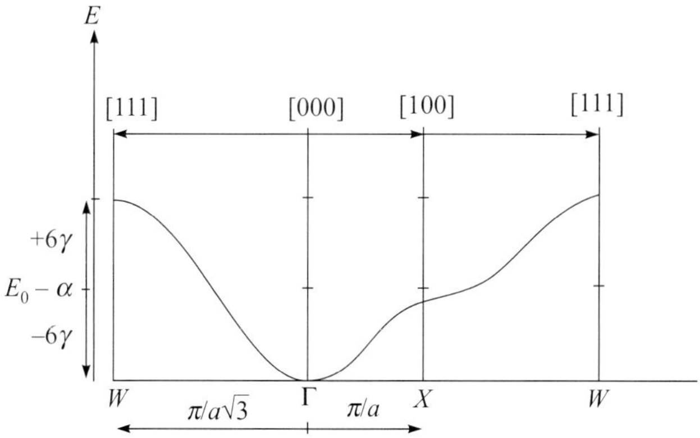

Figure 15

In the plane $k_z = 0$, the lines of constant energy $E' = E_0 - \alpha - 2\gamma$ correspond to $\cos k_x a + \cos k_y a = 0$, which leads to straight lines of the equation $k_y = \pm k_x \pm \pi/a$. This forms a square with vertices X_1, X_2, X_3, and X_4 in the plane (see Ex. 2a).

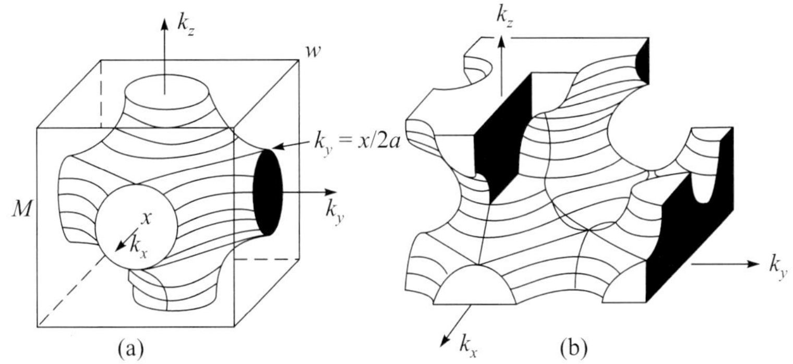

Figure 16

This reasoning also applies successively to the $k_x = 0$ and then the $k_y = 0$ planes.

The constant energy surface $E' = E_0 - \alpha - 2\gamma$ is therefore the cube at corners at the points X_1 to X_6 such that $\left(\pm\frac{\pi}{a},0,0\right)$, $\left(0,\pm\frac{\pi}{a},0\right)$, and $\left(0,0,\pm\frac{\pi}{a}\right)$.

Contrary to a 2D lattice, it will not be the Fermi surface of a monovalent element because its volume $\left(\frac{\pi\sqrt{2}}{a}\right)^3$ is smaller than and not equal to $\left(\frac{1}{2}\right)\left(\frac{2\pi}{a}\right)^3$.

The lines of constant energy $E'' = E_0 - \alpha$ correspond to $\cos k_x a + \cos k_y a = -\cos k_z a$.

For $k_z = \pm\frac{\pi}{2a}$, we find again that the lines $k_x = \pm k_y \pm \frac{\pi}{a}$ delimit a square (bold line in Fig. 16a).

For values of $|k_z|$ increasing between $\pi/2a$ and π/a, we find decreasing values in $-\cos k_z a$ and the lines of constant energy become circular without strictly achieving perfect circularity (at $k_z = \pi/a$, the pseudo-radius is $|k| \approx \pi/2a$).

Inversely, the values of k_z such that $0 < |k_z| < \pi/2a$ correspond to increasing constant energy lines (beyond the square in the 2D lattice: see Fig. 8b, Ex. 2a). These are pseudo arcs with circumferences centered on M. Taking symmetry into account we find the characteristics of Fig. 16a, which correspond to the Fermi surface of a monovalent element. To see this, one can repeat the resulting structure (translations of $2\pi/a$ along successively k_x, k_y, and k_z) and next take a point X as the origin to find that the volume of the unoccupied hole states reproduces exactly the volume of occupied states (see Fig. 16b).

In principle, we can evaluate $g(E)$ starting from $\iint \frac{\mathrm{d}S}{\left|\overrightarrow{\nabla_k E}\right|}$, where $\overrightarrow{\nabla_k E}$ is along k_x such as $2\gamma \sin k_x a$, etc.

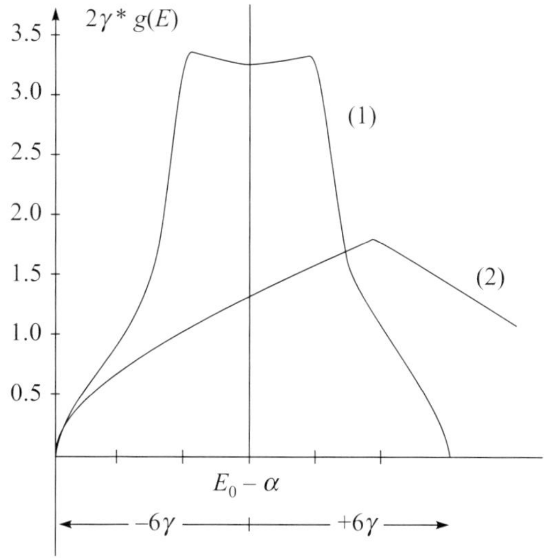

Figure 17

Despite the possibility of integration using a single variable, the final result is numerical. Thus, we may only note: (i) the symmetrical shape relative to energy $E_0 - \alpha$; (ii) the beginning evolves as $\sqrt{E}$, like free 3D electrons that have a mass m^*, which reaches a plateau when the constant energy surface E' comes into contact with the Brillouin zone (at the point X). This plateau corresponds to compensation between the decrease of the useful dS and the decrease of $\nabla_k E$. Figure 17 shows this evolution compared to free electrons of the same mass m^*, curve (2). Also note the differences with the curve $g(E)$ correspondingly obtained in 2D (Fig. 9, Ex. 2a).

(3) See Figs. 18a and 18b. Note that the curve $E(k_x)$ of p-electrons has a downward concavity while curves $E(k_y,k_z)$ have an upward concavity. The different constant energy curves contained in the space $E = f(k_x k_y)$ have a saddle-shaped surface and the effective mass of the p-electrons in the k_x direction is negative.

When the s- and p-bands overlap, the proposed relationships no longer hold and one must use hybrid orbitals that lead to a dispersion relation via the resolution of the secular equation (see Pb. 10 for the example of graphite and the classic works [1] [11] [25] for more details). It should be especially noted that to determine the allowed and forbidden bands one must start from a system with multiple atomic levels, that is, a system consisting of different atoms or if they are identical, with several orbitals or atoms with nonequivalent positions. The study of graphite concerns the latter two properties.

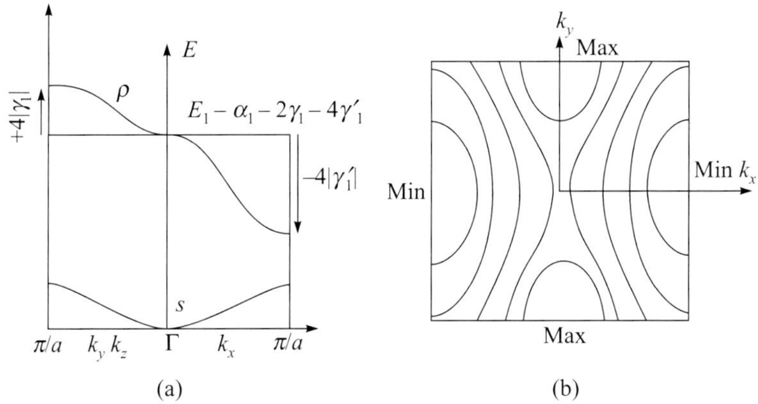

Figure 18

Exercise 3b: Tight bindings in the bcc and fcc lattices (variation of Ex. 3a)

We consider identical atoms located at points of a lattice that are body-centered cubic (bcc: i) and face-centered cubic (fcc: ii).

(a) Construct the first Brillouin zone for each of these lattices.

(b) For each case determine the dispersion relation for s-electrons, having the general form:

$$E = -\alpha - \gamma \sum_m e^{-\vec{k}\cdot\vec{\rho}_m}$$

where α and γ are the positive energies that can be calculated and $\vec{\rho}_m$ represents the vectors that connect an atom located at the origin to each of its nearest neighbors.

(c) Find the expressions for E at the points located at the intersection of the first Brillouin zone with the directions [100], [110], and [111] and thus the value $E(\Gamma)$ to the center of this zone.

Deduce the total width of the corresponding band, ΔE, and find the expression for the effective mass m^x in the vicinity of Γ as well as the shape of constant energy surfaces around this point.

Solution:

(a) See solution and Fig. 10 in Chapter I, Ex. 14.

(b & c) are a generalization of the previous exercises.

(i) Bcc: Eight nearest neighbors at $\rho = \left(\pm\frac{1}{2}, \pm\frac{1}{2}, \pm\frac{1}{2}\right)$; unit: a.

$$E = -\alpha - 8\gamma \cos\frac{k_x a}{2} \cdot \cos\frac{k_y a}{2} \cos\frac{k_z a}{2}$$

At Γ: $k_x = k_y = k_z = 0$. $E(\Gamma) = -\alpha - 8\gamma$

At N(100), $k_x = \frac{2\pi}{a}; k_y = 0, k_z = 0, E(H) = -\alpha + 8\gamma.$

At $P(100)$, $k_x = k_y = k_z = \frac{\pi}{a}.$ $E(P) = -\alpha.$

ΔE (band) = 16γ.

In the neighborhood of Γ, $E = -\alpha - 8\gamma + 4\gamma k^2 a^2$

The constant energy surfaces are spherical and occupied by electrons of mass $m^x = \frac{\hbar^2}{8\gamma a^2}$.

(ii) Fcc: Twelve nearest neighbors:

$$\rho = \left(\pm\frac{1}{2}, \pm\frac{1}{2}, 0\right); \left(\pm\frac{1}{2}, 0, \pm\frac{1}{2}\right); \text{and} \left(0, \pm\frac{1}{2}, \pm\frac{1}{2}\right).$$

$$E = -\alpha - 4\gamma\left(\cos\frac{k_x a}{2}\cdot\cos\frac{k_y a}{2} + \cos\frac{k_x a}{2}\cdot\cos\frac{k_z a}{2} + \cos\frac{k_y a}{2}\cdot\cos\frac{k_z a}{2}\right)$$

At Γ: $E(\Gamma) = -\alpha - 12\gamma$

At X(100): $k_x = \frac{2\pi}{a}, k_y = k_z = 0, E(X) = -\alpha + 4\gamma.$

At K(110): $k_x = k_y = \frac{3\pi}{2a}, k_z = 0, E(k) = -\alpha + 4\gamma\left(\sqrt{2} - \frac{1}{2}\right).$

At L(110): $k_x = k_y = k_z = \frac{\pi}{a}, E(L) = -\alpha.$

ΔE (band) = 16γ.

In the vicinity of Γ, $E = -\alpha - 12\gamma + 4\gamma k^2 a^2$

The constant energy surfaces are spherical and occupied by electrons with effective mass $m^x = \frac{\hbar^2}{8\gamma a^2}$.

In all cases, if the effective mass is isotropic and the constant energy surfaces are spherical around the Γ point, this no longer holds when developing the cosine term up to the second order (in k^4), which highlights the effects of the crystal symmetry.

A detailed study of the density of states would show that the g(E) curve is symmetric for the bcc lattice with a logarithmic singularity at the median energy $E_m = -\alpha$, whereas it is asymmetric for the fcc lattice with a logarithmic singularity at the maximal energy $E_M = -\alpha + 4\gamma$.

Exercise 4: Dimerization of a linear chain

Consider a linear chain of equally spaced atoms and of length Na. Each atom contributes an electron to the band that will be studied; this electron is localized in the orbital ϕ when the atom considered is alone. Describe this chain using the tight-binding method.

(1) It is assumed that the only elements of the single electron Hamiltonian matrix H_0 are

$< \ell|H_0|\ell \geq 0$

$< \ell|H_0|\ell + 1 \geq -\gamma\ (\gamma > 0)$

where $|\ell \geq$ represents the quantum state for which the wave function of the electron is $\phi_\ell(x)$, $(< \ell|\ell' = \delta_{\ell\ell'})$.

1.1: Observing that the chain is periodic (with period a), write the wave functions $\psi_k(x)$ which represent the stationary states of wave vector wave k.

1.2: Deduce the dispersion law $E(k)$ and show it graphically.

1.3: Determine the Fermi level (assuming that every atom contributes one electron).

1.4: Determine the contribution of this band to the cohesive energy.

(2) The atoms in even rows are held fixed, and the atoms in odd row undergo a slight movement u (Fig. 19).

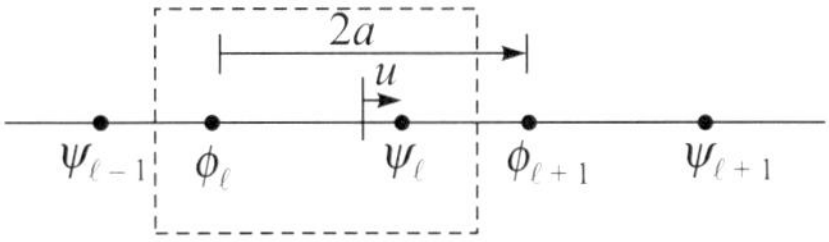

Figure 19

It is thus appropriate to choose a basis with two atoms (wave functions ϕ_ℓ and ψ_ℓ) with $a' = 2a$. The elements of the matrix of the new Hamiltonian H are thus (where $\alpha > 0$):

$$<\phi_\ell|H|\phi_\ell> = <\psi_\ell|H|\psi_\ell> = 0$$

$$<\phi_\ell|H|\psi_\ell> = -\gamma + \alpha u$$

$$<\psi_{\ell-1}|H|\phi_\ell> = <\psi_\ell|H|\phi_{\ell+1}> = -\gamma - \alpha u$$

2.1: Writing the wave functions in the form:

$$|\psi_k> = \sum_\ell e^{ik\ell a'}(A|\phi_\ell> + B|\psi_\ell>)$$

(a) Determine the new law of dispersion ξ (k).

(b) Calculate the width of the energy bandgap.

(c) What can you deduce from the cohesive energy? Calculate its change knowing that when z is small:

$$\int_0^{\pi/2}[1-(1-z^2)\sin^2 x]^{1/2}\,dx \approx 1 + \frac{1}{2}z^2\left[\ln\frac{4}{|z|} - \frac{1}{2}\right]$$

2.2: The atoms in the chain are in fact connected by "spring" constants with force β.

(a) Find the elastic energy of the chain as a function of u.
(b) Find the total energy (electronic and elastic) of the chain when it undergoes a distortion u.
(c) Establish the equilibrium expression of the distortion based on the other parameters (α, β, and γ).

Solution:

(1)

1.1 $\psi_k(x) = \sum_{\ell} e^{ik\ell a} \phi_\ell(x)$

1.2 $H_0|k> = \sum_{\ell} e^{ik\ell a} H_0|\ell> = E|k>$

$<\ell| - \rightarrow \qquad Ee^{ik\ell a} = \sum_{\ell'} e^{ik\ell' a} <\ell|H_0|\ell'>$

$$E = \sum_{\ell'} e^{ik\ell' a} <0|H_0|\ell'> = -2\gamma\cos ka;\ E - 2\gamma\cos ka.$$

The similarities between this exercise and the beginning of Ex. 1 should be noted. We establish here the dispersion relation that was suggested in the statement of Ex. 1. Up to the quantity $-\alpha$, the dispersion relation and the representative curve in Fig. 1 are the same.

1.3 $E_F = 0$ (see solution of Ex. 1 with $\alpha = 0$).

1.4 $\Delta E_c = 2\times 2\int_0^{\pi/2a} E\cdot\frac{dk}{2\pi/Na} = -N\frac{4\gamma}{\pi}$, where $\Delta E_c = -4\gamma/\pi$ per atom.

(2)

2.1 (a) $|k> = \sum_{\ell'} e^{ik\ell' a'}(A|\phi_{\ell'}> + B|\psi_{\ell'}>)$

$<\phi_0| et <\psi_0|$ which leads to

$$-\xi_k A + B[(-\gamma + \alpha u) - (\gamma + \alpha u)e^{-ika'}] = 0$$
$$A[(-\gamma + \alpha u) - (\gamma + \alpha u)e^{ika'}] - \xi_k B = 0$$

from which $\xi_k^2 = E^2 + \Delta_k^2$, where $\Delta_k = 2\alpha u\sin(ka'/2)$.
The dispersion curve, shown in Fig. 20b, is now limited to the interval $-\pi/2a, +\pi/2a$. Compared to the curve in 1 (Fig. 20a), there is an opening of a bandgap with amplitude $4\alpha|u|$ at $k = \pi/2a$.

(b) $E_g = 4\alpha|u|$. The lower band, which corresponds to $\xi_k = -(E^2 + \Delta^2)^{1/2}$, is going to be completely filled. The upper band $[\xi_k = (E^2 + \Delta^2)^{1/2}]$, will be completely empty (at 0 K).

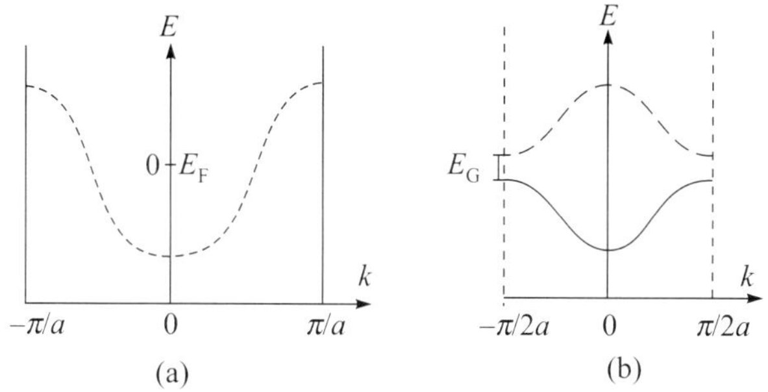

Figure 20

(c) Compared with 1, the appearance of the bandgap will increase the cohesive energy because the electron energy (particularly in the vicinity of $k = \pi/2a$) will decrease.

$$\Delta E_c = 2\times 2\int_0^{\pi/2a} \xi(k)\cdot\frac{dk}{2\pi/Na}$$

where $\xi(k) = -(4\gamma^2\cos^2 ka + 4\alpha^2u^2\sin^2 ka)^{1/2}$

$$= -2\gamma\left[1-\left(1-\frac{\alpha^2u^2}{\gamma^2}\right)\sin^2 ka\right]^{1/2}$$

Using the proposed relationship when $z = \dfrac{\alpha u}{\gamma}$ is small, we find

$$\Delta E_C = -\frac{4\gamma}{\pi}\{1+\frac{1}{2}\left(\frac{\alpha u}{\gamma}\right)^2\left[\log\frac{4}{\alpha u/\gamma}-\frac{1}{2}\right]\}$$

In this expression, the term $(1/2)(\alpha u/\gamma^2)[\ldots]$ represents the increase of the cohesive energy caused by dimerization.

2.2 (a) $E(\text{élast}) = N\left(\dfrac{1}{2}\beta u^2\right)$

(b) $\Delta E(\text{distortion}) = \dfrac{1}{2}\beta u^2 - \dfrac{2\alpha^2u^2}{\pi\gamma}\left[\log\dfrac{4}{|\alpha u/\gamma|}-\dfrac{1}{2}\right]$

(c) In [...], we can neglect the 1/2 because $\alpha u/\gamma$ is small.
ΔE takes the form: $\Delta E(\text{dist.}) = Au^2 + Bu^2\log\overline{Cu}$.

In equilibrium $\dfrac{\partial \Delta E(\text{dist.})}{\partial u}=0$ or $u=\dfrac{4\gamma}{a\sqrt{e}}e^{-\beta\pi\gamma/4\alpha^2}$

When α tends to zero, u tends to zero, and the dimerization disappears.

This exercise (kindly suggested by Prof. Gerl; Nancy University) illustrates how electrons can stabilize the dimerization resulting in a bandgap. For further information on dimerization, see Chapter III, Exs. 3 and 5; on Peierls instabilities in 1D conductors, see Chapter IV, Ex. 13; on their correlation with the Kohn anomaly, see Chapter III, Ex. 6; and on charge density waves, see Chapter I, Pb. 8, and for example, C. Noguera; *J. Phys. C* **19**, 1986, 2161.

Exercise 5a: Conductors and insulators

(a) In a linear chain of identical atoms equidistant by a, show that the solid considered is electrically insulating at 0 K if it has an even atomic valence electron number and if one takes into account the potential energy created by the ions of the network.

(b) In the context, the theory of nearly free electrons at 0 K considers a 2D square lattice (with basis a) composed of identical divalent atoms. Explore the nature of its electrical conductivity by representing the dispersion curves of electrons in the $\overrightarrow{\Gamma X}$ and the $\overrightarrow{\Gamma M}$ directions ([10] and [11] respectively). In the latter case, draw the Fermi surface and the first Brillouin zone.

(c) *Numerical application*: In the square lattice above, the width of the bandgap between the valence band, VB, and the conduction band, CB, $E_c - E_v$ is 4 eV in X in the [10] direction and 2 eV in M in the [11] direction. In addition: $\dfrac{\hbar^2}{2m}\cdot\dfrac{\pi^2}{a^2}=5\,\text{eV}$. Is it a conductor?

Solution:

(a) The dispersion curves $E = f(k)$ show the discontinuities (forbidden bands) for $k = n(\pi/a)$. Each allowed energy band spans over $2\pi/a$ and contains $2N$ electrons (N is the number of atoms in the row) because in the k-space the unit cells have a length of $2\pi/L = 2\pi/Na$, and it can accept two electrons each ($\uparrow\downarrow$).

In 1D if there is an even number (2, 4, or 6) of valence electrons, the bands (1, 2, or 3) will be completely filled. Electrons occupying these full bands cannot carry electrical current and thus the body will be electrically insulating (see Course Summary). This rule has exceptions, in particular, the transition metals oxides or ITO: $InTiO_2$.

(b) The first Brillouin zone contains N cells $\left[\frac{(2\pi/a)^2}{(2\pi/Na)^2} = L^2/a^2 = N = \text{number of lattice atoms}\right]$ and can receive $2N$ electrons. This filling must be carried out from the lowest energy levels taking into account the dispersion relation of electrons. The latter introduces energy discontinuities when the wave vector approaches the limits of the first Brillouin zone. Figure 21a shows the characteristics of the dispersion of nearly free electrons following the two principle directions [10] and [11]. The forbidden energy bands at point X (i.e., X_v and X_c) and at point M (M_V and M_C) have respective widths of $2V_{10}$ and $2V_{11}$ in the theory of almost free electrons.

Two types of figures can be obtained:

(1) The top of the first allowed band at point M (or M_V) has a smaller energy than the bottom of the second at the point X (or X_C): the first Brillouin zone accepts $2N$ valence electrons and is completely full; the concerned solid will thus be insulating.

(2) The energy $E(M_V)$ is greater than the energy $E(X_C)$ (see Fig. 21a): electrons occupy the lowest energy levels in the second band along the [10] direction before reaching the top of the first band in the [11] direction. The number of electrons contained in the second band will be the number of states left vacant in the first band: the two overlapping bands are incompletely filled and the body is conductive. The Fermi surface is shown in Fig. 21b.

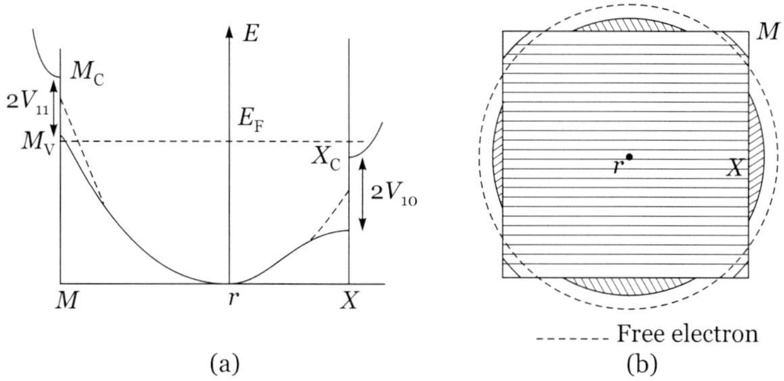

Figure 21

It should be noted that in the nearly free electron theory inequality (2) is written:

$$\frac{\hbar^2}{2m}\frac{2\pi^2}{a^2} - V_{11} > \frac{\hbar^2}{2m}\frac{\pi^2}{a^2} + V_{10} \text{ or } \frac{\hbar^2}{2m}\cdot\frac{\pi^2}{a^2} > V_{10} + V_{11}.$$

These results can be generalized to a 3D solid and can explain the fact that divalent elements such as Mg, Be, and Pb. have a metallic character while Si, Ge, Se, and Te are semiconductors. In the latter, the strong attraction between electrons and the lattice (covalent bonds), as in the case of the alkali halides, results in large or very large bandgaps and the overlap of the conduction and valence bands.

(c) $E(M_v)$ = 9 eV, $E(X_c)$ = 7 eV: it is thus a conductor.

Exercise 5b: Nearly free electrons in a rectangular lattice

We consider a rectangular lattice with parameters a= 3 Å and b = 4 Å with an atom of species A located at (0, 0) and an atom of species B at (1/2, 1/2) (see Chapter I, Ex. 3a, Question (2) and Chapter III, Ex. 14).

The valence electron dispersion relation obeys the nearly free electron theory that is to say that in a given $[m, n]$ direction, it follows the theory of free electrons except when the wave vector k approaches the limit of the Brillouin zone (here the first). In this latter situation, the dispersion curve deviates symmetrically from the value of $E^0(m, n)$ given by the theory of free electrons and the total amplitude of this deviation is U(m, n) [distance separating $E_1(m, n)$ from $E_2(m, n)$].

(1) Sketch the dispersion curves in the directions [00] $\rightarrow$ [10]; [00] $\rightarrow$ [01]; and [00] $\rightarrow$ [11].

(2) There are two almost free electrons per "molecule" AB. What relationship (inequality) must exist between U(m, n), a and b so that the material is insulating?

In the reverse assumption (the material is conductive), find the characteristics of the Fermi surface of the first Brillouin zone and specify the location of electrons and holes.

(3) $U(1,0) = 2$ eV; $U(0,1) = 1$ eV; $U(11) = 1.5$ eV. Is the material a conductor or an insulator?

(4) The atoms A and B are chemically identical. What is the shape of the new Brillouin and the modulus of k(m, n) for which discontinuities occur? Reconsider the conductive or insulating character of the material according to the amplitude of the discontinuities in the Brillouin zone. Successively considered the case where each atom A gives 1 and then 2 almost free electrons. ($\hbar^2/2m \approx 3.8$ eVÅ^2)

Solution:

(1) The reciprocal lattice is a simple rectangular lattice where $A = 2\pi/a$; $B = 2\pi/b$. (see Chapter I, Ex. 18, Fig. 18 for its construction).

In the [10] direction, the boundary of the Brillouin zone is $k = \pi/a$; $E^0(10) = (\hbar^2/2m)(\pi/a)^2$.

In the [01] direction: $k = \pi/b$; E^0 (01) = $(\hbar^2/2m)$ $(\pi/b)^2$.

In the [11] direction: $k = \pi(1/a^2 + 1/b^2)^{1/2}$.

and E^0 (11) = $(\hbar^2/2m)$ π^2 $(1/a^2 + 1/b^2)$.

The characteristics of the curves are shown in Fig. 22 where the dashed lines indicate free electrons dispersion: $E = \hbar^2k^2/2m$.

(2) The first Brillouin zone can accept two electrons per molecule. The material will be insulating when this zone is complete (a full band does not participate in the electron transport).

In order for this to occur, the bands must not overlap or, here, $E_1(11) < E_2(01)$, that is to say

$$E^0(11) - \frac{U_{11}}{2} < E^0(01) + \frac{U_{01}}{2} \text{ or } U_{01} + U_{11} > \frac{\hbar^2}{m}\frac{\pi^2}{a^2}$$

For the reverse assumption (shown in Fig. 22), the Fermi level will intercept the second band in the [01] direction, $E_F > E_2(01)$, and the first band in the direction [11], $E_F < E_1(11)$. The shapes of the Fermi surface are shown in Fig. 23. Note that the area covered by the nearly free electrons is identical to that covered by free electrons: $\pi k_F^{02} = 4\pi^2/ab$.

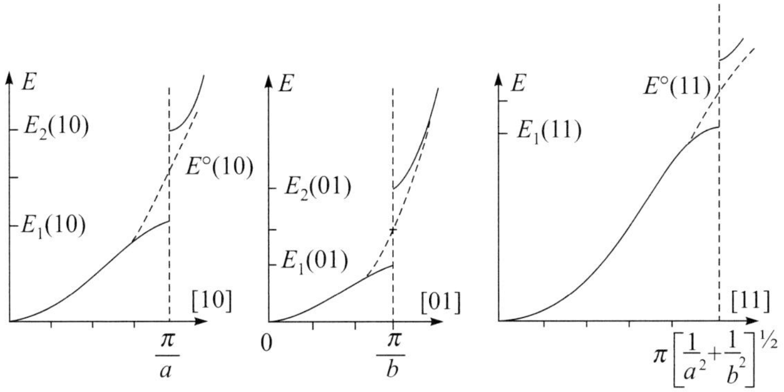

Figure 22

(3) $U_{01} + U_{11} = 2.5$ eV $< \hbar^2\pi^2/ma^2 = 8.3$ eV. Given the shape of the Fermi surface, the material is a conductive and will probably be semi-metallic, such as Sb or Bi.

(4) The reciprocal lattice of a rectangular centered lattice is itself rectangular centered. (In Fig. 23, remove the odd $h + k$ points such that (0, 1), (10); see also Chapter I, Ex. 19, Fig. 21.

The first Brillouin zone has a truncated diamond shape and corresponds to the first two Brillouin zones of the previous questions (constructed from the points (1,1); $(\bar{1},1)$; $(1,\bar{1})$; and $(\bar{1},\bar{1})$. If A is monovalent, the Fermi sphere corresponds to that shown in Fig. 23 for free electrons (radius k_f) because it has the same number of electrons as above. It is therefore a conductor.

However, if there are two electrons per atom (i.e., 4 per area $a \times$ b), its radius will be $\sqrt{2}$ times larger and it will overlap partly the second Brillouin zone in the [11] direction (see Fig. 12b in Chapter III, Ex. 14, with $k_D = k_{F^-}$).

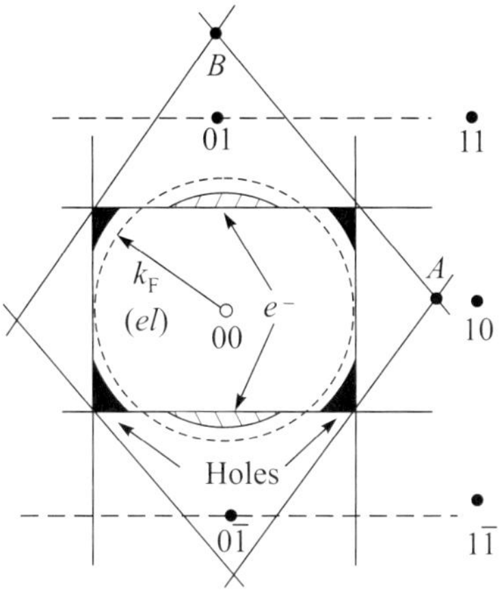

Figure 23

If the dispersion curve is unchanged (see Fig. 22) for the [11] direction, the discontinuities in the [10] and [01] directions are brought respectively to point A (Fig. 23) and to point [01] ($2\pi/b$) and not π/a and π/b. Following the same reasoning as above, the amplitude of these discontinuities (E_2 (A), E_2 [01] < or >E_1 relative to the point located at the mid-point between 00 and 11 which will determine the conductive or insulating nature of the material.

Exercise 6: Phase transition in the substitution alloys: application to CuZn alloys

Copper crystallizes in the fcc lattice with parameter a. The progressive substitution of copper atoms (one free electron per atom) by zinc atoms (two electrons per atom) causes an increase in the radius of the Fermi sphere without altering the crystallographic structure (phase α) of the alloy, until the Fermi sphere comes into contact with the first Brillouin zone, which results in the appearance of the β-phase (bcc). In this problem, we determine the relative concentration ρ_{Zn} of zinc atoms $\left(\rho_{\text{Zn}} = \dfrac{N_{\text{Zn}}}{N_{\text{Cu}} + N_{\text{Zn}}}\right)$, which corresponds to the phase transition ($\alpha \rightarrow \beta$) as well as the phase change $\beta \rightarrow \gamma$.

(a) Find the general expression for the Fermi wave vector k_F as a function of electronic concentration n ($n = N/V$) of free electrons.

(b) Specify the numerical value of k_F for pure copper with $a = 3.6$ Å.

(c) For an fcc structure, what is in the distance k_M in reciprocal space that separates the center of the Brillouin zone from its nearest face?

(d) Deduce the expression for ρ, that is ρ_0, for which $k_F = k_M$, which corresponds to the phase transition between α and β. One must first find k_F (evaluated above) as a function of ρ and a.

(e) When ρ becomes ρ_0, the lattice becomes bcc (phase β) with lattice parameter a_l. What is the new expression for k_F, denoted k'_F as a function of ρ and a_l? What is the new minimal distance k'_M that separates the origin of the new first Brillouin zone? Compare k'_M to k'_F when $\rho = \rho_0$ and deduce, see previous approach, the expression for $\rho = \rho'_0$, for which $k'_F = k'_M$ (phase transition $\beta \rightarrow Y$).

(f) It is assumed that copper and zinc atoms behave as hard spheres with approximately the same radius. This means that in a given phase, the substitution of copper by zinc does not change the crystalline parameter. Can you deduce the relationship between a and a_1 in the phase change $\alpha \rightarrow \beta$?

Find the energy $U(e)$ of the free electron gas (at 0 K) relative to N atoms in the α-phase and next in the β-phase. Comment on the resultant curve. Explain qualitatively the causes of the phase transition.

Solution:

(a) $k_F = (3\pi^2 n)^{1/3}$, where $n = \frac{N}{V} = \frac{4}{a^3}$ (for monovalent fcc elements)

(b) $k_F = (12\pi^2)^{1/3} \cdot (1/a) = 1.36$ Å^{-1}

(c) The reciprocal lattice of an fcc lattice is a bcc lattice and the distance between the center of the first Brillouin zone to the nearest side is equal to the half of the distance between the origin and the (111) point:

$$k_M = \frac{1}{2}\frac{2\pi}{a}(h^2 + k^2 + \ell^2)^{1/2} = \frac{\pi}{a}\sqrt{3}.$$

This result can also be obtained by considering that the first allowed reflection of the fcc lattice is the (111) reflection and the distance that separates two successive reflection planes is $d_{111} = \dfrac{a}{(h^2+k^2+\ell^2)^{1/2}} = \dfrac{a}{\sqrt{3}}$. At normal incidence, the Bragg's law gives $2d_{111} = \lambda_m$ or $\lambda_m = \dfrac{2a}{\sqrt{3}}$ and $k_M = \dfrac{2\pi}{\lambda_m} = \dfrac{\pi\sqrt{3}}{a}$.

(d) $k_F = (3\pi^2 n)^{1/3}$, where $n = 4(1+\rho)a^3$.

$\alpha \to \beta,\ k_F = k_M \to 1+\rho = \dfrac{\pi\sqrt{3}}{4} = 1.36$

or for $\alpha \to \beta$, $Zn_{0.36}Cu_{0.64}$.

(e) There are only two atoms per bcc cell from which $k'_F = (3\pi^2 n')^{1/3}$, where $n' = 2(1+\rho)/a_1^3$.

The new reciprocal lattice points are such that $h + k + l$ are even. The point nearest to the origin is the (110) point. Thus $k'_M = \pi\sqrt{2}/a_1$ and $k'_M < k'_F\,(\rho_0)$.

The transition $\beta \to \gamma$ occurs when $\rho = \rho'_0$, where $1+\rho'_0 = \dfrac{\pi\sqrt{2}}{3} = 1.48$, thus $Zn_{0.48}Cu_{0.52}$.

(f) The radius of atoms is such that $r = \dfrac{a\sqrt{2}}{4} = \dfrac{a_1\sqrt{3}}{4}$ or $a/a_1 = \sqrt{3/2}$, assuming that the atoms are in contact either along the diagonal relative to a face (fcc) or along the diagonal of the body-centered cube.

$$U(e) = \frac{3}{5} N E_F$$

$$\text{Phase } \alpha: \quad U_e = \frac{3}{5} N \frac{\hbar^2}{2m} \frac{[12\pi^2(1+\rho)]^{2/3}}{a^2}$$

$$\text{Phase } \beta: \quad U_e = \frac{3}{5} N \frac{\hbar^2}{2m} \frac{[6\pi^2(1+\rho)]^{2/3}}{a_1^2}$$

In both cases, the energy of the electron gas increases as $(1 + \rho)^{2/3}$ but, after simplification, the coefficient in front of $1 + \rho$ must be proportional to $\dfrac{2^{2/3}}{a^2} \approx \dfrac{1.587}{a^2}$ (fcc) or proportional to $3/2a^2 = 1.5a^2$(bcc).

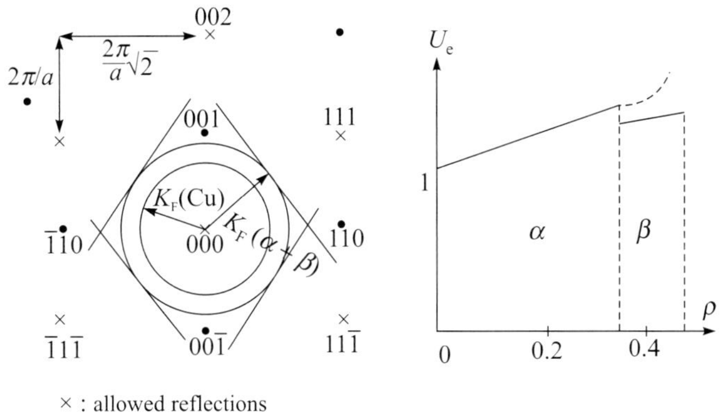

Figure 24

In the light of Fig. 24, we understand the reasons for the phase changes. The addition of Zn atoms provokes an increase in k_F. When the Fermi surface touches the first Brillouin zone (Fig. 24, left), this growth can no longer be isotropic because the extra electrons are forced to stay in the corners which would result in an increased slope in the energy evolution (Fig. 24, right: dotted line). The system recrystallizes in a structure where the Fermi surface will not be in contact with the first Brillouin zone and where the slope in the energy evolution will be smoother, until this mechanism starts over. Hume-Rothery was the first to draw attention to the strong influence of the electron concentration on the phase transitions in substitution alloys. Even though the reality is more complex (curvature of the bands: see preceding example, coexistence of phases $\alpha + \beta$, existence of ordered and disordered phases of β (see Chapter I, Pb. 5), it is remarkable that the simplified calculation above gives almost the exact result ($\alpha \rightarrow \beta : 1.38$ instead of 1.36; $\beta \rightarrow \gamma : 1.48$).

Exercise 7: Why nickel is ferromagnetic and copper is not

Figure 25 shows a naïve sketch of the density of electronic states (outer shells) of metals such as copper and nickel above the Curie point. The bands result from the 3d levels and are divided into two subbands for spins ↑ and ↓ and are juxtaposed next to the 4s levels. To simplify the problem, the densities of states are assumed to be constant in the various bands:

$g(E) = C_1$ (in the interval $0 < E < \infty$) for the band originating from the 4s levels

$g(E) = C_2$ (in the interval $0 < E < E_d$) for the band originating from the 3d ↑ and 3d ↓ levels.

In addition we recall that there are 10 electrons per atom (5 ↑ and 5 ↓) as can normally fill the d band.

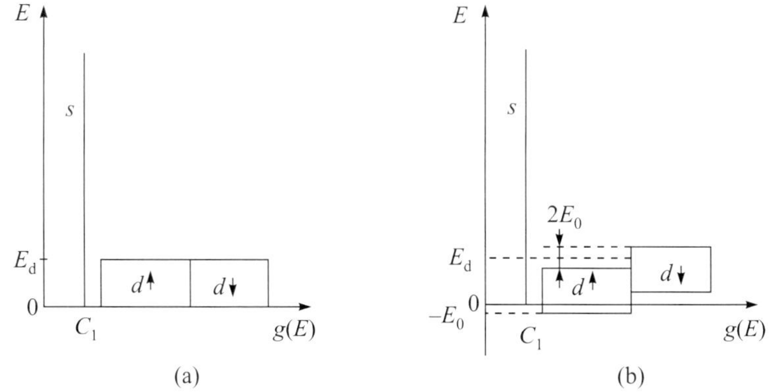

Figure 25

(a) Using the following values: E_d = 5 eV, C_1 = 0.14 e^-/eV and knowing that one must fill all bands with 11 and 10 electrons per atom respectively, find the position of the Fermi level of copper (Z = 29 and 11 electrons 4s + 3d per atom) and of nickel (Z = 28 and 10 electrons 4s + 3d per atom).

(b) How many available states remain in the d bands of nickel and of copper?

(c) Below the Curie point, the exchange interaction between spins shifts the d subbands by $\pm E_e$ from their original position (see Fig. 25b).
In the case of nickel, find the overpopulation of ↑ states compared to ↓ states and deduce the number of Bohr magneton assigned to each atom when E_e = 0.27 eV.
To simplify the calculation take $f(E) = 1$ when $E < E_F$ and $f(E) = 0$ when $E > E_F$.

Solution:

(a) For N atoms having z valence electrons, we have:

$$\int_0^{\infty} f(E)\, g(E)\mathrm{d}E = Nz.$$

With the notations and simplifications in the statement, reduced to an atom this relation becomes

$$\int_0^{E_\mathrm{F}} C_1 \mathrm{d}E + \int_0^{E_\mathrm{F},\, E_\mathrm{d}} C_2 \mathrm{d}E + \int_0^{E_\mathrm{F},\, E_\mathrm{d}} C_2 \mathrm{d}E = z.$$

The filling of the d subbands will also not exceed five electrons per atom if

(α) $E_\mathrm{F} < E_\mathrm{d}$—upper bound of the second and third integrals will be E_F (case for Ni).

(β) $E_\mathrm{F} > E_\mathrm{d}$—upper bound of the second and third integrals will be E_d (case for Cu) and each integral corresponds to five electron states per atom so that $C_2 = 5/E_\mathrm{d}$.

The Fermi energy of Cu reduces to $C_1 E_\mathrm{F} = 1$ or $E_\mathrm{F} = 7.17$ eV.

The Fermi energy of Ni reduces to $(C_1 + 2C_2)\, E_\mathrm{F} = 10$ or $E_\mathrm{F} =$ 4.67 eV.

(b) The number of non-occupied states in each d band obeys $P = \int_{E_\mathrm{F}}^{E_\mathrm{d}} C_2(E)\mathrm{d}E$ = 0.33 electrons/atom for Ni even though in Cu, $p = 0$ because the d subbands are filled.

(c) Taking into account the exchange energy the relation in (a) becomes

$$\int_0^{E_\mathrm{F}} C_1 \mathrm{d}E + \int_{-E_\mathrm{e}}^{E_\mathrm{F},\, E_\mathrm{d}-E_\mathrm{e}} C_2 \mathrm{d}E + \int_{E_\mathrm{e}}^{E_\mathrm{F},\, E_\mathrm{d}+E_\mathrm{e}} C_2 \mathrm{d}E = z.$$

The Cu Fermi energy remains unchanged and the resulting magnetic moment is zero. The Fermi energy in Ni is also unchanged, but the difference in spin population is

$$n_\mathrm{B} = \int_{-E_\mathrm{e}}^{E_\mathrm{F},\, E_\mathrm{d}-E_\mathrm{e}} C_2 \mathrm{d}E - \int_{E_\mathrm{e}}^{E_\mathrm{F},\, E_\mathrm{d}+E_\mathrm{e}} C_2 \mathrm{d}E.$$

For Ni the number of Bohr magneton is $n_\mathrm{B} = 2E_\mathrm{d}E_\mathrm{e}/5 = 0.54$.

Figures 25c and 25d represent the energy diagrams corresponding to Ni above and below the Curie point. The position of the Fermi level of copper, E_F(Cu) is also indicated for a comparison.

- The above-mentioned discussion is provided to explain that the ferromagnetism of the nickel and the non-ferromagnetism of copper are qualitatively correct. For the exact numerical values and for more details, see corresponding chapter in Ref. [15] to check a, b, or c.

This explanation, however, is incomplete because it does not consider in details the reason for which the two bands of the d-electrons are energetically shifted. This shift, due to the coupling energy between the spins of neighboring atoms (exchange energy), is very sensitive to the interatomic distances r_0 and the exchange energy is positive for Fe, Co, and Ni (as well as Cu), while it is negative for Cr and Mn (which are anti-ferromagnetic).

- Under the action of an external induction the s band will be responsible for the paramagnetism of free electrons (see Chapter IV, Ex. 22).

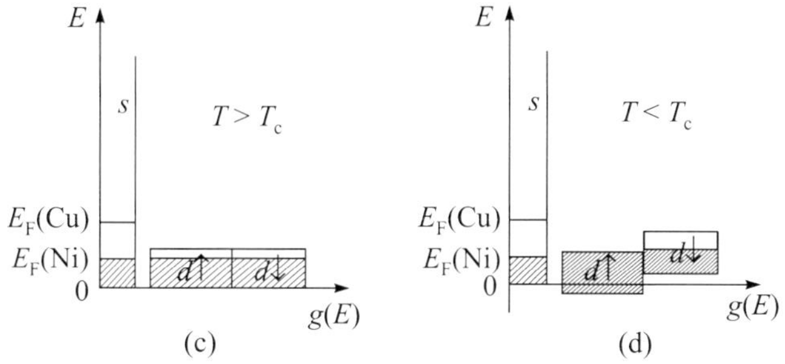

Figure 25

Exercise 8: Cohesive energy of transition metals

The atoms of the transition elements are characterized by an outer electron shell of incomplete d electrons with an energy E_d close in energy to the subsequent s-electrons. In the solid state, d levels form a band of width W, which is centered on E_d, and from which the density of states is supposed constant: $g(E) = C$, and which can accommodate 10 electrons per atom (see Fig. 26a).

(a) Consider a transition element with n_d d electrons per atom. Find the expression giving the total energy of n_d electrons (denoted E_M) and compare it with their initial energy in the atomic state (denoted E_A) in order to deduce the cohesive energy, $E_C = E_A - E_M$ for the element, assumed for simplicity to be at 0 K. How does E_c vary as a function from n_d?

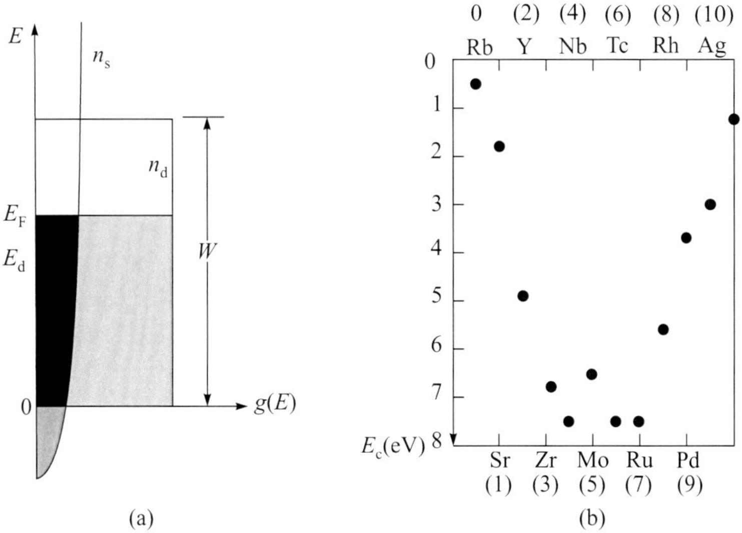

Figure 26

(b) The cohesive energy of the elements from $Z = 37$ (Rb) to $Z = 47$ (Ag) is shown in Fig. 26.
Compare these experimental results with those deduced from the above model. Assume $T = 0$ K and start from Rb ($n_d = 0$), the additional electrons occupying exclusively the 4d band: the n_d values are indicated between () in Fig. 26b. Take $W = 6$ eV and $E_D = \frac{W}{2}$. Discuss these results.

Solution:

(a) Taking the origin at the bottom of the d band, we find

$$E_A = n_d E_D; \quad E_M = \int_0^{E_F} E_g(E) \cdot dE = C\frac{E_F^{\,2}}{2}$$

We can write $n_d = \int_0^{E_F} g(E) \cdot dE = C \cdot E_F$ and

$$10 = \int_0^W g(E)dE = CW.$$

Basic operations lead to

$$E_C = n_d E_D - \frac{W}{20} n_d^{\,2} \text{ or when } E_D = \frac{W}{2} \text{ to } E_C = \frac{W}{2}\left(n_d - \frac{n_d^{\,2}}{10}\right).$$

The cohesive energy follows a parabolic law as a function of n_d electrons (per atom) occupying the d band.

This cohesive energy will be maximum at $n_d = 5$.

(b) See Fig. 26c.

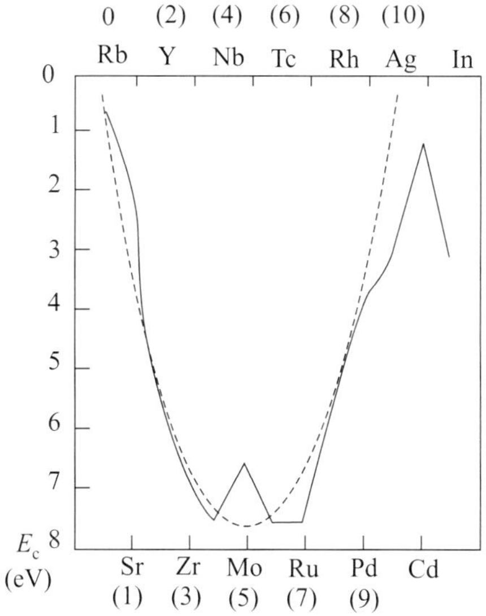

Figure 26 (c)

Comments: *Cohesive energy of transition metals*

The constant density of states model for the d bands of transition metals was originally proposed by J. Friedel [32]. This model captures well the high cohesion energy of these metals (also inducing a strong bulk modulus, *B*, and high melting temperature), especially for those with a half-full d band such as tungsten and molybdenum.

The excellent agreement is due to the adequate choice of W. In reality, the model could be improved by taking into account:

(i) the energy shift, α, between atomic levels and the centre of the d band.

(ii) the $\sqrt{E}$ evolution of the density of states s levels—with partial occupation from 0.5 to 1 electron per atom, as indicated in Fig. 26 with respect to osmium.

(iii) the d bandwidth, *W*, which varies from one element to another depending on the distance r_0 between nearest neighbors ($W \propto r_0^{-5}$, see W. A. Harrison [12] page 487).

These modifications result in a better agreement with the experimental results without changing the explanation of the essential physical phenomenon, high cohesion, and its evolution.

In addition, note that the transition metals are also characterized by an electronic specific heat C_e that can be 10 times greater than normal metals and thus by an effective d-electron mass that is very large:

$$\frac{m_e^x}{m_0} \approx 30.$$

These two phenomena are explained by the narrowness of the d band, which leads to a large density of states, and C_e is related to the density of states at the Fermi level, $g(E_F)$ by (see Chapter IV, Ex. 19):

$$C_e \approx \left(\frac{\pi^2 k_B{}^2 T}{3}\right)\frac{g(E_F)}{N}$$

We thus find the same trend for m^x by considering that $g(E) \propto m^{x\frac{3}{2}} \sqrt{E}$ and that the more one band is flat, the more m^x is large: $m^x \propto 1/W$ (Ex. 1 on tight binding in the present chapter).

Exercise 9: Semi-metals

The band structure of semi-metals (As, Sb, Bi) is characterized by the fact that the valence and conduction bands overlap in such a manner that the corresponding dispersion relations are described by:

$$E(V) = -(\hbar^2/2m_h{}^x)k^2 + E_0 \text{ and } E(C) = (\hbar^2/2m_e{}^x)k^2$$

where $E_0 > 0$, $m_e{}^x$ and $m_h{}^x$ are the effective masses of electrons and holes such that $m_e{}^x < m_h{}^x$.

(a) Draw the band diagram for a semi-metal and compare it with that of an alkali metal and a semiconductor.

(b) Starting from the density of states for a free electron gas, and including the corrections induced by the effective masses $m_e{}^x$ and $m_h{}^x$, find the expression for the density of conduction electrons n_e and the density of holes n_h, at 0 K.

Knowing that in semi-metals $n_e = n_h$, express the Fermi energy E_F as a function of E_0, $m_e{}^x$, and $m_h{}^x$. Can one reach this result more quickly?

From the data, express the common density $n_0 = n_e = n_h$.

(c) *Numerical application*: E_0 = 0.4 eV:
$m_e^x/m_0 = 0.05$; $m_h^x/m_0 = 0.1$.
Find E_F and n_0 using $\hbar^2/2m_0 \approx 3.8\ \text{eV}\text{Å}^2$
Find the electrical conductivity σ for this type of material taking the value for the mean free time for both electrons and holes to be $\tau = 2 \times 10^{-14}$ s.
Compare the values of n_0 and σ to typical values in good metals.

(d) What are the predictable electronic properties of semi-metals including the sign of the Hall effect, the effect of temperature on σ, and the electronic specific heat, etc.?
Compare the corresponding evolutions with those of metals and semiconductors. $(\hbar, m_0, e)$

Solution:

(a) See Fig. 27. Note the sharper curvature of E_c compared with E_v due to $m_e^x < m_h^x$.

(b) We use the expression for the density of states for 3D free electrons: $g(E) = \frac{V}{2\pi^2}\left(\frac{2m_0}{\hbar^2}\right)^{3/2}\sqrt{E}$; substitute m_e^x (and then m_h^x) to m_0 in order to take into account the bending of the different bands, and choosing the origin of energies at the bottom of E_c. Considering a unitary volume for the evaluation of n_e and n_h, we find

$$n_e = \int_0^{E_F} g_e(E)dE, \text{ where } g_e(E) = \frac{1}{2\pi^2}\left(\frac{2m_e^x}{\hbar^2}\right)^{3/2}\sqrt{E}$$

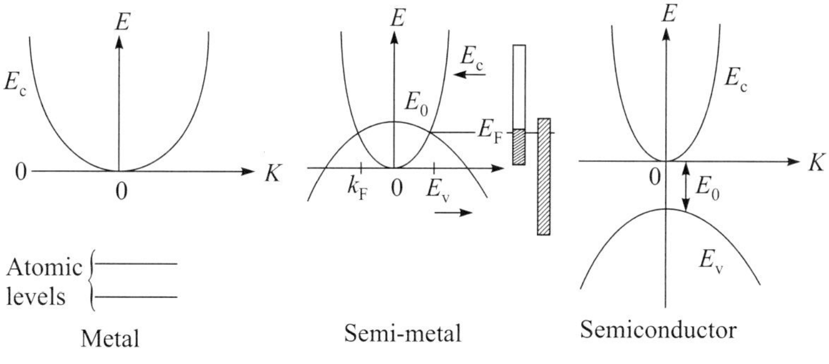

Figure 27

and $n_h = \int_{E_0}^{E_F} g_h(E)\mathrm{d}E$, where $g_h(E) = \frac{1}{2\pi^2}\left(\frac{2m_h^x}{\hbar^2}\right)^{3/2}\sqrt{E_0 - E}$,

which after integration becomes $n_e = \frac{1}{3\pi^2}\left(\frac{2m_e^x}{\hbar^2}\right)^{3/2} E_F^{3/2}$ and

$$n_h = \frac{1}{3\pi^2}\left(\frac{2m_h^x}{\hbar^2}\right)^{3/2}(E_0 - E_F)^{3/2}$$

The equality $n_e = n_h$ leads to $E_F = E_0 \frac{m_h^x}{m_e^x + m_h^x}$.

One can foresee this result by observing that the intrinsic nature of this type of material leads to an equality between the volume in k-space occupied by electrons to that occupied by holes, leading to a Fermi level located at the intersection of the two parabolas, as naively shown in Fig. 27b (even if the holes and electrons are not localized at the same point in this space).

$E_F = \left(\frac{\hbar^2}{2m_e^x}\right)k_F^2$; $E_F = -\left(\frac{\hbar^2}{2m_h^x}\right)k_F^2 + E_0$ or effectively

$$E_F = \frac{E_0 m_h^x}{(m_e^x + m_h^x)}$$

$$n_0 = \frac{1}{3\pi^2}\left(\frac{2}{\hbar^2}\frac{m_e^x \cdot m_h^x}{m_e^x + m_h^x}\right)^{3/2} E_0^{3/2}.$$

The latter expression is homogenous in the density of particles per $Å^{-3}$ if we take into account $\hbar^2/2m_0 = 3.8\ \mathrm{eV}Å^2$

(c) $n_0 = 7 \times 10^{-6}\,\mathrm{part}\cdot Å^{-3} = 7 \times 10^{24}\ \mathrm{part}\cdot \mathrm{m}^{-3}$

$E_F = 0.266\ \mathrm{eV}$

$\sigma = n_e e|\mu_c| + n_h e\mu_h$, where $\mu_e = e\tau/m_e^x$

$$\sigma = n_0 e^2 \tau\left(\frac{1}{m_h^x} + \frac{1}{m_h^x}\right) = 0.12 \times 10^6\,\Omega^{-1}\mathrm{m}^{-1}.$$

The electronic density here is four orders of magnitude smaller than that of alkali metals ($n_0 = 5 \times 10^{28}\ e^-\ \mathrm{m}^{-3}$) while the conductivity will only be two orders of magnitude smaller due to the lower effective masses. The mobility is thus larger: $\rho(\text{Na}) = 5 \times 10^{-6}\,\Omega$ cm and $\sigma(\text{Na}) = 2 \times 10^{7}\,\Omega^{-1}\,\text{m}^{-1}$.

(d) *Comments: Semi-metals*

Arsenic, bismuth, and antimony are pentavalent elements that crystallize with two nonequivalent atomic positions in the elementary lattice cell. Of the 10 valence electrons per elementary lattice, eight completely fill four valence bands and the two others partially fill the fifth valence band and very partially the conduction band.This explains, as for intrinsic semiconductors, the equality between the densities of electrons and holes, whereas in normal metals only conduction electrons have a significant density of states. This equality exists at 0 K and the value of n_0 is not significantly affected as temperature increases, contrary to the semiconductors which are theoretically insulators at 0 K. Despite this equality in the density of carriers, the sign of the Hall voltage is given by that of electrons due to their larger mobility. The magnitude of the Hall constant in bismuth is the largest of all metals due to the small value of n. The electronic specific heat of semi-metals follows a γT law as for normal metals (the inverse of semiconductors), but the coefficient γ varies between 5% (Bi) and 30% (Sb) compared with that determined from the theory of free electrons.

Excluding the arbitrary choice of the effective mass, the numerical data obtained here agree with reality:

$n_0(\text{Bi}) = 3 \times 10^{17}\ \text{cm}^{-3}$; $n_0(\text{As}) = 5 \times 10^{19}\ \text{cm}^{-3}$; $n_0(\text{Sb}) = 2 \times 10^{20}\ \text{cm}^{-3}$.

The fact that the largest carrier mobility in this material partially compensates for their small density (compared to normal metals) explains why the resistivity is only 10 to 100 times smaller for antimony (ρ = 39 μΩcm) and for arsenic (ρ = 33 μΩcm). As for normal metals, one can expect that the resistivity obeys the Matthiessen law and increases with temperature as well as with impurity concentration (see Chapter IV, Ex. 16).

In addition to the existence of two types of carriers, the analogy with semiconductors lies in the analogous approach leading to the law of mass action as well as the fact that, to go from one band structure to the other one, one simply moves the conduction band down below the valence band at $E_0 > 0$ for semi-metals while $E_g < 0$ for semiconductors, the origin of energy being chosen as in Fig. 27.

Exercise 10: Elementary study of an intrinsic semiconductor

Consider a semiconductor characterized by a constant density of states and equal to C in both the valence and conduction band. These bands, separated by E_g, have a respective width of E_v and E_c (see Fig. 28).

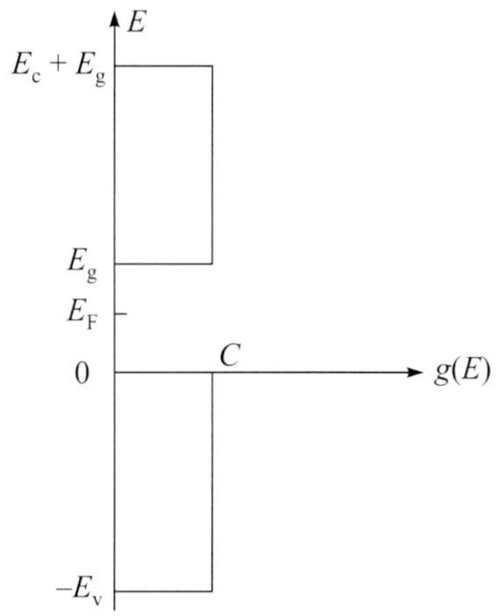

Figure 28

(a) Knowing that the Fermi level E_F is in the bandgap ($5k_BT < E_F < E_g - 5k_BT$), leading to the conventional simplification for $f(E)$, find the expression for the number of electrons n in the conduction band and the number of holes h in the valence band respectively as a function of the data and absolute temperature.

(b) Deduce the expression for the volumetric density of intrinsic carriers n_i at ambient temperature ($k_BT = 25$ meV) as well as the position of the Fermi level $E_F(i)$.
Numerical application: $E_g = 0.7$ eV, $E_c = E_v = 5$ eV, and $C = 2 \times 10^{21}$ electrons/cm^3/eV.

(c) At the same temperature, what is the intrinsic electronic conductivity of this material? Take $\mu_h = \mu_e = 1000$ cm^2/V·s.

(d) The material is doped with N_0 boron impurities. Find the expressions and the new values of n(d), h(d), σ(d), and E_F(d) with N_0 = 10^{12} impurities per cm^3 and next with N_0 = 10^{14} impurities per cm^3.

Solution:

(a) • $n = \int_{E_g}^{E_g+E_C} f_e(E)\cdot g(E)\cdot \mathrm{d}E$, where

$$f_e(E) = \frac{1}{e^{\frac{E-E_F}{k_B T}}+1} \approx e^{(E_F-E)/k_B T}$$

from which we find

$$n \approx \int_{E_g}^{E_g+E_c} Ce^{(E_F-E)/k_B T}\cdot \mathrm{d}E \cong CkTe^{(E_F-E_g)/k_B T}[1-e^{-E_e/k_B T}]$$

• $h = \int_0^{-E_v} g(E)\cdot f_h(E)\mathrm{d}E$, where $f_h(E) = 1 - f_e(E) \approx e^{(E-E_F)/k_B T}$

so that $h = CkTe^{-E_F/k_B T}\cdot [1 - e^{-E_F/k_B T}]$.

(b) $n_i^2 = (CkT)^2 e^{-E_g/k_B T}\,[1 - e^{-E_V/k_B T}]$.

This expression is very similar to the law of mass action (analyzed in detail in Pb. 4 for 3D semiconductors), especially if one notices that numerically the terms associated with the bandwidth are negligible.

For $k_B T \approx 0.025$ eV; $n_i \approx 1.5 \times 10^{13}$ p$\cdot$ cm^{-3}

$E_F(i)$ can be deduced from the two values of h (or n) determined in (a) and from $h = n_i$, determined in (b). Neglecting the terms in [...], this leads to $E_F(i) = E_g/2 = 0.375$ eV.

(c) $\sigma = n_i e(\mu + \mu_e) = 4.8 \times 10^{-3}\,\Omega^{-1}\,\mathrm{cm}^{-1}$

(d) Boron belongs to the third column of the periodic table and thus it acts as a 'p' dopant if (and only if) N_0 (ionized) > n_i. Otherwise (N_0 = 10^{12} impurities per cm^3) the material will remain essentially intrinsic and the results from (b) and (c) will be unchanged. When N_0 = 10^{14} impurities per cm^3, $t(d) = N_0$; $n(d) = n_1^2/N_0 = 2.25 \times 10^{12}\, e$ cm^{-3} (compensation); $\sigma(d) = t(d)\,\mu_t e\ 1.6 \times 10^{-2}\,\Omega^{-1}\,\mathrm{cm}^{-1}$.

E_F can be deduced from

(a) $h = Ck_B Te^{-E_F/k_B T} = N_0$

(b) $E_F = -k_B T \log(N_0/CkT) = 0.328$ eV

When a semiconductor is p-doped, the Fermi level is lowered. When it is n-doped it is raised.

Exercise 11: Density of states and bandgap

Figure 29 shows the evolution of the majority carrier concentration (in cm^{-3}) in a semiconductor as a function of temperature.

Analyze this evolution by explaining the reason of each of the three regimes represented.

Find graphically the width of the bandgap E_g and the ionization energy of the impurities E_i in eV, thus the density (in cm^{-3}) of impurities N_0. Assuming that these are due to residual *p*-type impurities, find the electrical conductivity σ when the concentrations of Fig. 29 are respectively equal to 10^{16}, 10^{15}, and 10^{14} cm^{-3} where $\mu_e = 1400$ $cm^2/V{\cdot}s$; $\mu_h = 500$ $cm^2/V{\cdot}s$. (k_B, e)

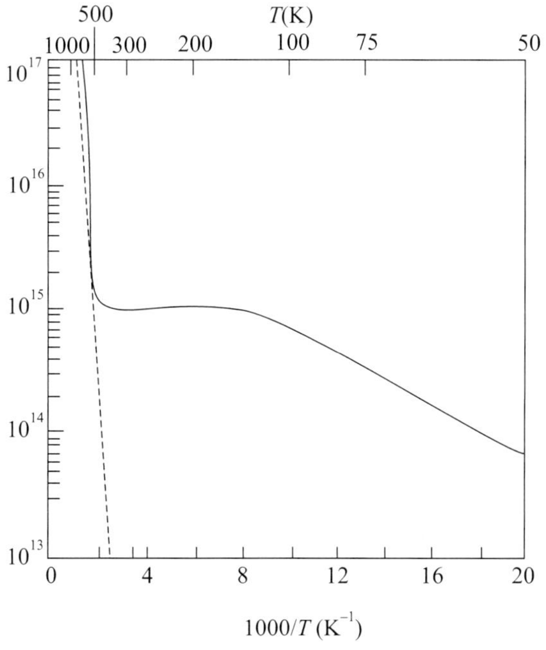

Figure 29

Solution:

Inspired from S. M. Sze [13], the curve shows three expected regimes (see Pb. 4 for more details).

- The intrinsic regime is observed at high temperatures: $n_i = \sqrt{N_c V_v} \exp-(E_g/2k_B T)$.
  In the logarithmic scale, the slope of the dashed lines (extrapolation from 10^{17} to 10^{13}) is equal to $-(E_g/2k_B)\times\frac{1}{T}$.
  Thus, we have $E_g = \frac{2k_B}{11.25\times10^{-3}}\frac{2.3[17-13]}{e} \approx 1.2\,\text{eV}$ with a relatively poor precision of 10%.
- The plateau corresponds to the regime where all the impurities are ionized. It is the case at room temperature. Their density corresponds to $N_0 = 10^{15}\text{ cm}^{-3}$.
- As temperature is lowered the freeze out regime of a part of impurities is reached and the density of carrier obeys $n = \sqrt{N_C N_0} \exp-E_i/2k_B T$.
- By evaluating the slope of the line as for E_g, we find that $E_i = 44$ meV.

We are thus exploring silicon ($E_g \approx 1.1$ eV) doped with either n-type impurities such as phosphorus or p-type impurities such as boron (where $E_i = 45$ meV for both cases). In the intrinsic regime $n_e = n_h = n_i = 10^{16}\text{ cm}^{-3}$ from which $\sigma = n_i e\,(\mu_h + \mu_e) = 3\Omega^{-1}\text{ cm}^{-1}$.

On the contrary, note that in the other two regimes we have $n_h << n_e$ from which $\sigma = n_h e \mu_h$ so that $\sigma = 8\times10^{-2}\ \Omega^{-1}\text{ cm}^{-1}$ and $8\times10^{-3}\ \Omega^{-1}\text{ cm}^{-1}$ for n respectively 10^{15} and 10^{14} cm^{-3}.

Exercise 12: Conductivity of semiconductors in the degenerate limit

The application of the law of mass action is limited to $5k_B T < E_F < E_g - 5\,k_B T$ when the semiconductor is nondegenerate. Show that the electrical conductivity σ of an n-doped degenerate semiconductor, with $E_F = E_g - 5\,k_B T$, is virtually independent of the nature of this semiconductor. It will be admitted, however, that the mobilities vary inversely to the effective masses and are inversely proportional to temperature.

Numerical application: Find σ_1 (limit) for Si ($\mu_c = 0.13\text{ m}^2/\text{Vs}$; $m_e^x = 0.2\,m_0$) and for GaAs ($\mu_c = 0.88\text{ m}^2/\text{Vs}$; $m_e^x = 0.07\,m_0$) at 77 K. Use the formula F1.

Solution:

Starting from the expression n_e given in the Course Summary, Section 7, we find

$$n_c/N_c = e^{(E_F - E_g)/k_BT} = e^{-5} = 6.7 \times 10^{-3},$$

$$\text{where } N_c = 2\left(\frac{2\pi m_e^x k_B T}{h^2}\right)^{3/2}.$$

$$\sigma = n_e e \mu_e \text{ and } \mu_e \approx A/m_e^x T, \text{ thus } \sigma \approx B\sqrt{m_e^x T}.$$

The result is independent of the bandgap and varies slightly with m^x and T.

Numerical application:

$N_c = 2.5(m_e^x/m_0)^{3/2}(T/300)^{3/2}10^{19}\ \text{cm}^{-3}$

$n_e(\text{Si},77) = 2 \times 10^{15}\ \text{cm}^{-3}$; $n_e\ (\text{GaAs}) = 4 \times 10^{14}\ \text{cm}^{-3}$

These densities correspond to the density of ionized impurities that allow one to obtain the degeneracy limit by raising the Fermi level to $E_g - 5k_BT$.

$$\sigma_1\ (\text{Si}) = 0.4\ \Omega^{-1}\ \text{cm}^{-1}$$

$$\sigma_1\ (\text{GaAs}) = 0.56\ \Omega^{-1}\ \text{cm}^{-1}$$

When the measured conductivity of the semiconductor is less than 0.1–1 Ω^{-1} cm^{-1}, one may assume that the semiconductor is nondegenerated.

Exercise 13: Carrier density of a degenerated semiconductor

Find the relation between the charge carriers n_e and the Fermi energy E_F in a semiconductor that is fully degenerate, starting from the following dispersion relation:

$$E = -\frac{E_g}{2} \pm \left(\frac{E_g^{\ 2}}{4} + \frac{2\hbar^2 k^2}{m} E_g\right)^{1/2}$$

Hint: Find n_e with the help of the density of states g(k), neglecting the influence of temperature ($f(E) = 1$ for $E \leq E_F$).

Solution:

The semiconductor is n-type and degenerate. Its Fermi level is in the conduction band and this level is quite insensitive to T, as for metals.

We retain therefore the + sign and observe that the energy origin is taken at the bottom of the conduction band.

$$n_e = \int_0^{E_F} g(E)dE = \int_0^{k_F} g(k)\,dk$$

The integration with respect to k is easier than in E and is allowed because the dispersion relation is isotropic. For a unitary volume, we evaluate

$n_e = \frac{1}{\pi^2}\int_0^{k_F} k^2 dk = k_F^3/3\pi^2$ (note that this is the same result as for free electrons), using the proposed dispersion relation we have

$$n_e = \frac{1}{3\pi^2}\left(\frac{m}{2\hbar^2 E_g}\right)^{3/2} (E_F{}^2 + E_F E_g)^{3/2}$$

Exercise 14: Semi-insulating gallium arsenide

The following data correspond to GaAs and are used in this exercise:

E_g = 1.4 eV; $m_e{}^x \approx 0.07m_0$; $m_h{}^x = 0.5m_0 \approx 0.5m_0$; μ_e = 8500 cm^2/V·s; $\mu_h \approx 400$ cm^2/V·s

In addition, it is useful to note that

$$N_0 = 2\left(\frac{2\pi m_0 k_B T}{h^2}\right)^{3/2} = 2\times 10^{25}\,\mathrm{m}^{-3}$$ at T = 300 K or $k_B T \approx 25$ meV

(a temperature remaining constant throughout the exercise). Also note that the law of mass action is always satisfied here.

(1) Assume that the semiconductor is perfectly pure.
Find the density of intrinsic carriers n_i, the electrical conductivity σ_i, and the position of the intrinsic Fermi level E_{Fi} measured relative to the top of the valence band.

(2) In practice, it is impossible to purify a material to better than N_d = 10^{14} imp.cm^{-3}. These residual impurities are of type n and their ionization energy E_d is small (E_d < 10^{-2}), which can reasonably be considered ionized at 300 K. Find the density of the different carriers $n_e(d)$, $n_h(d)$, the electrical conductivity $\sigma(d)$, and the position of the Fermi level $E_F(d)$.

(3) Chromium atoms are introduced into GaAs with a density N_{Cr} >10^{17} cm^{-3} and the non-ionized electronic levels are located

in the bandgap. This addition of such monovalent deep level impurities results in a decrease in the Fermi energy to that of these levels: E_F (Cr) = 0.7 eV (the hydrogenic model is not applicable for this type of deep impurity level).

Find the new density of carriers n_e(Cr) and n_h(Cr) and the new electrical conductivity σ(Cr). Discuss the effect of chromium doping on the conductivity of GaAs.

Solution:

The fundamental formulas that lead to the law of mass action are:
$n_e = N_c e^{(E_F - E_g)/k_B T}$ and $n_h = N_v e^{-E_F/k_B T}$
where the expressions for N_c and N_v are taken from the Course Summary. We find the general result:

(1) $$E_F = \frac{E_g}{2} + \frac{k_B T}{2} \log\left(\frac{n_e}{n_h} \cdot \frac{N_v}{N_c}\right)$$

$$n_i = \sqrt{n_e n_h} = 1.1 \times 10^{12}\ \mathrm{m}^{-3}$$

$$\sigma_i = n_i e(\mu_e + \mu_h) = 1.6 \times 10^{-7}\,\Omega^{-1}\mathrm{m}^{-1}$$

$$E_{Fi} = \frac{E_g}{2} + \frac{3k_B T}{4} \log(\mathrm{m_h}/m_e) = 0.737\ \mathrm{eV}$$

(2) $$n_e = N_d = 10^{14}\,\mathrm{cm}^{-3};\ n_h = (n_i^2/N_d) < 10^2\,\mathrm{cm}^{-3}$$

$$\sigma(d) \approx n_e e \mu_e \approx 13\,\Omega^{-1}\mathrm{m}^{-1}$$

$$E_F(d) \approx 1.15\,\mathrm{eV}$$

corresponding to a spectacular increase but logically of $\sigma(d)$ and of E_F.

(3) Starting from the expressions for n_e and n_h in which we use the new values for E_F, we find:

$$n_e(\mathrm{Cr}) = 22 \times 10^4 e \cdot \mathrm{cm}^{-3};\quad n_h(\mathrm{Cr}) = 4.2 \times 10^6\,\mathrm{h} \cdot \mathrm{cm}^{-3}$$

$$\sigma(\mathrm{Cr}) = 6 \times 10^{-8}\,\Omega^{-1}\mathrm{m}^{-1}$$

We will observe that the addition of sufficient amounts of chromium to GaAs freezes the electrical activity of the residual impurities that cannot be eliminated chemically.

This addition allows one to obtain a higher electrical resistivity than a chemically pure material impossible to prepare. We are thus justified in calling such a doped GaAs semi-insulating.

The latter result can be explained by the fact that in pure GaAs, the Fermi level is located above $E_g/2$ due to the large inequality between the effective mass of electrons and that of holes. With the addition of chromium the density of slow holes is increased to the detriment of the density of more mobile electrons.

Technologically the preceding result is important because it allows the electrical isolation of active GaAs electronic components by growth on substrates doped with deep impurities that are therefore highly resistive (see Sapoval and Hermann [23] p. 112, the inspiration for this exercise, for the justification of the pinning of E_F to that of the chromium level.)

For the conceptual point of view, one must keep in mind that the addition of impurities does not lead always to the increase of the electrical conductivity.

Exercise 15: Intrinsic and extrinsic electrical conductivity of some semiconductors

Assuming that it is not technologically possible to reduce the relative atomic concentration of impurity atoms below 10^{-10} for elements and 10^{-8} for binary compounds, find, at ambient temperature (k_BT = 25 meV):

(a) The intrinsic or extrinsic nature of the electrical conductivity σ for the following semiconductors: Ge, Si, InAs, and GaAs

(b) The order of magnitude of σ (assuming that the residual impurities are all n-type). Use the numerical data presented in the table at the beginning of this chapter and recall the law of mass action:

$n_i^2 = N_v N_c e^{-E_g/k_BT}$ in which $N_c = N_v = 5 \times 10^{19}$ cm^{-3}. To simplify, consider that the effective mass of electrons and holes are equal to the free electron mass and assume, in addition, that the average atomic density of the considered solids is ≈5 × 10^{22} cm^{-3} and that they are nondegenerate.

Solution:

(a) To know whether the semiconductors are extrinsic or intrinsic at ambient temperature, one must compare the thermally excited density of electrons (n_i) determined from the law of mass action (see Course Summary) to the density of residual impurities (assumed to be ionized). We thus obtain:

Ge: $n_i = 7.5 \times 10^{13}$ cm^{-3}; n(extrinsic) ~5×10^{12} cm^{-3}

Si: $n_i = 6 \times 10^{13}$ cm^{-3}; n(extrinsic) ~5×10^{12} cm^{-3}

InAs: $n_i = 5 \times 10^{16}$ cm^{-3}; n(extrinsic) ~5×10^{14} cm^{-3}

GaAs: $n_i = 2 \times 10^{7}$ cm^{-3}; n(extrinsic) ~5×10^{14} cm^{-3}

Inversely, we could also determine E_g such that the intrinsic concentration coincides with the extrinsic concentration. We thus obtain:

$E_g = 2k_BT \times \log\,[N_c/n_{(\text{impurities})}]$

or

E_g(limit) ≈ 0.8 eV for elements and E_g(limit) ≈ 0.6 eV for binary compounds Ge and InAs have an intrinsic conductivity at ambient temperature; Si and GaAs are extrinsic.

(b) σ(intrinsic) = $n_i e(\mu_e + \mu_h)$(see Course Summary)

σ (Ge) = 0.1 Ω^{-1} cm^{-1}; σ[(InAs) ≈ 250 Ω^{-1} cm^{-1}]

σ(extrinsic) = n(ext) × $e\mu_e$

σ(Si) = 10^{-3} Ω^{-1} cm^{-1}; σ[(GaAs) ≈ 1 Ω^{-1} cm^{-1}]

One can also refer to the next exercise concerning impurity bands in binary semiconductors. GaAs is treated more rigorously by first estimating if it is degenerate or not (see Ex. 12).

Exercise 16: Impurity orbitals

The physical characteristics of indium antimonide, InSb, are as follows: E_g = 0.23 eV; ε_r (dielectric constant) = 17; effective mass of an electron in the conduction band m_e^* = 0.014 m. Using the Bohr model calculate:

(a) The ionization energy of donors E_d

(b) Their orbital radius of the ground state r_i

(c) What is the minimum concentration of donors to have orbital overlaps between adjacent impurities? These overlaps tend to produce an impurity band, in which electron conduction can occur by electrons hopping from one impurity site to its ionized neighbor.

Solution:

(a) Using the reasoning and notation from Pb. 4, Question (3), we find:

$(E_d/E_H)=\frac{m_g{}^*/m}{\varepsilon_r{}^2}$, so that here we find E_d = 0.66 meV.

(b) $(r_i/r_B)=\frac{\varepsilon_r}{m_g{}^*/m}$, here we find r_i = 640 Å.

(c) The overlapping of impurity orbitals occurs when the concentration exceeds the value $\left(\frac{4\pi}{3}r_i^3\right)^{-1}$, here ≈ 10^{15} cm^{-3}.

In the III–V semiconductors, any deviation from their stoichiometry leads to the creation of acceptors, p (if there is an excess of III atoms), or donors, n (if there is an excess of V atoms). Here the relative surplus of antimony concentration relative to indium, of order 10^{-7} leads to the appearance of impurity bands and it is difficult, in fact technologically impossible, to approach the ideal stoichiometry (see Ex.15).

Exercise 17: Donor ionization

In a certain semiconductor there are 10^{13} donors/cm^3 with an ionization energy E_d of 10^{-3} eV and an effective mass of 10^{-2} m.

(a) What is the concentration of conduction electrons at 4 K?

(b) What is the value of the Hall constant, R_H?

Note: Assume that there are no acceptor atoms.

Solution:

(a) At low temperature, when the concentration of the impurities N_d is significant, the concentration of electrons n in the conduction band obeys (see Pb. 4, Question (3b)):

$$n=\left(\frac{N_c\cdot N_d}{2}\right)^{1/2}\exp-(E_d/(2k_BT))$$

where $N_c = 2(m_e{}^*\cdot k_BT/2\pi\hbar^2)^{3/2}$

At 4 K, k_BT = 0.345 meV and $N_c = 4.4\times10^{13}$ cm^{-3}, which leads to n =3.5 × 10^{12} cm^{-3} (50% atomic impurities are ionized).

(b) $R_H=-\frac{1}{n_e}$ (see Chapter IV, Pb. 2) here we have R_H = –1.8 m^3/C.

Exercise 18: Hall effect in a semiconductor with two types of carriers

Consider a rectangular parallelepiped semiconductor with sides parallel to Ox, Oy, and Oz. This sample is placed in a uniform magnetic induction $\vec{B}_0$ directed along Oz while a uniform constant current density j_x flows parallel to Ox.

(a) In the xOy plane, find the vector equation of motion for a charged particle (q) with mass m when it is subject (in addition to $\vec{B}_0$) to a constant electric field $\vec{E}(E_x, E_y)$ and a damping force of form $-m\vec{v}/\tau$, where v is the velocity of the particle, and τ is the relaxation time. Simplify this result in the steady state $[(d/dt) = 0]$.

(b) In the latter case (steady-state regime), find the expression for the v_x and v_y components of velocity as a function of E_x and E_y of $\vec{E}$ and of the algebraic mobility μ of the particle ($\mu = q\tau/m$). Simplify the results by neglecting the terms in $\mu^2 B_0{}^2$.

(c) The electrical conductivity of the semiconductor is assured by two types of carriers of charge q_1 and q_2 with respective densities per unit volume n_1 and n_2, with (algebraic) mobility μ_1 and μ_2.

As a function of this data and the components E_x and E_y of $\vec{E}$, find the j_x and j_y of the current density vector. Find the electrical conductivity σ_0 when $B_0 = 0$.

(d) The current is constrained to flow parallel to the $x'O_x$ axis ($j_y = 0$). What is the expression for the Hall electric field E_y? What is the expression of the Hall constant R_H, where $E_y = R_H . B_0 j_x$? Find R_H for

(i) two carrier types of the same sign (e.g., electrons)

(ii) two carrier types with different signs (e.g., electrons and holes)

(iii) only one carrier type

(e) *Numerical application*: The thickness c (along Oz) of the sample is c = 0.1 mm, the current flowing is I_0 = 1 mA. For B_0 = 1 T, find the Hall constant R_H and the Hall voltage V_H for intrinsic Ge at 290 K and for GaAs doped n using the following data: Ge: $n_e = 3 \times 10^{13}\ e/\text{cm}^3$, μ_e = 3,600 cm^2/V·s and μ_h = 1,700 cm^2/V·s.

- GaAs (n) has two types, light and heavy, electrons of mass:
  –light electrons: $m_l = 0.07$ m, $n_l = 10^{15}$ e/cm^3
  –heavy electrons: $m_L = 0.2$ m, $n_L = 10^{14}$ e/cm^3
  Assume that these two populations of electrons have the same relaxation time $\tau = 4 \times 10^{-14}$ s. (e,m)

Solution:

(a) $m(d\vec{v}/dt) = q \cdot \vec{E} + q(\vec{v} \wedge \vec{B_0}) - m\vec{v}/\tau$. In the steady-state regime, $(d/dt) = 0$ and we find: $\vec{E} = \dfrac{\vec{v}}{\mu} - (\vec{v} \wedge \vec{B_0})$.

(b) $E_x = (mv_x/q\tau) - v_y B_0$; $E_y = (mv_y/q\tau) + v_x B_0$.
From which $v_x\ (1 + \mu^2 B_0^2) = \mu E_x + \mu^2 B_0 E_y \approx v_x$ and $v_y(1 + \mu^2 B_0^2) = -B_0 \mu^2 E_x + \mu E_y \approx v_y$ (taking into account the algebraic expression $\mu = q\tau/m$ and not the usual positive $|\mu| = \dfrac{|q|\tau}{m}$).

(c) $\vec{j} = \sum_i n_i q_i \vec{v}_i$

$$j_x = n_1 q_1 v_{1x} + n_2 q_2 v_{2x} = (n_1 q_1 \mu_1 + n_2 q_2 \mu_2) E_x + B_0 E_y (n_1 q_1 \mu_1^2 + n_2 q_2 \mu_2^2)$$

$$j_y = n_1 q_1 v_{1y} + n_2 q_2 v_{2y} = (n_1 q_1 \mu_1 + n_2 q_2 \mu_2) E_y - B_0 E_x (n_1 q_1 \mu_1^2 + n_2 q_2 \mu_2^2)$$

$$\sigma_0 = n_1 q_1 \mu_1 + n_2 q_2 \mu_2.$$

(d) In equilibrium, the vector $\vec{j}$ is parallel to 0_x so that $j_y = 0$ from which we find

$$E_y = B_0 \cdot E_x \frac{n_1 q_1 \mu_1^2 + n_2 q_2 \mu_2^2}{n_1 q_1 \mu_1 + n_2 q_2 \mu_2} = \frac{B_0 E_x}{\sigma_0}(n_1 q_1 \mu_1^2 + n_2 q_2 \mu_2^2);$$

$$j_x \approx \sigma_0 \cdot E_x \text{ therefore, } E_y = \frac{B_0 j_x}{\sigma_0}(n_1 q_1 \mu_1^2 + n_2 q_2 \mu_2^2)$$

which leads to $R_H = (n_1 q_1 \mu_1^2 + n_2 q_2 \mu_2^2)/(n_1 q_1 \mu_1 + n_2 q_2 \mu_2)^2$.

(i) For two types of carriers with the same sign electrons for example $q_1 = q_2 = -q$:
$R_H = -(n_1 \mu_1^2 + n_2 \mu_2^2)/q(n_1 \mu_1 + n_2 \mu_2)^2$

(ii) For two types of carriers with different signs (holes and electrons):

$q_1 = -q_2 = q$

$R_H = -(n_h\mu_h^2 - n_e\mu_e^2)/q(n_h\mu_h - n_e\mu_e)^2$

with the sign convention in (b) $(\mu_e = -q\tau/m_e)$ or

$R_H = -\frac{1}{q}\frac{n_h\mu_h^2 - n_e\mu_e^2}{(n_h|\mu_h| + n_e|\mu_e|)^2}$ with the usual convention.

(iii) For one type of carrier $n_2 = 0$, $R_H = 1/n_1q_1$; a result that is in agreement with that from Pb. 2 of Chapter IV. As in Chapter IV, Pb. 2, and Ex. 19 of this chapter one may also explicitly write the electrical conductivity tensor.

(e)
- Intrinsic Ge: $n_e = n_h = n_i$ from which $R_H = -\frac{1}{n_i q}\frac{|\mu_h| - |\mu_e|}{|\mu_h| + |\mu_e|}$.

  $R_H = 7.5 \times 10^4\,\text{cm}^3\cdot\text{C}^{-1}; v_H = E_y \cdot b = b \cdot R_H \cdot j_x \cdot B_0 = R_H \cdot \frac{1}{C} \cdot B_0$

  $= 0.75\,\text{volt}.$

- GaAs (n): $\mu_l = q\tau/m_l = 3.5 \times 10^{-2}\ \text{m}^2/\text{V·s}$,

  $\mu_l = q\tau/m_L = 0.1\ \text{m}^2/\text{V·s}$.

  $R_H \approx 6.9 \times 10^{-3}\ \text{m}^{-3}\cdot\text{C}^{-1}$; $V_H = 69$ mV.

Exercise 19: Transverse magnetoresistance in a semiconductor with two types of carriers

We reconsider the semiconductor from the preceding exercise with the same geometry.

(a) Same question as Question (a) in the preceding exercise.

(b) Same question as Question (b) in the preceding exercise, still assuming that for the magnetic induction B_0 is such that the terms in $\mu^2B_0^2$ are small, << 1, and the terms in B_0^3 are negligible.

(c) Find the components of the tensor σ_{ij} that connect the current vector density and the electric field:

$j_x = \sigma_{xx}E_x + \sigma_{xy}E_y$

$j_y = \sigma_{yx}E_x + \sigma_{yy}E_y$

(d) The sample has finite dimensions (j_y is zero): find the longitudinal conductivity $\sigma_{||}(B)$ defined by $\sigma_{||}(B) = j_x/E_x$ and also the variation of the ratio $[\sigma_{||}(\text{B}) - \sigma(0)]/\sigma(0) = \Delta\sigma_{||}/\sigma(0)$

as a function of B_0 in which $\sigma(0)$ is the electrical conductivity of the material when the magnetic induction is zero.

(e) Find $\Delta\sigma_{||}/\sigma(0)$ for:

(i) Two types of carriers with the same sign

(ii) Two types of carriers with different signs

(iii) One type of carrier

(For two types of carriers with the same sign, see the explanation on light and heavy holes in the exercise later on the band structure of III–V compounds)

(f) With the numerical data from the preceding exercise, find the relative variation of the electrical resistance $\frac{R(B)-R(0)}{R(0)}$ for intrinsic Ge and for n-doped GaAs when the magnetic induction B_0 goes from 1 T to 0 T.

What DC voltage along O_x must be applied to maintain 1 mA of current across the sample when $B_0 = 0$? Take a (along x) = b (along y) = 1 mm.

(g) The sample is now limited in such a way that E_y (Hall) = 0, what is the new ratio E_y (Hall) = 0 for a single carrier type? Qualitatively find this result starting from basic considerations relative to the effect of B_0 on a mean free path Λ for charged carriers.

Solution:

(a) See answer to Question (a) of the preceding exercise.

(b) $v_{ix}(1+\mu_i^2 B_0^2) = \mu_i E_x + \mu_i^2 B_0 E_y;\;\; v_{ix} = \mu_i(1-\mu_i^2 B_0^2)E_x + \mu_i^2 B_0 E_y;$

$v_{iy}(1+\mu_i^2 B_0^2) = -B_0\mu_i^2 E_x + \mu_i E_y; v_{iy} = -B_0\mu_i^2 E_x + \mu_i(1-\mu_i^2 B_0^2)E_y.$

(c) $j_x = n_1 q_1 v_{1x} + n_2 q_2 v_{2x} = \sigma_{xx}E_x + \sigma_{xy}E_y$

$j_y = n_1 q_1 v_{1y} + n_2 q_2 v_{2y} = \sigma_{yx}E_x + \sigma_{yy}E_y$

$\sigma_{xx} = \sigma_{yy} = n_1 q_1 \mu_1(1-\mu_1^2 B_0^2) + n_2 q_2 \mu_2(1-\mu_2^2 B_0^2)$

$\sigma_{xy} = -\sigma_{yx} = (n_1 q_1 \mu_1^2 + n_2 q_2 \mu_2^2)B_0$

(d) $j_y = 0; E_y = (\sigma_{xy}/\sigma_{yy})E_x;\; j_x = [(\sigma_{xx}^2 + \sigma_{xy}^2)/(\sigma_{xx})]E_x$

$\sigma_{||}(B) = (n_1 q_1 \mu_1 + n_2 q_2 \mu_2) - [(n_1 n_2 q_1 q_2 \mu_1 \mu_2(\mu_1-\mu_2)^2)$

$/(n_1 q_1 \mu_1 + n_2 q_2 \mu_2)]\cdot B_0^2$

$$\Delta\sigma_{||}/\sigma(0) = -\frac{n_1 n_2 \mu_1 \mu_2 (\mu_1 - \mu_2)^2}{\sigma(0)^2} B_0^{\ 2},$$

where $\sigma(0) = n_1 q_1 \mu_1 + n_2 q_2 \mu_2$.

(e) (i) For two types of electrons:

$$q_1 = q_2 = -q;\ \Delta\sigma_{||}/\sigma(0) = -\frac{n_1 n_2 (\mu_1 - \mu_2)^2 \mu_1 \mu_2}{(n_1 \mu_1 + n_2 \mu_2)^2} B_0^{\ 2}$$

The result is identical for two types of holes.

(ii) For two different types of carriers with different sign: $q_1 = -q_2 = q$;

$$\Delta\sigma_{||}/\sigma(0) = -\frac{n_1 n_2 |\mu_1||\mu_2|(|\mu_1| + |\mu_2|)^2}{(n_1|\mu_1| + n_2|\mu_2|)^2} B_0^{\ 2}.$$

As in the preceding hypothesis, the resistance of a material is an increasing function and proportional to the square of B_0.

(iii) If $n_2 = 0 \rightarrow \dfrac{\Delta\sigma}{\sigma} = 0$, a result which is consistent with that of Chapter IV, Pb. 2.

(f) • Intrinsic Ge:

$$\Delta R/R = [\sigma(0) - \sigma(B)]/\sigma(0) = |\mu_h||\mu_e| \cdot B_0^{\ 2} = 6\% : V_0 = 3.9\ \text{V}.$$

• GaAs: $\dfrac{\Delta R}{R} = 0.7 \times 10^{-3}, V_0 = 1.38\ \text{V}.$

(g) If the semiconductor is infinite, the Hall field does not appear; however, $j_y \neq 0$. This is what happens in a different geometry: the Corbino disc.

For one type of carrier, the magnetoresistance is nonzero:

$\sigma(B) = \sigma_{xx} = \sigma_0 (1 - \mu^2 B_0^2)$.

One can qualitatively find this result by considering that charge carriers will follow a trajectory along a cycloid segment of the length Λ instead of a line segment of length Λ (in absence of B_0). The distance measured along x will be

$$\Lambda \cos\varphi \approx \Lambda\left(1 - \frac{\varphi^2}{2}\right) \approx \Lambda\left(1 - \frac{\mu^2 B_0^{\ 2}}{2}\right),$$

where Δ is the mean free path.

The resistance of the semiconductor increases and we find

$$\frac{\Delta\sigma}{\sigma(0)} = -\frac{\sigma^2 B_0^{\ 2}}{2}.$$

Taking into account the statistics nature of the mean free time and path, the ½ coefficient disappears and we find the same result as above.

Comment

The Hall effect is odd in B_0 and therefore it changes sign if B_0 changes direction. The magnetoresistance is even in B_0, and thus regardless of the sign of voltage V, there is an increase in the transverse resistance (that is to say $j_x \perp B_0$) for weak applied magnetic induction. For larger values of B_0, the magnetoresistance can be negative, see [14] p. 321.

Comment: Giant magnetoresistance (GMR); Nobel Prize in physics in 2007

In the present exercise, the explanation of magnetoresistance is based on classical arguments and it leads to change in the resistance of a few percent for semiconductors. Quite different is the giant magnetoresistance (GMR) effect that has been observed in thin-film structures composed of alternating ferromagnetic and nonmagnetic layers. Similarly to the quantum Hall effect of Chapter IV, Ex. 21, this effect can only be explained from quantum mechanical arguments involving the spin orientation and it leads to a large change in resistance (typically 10% to 20%) when the corresponding nanotechnology devices are subjected to a magnetic field. In these devices, the bit of information is carried by the spin of the electrons rather than the presence or absence of an electron and the corresponding technology is named "spintronics" instead of "microelectronic." In a GMR spintronic device, the first magnetic layer polarizes the electron spins. The second layer scatters the spins strongly if its moment is not aligned with the polarizer's moment. If the second layer's moment is aligned, it allows the spins to pass. The resistance therefore changes depending on whether the moments of the magnetic layers are parallel (low resistance) or antiparallel (high resistance). The overall resistance is relatively low for parallel alignment and relatively high for antiparallel alignment. GMR is used commercially by hard disk drive manufacturers. The Nobel Prize in physics in 2007 was awarded jointly to Albert Fert and Peter Grünberg "for the discovery of Giant Magnetoresistance." For additional details, see P. Grünberg, *Rev. Mod. Phys.*, **80**, 2008, 1531.

Exercise 20: Excitons

In semiconductors and pure insulators, the minimum energy E_T for which a photon can be absorbed may be slightly less than the bandgap E_g, when the excited electron from the valence band remains bounded to its hole by a Coulomb attraction. This bound electron–hole pair is called an "exciton."

We start from the Bohr model of atomic hydrogen applied to two particles of mass m_e^* and m_h^* of charge ±e and separated in material with relative dielectric constant ε_r. Find the quantum energies $E_L = E_1, E_2, \ldots E_n$ and the radii $r_{1,2,\ldots n}$ of an exciton with a weak bond (called Mott exciton).

Find the corresponding energetic positions relative to the conduction and valence bands as well as the minimum energy E_T of a photon leading to the creation of an exciton in its ground state.

Numerical application for CdTe:

$$E_g = 1.6 \text{ eV};\ \varepsilon_r = 10.2;\ m_e^x = 0.09m_0;\ m_h^x = 1.38m_0$$

Find E_1, r_1, and E_T known that $E_H = 13.6$ eV and $r_B = 0.53$ Å.

Solution:

The procedure is formally the same as that used for the calculation of the ionization of impurities in a semiconductor. In the Bohr model of an atom, we substitute ε_0 by $\varepsilon_0\varepsilon_r$ and the real mass by the effective mass in the expressions for E_H and r_B. Nevertheless here, the electron mass and that of holes are comparable and it is necessary to use the effective reduced mass of these two particles:

$$\mu^{-1} = m_e^{x-1} + m_h x^{-1}$$

We thus obtain: $E_L(n) = \frac{E_H}{n^2} \cdot \frac{\mu/m_0}{\varepsilon_r^2}$ and $r_n = \frac{r_B \cdot \varepsilon_r}{\mu / m_0} n^2$.

When $n = \infty$, the electron and hole are completely disassociated: one is in the conduction band and the other in the valence band.

When $n = 1$, the exciton is in its ground state. This corresponds to a bound electronic state, in the bandgap at a distance $E_L(1)$ below the bottom of the conduction band—see Fig. 30a. Then,

$$E_T = E_g - E_1(2, n), E_1 = 11 \text{ meV}; r_1 = 64 \text{ Å}; E_T = 1.589 \text{ eV}.$$

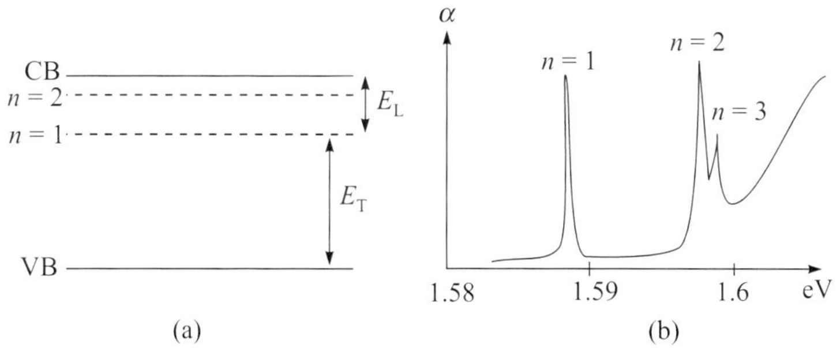

Figure 30

The characteristics of the optical absorption spectra can be obtained and are shown in Fig. 30b.

We note that the energy of excitons is such that $E_1 << E_g$ so that $E \approx E_g$, which does not significantly change the results from Ex. 28 concerning the color and transparency of semiconductors and insulators.

Remark: The electron–hole systems are more or less bonded.

Here we consider weakly bonded excitons (Mott–Wannier type) having a radius that is large compared to interatomic distances, which justifies the use of a macroscopic dielectric constant, as used for shallow impurity levels in semiconductors.

In the opposite case, Frenkel excitons result essentially from the excitation of a valence electron (excited state but not fully ionized), which leads to an electron–hole pair with a small radius and thus very localized on an atom.

Exercise 21: III–V compounds with a direct bandgap: light and heavy holes

Through some approximations (the Kane model), the band structure of III–V compounds (InSb, InAs, InP, GaAs, GaSb) having a direct bandgap E_g can be described by the following relations:

- CB: $E = E_g + \frac{\hbar^2 k^2}{2m} + \frac{1}{2}\left(\sqrt{E_g^2 + Qk^2} - E_g\right)$

- VB_1: $E = -\hbar^2 k^2 / 2m$

- VB_2: $E = -\frac{\hbar^2 k^2}{2m} - \frac{1}{2}(\sqrt{E_g^2 + Qk^2} - E_g)$

(1) For each band, find the effective mass of the particles situated at $k \approx 0$ with $Q > 0$.

(2) Knowing that the optical absorption starts at $\lambda_0 = 8670$ Å and that the effective mass of electrons (determined electrically) is $m^x \approx 0.07$ m in GaAs, find the numerical values of E_g (in eV) and of Q. Sketch the band structure of GaAs and indicate the light and heavy hole bands.

(3) Assuming that Q has the same value for all the III–V compounds considered, find the evolution of the effective mass of conduction electrons, their mobility as a function of the E_g (see Table 1 in of the Course Summary). Find μ_e of InSb, InAs, and InP knowing that μ_e(GaAs) = 8800 cm^2/V·s. Starting from the band structure, explain qualitatively why the electrical conductivity of these materials are essentially given by electrons even when they are intrinsic. $(\hbar, m)$

Solution:

(1) For CB: $\frac{\partial^2}{\partial k^2}(\sqrt{E_g^2 + Qk^2} - E_g) = QE_g^2(E_g^2 + Qk^2)^{-3/2}$

from which $\left(1/m^x = \frac{1}{\hbar^2}\frac{\partial^2 E}{\partial k^2}\right)$

- CB: $\frac{1}{m^x} = \frac{1}{m} + \frac{Q}{2\hbar^2 E_g}$: electrons $(m^x > 0)$

- VB_1: $m^x = -m$: heavy holes

- VB_2: $\dfrac{1}{m^x} = -\left(\dfrac{1}{m} + \dfrac{Q}{2\hbar^2 E_g}\right)$: light holes

(2) $E_g = hc/\lambda \cong 1.43$ eV; $Q = 290$ (eV.Å)^2
The characteristics of the corresponding band structure (including the spin-orbit band with equation $E = -\Delta - \dfrac{\hbar^2 k^2}{2m} - \dfrac{Q}{8}\dfrac{k^2}{(E_g + \Delta)}$) is shown in Fig. 31.

(3) $m/m_e^{\,x} = 1 + \dfrac{Qm}{2\hbar^2 E_g}$

As the bandgap increases, the electron effective mass increases ($m_e^x = 0$ for $E_g = 0$, $m_e^x = m$ for $E_g = \infty$). The electron mobility ($\mu_e \approx -e\tau/m^x$) increases when the bandgap decreases. One expects that the ratio of mobilities varies as the inverse of the effective mass which is inversely proportional to E_g. From this we find, μ_e(InSb) ≈ 70,000 cm^2/V·s; μ_e(InAs) ≈ 36,000 cm^2/V·s; and μ_e(InP) ≈ 10,000 cm^2/V·s.

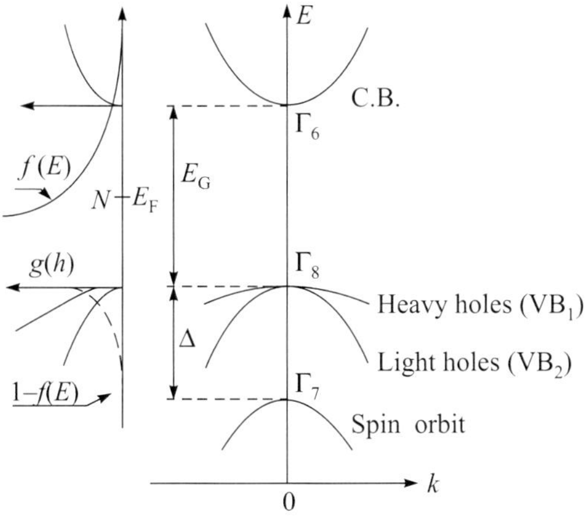

Figure 31

Comparing the calculated values with the experimental results, 77,000 cm^2/V·s; 33,000 cm^2/V·s; and 4,600 cm^2/V·s, respectively (see Table 1 in the Course Summary), we note that there is an excellent agreement for all compounds except InP.

In fact the semiconductor mobility also depends on τ by the Matthiessen's rule (see Chapter IV, Ex. 16): $\tau^{-1} = \tau^{-1}$(impurities) + τ^{-1}(temperature) + τ^{-1}(deformation).

In addition to the electronic mobility evolution in the III–V compounds, the band structure of these compounds (from the Kane model) can explain why their conductivity is essentially electronic when they are intrinsic.

The electronic conductivity σ obeys $\sigma = n_e\mu_e + n_{lh}\mu_{lh} + n_{hh}\mu_{hh}$ and for an intrinsic material: $n_e = n_{lh} + n_{hh}$. Nevertheless, the density of heavy holes (n_{hh}) is much larger than that of light holes (n_{lh}) because the density of states at the common maximum of the two valence bands is larger for heavy holes than for light holes ($g(E) \propto m^{x3/2}$). In addition, the Fermi–Dirac (or Maxwell–Boltzmann) statistics favor the occupation with holes on the top of the valence band with the smallest curvature that of heavy holes (see Fig. 31). From this it results that the mean effective mass of holes is nearly that of heavy holes and the average mobility of holes $\bar{\mu}_h$ is close to that of heavy holes. Thus, the contribution of holes (which is an order of magnitude smaller than that of electrons) will be negligible compared with electrons so that $\sigma = n_e e\mu_e$. This high electron mobility (compared with that of Si) explains in part the attention in III–V compounds for electronic applications.

Comment: Band structure of semiconductors

As in IV–IV semiconductors (Si–Ge), the chemical bonds between elements of type III (ns^2p^1) and of type V (ns^2p^3) leads to hybrid tetrahedral sp^3 orbitals taking into account the crystalline geometry. Knowing that there are two atoms per lattice, there are eight valence electrons per unit cell and if there are N unit cells in the crystal, the $8N$ valence electrons will be divided into two valence bands:

- A binding s-band with $2N$ states sufficiently deep so that it does not interfere with the principle properties of the material (not shown in the figure).
- A binding p-band with $6N$ states giving rise to three subbands (each with $2N$ states), which in the absence of spin–orbit coupling, that degenerate at $k = 0$ (VB_1, VB_2, and spin–orbit).

The spin–orbit coupling lifts this degeneracy. For the valence band at $k = 0$, we thus obtain, two bands Γ_8 ($j = 3/2$) and Γ_7 ($j = 1/2$) separated in energy by Δ. The first unoccupied band (E_c) corresponds to the s antibonding band containing $2N$ states: it is the conduction band, CB. The difference in energy at $k = 0$ between the conduction band (Γ_6 band) and the valence band (Γ_8) gives the bandgap of the material (E_g).

Qualitatively, the energy E_g represents the bonding/antibonding splitting of the IV–IV, III–V, or II–VI chemical bond. For the homopolar bonds of the IV column, the amplitude of this splitting is inversely proportional to the interatomic distance. The bandgap decreases (from diamond via Si to Ge) when the corresponding elemental atomic number increases (because the atomic radius and therefore the distance separating them increase with Z). For a given row of the periodic table, this splitting also increases with the ionic nature of the bond (for the same row E_g(IV–IV) < E_g(III–V) < E_g(II–VI), for instance:

E_g (Si) < E_g (AlP); E_g (Ge) < E_g (GaAs).

By combining these two rules, it is therefore easy to foresee that E_g (GaP) > E_g (GaAs) > E_g (InAs) > E_g (InSb) (simple knowledge of the periodic table without looking at a table giving E_g).

The objective of the present exercise is to correlate the values of E_g with the values of mobility that are an important part of the electric properties of these materials. Reciprocally, the electrical and optical measurements allow the experimental determination of the parameters E_g and Q, the deduction of the dispersion curves (VB_1, VB_2, and CB), and thus the corresponding density of states (out of the present exercise).

Ex. 25 will explore empirical expressions for the bandgap in ionic crystals and study their influence on the static dielectric constant.

Exercise 22: Electronic specific heat of intrinsic semiconductors

To simply evaluate the electronic specific heat of an intrinsic semiconductor with bandgap E_g, one can take into account (i) the increase of electronic energy of electrons Δn_i passing from the valence band to the conduction band for a temperature increase ΔT and (ii) the increase in kinetic energy of n_i conduction electrons associated with the same temperature variation.

(a) Find the corresponding expression. Precise the characteristics of the function C_e $(s{\cdot}c) = f\,(T)$. Express the result in the form $\frac{C_e(s\cdot c)}{n_i}$ and indicate the dominant mechanism (i) or (ii). Take $k_B T << E_g$ and in the law of mass action neglect the T dependence of the pre-exponential terms N_c and N_v. Indicate in integral form but do not try to resolve the expression that leads to a more exact result.

(b) Numerically compare this specific heat at room temperature with that of a metal with the same atomic density. The semiconductor is germanium $E_g \approx 0.7$ eV and $n_i \approx 10^{13}$ cm^{-3}, the metal is characterized by $E_F \approx 5$ eV, and $n = 5 \times 10^{22}$ cm^{-3}. Comment on the results.

Solution:

(a) (i) $\Delta U_e^{(i)} = E_g \cdot \Delta n_i$, where $n_i = \sqrt{N_c N_v}\,\exp-\left(\frac{E_g}{2k_B T}\right)$

(ii) $\Delta U_e^{(ii)} \approx n_i \cdot \Delta(k_B T)$

$$C_e = (\Delta U_e^{(i)} + \Delta U_e^{(ii)})/\Delta T$$

$$C_e = k_B\left[\frac{1}{2}\left(\frac{E_g}{k_B T}\right)^2 + 1\right]\sqrt{N_c N_v}\,\exp-(E_g/2k_B T)$$

The function C_e $(s{\cdot}c) = f\,(T)$ is dependent on the exponential term which is $\propto n_i$. The interband electronic transitions (i) are the dominant contribution and $\frac{C_e(s\cdot c)}{n_i} \approx \left(\frac{E_g}{k_B T}\right)^2\left(\frac{k_B}{2}\right)$.

The exact expression can be deduced from $U_e = \int_{E_g}^{\infty} E \cdot g(E)\,f(E)\,dE$

plus an analogous expression describing the energy of holes. The integral form to solve $\int x^{3/2} e^{-ax}\,dx$ is similar to that leading to the evolution of n, $\int x^{1/2} e^{-ax}\,dx$ with the same condition: $5k_B T < E_F < E_g - 5k_B T$.

(b) In the case of Ge at ambient temperature, we have $C_e\,(\text{Ge}) \approx 4 \times 10^{15} k_B$ (cm^{-3}).

For a metal at the same temperature, we find

$$C_e(\text{metal}) = n \cdot \frac{\pi^2}{2} k_B \left(\frac{k_B T}{E_F} \right) \approx 12 \times 10^{21} k_B (\text{cm}^{-3}).$$

These order of magnitudes show that the electronic specific heat of an intrinsic semiconductor is, at ambient, considerably smaller than that of a metal and it is inaccessible to experiment due to the lattice contribution. This explains why in the literature we only find experiments between 0.5 K and 4 K on heavily doped semiconductors (>10^{18} cm^{-3}) to determine the effective mass of the density of states (Baranski et al. [2], p. 106).

Nevertheless, the exponential behavior of C_e seen here is characteristic of electronic systems that have a bandgap and is very different from the linearity seen in a free electron gas. Such behavior was also seen in metallic superconductors at very low temperatures and it evidences a bandgap in these materials (see Ex. 23).

Exercise 23: Specific heat and the bandgap in metallic superconductors

The specific heat measurements in metallic superconductors lead to the following data in the superconducting state:

T(K)	2.48	1.91	1.62	1.46	1.24	1.08
C (mJ/mol·K)	10.33	5.01	3.03	2.1	1.14	0.65

An applied magnetic induction larger than a critical value results in the metal becoming normal and the evolution of C, then follows the conventional law of form: $C/T = \gamma + \alpha T^2$, where $\gamma = 1.65$ and $\alpha = 0.256$.

Graphically compare the evolution of the specific heat for normal electrons $C_{e.n}$ with that of superconducting electrons $C_{e.s}$ as a function of T. Deduce the bandgap E_g of the superconductor. (k_B, e)

Solution:

In the normal state, C includes an electronic term $C_e = \gamma T$ and a term related to the vibrations of the lattice αT^3.

In the superconducting state, one must subtract the lattice term in αT^3 from the measured values of C in order to extract only the contribution of the superconducting electrons $C_{e.s}$.

Figure 32 shows the evolution of $C_{e.s}$ compared with the variation in γT of $C_{e.n}$.

As bandgap materials have a specific heat that varies as exp – (E_g/k_BT) (see preceding exercise), it suffices to graphically determine the slope of the line following log $C_{e.s} = f(T^{-1})$ to determine E_g (see inset): $E_g \approx 4.9 \times 10^{-4}$ eV.

In fact the width of a superconducting bandgap decreases as the temperature increases and goes to zero at $T = T_c$. For simplicity, this problem (inspired by J. Giber [7]) ignores this difficulty.

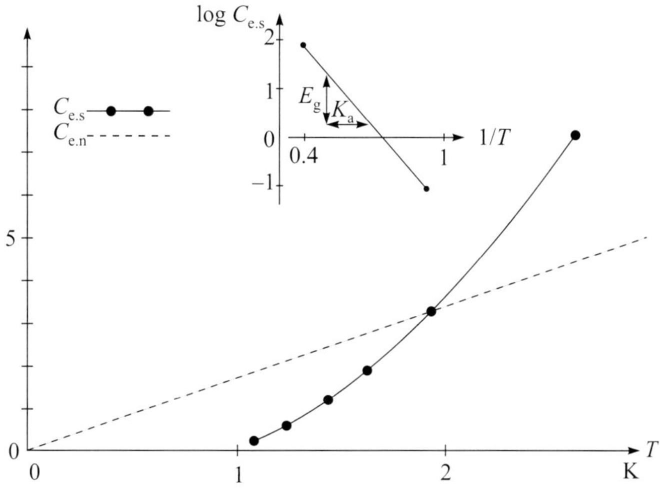

Figure 32

Exercise 24: The Burntein–Moss effect

In III–V semiconductors of direct bandgap that are completely degenerated, the Burstein–Moss effect results in an increase in the threshold energy for the optical absorption with n-doping of III–V compounds. Explain simply this effect by evaluating the filled level ΔE of the conduction band as a function of the impurity concentration N_D, assuming that all impurities are ionized. Use the expression for the density of states g(E) in the conduction band in 3D weighted by the effective mass m_e^x but neglecting the effect of temperature.

Numerical application: Evaluate the corresponding optical absorption threshold energies and wavelengths for InSb where $N_D = 10^{14}$ cm^{-3} and next 10^{17} cm^{-3}; $m_e^x = 0.017 m_0$ and $E_g = 240$ meV.

Recall that for free electrons $g(E)=\frac{V}{2\pi^2}\left(\frac{2m_0}{\hbar^2}\right)^{3/2}\sqrt{E}$ and that $\frac{\hbar^2}{2m_0}=3.8\,\text{eV}\text{Å}^{-2}$.

Solution:

The Fermi level is in the conduction band and the optical absorption begins when $h\nu = E_g + \Delta E$ (see Fig. 33).

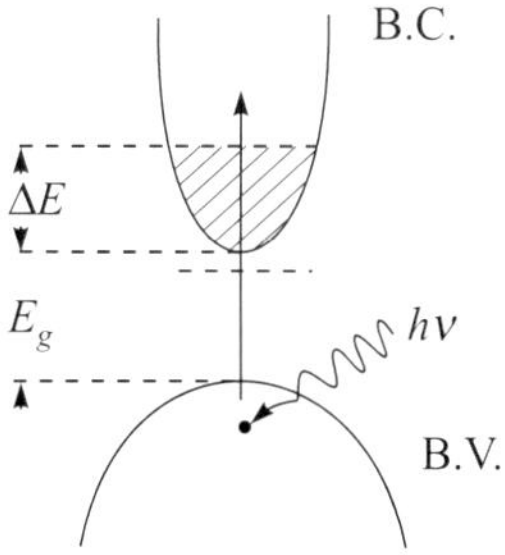

Figure 33

If all the impurities are ionized, we have, assuming a unitary volume:

$$n = N_D = \int_0^{\Delta E} g(E)\,dE = \frac{1}{2\pi^2}\left(\frac{2m_e^x}{\hbar^2}\right)^{3/2} \cdot \int_0^{\Delta E} \sqrt{E}\cdot dE,$$

$$\Delta E = \left(\frac{\hbar^2}{2m_e^x}\right)(3\pi^2 N_D)^{2/3},$$

where $\Delta E = 0.461$ meV for $N_D = 10^{14}$ cm^{-3}, but $\Delta E = 46.1$ meV for $N_D = 10^{17}$ cm^{-3}.

The optical absorption is shifted from $h\nu \cong 240.5$ meV to 286 meV. The phenomenon is observed in the infrared. The wavelength of optical absorption is 5.156 μm ($N_D = 10^{14}$ cm^{-3}) to 4.335 μm ($N_D = 10^{17}$ cm^{-3}).

***Remark*:** This effect, observed in 1954, is very noticeable in InSb because the conduction band fills quickly with doping due to the small value of the effective mass involved in g(E) that is proportional to $m_e^{x\,3/2}$.

In addition, the ionization energy of impurities is very small in this material (0.7 meV), taking into account its low effective mass and its large dielectric constant ($\varepsilon_r \cong 17.7$), which results in impurities being effectively ionized even at relatively low temperatures.

Exercise 25: Bandgap, transparency, and dielectric constant of ionic crystals

The bandgap E_g of compounds of type $A^N B^{8-N}$ crystallizing in the NaCl type structure (fcc structure, see Chapter I, Fig. 1e), is given by: $E_g = \eta_g(\hbar^2/mr_0^{\,2})$. The width of the valence band ΔE_V associated with p-states is $\Delta E_v = \eta_v(\hbar^2/mr_0^{\,2})$ where r_0 is the distance between nearest neighbors ($r_0 = a/2$) where a is the lattice parameter) and the coefficients η_g and η_v are such that $\eta_g = 12.9 - 3.8N$; $\eta_v = 2.1 + N$ where $N = 1$ for the alkali halides and $N = 2$ for the oxides, sulfides and selenides of Mg, Ca, Sr, or Ba.

Numerically determine E_g, $\varepsilon_1(0)$, and the energy $\hbar\omega'_p$ at which the dielectric constant $\varepsilon_1(\omega)$ is zero in the ultraviolet, for the compounds below. Also evaluate the energy interval at which $\varepsilon_2(\omega)$ and the energy loss function are zero.

LiF ($a = 4.02$ Å), NaCl (5.63 Å), KI (7.06 Å), MgO (4.2 Å),

and BaO ($a = 5.52$ Å).

For $\varepsilon_1(0)$, the evaluation is limited to the electronic contribution to the static dielectric constant using the simplified Lorentz model with eight electrons per molecule making transitions with an average energy of $\hbar\overline{\omega_T} = E_g + \Delta E_v$.

Compare $\hbar\omega'_p$ with $\hbar\omega_p$, the energy of plasma oscillations for a free electron gas (with eight molecules AB).

Neglecting the effect of impurities, which crystals are completely transparent in the visible, qualitatively describe the effect of temperature and pressure on the bandgap energy.

($\hbar^2/m \approx 7.6$ eVÅ^{-2}; ε_0; m)

Solution:

Taking into account that $r_0 = a/2$, we construct Table 2 where the experimental values of E_g are also indicated for comparison.

Table 2

	I-VII : $\eta_g = 9.1$ $\eta_v = 3.1$			II-VI : $\eta_g = 5.3$ $\eta_v = 4.1$	
	LiF	**NaCl**	**KI**	**MgO**	**BaO**
E_g (eV) Cal.	17	8.7	5.55	9.1	5.3
E_g (eV) Exp.	13.7	8.75	6.3	8	–
ΔE_v Cal.	5.8	3	1.9	7	4.1

The agreement between E_g (Cal) and E_g (Exp) is satisfactory.

The Lorentz model is of the form: $\varepsilon_1(\omega) = 1 + \dfrac{\omega_p^2}{\bar{\omega}_T^2 - \omega^2}$ (see Pb. 6), where $\hbar\bar{\omega}_T = E_g + \Delta E_v$. We have $\omega_p^2 = ne^2/m\varepsilon_0$, where $n = 8 \times 4/a^3$.

We thus obtain $\varepsilon_1(0) = 1 + \dfrac{\omega_p^2}{\bar{\omega}_T^2}$, while $\varepsilon_1(\omega) = 0$ for $\omega'^2_p = \omega_p^2 + \bar{\omega}_p^2$.

The numerical applications result are provided in Table 3 in which the experimental values of $\varepsilon_1(\omega)$ are also included.

Table 3

	LiF	**NaCl**	**KI**	**MgO**	**BaO**
$\hbar\omega_p$ (eV)	26.2	15.7	11.2	24.5	16.3
ε_1 (0) Cal.	2.3	2.8	3.25	3.3	4
ε_1 (0) Exp.	1.9 (9)	2.25 (5.9)	2.7 (5.1)	3 (9.8)	-
$\hbar\omega'_p$ (eV)	32.7	18.6	13	27.5	17.9

As noted in the statement of the problem, the evaluation of $\varepsilon_1(0)$ only concerns the electronic contribution to the static dielectric constant. The experiments cited as a comparison were done at frequencies much larger than the resonant frequency of ions and therefore correspond to a measure of $\varepsilon(\infty)$ in the Lyddane–Sachs–Teller (LST) relation relative to phonons (see Chapter III, Pb. 1). The true static dielectric constant mentioned in parenthesis in the table results from the sum of the electronic and ionic contributions. Finally, note that $\hbar\omega_p'$ is also slightly larger than $\hbar\omega_p$ (see comment on the bandwidth).

Since $E_g > 3$ eV, all the materials here are transparent in the visible spectrum (1.5 eV < $h\nu$ < 3 eV; see Ex. 28). The absence of optical absorption in the energetic interval extends from the beginning of the visible spectra up to E_g, corresponds to the zero of $\varepsilon_2(\omega)$ and therefore to the energy loss function $\varepsilon_2/(\varepsilon_1^2 + \varepsilon_2^2)$ in this interval.

Given that the r_0^{-2} law governs E_g, hydrostatic pressure, which brings atoms closer together, will result in an increase in the bandgap while an increase in temperature results in a thermal expansion and leads to a decrease in the E_g.

Comment: Width of the bandgap

The expressions given in this problem were proposed by Pantelides (*Phys. Rev. B* **11**, 1975, 5082) and can also be applied to certain metallic nitrides (III–V compounds) that crystallize in the NaCl structure. With $\eta_g = 2.8$, they can also be correctly applied to the bandgap of CaF_2 type crystals and with $\eta_g = 12.9$ can be used for rare gas crystals (Ne, Ar, Kr, and Xe).

The study of Pantelides is based on the tight-binding approximation and results in the complete valence band structure of 51 corresponding crystals using simply two empirically determined parameters. The expression obtained here for E_g precise the qualitative remarks made in Ex. 21 for diamond (C, Si, and Ge) and zincblende (GaAs) crystals: the bandgap increases when the distance between nearest neighbors decreases and when the ionic nature of the compound increases. This allows us to generalize that the bandgap of all semiconductors and insulators increases as a function of applied pressure and decreasing temperature.

Comment: $\varepsilon_1(0)$

The choice of the average transition energy as $\hbar\overline{\omega}_T = E_g + \Delta E_v$, as proposed in this exercise, is relative arbitrary. This explains the non-negligible difference between the calculated values and the measured values of $\varepsilon_1(0)$. The interest in this assumption lies in that it qualitatively explains the quasi-general tendency of the decrease in the static dielectric constant as the bandgap increases, a trend found in the semiconductors:

C(diamond): $E_g = 5.5$ eV, $\varepsilon_1 = 5.7$

C(Si): $E_g = 1.12$ eV, $\varepsilon_1 = 12$
C(Ge): $E_g = 0.7$ eV, $\varepsilon_1 = 16$
C(PbTe): $E_g = 0.31$ eV, $\varepsilon_1 = 30$

Comment: "Plasmons" in semiconductors and insulators

In the 1960s, there were many questions on the following experimental facts: the energy loss function of semiconductors and insulators often exhibited a maximum at $\hbar\omega'_p$, very near the calculated value of $\hbar\omega_p$, where all valence electrons are assumed to be free even though this was not the case. For example, in diamond we find $\hbar\omega'_p = 33$ eV and $\hbar\omega_p$ (free) = 31 eV, while E_g = 5.5 eV (J. Cazaux, *Surf. Sci.*, **29**, 1972, 114). Thus, the word "plasmon" was more or less associated to "free electrons" instead of a better spelling: collective excitations. The Lorentz model developed here provides a simple explanation: the loss function $\varepsilon_2/(\varepsilon_1^2 + \varepsilon_2^2)$ is maximum when $\varepsilon_1(\omega) = 0$ and the zero of ε_1 corresponds to

$$\hbar\omega'_p = [(\hbar\omega_p)^2 + (\hbar\overline{\omega}_T)^2]^{1/2} \approx \hbar\omega_p\left[1 + \frac{1}{2}\frac{(\hbar\overline{\omega}_T)^2}{(\hbar\omega_p)^2}\right] \text{ because}$$

$(\hbar\omega_p)^2 >> (\hbar\overline{\omega}_T)^2$.

We observe that the zeros of $\varepsilon_1(\omega)$ correspond to natural frequencies of oscillations for a longitudinal excitation (see Chapter IV, Pb. 5) and that the evolution of ε_1 in the UV (with ω_T, ω_p, plasmon) and in the infrared (with ω_T, ω_L, polariton) are very similar if we compare Fig. 17 of Chapter III, Pb. 1, and Fig. 34 in Ex. 26.

In addition the approximation made to determine the zeros of $\varepsilon_1(\omega)$ is the same type as that used to evaluate the optical index of X-rays in matter. In the latter, the Z atomic electrons were considered to be oscillating quasi-freely and these oscillations were the origin of the coherent diffusion of X-rays (see Chapter I, Ex. 23; Chapter IV, Ex. 30; and Ex. 27 of this chapter).

Finally, we note that if the Lorentz model allows us to explain the general trends of ε_1 and ε_2 at large values of ω, a detailed study of these evolutions must take into account quantum mechanics based on interband transitions at critical points in the Brillouin zone, especially when the energy interval is located near E_g (see Pb. 6). In particular, the bandgap corresponds to the energetic position of the optical absorption threshold, where the energy loss function begins: a fact not taken into account in the Lorentz model.

Exercise 26: Dispersion of light: Sellmeier formula

Transparent optical medium in the visible (0.4 μm < λ < 0.8 μm) have an optical index n which obeys the Sellmeier relation: $n^2 = \mathrm{A} + \dfrac{\mathrm{B}}{\lambda^2}$.

Show that this relation can be deduced from the evolution of the complex dielectric constant as described by the Lorentz model in the form $\tilde{\varepsilon}_\mathrm{r} = 1 + \dfrac{\omega_\mathrm{p}{}^2}{\omega_\mathrm{T}{}^2 - \omega^2 + i\gamma\omega}$, where $\omega_\mathrm{p}{}^2 = \dfrac{ne^2}{m\varepsilon_0}$ in which the n electrons per unit volume make an average energy transition $\hbar\omega_\mathrm{T}$ and γ is the damping constant.

Simplify the expression for $\tilde{\varepsilon}_\mathrm{r}$ using physical justifications and then express A and B as a function of the remaining parameters. Using these results, describe the microscopic causes of the dispersion of white light by a glass prism.

Solution:

In the spectral domain considered, the medium is transparent. This implies that the optical absorption conditioned by the damping coefficient is negligible (ε_2 = 0 because $\gamma \approx 0$) and that the visible photon energy is smaller than the bandgap energy of the material: $\hbar\omega < E_\mathrm{g} < \hbar\omega_\mathrm{T}$.

From this we have $\varepsilon_1 = 1 + \dfrac{\omega_\mathrm{p}^2}{\omega_\mathrm{T}^2\left(1 - \dfrac{\omega^2}{\omega_\mathrm{T}^2}\right)}$

$\varepsilon_1 \approx 1 + \dfrac{\omega_\mathrm{p}^2}{\omega_\mathrm{T}^2}\left(1 + \dfrac{\omega^2}{\omega_\mathrm{T}^2}\right)$ to first order.

Since $n = \sqrt{\varepsilon_\mathrm{r}}$, we find $A = 1 + \dfrac{\omega_\mathrm{p}^2}{\omega_\mathrm{T}^2}$ and $B = \dfrac{\omega_\mathrm{p}^2}{\omega_\mathrm{T}^4}(2\pi c)^2$, where $\lambda = \dfrac{2\pi c}{\omega}$.

The evolution of ε_1 in the spectral interval from the infrared to the ultraviolet is shown in Fig. 34.

In the visible this evolution is conditioned by the presence of the resonance, which is located in the UV (at $\omega = \omega_\mathrm{T}$). Physically, when going from the red to the blue, the dipolar moment $\vec{\mathrm{p}} = -e\vec{\mathrm{x}}$

induced by the electric field of the EM wave and associated with the polarization of each atom will increase because x increases as this resonance is approached (see top insert in Fig. 34). $\vec{p} = N\vec{P}$ and $\vec{p} = \varepsilon_0 \chi \vec{E} = \varepsilon_0(\varepsilon_r - 1)\vec{E}$, a microscopic increase of $\vec{p}$ explains the macroscopic increase of ε_r and therefore of n, which leads to an index n slightly larger for the blue than the red. We thus can explain the decomposition of light by a transparent prism.

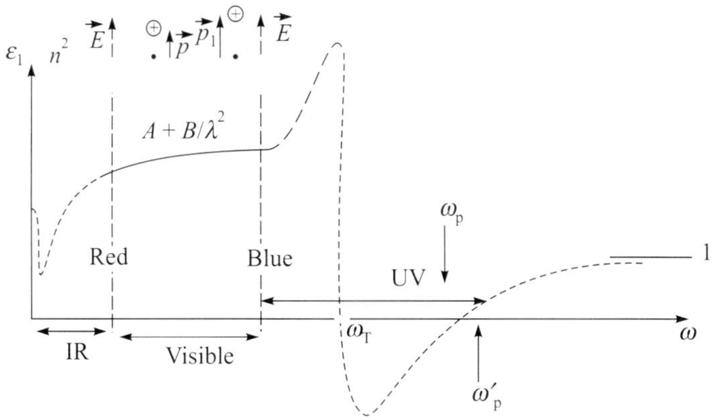

Figure 34

It is clear that to express $\tilde{\varepsilon}$ more rigorously than the Lorentz model used here, a more accurate summation of all the possible electronic transitions weighted by the corresponding oscillator strength is necessary. This procedure complicates the intermediate results without profoundly changing the physical description of the mechanism.

Exercise 27: Back to the optical index and absorption coefficient of X-rays

Consider an element irradiated with a beam of monochromatic X-rays with frequency ω. The different electronic levels (1s, 2s, 2p, etc.) of an atom of this element are characterized by their binding energy $\hbar\omega_{jT}$ and their population by z_j, in such a manner that the complex dielectric constant of this material can be described by an expression of the form:

$$\tilde{\varepsilon}_r = 1 + \sum_j \frac{\omega_{jp}^{\ 2}}{\omega_{jT}^{\ 2} - \omega^2 + i\gamma_j \omega}$$

where $\omega_{jp}{}^2 = N(\text{at})z_j \dfrac{e^2}{m\varepsilon_0}$, $N(\text{at})$ is atomic density, and γ_j is the damping constant.

(1) Consider the case of metallic sodium (Z = 11) radiated with X-ray photons of energy $\hbar\omega = \hbar\nu = 10\,\text{keV}$ (see Chapter IV, Ex. 30). Show that the proposed expression can be identified with that obtained by considering that all electrons, including atomic electrons, are free therefore that $(\hbar\omega)^2 >> (\hbar\omega_{j\text{T}})^2, (\hbar\omega_{jp})^2$.

Neglecting the term γ (because $\gamma_j << \omega_{j\text{T}}$), put ε_1 in the form $\varepsilon_1 = 1 - \Delta\varepsilon_1$ and find numerically $\Delta\varepsilon_1$ assuming that atomic electrons are free. Estimate the error introduced by such a simplification.

Na: $\hbar\omega_{j\text{T}}(1\text{s}) = 1070\,\text{eV}$,

$\hbar\omega_{2\text{T}}(2\text{s} + 2\text{p}) \approx 50\,\text{eV}$,

$\hbar\omega_{3\text{T}}(e^{-}\text{cond.}) = 0$,

$\hbar\omega_{\text{p}0} = 6\,\text{eV}$, where $\omega_{\text{p}0}{}^2 = N(\text{at}) \cdot \dfrac{e^2}{m\varepsilon_0}$.

(2) Find $\Delta\varepsilon_1$ and $\Delta\varepsilon_2$, the expressions for the complex index $\tilde{N} = n - ik$, the coefficient of attenuation with amplitude α, and the coefficient of attenuation in intensity μ.

(3) Now the goal is to correlate the factors governing the X-ray diffusion with those conditioning their propagation.

Can you express ε_1, ε_2, and μ as a function of the summation $\sum f_j$ of the electronic diffusion factors of the considered atoms? Recall the anomalous diffusion factor $\tilde{f}$ for a bound electron is $\tilde{f} = \dfrac{\omega^2}{(\omega^2 - \omega_\text{T}{}^2) - i\gamma\omega}$ (see Chapter I, Ex. 23), which leads to a comparison between a bound electron and a free electron.

Solution:

(1) The proposed simplification (also done in Chapter IV, Ex. 30) consists of identifying the expression:

$$1 + \frac{z_1(\hbar\omega_{0\text{p}})^2}{(\hbar\omega_{1\text{T}})^2 - (\hbar\omega)^2} + \frac{z_2(\hbar\omega_{0\text{p}})^2}{(\hbar\omega_{2\text{T}})^2 - (\hbar\omega)^2} - \frac{z_3(\hbar\omega_{0\text{p}})^2}{(\hbar\omega)^2} \quad \text{with}$$

$1-\frac{Z(\hbar\omega_{0p})^2}{(\hbar\omega)^2}$ where $z_1 = 2(1s)$; $z_2 = 8(2s + 2p)$; and $z_3 = 1(3s = \text{conduction e})$.

Taking into account that $(\hbar\omega)^2 >> (\hbar\omega_{jT})^2$, we note that the error will essentially only affect the term $\hbar\omega_{1T}$ relative to the most tightly bound 1s electrons. It will be of order of $z_1(\hbar\omega_{0p})^2\frac{(\hbar\omega_{1T})^2}{(\hbar\omega)^4}$, or 0.8×10^{-8} for a $\varepsilon_1 \approx 1 - 0.4 \times 10^{-5}$.

(2) $\tilde{N} = n - ik = \sqrt{\varepsilon_1 - i\varepsilon_2}$ or $n \approx 1 - \left(\frac{\Delta\varepsilon_1}{2}\right)$ and $k \approx \frac{\varepsilon_2}{2}$

$E = E_0 \exp i\omega\left(t - \tilde{N}\frac{z}{c}\right)$, where $\tilde{N} = \frac{c}{\tilde{v}}$

$E = E_0 \exp -\alpha z \cdot \exp i\omega\left(t - n\frac{z}{c}\right)$, where $\alpha = \omega\frac{k}{c} = \omega\frac{\varepsilon_2}{2c}$

$I = I_0 \exp -\mu z$ with $\mu = 2\alpha = \omega\frac{\varepsilon_2}{c}$.

(3) $\tilde{\varepsilon} = 1 - \sum_i \tilde{f}\frac{\omega_{jp}^2}{\omega^2} \rightarrow \tilde{f}_j = f_{1j} + if_{2j}$, where f_{ij} and $f_{2i} > 0$

$\tilde{\varepsilon}_r = \varepsilon_1 - i\varepsilon_2$, where ε_1, $\varepsilon_2 > 0$ (see note Chapter IV, Ex. 29),

so that $\mu = \sum_j \frac{f_{2j}}{c} \cdot \frac{\omega^2}{\omega} jp$.

Exercise 28: Optical absorption and colors of semiconductors and insulators

Consider a semiconductor with a direct bandgap $E_g \cong 2.3\,\text{eV}$ (GaP, for example).

(a) What is its color, as observed by transmission when pure?

(b) Same question for rutile; TiO_2 ($E_g \approx 3$eV); ZnS ($E_g \approx 3.6$ eV); and CdS ($E_g = 2.5$ eV).

(c) Suppose that GaP has an impurity level located at 0.7 eV above the valence band, what is its new color as viewed by transmission?

(d) Can you generalize the above results and explain why the majority of oxides and alkali halides are transparent when they are pure? (h)

Solution:

(a & b) Neglecting the possible creation of excitons (see Ex. 20), photons $h\nu$ can only be absorbed by electrons from the valence band if the photon energy allows these electrons to reach to the conduction band.

The visible spectra are 4000 Å ≤ λ ≤ 8000 Å or equivalently 3 eV ≥ $h\nu$ ≥ 1.5 eV and thus no absorption is possible in pure materials such that $E_g \geq 3$ eV. These materials (TiO_2, ZnS) will thus be transparent.

- GaP absorbs in the green, blue, and violet (see Fig. 35) and allows the yellow, orange, and red to pass. It will thus be orange by transmission.
- In addition, CdS allows the green to pass and will thus be yellow by transmission.

(c) If the material contains an impurity level "*n*," the transitions between this level and the valence band will occur, resulting in selective absorption of photons $h\nu = E_g - E_D$. In the case of GaP, the red will thus be absorbed and to a large extent change toward the yellow, which is the transmission color.

(d) The bandgap reflects the bonding energy of valence electrons. Its value will be larger if electrons are more tightly bound. Without knowing the exact value of the bandgap (but using the comment from Exs. 21 and 25), we expect that the oxides (Al_2O_3, BeO, MgO) and the alkali halides (LiF, NaCl) will have bandgaps greater than 3 eV. They will therefore be transparent (without color) if they are pure (without impurities; see comments below).

Most of the transparent materials are also electrical insulators with a noticeable exception: indium tin oxide (ITO, or tin-doped indium oxide) has a large bandgap of around 4 eV and, thus, it is mostly transparent in the visible part of the spectrum in thin layers, 200–850 nm, but it is also a conducting oxide via a large concentration of dopants such as Sn. The mechanism of DC conductivity is detailed in H. Peelaers et al., *Appl. Phys. Lett.* **100**, 2012, 011914.

These specific properties makes that it is widely used to make transparent conductive coatings for displays such as liquid crystal displays, flat panel displays, plasma displays, touch panels, and electronic ink applications. Its elaboration is increasing the thickness and increasing the concentration of charge carriers, which will increase the material's conductivity, but decrease its transparency.

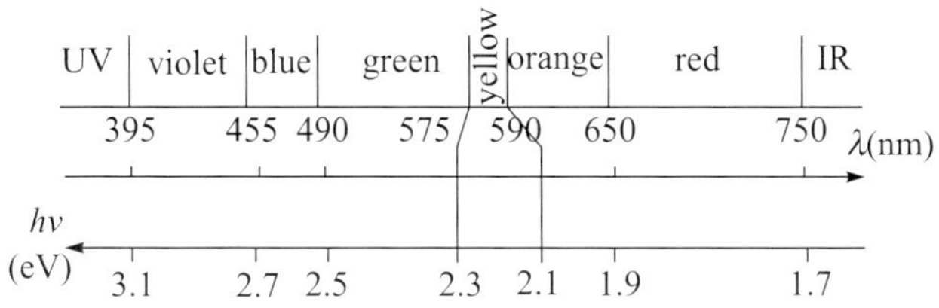

Figure 35

Comment: Color of insulators and semiconductors

The electrical insulators ($E_g \geq 3$ eV), when they are pure, are transparent for all radiations in the visible spectra but inescapably absorb photons in the ultraviolet when $h\nu$ reaches $h\nu \approx E_g$. In virtue of the law on the value of E_g developed in Exs. 21 and 25 (see comment), the largest bandgap is of LiF ($E_g \approx 12$ eV), if we exclude solidified rare gases E_g (Ne) = 18 eV). This implies that we can only build transparent lenses for UV optics below $\lambda \approx 100$ nm. It is more convenient to use quartz for the near UV but for the far UV the use of LiF (up to the limit mentioned above) is needed and its use requires operations in the vacuum in order to limit absorption by gas molecules.

For semiconductors, their reflecting color will be complimentary to that observed by transmission because the absorbed wavelengths will be the most efficiently reflected and rediffused (see for instance Pb. 6, Question 1b(ii)—R increases with ε_2 which itself starts at $h\nu = E_g$). With exceptions, they appear therefore blue by reflection.

The presence of impurities modifies the color of insulators because the absorption process (between the impurity levels and E_c or E_v) is often located in the visible. This explains the color of precious stones (ruby, sapphire, emerald, etc.), which are initially transparent metallic oxides that are accidentally or not doped with metallic impurities (for instance ruby is corundum, a variety of Al_2O_3 doped with chrome).

Comment: Color of insulators and semiconductors

The detailed mechanism of optical absorption differs depending on whether the transitions are direct or indirect.

If the forbidden band is direct (see Fig. 36a), the absorption coefficient μ is of the form (near the absorption threshold):

$$\mu \propto (h\nu - E_{gd})^{1/2}$$

This is quantum result that one can also deduce from $\mu = (\omega/c)\sqrt{2\varepsilon_2}$ (Pb. 6, Question 1b(ii)) where $\varepsilon_2 \propto 1/\omega$ (Question (2b) where $\omega \approx \omega_T$).

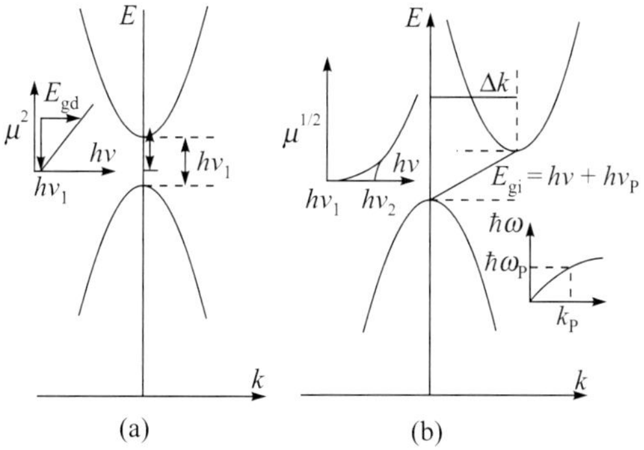

Figure 36

If the bandgap is indirect, phonon transitions (creation or absorption) are necessary to ensure wave vector conservation:

$$\hbar k_i + \hbar(2\pi/\lambda) + \hbar k_p = \hbar k_f;\ E_f = E_i + h\nu + h\nu_p.$$

Compared with direct transitions, conservation of energy is not modified significantly since $h\nu_p$(phonons) $\leq$ a few meV, whereas the change in momentum $\Delta k \approx k_p$ is essentially due to phonons, which modifies the characteristics of μ near the absorption threshold: $\mu^{1/2}$ will contain two linear parts (Fig. 36b) instead of being linear in μ^2 as for a direct transition.

Exercise 29: Optoelectronic properties of III–V compounds

In order to realize emissive optoelectronic devices (LEDs and lasers), the following constraints are imposed on the active materials:

(a) Its bandgap E_g must be direct and its value must be exactly adjusted to the wavelength of emission.

(b) The active material must be sandwiched by an adjacent material (substrate) having a bandgap E_g' greater than E_g.

(c) The active material and the substrate, both single crystals, must have a lattice parameter "a" (of the fcc lattice) relatively close to one another.

As a result, if the substrate is a simple binary compound $A'B'$ from the III–V family (III: Al, Ga, In; V: As, P, Sb), one must often use a tertiary compound of type $A_X{}^1, A_{1-x}{}^2B^1$, or $A^1B_y{}^1B_{1-y}{}^2$ $(0 \le x, y \le 1)$ to realize the active material.

Using the table showing the bandgap of III–V compounds as a function of their lattice parameter a (Table 1 in Course Summary), find the explicit composition ($A^{(1)}$, $A^{(2)}$, $B^{(1)}$, $B^{(2)}$, x, y?)of the active material and substrate necessary for the two following realizations:

(1) A laser emitting in the visible at λ = 7500 Å.

(2) An active device in the infrared at λ = 1.6 µm (minimum absorption of fiber optics). In each case, state the value of the crystal lattice a.

(3) What is the emission wavelength for a quaternary active material: $Ga_xIn_{1-x}As_yP_{1-y}$, where x = 0.29 and y = 0.63?

Note: In the transition $A^{(1)}, B \rightarrow A^{(1)}_{1-x}A^{(2)}_xB \rightarrow A^{(2)}B$ assume that the bandgap and the crystal parameter vary linearly with x (Vegard law).

Solution:

In each case, it is sufficient to trace a horizontal line from the y-axis corresponding to $E_g = hc/\lambda$ (constraint: a) and find the intersection of this line with the curves connecting the different semiconductors. To find the substrate for the constraints imposed by (b) and (c), one looks perpendicular and directly above this intersection for a III–V semiconductor, if it exists.

(1) $\lambda = 7500\ \text{Å} \rightarrow E_g \approx 1.65\ \text{eV}$. The first intersection connects AlAs and GaAs, with the substrate being AlAs.

The active compound will be $Al_xGa_{1-x}As$ where x is such that $xE_g(\text{AlAs}) + (1 - x)\, E_g\,(\text{GaAs}) = 1.65$ eV or $x \approx 0.30$.

The lattice parameters will be (Vegard law):

$xa(\text{AlAs}) + (1 - x)\ a(\text{GaAs}) = a(Al_xGa_{1-x}As)$, which gives a = 5.64 Å.

More precise values are tabulated in handbook on semiconductor such as that by Baranski et al. [Ref. 2].

(2) $\lambda = 1.6\,\mu\text{m} \rightarrow E_g = 0.775$ eV. The first intersection connects GaAs to InAs and is below InP. The substrate will thus be InP (a_0 = 5.87 Å; E_g = 1.27 eV). The active compound will be $Ga_{1-x}In_xAs$ where x = 0.61 and a = 5.90 Å (in reality the composition used is $Ga_{0.47}In_{0.53}As$, which has a lattice parameter closer to InP).

(3) $a, E_g = x[y(\text{GaAs}) + (1-y)(\text{GaP})] + (1-x)[y(\text{InAs}) + (1-y)(\text{InP})]$

$$E_g = 0.997\text{ eV};\ \lambda = 1.243\,\mu\text{m}.$$

This material is effectively used in the infrared (1.3 μm) on a InP substrate.

Comment: Optoelectronic properties of the III–V compounds

(a) The bandgap must be direct so that electronic transitions ($E_c \rightarrow E_v$) are vertical ($k_i = k_f + k(\text{photon}) \approx k_f$).
If the minimum of E_c (in k-space) does not correspond to the maximum E_v, the corresponding transitions will be oblique and the conservation of k will favor nonradiative transitions (Auger type—see Chapter IV, Pb. 4 and comment).
For direct bandgaps, the number of emitted photons will be proportional to the convolution of the density of electrons occupying the bottom of the conduction band ($g_{Ev}(E) \cdot f_h(E)$ where $f_h(E) = 1 - f_e$ and $f_e(E)$ is the Fermi–Dirac distribution function).

(b) The adjacent material (substrate) must have a bandgap that is larger than that of the active material so that electrons emitted from n-impurities in the substrate are trapped by the potential well of the active material (injection of carriers; see Pb. 10 for the electronic states in quantum wells and the band diagram).

(c) The crystallization of the adjacent and active materials must be continuous so that interfacial problems are avoided. This epitaxy (see Chapter I, Pb. 4) can only be done correctly if there is a small difference in lattice parameter a otherwise strains created will relax by forming dislocations (see Chapter II, Pb. 5).

All of these delicate technologies are well mastered and the characterization of the material must be carried out at each step (see Chapter I, Pbs. 9 and 10, for example).

Exercise 30: The Gunn diode

Consider a rectangular parallelepiped, of length c and cross-section a^2 of gallium arsenide (GaAs), which contains an excess of N arsenic atoms, all ionized at ambient temperature.

(a) When the applied bias U_0 at the ends of the bar is weak, the electrical conductivity σ is carried by a single type of (light) electrons with mass m_l = 0.07 m and mobility $\mu_l \approx 7500$ cm^2/V·s. Find the expression and the numerical values of σ and the resistance R_b with the following parameters' values: $N = 10^{14}$ cm^{-3}, a = 1 mm, and $c = 10^{-2}$ cm.

(b) The light electrons are located at the bottom of the conduction band which has a second minimum (see Fig. 37) corresponding to heavy electrons with an electron mass that nearly corresponds to that of the free electrons. These heavy electrons may participate to the electric conduction.

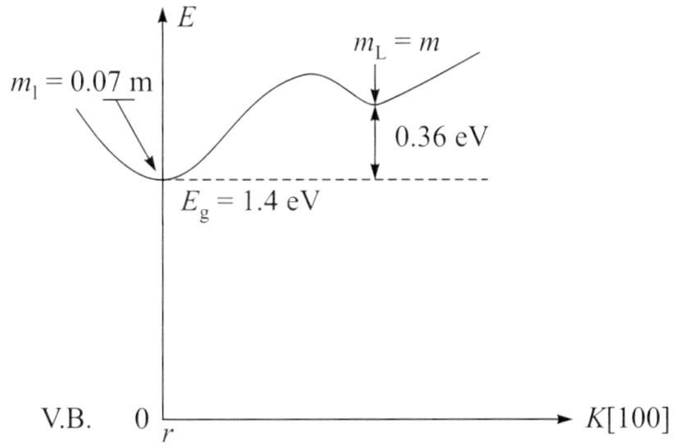

Figure 37 Conduction band of GaAs.

Find the expression of σ as a function of μ_l and densities n_l and n_h for charge carriers (light electrons and heavy electrons) by assuming that the relaxation time τ is the same for both carriers.

(c) Assume that the density n_h of heavy electrons increases proportionally to applied electric field E following the law $n_h = N\dfrac{E}{E_0}$ and that this increase occurs to the detriment of the density of n_l of light electrons until the latter has disappeared (for $E = E_0$). Thus, we have $n_h + n_l = N$.

Find the expression for σ, $\vec{j}$ (current density), and $\vec{v}$ (average drift velocity of electrons) for $E < E_0$ and next for $E > E_0$. Find the value of E_M, for which j is maximum in the interval $0 < E < E_0$.

(d) Sketch the characteristic curves $v = f(E)$ and $I = f(U_0)$ indicating the numerical values of v and I at the important points M and C ($E = E_M$ in M and $E = E_0$ in C) where $E_0 = 6$ kV/cm.

(e) A DC current is applied to the bar that is inserted in series in a circuit containing a resistance R and a source e_0. Using a graphical construction, find the maximum value of e_0 so that the Gunn diode functions at the average electric field $(E_M + E_0)/2$. What is the value of R?

(f) In fact in the latter hypothesis, the electric field and the charge distribution are not uniform in the diode: a domain formed from the large density of charge appears at the –pole, propagates in the crystal at the drift speed v and when it attains the +pole is instantly replaced by another domain (which appears at the –pole).

What is the frequency of the waves thus created?

Solution:

(a) $\sigma = Ne\mu_l = 0.12\,\Omega^{-1}\text{cm}^{-1}; R_b = \dfrac{1}{\sigma}\dfrac{c}{a^2} = 8.3\,\Omega.$

(b) $\mu_l = \dfrac{e\tau}{m_l}, \mu_h = \dfrac{e\tau}{m_h}; \sigma = e(n_h\mu_h + n_l\mu_l) = e\mu_l\left(n_l + n_h\dfrac{m_l}{m_h}\right)$

and $n_l + n_h = N$.

(c) $E < E_0; n_h = N\dfrac{E}{E_0}; n_l = N\left(1 - \dfrac{E}{E_0}\right); \sigma = Ne\mu_l\left(\dfrac{E}{E_0}\cdot\dfrac{m_l}{m_h} + 1 - \dfrac{E}{E_0}\right).$

$$j = Ne\mu_l\left[\left(\frac{m_l}{m_h} - 1\right)\frac{E}{E_0} + 1\right]E\,;\, v = \frac{j}{Ne} = \mu_l E\left(\frac{E}{E_0}\cdot\frac{m_l}{m_h} + 1 - \frac{E}{E_0}\right);$$

$$\frac{\mathrm{d}j}{\mathrm{d}E} = 0 \text{ for } E_M = \frac{E_0}{2}\frac{m_l}{m_h - m_l}.$$

$$E > E_0; n_l = 0; n_h = N; \sigma = Ne\mu_l\frac{m_l}{m_h}; j = Ne\mu_l\cdot\frac{m_l}{m_h}\cdot E; v = \mu_l\frac{m_l}{m_h}E.$$

(d) When $E = E_M; j_M = \frac{Ne\mu_l}{2} \cdot \frac{m_l}{m_h - m_l} E_0 = 360\ \text{A/cm}^2, I_M = 3.6\ \text{Å}$,

$$v_M = \frac{j}{Ne} = 2.25 \times 10^7\ \text{cm/s}.$$

When $E = E_0 : j_C = Ne\mu_l \cdot \frac{m_l}{m_h} E_0 = 50\ \text{A/cm}^2, I_C = 0.5\ \text{Å}$,

$v_c = 0.315 \times 10^7$ cm/s.

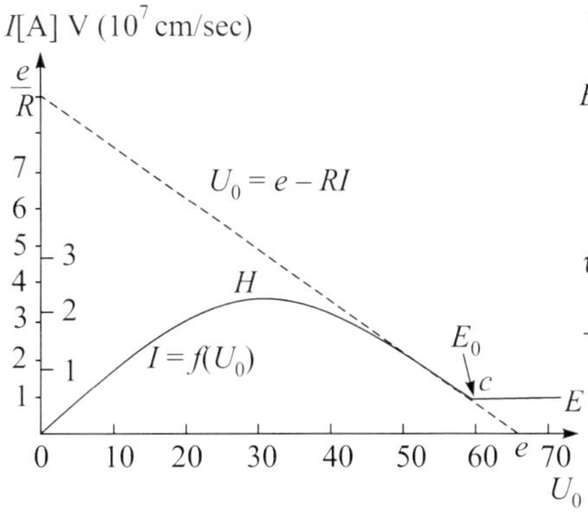

Figure 38

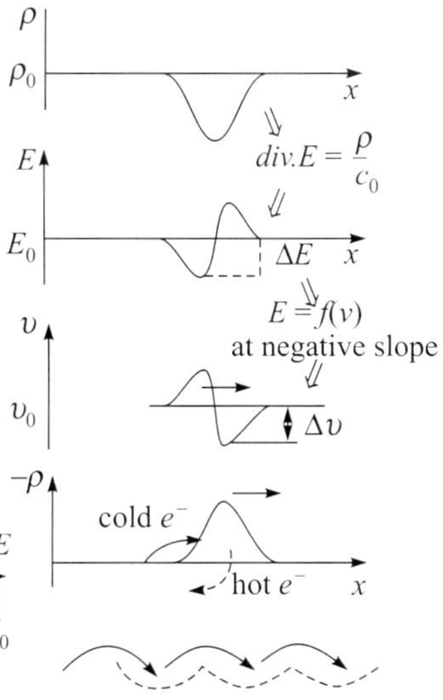

Figure 39

(e) The operating point corresponds to the resolution of the systems formed by: $U_0 = e - RI$ (1) (Ohm's law) and by $I = f(U_0)$ (2) (characteristic of the diode).

Knowing the analytic expression of (2) this system can be solved algebraically, but the statement of the problem suggests to solve it graphically, as usually in electronics when the characteristics are given graphically. We thus obtain from Fig. 38 that $e = 61$ V and $R \approx 6.7\ \Omega$.

(g) The periodicity T of the waves is such that $T = c/v$ from which

$$f = \frac{v}{c} = \frac{1.56 \times 10^7}{10^{-2}} \cong 1.6\ \text{GHz}.$$

Note: Gunn diodes

In partially ionic semiconductors such as GaAs or InP from which the electronic density N is such that $N \cdot c \approx 10^{12}$ cm^{-2}, the application of an intense electric field provokes the appearance of waves with frequencies between 1 and 10 GHz and inversely proportional to the length c of the bar.

Figure 39 schematically depicts the propagation mechanism of a strong electron concentration domain along the bar (the domain generating the wave). Hot and heavy electrons located in the front of the wave will abandon the wave which moves too fast for them and became light electrons. During this the cold and light electrons initially left by the impulse appearing from of the impulse in high-field region, etc.

These diodes serve as high-frequency generators for radars that are used in commercial airlines, as well as radars found on highways. With their light weight and low cost they can also be used to advantage in applications normally using the photoelectric effect such as alarms and automatic doors. In the LSA (limited space charge accumulation) mode, the charge domains do not have time to form and the amplification of electromagnetic waves is assured by the fact that the speed of charge carriers decreases as the electric field increases, they thus transfer the energy of the continuous electric field toward the high-frequency field.

Problems

Problem 1: Krönig–Penney model: periodic potential in 1D

In order to establish the existence of energy bands that are alternatively allowed and forbidden when electrons are subject to a periodic potential, we explore the behavior of an electron subject to a periodic potential created by N equidistant ions separated by a in a 1D lattice.

The potential energy $V(x)$, shown in Fig. 40, in the periodic interval $-b \le x \le c$ takes the successive values: Region I $(0 \le x \le c)$, $V(x) = 0$; region II $(-b \le x \le 0)$, $V(x) = V_0$ where $a = b + c$.

(1) Find the solutions to the Schrödinger equation in region I and in region II (use A, B, C, and D as integration constants).

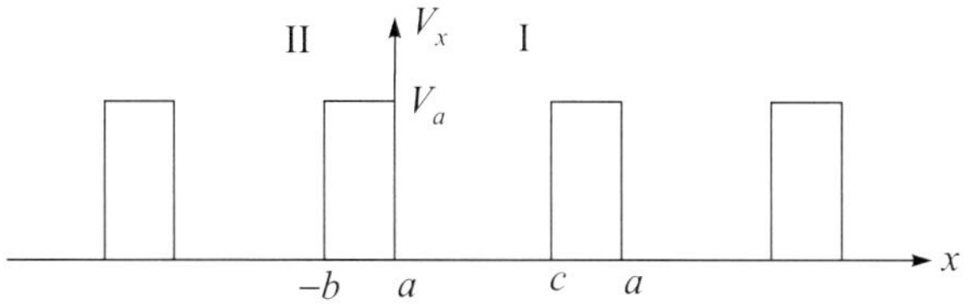

Figure 40

(2) When an electron is subject to a periodic potential, its wave function obeys cyclic boundary conditions $\psi(x) = \psi(x + Na)$ (Bloch's theorem), and must take the form of a Bloch wave: where $u(x)= u(x + a)$, (a is $\psi(x) = u_k(x)\cdot e^{ikx}$ the period between regions).

Find the wave function as a function of C and D in region III ($c \le x \le a$) and state the sequence of discrete values of the wave vector k.

(3) Find the integration constants $A, B, C,$ and D using the continuity of the wave functions and of their first derivative at $x = 0$ and $x =$ c. Show that the systems of equations thus obtained gives a trivial solution only when

$$\cos ka = \cos\alpha c \cdot \cos\beta b - \frac{\alpha^2 + \beta^2}{2\alpha\beta}\sin\alpha c \cdot \sin\beta b\,,$$

where $\alpha = \dfrac{(2mE)^{1/2}}{\hbar} \cdot \beta = \dfrac{[2m(E - V_0)]^{1/2}}{\hbar}$ is satisfied.

(4) Verify that we recover the relation $E = f(k)$ of free electrons when the potential energy everywhere is zero.

(5) Assume that the following inequalities are satisfied:

$qb << 1$(where $iq = \beta$); $E < V_0$.

Show that the condition established in (3) can be reduced to the form:

$$\cos ka = p\frac{\sin\alpha a}{\alpha a} + \cos\alpha a.$$ Find P.

Show the graphical evolution of the 2° term of the equality when $P = 3\pi/2$. Deduce the existence of energy bands alternating between allowed and forbidden. How many electronic states are in the allowed bands? Comment on the result.

What happens when P tends to infinite?

(6) *Numerical application*: By successive operations, find the energy width of the first allowed band and the forbidden band that follows it using $P = 3\pi/2$ and $a = 3$ Å.

(7) Find the symbolic expression and then the numerical value of the effective mass m^* for particles situated at the maximum of the first allowed band:

$$\frac{1}{m^*} = \frac{1}{\hbar^2}\frac{\partial^2 E}{\partial k^2}. \qquad (e, m, \hbar)$$

Solution:

(1) $\frac{\hbar^2}{2m}\frac{d^2\psi}{dx^2} + V\psi = E\psi$ or $\frac{d^2\psi}{dx^2} + \frac{2m}{\hbar^2}(E - V)\psi = 0$

Region I: $V = 0$; $\psi_1 = A \exp(i\alpha x) + B \exp(-1\alpha x)$, where

$$\alpha = \frac{(2mE)^{1/2}}{\hbar}.$$

Region II: $V = V_0$; $\psi_{II} = C\exp(i\beta x) + D\exp(i\beta x)$, where

$$\beta = \frac{[2m(E - V_0)]^{1/2}}{\hbar}$$

β is real if $E > V_0$, β is purely imaginary if $E < V_0$.

(2) The wave function in region III is deduced from that obtained in region II by a translation of a:

$\psi_{III}(x) = \psi_{II}(x + a) = u(x)e^{ik(x+a)} = \psi_{II}(x)\, e^{ika} = e^{ika}$ [Cexp $(i\beta x)$ + Dexp $(-i\beta x)$].

Cyclic boundary conditions result in $\psi(x + Na) = \psi(x)$ or $\exp(ikNa) = 1$ from which $k = \frac{2\pi n}{Na}$, where n is a positive or negative integer.

As for phonons (see Chapter III, Course Summary) or for free electrons (see Chapter IV, Course Summary), the electronic states are characterized by a wave vector k from which the discrete steps are $\hbar\, \frac{2\pi}{L} = \frac{2\pi}{Na}$ (periodic conditions).

(3) $\psi_{\mathrm{I}}(x=0)=\psi_{\mathrm{II}}(x=0)\Rightarrow$ A + B = C + D

$\psi'_{\mathrm{I}}(x=0)=\psi'_{\mathrm{II}}(x=0)\Rightarrow \alpha(\mathrm{A}-\mathrm{B}) = \beta(\mathrm{C}-\mathrm{D})$

$\psi_{\mathrm{I}}(x=c=\psi_{\mathrm{III}}(x=c)\Rightarrow$ A $\exp(i\alpha c)$ + B $\exp(-i\alpha c)$

$= \exp ika[C\exp(i\beta c)+D\exp(i\beta c)]$

$\psi'_{\mathrm{I}}(x=c=\psi'_{\mathrm{III}}(x=c)\Rightarrow \alpha[\mathrm{A}\exp(i\alpha c)-\mathrm{B}\exp(-i\alpha c)]$

$= \beta\exp ika[C\exp(i\beta c)-D\exp(i\beta c)]$

This set of linear and homogeneous equations has a nontrivial solution such that if the determinant of the coefficients is canceled, then

$$\cos ka=\cos\alpha c\times\cos\beta b-\frac{\alpha^2+\beta^2}{2\alpha\beta}\sin\alpha c\times\sin\beta b.$$

(4) When V_0 = 0 and b = 0, then c = a and coska = cosαa or $k=\frac{2me^{1/2}}{\hbar}$, such that $E=\frac{\hbar^2k^2}{2m}$ (see the dashed curve in Fig. 42).

(5) If $E < V_{0,}$ β is purely imaginary: $\beta = iq$, and the equality [given in (3) above] becomes

$$\cos ka=\cos\alpha c\times \mathrm{ch}qb+\frac{q^2-\alpha^2}{2q\alpha}\sin\alpha c\times \mathrm{sh}qb.$$

The bands prohibited are determined by the double inequality:

$$-1\le\cos\alpha c\times \mathrm{ch}qb+\frac{q^2-\alpha^2}{2q\alpha}\sin\alpha c\times \mathrm{sh}qb\le 1 \quad .$$

In the case where $qb \approx \varepsilon$ (ch$qb \approx 1$; sh$qb \approx qb$), equality (3) above takes the form

$$\cos ka = \cos\alpha a + \frac{mV_0}{h^2\alpha}\times b\sin\alpha a = \cos\alpha a + P\frac{\sin\alpha a}{\alpha a}$$

where $P=\frac{\mathrm{abm}V_0}{\hbar^2}$.

This transcendent equation must have a solution for α so that the wave functions of the form ψ_{I} exist.

Figure 41 represents the evolution of $P\frac{\sin\alpha a}{\alpha a}$ + cosαa as a function of αa for $P = \frac{3\pi}{2}$. As the cosine term of the first member can take only values between +1 and –1, only those values of αa will be permitted for which the second member of the equality is located in this interval. These intervals allowed for αa, drawn in thick lines in Fig. 41, ($\alpha = (2mE)^{1/2}/h$) correspond to the values of k equal to $\frac{n\pi}{a}$. The variation of E as a function of k is represented in Fig. 42.

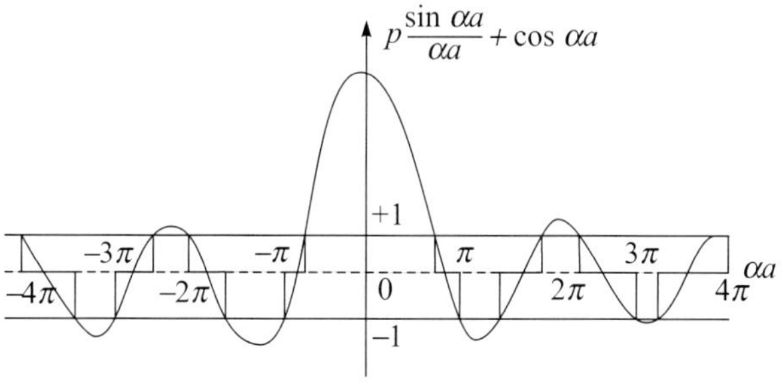

Figure 41

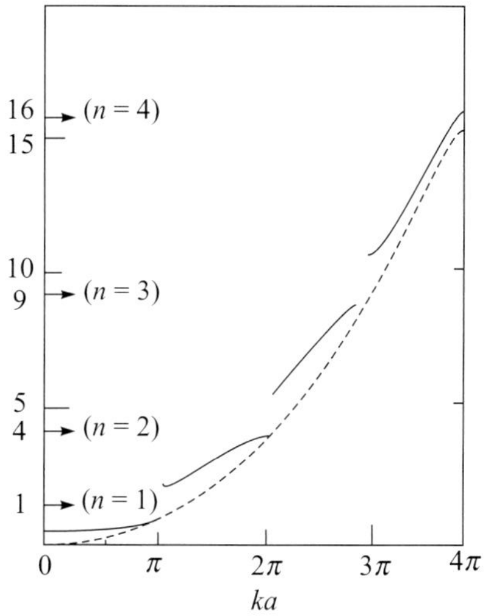

Figure 42 Energy as a function of wave vector for a Krönig-Penny potential (P = $3\pi/2$).

Has the interior of a band allowed, coska varies between –1 and +1 is $-\pi \le ka \le \pi$ or else $-\frac{N}{2} \le n \le \frac{N}{2}$ (see Question

(2)). If N is the number of faces of the crystal, there will be N discrete values of the vector waveform per band, which gives an opportunity to logically fill the first strip with 2 electron (↑ and ↓).

If P tends towards infinity, $\sin\alpha a \to 0$ and $\alpha a = \frac{(2mE)^{1/2}}{\hbar} \cdot a = n\,\pi$.

The energy spectrum becomes discrete, values itself are those of an electron trapped in a segment of side a (see the order in Fig. 42 which specifies the succession of these discrete values, $E = n^2 \frac{\hbar^2\pi^2}{2ma^2}$.

(6) The function $\frac{3\pi}{2}\frac{\sin\alpha a}{\alpha a} + \cos\alpha a$ is greater than 1 when $\alpha a = \frac{2\pi}{3}$ and is slightly smaller than 1 for $\alpha a = \frac{3\pi}{4}$;<< $\alpha_m a$ >> is therefore of the order of 2.2 radians, and $E_{\mathrm{m}} = \frac{\hbar^2}{2m}\alpha_{\mathrm{m}}^2 = 2\,\mathrm{eV}$.

The maximum of the allowed band corresponds to $E_{\mathrm{m}} = \frac{\hbar^2}{2ma^2}\pi^2 = 4.1\,\mathrm{eV}$.

The width of the first allowed band is thus 2.1 eV.
As the same function is equal to –1 for αa of order $\alpha_{2\mathrm{m}}a \approx 3.5$ radians, the width of the bandgap which follows is $E_{\mathrm{i}} = E_{2\mathrm{m}} - E_M \approx 1$ eV, where $E_{2\mathrm{m}} = 5.1$eV.

(7) $E = \frac{\hbar^2\alpha^2}{2m}$. At the top of the first allowed band $\alpha a = \pi$,

$\cos ka = -1$ and $\left(\frac{\mathrm{d}\alpha}{\mathrm{d}k}\right)_{\alpha a=\pi} = 0$.

At this point $\frac{\mathrm{d}^2E}{\mathrm{d}k^2} = \frac{\hbar^2}{m}\alpha\left(\frac{\mathrm{d}^2\alpha}{\mathrm{d}k^2}\right)_{\alpha a=\pi}$.

Differentiating $P\frac{\sin\alpha a}{\alpha a} + \cos\alpha = \cos ka$ two times with respect to k, we find $m^* = -\frac{mP}{\pi^2} = -\frac{3m}{2\pi} = -0.48\,m = -4.34\times10^{-31}\,\mathrm{kg}$.

As in other exercises and problems of this chapter, one finds that the particles located at the top of the band have a negative effective mass: they are holes.

Problem 2: Nearly free electrons in a 1D lattice

Consider a row of identical equidistant atoms separated by a along the axis Ox. A fraction of electrons of each atom can propagate along the row and are subject to a periodic potential of ions, which result in potential energy of $U = U_1\cos gx$ (where $g = 2\pi/a$) and U_1 is sufficiently small that the Schrödinger equation can be resolved by successive approximations.

(1) Find the wave function ψ_0 and energy E^0 of electrons as a function of wave vector k when $U_1 = 0$.

(2) Now assume that U_1 differs from zero and find the solutions for ψ in form of Bloch waves: $[\psi = \psi_0 \cdot u(x)]$, where such that

$$u(x) = 1 + \sum_{\substack{-\infty \\ n\neq 0}}^{+\infty} A_n e^{-ignx}.$$

Find the expression of A_n by assuming that U_1 is small, the energy of the Bloch waves is nearly equal to that of free electrons E^0, the first-order approximation.

(3) Starting from A_n calculated previously, re-evaluate the energy E of the corresponding waves, the second-order approximation.

(4) Show that the results thus obtained are in general compatible with the hypotheses made, specifically:

$$|A_n| << 1 \text{ and } \frac{|E - E^0|}{E^0} << 1 \text{ except when } k \approx \pm\frac{\pi}{a}.$$

Numerical application: a = 3 Å; U_1 = 2 eV. Find A_1 and A_{-1} as well as $|E - E^0|$ for $k = \pi/2a$ and $k = (\pi/a)\ (1 + \varepsilon)$, where $\varepsilon = -2\%$.

(5) To study the energy when the wave vector approaches $g/2$, $k \approx \pi/\text{a}$, we now consider that only the coefficient A_1 is not small compared to 1 in the expression of $u(x)$. Show that A_1 appears simultaneously in the two linear relations. Deduce the expression for energy.

Taking as a numerical example $k = \pi/2a$, show that when k is very different from π/a, E thus obtained is equal to the result evaluated in Question (4) and approximately corresponds to the energy of free electrons with the same wave vector.

(6) Find the simplified expression of E when $k = (g/2) - \Delta k$ (Δk is small compared to $g/2$). Deduce the existence of a discontinuity for $k = \pi/a = g/2$. What is the amplitude of this discontinuity and the value taken by A_1? Comment on the results.

(7) After having studied the evolution of E relative to the wavevector $k = -(g/2) + \Delta k$, show the shape of the curve $E = f(k)$ in the interval $(2\pi/a) < k < (2\pi/a)$ in the extended zone scheme and next in the reduced zone scheme.

(8) Show the shape of the density of states $g(E)$ relative to the two first allowed energy bands after having evaluating $g(k)$ for k neighboring π/a.

(9) Find the expression of the effective mass m^* for an electron whose wave vector neighbors π/a. Evaluate numerically the ratio m^*/m using the results from Question (4), and then with

$$k \approx \frac{\pi}{a} - \Delta k(|\Delta k| << \frac{\pi}{a} \text{ and } \Delta k \geq 0).$$

(10) The element considered is monovalent. Evaluate symbolically and next numerically the Fermi energy at 0 K.
The element considered is now divalent. Why it will not be a good conductor of electricity? $(e, \hbar, m)$

Note: In order to be accessible to the largest audience, the resolution of this problem does not use perturbation theory or calculus of variations which would result in a more elegant and faster solution.

Solution:

(1) When $U_1 = 0$, electrons are free and the solutions to the Schrödinger equation $-\frac{\hbar^2}{2m}\frac{d^2\psi_0}{dx^2} = E^0\psi_0$ correspond to plane waves $\psi_0 = \frac{1}{L^{1/2}}e^{ikx}$, where $E^0 = \frac{\hbar^2 k^2}{2m}$.

(2) We introduce Bloch waves of form $\psi(x) = \frac{1}{L^{1/2}}\left(e^{ikx} + \sum_{n \neq 0} A_n c^{i(k-gn)x}\right)$ in the Schrödinger equation

$$-\frac{\hbar^2}{2m}\frac{d^2\psi}{dx^2} + \psi U_1 \cos gx = E\psi \text{ and we obtain (1):}$$

$$\sum_{n\neq 0}[k^2-(k-ng)^2]A_n e^{i(k-ng)x}+\frac{2m}{\hbar^2}(E-E^0-U_1\cos gx)e^{ikx}$$

$$+\frac{2m}{\hbar^2}\sum_{n\neq 0}(E-E^0-U_1\cos gx)A_n e^{i(k-ng)x}=0.$$

In the first-order approximation, we neglect the last summation because the coefficients $A_n U_1$ are of second order, and $E \approx E^0$ so that, after developing the cosine, the equation simplifies to:

$$\sum_{n\neq 0}[k^2-(k-ng)^2]A_n e^{i(k-ng)x}=\frac{m}{\hbar^2}U_1[e^{i(k-g)x}+e^{i(k-g)x}].$$

By identifying term by term, the terms having the same exponentials, we find:

$$A_1=\frac{m}{\hbar^2}\frac{U_1}{g(2k-g)},\ A_{-1}=-\frac{m}{\hbar^2}\frac{U_1}{g(2k+g)},\text{ and } A_n(n\neq 0,1,-1)=0$$

In the first-order approximation, the solutions to the Schrödinger equation (taking the form of a Mathieu equation in the case of a sinusoidal potential) are

$$\psi=\frac{1}{(L)^{1/2}}e^{ikx}\left[1+\frac{amU_1}{\hbar^2.4\pi}\left(\frac{e^{-i2\pi\frac{x}{a}}}{k-\frac{\pi}{a}}-\frac{e^{i2\pi\frac{x}{a}}}{k+\frac{\pi}{a}}\right)\right] \qquad (2)$$

(3) In the second-order approximation, we evaluate the unknown E starting from the A_n previously calculated and identify the terms between them in factor of exp ikx.

We thus find $E-E^0=\frac{U_1}{2}(A_1+A_{-1})=\frac{m}{\hbar_2}\frac{U_1^2}{(4k^2-g^2)}$.

(4) The coefficients A_1 and A_{-1} are in general very small to first order (in U_1) and E – E^0 is infinitely small in second order (in $U_1{}^2$), except when $k\approx\pm\frac{g}{2}\approx\pm\pi/a$.

We thus obtain numerically

For $k=\frac{\pi}{2a}$,

$$A_1=-0.12;\ A_{-1}=-0.04;\ E-E^0=0.16\text{ eV};$$

$$E^0=1.04\text{ eV; and }\frac{|E-E^0|}{E^0}\approx 0.15<<1$$

For $k = \frac{\pi}{a}(1+\varepsilon)$, where $\varepsilon = -2\%$.

$A_1 = -3$; $A_{-1} = 3\%: E - E_0 = -3.03\ \text{eV}; E^0 \approx 4.16\ \text{eV}$; and

$$\frac{\left|E - E^0\right|}{E^0} = 0.72.$$

(5) Setting $\psi = \frac{1}{\sqrt{L}}(e^{ikx} + A_1 e^{i(k-g)x})$ into the Schrödinger equation, we find the expression

$$\left[\frac{\hbar^2 k^2}{2m} + \frac{A_1 U_1}{2} - E\right]e^{ikx} + \left[\frac{\hbar^2}{2m}(k-g)^2 A_1 + \frac{U_1}{2} - EA_1\right]e^{i(k-g)x} = 0$$

which reduces to system of two relations:

$$\frac{\hbar^2 k^2}{2m} - E + \frac{U_1}{2}A_1 = 0$$

$$\frac{U_1}{2} + \left[\frac{\hbar^2}{2m}(k-g)^2 - E\right]A_1 = 0$$

They are only compatible when E is the root of

$(E_k^0 - E)(E_{k-g}^0 - E) - \frac{U_1^2}{4} = 0$, where $E_k^0 = \frac{\hbar^2 k^2}{2m}$

and $E_{k-g}^0 = \frac{\hbar^2}{2m}(k-g)^2$.

We deduce that the energy is

$E = \frac{1}{2}\left[E_k^0 + E_{k-g}^0 \pm \sqrt{(E_k^0 - E_{k-g}^0)^2 + U_1^2}\right]$, which leads to

$E \approx E_k^0 + \frac{U_1^2}{4(E_k^\circ - E^\circ_{k-g})} \simeq \frac{\hbar^2 k^2}{2m}$ (when k is very different from π/a).

Numerical application: $E\left(k = \frac{\pi}{a}\right) = 1.04\ \text{eV} - 0.12\ \text{eV}$.

The correction term has the same magnitude as that evaluated in Question (4), and it will be negligible when the potential created by the ions of the lattice is weak.

We have retained the (+) solution for E because $|k| \le \frac{\pi}{a}$; the (−) solution corresponds to $\frac{\pi}{a} \le |k| \le \frac{2\pi}{a}$ and in the last case the correction term is positive.

(6) For $k = g/2 - \Delta k$, we find

$$E_k^0 = \frac{\hbar^2}{2m}\left(\frac{g^2}{4} - g\Delta k + \Delta k^2\right) \text{ and}$$

$$E_{k-g}^0 = \frac{\hbar^2}{2m}\left(\frac{g^2}{4} + g\Delta k + \Delta k^2\right), \text{ from which}$$

$$E = \frac{1}{2}\left\{\frac{\hbar^2}{2m}\left(\frac{g^2}{2} + 2\Delta k^2\right) \pm U_1\left[1 + \left(\frac{\hbar^2}{2m}\right)^2 \frac{4g^2 \cdot \Delta k^2}{U_1^2}\right]^{1/2}\right\}.$$

Using $E_B^0 = \frac{\hbar^2}{2m}\frac{g^2}{4}$:

- $E - E_B^0 = \frac{\hbar^2}{2m}\Delta k^2\left(1 - \frac{4E^0_B}{U_1}\right) - \frac{U_1}{2}$ for $\Delta k > 0$: parabolic arc where the maximum is at point $[\pi/a, E^0_B - (U_1/2)]$or
- $E - E_B^0 = \frac{\hbar^2}{2m}\Delta k^2\left(1 + \frac{4E_B^0}{U_1}\right) + \frac{U_1}{2}$ for $\Delta k < 0$: parabolic arc, where the minimum is at point $(\pi/a, E^0_B + (U_1/2))$.

If $\Delta k = 0$: $k = -(k - g)$, $E^0_k - E^0_{k-g}$, the equation for E results in a double root: $E = E_k^0 \pm \frac{U_1}{2}$.

We note that E is discontinuous when the wave vector k approaches the first Brillouin zone. The amplitude of the discontinuity is equal to U_1: it is the width of the forbidden bandgap in the model for nearly free electrons.

In this case, $A_1 = -1$, the electronic waves propagating in the direction of increasing x with a wavelength of $a/2$ are subject to Bragg reflections on equidistant ions; the reflected waves have, in absolute value, the same amplitude as the incident waves and the resulting is a standing wave. If the potential was periodic without being purely sinusoidal, its decomposition in Fourier series would result in discontinuities of amplitude V_n for k values such that $k = ng/2$.

(7) The energy is an even function of k; the dispersion curves are shown in Figs. 43a (extended zone scheme) and 43b (reduced zone scheme).

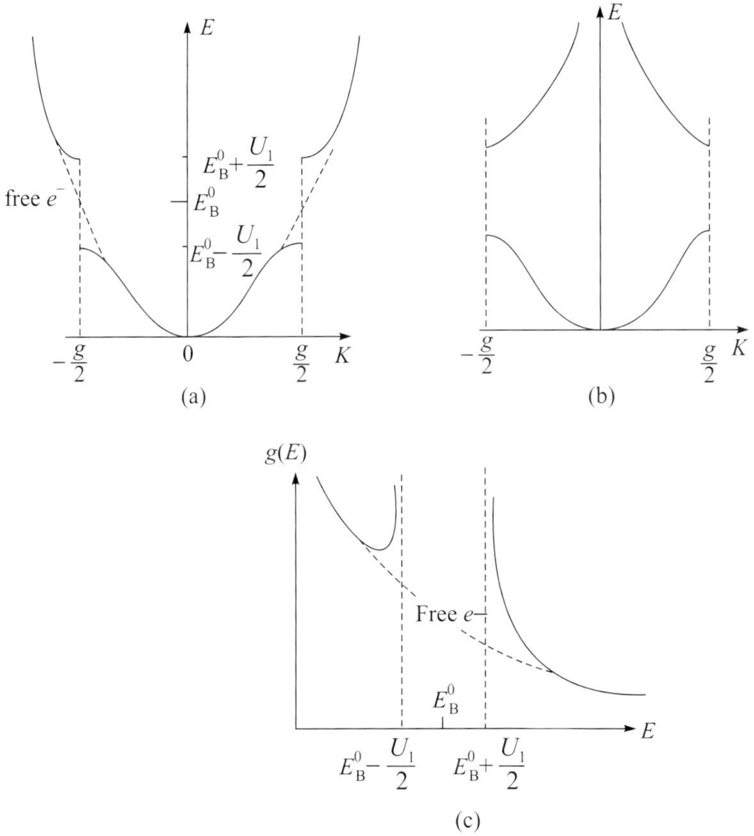

Figure 43

(8) $g(E)\cdot dE = 2g(|k|)\cdot dk = 4\frac{Na}{2\pi}\cdot dk.$

When $|k| << \frac{g}{2}.E \approx \frac{\hbar^2 k^2}{2m}$ and $g(E) = \frac{Na}{\pi}\left(\frac{2m}{\hbar^2}\right)^{1/2}\frac{1}{\sqrt{E}}$

When $k = \frac{g}{2} - \Delta k; \frac{\mathrm{d}(\Delta k)}{\mathrm{d}E} = \left[\frac{\hbar^2}{m}\Delta k\left(\frac{\mp 4E_B^\circ}{U}\right)\right]^{-1}$ from which

(a) For $\Delta k > 0, \frac{E_B^0}{U_1}$ is larger than 1:

$$g(E) = \frac{Na}{\pi}\left(\frac{2m}{\hbar^2}\right)^{1/2}\left(\frac{4E_B^0}{U_1} - 1\right)^{-1/2}\left(-E + E_B^0 - \frac{U_1}{2}\right)^{-1/2}; E \le E_B^0 - \frac{U_1}{2}.$$

(b) For $\Delta k < 0$:

$$g(E) = \mathrm{Na}/\pi \left(\frac{2m}{\hbar^2}\right)^{1/2} \left(1 + \frac{4E_B^0}{U_1}\right)^{-1/2} \left(E - E_B^0 - \frac{U_1}{2}\right)^{-1/2};$$

$$E \geq E_B^0 + \frac{U_1}{2}$$

$$g(E) \to \infty \text{ when } E \to E_B^0 - \frac{U_1}{2} \text{ and } E \to E_B^0 + \frac{U_1}{2}.$$

The synthesis of these results is shown in Fig. 43c.

(9) $$\frac{1}{m^*} = \frac{1}{\hbar^2}\frac{\partial^2 E}{\partial k^2}, \text{ if } E = \frac{\hbar^2 k^2}{2m} m^* = m,$$

$$\frac{1}{m^*} = \frac{1}{m}\left(1 - \frac{4E_B^0}{U_1}\right) \text{ for } \Delta k \geq 0 \text{ so that } \frac{m^*}{m} = -0.137,$$

$$\frac{1}{m^*} = \frac{1}{m}\left(1 + \frac{4E_B^0}{U_1}\right) \text{ for } \Delta k \leq 0 \text{ so that } \frac{m^*}{m} = +0.107.$$

Under the action of an applied field, the electrons located at the maximum of the first band, near the energy discontinuity, behave like free electrons with a negative effective mass, independent of E; the equation of motion only involve the ratio q/m. We can thus treat the particles as holes with charge $+e$ and mass $|m^*|$.

Electrons located at the bottom of the second band near the energy discontinuity behave as free electrons with a positive effective mass, $m^* < m$. This is a general result. We also note that the larger the crystalline potential U, the flatter the allowed bands and the higher the effective mass.

(10) In each energy band, we can place two electrons per basis. If the element is monovalent, only half the first energy band will be complete and the second band will be empty (at 0 K). Under the action of an applied electric field, the electrons located in the full band cannot increase their kinetic energy and therefore cannot transport an electrical current: the material is an insulator. At $T \neq 0$ K, some electrons are thermally excited to the conduction band leaving holes in the valence band: electrical conductivity of the material increases with T. We

observe that, as at the extremities of the band the electrons are no longer free, the Fermi energy is not equal to $E_B^0 - \frac{U_1}{2}$, but it must be evaluated using the method developed in the next problem.

Problem 3: 1D semiconductor: electronic specific heat

In the nearly free electron approximation, conduction electrons subject to a periodic potential energy of form $U = U_1 \cos gx$ (where $g = \frac{2\pi}{a}$ and a is the lattice parameter of the linear lattice), the energy E of electrons with wave vector $k = \frac{\pi}{a} - \Delta k$ is of the form (see Pb. 2)

$$E - E_B^0 = \frac{\hbar^2}{2m}\Delta k^2 \left(1 - \frac{4E_B^0}{U_1}\right) - \frac{U_1}{2} \text{ for } \Delta k > 0$$

$$E - E_B^0 = \frac{\hbar^2}{2m}\Delta k^2 \left(1 + \frac{4E_B^0}{U_1}\right) + \frac{U_1}{2} \text{ for } \Delta k < 0$$

where $U_1 > 0$, $E_B^0 = \frac{\hbar^2 \pi^2}{2ma^2} > U_1$.

(a) Sketch the curves $E = f(k)$ in the vicinity of $k = \pi/a$.

(b) Find the expression for the effective mass m^* of particles situated near from the maximum of the lower band [$E < E^0_B - (U_1)/2$] and that of particles located at the minimum of the upper band $E > E^0_B + (U_1)/2$. Comment on the results. Numerical application with $U_1 = 0.2$ eV, $E^0_B = 1$ eV.

(c) After recalling the expression for the density of states in k-space ($k \geq 0$) for a 1D lattice, $g(|k|)$, find the density of state $g(E)$ at the maximum of the lower band as well as at the bottom of the upper band. Compare, in the same energy domain, this result to the density of states of a 1D gas of free particles (affected by the effective mass calculated in (b). Comment on the results.

(d) The solid considered consists of two electrons per basis which, at non-zero temperature, occupy nearly all the states of the lower (valence) band and very partially the states of the upper (conduction) band.

Taking the maximum of the valence band as the new origin of energy, find the integral expression giving the density of n per unit length of electrons located in the conduction band. Same question for the density p (per unit length) of holes located in the valence band.

(e) Suppose that the bands are semi-infinite and that E_F is located in the bandgap (which allows us to relax the Fermi distribution into a Boltzmann distribution), find the linear densities, n and p (for a nondegenerate semiconductor) and thus the product np. Recall that $\int_0^{\infty} x^{-1/2}e^{-\alpha x} = \left(\frac{\pi}{\alpha}\right)^{1/2}$

(f) When the semiconductor is intrinsic, find the expression for E_F and describe the evolution of n and p with temperature. What are the numerical values taken by n, p, and E_F at T =290 K?

(g) Find the increase in the number of holes and electrons resulting from an increase in temperature dT. Deduce the approximate expression for the corresponding increase of electronic energy dU. Establish the expression for the specific electronic heat C_v of the semiconductor. Using the preceding numerical data, find C_v at 290 K. $(e, m, \hbar, k_B)$

Solution:

(a) See Pb. 2 and Fig. 44.

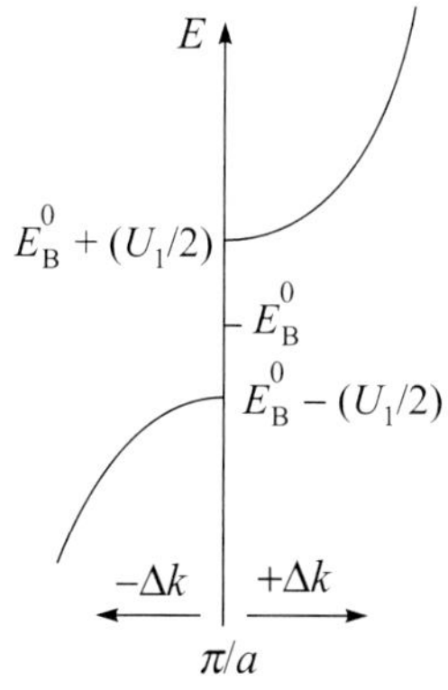

Figure 44

(b) $\frac{1}{m^*}=\frac{1}{\hbar^2}\frac{\partial^2 E}{\partial k^2};\frac{1}{m^*}=\frac{1}{m}\left(1-\frac{4E_{\mathrm{B}}^0}{U_1}\right)$ for $E\leq E_{\mathrm{B}}^0-\frac{U_1}{2}$.

$$\frac{1}{m^*}=\frac{1}{m}\left(1+\frac{4E_{\mathrm{B}}^0}{U_1}\right) \text{ for } E\geq E_{\mathrm{B}}^0+\frac{U_1}{2}.$$

The particles occupying the maximum of the lower band behave as holes with mass m_{h}^* such that

$$\frac{1}{m_{\mathrm{h}}^*}=\frac{1}{m}\left(\frac{4E_{\mathrm{B}}^0}{U_1}-1\right)$$

Numerical application:

$$m_{\mathrm{h}}^*=\frac{m}{19}=0.0526m, m_{\mathrm{e}}^*=\frac{m}{21}=0.0476m.$$

(c) $g(|k|)=\frac{Na}{\pi}; g(E)dE=2g(|k|)\cdot dk=2\cdot\frac{Na}{\pi}\cdot dk,$

$$g(E)=\frac{Na}{\pi}\left(\frac{2m}{\hbar^2}\right)^{\frac{1}{2}}\cdot\left(\frac{4E_{\mathrm{B}}^0}{U_1}-1\right)^{-1/2}\cdot\left(-E+E_{\mathrm{B}}^0-\frac{U_1}{2}\right)^{-1/2}$$

So for $E\leq E_{\mathrm{B}}^0-\frac{U_1}{2}$;

$$g(E)=\frac{\mathrm{Na}}{\pi}\left(\frac{2m}{\hbar^2}\right)^{1/2}\cdot\left(1+\frac{4E_{\mathrm{B}}^0}{U_1}\right)^{-1/2}\cdot\left(E-E_{\mathrm{B}}^0-\frac{U_1}{2}\right)^{-1/2}$$

and for $E\geq E_{\mathrm{B}}^0+\frac{U_1}{2}$

The density of states of a 1D free electron gas is

$$g(E)=\frac{Na}{\pi}\left(\frac{2m}{\hbar^2}\right)^{1/2}\cdot E^{-1/2}.$$

When the origin of energy coincides with the bottom of the parabola, $E=\frac{\hbar^2}{2m}\cdot k^2$, the density of states takes the form

$$g(E)=\frac{Na}{\pi}\left(\frac{2m}{\hbar^2}\right)^{1/2}\left(-\frac{U_1}{2}\pm E_{\mathrm{B}}^0\mp E\right)^{-1/2}$$ respectively for $E\leq E_{\mathrm{B}}^0-\frac{U_1}{2}$

and $E\geq E_{\mathrm{B}}^0+\frac{U_1}{2}$ when taking these limiting values as extremes of the parabolas.

By replacing m with the expression for the effective mass evaluated in (b), we find the results above. We verify that it is possible to evaluate $g(E)$ in the general case starting from the expression relative to free electrons and substituting it into the effective mass, which takes into account the particular form of the dispersion curves and therefore the effects of the lattice.

(d) Per unit length Na = 1:

$$n=\int_{U_1}^{\infty} f_e(E)\cdot g(E)\cdot dE=\frac{1}{\pi}\left(\frac{2m_{\mathrm{e}}{}^*}{\hbar^2}\right)^{1/2}\int_{U_1}^{\infty}\left(e^{\frac{E-E_{\mathrm{F}}}{k_{\mathrm{B}}T}}+1\right)^{-1}\cdot(E-U_1)^{-1/2}\cdot dE$$

$$p=\int_{-\infty}^{0} f_h(E)\cdot g_h(E)\cdot dE=\frac{1}{\pi}\left(\frac{2m_{\mathrm{h}}{}^*}{\hbar^2}\right)^{1/2}\int_{-\infty}^{0}\left(e^{\frac{E_{\mathrm{F}}-E}{k_{\mathrm{B}}T}}+1\right)^{-1}\cdot(-E)^{-1/2}\cdot dE.$$

(e) If $\dfrac{U_1-E_{\mathrm{F}}}{k_{\mathrm{B}}T}>>1;\ n\approx\dfrac{1}{\pi}\left(\dfrac{2m_{\mathrm{e}}{}^*}{\hbar^2}\right)^{1/2}\cdot e^{\frac{E_{\mathrm{F}}-U_1}{k_{\mathrm{B}}T}}\displaystyle\int_0^{\infty}x^{-1/2}e^{\frac{x}{k_{\mathrm{B}}T}}\cdot dx$

which gives

$$n\approx\left(\frac{2m_{\mathrm{e}}{}^*\cdot k_{\mathrm{B}}T}{\hbar^2\pi}\right)^{1/2}e^{(E_{\mathrm{F}}-U_1)/k_{\mathrm{B}}T},\ p=\left(\frac{2m_{\mathrm{h}}{}^*}{\hbar^2}\cdot\frac{k_{\mathrm{B}}T}{\pi}\right)^{1/2}\cdot e^{-E_{\mathrm{F}}/k_{\mathrm{B}}T}$$

from which we find $n\cdot p=\dfrac{2k_{\mathrm{B}}T}{\hbar^2\pi}(m_{\mathrm{h}}{}^*\cdot m_{\mathrm{e}}{}^*)^{1/2}e^{-U_1/k_{\mathrm{B}}T}$.

(f) If the semiconductor is intrinsic,

$$n=p=\left(\frac{2k_{\mathrm{B}}T}{\hbar^2\pi}\right)^{1/2}\cdot(m_{\mathrm{h}}{}^*\cdot m_{\mathrm{e}}{}^*)^{1/4}e^{-U_1/2k_{\mathrm{B}}T}=4.7\times10^6\,\mathrm{m}^{-1}$$

$$E_{\mathrm{F}}=\frac{U_1}{2}+\frac{k_{\mathrm{B}}T}{4}\log(m_{\mathrm{h}}{}^*/m_{\mathrm{e}}{}^*)=100.6\ \mathrm{meV}.$$

(g) n and p have the form $p = n = AT^{1/2}\cdot e^{-B/T}$ from which:

$$\frac{dp}{dT}=\frac{dn}{dT}=A\cdot T^{-1/2}\cdot e^{-B/T}\left(\frac{1}{2}+\frac{B}{T}\right)\ .$$

For an increase of temperature dT, each of the dn or dp particles gains energy equal to the width of the bandgap U_1 (the electrons jump to the conduction band, the holes fall in the valence band). Then

$$C_v \approx \frac{dU}{dT} = U_1 \cdot A \cdot T^{-1/2} \cdot e^{-B/T}\left(1+\frac{2B}{T}\right) \approx \frac{n \cdot U_1}{T}\left(\frac{U_1}{k_B T}+1\right).$$

Numerically, $C_v = 46 \times 10^{-16}$ J/deg.m = 29×10^3 eV/deg·m

We find that C_v does not obey the linear T dependence of free electrons but a law dependent in exp – $(E_g/2kT)$, which governs the generation of the electron/hole pairs for intrinsic semiconductors. See Ex. 22 for another calculation of C_v.

Problem 4: DC conductivity of intrinsic and doped Ge and Si

We consider a 3D semiconductor in which the Fermi energy E_F is located in the bandgap of width E_g such that taking the origin of energy to be at the valence band maximum, the inequality $5k_BT < E_F < E_g - 5k_BT$ is satisfied (nondegenerate semiconductor).

(1) Law of mass action

(a) Recall the density of states of a free electron gas for unitary volume. Taking into account the fact that the conduction electrons behave as free electrons with effective mass m_e^*, express their number, n, at temperature T, a function of E_F, E_g, and m_e^* (assuming that the conduction band is semi-infinite).

Show that the result can be expressed in the form $n = N_c \exp-\left(\frac{E_g - E_F}{k_B T}\right)$ and find N_c.

Recall that $\int_0^\infty (x)^{1/2} e^{-\beta x} dx = \frac{1}{2\beta}\left(\frac{\pi}{\beta}\right)^{1/2}$.

(b) What is the probability of electron occupation $f(E)$ in the valence band? Deduce, for this same band, the probability of holes, $f_h(E)$. Simplify this expression assuming that $E_F/k_BT >> 1$ is satisfied. Deduce the number of holes p located in the valence band, assumed to be semi-infinite $(-\infty < E_v < 0)$ taking into account that holes behave as positive free charges with an effective mass m_h^*. Show that the result can be presented in the following form:

$p = N_v \exp - (E_F/k_BT)$ and find N_v.

(c) Evaluate the product np (law of mass action) and show that it is independent of E_F.

(2) Intrinsic semiconductors

The semiconductor is absolutely pure and perfectly regular (intrinsic) and the conduction electrons (in the band) at temperature T can only originate from the breaking of certain valence bonds.

(a) In this case compare the concentration n and the concentration p.

(b) Express the law of variation of n as a function of T and E_g.

(c) Deduce E_F as a function of E_g, m_e^*, and m_h^*.

(d) Find E_F and n at 290 K, using the following numerical data:
Germanium: $E_g \approx 0.6$ eV, $m_e^* = m_h^* = 0.1\, m_0$
Silicon: $E_g \approx 1$ eV, $m_e^* = m_h^* = 0.2\, m_0$, where m_0 is the mass of a free electron.
What note suggests this result at room temperature, knowing that it is technologically impossible to obtain a crystal containing less than 10^{11} impurities.cm^{-3} (or a relative concentration of 10^{-11}).

(e) We denote by μ_n and μ_p the respective mobility of the two types of charge carriers (electrons and holes). Find the expression for the electrical conductivity σ of an intrinsic semiconductor. Express the temperature variation of σ knowing that μ_n and μ_p vary as $T^{-3/2}$. Show the curve $\log \sigma = f(1/T)$ in the case of silicon and germanium after having determined σ at 290 K. Use
μ_n (Si) = 1200 cm^2/V·s, μ_h (Si) = 450 cm^2/V·s,
μ_n (Ge) = 3600 cm^2 /V·s; μ_h (Ge) = 1700 cm^2/V·s
Specify what temperature increase, ΔT, around ambient leads to a relative increase of 100% in their conductivity.

(3) Influence of impurities:doped semiconductors

During the elaboration of samples, pentavalent impurities (arsenic) are added to the sample with a concentration N_d. Each impurity atom substitutes into the crystalline lattice in the place of one germanium or silicon atom such that four electrons make covalent bonds with their nearest neighbors and the fifth (with charge $-e$ and mass m_e^*) rotates around the impurity of charge $+e$.

(a) Assume that this behavior can be described by the Bohr model for a hydrogen atom in order to establish the expression for r_i (radius of the ground state circular orbit described by an excess electron). Hint: Take into account the relative dielectric constant of the medium ε_r and the effective mass of an electron in a periodic potential.

Evaluate r_i numerically using $\varepsilon_r(\text{Si}) = 11.7$ and $\varepsilon_r(\text{Ge}) \approx 15.7$ and find the ratio r_i/r_B in which r_B is the Bohr radius of the hydrogen atom ($r_B = 0.53$ Å).

Each impurity atom can be thermally ionized and the electron can thus escape the attraction of the impurity and move freely in the lattice (always with the mass m_e^*). Find the expression for the ionization energy E_d of impurities and determine the ration E_d/E_H in which E_H is the ionization energy of atomic hydrogen. *Numerical application*: $E_H = 13.6$ eV. In an energy diagram show the donor level relative to the unionized impurity.

(b) Find the density of electrons occupying the donor level of the unionized impurity at temperature T as a function of E_F, E_g, E_d, and N_d and taking into account the distribution function of electrons for impurity states of energy E is of the form $f(E)=\left[\frac{1}{g_i}e^{\frac{E-E_F}{k_BT}}+1\right]^{-1}$, where $g_i = 2$ because each state can be occupied by one electron with spin $\uparrow$ or $\downarrow$, see Ref. [14] p. 206.

Express the density n of electrons that is found in the conduction band.

(c) Identify the preceding result with that established more generally in Question (1a), in order to give the equation that must satisfy the Fermi energy E_F.

Find explicitly the expressions for E_F in the two following limiting cases:

(i) $8(N_d/N_c)\exp E_d/k_BT << 1$, weak doping

(ii) $8(N_d/N_c)\exp E_d/k_BT >> 1$, low temperature

Deduce the corresponding expressions for n such that the unknown E_F does not appear.

In the case of silicon, numerically determine E_F and n for $N_d = 10^{14}$ imp.cm^{-3} and $T = 290$ K; then for N_d =

10^{17} imp.cm^{-3} at 245 K. Compare these results with those of intrinsic Si. What are the values taken for the density of holes p?

(d) Find the expression of electrical conductivity σ for a n-doped semiconductor. Describe the evolution of σ as a function of $1/T$ when N_d is large and show the result on the curve Log $\sigma = f(1/T)$ established in Question (2b). Find the numerical value for Si at 145 K where $N_d = 10^{18}$ impurity·cm^{-3}. $(e, m, k_B, \hbar)$

Solution:

(1) (a) $g(E) = \dfrac{m^{3/2}}{\pi^2\hbar^3}\sqrt{2E}$.

Taking into account the choice of origin for the energies and replacing m by m_e* to account for the effects of the lattice potential, we find

$$n = \int_{\min}^{\max} g(E)\cdot f(E)\cdot dE = \frac{\sqrt{2}}{\pi^2\hbar^3} m_e{}^{*3/2} \int_{E_g}^{\infty} (E - E_g)^{1/2} \frac{dE}{e^{\frac{E-E_F}{k_B T}} + 1}$$

$$\approx \frac{\sqrt{2}m_e{}^{*3/2}}{\pi^2\hbar^3} \int_{E_g}^{\infty} \left(E - E_g\right)^{1/2} \cdot \exp\frac{E_F - E}{k_B T} dE.$$

$$n = \frac{\sqrt{2}m_e{}^{*3/2}}{\pi^2\hbar^3} \exp\left(\frac{E_F - E_g}{k_B T}\right) \int_0^{\infty} x^{1/2} e^{-\beta x} dx,$$ where we have substituted $x = E - E_g$.

$$n = N_c \exp\frac{E_F - E_g}{k_B T}, \text{ where } N_c = 2\left[\frac{m_e{}^* k_B T}{2\pi\hbar^2}\right]^{3/2}.$$

(b) In the case of a nondegenerate semiconductor, the Fermi function can be approximated by the Boltzmann distribution.

$$f_e = \frac{1}{e^{\frac{E-E_F}{k_B T}} + 1}; f_h = 1 - f_e = \frac{1}{e^{\frac{E_F-E}{k_B T}} + 1} \approx e^{\frac{E-E_F}{k_B T}}, \text{ where } E \le 0.$$

$$p = \int_{\min}^{\max} g_h(E)\cdot f_h(E)\cdot dE = \int_0^{\infty} \frac{m_h{}^{*3/2}}{\pi^2\hbar^3}(-2E)^{1/2} \exp\left(\frac{E - E_F}{k_B T}\right)\cdot dE$$

$$p = N_v e^{-\frac{E_F}{k_B T}}, \text{ where } N_v = 2\left[\frac{m_h^* \cdot k_B T}{2\pi\hbar^2}\right]^{3/2}.$$

(c) $n.p = N_v \cdot N_c \cdot e^{-E_g/k_B T}$: Law of mass action.

(2) (a) If the semiconductor is intrinsic $p = n = n_i$ from which we have

(b) $n_i = (N_v N_c)^{1/2} \exp -\dfrac{E_g}{2k_B T}$.

(c) Comparing this expression with the result obtained in Question (1a) or (1b), we find

$$E^{-\frac{E_F}{k_B T}} = \left(\frac{N_c}{N_v}\right)^{1/2} \exp -\frac{E_g}{2k_B T} \text{ or } E_F = \frac{E_g}{2} + \frac{3}{4} k_B T \log \frac{m_h^*}{m_e^*}.$$

$$\text{Ge}: E_F = \frac{E_g}{2} = 0.30\,\text{eV}, N_v = N_c = 0.75 \times 10^{24}\,\text{m}^{-3},$$

$$n(\text{p}) = 4.6 \times 10^{12} e(t)\,\text{cm}^{-3}$$

$$\text{Si}: E_F = 0.5\,\text{eV}, N_v = N_c = 2.1 \times 10^{24}\,\text{m}^{-3},$$

$$n(\text{p}) = 4.3 \times 10^{9}\, e(t)\,\text{cm}^{-3}.$$

(d) At ambient temperature, silicon, contrarily to germanium cannot be an intrinsic semiconductor because residual impurities play a non-negligible role and the calculations of σ(Si) done in Question (2e) are purely formal.
$\vec{j} = -n_i e\,\vec{v} + n_+ e\vec{v}_+ = e(n|\mu| + n_+\mu_+)\vec{E} = \sigma\vec{E}$ (see Course Summary)
Here $n_- = n_+ = n = p$ from which $\sigma = en_i(\mu_p + |\mu_n|)$
$\sigma(\text{Ge}) \approx 0.4 \times 10^{-2}\,\Omega^{-1}\cdot\text{cm}^{-1}$, $\sigma(\text{Si}) \approx 1.1 \times 10^{-6}\,\Omega^{-1}\cdot\text{cm}^{-1}$.

(e) We observe that the conductivity of germanium exceeds that of silicon by several orders of magnitude because of the larger bandgap of Si (E_g plays a role for n and p in an exponent with a negative sign).
If we assume that the mobilities vary as $A \cdot T^{-3/2}$, this evolution compensates for the $T^{3/2}$ factor of N_v and N_p and $\sigma = C \cdot e^{-E_g/2k_B T}$, where C is independent of T;
$\log\sigma = \log C - \dfrac{E_g}{2k_B T}$; the curve $\log \sigma = f(1/T)$ is a line with slope $- E_g/2k_B$. We can deduce the double conductiv-

ity for an increase of ΔT around ambient such that $\Delta T \approx \dfrac{2k_B T^2}{E_g}\log 2$; σ(Ge) doubles every 15°, σ(Si) doubles approximately every 9°.

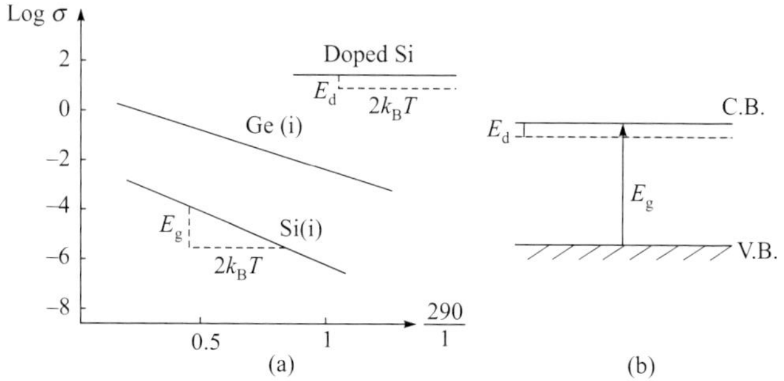

Figure 45

(3) (a) The Bohr radius of the ground state of hydrogen is $r_H = \dfrac{4\pi\varepsilon_0\hbar^2}{me^2} = 0.53\,\text{Å}$, and the corresponding energy is

$$E_H = \frac{1}{8\pi\varepsilon_o}\frac{e^2}{r_H} = \frac{me^4}{2(4\pi\varepsilon_o)^2\hbar^2} = 13.6\,\text{eV}.$$

$r_i = \dfrac{4\pi\varepsilon_0\varepsilon_r\hbar^2}{m_e^*.e^2} = \varepsilon_r\cdot\dfrac{m}{m_e^*}\cdot r_H$. Applied to an impurity, this model must be modified by taking into account the potential effects (m^*) and the relative dielectric constant of the medium, ε_r. Thus, we have:

$$E_d = \frac{m_e^* c^4}{2(4\pi\varepsilon_0\varepsilon_r)^2\hbar^2} = \frac{m_e^*}{m}\cdot\frac{E_H}{\varepsilon_r^2}$$

and from which we find for silicon: r_i = 31 Å, E_d = 20 meV and for germanium r_i = 83 Å, E_d = 6 meV.

When the impurity is ionized, an electron escape from the atomic attraction to and it propagates freely in the crystal. The free energetic state corresponds to the bottom of the conduction band. Thus, the bound state corresponds to an energy smaller than E_d of the previous state and is located in the forbidden gap at a distance E_d below the bottom of the E_c (see Fig. 45b, Course Summary and Pb. 6). The combined effects of light effective masses and of relative dielectric constants larger than unity may facilitate the ionization of impurities and the values of r_i far larger than the interatomic distances to justify the use of a macroscopic dielectric constant.

(b) The density of bounded electrons is

$$n_e = N_d \cdot f(E) = \frac{N_d}{\frac{1}{2}e^{\alpha}+1} \text{ where } \alpha = \frac{E_g - E_d - E_F}{k_B T}.$$

We thus deduce the density of electrons in the conduction band:

$n = N_d - n_e = N_d/(2e^{-\alpha} + 1)$.

(c) The result of Question (1a) was established without making a hypothesis on the nature of the semiconductor (intrinsic or extrinsic) and thus it remains valid.

Equating the two expressions for n and substituting

$$g = \exp\frac{E_g}{k_B T};\ d = \exp\frac{E_d}{k_B T}, \text{ we find } x^2 + \frac{1}{2}\frac{g}{d}x - \frac{1}{2}\frac{N_d}{N_c}\frac{g^2}{d} = 0.$$

The only positive solution is $x = \frac{g}{4d}\{-1+\left(1+8d\frac{N_d}{N_c}\right)^{1/2}\}$,

where $x = \exp\frac{E_F}{k_B T}$.

(i) When $8\frac{N_d}{N_c}e^{E_d/k_B T} << 1$, weak doping and large T

$$x \cong g\cdot\left(\frac{N_d}{N_c}\right) \text{ so that } E_F = E_g + k_B T \log\frac{N_d}{N_c}.$$

Using this result in the equation 1a, we find $n \approx N_d$: all the impurities are ionized.

(ii) When $8\frac{N_d}{N_c}e^{\frac{E_d}{k_B T}} >> 1$, large doping and small T

$$x \approx \frac{g}{(d)^{1/2}}\cdot\left(\frac{N_d}{2N_c}\right)^{1/2} \text{ so that}$$

$$E_F = E_g - \frac{E_d}{2} + \frac{k_B T}{2}\log\frac{N_d}{2N_c},$$

and $n \approx \left(\frac{N_c \cdot N_d}{2}\right)^{1/2} \exp - \frac{E_d}{2k_B T}$.

- For silicon at 300 K where $N_d = 10^{14}$ impurities·cm^{-3}, we are in situation α, $E_F = 0.79$ eV
  or $n \approx 10^{14}$ e·cm^{-3} and from the law of mass action:
  $p \approx \frac{n_i^2}{n} = 2 \times 10^5\,\text{cm}^{-3}$ (negligible).
- At 150 K with $N_d = 10^{17}$ impurities.cm^{-3},
  $N_c = \frac{2.1 \times 10^{24}\,\text{e} \cdot \text{m}^{-3}}{2^{3/2}} = 7.5 \times 10^{17}\,\text{e/cm}^3$
  $8(N_d/N_c)\exp(E_d/k_B T) \approx 5 > 1$ we are in situation β:
  $E_F = 0.973$ eV, $n = 8.6 \times 10^{16}$ e/cm^3, p is negligible.
  Nevertheless the doping is insufficient for silicon to be degenerate and the approximation 1a is justified (see Pb. 5). When the concentration of donors increases, the Fermi level increases while np remains constant: The concentration of holes becomes negligible.
  The reverse conclusion holds for a p-doping.

(d) There is only one type of charge carrier $\sigma = n_e e \mu_e$ so that $\alpha = e \cdot A \cdot T^{-3/2} \cdot \left(\frac{N_d N_c}{2}\right)^{1/2} \exp - \frac{E_d}{2k_B T}$; when the doping is important the curve $\log\sigma = f(1/T)$ is a line with slope $-(E_d/(2k_B T))$ instead of slope $-E_g/(2k_B T))$ obtained when the semiconductor is intrinsic.
We find $\mu_n(145\text{ K}) = 3.400\,\text{cm}^2/\text{V} \cdot \text{s}; \sigma = 47\,\Omega^{-1} \cdot \text{cm}^{-1}$

Problem 5: Degenerated and nondegenerated semiconductors

We consider a 3D semiconductor in which the n-type impurity atoms with density N_d have an ionization energy E_d.

(a) Assume that the impurity states with energy E has a distribution function of form $f(E) = \left[\frac{1}{g_i} e^{(E-E_F)/(k_B T)} + 1\right]^{-1}$ where $g_i = 2$, and find the density n_e of electrons that occupy the donor levels at temperature T. Deduce the density n of electrons, originating from donor levels, that occupy the conduction band. Express the result as a function of N_d, E_d, E_g (bandgap), and E_F (Fermi energy). Take the energy origin at the top of the valence band.

(b) After recalling the expression for the density of states $g(E)$ for a free electron gas enclosed in a unitary volume, find in the form of an integral the general expression of the density of electrons n in a conduction band assumed to be semi-infinite. The result should be expressed as a function of E_F, E_g, and m_e^*, the effective mass of electrons in this band.

The expression thus established is independent of the nature of the semiconductor (doped or not) but the resulting integration depends on the position of the Fermi energy. We can thus envisage two limiting cases:

(i) The Fermi level is in the bandgap and the Boltzmann distribution can be used as the distribution function $\int_0^\infty \frac{x^{1/2}\cdot dx}{e^{x-\xi}+1} = \frac{\sqrt{\pi}}{2}e^{\xi}$ when $-\infty < \xi \leq 0$; the semiconductor is nondegenerated

(ii) The Fermi level is located at an energy greater than $5k_BT$ above the bottom of the conduction band and the Fermi integral obeys

$\Phi^{1/2}(\xi) = \int_0^\infty \frac{x^{1/2}\cdot dx}{e^{x-\xi}+1} = \frac{2}{3}\xi^{3/2}$ when $5 \leq \xi < \infty$; the semiconductor is completely degenerated.

(c) In the case of a nondegenerated semiconductor, show that the density of conduction electrons can be put in the form:

$n = N_C \exp - (E_g - E_F)/k_BT$. Express N_c.

Comparing this result with that established in (a), find the expression for E_F assuming that $8\frac{N_d}{N_c}\cdot e^{E_d/k_BT} >> 1$.

For which temperature T(max) is E_F maximum? What is the value of E_F(max)?

At which critical concentration N_d(critical) that the semiconductor becomes partially degenerated, E_F reaching the conduction band minimum?

Numerically determine N(critical) for silicon using

$E_d = 0.02$ meV, $m_e^* = 0.2m$ and for InSb using $E_d = 10^{-4}$ eV and $m_e^* = 10^{-2}m$.

(d) Find the density of conduction electrons for a completely degenerated semiconductor (hypothesis (ii) above) as a function of E_F and E_g. Comment on the result. Which equation does E_F satisfy? Numerically find the impurity concentration Nd for Si and InSb at room temperature (T = 290 K) for the Fermi energy to be located at $5k_BT$ above the bottom of the conduction band. Comment on the result. $(e, m, \hbar, k)$

Solution:

(a) $n_e = N_d \cdot f(E) = \dfrac{N_d}{\frac{1}{2}e^{\alpha}+1}$ and $n = \dfrac{N_d}{2e^{-\alpha}+1}$ where

$$\alpha = \frac{E_g - E_d - E_F}{k_B T} \quad \text{(see Pb. 4, Question (3b))}$$

(b) $g_{el}(E) = \dfrac{m^{3/2}}{\pi^2 \hbar^3}\sqrt{2E}$ and

$$n = \int_{E_g}^{\infty} g(E)\cdot f(E)\cdot dE = \frac{\sqrt{2}m_e^{*3/2}}{\pi^2\hbar^3}\int_{E_g}^{\infty}\frac{\sqrt{E-E_g}\,dE}{e^{\frac{E-E_F}{k_BT}}+1}$$

(c) Nondegenerated semiconductor:
Using $(E - E_g)/ k_BT = x$, $(E_g - E_F)/k_BT = \xi$.

We find $n = N_e \exp(E_F - E_g)/kT$, where $N_c = 2\left[\dfrac{m_e^*\cdot k_BT}{2\pi\hbar^2}\right]^{3/2}$.

Identifying the values of n obtained in (a) and (c), E_F is given by $y^2 + \dfrac{1}{2}\dfrac{g}{d}y - \dfrac{1}{2}\dfrac{N_d}{N_c}\dfrac{g^2}{d} = 0$ where

$y = \exp(E_F/k_BT)$, $g = \exp(E_g/k_BT)$, and $d = \exp(E_d/k_BT)$.

This result conforms to that already established in Pb. 4 which used the same notations. In the hypothesis that

$$8\frac{N_d}{N_c}e^{E_d/k_BT} >> 1, E_F = E_g - \frac{E_d}{2} + \frac{k_BT}{2}\log\frac{N_d}{2N_c}.$$

E_F has a maximum at

$$\frac{dE_F}{dT}=0;\ \frac{k_B}{2}\log\frac{N_d}{2N_c}-\frac{k_BT}{2}\cdot\frac{N_d}{2N_c}\cdot\frac{2}{N_d}\cdot\frac{dN_c}{dT}=0,\ \text{which gives}$$

$$\log(N_d/2N_c)=(T/N_c)\cdot(dN_c\, dT),$$

where $(dN_c/dT) = (3/2)\cdot(N_c/T)$, resulting in

$$\log(N_d/2N_c) = 3/2;\ (N_d/2N_c) = e^{3/2}.$$

$$T(\text{Max})=N_d^{2/3}\cdot\frac{e^1}{2^{1/3}}\cdot\frac{\pi\hbar^3}{m_e{}^*k_B}=8.15\left(\frac{m}{m_e{}^*}\right)\cdot\left(\frac{N_d}{10^{18}}\right)^{2/3},\ \text{where } N_d$$

is in cm^{-3}.

$$E_F(\text{Max})=E_g-\left(\frac{E_d}{2}\right)+\left(\frac{3}{4}\right)\cdot k_BT(\text{Max})$$

$$=E_g-\left(\frac{E_d}{2}\right)+\left(3\frac{e^1}{2^{7/3}}\right)\cdot\left(\frac{\pi\hbar^3}{m_e{}^*}\right)N_d^{2/3}$$

$$E_F(\text{Max})=E_g-\left(\frac{E_d}{2}\right)+5.3\times10^{-4}\left(\frac{m_e{}^*}{m}\right)\cdot\left(\frac{n_d}{10^{18}}\right)^{2/3}$$

$E_F(\text{Max})=E_g$ when N_d (critical)

$$=2\left(\frac{m_e^*}{\pi\hbar^3}\cdot\frac{E_d}{3e^1}\right)^{3/2}\cong10^{22}\times5\left(\frac{m_e{}^*}{m}\right)^{3/2}\cdot E_d{}^{3/2}.$$

Si: N_d (critical) = 8×10^{18} cm^{-3}; T(Max) = 160 K
InSb: N_d (critical) = 3×10^{13} cm^{-3}; T(Max)$\approx$ 1 K
We note that in the III–V semiconductor group, InSb has a very low critical concentration, related to its small effective mass and ionization energy. Taking into account the difficulties to obtain a *stoichiometric* alloy, it is almost always a partially degenerate semiconductor.

(d) Substituting $(E - E_g/k_BT) = x$ and $(E_F - E_g/(K_BT) = \xi$, the electronic concentration in a completely degenerate semiconductor is described by

$$n=\int_0^\infty g(E)\cdot f(E)\cdot dE$$

$$=\frac{\sqrt{2}m^{*3/2}}{\pi^2\hbar^3}(k_BT)^{3/2}\int_0^\infty\frac{x^{1/2}\cdot dx}{e^{x-\xi}+1}=\frac{1}{3\pi^2}\left[\frac{2m_c{}^*k_BT}{\hbar^2}\right]^{3/2}\cdot\left[\frac{E_F-E_g}{k_BT}\right]^{3/2}$$

$$n = \frac{4}{3\sqrt{\pi}} \cdot N_c \cdot \left(\frac{E_F - E_g}{k_B T}\right)^{3/2} \approx \frac{1}{3\pi^2}\left(\frac{2m_e{}^*}{\hbar^2}\right)^{3/2} (E_F - E_g)^{3/2}, \text{ from}$$

which we obtain $E_F = \dfrac{\hbar^2}{2m_e{}^*}(3\pi^2 n)^{2/3}$.

Note that the electronic concentration does not explicitly depend on temperature and that there exists an identical relation between n and E_F as that established for metals (see also Exs. 13 and 24): assuming that the energy origin is at the bottom of the conduction band.

Identifying this concentration with that established in (a) and taking into account $-\alpha \geq (E_d/kT)+5$, we find

$$n \approx \frac{N_d}{2} \exp \frac{E_g - E_d - E_F}{k_B T} \approx \frac{4}{3\sqrt{\pi}} N_c \left(\frac{E_F - E_g}{k_B T}\right)^{3/2},$$

which leads to the following equality and allows, in principle, the calculation of E_F: $\xi^{3/2} e\xi = \dfrac{3\sqrt{\pi}}{8} e^{-\frac{E_d}{kT}} \cdot \dfrac{N_d}{N_c}$, when

$$E_F = E_g + 5k_B T;\, N_d = N_c \cdot \frac{5^{3/2} \cdot e^5 .8}{3\sqrt{\pi}} \cdot e^{\frac{E_d}{k_B T}} = 2 \cdot 5 \times 10^3 e^{E_d/k_B T} \cdot N_c\, .$$

At 290 K, for Si: N_d = 1.2 × 10^{22} cm^{-3}; for InSb : N_d = 6 × 10^{19} cm^{-3}.

As the atomic concentration of pure silicon is of order 6 × 10^{22} cm^{-3}, a doping change of 20% will radically change the physical properties of the element and the resulting behavior will be that of an alloy, whose electronic properties cannot be described by the method used here. This shows that silicon is never completely degenerated at ambient temperature.

The critical concentration N_d in InSb varies with temperature as N_c, thus has a $T^{3/2}$ dependence because $e^{E_d/k_B T} \approx 1$ and this semiconductor can be completely degenerate. But at the very large impurity concentrations needed to ensure its degeneracy, the impurity levels broaden and an impurity band (formed by the overlap between electrons placed in orbits r_i becomes important (see Ex. 16). This band will also participate in the electrical conductivity of the material. The theoretical study of heavily doped semiconductors thus requires special techniques (see Ref. [14], p. 232).

Problem 6: Electron transitions: optical properties of semiconductors and insulators

(1) General relations between optical constants:

In a vacuum (ε_0, μ_0), which occupy the half space of negative z, a monochromatic electromagnetic wave with angular frequency ω and electric field amplitude E_i is polarized along Ox, and propagates along increasing z. It hits a medium occupied along the half space $z > 0$ that is characterized by μ_0 and by a complex relative dielectric constant $\tilde{\varepsilon}_r = \varepsilon_1 + i\varepsilon_2$.

(a) Starting from Maxwell's equations, show that the transmitted wave in the dielectric is of the form: $E_T(z,t) = \overline{E}_T \cdot e^{-\frac{\omega}{c}\chi z} \exp i\left(n\frac{z}{c} - \omega t\right)$ where c is the speed of light in a vacuum.

Find explicitly the index of refraction n and the extinction coefficient χ as a function of ε_1 and ε_2 after having determined the phase velocity v_p and the wavelength λ in the medium of interest.

Show that the attenuation of the wave intensity is of form $I = I_0 e^{-\mu z}$ and find the optical linear absorption co-efficient μ as a function of ω, χ, and c.

(b) Using the boundary conditions and after having found the impedances $Z = E/H$ of the incident, reflected, and transmitted waves, find the expression for the intensity of the reflection coefficient at normal incidence, R, as a function of ε_1 and ε_2 and then as a function of n and χ.

Specify the values taken by n, χ, μ, R, and $T = 1 - R$ with the following assumptions:

$\alpha : \varepsilon_1 > 1, \varepsilon_2 = 0; \beta : \varepsilon_1 < 0, \varepsilon_2 >> |\varepsilon_1|; \gamma : \varepsilon_1 \approx 1, \varepsilon_2 << \varepsilon_1$

(2) Classical model of the dielectric function

We assume that the dielectric medium is an assembly of N atoms per unit volume in a volume immersed in a vacuum (ε_0). Subject to the action of the electric field wave, $E_0 e^{-i\omega t}$, a fraction n of electrons from an atom are displaced from their equilibrium position, each one remaining linked to a central ion ($+e$) due to a restoring force of form: $F_r = - m\omega_T^2 x$ (where x is the elongation of an electron, m is its mass, and ω_T is a constant) and is subject to a frictional force proportional to its velocity v: $F_F = -m\gamma v$.

(a) Write the differential equation of motion for an electron under the action of an electric field wave. Find the steady-state solution of its elongation, deduce the expression for the polarization (per unit volume) P of the medium and its relative dielectric constant $\tilde{\varepsilon}_r$.
Simplify the problem by using $Ne^2/m\varepsilon_0 = \omega_p^2$, where $N = nN$.

(b) Find the expressions for ε_1 and ε_2. Find the limiting values when $\omega = 0$ and $\omega = \infty$.

(c) Now assume that the damping constant γ is small such that $(\gamma/\omega_T) << 1$. How does this simplify the expressions for $\varepsilon_1 - 1$ and for ε_2 when the excitation frequency ω is near the resonance frequency ω_T?
Sketch the representative curves and specify the extreme value and the full width at half maximum of ε_2.
How are the expressions for $\varepsilon_1 - 1$ and ε_2 simplified when far from the resonance: $|\omega_T - \omega| >> \gamma$?

(d) Starting from the previous results in the limiting case where $\gamma = 0$, show graphically the evolution of $\varepsilon_1 - 1$ and ε_2 and that of n and χ.

(3) Elementary quantum approach

(a) The considered medium is in fact a semiconductor without impurities having a horizontal valence and conduction band: $E_v(k) = C$, $E_c(k) = C'$ (C and C' are the constants independent of the electron wave vectors k and we can write $E_g = C' - C$). A flux of photons with energy $h\nu = \hbar\omega$ irradiates this material.
Write the equations of energy momentum conservation when a photon can be absorbed by an electron.
Is optical absorption possible when $\hbar\omega < E_g$?
What is the physical meaning (and value) of the parameter ω_T used in the first two parts of this problem?
Qualitatively describe the optical behavior of the material considered.
Show that the electronic transitions leave the wave vector practically unchanged (vertical transitions) if the modulus

of the electron wave vector taken into account is of order $\pi/a \approx 1\,\text{Å}^{-1}$, if $E_g \approx 1\,\text{eV}$ and if there is a direct forbidden bandgap.

(b) The energy bands of the semiconductor considered are no longer flat and the density of electronic states in the valence and conduction bands are shown schematically in Fig. 46.

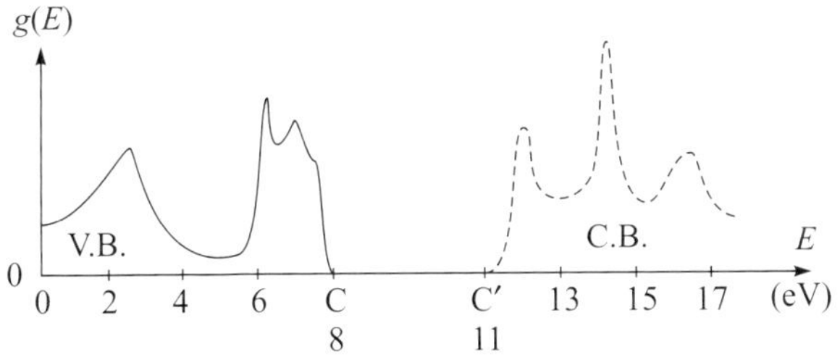

Figure 46

Sketch the optical absorption curve of this material [or the curve $\varepsilon_2 = f(\omega)$] as a function of the angular frequency of the incident electromagnetic waves.

What is the maximal wavelength λ_0 at which optical absorption begins?

Deduce the color of the semiconductor.

What is the approximate value of the static dielectric constant $\varepsilon_r(0)$ knowing that the atomic density is of order 5×10^{22} cm^{-3}, that there are four valence electrons per atom to consider and neglecting the other possible contributions to ε_r. $(e, \hbar, m, \varepsilon_0)$

Solution:

(1) (a) $\nabla^2 \vec{E} + \frac{\omega^2}{c^2}\tilde{\varepsilon}_r \vec{E} = 0$; $\frac{d^2 E_x}{dz^2} + \frac{\omega^2}{c^2}\tilde{\varepsilon}_r E_x = 0.$

From which we have

Substituting $(\tilde{\varepsilon}_r)^{1/2} = n + i\chi$, we find (not retaining the positive solution that represents the propagation for increasing z):

$$E_T(z,t) = \bar{E}_T \exp -\frac{\omega\chi}{c} z \exp i\omega(\frac{nz}{c} - t)$$

$$= \bar{E}_T \exp -\frac{\omega\chi}{c} z \exp i\omega\left(\frac{z}{v_p} - t\right)$$

$$= \bar{E}_T \exp -\frac{\omega\chi}{c} z \exp i2\pi\left(\frac{z}{\lambda} - \frac{t}{T}\right)$$

where we have $v_p = \frac{c}{n}$, and $\lambda = \frac{c}{nv}$, and $(\tilde{\varepsilon}_r)^{1/2} = \tilde{N} = n + i\chi$, where

$$n = \sqrt{\frac{1}{2}[\varepsilon_1 + (\varepsilon_1^2 + \varepsilon_2^2)^{1/2}]} \text{ and } \chi = \sqrt{\frac{1}{2}[-\varepsilon_1 + (\varepsilon_1^2 + \varepsilon_2^2)^{1/2}]},$$

or equivalently $\varepsilon_1 = n^2 - \chi 2$ and $\varepsilon_2 = 2n\chi$.
In addition, since the intensity is proportional to the square of the amplitude of the electric field: $\mu = \frac{2\omega\chi}{c}$.

(b) By introducing $E_T = \bar{E}_T \exp i\omega\left(\frac{\sqrt{\tilde{\varepsilon}_r}}{c} z - t\right)$ into one Maxwell's equations, we find for the wave transmitted in the dielectric; $\bar{E}_T/\bar{H}_T = z_T = (\mu_0/\varepsilon_0\tilde{\varepsilon}_r^{\ 1/2})$ and for the incident and reflected waves (in vacuum):
$\bar{E}_i/\bar{H}_i = Z_0 = (\mu_0/\varepsilon_0)^{1/2}, \bar{E}_r/\bar{H}_r = -Z_0 = -(\mu_0/\varepsilon_0)^{1/2}$.
At $z = 0$, the boundary conditions can be written as $\bar{E}_i + \bar{E}_r = \bar{E}_t$ and $\bar{H}_i + \bar{H}_r = \bar{H}_t$ so that

$$r = E_t/E_i = \frac{1-(Z_0/Z_T)}{1+(Z_0/Z_T)} = \frac{1-(\tilde{\varepsilon}_r)^{1/2}}{1+(\tilde{\varepsilon}_r)^{1/2}} = \frac{1-n-i\chi}{1+n+i\chi}$$

$$R = r \cdot r^x = \frac{(n-1)^2 + \chi^2}{(n+1)^2 + \chi^2}.$$

(i) $\varepsilon_2 = 0$ and $\chi = \mu = 0$: the medium is not absorbing

$n = (\varepsilon_1)^{1/2}$ and $R = \left(\frac{n-1}{n+1}\right)^2$: the reflection is weak and

$$T = \frac{4n}{(n+1)^2}$$

(ii) $\varepsilon_2 >> \varepsilon_1, n \approx \chi \approx (\varepsilon_2/2)^{1/2}$ and $\mu \approx \frac{\omega}{c}(2\omega_2)^{1/2}$:

$$T = 1 - R = \frac{2(2\varepsilon_2)^{1/2}}{1 + \varepsilon_2 + (2\varepsilon_2)^{1/2}}.$$

(iii) As in (a) χ and μ are small but $n \approx 1$, the reflection is weak and transmission is nearly complete, then

(2) (a) $$\mathrm{m}\frac{d^2x}{dt^2} = -eE_0e^{-i\omega t} - m\gamma v - m\omega_{\mathrm{T}}^2 x.$$

The steady-state solution is in the form $x = x_0e^{-i\omega t}$ from which we have $(-m\omega^2 - im\omega\gamma + m\omega_{\mathrm{T}}^2)x_0 = -eE_0$ so that

$$x_0 = \frac{-eE_0}{m(\omega_{\mathrm{T}}^2 - \omega^2 - i\gamma\omega)}$$

We find $\vec{P} = \frac{Ne^2\vec{E}_0}{m(\omega_{\mathrm{T}}^2 - \omega^2 - i\gamma\omega)}$; or $\vec{D} = \varepsilon_0\vec{E}_0 + \vec{P} = \varepsilon_0\tilde{\varepsilon}_{\mathrm{r}}\vec{E}_0$,

from which we obtain $\varepsilon_{\mathrm{r}} = 1 + \frac{\omega_{\mathrm{p}}^2}{\omega_{\mathrm{T}}^2 - \omega^2 - i\gamma\omega} = \varepsilon_1 + i\varepsilon_2$.

(b) $\varepsilon_1(\omega) = 1 + \frac{\omega_{\mathrm{p}}^2(\omega_{\mathrm{T}}^2 - \omega^2)}{(\omega_{\mathrm{T}}^2 - \omega^2)^2 + \gamma^2\omega^2}, \varepsilon_2(\omega) = \frac{\gamma\omega\omega_{\mathrm{p}}^2}{(\omega_{\mathrm{T}}^2 - \omega^2)^2 + \gamma^2\omega^2}$:

- $\varepsilon_1(0) = 1 + \frac{\omega_{\mathrm{p}}^2}{\omega_{\mathrm{T}}^2}, \varepsilon_2(0) = 0$, except when $\omega_{\mathrm{T}} = 0$.
- $\varepsilon_1(\infty) = 1, \varepsilon_2(\infty) = 0$.

(c) When ω is in the vicinity of ω_{T}, the approximation $\omega^2_{\mathrm{T}} - \omega^2 = (\omega_{\mathrm{T}} + \omega)(\omega_{\mathrm{T}} - \omega) \cong 2\omega_{\mathrm{T}}(\omega_{\mathrm{T}} - \omega)$ is valid and we find

$$\varepsilon_1 - 1 \approx \frac{\omega_{\mathrm{p}}^2}{2\omega_{\mathrm{T}}}\cdot\frac{\omega_{\mathrm{T}} - \omega}{(\omega_{\mathrm{T}} - \omega)^2 + (\gamma/2)^2}$$

$$\varepsilon_2 \approx \frac{\omega_{\mathrm{p}}^2\gamma}{4\omega_{\mathrm{T}}}\cdot\frac{1}{(\omega_{\mathrm{T}} - \omega)^2 + (\gamma/2)^2}$$

The corresponding curves, shown in Fig. 47, are Lorentzian: the function ε_2 is even compared to $\omega_{\mathrm{T}} - \omega$ and $\varepsilon_1 - 1$ is odd; the full width at half maximum of ε_2 is equal to γ at the points where the extremes of $\varepsilon_1 - 1$ are located. Finally ε_2 is maximum on resonance ($\omega = \omega_T$) and the amplitude M of this maximum is $M = \omega_p^2/\omega_T\gamma$, it can be very sharp. Far from resonance, γ can be neglected in the denominator and $\varepsilon_1 - 1 = \dfrac{\omega_p^2}{\omega_T^2 - \omega^2}$, $\varepsilon_2 = \dfrac{\omega_p^2\gamma\omega}{(\omega_T^2 - \omega^2)^2}$.

In this approximation, ε_2 is very small and corresponds most often to $\varepsilon_2 << |\varepsilon_1|$.

(d) In the limit when $\gamma = 0$, ε is zero everywhere except when $\omega = \omega_T$: $\varepsilon_2(\omega) = A\dfrac{\omega_p^2}{\omega^2}\delta(\omega - \omega_T)$.

(e) Using the Kramers–Krönig relations (Chapter IV, Pb. 5), we can show that the constant A is equal to $\pi/2$. The evolutions of $\tilde{\varepsilon}(\omega)$ and of $\tilde{N} = n + i\chi$ are shown in Fig. 47.

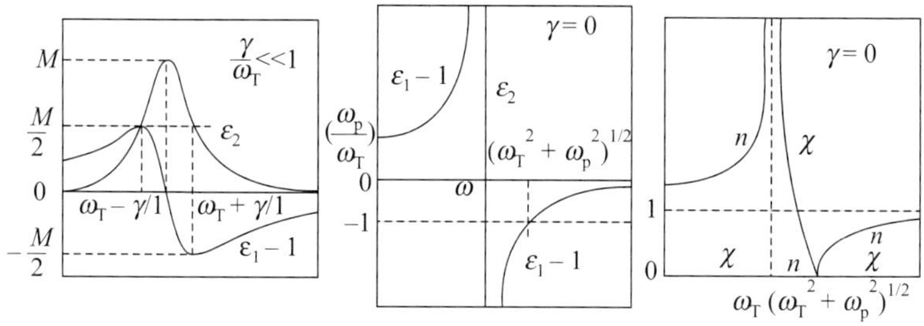

Figure 47

(3) (a) Denoting the electron energy E_i and wave vector $\vec{k}_i$ before absorption of a photon, and E_F and $\vec{k}_f$ as the values after absorption by a photon of energy $h\nu$ and with wave vector $\dfrac{h\nu}{c} = \dfrac{\hbar 2\pi}{\lambda}$. Conservation of energy and momentum impose that $E_f = E_i + h\nu$; $\hbar\vec{k}_f = \hbar\vec{k}_i + \hbar\left(\dfrac{2\pi}{\lambda}\right)\vec{u}$.

These relations can only be satisfied when the final state is allowed. When $\hbar\omega r$ is less than E_g, the final energy of the electron is located in the bandgap and absorption cannot occur: the solid is considered to be transparent to all corresponding electromagnetic radiation: ($\mu = \varepsilon_2 = 0$).

Absorption can only start when $h\nu = E_g$ or equivalently $\hbar\omega_T = E_g$. We thus are in the case 2d studied above where ε_2 takes the form of a delta function: reflection is weak for ω located near $\omega_T = E_g/\hbar$ and total transmission occurs for ω above $(\omega_T + \omega_p^2)^{1/2}$.

When $h\nu$ = 1 eV, the photon wave vector $|k_p| = 2\pi/\lambda = \frac{2\pi}{12400}\text{Å}^{-1} = 5\times10^{-4}\text{Å}^{-1}$ is much smaller than the electron wave vector $(0 < |k| < 1\text{ Å}^{-1})$: in the band diagram $E = f(\vec{k})$ the electronic transitions are vertical because they occur without notable change in the wave vector $|k_f| \approx |k_i|$. For indirect transitions, see Ex. 28.

(b) The initial electronic levels (in the valence band) and final ones (in the conduction band) are not discrete and one must observe an absorption band which begins around $h\nu = E_g = E_{c'} - E_c = 3$ eV. Neglecting the selection rules, $\mu(\omega)$ and $\varepsilon_2(\omega)$ must reflect the evolution of the convolution product of $g_{vb}(E)dE$ and $g_{CB}(E')dE'$, giving in particular the first maximum of energy E that separates the last maximum of the valence band from the first maximum of the conduction band. As in this region $\varepsilon_2 >> \varepsilon_1 : \mu \propto \omega\sqrt{\varepsilon_2}$, the features of ε_2 show up in μ but with a weight that varies with $h\nu$, see comment 2 for details. Figure 48 gives the characteristics of $\varepsilon_2(\omega)$ and $\varepsilon_1(\omega)$ taking into account the most probably transitions located at $h\nu$ = 5 eV (from the VB: 7 eV toward the CB: 12 eV); at $h\nu$ = 7–8 eV (from the VB: 7 eV toward the CB: 14 eV), as well as at $h\nu \approx 10$ eV (from the VB: 2 eV toward the CB: 12 eV and from the VB: 7 eV toward the CB: 17 eV).

- $\lambda_0 = \frac{\hbar c}{E} = 4100\text{ Å}$: This corresponds to the end of the violet at the limit of the near UV. The solid is transparent for all radiation in the visible spectra and is therefore colorless.
- $\varepsilon_r(0) = 1 + (\omega_p^2/\omega_T^2)$ (see Solution 2b). Taking into account the electronic density $\hbar\omega_p$ is equal to 16 eV and that, taking the average transition energy $\hbar\omega_T = 6$ eV, we find that $\varepsilon_1(0) \approx 8$. We are thus

concerned with the electronic contribution to $\varepsilon_1(0)$, see Ex. 25.

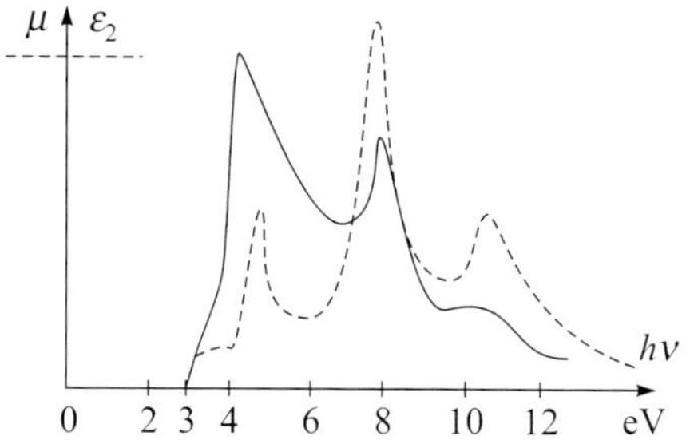

Figure 48

Comments

(1) Setting $\omega_T = 0$ the results obtained in the Question (2) of this exercise can be identified with those of Chapter IV, Pb. 5, relative to a plasma of free electrons. In particular, they show that, when $\gamma = 0$, ε_2 is not zero everywhere (which would result in $\varepsilon_1(\omega) = 1$; see comments in Chapter IV, Pb. 5) because it is not zero at the origin; the limiting value is $\frac{\pi}{2}\frac{\omega_p^2}{\omega_T}\delta(\omega-\omega_T)$ when $\omega_T \to 0$.

(2) In the third part, we have seen that probabilities of the electronic transitions are not uniform but that they depend on the density of states in the valence and conduction bands. Using quantum mechanics, we find (see Greenaway and Harbeke, citation in Chapter IV, Pb. 7):

$$\varepsilon_2(\omega) = \frac{\pi e^2}{m^2 \varepsilon_0 \omega^2} \int_{BZ} \frac{2dk}{(2\pi)^3} \left| \vec{e} \cdot \vec{M}_{vc} \right|^2 \delta[E_c(\vec{k}) - E_v(\vec{k}) - \hbar\omega] \, .$$

Averaging certain approximations concerning the matrix elements $|M_{vc}|$ which describe the electronic transition probabilities between the initial and final state, we can show that ε_2 is proportional to the surface integral

$$J_{vc} = \frac{2}{(2\pi)^3} \int \int_{E_c - E_v = cst} \frac{dS}{\left| \nabla_k (E_c - E_v) \right|} \, .$$

involving the joint-density of states between the valence and conduction bands. The most probable transitions are located at critical points in the Brillouin zone (known as Van Hove singularities) for which $\nabla_k (E_c - E_v) = 0$ and which often correspond to points where the dispersion curves $E = f(k)$ are in contact with the Brillouin zone (for which $\nabla_k E = 0$).

The theoretical calculation of the semiconductor or insulator band structure thus allows one to determine the position of these critical points and therefore foresee the optical properties in the visible and ultraviolet. Reciprocally obtained from the reflection spectra or characteristic energy losses, the experimental evolution of $\varepsilon_2(\omega)$ can be used to verify the validity of band structure calculations found in the literature and to interpret their results. (Pb. 10 illustrates this point for graphite). Such a procedure will be facilitated by taking into account symmetry considerations and the use of summation rules (deduced by quantum considerations and from the Kramers–Krönig relations) such as:

$$\int_0^\infty \omega\varepsilon_2(\omega)\mathrm{d}\omega = \frac{\pi}{2}\omega_\mathrm{p}^2 \text{ or } \varepsilon_1(0) = 1 + \frac{2}{\pi}\int_0^\infty \frac{\varepsilon^2(\omega)}{\omega}\mathrm{d}\omega.$$

Technical Note: See the corresponding technical note of Chapter IV, Ex. 29, for the present use of $\varepsilon_1 + i\varepsilon_2$ and $n + i\chi$. Also, the imaginary part of the index is represented here by χ and not k as previously (in order to avoid confusion with the wave vector k).

Problem 7: The p–n junction

At ambient temperature, a p–n junction diode of germanium is doped by p_1 acceptors (per unit volume) in the p-region and by n_2 donors (per unit volume) in the n-region. The indices 1 and 2 designate respectively the regions p and n. The concentration of impurity atoms is assumed to be uniform in each of these two regions and all impurities are ionized. In addition, the p–n interface is plane and abrupt. The potential $V(x)$ at a point M is a unique function of the x-axis relative to the perpendicular axis of the junction plane. The origin 0 is located at the interface between region A ($x < 0$) and region 2 ($x > 0$).

(1) Diffusion current and potential barrier

(a) Discuss the nature of the currents that appear when the p-semiconductor is set in contact with the n-semiconductor. Justify qualitatively the fact that the electrostatic distribution of charges can, in equilibrium, be represented by the curve in Fig. 49.

Using the Einstein relation $\frac{D}{\mu} = \frac{kT}{e}$, where D is the diffusion coefficient, μ is the mobility, and e is the absolute value of electron charge), find the expression for the contact potential U_0 [$U_0 = V(x_2) - V(-x_1)$] which arises between the p region and the n region in the absence of an external applied field.

Express the result first as a function of n_2, p_1, and n_i ($n_i^2 = n_i p_i$) and next as a function of the electrical conductivity σ_p and σ_n of regions 1 and 2.

(b) Numerically determine U_0, n_2, and p_1 using

$\sigma_p = 100\,\Omega^{-1}\text{cm}^{-1}, \sigma_n = 1\,\Omega^{-1}\text{cm}^{-1}$

$n_1 = 2.5 \times 10^{13}$ cm^{-3},

$\mu_p = 1700$ cm^2/V·s,

$\mu_n = 3600$ cm^2/V·s.

(c) From purely electrostatic considerations, find the evolution of the internal field $E(x)$ and then the potential function $V(x)$. Draw the representative curves (take $V = 0$ at $x = 0$) and find the expression for U_0.

(d) *Numerical application*: For Ge, ε_r (relative dielectric constant) = 16. Find the width $d = x_2 + x_1 \cong x_2$ of the transition region and the value E_m of the maximal electric field.

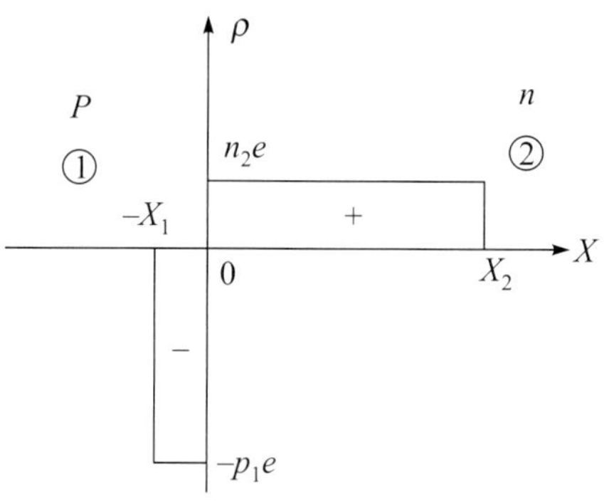

Figure 49

(2) Reverse bias: capacitance of the junction

A DC voltage $-U$ is applied between the negative pole at the p-side and positive pole at the n-side:

(a) What is the new width of the transition zone (assuming that the electrostatic concentration remains the same as shown in Fig. 49)?

(b) S being the area of the section of the junction, find the charge $+Q$ that is stored on the n-side, $0 \le x \le x_2$.

(c) The voltage $-U$ fluctuates with an amplitude u around the value U_1, $-U = -U_1 \pm u$ where $u << U_1$. Find the expression for the differential capacitance $C_j = \dfrac{dQ}{dU}$ for $U = U_1$.

(d) Using the previous numerical data, find the capacitance C_j for $S = 1\ mm^2$, $-U_1 = -5$ V and $U_1 = 0$. Sketch the curve $C_j = f(-U_1)$.

(e) The voltage u is sinusoidal with angular frequency ω. Show that the current i flowing through the diode contains terms with frequency 2ω. Find the weight of this harmonic compared to the fundamental one. Comment on the result.

(3) Forward bias

(a) Using the result of Question (1c), show schematically the energetic position of the valence and conduction bands of both p and n sides, as well as the Fermi level when 0 voltage is applied (Hint: The Fermi level E_F is in the forbidden bandgap).

(b) A forward bias is now applied to the diode. What are the modifications to the band diagram above? Starting from this diagram and using simplified Fermi–Dirac statistics, justify the following expression relating the current in the diode to the applied voltage: $i = i_s\ (e^{eU/k_BT} - 1)$. What happens when U is negative?

Find the characteristics of the curve $i = f(U)$ and find the values of current at ambient temperature when $U = 0.5$ V and $U = -0.5$ V (take $i_s = 10\ \mu A$).

Why is the result obtained when $U = 0.5$ V unrealistic?

(c) $U = u_0 \cos\omega t$ where $U_0 << k_BT/e$. In the current flowing through the diode, find the weight of the 2ω-harmonic relative to the first one in ω. (e, ε_0, k_B)

Solution:

(1) (a) When a p-doped semiconductor is put into contact with a n-doped semiconductor, the strong concentration gradient of mobile carriers that exists at the interface gives rise to a diffusion current $\vec{i}_{\mathrm{p}} = -eD_{\mathrm{p}}\overrightarrow{\nabla p}$ that circulates from 1 toward 2 and leads to a decrease in the hole concentration in the p region near the interface.

In the same way, a flux of electron will diffuse in the opposite direction, giving rise to a current $\vec{i}_{\mathrm{n}} = eD_{\mathrm{n}}\overrightarrow{\nabla n}$ and a decrease in the electron concentration on the n-side. The two parts of the junction, initially neutral electrostatically, will thus be charged negatively in the p region and positively in the n region. In equilibrium a double layer −/+ will block any further diffusion by the creation of an internal electric field $\vec{E}$ (from which a contact potential arises); the resulting currents are zero.

$i_{\mathrm{n}} = ne|\mu_{\mathrm{n}}|\vec{E} + eD_{\mathrm{n}}\vec{\nabla} n = 0, i_{\mathrm{p}} = pe\mu_{\mathrm{p}}\vec{E} - eD_{\mathrm{p}}\vec{\nabla} p = 0.$

Using the Einstein relation, we find

$$E_x = -\frac{k_{\mathrm{B}}T}{e}\cdot\frac{dn}{dx}\cdot\frac{1}{n};\ V = \frac{k_{\mathrm{B}}T}{e}\log n \text{ or}$$

$$U_0 = \frac{k_{\mathrm{B}}T}{e}\log\frac{n_2}{n_1} = \frac{k_{\mathrm{B}}T}{e}\log\frac{n_2 p_1}{n_1^2}.$$

The regions 1 and 2 are doped so that their electrical conductivity is due to one type of carrier:

$\sigma_{\mathrm{n}} = n_2 e\mu_{\mathrm{n}}$, $\sigma_{\mathrm{p}} = p_1 e\mu_{\mathrm{p}}$ from which we find

$$U_0 = \frac{k_{\mathrm{B}}T}{e}\log\frac{\sigma_{\mathrm{n}}\cdot\sigma_{\mathrm{p}}}{n_{\mathrm{i}}^2\cdot e^2\cdot\mu_{\mathrm{p}}\mu_{\mathrm{n}}}$$

(b) $U_0 = 0.3$ volt, $p_1 = 3.6\times 10^{17}$ cm^{-3}, $n_2 = 1.7\times 10^{15}$ cm^{-3}.

(c) $$\nabla E = \frac{\rho}{\varepsilon_0\varepsilon_{\mathrm{r}}}\left|\begin{array}{l}\dfrac{dE}{dx} = -\dfrac{p_1 e}{\varepsilon_0\varepsilon_{\mathrm{r}}} \text{ where } E = -\dfrac{p_1 e}{\varepsilon_0\varepsilon_{\mathrm{r}}}x + b_1 \text{ for } -x_1 < x < 0 \\ \dfrac{dE}{dx} = -\dfrac{n_2 e}{\varepsilon_0\varepsilon_{\mathrm{r}}} \text{ where } E = -\dfrac{n_2 e}{\varepsilon_0\varepsilon_{\mathrm{r}}}x + b_2 \text{ for } 0 < x < x_2\end{array}\right.$$

- $E = 0$ for $x < -x_1$ and $x > x_2$, it is continuous at $x = 0$ ($b_1 = b_2$). At the latter point, it reaches its absolute maximum:

$$|E_M| = -b_1 = -b_2 = \frac{p_1 e}{\varepsilon_0 \varepsilon_r}|x_1| = \frac{n_2 e}{\varepsilon_0 \varepsilon_r} x_2.$$

We observe that the system is globally neutral, that is, $p_1 x_1 = n_2 x_2$

$$\vec{E} = -\overrightarrow{\nabla V} \left| \begin{array}{l} V = \dfrac{p_1 e}{\varepsilon_0 \varepsilon_r}\dfrac{x^2}{2} + \dfrac{p_1 e}{\varepsilon_0 \varepsilon_r} x_1 x + c \quad (-x_1 < x < 0) \\ V = \dfrac{n_2 e}{\varepsilon_0 \varepsilon_r}\dfrac{x^2}{2} + \dfrac{n_2 e}{\varepsilon_0 \varepsilon_r} x_2 x + c \quad (0 < x < x_2) \end{array} \right.$$

From which we obtain the graphs shown in Fig. 50.

$$U_0 = V(x_2) - V(-x_1) = \frac{e}{2\varepsilon_0 \varepsilon_r}[n_2 x_2^2 + p_1 x_1^2]$$

$$= \frac{n_2 x_2}{2\varepsilon_0 \varepsilon_r} d \approx \frac{n_2 e}{2\varepsilon_0 \varepsilon_r} d^2.$$

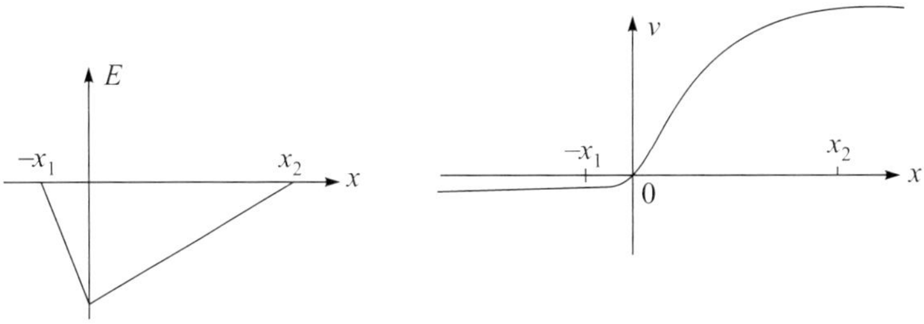

Figure 50

(d) $$d = \left[\frac{2\varepsilon_0 \varepsilon_r U_0}{n_2 e}\right]^{1/2} = 0.56\ \mu\text{m}$$

$$E_M \approx \frac{n_2 e}{\varepsilon_0 \varepsilon_r} d \approx \frac{2U_0}{d} \approx 10^4\ \text{V/cm}$$

(2) (a) The charge distribution gets larger without an increase in density. It is thus sufficient to consider the result from Question (1c) and replace U_0 by $U_0 + U$ from which we find

$$d = \left[\frac{2\varepsilon_0 \varepsilon_r}{n_2 e}(U_0 + U)\right]^{1/2}$$

(b) $Q = \int_0^d n_2 e \cdot S \cdot \mathrm{d}x = n_2 eSd = S[2\varepsilon_0 \varepsilon_r n_2 e(U_0 + U)]^{1/2}$

(c) $C_j = \dfrac{\mathrm{d}Q}{\mathrm{d}U} = \left[\dfrac{\varepsilon_0 \varepsilon_r n_2 e}{2(U_0 + U_1)}\right]^{1/2} S$

(d) $U_1 = 0, C_j = 320\ pF$, $U_1 = 5v$, $C_j = 60\ pF$

(e) From 2b, Q is of the form $\alpha(U_1 + u)^{1/2}$

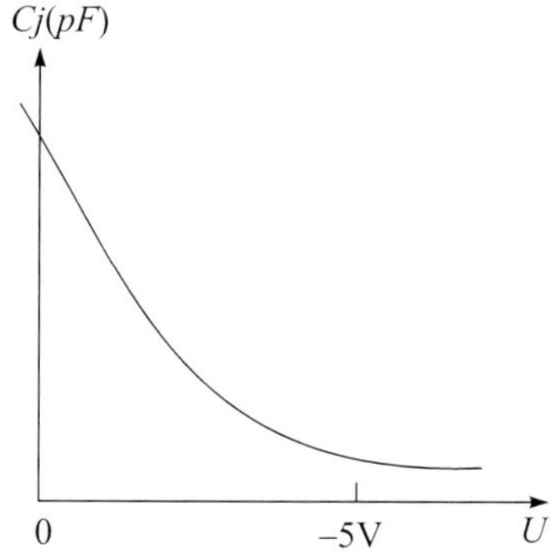

Figure 51

With limited development

$$Q = \alpha U_1^{1/2}\left(1 + \frac{u}{U_1}\right)^{1/2} \approx \alpha U_1^{1/2}\left[1 + \frac{u}{2U_1} - \frac{1}{8}\frac{u^2}{U_1^2} + \cdots\right]$$

$$i = \frac{\mathrm{d}Q}{\mathrm{d}t}\alpha U_1^{1/2}\left[\frac{u_0\omega}{2U_1}\cos\omega t - \frac{u_0}{4U_1^2}\sin\omega t\cos\omega t + \cdots\right]$$

where $u = u_0 \sin \omega t$.

Taking into account $\sin \omega t \cdot \cos \omega t = (1/2) \cdot \sin 2\omega t$, we find that $\dfrac{i(2\omega)}{i(\omega)} = \dfrac{1}{4}\dfrac{u_0}{U_1}$.

The diodes with a variable capacitance can serve as harmonic generators in telecommunications.

(3) (a) Figure 52 shows the electronic potential energy diagram: (a) Before p and n regions are connected and (b) after contact [U(ext) = 0].

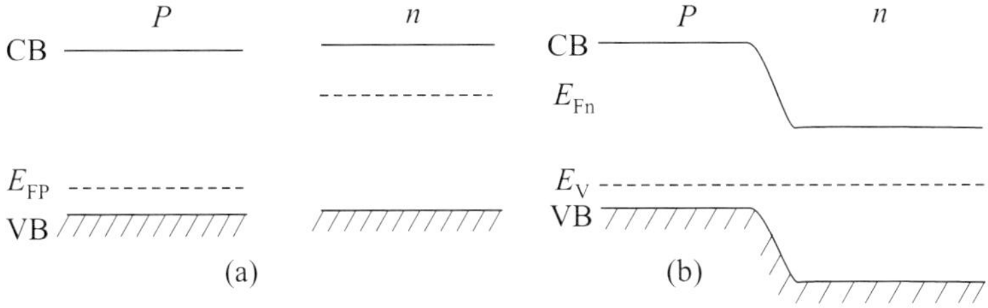

Figure 52

After contact, the Fermi levels align and $E_{F_p} - E_{F_n} = -eU_0$ in which U_0 represents the contact potential difference calculated in solutions 1a and b (note the sign change when going from potential to potential energy).

Starting from the expression $E_{F_p} = \frac{E_g}{2} + \frac{k_B T}{2} \log \frac{N_v}{N_c} \frac{n_1}{p_1}$ (see Ex. 14) and the equivalent expression for E_{F_n}, we find again the expression for U_0 obtained in 1a.

(b) See Fig. 53.

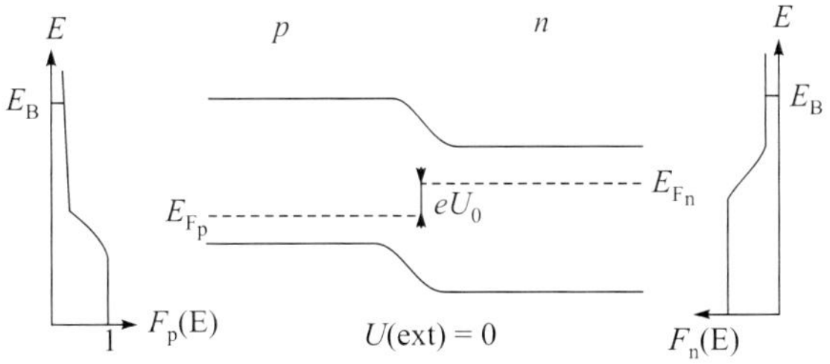

Figure 53

When the diode is forward biased, the level E_{F_n} is shifted upward on the y-axis by a quantity of eU_0 relative to E_{F_p}. The number of electrons that can diffuse from n toward p is proportional to $f_n(E_B) = \frac{1}{e^{(E_B - E_{F_n})/(k_B T)} + 1} \approx e^{(E_{F_n} - E_B)/k_B T}$.

Analogously, the number of electrons that can diffuse from p to n is proportional to $f_p(E_B) = \frac{1}{e^{(E_B - E_{F_p})/(k_B T)} + 1} \approx e^{(E_{F_p} - E_B)/k_B T}$.

The resulting electrical current is thus proportional to

$$f_n(E_B)-f_p(E_B)=e^{(E_{F_p}-E_B)/k_BT}(e^{eU/k_BT}-1) \text{ or } i=i_s(e^{eU/k_BT}-1).$$

We must also add the contribution of holes but the final relation will not be modified.

The relation obtained is algebraic and valid when U is positive as well as negative.

For U = 0.5 eV, $i(0.5\text{v}) \approx I \times (e^{20} - 1) \approx I_g e^{20} \approx 5.000$ A.

This result is not realistic because one must take into account the voltage drop in the p and n regions (here especially the n region). If for example, S = 1 mm^2 and e = 0.1 mm, $R \approx 1\ \Omega$, and $i(0.5\text{ V}) = 0.25$ A. As for a Gunn diode (Fig. 32, Ex. 30), the operating point will be determined using a graphical construction where the result of an intersection of the diode characteristics $i = f(U)$ with the line $U = E - Ri$ (see Fig. 54).

$U = -0.5$ eV; $i(0.5\text{V}) = i_s(e^{20} - 1) \approx -i_s = 10\ \mu$A.

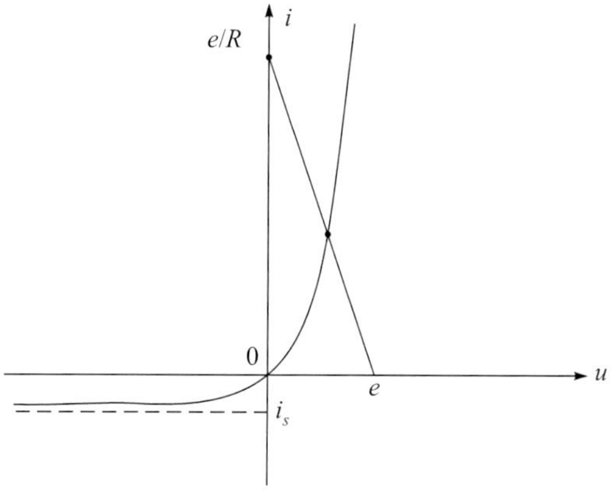

Figure 54

(c) $\dfrac{i}{i_s}=e^{u/U_T}-1\approx\dfrac{u}{U_T}+\dfrac{1}{2}\left(\dfrac{u}{U_T}\right)^2+\cdots$, where $U_T=k_BT/e$ and $u=u_0\cos\omega t$.

Expanding the quadratic term, we find: $\dfrac{1}{2}\cdot\dfrac{u_0^2}{U_T^2}\cdot\dfrac{1}{2}\cos(2\omega t+1)$ from which $\dfrac{i(2\omega)}{i(\omega)}=\dfrac{u_0}{4U_T}$.

The characteristics $I = f(V)$ are strongly nonlinear and this property may be used for the generation of harmonics 2ω, 3ω, etc.

Problem 8: The transistor

As shown below, a transistor consists schematically of a base with thickness a, which lacks electrons and in which diffusion phenomena of electric charges will occur. It is placed between an emitter, E, and a collector, C, both rich in electrons. In all that follows, we consider only 1D, x, and are only interested in the physical phenomena that occur in the base.

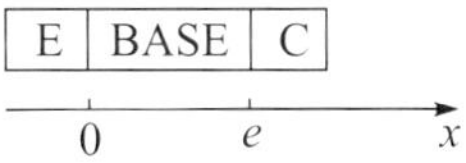

The electric current flowing through the junction results almost exclusively of electrons whose number n per unit length (carrier density) is controlled by two phenomena:

- Electrons disappear from their recombination of positive ions and their number per unit length and per unit time is $\dfrac{n-n_0}{\tau}$ where n_0 and τ are two constants that characterize the base;
- Diffusion at each point of the x-axis the intensity I of the electrical current is related to the derivative of the carrier density per unit length $\dfrac{\partial n}{\partial x}$ via the relation: $I = qD\dfrac{\partial n}{\partial x}$ where D is the diffusion coefficient and q is the absolute value of the charge of an electron.

(1) Show that the general equation for the density of carriers in the base obeys to $\dfrac{\partial n}{\partial t} = \dfrac{n_0 - n}{\tau} + D\dfrac{\partial^2 n}{\partial x^2}$.

(2) In thermodynamic equilibrium, the density of carriers n per unit length is given by $n(x,t) = N\exp\left(-\dfrac{q\varphi(x,t)}{k_B T}\right)$, where N is a constant k_B is the Boltzmann constant, T is absolute temperature, and $\phi(x,t)$ is the suitable electric potential such that:

- When the base is at a constant and uniform electric potential (electrical equilibrium) ϕ takes a constant value ϕ_0 and n is n_0.

- When a potential difference V_E is applied between the emitter, E, and the base (at $x = 0$), the effective voltage the value of ϕ is given by $\phi = \phi_0 - V_E$.
- When a potential difference V_C is applied between the collector, C, and the base (at x = a), the above expression applies with $+V_C$, instead of $-V_E$, so that the value of ϕ is very large and positive and the density of carriers n per unit length is thus equal to zero at x = a.

  Express the number of carriers at $x = 0$ as a function of n_0, V_E, and T as well as at $x = a$ (as a function of V_c).

(3) We investigate now the mechanisms of the transistor in a DC state regime where, thus, V_E is constant potential.

  (a) Write the equation for n, using $L=(\tau D)^{1/2}$. Find the expression for $n(x)$ as a function of the parameters n_0, V_E, T, a, and L. What is the physical interpretation of L?

  (b) Calculate the electrical current $I(x)$ in the base. Denote I_E and I_C respectively as the current at $x = 0$ and x = a. Calculate the static gain $\sigma_0 = I_C/I_E$ as a function of a and L by assuming that a is very small compared to L and that qV_E is very large compared to k_BT.

(4) A small sinusoidal potential v is now superimposed on the potential difference V_E: $v = v_E \exp(i\omega t)$.

  (a) Write the boundary conditions at $x = 0$ and at $x = a$ of the carrier density n assuming that qv_e is very small compared to k_BT.

  (b) Calculate the term dependent on time $n_1(x, t)$ of the carrier density. To simplify use $\beta^2 = \dfrac{1+i\omega\tau}{L^2}$.

  (c) As previously, calculate the alternating currents i_E at $x = 0$ and i_c at x = a. Calculate the dynamic gain defined by $\sigma = i_c/i_E$. If a is very small compared to L and for low frequencies ($\omega\tau \ll 1$), how can one simplify α? What is the physical interpretation of the imaginary part of α?

  (d) Determine the frequency ω_1 for which the modulus of the dynamic gain changes by 10% compared to the static gain α_0.

Solution:

(1) We consider an element of length between x and $x + dx$. Carried by the diffusion current, the flux of particles (per unit surface) going through the plane at x during a unit interval of time is $D \cdot \partial n(x)/\partial x$. The exiting flux at $x + dx$ is $D\frac{\partial}{\partial x}\left[n+\left(\frac{\partial n}{\partial x}\right)dx\right]$. The density of carriers that recombine between x and dx is $\frac{n-n_0}{\tau}dx$. Combining these elements we have $\frac{\partial n}{\partial t}=D\frac{\partial^2 n}{\partial x^2}+\frac{n_0-n}{\tau}$. We could also obtain this result directly using charge conservation equation: $div\ j+\frac{\partial \rho}{\partial t}=0$.

(2) The boundary conditions imposed successively are
$n_0 = N \exp - (q\phi_0)/(k_B T)$ from which $N = n_0 \exp(q\phi_0/k_B T)$.
- At $x = 0$, the density $n(0)$ is such that
$n(0) = N \exp - q(\phi_0 - V_E)/k_B T = n_0 \exp + (qV_E/k_B T)$
- At $x = a$, the density $n(a)$ is such that
$n(a) = N \exp - q(\phi_0 + V_c)/(k_B T) \approx 0$

(3) (a) In the steady-state regime, $\frac{\partial n}{\partial t}=0$ and the expression obtained in (1) becomes $\partial^2 n/\partial x^2 - n/L^2 = -n_0/L^2$ which gives $n(x)=A\exp-\frac{x}{L}+B\exp\frac{x}{L}+n_0$.

When $x = 0$, $n(0)=n_0\exp\frac{qV_E}{k_B T}=A+B+n_0$ or

$$A+B=n_0\left[\exp\left(\frac{qV_E}{k_B T}\right)-1\right].$$

When $x=a, n(a)=0=A\exp-\frac{a}{L}+B\exp\frac{a}{L}+n_0$.

The integration constants A and B are given by

$$A=\frac{n_0[\exp(qV_E/k_B T)-1-\exp-(a/L)]}{1+\exp-\frac{2a}{L}} \text{ and}$$

$$B=\frac{n_0[\exp(qV_E/k_B T)-1-\exp(a/L)]}{1+\exp\frac{2a}{L}},$$

where L is the diffusion length of the particles.

(b) $I(x)=qD\dfrac{\partial n}{\partial x}=\dfrac{qD}{L}\left(-A\exp-\dfrac{x}{L}+B\exp\dfrac{x}{L}\right)$

where $I_{\mathrm{E}}=\dfrac{qD}{L}(-A+B)$ and

$$I_{\mathrm{C}}=\frac{qD}{L}\left[-A\exp\left(-\frac{a}{L}\right)+B\exp\left(\frac{a}{L}\right)\right].$$

With the help of the expressions and integration constants obtained above, we have $\alpha=\dfrac{e^{qV_{\mathrm{E}}/k_{\mathrm{B}}T}-1+ch(a/L)}{(e^{qV_{\mathrm{E}}/k_{\mathrm{B}}T}-1)\,ch(a/L)+1}$.

Taking into account the suggested approximations, this simplifies to $\alpha\approx\dfrac{1}{ch\dfrac{a}{L}}\approx\dfrac{1}{1+\dfrac{a^2}{2L^2}}$.

(4) (a) Replacing V_{E} by $V_{\mathrm{E}}+v$, the condition (2) at $x=0$ becomes

$$n(0)=n_0\exp\frac{q(V_{\mathrm{E}}+v)}{k_{\mathrm{B}}T}\approx n_0[(1+qv/k_{\mathrm{B}}T)]\exp(qV_{\mathrm{E}})/(k_{\mathrm{B}}T).$$

(b) The term dependent on time will be of the form $n_1(x,\ t)=f(x).\exp\ i\omega t$ and must satisfy the continuity equation (Question (1)) which becomes

$$i\omega t=-\frac{f}{\tau}+D\frac{\mathrm{d}^2f}{\mathrm{d}x^2}\ \text{or}\ \frac{\mathrm{d}^2f}{\mathrm{d}x^2}-\frac{i\omega\tau+1}{L^2}f=0\ \text{or}\ \frac{\mathrm{d}^2f}{\mathrm{d}x^2}-\beta^2f=0.$$

We thus have that $f=C\exp-\beta x+E\exp\beta x$ and
$n_1(x,t)=C\exp(i\omega t-\beta x)+E\exp(i\omega t+\beta x)$.
This solution is dependent on time and must satisfy the boundary conditions

$$n_1(0,t)=n_0\frac{qve^{i\omega t}}{k_{\mathrm{B}}T}\cdot\exp\frac{qV_{\mathrm{E}}}{k_{\mathrm{B}}T}\ \text{and}\ n_i(a,t)=0.$$

(c) The expression for the sinusoidal current $i(x)$ will be given by $i(x)=qD\dfrac{\partial n_1}{\partial x}=qD[-\beta C\exp(i\omega t-\beta x)+\beta E\exp(i\omega t+\beta x)]$

or $i(E)=i(0)=qD\beta[E-C]\exp(i\omega t)$,
$i_c=i(a)=qD\beta[-C\exp-\beta a+E\exp(\beta a)]\exp(i\omega t)$.

$$\frac{i_C}{i_E} = \alpha = \frac{-C\exp{-\beta a} + \exp\beta a}{E - C} \text{ with (boundary conditions)}$$

$$C + E = n_0 \frac{qv}{k_B T}\exp\frac{qV_E}{k_B T} \text{ and } Ce^{-\beta a} + Ee^{\beta a} = 0.$$

We thus obtain

$$\alpha(\omega) = 2\frac{sh\,\beta a}{sh\,2\beta a} \cong \frac{1 + (a^2/6L^2)(1 + i\omega\tau)}{1 + \frac{2}{3}\frac{a}{L^2}(1 + i\omega\tau)}.$$

This result can be identified with that established in condition 3 when ω is zero.

The imaginary part of α represents the delay between I_c and I_E due to the recombination time of minority carriers. One may treat this phenomenon by introducing the notion of emitter-base (diffusion) capacitance.

(d) $\frac{|\alpha(\omega_I)|^2}{\alpha^2(0)} = 0.8$ when $\omega_I\tau = 0.75\frac{L^2}{a^2}$

Note: Nobel Prize in physics in 1956: J. Bardeen, W. Shockley, and W. Brattain won the Nobel Prize in physics in 1956 for the invention of the *transistor*.

Problem 9a: Electronic states in semiconductor quantum wells and superlattices

Consider a 3D gas of electrons that can propagate freely (with an effective mass m^x) in the xOy plane but are limited in the interval $0 \le z \le d$ by the existence of two potential barriers (assumed to be infinite) at $z = 0$ and $z = d$.

(1) By applying periodic boundary conditions of period L along x and y, and fixed conditions at $z = 0$ and $z = d$, find the wave function for these electrons and characterize the quantization of the wave-vector components, k_x, k_y, and k_z, by the quantum numbers n_x, n_y, and n_z. What are the conditions imposed on n_z?

Find the expression for the quantized energy of these electrons and specify the minimal value E_1 of this energy.

(2) Looking temporarily to the electron motion along z ($k_x = k_y = 0$), find the symbolic expression and the numerical value

(in eV) of the first three allowed energy levels E_1, E_2, and E_3 where d = 100 Å and m^x (As Ga) = 0.066 m.

(3) Consider that $d << L$, sketch the dispersion curve along the parallel direction $k^{||}$ and perpendicular k_z to the xOy plane.
Describe the progressive filling of the allowed levels. Specify, in particular, the conditions on n_x and n_y when going from n_z = 1 to n_z = 2. Deduce the characteristics of the density of states $g(E)$ curve.

(4) In reality, the electrons studied above are electrons in the conduction band of a semiconductor SC_1 with bandgap E_{g1} and thickness d which is surrounded (quantum well) by a semiconductor SC_1 with bandgap E_{g2}. The two semiconductors in contact have the same reference potential (vacuum level) relative to which (with the help of their electron affinity respectively $e\chi_1$ and $e\chi_2$) is located the bottom of their conduction bands.
Sketch the position of the conduction band minima and the valence band maxima when going through the structure $SC_2/SC_1/SC_2$ along z. Choose the example of the structure AlAs/GaAs/AlAs where (in eV): E_g(GaAs) = 1.43; E_g(AlAs) = 2.16; $e\chi$(GaAs) = 4.07; and $e\chi$(AlAs) = 3.5.
Find in particular the height of the potential barrier at the SC_2/SC_1 interface. Indicate them on the diagram of the energy states E_1, E_2, and E_3, evaluated previously (assuming that the finite height of the barrier does not significantly influence the positions established in 2).
Also indicate the position $E_1(t)$, of holes in GaAs, taking into account their effective mass m^x(h) = 0.68 m. Deduce the radiation that may be emitted from GaAs.

(5) Show the diagram of energy levels through the Oz axis of a heterostructure of type GaSb/InAs/GaSb. Take, in eV, E_g (GaSb) = 0.81; E_g(InAs) = 0.42; eχ(GaSb) = 4.06 and $e\chi$(InAs) = 4.9.
Specify the regions in space where electrons as well as holes will be localized in such a structure.

(6) Now consider a succession of quantum wells with width L_1, distant from L_2 resulting in a juxtaposition of alternating multilayers with identical composition to that studied in

Question (4): AlAs/GaAs/AlAs/GaAs/ where $L_1 = L_2 = 100$ Å. Can you sketch the approximate dispersion relation of conduction electrons of GaAs that propagate along the Oz axis and give the order of magnitude ΔE_m of the first allowed energy subband taking into account the limits of the Brillouin zone of the superlattice. Note that L_1 and $L_2 >> a$ because a(GaAs) = 5.65 Å.

Solution:

(1) See Chapter IV, Exs. 1 to 4 and especially Ex. 7 (object 4):
$\phi = A\sin k_z z.\exp i(k_x x + k_y y)$; $k_x = n_x 2\pi/L$; $k_y = n_y 2\pi/L$.
$n_x.n_y$ integers > 0 or < 0 or = 0.
$k_z = n_z\pi/d$ where $n_z \geq 1$

$$E = \frac{\hbar^2 k^2}{2m^x} = \frac{\hbar^2\pi^2}{2m^x}\left[\frac{4}{L^2}(n_x^2 + n_y^2) + \frac{n_z^2}{d^2}\right]$$

$$E_1 = \frac{\hbar^2\pi^2}{2m^x d^2}$$ with $n_x = n_y = 0$ and $n_z = 1$.

(2) E_1 = (see the solution of Question (1)); $E_2 = 4E_1$: $E_3 = 9E_1$ (which corresponds respectively to n_z = 1.2 and 3 with n_x, n_y = 0). $E_1 = 56$ meV; $E_2 = 0.226$ eV; $E_3 = 0.51$ eV.

(3) See Fig. 55.
The first occupied state is $n_z = 1$; $n_x = n_y = 0$. Starting from this state, the filling occurs along k_x and k_y (n_x and n_y increasing in absolute value). n_z remains equal to 1 until

$$E_1 + \frac{\hbar^2}{2m^x}(k_x^2 + k_y^2) = E_2$$ or $n_x^2(1) + n_y^2(1) = (3/4)L^2/d^2$.

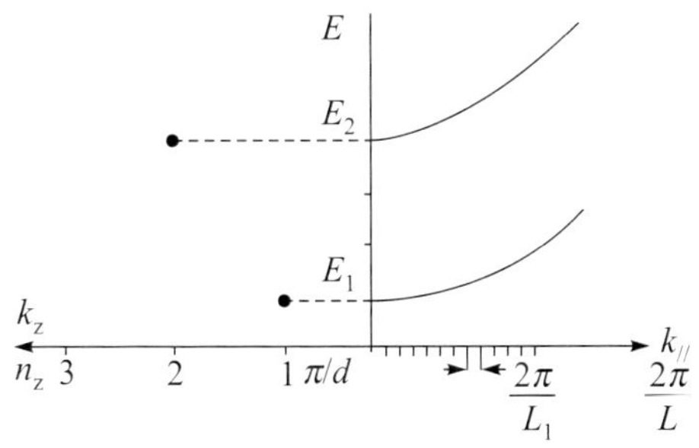

Figure 55

After this, the occupied states occur simultaneously for
$n_z = 1; |n_x|, |n_y| \geq |n_x(1)|, |n_y(1)|$ and n_z = 2; $n_x, n_y = 0, \ \pm 1, \ \pm 2, \cdots$
The density of states $g(E)$ will be 2D and therefore constant [$g(E) = L^2 m/\pi\hbar^2$, see Chapter IV, Ex. 14] between E_1 and E_2, E_2 and E_3, etc.
It will change suddenly when going from E_2 to E_3, as seen in Fig. 56.

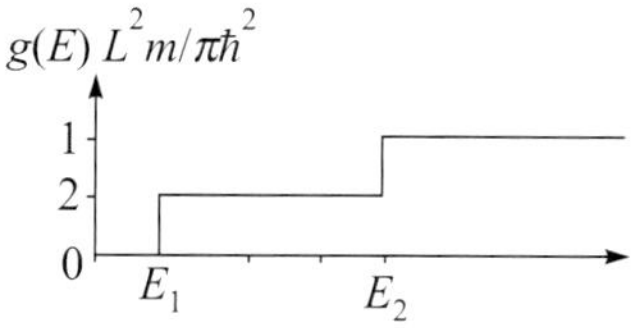

Figure 56

(4 & 5) The height of the potential barrier in the conduction bands is $e\chi_1 - e\chi_2$, or (AlAs/GaAs) = 0.57 eV. This value makes the infinite height approximation acceptable for level E_1 but not for level E_3. For holes, the potential barrier will be $\Delta(\text{VB}) = e\chi_2 + E_{g2} - e\chi_1 - E_{g1}$ (or 0.11 eV in the case considered here). Measured starting from the maximum of the valence band, the energy $E_1(t)$ of holes will be
$E_1\,(t) = \hbar^2\pi^2/2m_t^x d^2 \approx 5.6$ meV·
$h\nu = E_1 + E_g + E_1\,(t) \approx 1.5$ eV.
In the case of a GaSb/InAs/GaSb structure, the same analysis leads to $e\chi_1 - e\chi_2 = 0.84$ eV .
Contrarily, the maximum of the valence band of SC_2 is now higher than the bottom of the conduction band SC_1. The holes will be localized in SC_2 (because Δ (VB) = −0.45 eV), while the electrons remain localized in the quantum wells of SC_1.
The results of this analysis are shown in Fig. 57.

(6) The dispersion relation will be of the form $E = E_1 + \hbar^2 k^2/2m^x$. The imposed discontinuities of the lattice (at $k = \pi/a$) will be superposed with the discontinuities (which are much compressed) imposed by the periodicity of the superlattice (in $k = n\pi/d$). Neglecting the amplitude of the latter discontinuities, we find $\Delta E_m = (\hbar^2/2m^x)(\pi/d)^2$ of order E_1, see Fig. 58.

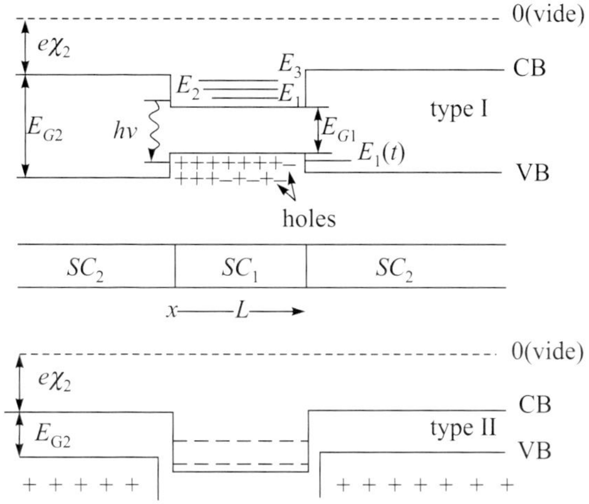

Figure 57

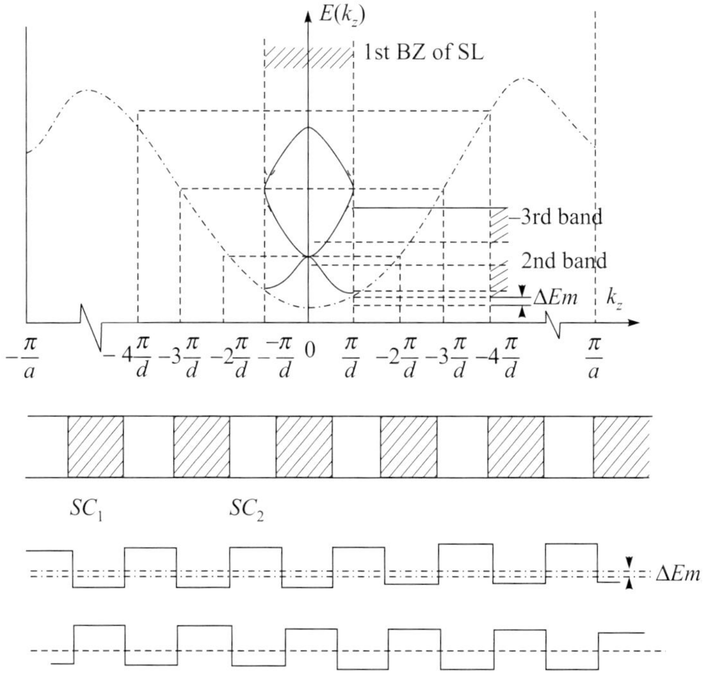

Figure 58

Comment: Semiconductor heterostructures; Nobel Prizes in physics in 1973 and 2000

Submitted by Leo (or Reona) Esaki, the original version of a paper on semiconductor *superlattices* was rejected for publication by *Physical Review* on the referee's assertion that it was "too speculative" and involved "no new physics." Nevertheless Esaki shared the Nobel Prize in physics in 1973 with Ivar Giaever and Brian David Josephson for his discovery of the phenomenon of electron tunneling. He is known for his invention of the Esaki diode, which exploited that phenomenon. Proposed by L. Esaki and Tsu (see *IBM Journal of Res and Dev* **14**, 1970, 61) the present heterostructures can be classified as either type I: AlAs/GaAs (Question (4)) or type II, GaSb/ InAs (Question (5)).

In the first case, heterostructures of type I, the mobile electrons and holes are localized in the semiconductor that having a smaller bandgap (SC_1). The probability of electron–hole radiative recombination of electron–hole will be large. Therefore, these heterostructures are used in optoelectronic devices (lasers) due to their efficient emission of photons. Electrons can be injected by doping the semiconductor SC_2. They will occupy, nevertheless the states E_1, E_2, and E_3 of SC_1. The wavelength of emission is in part adjustable by an adequate choice of quantum well width (Question (4)).

In heterostructures of type II, mobile electrons (in SC_1) and mobile holes (in SC_2) are spatially delocalized. The recombination probability will be reduced to a minimum, especially if the semiconductor SC_1 does not contain impurities. Taking into account the relative position of the bottom of the conduction band CB_1, compared with the valence band VB_2, the electrons issued from SC_2 will occupy the states E_1, E_2, and E_3 of SC_1. The structure presents a semi-metallic character related to the overlapping of the VB_2–CB_1 (see Ex. 9). This property will be useful for realizing very fast devices due to the high mobility (large time of flight) of electrons in this type of heterostructure.

In superlattices, the coupling can be adjusted by a careful choice of L_2. A more detailed study (of Question (6)) can use the Krönig–Penney model (Ex. 1) but the exercise would lose in clarity what it gains in rigor. For further information see Ref. [16] Ch. 10.2 and 10.3. From the point of view of elaboration of quantum wells a decisive step has been reached from the use of molecular beam epitaxy

(Dingle, Wiegmann, and Henry, *Phys. Rev. Lett.* **33**, 1974, 827) *and* their characterization in situ is detailed in Chapter I, Pbs. 4 and 10.

As in 3D homogenous semiconductors, the Coulomb association of an electron and a hole in quantum wells can lead to the formation of an exciton, see Ex. 20. If the energy of the exciton thus formed is relatively weak, the quantization of an electron and a hole will be independent and the electron–hole coupling will lead to a 2D gas of excitons that can move freely parallel to the plane of the layer.

In CdS, CdSe, and CdTe quantum wells, another possibility was demonstrated by luminescence of semiconductors where strongly bound excitons form a hydrogenic exciton whose center of mass motion is quantized when it is perpendicular to the plane of the layer. These quantized states are easy to calculate because they only concern particles with mass $m_e^x + m_h^x$ enclosed in a well with width corresponding to the thickness of the layer and the height determined by the value of the bandgap. We also know how to fabricate heterostructures whose properties are novel such as CdZnTe/CdTe where the opposing constraints in the wells of CdTe and the barriers of CdZnTe effectively spatially delocalize light holes from heavy holes. Thus, we obtain heterostructures of type I for heavy holes and heterostructures of type II for light holes. As a result, heavy excitons of type I and light excitons of type II can exist in such structures.

We note that understanding these phenomena require only the knowledge acquired in Chapter I, Pbs. 4 and 10 (elaboration and characterization of superlattices) and Ex. 21 (heavy and light holes), Ex. 20 (excitons), and thus the quantization of electronic states in potential wells such as seen in the present exercise.

The Nobel Prize in physics in 2000 was awarded "for basic work on information and communication technology" with one half jointly to J. Zhores, I. Alferov, and H. Kroemer "for developing semiconductor heterostructures used in high-speed- and opto-electronics" and the other half to Jack S. Kilby "for his part in the invention of the integrated circuit

Problem 9b: Electronic states in 2D quantum wells (variation of Problem 9a)

We consider an electron gas (with effective mass m^x) that can propagate freely along the Oz axis but is limited in the interval

$0 \le x \le d$ and $0 \le y \le d$ by the existence in the xOy plane of the boundaries of potential barriers assumed to be infinite. In reality, we are studying the electronic states (of conduction electrons) for a semiconductor SC_1 forming a parallelepiped rod of length L with a square cross-section ($d \times d$) surrounded by a semiconductor SC_2 in such a way that the electrons of SC_1 are localized into wells (the bottom of the conduction band SC_2 is such that $BC_2 \ge BC_1$).

(1) Applying periodic boundary conditions with period L along z and fixed boundary conditions along x and y, find the wave function for these electrons and characterize the quantization of the wave vector k by the quantum numbers n_x, n_y, and n_z. What are the conditions imposed on n_x and n_y?
Deduce the expressions for quantized energies and specify the minimal value E_1.

(2) Knowing that $L >> d$, find the symbolic expression for the first three distinct energy levels for variable n_z first (with n_x and n_y fixed and = 1), and next for n_z (n_x and n_y varying).
Numerical application: d = 100 Å; L = 10 μm; m^x = 0.066 m (GaAs). Deduce the characteristics of the dispersion curve along the directions $k_{||}$ parallel to the xOy plane and next along the direction k_z perpendicular to the xOy plane.

(3) Describe the progressive filling of the energy levels and sketch the density of states. $(\hbar^2/2m \approx 3.8\ \text{eVÅ}^2)$

Solution:

(1) See Chapter IV, Exs. 1 to 4 and especially Ex. 7 as well as the preceding problem.

$\psi = A \sin k_x x \sin k_y y \exp i(k_z z)$

$k_x = n_x\pi/d;\ k_y = n_y\pi/d$, where n_x, n_y are integers ≥ 1

$k_z = n_z 2\pi/L$, where n_z is an integer > or <0 or zero

$$E = \frac{\hbar^2 k^2}{2m^x} = \frac{\hbar^2\pi^2}{2m^x}\left[\left(\frac{n_x^2 + n_y^2}{d^2}\right) + \frac{4n_z^2}{L^2}\right]$$

$$E_1 = \frac{\hbar^2\pi^2}{m^x d^2},$$ where $n_x = n_y = 1$, $n_z = 0$.

(2) Starting from the level $E_1(110)$, the energy variation along z will be very small: $n_z = \pm 1$; $\delta E_1 = 2\hbar^2\pi^2/m^x L^2$

$$n_z = \pm 2;\ \delta E_2 = 8\hbar^2\pi^2/m^x L^2, \text{ etc.}$$

On the contrary, starting from the same level $E_1(110)$, the energy level $E_2(210)$ or $E_2(120)$ will be such that $E_2 = \frac{5}{2}\frac{\hbar^2\pi^2}{md^2}$ and $E_3(220) = \frac{4\hbar^2\pi^2}{md^2}$. Taking into account that $L >> d$, the parabolic dispersion curve along k_z will be quasi-continuous and increase very slightly when n_z increases even though it will be formed from discrete points along $k_{||}$. Permuting k_z and $k_{||}$, its characteristics will be analogous to that found in Fig. 55 from the preceding problem with $E_2 = \frac{5}{2}E_1$ and not $E_2 = 4E_1$ as in Fig. 55.

$$E_1 \approx 113 \text{ meV};\ E_2 = 284 \text{ meV};\ E_3 = 454 \text{ meV}$$
$$\delta E_1 \approx 0.23\ \mu\text{eV};\ \delta E_2 = 0.9\ \mu\text{eV};\ \delta E_3 \approx 2\ \mu\text{eV}$$

The filling of the states will first start in the k_z direction. Aside from the (110) level, E_0 occupied by $2e^-$ (↑ ↓), all the other levels will be occupied by $4e$ $(1, 1, \pm n_z)$ and the energetic difference between these levels increases (see Fig. 59).

The result is that the total number of electrons between E_1 and E (where $E - E_1 >> \delta E$) will increase as $\sqrt{E}$ and the density of states, which is the derivative of this quantity, will evolve as $1/\sqrt{E}$.

This result is expected because this is just the evolution of the density of free electron states in 1D when considering the intervals ΔE such that $\Delta E >> \delta E$ (see Chapter IV, Ex. 13, Fig. 12).

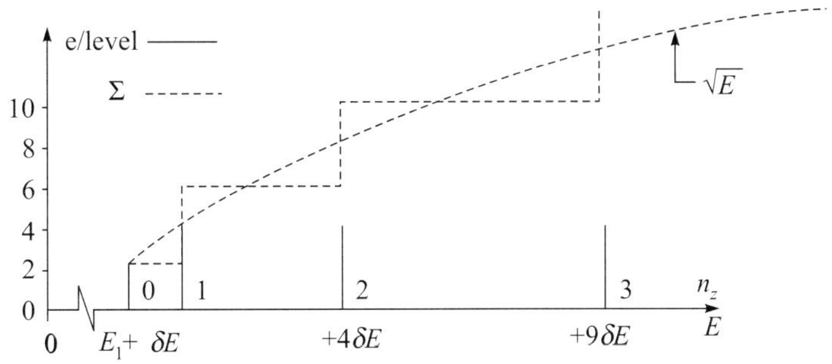

Figure 59

Having attained the energy E_2 along k_z, the filling of states now proceeds starting from the energy levels $E(210)$ and $E(120)$ by an analogous operation to that described above for the initial (110) state. The characteristics of the density of states curve, shown in Fig. 60, clearly illustrate the essential phenomenon: the discontinuities in E_1, E_2, and E_3 are related to the quantization of energy levels in the k_x and k_y plane.

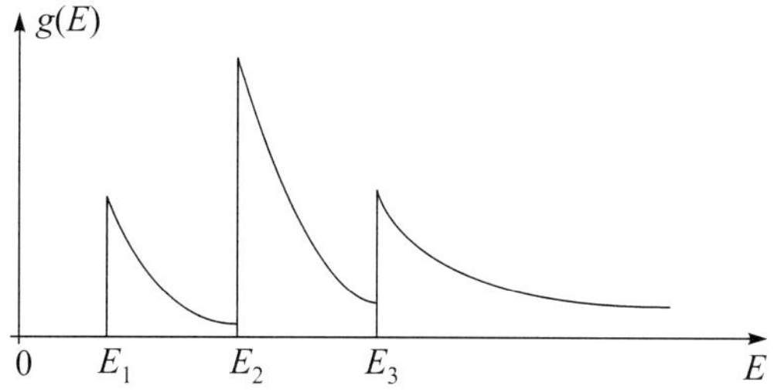

Figure 60

Taking into account the characteristics of this curve, one must expect more marked quantum effects than in the case of a well limited in a single direction (compare Fig. 56 in the preceding problem with Fig. 60).

Due to recent progress in nanolithography of semiconductors, we now know how to realize superlattices in 2D made from a matrix formed from regularly distributed quantum wells in the xOy plane and coupled between them (see for example, Marzin et al. *Phys. Rev. Lett.* **73**, 1994, 716).

Problem 10: Band structure and optical properties of graphite in the ultraviolet

The calculated band structure diagram for graphite (using the tight-binding approximation for a 2D lattice, see Remark) is shown in Fig. 61. We propose to analyze this structure and deduce the optical properties of the crystal.

In the direct space, the 2D structure of graphite is shown in Fig. 14 of Chapter I, Ex. 17 (see also Fig. 48 of Chapter I, Pb. 12) in which each carbon atom ($Z = 6$) has four valence electrons where three of these electrons, σ, make covalent bonds with three nearest neighbors and the forth, π, is more or less free.

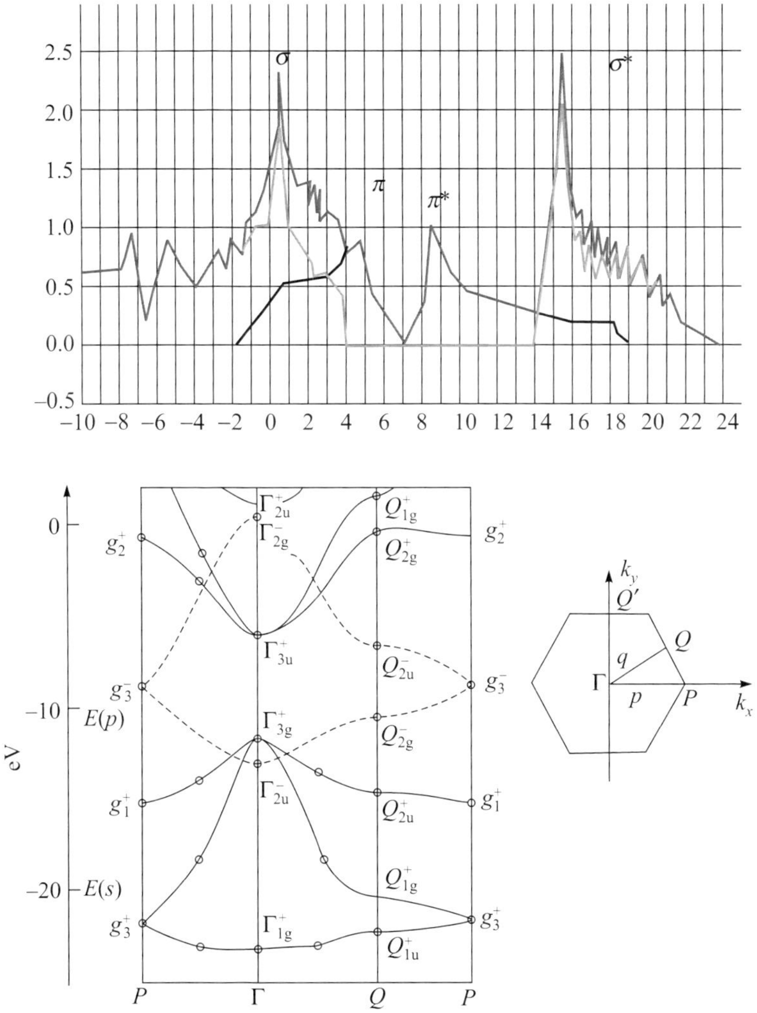

Figure 61 In the middle, calculated electron band structure of graphite in the 2D and tight-binding approximations. Above, experimental density of states. Below, first Brillouin zone.

(1) It is well known that graphite conducts electricity. Where is the conduction band? Does its dispersion relation follow a parabolic law? Where is the Fermi level found? Does the Fermi energy E_F obtained correspond to the value deduced from free electron theory ($E_F \approx 9$ eV, see Chapter IV, Ex. 14b)?

(2) When electromagnetic waves in the UV range irradiated a graphite crystal with its electric field contained in the plane of the layer ($\vec{E} \perp \vec{c}$), the photon absorption results from the allowed electronic transitions between states of the same parity, that is to say between occupied levels and empty levels represented either by dashed lines (π states) or full lines (σ states). Eventually with the help of the experimental evolution of the imaginary part $\varepsilon_2(h\nu)$ of $\tilde{\varepsilon}(\omega)$, Fig. 62, can you localize the regions where vertical electronic transitions are located in Fig. 61?

Compare the evolution of $\varepsilon_1(h\nu)$ and $\varepsilon_2(h\nu)$ to that deduced from the Lorentz model of bound electrons by considering that the average energy of transitions experienced by one type of electrons is $\hbar\omega_{\text{IT}}$, while the average energy of transitions experienced by the other type of electrons is $\hbar\omega_{2\text{T}}$.

The numerical values of these energies will be identified at the max of $\varepsilon_2(\omega)$, the electronic densities that come into play are such that $\hbar\omega_{1\text{p}} = 12.5\text{ eV}$, where $\omega_{1\text{p}}^2 = n_1 \dfrac{e^2}{m\varepsilon_0}$ and $n_1 = 1$ π electron (per C atom) and $\hbar\omega_{2\text{p}} = \sqrt{3}\hbar\omega_{1\text{p}}$, corresponding to three σ electrons per C atom while the damping terms, chosen arbitrarily are

$$\gamma_1 = \tau_1^{-1} = \frac{\omega_{1\text{p}}}{5} \text{ and } \gamma_2 = \tau_2^{-1} = \frac{\omega_{2\text{p}}}{5}.$$

Find the energetic position of the maxima of the energy loss function.

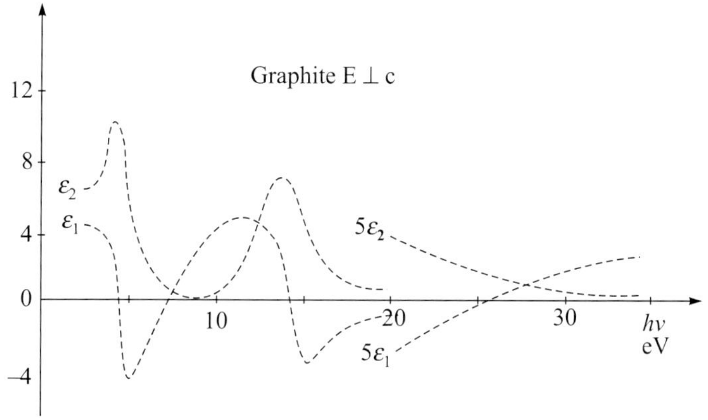

Figure 62

(3) The electric field of the electromagnetic wave is now polarized parallel to the c-axis. Describe qualitatively the expected consequences on the evolution of $\tilde{\varepsilon}(\omega)$ from the selection rules for vertical transitions (which take place between bands of opposite parity $\pi \to \sigma^x$ and $\sigma \to \pi^x$ when $E||c$.

What comments can be made concerning the anisotropy of the optical constants and the energy loss function of graphite in the far UV?

***Remark*:** The band structure presented (Fig. 61) corresponds to that calculated by Bassani and Parrivicini *(Nuovo Cimento* **50B**, 1967, 95) in which the beginning of the branch passing by Γ_{1u^+} has been added. The electron energies of the 2s and 2p valence of the carbon atom are indicated by E(s) and E(p) with the origin of energies corresponding to the vacuum level (habitual choice for calculations using the LCAO method). We will also note the equality between the length of the ΓP and ΓQ segments in the representation, which in reality is unequal (see the Brillouin zone to the right).

Solution:

(1) Graphite consists of two carbon atoms C_1 and C_2 at nonequivalent positions (see Chapter I, Ex. 17) where each atom has four valence electrons.

We expect that four valence bands occupied by eight atoms per basis and that the atomic wave functions are of type 2s, $2p_x$, $2p_y$, and $2p_z$. For atoms distributed along a hexagonal lattice, three of the four valence electrons for each C form covalent bonds with their neighbor atoms. These σ bonds are described by hybrid orbitals and the fourth electron per atom is described by an orbital of type $2p_z$ (it is a π electron). The Hamiltonian of graphite will be invariant under reflection, the σ wave functions are even and the π wave functions are odd in this reflection.

We therefore find eight bands of which the first four are occupied following the increasing energy sequence: $3\sigma \to \pi \to \pi^x \to 3\sigma^x$ where the symbol x represents the anti-bonding orbitals (CB) while the absence of the symbol represents bonding orbitals. Thus, the first band σ, in full lines, which passes through the point $\Gamma_{1G}{}^{+}$ results from combinations of type $p_x(C_1) - p_x(C_2)$ and $p_y(C_1) - p_y(C_2)$. In addition the band π pass through $\Gamma_{2u}{}^{-}$ and corresponds to combinations of the type $p_z(C_1)+p_z(C_2)$. Finally the anti-bonding orbitals (x) result from a sign change in the combinations and occur in the inverse order.

The crystal cohesion is assured by the fact that the average energy of the four valence bands is smaller than the electronic energy of the initial C–C molecule, $4E(\mathrm{s}) + 4E(\mathrm{p})$, because the density of states is maximal at the bottom of each band (regions where the bands are quite horizontal: a $\partial E/\partial k$ small leads to a $g(E)$ max).

We note in particular that the first band $\sigma(\Gamma_{1g}{}^{+})$ is nearly flat and very narrow (quasi-atomic state). The π electron bands in dotted lines have the nearly expected parabolic shape (see Chapter IV, Ex.14b) along ΓP direction (Γ_{2u}^{-}-P_3^{-} followed by P_3^{-} Γ_{2g}^{-}, in the reduced zone scheme). As foreseen also, along the ΓQ direction deviations from parabolic behavior are observed in the vicinity of the Brillouin zone (near Q_{2g}^{-}). The Fermi level is found at $E(\mathrm{p})$ (i.e., P_3^{-}) between four full bands and four empty bands. The width of the bandgap is zero and the electrical conductivity results in the continuity of the states at point $P_3{}^{-}$. Measured from the bottom of the π band the Fermi energy is $E_F \sim 4.5$ eV which is half the result expected from the theory of free electrons.

(2) The most probable vertical electronic transitions $\left(k_f - k_i << \dfrac{2\pi}{a}\right)$ are found at the points where $\nabla_k(E_c - E_v) = 0$ in reciprocal space and ,here, around regions that have the larger density of states in the valence and conduction bands (see Pb. 6 for additional details). The most probable transitions of type $\pi \rightarrow \pi^x$ take place essentially between Q_{2u}^{-} and Q_{2g}^{-} (and to a lesser degree between Γ_{2u}^{-} and Γ_{2g}^{-}), while transitions of type $\sigma \rightarrow \sigma^x$ take place essentially between Q_{2u}^{+} and Q_{2g}^{+} , P_1^{+} and P_2^{+}, and to a lesser degree between Γ_{3g}^{+} and Γ_{3u}^{+}.

The average energy of the first transitions ($\varepsilon \rightarrow \pi^{*}$) is of the order of 4.5 eV while the average of the seconds is of the order of 14 eV, which confirms the analysis of the evolution of ε_2, which exhibits maxima at these energies.

A Lorentz model, applied to two types of bound electrons (the π and the σ), results in transitions of type $\pi \to \pi^*$ at $h\omega_{1T} \approx 4.5$ eV and the others of type $\sigma \to \sigma^*$ at $h\omega_{2T} \approx 14$ eV leads to

$$\tilde{\varepsilon}_r(\omega) = \varepsilon_1 + i\varepsilon_2 = 1 + \frac{\omega_{1p}^2}{\omega_{1T}^2 - \omega^2 - i\omega\gamma_1} + \frac{\omega_{2p}^2}{\omega_{2T}^2 - \omega^2 - i\omega\gamma_2}.$$

This is just a simple extension of the analytical results obtained in Ex. 25. Separating the real and imaginary parts we obtain the curves with dashed lines in Fig. 63.

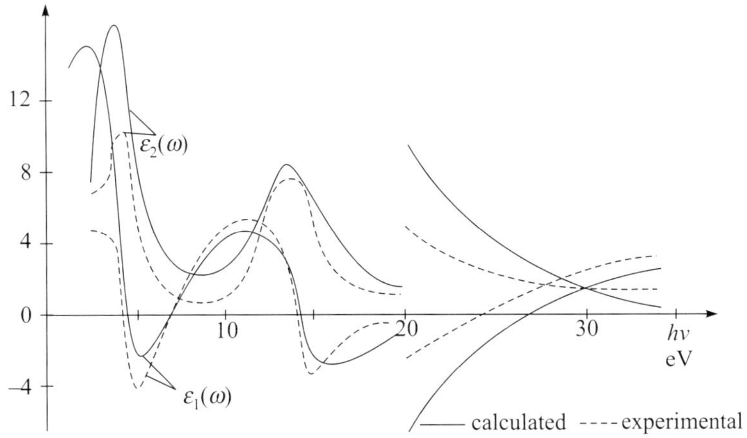

Figure 63

Taking into account the simplicity of this model and the very arbitrary choice of damping coefficients γ_1 and γ_2, the agreement with experiment is satisfactory.

The energy loss function $\dfrac{\varepsilon_2(\omega)}{[\varepsilon_1^2(\omega) + \varepsilon_2^2(\omega)]}$ is maximal when $\varepsilon_1(\omega)$ is zero which corresponds to $\hbar\omega_1' \approx 7$ eV and $\hbar\omega'' \approx 27$ eV in Fig. 62. We can also analytically find these maxima by neglecting the damping terms γ_1 and γ_2 and then determining the zeros of ε_1 in $\tilde{\varepsilon}_r(\omega)$ above, which leads to a simple solution of the double-squared equation (see J. Cazaux, *Optics Com.* **3**, 1971, 221 and 225).

The comparison between this function deduced from experimental values of $\tilde{\varepsilon}(\omega)$ and that deduced from the model (J. Cazaux *Solid State Comm.* **8**, 1970, 545) is very good concerning the position of the maxima, but it is much worse for the amplitudes. The latter result is explained by the amplification of the differences due to the inversion of $\tilde{\varepsilon}^{-1}$ in the spectral regions where ε_1 and ε_2 are small and consequently $\tilde{\varepsilon}^{-1}$ is large.

(3) When the electric field is such that $\vec{E}||\vec{c}$ selection rules impose transitions of opposite parity of type

$\pi \to \sigma^x (Q^+_{1u} \to Q^-_{2g})$ and $\sigma \to \pi(Q^+_{1g} \to Q^-_{2u})$.

The mean energy of the first is of the order of 16 eV and that of the second is of order 11 eV. One may expect essentially a completely different evolution of $\tilde{\varepsilon}(\omega)$ compared to that obtained previously. Figure 64 shows the experimental evolution of $\varepsilon_1(\omega)$ and $\varepsilon_2(\omega)$—in lines and with points—compared with the results from the Lorentz model, analogous to that developed for $E \perp c$ but taking into account the new transitions and the effective number of electrons being involved (For additional details see J. Cazaux, *Optics Comm.* **2**, 1970, 173 and **3**, 1971, 221 and 225).

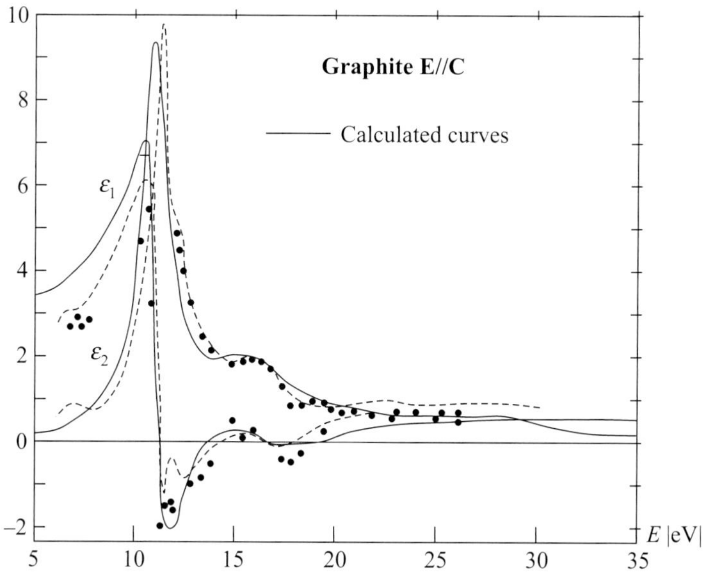

Figure 64

Comments

The anisotropy of graphite and of a number of lamellar crystals such as boron nitride in the UV is a consequence of the electronic selection rules (VB→CB) which leads to a different evolution of $\varepsilon_2(\omega)$ for $\vec{E} \perp \vec{c}$ and $\vec{E} \| \vec{c}$. Via the Kramers–Krönig relations, it also results a change for $\varepsilon_1(\omega)$ with different polarizations. Taking into account the relations between the complex optical index $\tilde{N}(\omega)$ and the complex dielectric function $\tilde{\varepsilon}_r(\omega)$: $\tilde{N}(\omega) = \sqrt{\tilde{\varepsilon}(\omega)}$, the initial anisotropy of $\tilde{\varepsilon}_r$ leads to the anisotropy of $\tilde{N}(\omega)$, and thus of the reflecting power and of its energy loss function.

Starting from numerical values deduced from the band structure curves in Fig. 61, it is possible to directly calculate $\varepsilon_2(\omega)$ from the evaluation of the joint-density of states (see comment of Pb. 6) and the result can then be compared withthose of optical measurements. For the limits of the Lorentz model and plasmons, see comment of Ex. 25.

Problem 11: π–π^* band structure of graphene

Graphene is a single layer of carbon atoms arranged in a honeycomb structure as shown in Fig. 65. The distance between two neighboring atoms is d = 1.42 Å.

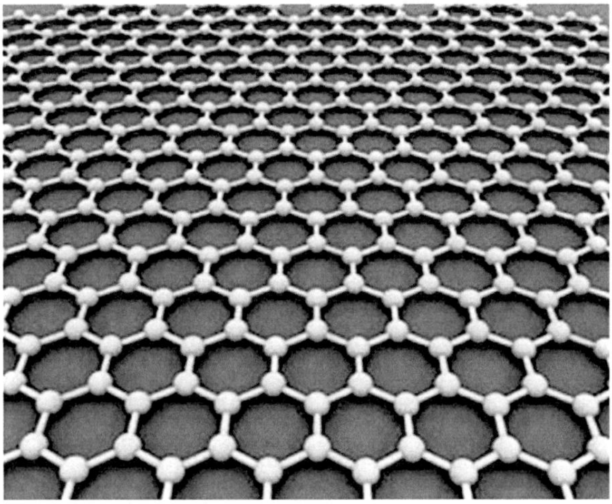

Figure 65

The band structure of graphene has been calculated using the tight-binding approximation by taking into account the $2p_z$ orbital only for the two neighboring atoms of the primitive cells. Ignoring the interaction between second nearest neighboring atoms the energy dispersion of $\pi(-)$ and $\pi^*(+)$ bands is given by

$$E = \pm\sqrt{\gamma_0^2\left(1+4\cos^2\frac{k_y a}{2}+4\cos\frac{k_y a}{2}\cdot\cos\frac{k_x\sqrt{3}a}{2}\right)} \tag{1}$$

Γ_0 is the nearest-neighbor hopping energy $\gamma_0 \approx 2.8$ eV; "a" is the lattice constant; **k** is the wave vector of orthogonal components k_x and k_y.

(1) Represent the **A** and **B** vectors of the reciprocal lattice and the first Brillouin zone (BZ) with its central point, Γ at (0,0), and special points M, M_1, and K at the BZ surface where ΓM is in the [11] direction like k_x while ΓM_1 is in the [01] direction, K being the intersection point with k_y.
Evaluate the electron energies at these special points, $E(\Gamma)$, $E(M_1)$, and $E(K)$. Draw the band structure scheme, $E(ka)$, along ΓM_1, ΓK (simplify eventually the energy expression in this direction) and KM_1; compare the calculated structure to that shown in Pb. 10, Fig. 61. Indicate the position of the Fermi level E_F.

(2) The next step is focused on some properties of graphene as deduced from its band structure at around point K.
Express the group velocity, $v_g = (1/\hbar)\partial E/\partial k$, and the effective mass of the particles at the top and bottom of the π and π^* band respectively. Give the order of magnitude of their numerical values and remarks. With $k_y = \Gamma K \pm \Delta k_y$ and $k_x = \pm\Delta k_x$ with $\Delta k_y << \Gamma K$, show that the constant energy surfaces are circles at around M. Express the density of states, $g(E)dE$, around K. Compare the corresponding expressions for a free electron gas in 2D. ($\hbar$, e)

Solution:

(1) In the primitive unit cell, there are two carbon atoms in nonequivalent position $a = d\sqrt{3} = 2.46$ Å; $A\ a\cos 30° = 2\pi$;

$|\mathbf{A}|=|\mathbf{B}|=(2\pi/a\ \cos30°)=4\pi/3d$ (see Chapter I, Ex. 17, and Chapter IV, Ex. 14b).

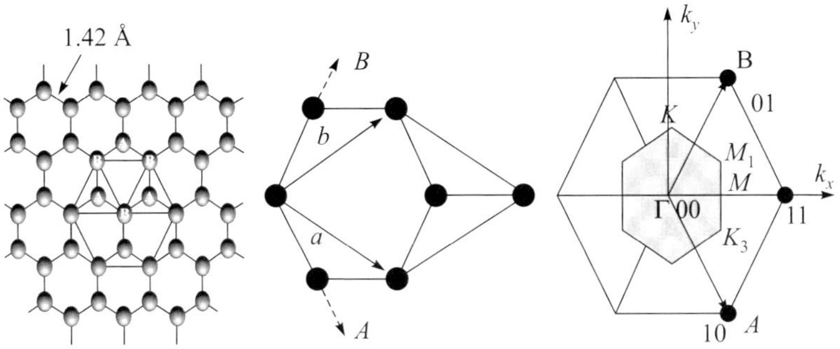

Figure 66

$\Gamma M=|A|/2=2\pi/a\sqrt{3}$; $\Gamma K=\Gamma M/\cos30°=4\pi/3a$; $KM_1=\Gamma M_1\ \sin30°=\pi/a\sqrt{3}$.

$E(\Gamma)=\pm3\gamma_0=8.4$ eV, $E(M)=\pm\gamma_0=\pm2.8$ eV; $E(K)=0$ eV.

Note that Eq. 1 takes a simplified form along the k_y axis:

$$E=\pm\gamma_0[1+2\cos(k_y a/2)] \qquad (2)$$

There are two carbon atoms and then two π electrons per primitive cell. Therefore, for a perfect undoped graphene sheet the π^* band is empty and the Fermi level is situated between the two symmetrical bands at the points where the π band touches the π^* band: E_F is at $E=0$ in the suggested energy scale.Then a zero excitation energy is needed to excite an electron from just below the Fermi energy to just above at the K points in the corners of the Brillouin zone. The calculated band structure scheme is similar to that of graphite derived also from tight-binding calculations but including the dispersion of the σ electrons (Fig. 61, Pb. 10). In particular there is the zero gap between π and π^* bands at point labeled K here and labeled P in Fig. 61). This zero gap energy at this point explains the good electric conductivity of graphite. The key difference is that the energy dispersion is electron–hole symmetric here.

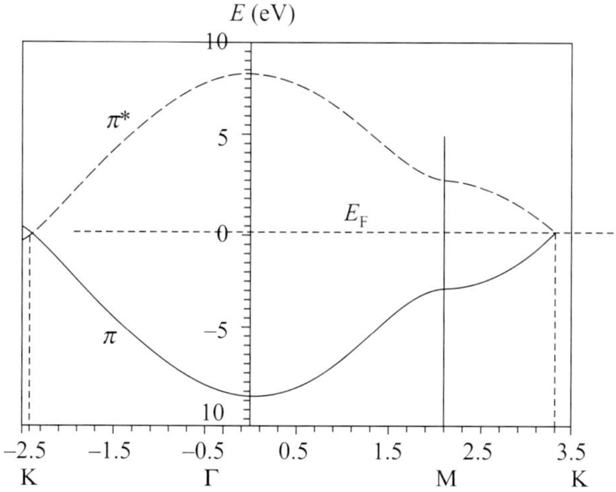

Figure 67

(2) From Equation 2, one obtains $v_g = 1/\hbar\ (\partial E/\partial k_y) = \pm(1/\hbar)\ \gamma_0 a \sin(k_y a/2) = \pm\gamma_0 a\sqrt{3}/2\hbar$.

The effective mass obeys to $1/m^* = (1/\hbar^2)(\partial^2 E/\partial k^2) = (-/+)\ \gamma_0 a^2 \cos(k_y a/2)/2\hbar^2 = (-/+)\gamma_0 a^2/4\hbar^2$.

The numerical values are $v_g = 5.7 \times 10^5$ m/s and $1/m^* = (-/+)\ 6.15 \times 10^{29}$ kg^{-1}. The particles are holes on the top of the π band and are electrons on the bottom of the π^* band; their mass is ~0.

At around point K and along the Kk_y direction the change of E, ΔE, with Δk_y may be dired from $\partial E/\partial k_y$ of Equation 2, one obtains $\Delta E = \pm(\sqrt{3}/2)\gamma_0\ (\Delta k_y a)$ (3)

A similar expression, $E = \pm(\sqrt{3}/2)\gamma_0\ (\Delta k_x a)$, is obtained along the ΓM_1 direction from developments of the form $\cos(\alpha + \varepsilon) = \cos\alpha \cos\varepsilon - \sin\alpha \sin\varepsilon \sim \cos\alpha - \varepsilon \sin\alpha$ where $\varepsilon << \alpha$.

The surfaces of constant energies are concentric circles in the k_x, k_y plane or conical surfaces in a 3D representation: E; k_x; and k_y (see Fig. 37). At points K the energy dispersion is $E = \pm\hbar v_g|k|$ and it is distinct from the energy dispersion of free electrons $E = \hbar^2k^2/2m$ (Chapter IV, Ex. 14b). The density of states is $g(k)\mathrm{d}k = 2\pi k\mathrm{d}k/(2\pi)^2/L_xL_y = g(E)\mathrm{d}E/2$ with 1/2 for the spin degenerency. Then for a unit surface:

$g(E) = \pi^{-1}k\mathrm{d}k/\mathrm{d}E = 4E/3\pi a^2\gamma_0{}^2$.

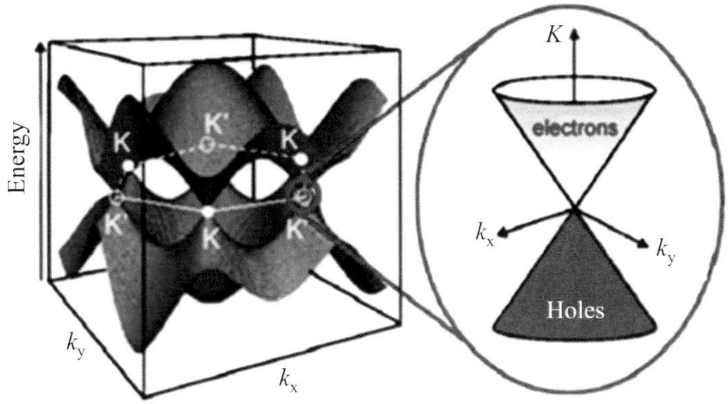

Figure 68

In the k-space, the density of states is the same for all the 2D objects but $g(E)$ differs in function of each specific dispersion relation (Chapter IV, Ex. 14). $g(E)$ is a constant for a free electron gas but it increases linearly with E in the present example.

Comments: *Graphene; Nobel Prize in physics in 2010*

The term graphene first appeared in 1987 to describe single (2D) sheets of graphite solids conceptually different from the surface layer on the top 3D graphite. Sometimes the term is used in descriptions of carbon nanotubes. The Nobel Prize in physics in 2010 was awarded to A. Geim and K. Novoselov at the University of Manchester "for groundbreaking experiments regarding the 2D material graphene."

The present exercise is just a simple introduction to the theoretical aspect of this fascinating material in showing that graphene differs from most conventional 3D materials. Intrinsic graphene is a semi-metal or zero-gap semiconductor. Its band structure is characterized by an E–k relation linear for low energies near the six corners of the 2D hexagonal Brillouin zone, leading to zero effective mass for electrons and holes. Due to this linear (or "conical") dispersion relation at low energies, electrons and holes near these six points, two of which are inequivalent, behave like relativistic particles described by the Dirac equation for spin 1/2 particles. Hence, the electrons and holes are called Dirac fermions, and the six corners of the Brillouin zone are called the Dirac points. Moreover, in the present case this pseudo-relativistic description restricted to vanishing rest mass leads to interesting additional features such as the unimpeded penetration of relativistic particles through high and wide potential barriers: one of the most exotic and counterintuitive phenomena.

Experimental results from transport measurements show that electrons and holes of graphene have high mobilities >15,000 $cm^2V^{-1}s^{-1}$ presently limited by the scattering on defects while the intrinsic limit, scattering by acoustic phonons, would be to 200,000 $cm^2V^{-1}s^{-1}$ at room temperature. At a carrier density of 10^{12} cm^{-2}, the corresponding resistivity of the graphene sheet would be 10^{-6} Ω·cm that is less than the resistivity of silver, the lowest resistivity substance known at room temperature. Despite the zero carrier density near the Dirac points, graphene exhibits a minimum conductivity of the order of $4e^2/h$. In the presence of a magnetic field, graphene displays an interesting quantum Hall effect (see Chapter IV, Ex. 23), which can even be measured at room temperature. This anomalous behavior is a direct result of the emergent massless Dirac electrons in graphene in a magnetic field leading to Landau level with energy precisely at the Dirac points.

Epitaxial graphene on SiC can be patterned using standard microelectronics methods and very high-frequency transistors were produced on monolayer graphene on SiC. Bandgap of the epitaxial graphene can be tuned from 0 to 0.25 eV by applying voltage to a dual-gate bilayer graphene field-effect transistor (FET) at room temperature or by irradiating with laser beams.

As an alternative to microelectronics where the information is carried by the electric charge graphene is thought to be an ideal material for spintronics (information carried by the spin) due to small spin–orbit interaction and near absence of nuclear magnetic moments in carbon. Electrical spin-current injection and detection in graphene was recently demonstrated up to room temperature with a spin coherence length above 1 μm at room temperature.

From this brief overview it appears that the potential applications of graphene are very large. These are concerned with spintronics, microelectronics, and optoelectronics with graphene ballistic transistors; graphene optical modulators; and its use as transparent conducting electrodes and in solar cells. They are also concernedwith nonelectronic applications such as single-molecule gas detection, room temperature distillation of ethanol for fuel and human consumption and graphene biodevices including the most ambitious biological application of graphene for rapid, inexpensive electronic DNA sequencing. The interested readers are referred to excellent review articles from Wu, Yu, and Shen with its 405 references in bibliography (*J. Appl. Phys.* **108**, 2010, 071301) or from Abergel et al. (*Advances in Physics* **59**, 2010, 261) from which the present comments are inspired.

Problem 12: Single-wall carbon nanotubes (SWCNTs)

Conceptually, single-wall carbon nanotubes (SWCNTs) can be considered to be formed by the rolling of a graphene layer (single layer of graphite) into a seamless cylinder. Hence, nanotubes are 1D objects with a well-defined direction along the nanotube axis, *T*. The way the graphene ribbon is wrapped is represented by a chiral vector, $\vec{c} = n\vec{a} + m\vec{b}$ where the integers *n* and *m* denote the number of unit vectors along the two directions in the in-plane directions (*n*, *m*) of graphene relative to its unit vectors, *a* and *b*, while $|c|$ is the length of the circumference of the tube perpendicular to *T* (corresponding to the minimal lattice vector denoted with m' and n'). If $m = 0$, the nanotubes are called zigzag nanotubes, and if $n = m$, the nanotubes are called armchair nanotubes. Otherwise, they are called chiral, E^1. Because of the curvature, *c* corresponds to a few hexagonal cells.

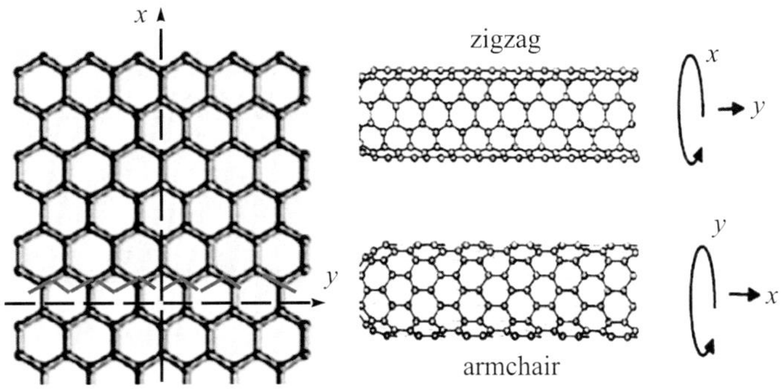

Figure 69

(1) Represent the chiral vector (8, 4) and the corresponding nanotube axis T in the graphene plane. Express the diameter d_t of a nanotube (m, n) as a function of the lattice parameter a. Numerical application for an ideal (8, 4) nanotube with d (nearest neighbor C–C distance) =1.42 Å.

(2) In an SWNT, the tube is infinitely long, k_z wave vector is continuous in the interval $-\pi/a$, π/a. Along the circumference, the wave vector, $k_\downarrow$ of the electrons moving around the circumference of the tube is quantized reducing the available states to slices through the 2D band structure of π and π^* electron states of graphene. Then, with –for π and + for π^* states, this band structure is composed of multiple 1D subbands obeying to the dispersion relation of the form:

$$E(k) = \pm[(\hbar \mathbf{v}_F k)^2 + (E^i{}_g/2)^2]^{1/2} \qquad (1)$$

For nanotubes of type $n-m=3l$, where l is zero or any positive integer with the fundamental gap, E^1 is very small ~0.0 eV. For other nanotubes, the fundamental gap obeys:

$$E^1{}_g = 2\gamma_0 d/d_t \qquad (2)$$

with γ_0 (nearest-neighbor hopping energy) ≈2.8 eV. Evaluate E^1 for a (8,4) SWNT and suggest a value for E^2. From the dispersion relation (1), express the density of states, $g(E)$, and represent the corresponding curves with $v_F \sim 6 \times 10^5$ m/s. Also represent the dispersion curves, $E(k_z)$ in the interval -0.2 Å$^{-1} \le k_z \le 0.2$ Å^{-1} and compare it to the dispersion curve of a (5,5) SWNT. Provide remarks concerning the electrical conductivity of the two. ($\hbar$, e)

Solution:

(1) From triangles such as OKL, one obtains $OL^2 = OK^2+KL^2 - 2OK{\cdot}KL\cos(\alpha)$ or $c^2 = (\pi d_t)^2 = (na)^2+(ma)^2-(ma{\cdot}na)\cos(120°)$. Then $d_t = (a/\pi)\ (n^2+nm+m^2)^{1/2}$. With $a = d\sqrt{3} = 2.46$ Å, $d_t(8,4) = 8.25$ Å: a value in agreement to the fact that most of SWNTs have a diameter of close to 1 nm, with a tube length L that can be many millions of times longer.

(2) $E^1{}_g \sim 0.96$eV and $E^2{}_g \sim 1.92$eV. Along a circumference, the electron waves are stationary with a wavelength λ obeying to $n\lambda = \pi d_t$. The bandgap energy for $n = 2$, $E^2{}_g$ is $2E^1{}_g$, also obtained from Eq. 2 with $d_t/2$ instead of d_t.

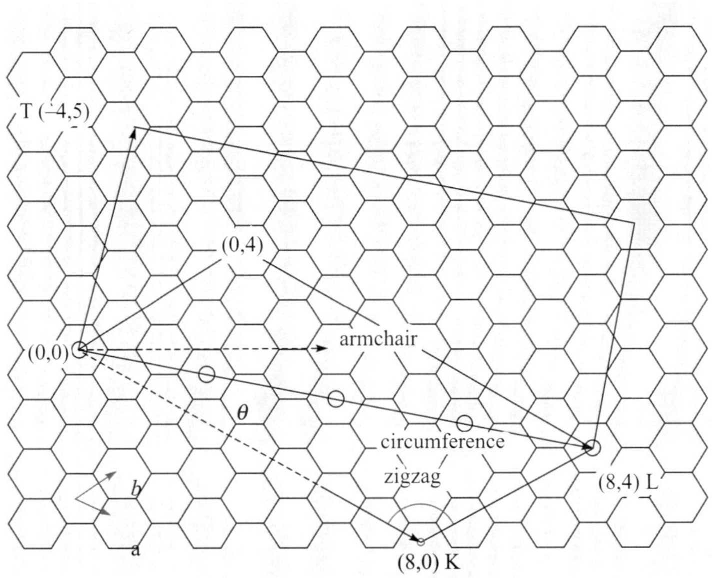

Figure 70

Along the tube axis, one deals with a 1D material: $g(E)dE = 2$ and $g(k)dk = 2L|dk|/\pi$–factor 2 for the spin degeneracy (see Chapter IV, Ex. 1). From Eq. 1, $dk = (1/\hbar v_F)EdE[E^2 - (E^1{}_g/2)^2]^{-1/2}$. Then per unit length:

$$g(E) = (2/\pi)\ (|\mathrm{d}k|/\mathrm{d}E) = (2/\pi\hbar v_F)E[E^2 - (E^1{}_g/2)^2]^{-1/2}.$$

Equation 1 is only valid at around point K of the Brillouin zone of graphene (see Pb. 11). This explains the restricted interval -0.2 Å$^{-1} \leq k_z \leq 0.2$ Å^{-1} within the validity of the above expression for $g(E)$. With two π electrons per unit cell, the π band is filled and the π^* band is empty at 0 K. Then the chiral SWNT (8,4) is a semiconductor and the SWNT (5,5), an armchair nanotube with $n = m$, is a conductor: $E_g = 0$. The corresponding dispersion curves are shown in Fig. 40, at left. The energy density of states of a conductor nanotube is a constant: $g(E)= 4/\hbar v_F$ such as for SWNT (5,5), where $E_g = 0$ but not for bandgap nanotubes for which the corresponding density of states are shown in Fig. 40, at right. $g(k)\mathrm{d}k$ is the same for all the 1D objects, but $g(E)$ differs in function of each specific dispersion relation, $E(k)$. For a 1D free electron gas, $g(E)$ is $g(E)= (2L/\hbar\pi)(2m/E)^{1/2}$.

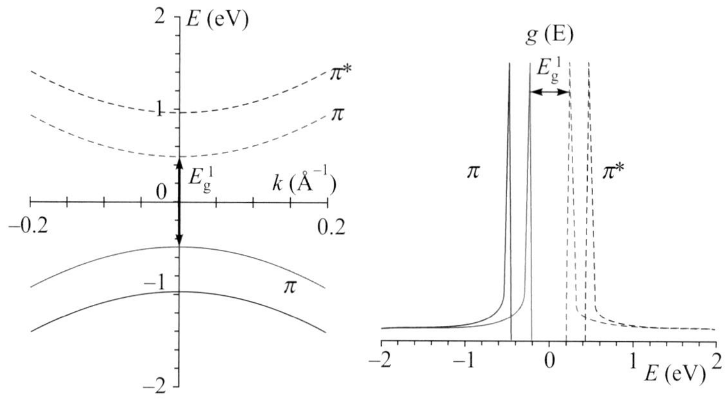

Figure 71 (Left) Energy dispersion curves of a (8,4) SWNT—full lines: π states; dashed lines π^* states compared to the energy dispersion curves of a (5,5) SWNT—dotted lines. (Right) Density of states of a (8,4) SWNT.

Comments

(a) *On SWNT*

Carbon nanotubes have been discovered in 1991 by S. Iijima (*Nature* **354**, 1991, 56) and this discovery has been followed by an increased number of graphene-related papers (>10,000 in 20011). This exercise is inspired from a review article from M. J. Biercuk et al. that appeared in *Carbon Nanotubes* **111**, 2008, 455. Like for the previous exercise on graphene, the present exercise is a simple introduction to these fascinating objects in illustrating, here, the difference of their electrical conductivity as a function of the way a graphene ribbon is wrapped.

(b) *On fullerenes: Nobel Prize in chemistry in 1996*

Fullerenes can be considered as large molecules of carbon in the form of a soccer ball. In this category, the most well-known molecule, C_{60}, is composed of 60 atoms of carbon regularly distributed on the spherical surface, which can be envisioned as a nanoscopic roll of a marble, as shown in Fig. 72. Each atom of carbon has only three nearest neighbors, the fourth valence electron of each carbon can therefore be considered to be free and the 60 free electrons are thus constrained to move on the spherical surface less than 10 Å in diameter.

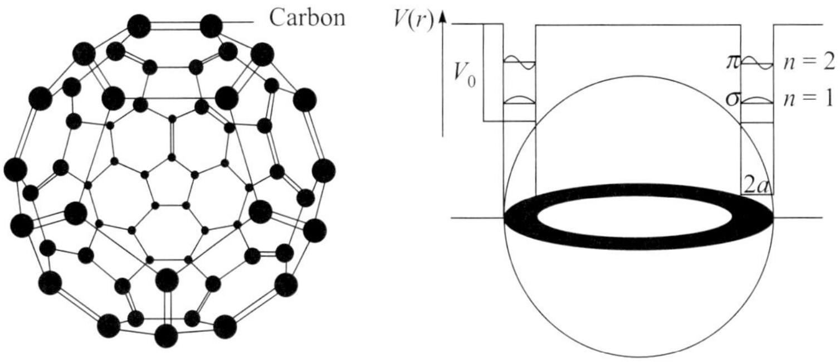

Figure 72 (Left) Representation of the molecular structure of C_{60}. (Right) free electron spherical shell model.

Because of the quasi-spherical shape molecule, it is very tempting to apply a free electron approach in order to obtain a simple description of the energy level diagram. Within this model, the π electrons of the C_{60} molecule are treated as if they were in a spherically symmetric potential $V(r)$. The solutions of the radial Schrödinger equation are the spherical harmonics leading to the radial quantum number n when matching the radial function to the well boundaries. In the limit of an infinitesimally narrow potential well, and an infinitely high potential barrier, a simple analytical expression can be derived for the energies of the eigenstates (see G. Gensterblum, *J. Elec. Spectrosc. Rel. Phen.* **81**, 1996, 89–223). Their optical properties in the UV range may be derived from an expression established for graphite when the electric field is parallel to the c-axis (see Pb. 10). Postulating that the damping terms are negligible in the dielectric function, this expression may be written in the form

$$\varepsilon(\omega) = 1 + \frac{1}{4}\frac{\omega_p^2}{\omega_\pi^2 - \omega^2} + \frac{3}{4}\frac{\omega_p^2}{\omega_\sigma^2 - \omega^2}$$

where ω_p is the free plasma angular frequency of $n\,(1\pi + 3\sigma)$ electrons per carbon atom in a fcc lattice of C_{60}, ω_π and ω_σ corresponding to the collective $\pi - \pi^*$ and $\sigma - \sigma^*$electronic transitions

In fact the electronic structure of such objects has not yet been fully elucidated. We know that they can be doped by alkaline metals and that atoms can be easily placed in the spherical cavity. The electrons from these dopants are entirely transferred toward the unoccupied states of the macromolecule but understanding the physical properties of these surprising objects is not complete. In particular why do K_3C_{60} and Rb_3C_{60} become superconducting below 19 and 20K respectively, while Na_3C_{60} and K_4C_{60} are insulators? In the field of nanotechnology, heat resistance, and superconductivity are some of the more heavily studied properties.There are other fullerenes such as C_{70}, but fullerenes with 72, 76, 84, and even up to 100 carbon atoms are commonly obtained. Nano "onions" are spherical particles based on multiple carbon layers surrounding a buckyball core. Silicon buckyballs have been created around metal ions and have been proposed for lubricants.

For the past decade, the chemical and physical properties of fullerenes have been a hot topic, fullerenes were under study for potential medicinal use: binding specific antibiotics to the structure to target resistant bacteria and even target certain cancer cells such as melanoma.

H. W. Kroto, R. F. Curl, and R. E. Smalley were awarded the Nobel Prize in Chemistry 1996 for the discovery of this new form of carbon, buckminsterfullerene ("buckyballs"). [See also their letter to *Nature* **318,** 1985, 162]. The name was homage to Buckminster Fuller, whose geodesic domes it resembles.

Questions

(Q.1) The law of mass action leads to $n_e \cdot n_h$ = const. for a given temperature. What physical arguments can you make to explain that an increase in n-doping of a semiconductor leads to a decrease in the density of holes?

(Q.2) The static dielectric constant of water is $\epsilon_r = 80$, its optical index in the visible is $n \approx 4/3$ or $n \approx \sqrt{\varepsilon_r}$. Where is the error?

(Q.3) What is the microscopic cause of the dispersion of white light by a glass prism?

(Q.4) Why do bismuth and antimony have an electrical resistivity greater than that of copper even though their atoms have each five valence electrons and those of copper have just one?

(Q.5) How does the semiconductor bandgap vary as a function of temperature? Why?

(Q.6) What is the general effect of adding impurities on the resistivity of a semiconductor? Are there exceptions?

(Q.7) With the help of the periodic table, order the following materials by increasing width of the bandgap: AlP, BeO, GaAs, LiF, MgO, and Ge.

(Q.8) Why silicon cannot intrinsic at ambient temperatures while germanium can be?

(Q.9) Why does the plasma frequency of semiconductor and insulators often correspond to the formula given by free electrons ($\omega^2_p = Ne^2/m\varepsilon_0$) even though the valence electrons are not free?

(Q.10) What characterizes the electronic specific heat of a material with a bandgap?

(Q.11) What is a heavy hole and a light hole?

(Q.12) Why do substitution alloys of the CuZn type of crystal structure change when they are enriched with one of the elements to the detriment of the other?

(Q.13) Why is the ionization energy of impurities in semiconductors smaller than in free atoms?

(Q.14) Why are the majority of oxides, alkali halides, and also diamond transparent when they are pure?

(Q.15) Why is ruby red?

(Q.16) What is the Gunn effect?

(Q.17) What is an exciton?

(Q.18) What is a superlattice of type I?

(Q.19) Why is molybdenum harder than silver?

(Q.20) Why is nickel ferromagnetic and not copper?

(Q.21) In which direction does the Fermi level of a semiconductor vary when it is p-doped? It is then placed in contact with an *n*-type material. Can you find a simple analogy to explain the equilibration of the Fermi level upon contact?

(Q.22) What are the analogies and differences between "plasmons" and "polaritons"?

(Q.23) Why can one approximately say for compounds of the type $A^N B^{8-N}$: the larger is the bandgap energy, the smaller is the static dielectric constant?

(Q.24) In optoelectronics, what is the interest to elaborate materials with an adjustable bandgap? How a superlattices be used to provide an alternative elaboration to homogenous materials of type $A_x A'_{1-x} B$?

(Q.25) The condition for a photon with energy $h\nu$ to be absorbed by a pure material with bandgap E_g is strictly when $h\nu \geq E_g$. Is this condition correct?

(Q.26) For semiconductors, give the order of magnitude of the limits for carrier density and resistivity. What is the order of magnitude for the resistivity of the best insulating solids?

(Q.27) What are the microscopic causes of the optical anisotropy of graphite in the UV?

(Q.28) Cite at least one method that allows the experimental determination of E_g in semiconductors and insulators.

(Q.29) Most of the transparent materials are also electrical insulators. Give a noticeable exception.

(Q.30) Compare normal magneto resistance in semiconductors to giant magnetoresistance (GMR) in alternating magnetic/nonmagnetic layers.

(Q.31) What is graphene? What are their specific properties?

(Q.32) What are the structural and electrical differences between a zigzag and an armchair nanotube?

ANSWERS AT THE END OF THE BOOK

Answers to Questions

Chapter I

(Q.1) $d \sim 3 - 4$ Å, from which we obtain $N_V \approx 1/d^3 \approx$ several 10^{28} atm^{-3}; $N_S = N_V^{\frac{2}{3}} \approx d^{-2} \approx 10^{19}$ at m^{-2}; $N_L = N_V^{\frac{1}{3}} = 1/d \approx 3 - 4 \times 10^9$ atm^{-1}.

(Q.2) (a) All of the atomic electrons (see Ex. 23).
(b) the electric potential (see Pbs. 3 and 4).
(c) the atomic nucleus and the magnetic moment (see Pbs. 7 and 11).

(Q.3) Taking the into account the strong interaction with crystalline potential, the slow electrons are only sensitive to the first atomic layers and the reciprocal lattice is formed of lines because the direct lattice is essentially 2D: the Ewald sphere is always intercepted by these lines (see Pb. 3). On the other hand, fast electron have a very short wavelength which results in a very large radius of the Ewald sphere (compared to $2\pi/a$; see Ex. 21).

(Q.4) Nothing, because the main interaction, photoelectric effectleads to the absorption of X-rays. The probability of the Bragg conditions to be satisfied is essentially zero.

(Q.5) This is specular reflection (see Exs. 18 and 18b and also Pb. 4).

(Q.6) No, it is the reverse via the expansion of the reciprocal lattice.

(Q.7) See comment in Pb. 4.

(Q.8) See comment in Pb. 3 and illustration.

(Q.9) 10^{-6} torr sec. See comment in Pb. 3.

(Q.10) See comment in Pb. 4.

(Q.11) See Pb. 5.

(Q.12) See Pb. 10.

(Q.13) See comment in Pb. 12.

(Q.14) None. The plane (300) is a virtual plane parallel to (100) and such that $d_{300} = d_{100}/3$. The first expression describes the reflection of order 1 on the (300) plane which is equivalent to describing the third reflections on the (100) plane. It is nevertheless preferable to use the notation of the type (300) when reasoning in reciprocal space because the points nh, nk, are nl are naturally found (e. g., 420, 330, 200, . . .).
One should not combine the two notations: $2d_{300}\sin\theta = \lambda$ represents the reflections of the ninth order on the (100) plane or that of the first order on the (900) plane.

(Q.15a) The diagrams are very similar because they have in common the lack of reflections that are forbidden by the fcc lattice. In addition, the forbidden reflections by the basis correspond to $2n + 2 = 2, 6, 10$ (see Pb. 1) for Ge and are very weak for GaAs because $f_{Ga} - f_{As} \approx 2$ (see Pb. 9) while the allowed reflections give: $f_{Ga} + f_{As} \approx 2f_{Ge}$. The situation is comparable to the quasi-extinction of certain reflections in the binary crystals of the KCl type where Z_1 and Z_2 are very nearly equal or quite identical if one takes into account the transfer of charge from K to Cl (see Pb. 2).

(Q.15b) h = odd because $f(K^+) = f(Cl^-)$ so that $Z(K^+) = Z(Cl^-) = 18$. See Pb. 7.

(Q.16) See Course Summary:
(a) $\lambda(\text{Å}) = 12.400/E\,(\text{eV})$ or $E \approx 8050$ eV.
(b) $\lambda(\text{Å}) = 12.26/\sqrt{V(\text{volts})}$ or $V \approx 63$ V.
(c) $\lambda(\text{Å}) = 0.28/\sqrt{E(\text{eV})}$ or $E \approx 33 \times 10^{-3}$ V. This result can be obtained directly from b) by considering M (neutron)/m (electrons) = 1840 so that $0.28 = 12.26/\sqrt{1840}$. Expressed in Kelvin, the neutron energy corresponds to 400 K.

(Q.17) The reciprocal lattice of a simple cubic is a simple cubic and the first Brillouin zone is cubic. The reciprocal lattice of a fcc lattice is a bcc lattice (see Fig. 10b). The basis does not play a role in these constructions so that they are the same for NaCl and GaAs.

(Q.18) The following techniques are named often by their acronyms, what are their full name and their main characteristics:

(a) LEED: Low-energy electron diffraction. See Pb. 3 for the characteristics.

(b) RHEED: Reflection high-energy electron diffraction. See Pb. 4 for the characteristics.

(c) EBSD: Electron backscattered diffraction. See Ex. 21b for the characteristics.

(Q.19) (a) The first Brillouin zone corresponds to the minimum volume limited by the symmetry planes between the origin of the reciprocal space and its various points.

(b) and (c) See Ex. 21b.

(Q.20) X-ray tubes and synchrotron radiation: See Summary of Course

(Q.21) The Laüe's experiments demonstrate the periodic structure of crystals that was previously suspected. These experiments demonstrate also the wave nature of the X-rays; see Ex. 20.

(Q.22) The Davisson and Germer experiments demonstrate the wave nature of incident electrons; see Pb. 3.

Chapter II

(Q.1) See Ex. 9.

(Q.2) $\in$ represents the cohesion energy of a pair of atoms (diatomic molecule). In the fcc structure each atom as 12 neighbors or $E_c = 6\in = 62.4$ meV. In addition, $r_0 = 1,12\sigma = \frac{a\sqrt{2}}{2}$ from which we find $a = 5.38$, see Ex. 10.

(Q.3) $\alpha(\text{MgO}) = 4\alpha(\text{NaCl})$, see Ex. 8.

(Q.4) The Madelung constant decreases with the number of neighbors. Thus, the binding energy is weaker for the ions located at the corners of a crystal, see Ex. 5.

When water molecules are interposed between the ion and the crystal the Coulomb attraction between the two is decreased by a factor equal to the relative dielectric constant of water: $\varepsilon_r \sim 80$.

(Q.5 and Q.6) See Ex. 11.

(Q.7) They are the multiplier coefficients of the deformations produced by a tension (C_{11}) or a shear (C_{44}). For a common value of the deformation slip (shear) requires less effort than traction (or compression). The transverse waves propagate slower than the longitudinal waves $V_T/V_L = \sqrt{C_{44}/C_{11}}$, see Ex. 16 of Chapter II and also Question (5) of Chapter III.

(Q.8) A tension produces a relative elongation $\Delta c/c$ along 1D associated and a relative decrease for the two other dimensions $(\Delta a/a) = (\Delta b/b) = -\sigma(\Delta c/c)$. If there is no change in volume during the tension: $(\Delta V/V) = (\Delta a/a) + (\Delta b/b) + (\Delta c/c) = 0$, so that $\sigma = 0.5$. In practice σ is of order $1/3$.

(Q.9) The compressibility β is such that $\beta = \dfrac{3(1-2\sigma)}{E}$, see Ex. 13.

$\sigma \approx 1/3$ so that $\beta \approx 1/E$. The numerical data for β indicated in the various exercises of this chapter lead to $E \approx 7 \times 10^{10}$ Pa (Al); 2.5×10^{10} Pa (alkali halides); $\approx 5 \times 10^8$ Pa (rare gas crystals). Some materials such as diamond are outside this range.

(Q.10) $E = 2.5 \times 10^{10}$ Pa (see above); $\Delta V/V = \beta \Delta p = 4 \times 10^{-6}$ for $\Delta p = 10^5$ Pa. We have dilation.

(Q.11) The Madelung constant measures the cohesive energy and it must be larger for an ion in a crystal than for the ion in a molecule. Otherwise the crystal would dissociate into molecules (see comment in Ex. 2b and Ex. 4).

Chapter III

(Q.1) The speed of sound is proportional to $\sqrt{\beta/M}$. The restoring constants in diamond are much larger (strong covalent binding) than in lead while the atomic mass is much lower: $V_s(Pb) \ll V_s$ (diamond).

(Q.2) The Debye temperature is proportional to v_s.
The speed of sound in Pb is small (see reply to Question (1)). For $T > \theta_D$ (which is the case): U (vibration) = $3 k_B T$ per atom, that is 75 meV at ambient temperature.

(Q.3) See reply to Question (1) taking into account that θ_D is proportional to v_S.

(Q.4) $C(50K)/C(5K) \approx 10^3$ because we are in the region where C_v varies as T^3 ($T << \theta_D$). $K(50K)/K(5K) \approx 10^3$ because $K = \frac{1}{3} C_v \Lambda$ and K varies as C while the mean free path of phonons Λ is limited by the size L of the sample.

(Q.5) See Pb. 7. The restoring constants β related to a shear are smaller than those related to a compression. These microscopic causes are also responsible for macroscopic effects evoked in Question (7) of Chapter II.

(Q.6) $K = (1/3)C_v\Lambda$ from which $\Lambda = 3.1 \times 10^{-8}$ m = 310 Å.

(Q.7) $3k_B$ per atom for T > θ_D. $C \propto T^3$, see Table I of Ex. 18.

(Q.8) k_B per atom for $T > \theta_D$. $C \propto T^3$, see Table I of Ex. 18.

(Q.9) No. Like the concept of photon, quantification in energy, applies in the free space and does implies that the EM wave is enclosed in a cavity, the concept of phonon, quantification in vibration energy, does not requires the use of boundary conditions: see Course Summary, paragraph 5 and comment in Ex. 11.

(Q.10) $\approx d/8$, see Ex. 21.

(Q.11) Due to a half quantum, see Ex. 21.

(Q.12) See Ex. 5.

(Q.13) The microscopic causes are due to a non-strictly parabolic potential to which atoms are subject, which induces a restoring force that is not exactly proportional to the elongation: deviation from the Hooke's law, see Pb. 6.

(Q.14) See Ex. 6.

(Q.15) There is an optical branch in the dispersion curve. It is related to the existence of two atoms per basis identical but in non-equivalent positions having therefore restoring forces different in direction (see Pb. 7 and Ex. 1).

(Q.16) Li F, NaCl, KBr, RbI, CsI. The greater the ion mass the lower the resonant frequency, see Ex. 10.

(Q.17) $\chi(\text{ions}) = \text{Ne}^2/\mu\omega_T^2 = \varepsilon_g - n^2$, see Pb. 1, paragraph 3

$\chi(\text{ions}) = 7(\text{LiF}); 3.65(\text{NaCl}); 8.3(\text{AgCl})$

The LST relation gives
$\omega_L/\omega_T = \sqrt{\varepsilon_5/\varepsilon_\infty}$, where $\varepsilon_\infty = n^2$
$\omega_L/\omega_T = 2.6(\text{LiF}); 1.6(\text{NaCl}); 1.75(\text{AgCl})$.

(Q.18) In general $\varepsilon_r < 1$ implies that $\chi < 0$, that is to say that the elementary dipolar moments are opposite the electric field that they created. This phase opposition is here due to the inertia of the ions when the electric frequency ω, in the IR, is above that of the resonance.

For free or bonded electrons, the same phenomena are observed in the UV, see Chapter IV, Ex. 29, and Chapter V, Ex. 26). The experimental consequence is an evanescent wave and a frequency interval $\omega_T < \omega < \omega_r$ where there is a total reflection because $\varepsilon_r < 0$: see Pb. 1.

(Q.19) Diamond is an electrical insulator with a band gap energy of ~5 eV and its thermal conductivity is only due to phonons $K = (1/3)C_v\Lambda$ but the velocity of sound, v is very large ~17,000 ms^{-1} (Chapter II, Pb. 4) because of the large value of $\sqrt{\beta/m}$ (see answer to Question (1)).

(Q.20) See Remark 2 at the end of the solution of Pb. 9.

Chapter IV

(Q.1) The alkali metals are transparent in the UV (for $\omega > \omega_p$). Below the plasma frequency, and therefore in the visible, they are totally reflective and thus with a metallic shine, see Ex. 29 and Pb. 5. Unfortunately, they oxidize in air and one must use other metals to realize mirrors. In the antiquity Cleopatra used polished silver mirrors.

(Q.2) After transfer into vacuum the sample is slightly expanded (see Chapter II, Question and Answer (10)). The average velocity of electrons and their Fermi energy are therefore lower: $E_F \propto (N/V)^{2/3}$. The relative increase in volume will be 1.6×10^{-5}, the relative decrease of the Fermi energy is of order 10^{-5}, around 50 μeV in absolute value; see Ex. 9.

(Q.3) The error results from the omission of the natural speed of electrons, $\approx v_F$ in the calculation of the transport distance: $\Lambda \approx v_F\tau = c\tau/200 \approx 150$ Å.

(Q.4) $\varepsilon_r - i\varepsilon_2$, if $E = E_0 e^{i\omega\tau}$; or $\varepsilon_1 - i\varepsilon_2$, if $E = E_0 e^{-i\omega\tau}$. See Technical Remark, Ex. 29.

(Q.5) $\rho = AT$; $\Delta\rho = A\Delta T$ from which $\alpha = 1/T \approx 3 \times 10^{-4}$, see Ex. 16.

(Q.6) $\sin i = n \sin r$ but for X-rays: $n \approx 1$, see Ex. 30. Same explanation for the Bragg law.

(Q.7) It is due to the Meissner effect: $B_{int} = \mu_0(1 + \chi) H = 0$ from which $\chi = -1$, see Pb. 7.

(Q.8) $\rho \approx 1 - 2$ μΩ·cm. This is a consequence of the Matthiessen law, see Ex. 16. During the elaboration one must eliminate the maximum number of impurities, elaborate it as a single crystalline and then use it at very low temperatures. For copper, one can thus win 5 orders of magnitude on τ and therefore, the mean free path can go from $\Lambda = 300$ Å to $\Lambda = 300$ cm, resulting in $\rho(\text{Cu}) \approx 10^{-11}$ Ω cm (at 4 K).

(Q.9) If we assume that $\varepsilon_2(\omega) = 0$ at all frequencies, the simple application of the KK relations leads to $\varepsilon_1 = 1$, see Pb. 5. The only material with an optical index of 1 over all the frequencies is the vacuum! On the contrary, for all others ε_1 and n vary with ω, most notably in the visible it is the variation of n with λ that is responsible for the dispersion of white light by a prism, see also Chapter V, Ex. 26).

(Q.10) Diffraction is used for the examination of periodic structures (ordered at long range distances), while EXAFS is used for short distances between first neighbors, which concerns dispersed materials, for instance copper sulfate in solution, see Pb. 4.

(Q.11) See Ex. 26.

(Q.12) Several hundreds of thousands see the Kubo criteria, Ex. 5.

(Q.13) Widemann–Franz law, K/σ = LT suggests that a good conductor of electricity is also a good conductor of heat, see Ex. 12 and Course Summary.

(Q.14) $t > \Lambda$, see Ex. 20.

(Q.15) The measurement of electrical resistivity at low temperature, see Ex. 17.

(Q.16) None in principle, in the free electron theory, the behavior of 'free' electrons is therefore, insensitive to the crystalline potential. In fact several exercises (no. 4b, 15 . . .) make a reference to crystalline effects but anecdotally.

(Q.17) By measuring the value of the Hall voltage, we deduce that n of $R_H = (n_e)^{-1}$, by measuring the conductivity we deduce μ in $\sigma = n_e \mu$, see Pb. 6.

(Q.18) The ejection of an Auger electron is much more likely that the emission of an X-ray photon when the initial binding energy of the atomic electron ejected is $E_L < 5$ keV, see Pb. 4.

(Q.19) The vertical resolution; see Pb. 5.

(Q.20) Photoemission experiments and angle resolved photo-electron spectroscopy, see Pb. 4.

(Q.21) The surface dipole layer formed by electron wave functions escaping a few into the vacuum and the fixed ions. See Pb. 2, Fig. 33, and comments.

Chapter V

(Q.1) A small portion of excess electrons will neutralize the majority of holes. If $n_i = 10^{13}$ cm^{-3} and if n_e becomes 10^{15} by doping, it is sufficient for 1% of the electrons provided by doping to neutralize (via electron/hole recombination) 99% of the initial holes. The final concentration will thus be $n_e = 10^{15}$ cm^{-3} and $n_h \approx 10^{11}$ cm^{-3}.

(Q.2) The dielectric constant is so unstable that we should call it the 'dielectric function'. If for water $\varepsilon_r(0) = 80$, its value in the visible spectral domain (0.4 μm < λ < 0.8 μm) is $\varepsilon_r(\text{visible}) = 16/9$.

(Q.3) See Ex. 29.

(Q.4) See Ex. 9.

(Q.5) It decreases when T increases because the atoms separate from each other due to thermal dilation, see comment in Ex. 25.

(Q.6) A very rapid decrease related to the addition of impurities if they are easily ionized, see Pb. 4. Nevertheless, for deep levels, they are not and this addition can induce a change in the Fermi level that favors less mobile carriers to the detriment of more mobile ones, so that ρ increases. One of the exceptions to the general tendency is therefore semi-insulating GaAs, see Ex. 20.

(Q.7) Ge, GaAs, AlP, MgO, BeO, LiF. The band gap energy increases as the size of atoms decreases and their ionic character increases, see Exs. 21 and 25.

(Q.8) Because n_i corresponds to 10^{10} cm^{-3}, which assumes that one can realize it with an impurity concentration lower than n_i/N^0 (where $N^0 \approx 5 \times 10^{22}$ cm^{-3}), that is to say, smaller than 10^{-12}, which is not possible. On the other hand, since E_g is small for Ge, $n_i \approx 10^{-13}$ cm^{-3}, its purification to 10^{-9} is possible.

The semiconductor industry owes a debt to chemists who were able to obtain this degree of perfection in purification.

(Q.9) See Ex. 25.

(Q.10) Instead of a monotonic increase as a function of temperature as for normal metals, it exhibits an exponential behavior, see Exs. 22 and 23.

(Q.11) See Ex. 21.

(Q.12) See Ex. 6.

(Q.13) The Coulomb attraction is reduced by the value of the dielectric constant and the particle mass by the effective mass, see Ex. 16 and Pb. 4.

(Q.14) Because the band gap is greater than 3 eV or in other words because the binding energy of valence electrons is greater than that of blue photons. For more details, see Ex. 25.

(Q.15) Due to chromium impurities, which introduce deep levels in the band gap, see comment in Ex. 28.

(Q.16) See Ex. 30.

(Q.17) See Ex. 19.

(Q.18) See Ex. 9.

(Q.19) Because the *d* electron bands are half full, see Ex. 8.

(Q.20) See Ex. 7.

(Q.21) Starting from an intrinsic semiconductor where E_F is in the middle of the band gap, the addition of electrons (*n*-doping) raises E_F, the neutralization of electrons (*p*-doping) lowers it. When these two parts come into contact, the electrons on the *n*-side diffuse to the *p*-side in the same way as two containers of a gas at different pressures.

(Q.22) Their respective excitation frequencies correspond, in each case to $\in_1(\omega) = 0$, see Chapter III, Pbs. 1 and 2; Chapter IV, Pb. 4; and Chapter V, Pb. 10. The only difference is the spectral domain concerned: the infrared for polaritons and the ultraviolet for plasmons, which is a consequence of the different particle mass involved (ions for one part and electrons for other).

(Q.23) If we restrict ourselves to the electronic contribution to the static dielectric constant, we find that $\chi_e(0) \approx \frac{\omega_p^2}{\bar{\omega}_T^2}$ (see Ex. 25), where $\bar{\omega}_T$ increases with the band gap energy.

(Q.24) This allows the adjustment of the emission wavelength of photons (laser diodes and photodiodes) or the absorption (photo-detector) where $h\nu = E_g$. In addition to the compositional change for homogenous materials, the choice of quantum well width (or period in a super-lattice) introduces an additional degree of liberty in the choice, see comment in Ex. 29 and in Pb. 9.

(Q.25) No if we think of excitons, see Ex. 19.

(Q.26) On the side of semi-metals: $n = 10^{18}$ cm^{-3}; $\rho \approx 10^{-2}\,\Omega$ cm
On the side of insulators: $n = 10^{10} - 10^{11}$ cm^{-3};
$\rho \approx 10^6 - 10^7\,\Omega$ cm
$\rho(Al_2O_3) \approx 10^6\,\Omega$ cm, ρ(sulfur) $\approx 10^{17}\,\Omega$ cm

(Q.27) The transition selection rules, see Pb. 10.

(Q.28) The optical measurement of the absorption threshold position is the most general method, see Ex. 28 and Pb. 6. Combined photoemission and BIS experiments permit also the measurement of E_G, see Fig. 55(left) in Pb. 10 and Chapter IV, Pb. 4. However, for direct band gap materials, we can also measure the radiation of fluorescence; for intrinsic semiconductors (E_g small), we can use electrical measurements where $\sigma = f(1/T)$, see Ex. 11 and Pb. 4, while for insulators (E_G large), we can locate the starting position from the function $\varepsilon_2/(\varepsilon_1^2 + \varepsilon_2^2)$ in electron loss spectroscopy, see Ex. 25.

(Q.29) See comment on ITO in Ex. 28.

(Q.30) See comments on GMR in Ex. 19.

(Q.31) See Pb. 11 and comments.

(Q.32) See Pb. 12 and comments.

Bibliography*

[1] ASHCROFT N. W. and MERMIN N. D. — Solid state physics — Saunders College (1976)

[2] BARANSKI R. et al. — Electronique des semiconducteurs — Editions de Moscou (1978)

[3] BENARD et al. — Métallurgie générate — 2nd Ed., Masson (1984)

[4] BUBE R. H. — Electrons in solids — Academic Press (1981)

[5] BURNS G. — Solid state physics — Academic Press (1985)

[6] EBERHART J. P. — Analyse structurale et chimique des matériaux — Dunod (1989)

[7] GIBER J. — Szilárdtestfizikai feladatok és számitásol — Múszaki Köryvkiadó (1982)

[8] GOLDMID H. S. — Problems in solid state physics — Academic Press (1968)

[9] GUINIER A. — Théorie et technique de la radiocristallographie — Dunod (1964)

[10] GUINIER G. — Eléments de physique moderne théorique — Bordas (1964)

[11] HARRISON W. A. — Solid state theory — Dover Publications (1979)

[12] HARRISON W. A. — Electronic structure and the properties of solids — Dover Publications (1989)

[13] JANOT C. et al. — Propridtés éléctronique des métaux et alliages — Masson (1971)

[14] KIREEV P. — La physique des semiconducteurs — MIR (1975)

*This bibliography lists only general books that have inspired exercises and problems of this book. References to scientific articles are indicated along the corresponding texts.

[15] KITTEL G. — Introduction á la physique de l'état solide — Dunod (a: 1958, lst Ed., Fr.; b: 1970, 2nd Ed. Fr.; c: 1972, 3rd Ed., Fr.; d: 1983, 5th Ed., Fr.)

[16] MATHIEU H. — Physique des semiconducteurs et composants éctroniques — Masson (1987)

[17] MOTT N. F. and JONES H. — The theory of the properties of metals and alloys — Clarendon Press (1936)

[18] NOGUERA C. — Physique et chimie des surfaces d'oxydes — Eyrolles (1995)

[19] NYE J. R. — Propriétés physiques des cristaux — Dunod (1961)

[20] PHILIBERT J. — Diffusion et transport de matiére dans les solides — Les editions de physique (1990)

[21] QUERE Y. — Physique des matériaux — Ellipses (1988)

[22] ROSENBERG H. M. — The solid state — Oxford Physics Series (1988)

[23] SAPOVAL B. and HERMANN C. — Physique des semiconducteurs — Ellipses (1990)

[24] SEEGER K. — Semi-conductor physics — Springer-Verlag (1982)

[25] SEITZ F. — Theorie moderne des solides — Masson (1949)

[26] DESJONQUÉRES M. C. and SPANJAARD D. — Concepts in surface physics — Springer (1998), ISBN 3-540-58622-9

[27] KING D. A. and WOODRUFF D. P. — The chemical physics of solid surfaces — Vol. 8, Elsevier (1997)

[28] OURA K. et al. — Surface science: an introduction — Springer-Verlag (2003)

[29] JULIAN CHEN C. — Introduction to scanning tunneling microscopy, 2nd Ed. — Oxford University Press (2008)

[30] MICHAELIDES A. and SCHEFFLER M. — An introduction to the theory of crystalline elemental solids and their surfaces — in Textbook of surface and interface science, WANDELT K., ed., pp. 13–72 — Springer (2012)

[31] HOFMANN P. — Surface physics: an introduction — self-published (2013), ebook ISBN 978-87-996090-0-0

[32] FRIEDEL J. — Physics of metals — Cambridge University Press (1969)

Further Readings

[1] BORN M. — Atomic physics — Hafner (1962)

[2] BORN M. and HUANG K. — Dynamical theory of crystal lattices — Clarendon Press (1954)

[3] BROUSSEAU M. — Physique du solide — Masson (1992)

[4] CHAHINE C. and DEVAUX P. — Problémes de thermodynamique statique — Dunod (1985)

[5] COHEN-TANNOUDJI C. et al. — Mécanique quantique — Hermann (1973)

[6] DAVIES P. — La nouvelle physique — Flammarion (1993)

[7] GUINIER A. and JULLIEN R. — La matière à l'état solide: des supraconducteurs aux superalliages — Hachette (1988)

[8] IBACH H. and LUTH H. — Solid state physics — 2nd Ed., Springer-Verlag (1993)

[9] KITTEL C. — Thèorie quantique du solide — Dunod (1967)

[10] PAULING L. — The chemical bond — Cornell University Press (1967)

[11] PAVLOV P. and KHOKHLOV V. — Physique du solide — MIR (1989)

[12] PEIERLS R. E. — Quantum theory of solids — Clarendon Press (1955)

[13] SZE S. M. — Semiconductor devices — John Wiley and Sons (1985)

[14] ZIMAN J. M. — Electrons and phonons — Clarendon Press (1962)

[15] PETTIFOR D. — Bonding and structure of molecules and solids — Oxford University Press (1995)

Index